# Brief Table of Integrals

**1.** $\displaystyle\int u^n \, du = \frac{u^{n+1}}{n+1} + C, \; n \neq -1$

**2.** $\displaystyle\int \frac{1}{u} \, du = \ln|u| + C$

**3.** $\displaystyle\int e^u \, du = e^u + C$

**4.** $\displaystyle\int a^u \, du = \frac{1}{\ln a} a^u + C$

**5.** $\displaystyle\int \sin u \, du = -\cos u + C$

**6.** $\displaystyle\int \cos u \, du = \sin u + C$

**7.** $\displaystyle\int \sec^2 u \, du = \tan u + C$

**8.** $\displaystyle\int \csc^2 u \, du = -\cot u + C$

**9.** $\displaystyle\int \sec u \tan u \, du = \sec u + C$

**10.** $\displaystyle\int \csc u \cot u \, du = -\csc u + C$

**11.** $\displaystyle\int \tan u \, du = -\ln|\cos u| + C$

**12.** $\displaystyle\int \cot u \, du = \ln|\sin u| + C$

**13.** $\displaystyle\int \sec u \, du = \ln|\sec u + \tan u| + C$

**14.** $\displaystyle\int \csc u \, du = \ln|\csc u - \cot u| + C$

**15.** $\displaystyle\int u \sin u \, du = \sin u - u \cos u + C$

**16.** $\displaystyle\int u \cos u \, du = \cos u + u \sin u + C$

**17.** $\displaystyle\int \sin^2 u \, du = \tfrac{1}{2}u - \tfrac{1}{4}\sin 2u + C$

**18.** $\displaystyle\int \cos^2 u \, du = \tfrac{1}{2}u + \tfrac{1}{4}\sin 2u + C$

**19.** $\displaystyle\int \tan^2 u \, du = \tan u - u + C$

**20.** $\displaystyle\int \cot^2 u \, du = -\cot u - u + C$

**21.** $\displaystyle\int \sin^3 u \, du = -\tfrac{1}{3}\left(2 + \sin^2 u\right)\cos u + C$

**22.** $\displaystyle\int \cos^3 u \, du = \tfrac{1}{3}\left(2 + \cos^2 u\right)\sin u + C$

**23.** $\displaystyle\int \tan^3 u \, du = \tfrac{1}{2}\tan^2 u + \ln|\cos u| + C$

**24.** $\displaystyle\int \cot^3 u \, du = -\tfrac{1}{2}\cot^2 u - \ln|\sin u| + C$

**25.** $\displaystyle\int \sec^3 u \, du = \tfrac{1}{2}\sec u \tan u + \tfrac{1}{2}\ln|\sec u + \tan u| + C$

**26.** $\displaystyle\int \csc^3 u \, du = -\tfrac{1}{2}\csc u \cot u + \tfrac{1}{2}\ln|\csc u - \cot u| + C$

**27.** $\displaystyle\int \sin au \sin bu \, du = \frac{\sin(a-b)u}{2(a-b)} - \frac{\sin(a+b)u}{2(a+b)} + C$

**28.** $\displaystyle\int \cos au \cos bu \, du = \frac{\sin(a-b)u}{2(a-b)} + \frac{\sin(a+b)u}{2(a+b)} + C$

**29.** $\displaystyle\int e^{au} \sin bu \, du = \frac{e^{au}}{a^2+b^2}(a\sin bu - b\cos bu) + C$

**30.** $\displaystyle\int e^{au} \cos bu \, du = \frac{e^{au}}{a^2+b^2}(a\cos bu + b\sin bu) + C$

**31.** $\displaystyle\int \sinh u \, du = \cosh u + C$

**32.** $\displaystyle\int \cosh u \, du = \sinh u + C$

**33.** $\displaystyle\int \operatorname{sech}^2 u \, du = \tanh u + C$

**34.** $\displaystyle\int \operatorname{csch}^2 u \, du = -\coth u + C$

**35.** $\displaystyle\int \tanh u \, du = \ln(\cosh u) + C$

**36.** $\displaystyle\int \coth u \, du = \ln|\sinh u| + C$

**37.** $\displaystyle\int \ln u \, du = u \ln u - u + C$

**38.** $\displaystyle\int u \ln u \, du = \tfrac{1}{2}u^2 \ln u - \tfrac{1}{4}u^2 + C$

**39.** $\displaystyle\int \frac{1}{\sqrt{a^2 - u^2}} \, du = \sin^{-1}\frac{u}{a} + C$

**40.** $\displaystyle\int \frac{1}{\sqrt{a^2 + u^2}} \, du = \ln\left|u + \sqrt{a^2 + u^2}\right| + C$

**41.** $\displaystyle\int \sqrt{a^2 - u^2} \, du = \frac{u}{2}\sqrt{a^2 - u^2} + \frac{a^2}{2}\sin^{-1}\frac{u}{a} + C$

**42.** $\displaystyle\int \sqrt{a^2 + u^2} \, du = \frac{u}{2}\sqrt{a^2 + u^2} + \frac{a^2}{2}\ln\left|u + \sqrt{a^2 + u^2}\right| + C$

**43.** $\displaystyle\int \frac{1}{a^2 + u^2} \, du = \frac{1}{a}\tan^{-1}\frac{u}{a} + C$

**44.** $\displaystyle\int \frac{1}{a^2 - u^2} \, du = \frac{1}{2a}\ln\left|\frac{a+u}{a-u}\right| + C$

**Note:** Some techniques of integration, such as integration by parts and partial fractions, are reviewed in the *Student Resource Manual* that accompanies this text.

# 11E

# A First Course in Differential Equations with Modeling Applications

## Dennis G. Zill
### *Loyola Marymount University*

CENGAGE
Learning

Australia • Brazil • Mexico • Singapore • United Kingdom • United States

**A First Course in Differential Equations with Modeling Applications, Eleventh Edition**
Dennis G. Zill

Product Director: Terry Boyle

Product Manager: Kate Schrumpf

Content Developer: Spencer Arritt

Product Assistant: Stephen Previte

Marketing Manager: Ana Albinson

Content Project Manager: Jennifer Risden

Art Director: Vernon Boes

Manufacturing Planner: Becky Cross

Production Service: MPS Limited

Photo Researcher: Lumina Datamatics

Text Researcher: Lumina Datamatics

Copy Editor: Steven Summerlight

Text Designer: Terri Wright

Cover Designer: Terri Wright

Cover Image: STS-114 Crew, ISS Expedition 11 Crew, NASA

Compositor: MPS Limited

Portions of this text appear in *Advanced Engineering Mathematics*, Sixth Edition, Copyright 2018, Jones & Bartlett Learning, Burlington, MA 01803 and are used with the permission of the publisher.

For product information and technology assistance, contact us at **Cengage Learning Customer & Sales Support, 1-800-354-9706**.

For permission to use material from this text or product, submit all requests online at **www.cengage.com/permissions**. Further permissions questions can be e-mailed to **permissionrequest@cengage.com**.

Library of Congress Control Number: 2016936077

Student Edition:

ISBN: 978-1-305-96572-0

Loose-leaf Edition:

ISBN: 978-1-337-29312-9

**Cengage Learning**
20 Channel Center Street
Boston, MA 02210
USA

Cengage Learning is a leading provider of customized learning solutions with employees residing in nearly 40 different countries and sales in more than 125 countries around the world. Find your local representative at **www.cengage.com**.

Cengage Learning products are represented in Canada by Nelson Education, Ltd.

To learn more about Cengage Learning Solutions, visit **www.cengage.com**.

Purchase any of our products at your local college store or at our preferred online store **www.cengagebrain.com**.

Printed in the United States of America

Print Number: 03      Print Year: 2017

# Contents

Paul B. Moore/Shutterstock .com

## 9    Numerical Solutions of Ordinary Differential Equations    368

### Appendices

## TO THE STUDENT

Authors of books live with the hope that someone actually *reads* them. Contrary to what you might believe, almost everything in a typical college-level mathematics text is written for you and not the instructor. True, the topics covered in the text are chosen to appeal to instructors because they make the decision on whether to use it in their classes, but everything written in it is aimed directly at you the student. So I want to encourage you—no, actually I want to *tell* you—to read this textbook! But do not read this text as you would a novel; you should not read it fast and you should not skip anything. Think of it as a *work*book. By this I mean that mathematics should always be read with pencil and paper at the ready because, most likely, you will have to *work* your way through the examples and the discussion. Before attempting any problems in the section exercise sets, work through *all* the examples in that section. The examples are constructed to illustrate what I consider the most important aspects of the section, and therefore, reflect the procedures necessary to work most of the problems. When reading an example, copy it down on a piece of paper and do not look at the solution in the book. Try working it, then compare your results against the solution given, and, if necessary resolve any differences. I have tried to include most of the important steps in each example, but if something is not clear you should always try—and here is where the pencil and paper come in again—to fill in the details or missing steps. This may not be easy, but it is part of the learning process. The accumulation of facts followed by the slow assimilation of understanding simply cannot be achieved without a struggle.

Specifically for you, a *Student Resource Manual* (*SRM*) is available as an optional supplement. In addition to containing solutions of selected problems from the exercises sets, the *SRM* contains hints for solving problems, extra examples, and a review of those areas of algebra and calculus that I feel are particularly important to the successful study of differential equations. Bear in mind you do not have to purchase the *SRM*; you can review the appropriate mathematics from your old pre-calculus or calculus texts.

In conclusion, I wish you good luck and success. I hope you enjoy the text and the course you are about to embark on—as an undergraduate math major it was one of my favorites because I liked mathematics that connected with the physical world. If you have any comments, or if you find any errors as you read/work your way through the text, or if you come up with a good idea for improving either it or the *SRM*, please feel free to contact me through Cengage Learning:

spencer.arritt@cengage.com.

## TO THE INSTRUCTOR

In case you are examining this text for the first time, *A First Course in Differential Equations with Modeling Applications,* Eleventh Edition, is intended for a one-semester or one-quarter course in ordinary differential equations. The longer version of the text, *Differential Equations with Boundary-Value Problems,* Ninth Edition, can be used for either a one- or two-semester course that covers ordinary and partial differential equations. This text contains six additional chapters. For a one-semester course, it is assumed that the students have successfully completed at least two semesters of calculus. Since you are reading this, undoubtedly you have already examined the table of contents for the topics that are covered. You will not find a "suggested syllabus" in this preface; I will not pretend to be so wise as to tell other teachers what to teach. I feel that there is plenty of material here to choose from and to form a course to your liking. The text strikes a reasonable balance between the analytical, qualitative, and quantitative approaches to the study of differential equations. As far as my "underlying philosophy" goes, it is this: An undergraduate text should be written with the students' understanding kept firmly in mind, which means to me that the material should be presented in a straightforward, readable, and helpful manner, while keeping the level of theory consistent with the notion of a "first course."

For those who are familiar with the previous editions, I would like to mention a few improvements made in this edition. Many exercise sets have been updated by the addition of new problems. Some of these problems involve new and, I think, interesting mathematical models. Additional examples, figures, and remarks have been added to many sections. Throughout the text I have given a greater emphasis to the concepts of piecewise-linear differential equations and solutions that involve nonelementary integrals. Finally, Appendix A, *Integral-Defined Functions*, is new to the text.

## Student Resources

- *Student Resource Manual (SRM)*, prepared by Warren S. Wright and Roberto Martinez (ISBN 978-1-305-96573-7, accompanies *A First Course in Differential Equations with Modeling Applications*, Eleventh Edition, and ISBN 978-1-305-96581-2 accompanies *Differential Equations with Boundary-Value Problems*, Ninth Edition), provides important review material from algebra and calculus, the solution of every third problem in each exercise set (with the exception of the Discussion Problems and Computer Lab Assignments), relevant command syntax for the computer algebra systems *Mathematica* and *Maple*, and lists of important concepts, as well as helpful hints on how to start certain problems.
- **MindTap for** *A First Course in Differential Equations with Modeling Applications*, Eleventh Edition, is a digital representation of your course that provides you with the tools you need to better manage your limited time, stay organized, and be successful. You can complete assignments whenever and wherever you are ready to learn with course material specially customized for you by your instructor and streamlined in one proven, easy-to-use interface. With an array of study tools, you'll get a true understanding of course concepts, achieve better grades, and set the groundwork for your future courses. Learn more at **www.cengage.com/mindtap.**

## Instructor Resources

- *Instructor's Solutions Manual (ISM)*, prepared by Warren S. Wright and Roberto Martinez, provides complete worked-out solutions for all problems in the text. It is available through the Instructor Companion website at **www .cengage.com.**
- *Cengage Learning Testing Powered by Cognero* is a flexible, online system that allows you to author, edit, and manage test bank content, create multiple test versions in an instant, and deliver tests from your learning management system (LMS), your classroom, or wherever you want. Cognero is available online at **www.cengage.com/login.**
- Turn the light on with **MindTap for** *A First Course in Differential Equations with Modeling Applications*, Eleventh Edition. Through personalized paths of dynamic assignments and applications, MindTap is a digital learning solution and representation of your course.

    **The Right Content:** With MindTap's carefully curated material, you get the precise content and groundbreaking tools you need for every course you teach.

**Personalization:** Customize every element of your course—from rearranging the Learning Path to inserting videos and activities.

**Improved Workflow:** Save time when planning lessons with all of the trusted, most current content you need in one place in MindTap.

**Tracking Students' Progress in Real Time:** Promote positive outcomes by tracking students in real time and tailoring your course as needed based on the analytics.

Learn more at **www.cengage.com/mindtap.**

## ACKNOWLEDGEMENTS

Compiling a mathematics textbook such as this and making sure that its thousands of symbols and hundreds of equations are accurate is an enormous task, but since I am called "the author" that is my job and responsibility. But many people besides myself have expended enormous amounts of time and energy in working towards its eventual publication. So I would like to take this opportunity to express my sincerest appreciation to everyone—most of them unknown to me—at Cengage Learning and at MPS North America who were involved in the publication of this edition. A special word of thanks goes to Spencer Arritt, Kathryn Schrumpf, Jennifer Risden, Vernon Boes, and Jill Traut for their guidance in the labyrinth of the production process.

Finally, over the years this text has been improved in a countless number of ways through the suggestions and criticisms of the reviewers. Thus it is fitting to conclude with an acknowledgement of my debt to the following generous people for sharing their expertise and experience.

### REVIEWERS OF PAST EDITIONS

William Atherton, *Cleveland State University*
Philip Bacon, *University of Florida*
Bruce Bayly, *University of Arizona*
William H. Beyer, *University of Akron*
R. G. Bradshaw, *Clarkson College*
Bernard Brooks, *Rochester Institute of Technology*
Allen Brown, *Wabash Valley College*
Dean R. Brown, *Youngstown State University*
David Buchthal, *University of Akron*
Nguyen P. Cac, *University of Iowa*
T. Chow, *California State University–Sacramento*
Dominic P. Clemence, *North Carolina Agricultural and Technical State University*
Pasquale Condo, *University of Massachusetts–Lowell*
Vincent Connolly, *Worcester Polytechnic Institute*
Philip S. Crooke, *Vanderbilt University*
Bruce E. Davis, *St. Louis Community College at Florissant Valley*
Paul W. Davis, *Worcester Polytechnic Institute*
Richard A. DiDio, *La Salle University*
James Draper, *University of Florida*
James M. Edmondson, *Santa Barbara City College*
John H. Ellison, *Grove City College*
Raymond Fabec, *Louisiana State University*
Donna Farrior, *University of Tulsa*
Robert E. Fennell, *Clemson University*
W. E. Fitzgibbon, *University of Houston*
Harvey J. Fletcher, *Brigham Young University*
Paul J. Gormley, *Villanova*
Layachi Hadji, *University of Alabama*
Ruben Hayrapetyan, *Kettering University*
Terry Herdman, *Virginia Polytechnic Institute and State University*
Zdzislaw Jackiewicz, *Arizona State University*

S. K. Jain, *Ohio University*
Anthony J. John, *Southeastern Massachusetts University*
David C. Johnson, *University of Kentucky–Lexington*
Harry L. Johnson, *Virginia Polytechnic Institute and State University*
Kenneth R. Johnson, *North Dakota State University*
Joseph Kazimir, *East Los Angeles College*
J. Keener, *University of Arizona*
Steve B. Khlief, *Tennessee Technological University*
Helmut Knaust, *The University of Texas at El Paso*
C. J. Knickerbocker, *Sensis Corporation*
Carlon A. Krantz, *Kean College of New Jersey*
Thomas G. Kudzma, *University of Lowell*
Alexandra Kurepa, *North Carolina A&T State University*
G. E. Latta, *University of Virginia*
Cecelia Laurie, *University of Alabama*
Mulatu Lemma, *Savannah State University*
James R. McKinney, *California Polytechnic State University*
James L. Meek, *University of Arkansas*
Gary H. Meisters, *University of Nebraska–Lincoln*
Stephen J. Merrill, *Marquette University*
Vivien Miller, *Mississippi State University*
George Moss, *Union University*
Gerald Mueller, *Columbus State Community College*
Philip S. Mulry, *Colgate University*
Martin Nakashima, *California State Polytechnic University–Pomona*
C. J. Neugebauer, *Purdue University*
Tyre A. Newton, *Washington State University*
Brian M. O'Connor, *Tennessee Technological University*
J. K. Oddson, *University of California–Riverside*
Carol S. O'Dell, *Ohio Northern University*
Bruce O'Neill, *Milwaukee School of Engineering*
A. Peressini, *University of Illinois, Urbana–Champaign*
J. Perryman, *University of Texas at Arlington*
Joseph H. Phillips, *Sacramento City College*
Jacek Polewczak, *California State University–Northridge*
Nancy J. Poxon, *California State University–Sacramento*
Robert Pruitt, *San Jose State University*
K. Rager, *Metropolitan State College*
F. B. Reis, *Northeastern University*
Brian Rodrigues, *California State Polytechnic University*
Tom Roe, *South Dakota State University*
Kimmo I. Rosenthal, *Union College*
Barbara Shabell, *California Polytechnic State University*
Seenith Sivasundaram, *Embry–Riddle Aeronautical University*
Don E. Soash, *Hillsborough Community College*
F. W. Stallard, *Georgia Institute of Technology*
Gregory Stein, *The Cooper Union*
M. B. Tamburro, *Georgia Institute of Technology*
Patrick Ward, *Illinois Central College*
Jianping Zhu, *University of Akron*
Jan Zijlstra, *Middle Tennessee State University*
Jay Zimmerman, *Towson University*

Dennis G. Zill
Los Angeles, CA

# A First Course in Differential Equations with Modeling Applications

# Introduction to Differential Equations

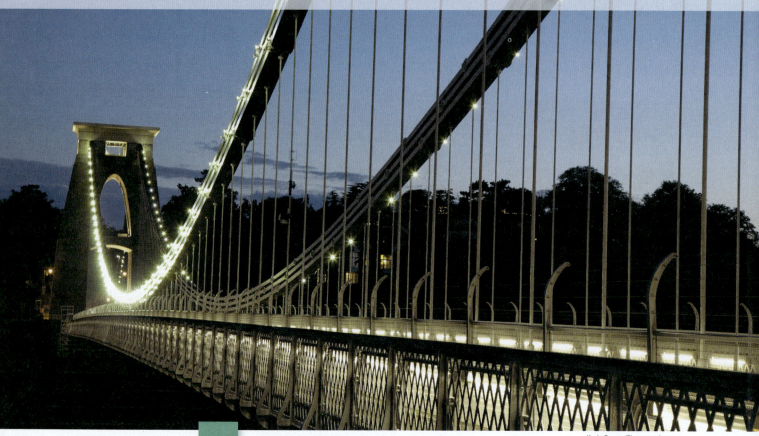

Kevin George/Shutterstock.com

The words *differential* and *equations* suggest solving some kind of equation that contains derivatives $y'$, $y''$, .... Analogous to a course in algebra, in which a good amount of time is spent solving equations such as $x^2 + 5x + 4 = 0$ for the unknown number $x$, in this course *one* of our tasks will be to solve differential equations such as $y'' + 2y' + y = 0$ for an unknown function $y = \phi(x)$. As the course unfolds, you will see there is more to the study of differential equations than just mastering methods that mathematicians over past centuries devised to solve them. But first things first. In order to read, study, and be conversant in a specialized subject you have to learn some of the terminology of that discipline. This is the thrust of the first two sections of this chapter. In the last section we briefly examine the link between differential equations and the real world.

## 1.1 Definitions and Terminology

**INTRODUCTION** The derivative $dy/dx$ of a function $y = \phi(x)$ is itself another function $\phi'(x)$ found by an appropriate rule. The exponential function $y = e^{0.1x^2}$ is differentiable on the interval $(-\infty, \infty)$ and by the Chain Rule its first derivative is $dy/dx = 0.2xe^{0.1x^2}$. If we replace $e^{0.1x^2}$ on the right-hand side of the last equation by the symbol $y$, the derivative becomes

$$\frac{dy}{dx} = 0.2xy. \tag{1}$$

Now imagine that a friend of yours simply hands you equation (1)—you have no idea how it was constructed—and asks, *What is the function represented by the symbol y?* You are now face to face with one of the basic problems in this course:

*How do you solve an equation such as* (1) *for the function $y = \phi(x)$?*

**A DEFINITION** The equation that we made up in (1) is called a **differential equation.** Before proceeding any further, let us consider a more precise definition of this concept.

> **DEFINITION 1.1.1 Differential Equation**
>
> An equation containing the derivatives of one or more unknown functions (or dependent variables), with respect to one or more independent variables, is said to be a **differential equation (DE).**

To talk about them, we shall classify differential equations according to **type, order,** and **linearity.**

**CLASSIFICATION BY TYPE** If a differential equation contains only ordinary derivatives of one or more unknown functions with respect to a *single* independent variable, it is said to be an **ordinary differential equation (ODE).** An equation involving partial derivatives of one or more unknown functions of two or more independent variables is called a **partial differential equation (PDE).** Our first example illustrates several of each type of differential equation.

**EXAMPLE 1** Types of Differential Equations

**(a)** The equations

*an ODE can contain more than one unknown function*

$$\frac{dy}{dx} + 5y = e^x, \quad \frac{d^2y}{dx^2} - \frac{dy}{dx} + 6y = 0, \quad \text{and} \quad \frac{dx}{dt} + \frac{dy}{dt} = 2x + y \tag{2}$$

are examples of ordinary differential equations.

**(b)** The following equations are partial differential equations:[*]

$$\frac{\partial^2 u}{\partial x^2} + \frac{\partial^2 u}{\partial y^2} = 0, \quad \frac{\partial^2 u}{\partial x^2} = \frac{\partial^2 u}{\partial t^2} - 2\frac{\partial u}{\partial t}, \quad \frac{\partial u}{\partial y} = -\frac{\partial v}{\partial x}. \tag{3}$$

---

[*]Except for this introductory section, only ordinary differential equations are considered in *A First Course in Differential Equations with Modeling Applications*, Eleventh Edition. In that text the word *equation* and the abbreviation DE refer only to ODEs. Partial differential equations or PDEs are considered in the expanded volume *Differential Equations with Boundary-Value Problems*, Ninth Edition.

Notice in the third equation that there are two unknown functions and two independent variables in the PDE. This means $u$ and $v$ must be functions of *two or more* independent variables.　■

**NOTATION**　Throughout this text ordinary derivatives will be written by using either the **Leibniz notation** $dy/dx$, $d^2y/dx^2$, $d^3y/dx^3$, ... or the **prime notation** $y'$, $y''$, $y'''$, .... By using the latter notation, the first two differential equations in (2) can be written a little more compactly as $y' + 5y = e^x$ and $y'' - y' + 6y = 0$. Actually, the prime notation is used to denote only the first three derivatives; the fourth derivative is written $y^{(4)}$ instead of $y''''$. In general, the $n$th derivative of $y$ is written $d^ny/dx^n$ or $y^{(n)}$. Although less convenient to write and to typeset, the Leibniz notation has an advantage over the prime notation in that it clearly displays both the dependent and independent variables. For example, in the equation

unknown function
or dependent variable

$$\frac{d^2x}{dt^2} + 16x = 0$$

independent variable

it is immediately seen that the symbol $x$ now represents a dependent variable, whereas the independent variable is $t$. You should also be aware that in physical sciences and engineering, Newton's **dot notation** (derogatorily referred to by some as the "flyspeck" notation) is sometimes used to denote derivatives with respect to time $t$. Thus the differential equation $d^2s/dt^2 = -32$ becomes $\ddot{s} = -32$. Partial derivatives are often denoted by a **subscript notation** indicating the independent variables. For example, with the subscript notation the second equation in (3) becomes $u_{xx} = u_{tt} - 2u_t$.

**CLASSIFICATION BY ORDER**　The **order of a differential equation** (either ODE or PDE) is the order of the highest derivative in the equation. For example,

second order　　　first order

$$\frac{d^2y}{dx^2} + 5\left(\frac{dy}{dx}\right)^3 - 4y = e^x$$

is a second-order ordinary differential equation. In Example 1, the first and third equations in (2) are first-order ODEs, whereas in (3) the first two equations are second-order PDEs. A first-order ordinary differential equation is sometimes written in the **differential form**

$$M(x, y)\,dx + N(x, y)\,dy = 0.$$

---

**EXAMPLE 2**　**Differential Form of a First-Order ODE**

If we assume that $y$ is the dependent variable in a first-order ODE, then recall from calculus that the differential $dy$ is defined to be $dy = y'\,dx$.

**(a)** By dividing by the differential $dx$ an alternative form of the equation $(y - x)\,dx + 4x\,dy = 0$ is given by

$$y - x + 4x\frac{dy}{dx} = 0 \quad \text{or equivalently} \quad 4x\frac{dy}{dx} + y = x.$$

**(b)** By multiplying the differential equation

$$6xy\frac{dy}{dx} + x^2 + y^2 = 0$$

by $dx$ we see that the equation has the alternative differential form

$$(x^2 + y^2)\,dx + 6xy\,dy = 0.$$ ■

In symbols we can express an $n$th-order ordinary differential equation in one dependent variable by the general form

$$F(x,\,y,\,y',\,\dots,\,y^{(n)}) = 0, \tag{4}$$

where $F$ is a real-valued function of $n + 2$ variables: $x, y, y', \dots, y^{(n)}$. For both practical and theoretical reasons we shall also make the assumption hereafter that it is possible to solve an ordinary differential equation in the form (4) uniquely for the highest derivative $y^{(n)}$ in terms of the remaining $n + 1$ variables. The differential equation

$$\frac{d^n y}{dx^n} = f(x, y, y', \dots, y^{(n-1)}), \tag{5}$$

where $f$ is a real-valued continuous function, is referred to as the **normal form** of (4). Thus when it suits our purposes, we shall use the normal forms

$$\frac{dy}{dx} = f(x, y) \qquad \text{and} \qquad \frac{d^2 y}{dx^2} = f(x, y, y')$$

to represent general first- and second-order ordinary differential equations.

---

### EXAMPLE 3   Normal Form of an ODE

**(a)** By solving for the derivative $dy/dx$ the normal form of the first-order differential equation

$$4x\frac{dy}{dx} + y = x \quad \text{is} \quad \frac{dy}{dx} = \frac{x - y}{4x}.$$

**(b)** By solving for the derivative $y''$ the normal form of the second-order differential equation

$$y'' - y' + 6 = 0 \quad \text{is} \quad y'' = y' - 6y.$$ ■

**CLASSIFICATION BY LINEARITY**   An $n$th-order ordinary differential equation (4) is said to be **linear** if $F$ is linear in $y, y', \dots, y^{(n)}$. This means that an $n$th-order ODE is linear when (4) is $a_n(x)y^{(n)} + a_{n-1}(x)y^{(n-1)} + \cdots + a_1(x)y' + a_0(x)y - g(x) = 0$ or

$$a_n(x)\frac{d^n y}{dx^n} + a_{n-1}(x)\frac{d^{n-1}y}{dx^{n-1}} + \cdots + a_1(x)\frac{dy}{dx} + a_0(x)y = g(x). \tag{6}$$

Two important special cases of (6) are linear first-order ($n = 1$) and linear second-order ($n = 2$) DEs:

$$a_1(x)\frac{dy}{dx} + a_0(x)y = g(x) \qquad \text{and} \qquad a_2(x)\frac{d^2 y}{dx^2} + a_1(x)\frac{dy}{dx} + a_0(x)y = g(x). \tag{7}$$

In the additive combination on the left-hand side of equation (6) we see that the characteristic two properties of a linear ODE are as follows:

- The dependent variable $y$ and all its derivatives $y', y'', \dots, y^{(n)}$ are of the first degree, that is, the power of each term involving $y$ is 1.
- The coefficients $a_0, a_1, \dots, a_n$ of $y, y', \dots, y^{(n)}$ depend at most on the independent variable $x$.

A **nonlinear** ordinary differential equation is simply one that is not linear. Nonlinear functions of the dependent variable or its derivatives, such as $\sin y$ or $e^{y'}$, cannot appear in a linear equation.

### EXAMPLE 4  Linear and Nonlinear ODEs

**(a)** The equations

$$(y - x)\,dx + 4x\,dy = 0, \quad y'' - 2y + y = 0, \quad x^3 \frac{d^3y}{dx^3} + x\frac{dy}{dx} - 5y = e^x$$

are, in turn, *linear* first-, second-, and third-order ordinary differential equations. We have just demonstrated in part (a) of Example 2 that the first equation is linear in the variable $y$ by writing it in the alternative form $4xy' + y = x$.

**(b)** The equations

$$\underset{\text{coefficient depends on } y}{(1 - y)y' + 2y = e^x}, \quad \frac{d^2y}{dx^2} + \sin y = 0, \quad \text{and} \quad \frac{d^4y}{dx^4} + y^2 = 0$$

are examples of *nonlinear* first-, second-, and fourth-order ordinary differential equations, respectively.

**SOLUTIONS**  As was stated on page 2, one of the goals in this course is to solve, or find solutions of, differential equations. In the next definition we consider the concept of a solution of an ordinary differential equation.

### DEFINITION 1.1.2  Solution of an ODE

Any function $\phi$, defined on an interval $I$ and possessing at least $n$ derivatives that are continuous on $I$, which when substituted into an $n$th-order ordinary differential equation reduces the equation to an identity, is said to be a **solution** of the equation on the interval.

In other words, a solution of an $n$th-order ordinary differential equation (4) is a function $\phi$ that possesses at least $n$ derivatives and for which

$$F(x, \phi(x), \phi'(x), \ldots, \phi^{(n)}(x)) = 0 \qquad \text{for all } x \text{ in } I.$$

We say that $\phi$ *satisfies* the differential equation on $I$. For our purposes we shall also assume that a solution $\phi$ is a real-valued function. In our introductory discussion we saw that $y = e^{0.1x^2}$ is a solution of $dy/dx = 0.2xy$ on the interval $(-\infty, \infty)$.

Occasionally, it will be convenient to denote a solution by the alternative symbol $y(x)$.

**INTERVAL OF DEFINITION**  You cannot think *solution* of an ordinary differential equation without simultaneously thinking *interval*. The interval $I$ in Definition 1.1.2 is variously called the **interval of definition,** the **interval of existence,** the **interval of validity,** or the **domain of the solution** and can be an open interval $(a, b)$, a closed interval $[a, b]$, an infinite interval $(a, \infty)$, and so on.

EXAMPLE 5    Verification of a Solution

Verify that the indicated function is a solution of the given differential equation on the interval $(-\infty, \infty)$.

(a) $\dfrac{dy}{dx} = xy^{1/2}; \quad y = \frac{1}{16}x^4$    (b) $y'' - 2y' + y = 0; \quad y = xe^x$

SOLUTION    One way of verifying that the given function is a solution is to see, after substituting, whether each side of the equation is the same for every $x$ in the interval.

(a) From

$$\textit{left-hand side:} \quad \dfrac{dy}{dx} = \dfrac{1}{16}(4 \cdot x^3) = \dfrac{1}{4}x^3,$$

$$\textit{right-hand side:} \quad xy^{1/2} = x \cdot \left(\dfrac{1}{16}x^4\right)^{1/2} = x \cdot \left(\dfrac{1}{4}x^2\right) = \dfrac{1}{4}x^3,$$

we see that each side of the equation is the same for every real number $x$. Note that $y^{1/2} = \frac{1}{4}x^2$ is, by definition, the nonnegative square root of $\frac{1}{16}x^4$.

(b) From the derivatives $y' = xe^x + e^x$ and $y'' = xe^x + 2e^x$ we have, for every real number $x$,

$$\textit{left-hand side:} \quad y'' - 2y' + y = (xe^x + 2e^x) - 2(xe^x + e^x) + xe^x = 0,$$

$$\textit{right-hand side:} \quad 0.$$    ■

Note, too, that each differential equation in Example 5 possesses the constant solution $y = 0$, $-\infty < x < \infty$. A solution of a differential equation that is identically zero on an interval $I$ is said to be a **trivial solution.**

SOLUTION CURVE    The graph of a solution $\phi$ of an ODE is called a **solution curve.** Since $\phi$ is a differentiable function, it is continuous on its interval $I$ of definition. Thus there may be a difference between the graph of the *function* $\phi$ and the graph of the *solution* $\phi$. Put another way, the domain of the function $\phi$ need not be the same as the interval $I$ of definition (or domain) of the solution $\phi$. Example 6 illustrates the difference.

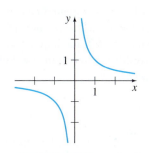

(a) function $y = 1/x$, $x \neq 0$

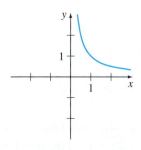

(b) solution $y = 1/x$, $(0, \infty)$

FIGURE 1.1.1 In Example 6 the function $y = 1/x$ is not the same as the solution $y = 1/x$

EXAMPLE 6    Function versus Solution

(a) The domain of $y = 1/x$, considered simply as a *function*, is the set of all real numbers $x$ except 0. When we graph $y = 1/x$, we plot points in the $xy$-plane corresponding to a judicious sampling of numbers taken from its domain. The rational function $y = 1/x$ is discontinuous at 0, and its graph, in a neighborhood of the origin, is given in Figure 1.1.1(a). The function $y = 1/x$ is not differentiable at $x = 0$, since the $y$-axis (whose equation is $x = 0$) is a vertical asymptote of the graph.

(b) Now $y = 1/x$ is also a solution of the linear first-order differential equation $xy' + y = 0$. (Verify.) But when we say that $y = 1/x$ is a *solution* of this DE, we mean that it is a function defined on an interval $I$ on which it is differentiable and satisfies the equation. In other words, $y = 1/x$ is a solution of the DE on *any* interval that does not contain 0, such as $(-3, -1)$, $\left(\frac{1}{2}, 10\right)$, $(-\infty, 0)$, or $(0, \infty)$. Because the solution curves defined by $y = 1/x$ for $-3 < x < -1$ and $\frac{1}{2} < x < 10$ are simply segments, or pieces, of the solution curves defined by $y = 1/x$ for $-\infty < x < 0$ and

$0 < x < \infty$, respectively, it makes sense to take the interval $I$ to be as large as possible. Thus we take $I$ to be either $(-\infty, 0)$ or $(0, \infty)$. The solution curve on $(0, \infty)$ is shown in Figure 1.1.1(b). ∎

**EXPLICIT AND IMPLICIT SOLUTIONS**    You should be familiar with the terms *explicit functions* and *implicit functions* from your study of calculus. A solution in which the dependent variable is expressed solely in terms of the independent variable and constants is said to be an **explicit solution.** For our purposes, let us think of an explicit solution as an explicit formula $y = \phi(x)$ that we can manipulate, evaluate, and differentiate using the standard rules. We have just seen in the last two examples that $y = \frac{1}{16}x^4$, $y = xe^x$, and $y = 1/x$ are, in turn, explicit solutions of $dy/dx = xy^{1/2}$, $y'' - 2y' + y = 0$, and $xy' + y = 0$. Moreover, the trivial solution $y = 0$ is an explicit solution of all three equations. When we get down to the business of actually solving some ordinary differential equations, you will see that methods of solution do not always lead directly to an explicit solution $y = \phi(x)$. This is particularly true when we attempt to solve nonlinear first-order differential equations. Often we have to be content with a relation or expression $G(x, y) = 0$ that defines a solution $\phi$ implicitly.

---

> **DEFINITION 1.1.3**    Implicit Solution of an ODE
>
> A relation $G(x, y) = 0$ is said to be an **implicit solution** of an ordinary differential equation (4) on an interval $I$, provided that there exists at least one function $\phi$ that satisfies the relation as well as the differential equation on $I$.

---

It is beyond the scope of this course to investigate the conditions under which a relation $G(x, y) = 0$ defines a differentiable function $\phi$. So we shall assume that if the formal implementation of a method of solution leads to a relation $G(x, y) = 0$, then there exists at least one function $\phi$ that satisfies both the relation (that is, $G(x, \phi(x)) = 0$) and the differential equation on an interval $I$. If the implicit solution $G(x, y) = 0$ is fairly simple, we may be able to solve for $y$ in terms of $x$ and obtain one or more explicit solutions. See (*iv*) in the *Remarks*.

**EXAMPLE 7**    Verification of an Implicit Solution

The relation $x^2 + y^2 = 25$ is an implicit solution of the differential equation

$$\frac{dy}{dx} = -\frac{x}{y} \tag{8}$$

on the open interval $(-5, 5)$. By implicit differentiation we obtain

$$\frac{d}{dx}x^2 + \frac{d}{dx}y^2 = \frac{d}{dx}25 \qquad \text{or} \qquad 2x + 2y\frac{dy}{dx} = 0. \tag{9}$$

Solving the last equation in (9) for the symbol $dy/dx$ gives (8). Moreover, solving $x^2 + y^2 = 25$ for $y$ in terms of $x$ yields $y = \pm\sqrt{25 - x^2}$. The two functions $y = \phi_1(x) = \sqrt{25 - x^2}$ and $y = \phi_2(x) = -\sqrt{25 - x^2}$ satisfy the relation (that is, $x^2 + \phi_1^2 = 25$ and $x^2 + \phi_2^2 = 25$) and are explicit solutions defined on the interval $(-5, 5)$. The solution curves given in Figures 1.1.2(b) and 1.1.2(c) are segments of the graph of the implicit solution in Figure 1.1.2(a).

**(a)** implicit solution

$x^2 + y^2 = 25$

**(b)** explicit solution

$y_1 = \sqrt{25 - x^2}, -5 < x < 5$

**(c)** explicit solution

$y_2 = -\sqrt{25 - x^2}, -5 < x < 5$

**FIGURE 1.1.2** An implicit solution and two explicit solutions of (8) in Example 7

Because the distinction between an explicit solution and an implicit solution should be intuitively clear, we will not belabor the issue by always saying, "Here is an explicit (implicit) solution."

**FAMILIES OF SOLUTIONS** The study of differential equations is similar to that of integral calculus. When evaluating an antiderivative or indefinite integral in calculus, we use a single constant $c$ of integration. Analogously, we shall see in Chapter 2 that when solving a first-order differential equation $F(x, y, y') = 0$ we *usually* obtain a solution containing a single constant or parameter $c$. A solution of $F(x, y, y') = 0$ containing a constant $c$ is a set of solutions $G(x, y, c) = 0$ called a **one-parameter family of solutions.** When solving an $n$th-order differential equation $F(x, y, y', \ldots, y^{(n)}) = 0$ we seek an **$n$-parameter family of solutions** $G(x, y, c_1, c_2, \ldots, c_n) = 0$. This means that a single differential equation can possess an infinite number of solutions corresponding to an unlimited number of choices for the parameter(s). A solution of a differential equation that is free of parameters is called a **particular solution.**

The parameters in a family of solutions such as $G(x, y, c_1, c_2, \ldots, c_n) = 0$ are *arbitrary* up to a point. For example, proceeding as in (9) a relation $x^2 + y^2 = c$ formally satisfies (8) for any constant $c$. However, it is understood that the relation should always make sense in the real number system; thus, if $c = -25$ we cannot say that $x^2 + y^2 = -25$ is an implicit solution of the differential equation.

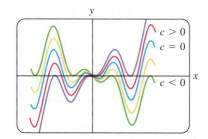

**FIGURE 1.1.3** Some solutions of DE in part (a) of Example 8

**EXAMPLE 8** Particular Solutions

**(a)** For all real values of $c$, the one-parameter family $y = cx - x \cos x$ is an explicit solution of the linear first-order equation

$$xy' - y = x^2 \sin x$$

on the interval $(-\infty, \infty)$. (Verify.) Figure 1.1.3 shows the graphs of some particular solutions in this family for various choices of $c$. The solution $y = -x \cos x$, the blue graph in the figure, is a particular solution corresponding to $c = 0$.

**(b)** The two-parameter family $y = c_1 e^x + c_2 x e^x$ is an explicit solution of the linear second-order equation

$$y'' - 2y' + y = 0$$

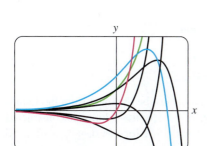

**FIGURE 1.1.4** Some solutions of DE in part (b) of Example 8

in part (b) of Example 5. (Verify.) In Figure 1.1.4 we have shown seven of the "double infinity" of solutions in the family. The solution curves in red, green, and blue are the graphs of the particular solutions $y = 5xe^x$ ($c_1 = 0, c_2 = 5$), $y = 3e^x$ ($c_1 = 3, c_2 = 0$), and $y = 5e^x - 2xe^x$ ($c_1 = 5, c_2 = 2$), respectively.

Sometimes a differential equation possesses a solution that is not a member of a family of solutions of the equation—that is, a solution that cannot be obtained by specializing *any* of the parameters in the family of solutions. Such an extra solution is called a **singular solution.** For example, we have seen that $y = \frac{1}{16}x^4$ and $y = 0$ are solutions of the differential equation $dy/dx = xy^{1/2}$ on $(-\infty, \infty)$. In Section 2.2 we shall demonstrate, by actually solving it, that the differential equation $dy/dx = xy^{1/2}$ possesses the one-parameter family of solutions $y = \left(\frac{1}{4}x^2 + c\right)^2, c \geq 0$. When $c = 0$, the resulting particular solution is $y = \frac{1}{16}x^4$. But notice that the trivial solution $y = 0$ is a singular solution since it is not a member of the family $y = \left(\frac{1}{4}x^2 + c\right)^2$; there is no way of assigning a value to the constant $c$ to obtain $y = 0$.

In all the preceding examples we used $x$ and $y$ to denote the independent and dependent variables, respectively. But you should become accustomed to seeing and working with other symbols to denote these variables. For example, we could denote the independent variable by $t$ and the dependent variable by $x$.

### EXAMPLE 9    Using Different Symbols

The functions $x = c_1 \cos 4t$ and $x = c_2 \sin 4t$, where $c_1$ and $c_2$ are arbitrary constants or parameters, are both solutions of the linear differential equation

$$x'' + 16x = 0.$$

For $x = c_1 \cos 4t$ the first two derivatives with respect to $t$ are $x' = -4c_1 \sin 4t$ and $x'' = -16c_1 \cos 4t$. Substituting $x''$ and $x$ then gives

$$x'' + 16x = -16c_1 \cos 4t + 16(c_1 \cos 4t) = 0.$$

In like manner, for $x = c_2 \sin 4t$ we have $x'' = -16c_2 \sin 4t$, and so

$$x'' + 16x = -16c_2 \sin 4t + 16(c_2 \sin 4t) = 0.$$

Finally, it is straightforward to verify that the linear combination of solutions, or the two-parameter family $x = c_1 \cos 4t + c_2 \sin 4t$, is also a solution of the differential equation. ■

The next example shows that a solution of a differential equation can be a piecewise-defined function.

### EXAMPLE 10    Piecewise-Defined Solution

The one-parameter family of quartic monomial functions $y = cx^4$ is an explicit solution of the linear first-order equation

$$xy' - 4y = 0$$

on the interval $(-\infty, \infty)$. (Verify.) The blue and red solution curves shown in Figure 1.1.5(a) are the graphs of $y = x^4$ and $y = -x^4$ and correspond to the choices $c = 1$ and $c = -1$, respectively.

The piecewise-defined differentiable function

$$y = \begin{cases} -x^4, & x < 0 \\ x^4, & x > 0 \end{cases}$$

is also a solution of the differential equation but cannot be obtained from the family $y = cx^4$ by a single choice of $c$. As seen in Figure 1.1.5(b) the solution is constructed from the family by choosing $c = -1$ for $x < 0$ and $c = 1$ for $x \geq 0$. ■

**(a)** two explicit solutions

**(b)** piecewise-defined solution

**FIGURE 1.1.5** Some solutions of DE in Example 10

**SYSTEMS OF DIFFERENTIAL EQUATIONS**    Up to this point we have been discussing single differential equations containing one unknown function. But often in theory, as well as in many applications, we must deal with systems of differential equations. A **system of ordinary differential equations** is two or more equations involving the derivatives of two or more unknown functions of a single independent variable. For example, if $x$ and $y$ denote dependent variables and $t$ denotes the independent variable, then a system of two first-order differential equations is given by

$$\frac{dx}{dt} = f(t, x, y)$$

$$\frac{dy}{dt} = g(t, x, y). \tag{10}$$

A **solution** of a system such as (10) is a pair of differentiable functions $x = \phi_1(t)$, $y = \phi_2(t)$, defined on a common interval $I$, that satisfy each equation of the system on this interval.

## REMARKS

(***i***) It might not be apparent whether a first-order ODE written in differential form $M(x, y)\,dx + N(x, y)\,dy = 0$ is linear or nonlinear because there is nothing in this form that tells us which symbol denotes the dependent variable. See Problems 9 and 10 in Exercises 1.1.

(***ii***) We will see in the chapters that follow that a solution of a differential equation may involve an **integral-defined function.** One way of defining a function $F$ of a single variable $x$ by means of a definite integral is:

$$F(x) = \int_a^x g(t)\,dt. \tag{11}$$

If the integrand $g$ in (11) is continuous on an interval $[a, b]$ and $a \le x \le b$, then the derivative form of the Fundamental Theorem of Calculus states that $F$ is differentiable on $(a, b)$ and

$$F'(x) = \frac{d}{dx} \int_a^x g(t)\,dt = g(x) \tag{12}$$

The integral in (11) is often **nonelementary,** that is, an integral of a function $g$ that does not have an elementary-function antiderivative. Elementary functions include the familiar functions studied in a typical precalculus course:

*constant, polynomial, rational, exponential, logarithmic, trigonometric, and inverse trigonometric functions,*

as well as rational powers of these functions; finite combinations of these functions using addition, subtraction, multiplication, division; and function compositions. For example, even though $e^{-t^2}, \sqrt{1 + t^3}$, and $\cos t^2$ are elementary functions, the integrals $\int e^{-t^2}\,dt, \int \sqrt{1 + t^3}\,dt$, and $\int \cos t^2\,dt$ are nonelementary. See Problems 25–28 in Exercises 1.1. Also see Appendix A.

(***iii***) Although the concept of a solution of a differential equation has been emphasized in this section, you should be aware that a DE does not necessarily have to possess a solution. See Problem 43 in Exercises 1.1. The question of whether a solution exists will be touched on in the next section.

*(continued)*

(*iv*) A few last words about implicit solutions of differential equations are in order. In Example 7 we were able to solve the relation $x^2 + y^2 = 25$ for $y$ in terms of $x$ to get two explicit solutions, $\phi_1(x) = \sqrt{25 - x^2}$ and $\phi_2(x) = -\sqrt{25 - x^2}$, of the differential equation (8). But don't read too much into this one example. Unless it is easy or important or you are instructed to, there is usually no need to try to solve an implicit solution $G(x, y) = 0$ for $y$ explicitly in terms of $x$. Also do not misinterpret the second sentence following Definition 1.1.3. An implicit solution $G(x, y) = 0$ can define a perfectly good differentiable function $\phi$ that is a solution of a DE, yet we might not be able to solve $G(x, y) = 0$ using analytical methods such as algebra. The solution curve of $\phi$ may be a segment or piece of the graph of $G(x, y) = 0$. See Problems 49 and 50 in Exercises 1.1. Also, read the discussion following Example 4 in Section 2.2.

(*v*) It might not seem like a big deal to assume that $F(x, y, y', \ldots, y^{(n)}) = 0$ can be solved for $y^{(n)}$, but one should be a little bit careful here. There are exceptions, and there certainly are some problems connected with this assumption. See Problems 56 and 57 in Exercises 1.1.

(*vi*) If *every* solution of an $n$th-order ODE $F(x, y, y', \ldots, y^{(n)}) = 0$ on an interval $I$ can be obtained from an $n$-parameter family $G(x, y, c_1, c_2, \ldots, c_n) = 0$ by appropriate choices of the parameters $c_i$, $i = 1, 2, \ldots, n$, we then say that the family is the **general solution** of the DE. In solving linear ODEs, we shall impose relatively simple restrictions on the coefficients of the equation; with these restrictions one can be assured that not only does a solution exist on an interval but also that a family of solutions yields all possible solutions. Nonlinear ODEs, with the exception of some first-order equations, are usually difficult or impossible to solve in terms of elementary functions. Furthermore, if we happen to obtain a family of solutions for a nonlinear equation, it is not obvious whether this family contains all solutions. On a practical level, then, the designation "general solution" is applied only to linear ODEs. Don't be concerned about this concept at this point, but store the words "general solution" in the back of your mind—we will come back to this notion in Section 2.3 and again in Chapter 4.

## EXERCISES 1.1

*Answers to selected odd-numbered problems begin on page ANS-1.*

In Problems 1–8 state the order of the given ordinary differential equation. Determine whether the equation is linear or nonlinear by matching it with (6).

**1.** $(1 - x)y'' - 4xy' + 5y = \cos x$

**2.** $x \dfrac{d^3y}{dx^3} - \left(\dfrac{dy}{dx}\right)^4 + y = 0$

**3.** $t^5 y^{(4)} - t^3 y'' + 6y = 0$

**4.** $\dfrac{d^2u}{dr^2} + \dfrac{du}{dr} + u = \cos(r + u)$

**5.** $\dfrac{d^2y}{dx^2} = \sqrt{1 + \left(\dfrac{dy}{dx}\right)^2}$

**6.** $\dfrac{d^2R}{dt^2} = -\dfrac{k}{R^2}$

**7.** $(\sin \theta)y''' - (\cos \theta)y' = 2$

**8.** $\ddot{x} - \left(1 - \dfrac{\dot{x}^2}{3}\right)\dot{x} + x = 0$

In Problems 9 and 10 determine whether the given first-order differential equation is linear in the indicated dependent variable by matching it with the first differential equation given in (7).

**9.** $(y^2 - 1)\,dx + x\,dy = 0$; in $y$; in $x$

**10.** $u\,dv + (v + uv - ue^u)\,du = 0$; in $v$; in $u$

In Problems 11–14 verify that the indicated function is an explicit solution of the given differential equation. Assume an appropriate interval $I$ of definition for each solution.

**11.** $2y' + y = 0$;   $y = e^{-x/2}$

**12.** $\dfrac{dy}{dt} + 20y = 24;\quad y = \dfrac{6}{5} - \dfrac{6}{5}e^{-20t}$

**13.** $y'' - 6y' + 13y = 0;\quad y = e^{3x}\cos 2x$

**14.** $y'' + y = \tan x;\quad y = -(\cos x)\ln(\sec x + \tan x)$

In Problems 15–18 verify that the indicated function $y = \phi(x)$ is an explicit solution of the given first-order differential equation. Proceed as in Example 6, by considering $\phi$ simply as a *function* and give its domain. Then by considering $\phi$ as a *solution* of the differential equation, give at least one interval $I$ of definition.

**15.** $(y - x)y' = y - x + 8;\quad y = x + 4\sqrt{x + 2}$

**16.** $y' = 25 + y^2;\quad y = 5\tan 5x$

**17.** $y' = 2xy^2;\quad y = 1/(4 - x^2)$

**18.** $2y' = y^3\cos x;\quad y = (1 - \sin x)^{-1/2}$

In Problems 19 and 20 verify that the indicated expression is an implicit solution of the given first-order differential equation. Find at least one explicit solution $y = \phi(x)$ in each case. Use a graphing utility to obtain the graph of an explicit solution. Give an interval $I$ of definition of each solution $\phi$.

**19.** $\dfrac{dX}{dt} = (X - 1)(1 - 2X);\quad \ln\left(\dfrac{2X - 1}{X - 1}\right) = t$

**20.** $2xy\,dx + (x^2 - y)\,dy = 0;\quad -2x^2y + y^2 = 1$

In Problems 21–24 verify that the indicated family of functions is a solution of the given differential equation. Assume an appropriate interval $I$ of definition for each solution.

**21.** $\dfrac{dP}{dt} = P(1 - P);\quad P = \dfrac{c_1 e^t}{1 + c_1 e^t}$

**22.** $\dfrac{dy}{dx} + 4xy = 8x^3;\quad y = 2x^2 - 1 + c_1 e^{-2x^2}$

**23.** $\dfrac{d^2y}{dx^2} - 4\dfrac{dy}{dx} + 4y = 0;\quad y = c_1 e^{2x} + c_2 x e^{2x}$

**24.** $x^3\dfrac{d^3y}{dx^3} + 2x^2\dfrac{d^2y}{dx^2} - x\dfrac{dy}{dx} + y = 12x^2;$

$y = c_1 x^{-1} + c_2 x + c_3 x\ln x + 4x^2$

In Problems 25–28 use (12) to verify that the indicated function is a solution of the given differential equation. Assume an appropriate interval $I$ of definition of each solution.

**25.** $x\dfrac{dy}{dx} - 3xy = 1;\quad y = e^{3x}\displaystyle\int_1^x \dfrac{e^{-3t}}{t}\,dt$

**26.** $2x\dfrac{dy}{dx} - y = 2x\cos x;\quad y = \sqrt{x}\displaystyle\int_4^x \dfrac{\cos t}{\sqrt{t}}\,dt$

**27.** $x^2\dfrac{dy}{dx} + xy = 10\sin x;\quad y = \dfrac{5}{x} + \dfrac{10}{x}\displaystyle\int_1^x \dfrac{\sin t}{t}\,dt$

**28.** $\dfrac{dy}{dx} + 2xy = 1;\quad y = e^{-x^2} + e^{-x^2}\displaystyle\int_0^x e^{t^2}\,dt$

**29.** Verify that the piecewise-defined function

$$y = \begin{cases} -x^2, & x < 0 \\ x^2, & x \ge 0 \end{cases}$$

is a solution of the differential equation $xy' - 2y = 0$ on $(-\infty, \infty)$.

**30.** In Example 7 we saw that $y = \phi_1(x) = \sqrt{25 - x^2}$ and $y = \phi_2(x) = -\sqrt{25 - x^2}$ are solutions of $dy/dx = -x/y$ on the interval $(-5, 5)$. Explain why the piecewise-defined function

$$y = \begin{cases} \sqrt{25 - x^2}, & -5 < x < 0 \\ -\sqrt{25 - x^2}, & 0 \le x < 5 \end{cases}$$

is *not* a solution of the differential equation on the interval $(-5, 5)$.

In Problems 31–34 find values of $m$ so that the function $y = e^{mx}$ is a solution of the given differential equation.

**31.** $y' + 2y = 0$  **32.** $5y' = 2y$

**33.** $y'' - 5y' + 6y = 0$  **34.** $2y'' + 7y' - 4y = 0$

In Problems 35 and 36 find values of $m$ so that the function $y = x^m$ is a solution of the given differential equation.

**35.** $xy'' + 2y' = 0$

**36.** $x^2y'' - 7xy' + 15y = 0$

In Problems 37–40 use the concept that $y = c$, $-\infty < x < \infty$, is a constant function if and only if $y' = 0$ to determine whether the given differential equation possesses constant solutions.

**37.** $3xy' + 5y = 10$  **38.** $y' = y^2 + 2y - 3$

**39.** $(y - 1)y' = 1$  **40.** $y'' + 4y' + 6y = 10$

In Problems 41 and 42 verify that the indicated pair of functions is a solution of the given system of differential equations on the interval $(-\infty, \infty)$.

**41.** $\dfrac{dx}{dt} = x + 3y$

$\dfrac{dy}{dt} = 5x + 3y;$

$x = e^{-2t} + 3e^{6t},$

$y = -e^{-2t} + 5e^{6t}$

**42.** $\dfrac{d^2x}{dt^2} = 4y + e^t$

$\dfrac{d^2y}{dt^2} = 4x - e^t;$

$x = \cos 2t + \sin 2t + \tfrac{1}{5}e^t,$

$y = -\cos 2t - \sin 2t - \tfrac{1}{5}e^t$

## Discussion Problems

**43.** Make up a differential equation that does not possess any real solutions.

**44.** Make up a differential equation that you feel confident possesses only the trivial solution $y = 0$. Explain your reasoning.

**45.** What function do you know from calculus is such that its first derivative is itself? Its first derivative is a constant multiple $k$ of itself? Write each answer in the form of a first-order differential equation with a solution.

**46.** What function (or functions) do you know from calculus is such that its second derivative is itself? Its second derivative is the negative of itself? Write each answer in the form of a second-order differential equation with a solution.

**47.** The function $y = \sin x$ is an explicit solution of the first-order differential equation $\dfrac{dy}{dx} = \sqrt{1 - y^2}$. Find an interval $I$ of definition. [*Hint*: $I$ is *not* the interval $(-\infty, \infty)$.]

**48.** Discuss why it makes intuitive sense to presume that the linear differential equation $y'' + 2y' + 4y = 5 \sin t$ has a solution of the form $y = A \sin t + B \cos t$, where $A$ and $B$ are constants. Then find specific constants $A$ and $B$ so that $y = A \sin t + B \cos t$ is a particular solution of the DE.

In Problems 49 and 50 the given figure represents the graph of an implicit solution $G(x, y) = 0$ of a differential equation $dy/dx = f(x, y)$. In each case the relation $G(x, y) = 0$ implicitly defines several solutions of the DE. Carefully reproduce each figure on a piece of paper. Use different colored pencils to mark off segments, or pieces, on each graph that correspond to graphs of solutions. Keep in mind that a solution $\phi$ must be a function and differentiable. Use the solution curve to estimate an interval $I$ of definition of each solution $\phi$.

**49.**

**FIGURE 1.1.6** Graph for Problem 49

**50.**

**FIGURE 1.1.7** Graph for Problem 50

**51.** The graphs of members of the one-parameter family $x^3 + y^3 = 3cxy$ are called **folia of Descartes**. Verify that this family is an implicit solution of the first-order differential equation

$$\frac{dy}{dx} = \frac{y(y^3 - 2x^3)}{x(2y^3 - x^3)}.$$

**52.** The graph in Figure 1.1.7 is the member of the family of folia in Problem 51 corresponding to $c = 1$. Discuss: How can the DE in Problem 51 help in finding points on the graph of $x^3 + y^3 = 3xy$ where the tangent line is vertical? How does knowing where a tangent line is vertical help in determining an interval $I$ of definition of a solution $\phi$ of the DE? Carry out your ideas and compare with your estimates of the intervals in Problem 50.

**53.** In Example 7 the largest interval $I$ over which the explicit solutions $y = \phi_1(x)$ and $y = \phi_2(x)$ are defined is the open interval $(-5, 5)$. Why can't the interval $I$ of definition be the closed interval $[-5, 5]$?

**54.** In Problem 21 a one-parameter family of solutions of the DE $P' = P(1 - P)$ is given. Does any solution curve pass through the point $(0, 3)$? Through the point $(0, 1)$?

**55.** Discuss, and illustrate with examples, how to solve differential equations of the forms $dy/dx = f(x)$ and $d^2y/dx^2 = f(x)$.

**56.** The differential equation $x(y')^2 - 4y' - 12x^3 = 0$ has the form given in (4). Determine whether the equation can be put into the normal form $dy/dx = f(x, y)$.

**57.** The normal form (5) of an $n$th-order differential equation is equivalent to (4) whenever both forms have exactly the same solutions. Make up a first-order differential equation for which $F(x, y, y') = 0$ is not equivalent to the normal form $dy/dx = f(x, y)$.

**58.** Find a linear second-order differential equation $F(x, y, y', y'') = 0$ for which $y = c_1 x + c_2 x^2$ is a two-parameter family of solutions. Make sure that your equation is free of the arbitrary parameters $c_1$ and $c_2$.

Qualitative information about a solution $y = \phi(x)$ of a differential equation can often be obtained from the equation itself. Before working Problems 59–62, recall the geometric significance of the derivatives $dy/dx$ and $d^2y/dx^2$.

**59.** Consider the differential equation $dy/dx = e^{-x^2}$.

    **(a)** Explain why a solution of the DE must be an increasing function on any interval of the $x$-axis.

    **(b)** What are $\lim\limits_{x \to -\infty} dy/dx$ and $\lim\limits_{x \to \infty} dy/dx$? What does this suggest about a solution curve as $x \to \pm\infty$?

    **(c)** Determine an interval over which a solution curve is concave down and an interval over which the curve is concave up.

    **(d)** Sketch the graph of a solution $y = \phi(x)$ of the differential equation whose shape is suggested by parts (a)–(c).

**60.** Consider the differential equation $dy/dx = 5 - y$.

    **(a)** Either by inspection or by the method suggested in Problems 37–40, find a constant solution of the DE.

    **(b)** Using only the differential equation, find intervals on the $y$-axis on which a nonconstant solution $y = \phi(x)$ is increasing. Find intervals on the $y$-axis on which $y = \phi(x)$ is decreasing.

**61.** Consider the differential equation $dy/dx = y(a - by)$, where $a$ and $b$ are positive constants.

    **(a)** Either by inspection or by the method suggested in Problems 37–40, find two constant solutions of the DE.

    **(b)** Using only the differential equation, find intervals on the $y$-axis on which a nonconstant solution $y = \phi(x)$ is increasing. Find intervals on which $y = \phi(x)$ is decreasing.

    **(c)** Using only the differential equation, explain why $y = a/2b$ is the $y$-coordinate of a point of inflection of the graph of a nonconstant solution $y = \phi(x)$.

    **(d)** On the same coordinate axes, sketch the graphs of the two constant solutions found in part (a). These constant solutions partition the $xy$-plane into three regions. In each region, sketch the graph of a nonconstant solution $y = \phi(x)$ whose shape is suggested by the results in parts (b) and (c).

**62.** Consider the differential equation $y' = y^2 + 4$.

    **(a)** Explain why there exist no constant solutions of the DE.

    **(b)** Describe the graph of a solution $y = \phi(x)$. For example, can a solution curve have any relative extrema?

(c) Explain why $y = 0$ is the $y$-coordinate of a point of inflection of a solution curve.

(d) Sketch the graph of a solution $y = \phi(x)$ of the differential equation whose shape is suggested by parts (a)–(c).

### Computer Lab Assignments

In Problems 63 and 64 use a CAS to compute all derivatives and to carry out the simplifications needed to verify that the indicated function is a particular solution of the given differential equation.

63. $y^{(4)} - 20y''' + 158y'' - 580y' + 841y = 0$;

$y = xe^{5x} \cos 2x$

64. $x^3 y''' + 2x^2 y'' + 20xy' - 78y = 0$;

$y = 20 \dfrac{\cos(5 \ln x)}{x} - 3 \dfrac{\sin(5 \ln x)}{x}$

---

## 1.2 Initial-Value Problems

**INTRODUCTION**  We are often interested in problems in which we seek a solution $y(x)$ of a differential equation so that $y(x)$ also satisfies certain prescribed side conditions, that is, conditions that are imposed on the unknown function $y(x)$ and its derivatives at a number $x_0$. On some interval $I$ containing $x_0$ the problem of solving an $n$th-order differential equation subject to $n$ side conditions specified at $x_0$:

$$\textbf{Solve:} \quad \frac{d^n y}{dx^n} = f(x, y, y', \ldots, y^{(n-1)})$$

$$\textbf{Subject to:} \quad y(x_0) = y_0, y'(x_0) = y_1, \ldots, y^{(n-1)}(x_0) = y_{n-1}, \tag{1}$$

where $y_0, y_1, \ldots, y_{n-1}$ are arbitrary constants, is called an **$n$th-order initial-value problem (IVP)**. The values of $y(x)$ and its first $n-1$ derivatives at $x_0$, $y(x_0) = y_0$, $y'(x_0) = y_1, \ldots, y^{(n-1)}(x_0) = y_{n-1}$ are called **initial conditions (IC)**.

Solving an $n$th-order initial-value problem such as (1) frequently entails first finding an $n$-parameter family of solutions of the differential equation and then using the initial conditions at $x_0$ to determine the $n$ constants in this family. The resulting particular solution is defined on some interval $I$ containing the number $x_0$.

**GEOMETRIC INTERPRETATION**  The cases $n = 1$ and $n = 2$ in (1),

$$\textbf{Solve:} \quad \frac{dy}{dx} = f(x, y)$$

$$\textbf{Subject to:} \quad y(x_0) = y_0 \tag{2}$$

and

$$\textbf{Solve:} \quad \frac{d^2 y}{dx^2} = f(x, y, y')$$

$$\textbf{Subject to:} \quad y(x_0) = y_0, \ y'(x_0) = y_1 \tag{3}$$

are examples of **first-** and **second-order** initial-value problems, respectively. These two problems are easy to interpret in geometric terms. For (2) we are seeking a solution $y(x)$ of the differential equation $y' = f(x, y)$ on an interval $I$ containing $x_0$ so that its graph passes through the specified point $(x_0, y_0)$. A solution curve is shown in blue in Figure 1.2.1. For (3) we want to find a solution $y(x)$ of the differential equation $y'' = f(x, y, y')$ on an interval $I$ containing $x_0$ so that its graph not only passes through $(x_0, y_0)$ but the slope of the curve at this point is the number $y_1$. A solution curve is shown in blue in Figure 1.2.2. The words *initial conditions* derive from physical systems where the independent variable is time $t$ and where $y(t_0) = y_0$

**FIGURE 1.2.1**  Solution curve of first-order IVP

**FIGURE 1.2.2**  Solution curve of second-order IVP

and $y'(t_0) = y_1$ represent the position and velocity, respectively, of an object at some beginning, or initial, time $t_0$.

### EXAMPLE 1    Two First-Order IVPs

**(a)** In Problem 45 in Exercises 1.1 you were asked to deduce that $y = ce^x$ is a one-parameter family of solutions of the simple first-order equation $y' = y$. All the solutions in this family are defined on the interval $(-\infty, \infty)$. If we impose an initial condition, say, $y(0) = 3$, then substituting $x = 0$, $y = 3$ in the family determines the constant $3 = ce^0 = c$. Thus $y = 3e^x$ is a solution of the IVP

$$y' = y, \quad y(0) = 3.$$

**(b)** Now if we demand that a solution curve pass through the point $(1, -2)$ rather than $(0, 3)$, then $y(1) = -2$ will yield $-2 = ce$ or $c = -2e^{-1}$. In this case $y = -2e^{x-1}$ is a solution of the IVP

$$y' = y, \quad y(1) = -2.$$

The two solution curves are shown in dark blue and dark red in Figure 1.2.3. ◼

The next example illustrates another first-order initial-value problem. In this example notice how the interval $I$ of definition of the solution $y(x)$ depends on the initial condition $y(x_0) = y_0$.

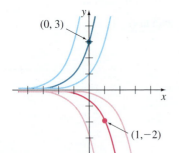

**FIGURE 1.2.3** Solution curves of two IVPs in Example 1

### EXAMPLE 2    Interval *I* of Definition of a Solution

In Problem 6 of Exercises 2.2 you will be asked to show that a one-parameter family of solutions of the first-order differential equation $y' + 2xy^2 = 0$ is $y = 1/(x^2 + c)$. If we impose the initial condition $y(0) = -1$, then substituting $x = 0$ and $y = -1$ into the family of solutions gives $-1 = 1/c$ or $c = -1$. Thus $y = 1/(x^2 - 1)$. We now emphasize the following three distinctions:

- Considered as a *function*, the domain of $y = 1/(x^2 - 1)$ is the set of real numbers $x$ for which $y(x)$ is defined; this is the set of all real numbers except $x = -1$ and $x = 1$. See Figure 1.2.4(a).
- Considered as a *solution of the differential equation* $y' + 2xy^2 = 0$, the interval $I$ of definition of $y = 1/(x^2 - 1)$ could be taken to be any interval over which $y(x)$ is defined and differentiable. As can be seen in Figure 1.2.4(a), the largest intervals on which $y = 1/(x^2 - 1)$ is a solution are $(-\infty, -1)$, $(-1, 1)$, and $(1, \infty)$.
- Considered as a *solution of the initial-value problem* $y' + 2xy^2 = 0$, $y(0) = -1$, the interval $I$ of definition of $y = 1/(x^2 - 1)$ could be taken to be any interval over which $y(x)$ is defined, differentiable, *and* contains the initial point $x = 0$; the largest interval for which this is true is $(-1, 1)$. See the red curve in Figure 1.2.4(b). ◼

See Problems 3–6 in Exercises 1.2 for a continuation of Example 2.

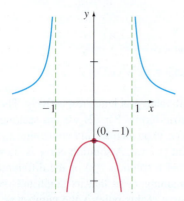

**(a)** function defined for all $x$ except $x = \pm 1$

**(b)** solution defined on interval containing $x = 0$

**FIGURE 1.2.4** Graphs of function and solution of IVP in Example 2

### EXAMPLE 3    Second-Order IVP

In Example 9 of Section 1.1 we saw that $x = c_1 \cos 4t + c_2 \sin 4t$ is a two-parameter family of solutions of $x'' + 16x = 0$. Find a solution of the initial-value problem

$$x'' + 16x = 0, \quad x\left(\frac{\pi}{2}\right) = -2, \quad x'\left(\frac{\pi}{2}\right) = 1. \tag{4}$$

**SOLUTION** We first apply $x(\pi/2) = -2$ to the given family of solutions: $c_1 \cos 2\pi + c_2 \sin 2\pi = -2$. Since $\cos 2\pi = 1$ and $\sin 2\pi = 0$, we find that $c_1 = -2$. We next apply $x'(\pi/2) = 1$ to the one-parameter family $x(t) = -2 \cos 4t + c_2 \sin 4t$. Differentiating and then setting $t = \pi/2$ and $x' = 1$ gives $8 \sin 2\pi + 4c_2 \cos 2\pi = 1$, from which we see that $c_2 = \frac{1}{4}$. Hence $x = -2 \cos 4t + \frac{1}{4} \sin 4t$ is a solution of (4). ■

**EXISTENCE AND UNIQUENESS**    Two fundamental questions arise in considering an initial-value problem:

*Does a solution of the problem exist? If a solution exists, is it unique?*

For the first-order initial-value problem (2) we ask:

**Existence** $\begin{cases} \textit{Does the differential equation } dy/dx = f(x, y) \textit{ possess solutions?} \\ \textit{Do any of the solution curves pass through the point } (x_0, y_0)? \end{cases}$

**Uniqueness** $\begin{cases} \textit{When can we be certain that there is precisely one solution curve} \\ \textit{passing through the point } (x_0, y_0)? \end{cases}$

Note that in Examples 1 and 3 the phrase "*a solution*" is used rather than "*the* solution" of the problem. The indefinite article "a" is used deliberately to suggest the possibility that other solutions may exist. At this point it has not been demonstrated that there is a single solution of each problem. The next example illustrates an initial-value problem with two solutions.

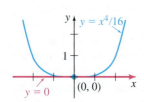

**FIGURE 1.2.5** Two solution curves of the same IVP in Example 4

<table>
<tr><td>**EXAMPLE 4**</td><td>**An IVP Can Have Several Solutions**</td></tr>
</table>

Each of the functions $y = 0$ and $y = \frac{1}{16} x^4$ satisfies the differential equation $dy/dx = xy^{1/2}$ and the initial condition $y(0) = 0$, so the initial-value problem

$$\frac{dy}{dx} = xy^{1/2}, \quad y(0) = 0$$

has at least two solutions. As illustrated in Figure 1.2.5, the graphs of both functions, shown in red and blue pass through the same point $(0, 0)$. ■

Within the safe confines of a formal course in differential equations one can be fairly confident that *most* differential equations will have solutions and that solutions of initial-value problems will *probably* be unique. Real life, however, is not so idyllic. Therefore it is desirable to know in advance of trying to solve an initial-value problem whether a solution exists and, when it does, whether it is the only solution of the problem. Since we are going to consider first-order differential equations in the next two chapters, we state here without proof a straightforward theorem that gives conditions that are sufficient to guarantee the existence and uniqueness of a solution of a first-order initial-value problem of the form given in (2). We shall wait until Chapter 4 to address the question of existence and uniqueness of a second-order initial-value problem.

**THEOREM 1.2.1    Existence of a Unique Solution**

Let $R$ be a rectangular region in the $xy$-plane defined by $a \leq x \leq b, c \leq y \leq d$ that contains the point $(x_0, y_0)$ in its interior. If $f(x, y)$ and $\partial f/\partial y$ are continuous on $R$, then there exists some interval $I_0 : (x_0 - h, x_0 + h), h > 0$, contained in $[a, b]$, and a unique function $y(x)$, defined on $I_0$, that is a solution of the initial-value problem (2).

**FIGURE 1.2.6** Rectangular region $R$

The foregoing result is one of the most popular existence and uniqueness theorems for first-order differential equations because the criteria of continuity of $f(x, y)$ and $\partial f / \partial y$ are relatively easy to check. The geometry of Theorem 1.2.1 is illustrated in Figure 1.2.6.

**EXAMPLE 5** Example 4 Revisited

We saw in Example 4 that the differential equation $dy/dx = xy^{1/2}$ possesses at least two solutions whose graphs pass through $(0, 0)$. Inspection of the functions

$$f(x, y) = xy^{1/2} \qquad \text{and} \qquad \frac{\partial f}{\partial y} = \frac{x}{2y^{1/2}}$$

shows that they are continuous in the upper half-plane defined by $y > 0$. Hence Theorem 1.2.1 enables us to conclude that through any point $(x_0, y_0)$, $y_0 > 0$ in the upper half-plane there is some interval centered at $x_0$ on which the given differential equation has a unique solution. Thus, for example, even without solving it, we know that there exists some interval centered at 2 on which the initial-value problem $dy/dx = xy^{1/2}$, $y(2) = 1$ has a unique solution. ■

In Example 1, Theorem 1.2.1 guarantees that there are no other solutions of the initial-value problems $y' = y$, $y(0) = 3$ and $y' = y$, $y(1) = -2$ other than $y = 3e^x$ and $y = -2e^{x-1}$, respectively. This follows from the fact that $f(x, y) = y$ and $\partial f / \partial y = 1$ are continuous throughout the entire $xy$-plane. It can be further shown that the interval $I$ on which each solution is defined is $(-\infty, \infty)$.

**INTERVAL OF EXISTENCE/UNIQUENESS** Suppose $y(x)$ represents a solution of the initial-value problem (2). The following three sets on the real $x$-axis may not be the same: the domain of the function $y(x)$, the interval $I$ over which the solution $y(x)$ is defined or exists, and the interval $I_0$ of existence *and* uniqueness. Example 6 of Section 1.1 illustrated the difference between the domain of a function and the interval $I$ of definition. Now suppose $(x_0, y_0)$ is a point in the interior of the rectangular region $R$ in Theorem 1.2.1. It turns out that the continuity of the function $f(x, y)$ on $R$ by itself is sufficient to guarantee the existence of at least one solution of $dy/dx = f(x, y)$, $y(x_0) = y_0$, defined on some interval $I$. The interval $I$ of definition for this initial-value problem is usually taken to be the largest interval containing $x_0$ over which the solution $y(x)$ is defined and differentiable. The interval $I$ depends on both $f(x, y)$ and the initial condition $y(x_0) = y_0$. See Problems 31–34 in Exercises 1.2. The extra condition of continuity of the first partial derivative $\partial f / \partial y$ on $R$ enables us to say that not only does a solution exist on some interval $I_0$ containing $x_0$, but it is the *only* solution satisfying $y(x_0) = y_0$. However, Theorem 1.2.1 does not give any indication of the sizes of intervals $I$ and $I_0$; *the interval $I$ of definition need not be as wide as the region $R$, and the interval $I_0$ of existence and uniqueness may not be as large as $I$.* The number $h > 0$ that defines the interval $I_0$: $(x_0 - h, x_0 + h)$ could be very small, so it is best to think that the solution $y(x)$ is *unique in a local sense*—that is, a solution defined near the point $(x_0, y_0)$. See Problem 51 in Exercises 1.2.

**REMARKS**

(*i*) The conditions in Theorem 1.2.1 are sufficient but not necessary. This means that when $f(x, y)$ and $\partial f / \partial y$ are continuous on a rectangular region $R$, it must always follow that a solution of (2) exists and is unique whenever $(x_0, y_0)$ is a point interior to $R$. However, if the conditions stated in the hypothesis of

Theorem 1.2.1 do not hold, then anything could happen: Problem (2) *may* still have a solution and this solution *may* be unique, or (2) may have several solutions, or it may have no solution at all. A rereading of Example 5 reveals that the hypotheses of Theorem 1.2.1 do not hold on the line $y = 0$ for the differential equation $dy/dx = xy^{1/2}$, so it is not surprising, as we saw in Example 4 of this section, that there are two solutions defined on a common interval $(-h, h)$ satisfying $y(0) = 0$. On the other hand, the hypotheses of Theorem 1.2.1 do not hold on the line $y = 1$ for the differential equation $dy/dx = |y - 1|$. Nevertheless it can be proved that the solution of the initial-value problem $dy/dx = |y - 1|$, $y(0) = 1$, is unique. Can you guess this solution?

(**ii**) You are encouraged to read, think about, work, and then keep in mind Problem 50 in Exercises 1.2.

(**iii**) Initial conditions are prescribed at a *single* point $x_0$. But we are also interested in solving differential equations that are subject to conditions specified on $y(x)$ or its derivative at *two* different points $x_0$ and $x_1$. Conditions such as

$$y(1) = 0, \quad y(5) = 0 \quad \text{or} \quad y(\pi/2) = 0, \quad y'(\pi) = 1$$

are called **boundary conditions (BC).** A differential equation together with boundary conditions is called a **boundary-value problem (BVP).** For example,

$$y'' + y = 0, \quad y'(0) = 0, \quad y'(\pi) = 0$$

is a boundary-value problem. See Problems 39–44 in Exercises 1.2.

When we start to solve differential equations in Chapter 2 we will solve only first-order equations and first-order initial-value problems. The mathematical description of many problems in science and engineering involve second-order IVPs or two-point BVPs. We will examine some of these problems in Chapters 4 and 5.

---

## EXERCISES 1.2

Answers to selected odd-numbered problems begin on page ANS-1.

In Problems 1 and 2, $y = 1/(1 + c_1 e^{-x})$ is a one-parameter family of solutions of the first-order DE $y' = y - y^2$. Find a solution of the first-order IVP consisting of this differential equation and the given initial condition.

**1.** $y(0) = -\frac{1}{3}$

**2.** $y(-1) = 2$

In Problems 3–6, $y = 1/(x^2 + c)$ is a one-parameter family of solutions of the first-order DE $y' + 2xy^2 = 0$. Find a solution of the first-order IVP consisting of this differential equation and the given initial condition. Give the largest interval $I$ over which the solution is defined.

**3.** $y(2) = \frac{1}{3}$

**4.** $y(-2) = \frac{1}{2}$

**5.** $y(0) = 1$

**6.** $y\left(\frac{1}{2}\right) = -4$

In Problems 7–10, $x = c_1 \cos t + c_2 \sin t$ is a two-parameter family of solutions of the second-order DE $x'' + x = 0$. Find a solution of the second-order IVP consisting of this differential equation and the given initial conditions.

**7.** $x(0) = -1, \quad x'(0) = 8$

**8.** $x(\pi/2) = 0, \quad x'(\pi/2) = 1$

**9.** $x(\pi/6) = \frac{1}{2}, \quad x'(\pi/6) = 0$

**10.** $x(\pi/4) = \sqrt{2}, \quad x'(\pi/4) = 2\sqrt{2}$

In Problems 11–14, $y = c_1 e^x + c_2 e^{-x}$ is a two-parameter family of solutions of the second-order DE $y'' - y = 0$. Find a solution of the second-order IVP consisting of this differential equation and the given initial conditions.

**11.** $y(0) = 1, \quad y'(0) = 2$

**12.** $y(1) = 0, \quad y'(1) = e$

**13.** $y(-1) = 5, \quad y'(-1) = -5$

**14.** $y(0) = 0, \quad y'(0) = 0$

In Problems 15 and 16 determine by inspection at least two solutions of the given first-order IVP.

**15.** $y' = 3y^{2/3}$,   $y(0) = 0$

**16.** $xy' = 2y$,   $y(0) = 0$

In Problems 17–24 determine a region of the $xy$-plane for which the given differential equation would have a unique solution whose graph passes through a point $(x_0, y_0)$ in the region.

**17.** $\dfrac{dy}{dx} = y^{2/3}$

**18.** $\dfrac{dy}{dx} = \sqrt{xy}$

**19.** $x\dfrac{dy}{dx} = y$

**20.** $\dfrac{dy}{dx} - y = x$

**21.** $(4 - y^2)y' = x^2$

**22.** $(1 + y^3)y' = x^2$

**23.** $(x^2 + y^2)y' = y^2$

**24.** $(y - x)y' = y + x$

In Problems 25–28 determine whether Theorem 1.2.1 guarantees that the differential equation $y' = \sqrt{y^2 - 9}$ possesses a unique solution through the given point.

**25.** $(1, 4)$

**26.** $(5, 3)$

**27.** $(2, -3)$

**28.** $(-1, 1)$

**29. (a)** By inspection find a one-parameter family of solutions of the differential equation $xy' = y$. Verify that each member of the family is a solution of the initial-value problem $xy' = y$, $y(0) = 0$.

**(b)** Explain part (a) by determining a region $R$ in the $xy$-plane for which the differential equation $xy' = y$ would have a unique solution through a point $(x_0, y_0)$ in $R$.

**(c)** Verify that the piecewise-defined function

$$y = \begin{cases} 0, & x < 0 \\ x, & x \geq 0 \end{cases}$$

satisfies the condition $y(0) = 0$. Determine whether this function is also a solution of the initial-value problem in part (a).

**30. (a)** Verify that $y = \tan(x + c)$ is a one-parameter family of solutions of the differential equation $y' = 1 + y^2$.

**(b)** Since $f(x, y) = 1 + y^2$ and $\partial f/\partial y = 2y$ are continuous everywhere, the region $R$ in Theorem 1.2.1 can be taken to be the entire $xy$-plane. Use the family of solutions in part (a) to find an explicit solution of the first-order initial-value problem $y' = 1 + y^2$, $y(0) = 0$. Even though $x_0 = 0$ is in the interval $(-2, 2)$, explain why the solution is not defined on this interval.

**(c)** Determine the largest interval $I$ of definition for the solution of the initial-value problem in part (b).

**31. (a)** Verify that $y = -1/(x + c)$ is a one-parameter family of solutions of the differential equation $y' = y^2$.

**(b)** Since $f(x, y) = y^2$ and $\partial f/\partial y = 2y$ are continuous everywhere, the region $R$ in Theorem 1.2.1 can be taken to be the entire $xy$-plane. Find a solution from the family in part (a) that satisfies $y(0) = 1$. Then find a solution from the family in part (a) that satisfies $y(0) = -1$. Determine

the largest interval $I$ of definition for the solution of each initial-value problem.

**(c)** Determine the largest interval $I$ of definition for the solution of the first-order initial-value problem $y' = y^2$, $y(0) = 0$. [*Hint*: The solution is not a member of the family of solutions in part (a).]

**32. (a)** Show that a solution from the family in part (a) of Problem 31 that satisfies $y' = y^2$, $y(1) = 1$, is $y = 1/(2 - x)$.

**(b)** Then show that a solution from the family in part (a) of Problem 31 that satisfies $y' = y^2$, $y(3) = -1$, is $y = 1/(2 - x)$.

**(c)** Are the solutions in parts (a) and (b) the same?

**33. (a)** Verify that $3x^2 - y^2 = c$ is a one-parameter family of solutions of the differential equation $y \, dy/dx = 3x$.

**(b)** By hand, sketch the graph of the implicit solution $3x^2 - y^2 = 3$. Find all explicit solutions $y = \phi(x)$ of the DE in part (a) defined by this relation. Give the interval $I$ of definition of each explicit solution.

**(c)** The point $(-2, 3)$ is on the graph of $3x^2 - y^2 = 3$, but which of the explicit solutions in part (b) satisfies $y(-2) = 3$?

**34. (a)** Use the family of solutions in part (a) of Problem 33 to find an implicit solution of the initial-value problem $y \, dy/dx = 3x$, $y(2) = -4$. Then, by hand, sketch the graph of the explicit solution of this problem and give its interval $I$ of definition.

**(b)** Are there any explicit solutions of $y \, dy/dx = 3x$ that pass through the origin?

In Problems 35–38 the graph of a member of a family of solutions of a second-order differential equation $d^2y/dx^2 = f(x, y, y')$ is given. Match the solution curve with at least one pair of the following initial conditions.

**(a)** $y(1) = 1$,   $y'(1) = -2$

**(b)** $y(-1) = 0$,   $y'(-1) = -4$

**(c)** $y(1) = 1$,   $y'(1) = 2$

**(d)** $y(0) = -1$,   $y'(0) = 2$

**(e)** $y(0) = -1$,   $y'(0) = 0$

**(f)** $y(0) = -4$,   $y'(0) = -2$

**35.**

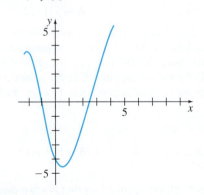

**FIGURE 1.2.7** Graph for Problem 35

**36.**

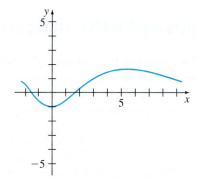

**FIGURE 1.2.8** Graph for Problem 36

**37.**

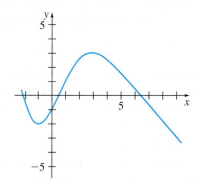

**FIGURE 1.2.9** Graph for Problem 37

**38.**

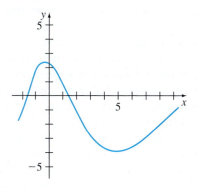

**FIGURE 1.2.10** Graph for Problem 38

In Problems 39–44, $y = c_1 \cos 2x + c_2 \sin 2x$ is a two-parameter family of solutions of the second-order DE $y'' + 4y = 0$. If possible, find a solution of the differential equation that satisfies the given side conditions. The conditions specified at two different points are called boundary conditions.

**39.** $y(0) = 0$, $y(\pi/4) = 3$

**40.** $y(0) = 0$, $y(\pi) = 0$

**41.** $y'(0) = 0$, $y'(\pi/6) = 0$

**42.** $y(0) = 1$, $y'(\pi) = 5$

**43.** $y(0) = 0$, $y(\pi) = 2$

**44.** $y'(\pi/2) = 1$, $y'(\pi) = 0$

### Discussion Problems

In Problems 45 and 46 use Problem 55 in Exercises 1.1 and (2) and (3) of this section.

**45.** Find a function whose graph at each point $(x, y)$ has the slope given by $8e^{2x} + 6x$ and has the $y$-intercept $(0, 9)$.

**46.** Find a function whose second derivative is $y'' = 12x - 2$ at each point $(x, y)$ on its graph and $y = -x + 5$ is tangent to the graph at the point corresponding to $x = 1$.

**47.** Consider the initial-value problem $y' = x - 2y$, $y(0) = \frac{1}{2}$. Determine which of the two curves shown in Figure 1.2.11 is the only plausible solution curve. Explain your reasoning.

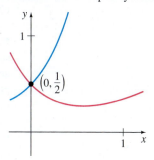

**FIGURE 1.2.11** Graphs for Problem 47

**48.** Show that

$$x = \int_0^y \frac{1}{\sqrt{t^3 + 1}}\, dt$$

is an implicit solution of the initial-value problem

$$2\frac{d^2y}{dx^2} - 3y^2 = 0, \quad y(0) = 0, \quad y'(0) = 1.$$

Assume that $y \geq 0$. [*Hint*: The integral is nonelementary. See (*ii*) in the *Remarks* at the end of Section 1.1.]

**49.** Determine a plausible value of $x_0$ for which the graph of the solution of the initial-value problem $y' + 2y = 3x - 6$, $y(x_0) = 0$ is tangent to the $x$-axis at $(x_0, 0)$. Explain your reasoning.

**50.** Suppose that the first-order differential equation $dy/dx = f(x, y)$ possesses a one-parameter family of solutions and that $f(x, y)$ satisfies the hypotheses of Theorem 1.2.1 in some rectangular region $R$ of the $xy$-plane. Explain why two different solution curves cannot intersect or be tangent to each other at a point $(x_0, y_0)$ in $R$.

**51.** The functions $y(x) = \frac{1}{16} x^4$, $-\infty < x < \infty$ and

$$y(x) = \begin{cases} 0, & x < 0 \\ \frac{1}{16} x^4, & x \geq 0 \end{cases}$$

have the same domain but are clearly different. See Figures 1.2.12(a) and 1.2.12(b), respectively. Show that both functions are solutions of the initial-value problem $dy/dx = xy^{1/2}$, $y(2) = 1$ on the interval $(-\infty, \infty)$. Resolve the apparent contradiction between this fact and the last sentence in Example 5.

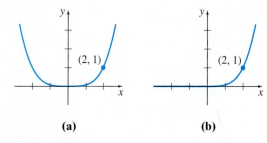

**(a)**          **(b)**

**FIGURE 1.2.12** Two solutions of the IVP in Problem 51

## 1.3    Differential Equations as Mathematical Models

**INTRODUCTION** In this section we introduce the notion of a differential equation as a mathematical model and discuss some specific models in biology, chemistry, and physics. Once we have studied some methods for solving differential equations in Chapters 2 and 4, we return to and solve some of these models in Chapters 3 and 5.

**MATHEMATICAL MODELS** It is often desirable to describe the behavior of some real-life system or a phenomenon—whether physical, sociological, or even economic—in mathematical terms. The mathematical description of a system or a phenomenon is called a **mathematical model** and is constructed with certain goals in mind. For example, we may wish to understand the mechanisms of a certain ecosystem by studying the growth of animal populations in that system, or we may wish to date fossils by means of analyzing the decay of a radioactive substance, either in the fossil or in the stratum in which it was discovered.

Construction of a mathematical model of a system starts with

(*i*)    identification of the variables that are responsible for changing the system. We may choose not to incorporate all these variables into the model at first. In this step we are specifying the **level of resolution** of the model.

Next

(*ii*)    we make a set of reasonable assumptions, or hypotheses, about the system we are trying to describe. These assumptions will also include any empirical laws that may be applicable to the system.

For some purposes it may be perfectly within reason to be content with low-resolution models. For example, you may already be aware that in beginning physics courses, the retarding force of air friction is sometimes ignored in modeling the motion of a body falling near the surface of the Earth, but if you are a scientist whose job it is to accurately predict the flight path of a long-range projectile, you have to take into account air resistance and other factors such as the curvature of the Earth.

Since the assumptions made about a system frequently involve *a rate of change* of one or more of the variables, the mathematical depiction of all these assumptions may be one or more equations involving *derivatives*. In other words, the mathematical model may be a differential equation or a system of differential equations.

Once we have formulated a mathematical model that is either a differential equation or a system of differential equations, we are faced with the not insignificant problem of trying to solve it. *If* we can solve it, then we deem the model to be reasonable if its solution is consistent with either experimental data or known facts about the behavior of the system. But if the predictions produced by the solution are poor, we can either increase the level of resolution of the model or make alternative assumptions about the mechanisms for change in the system. The steps of the modeling process are then repeated, as shown in the diagram in Figure 1.3.1.

**FIGURE 1.3.1**  Steps in the modeling process with differential equations

Of course, by increasing the resolution, we add to the complexity of the mathematical model and increase the likelihood that we cannot obtain an explicit solution of the differential equation.

A mathematical model of a physical system will often involve the variable time $t$. A solution of the model then gives the **state of the system;** in other words, the values of the dependent variable (or variables) for appropriate values of $t$ describe the system in the past, present, and future.

**POPULATION DYNAMICS**    One of the earliest attempts to model human **population growth** by means of mathematics was by the English clergyman and economist **Thomas Malthus** (1766–1834) in 1798. Basically, the idea behind the Malthusian model is the assumption that the rate at which the population of a country grows at a certain time is proportional* to the total population of the country at that time. In other words, the more people there are at time $t$, the more there are going to be in the future. In mathematical terms, if $P(t)$ denotes the total population at time $t$, then this assumption can be expressed as

$$\frac{dP}{dt} \propto P \qquad \text{or} \qquad \frac{dP}{dt} = kP, \tag{1}$$

where $k$ is a constant of proportionality. This simple model, which fails to take into account many factors that can influence human populations to either grow or decline (immigration and emigration, for example), nevertheless turned out to be fairly accurate in predicting the population of the United States during the years 1790–1860. Populations that grow at a rate described by (1) are rare; nevertheless, (1) is still used to model *growth of small populations over short intervals of time* (bacteria growing in a petri dish, for example).

**RADIOACTIVE DECAY**    The nucleus of an atom consists of combinations of protons and neutrons. Many of these combinations of protons and neutrons are unstable—that is, the atoms decay or transmute into atoms of another substance. Such nuclei are said to be radioactive. For example, over time the highly radioactive radium, Ra-226, transmutes into the radioactive gas radon, Rn-222. To model the phenomenon of **radioactive decay,** it is assumed that the rate $dA/dt$ at which the nuclei of a substance decay is proportional to the amount (more precisely, the number of nuclei) $A(t)$ of the substance remaining at time $t$:

$$\frac{dA}{dt} \propto A \qquad \text{or} \qquad \frac{dA}{dt} = kA. \tag{2}$$

Of course, equations (1) and (2) are exactly the same; the difference is only in the interpretation of the symbols and the constants of proportionality. For growth, as we expect in (1), $k > 0$, and for decay, as in (2), $k < 0$.

The model (1) for growth can also be seen as the equation $dS/dt = rS$, which describes the growth of capital $S$ when an annual rate of interest $r$ is compounded continuously. The model (2) for decay also occurs in biological applications such as determining the half-life of a drug—the time that it takes for 50% of a drug to be eliminated from a body by excretion or metabolism. In chemistry the decay model (2) appears in the mathematical description of a first-order chemical reaction. The point is this:

*A single differential equation can serve as a mathematical model for many different phenomena.*

---

*If two quantities $u$ and $v$ are proportional, we write $u \propto v$. This means that one quantity is a constant multiple of the other: $u = kv$.

Mathematical models are often accompanied by certain side conditions. For example, in (1) and (2) we would expect to know, in turn, the initial population $P_0$ and the initial amount of radioactive substance $A_0$ on hand. If the initial point in time is taken to be $t = 0$, then we know that $P(0) = P_0$ and $A(0) = A_0$. In other words, a mathematical model can consist of either an initial-value problem or, as we shall see later on in Section 5.2, a boundary-value problem.

**NEWTON'S LAW OF COOLING/WARMING**   According to Newton's empirical law of cooling/warming, the rate at which the temperature of a body changes is proportional to the difference between the temperature of the body and the temperature of the surrounding medium, the so-called ambient temperature. If $T(t)$ represents the temperature of a body at time $t$, $T_m$ the temperature of the surrounding medium, and $dT/dt$ the rate at which the temperature of the body changes, then Newton's law of cooling/warming translates into the mathematical statement

$$\frac{dT}{dt} \propto T - T_m \qquad \text{or} \qquad \frac{dT}{dt} = k(T - T_m), \qquad (3)$$

where $k$ is a constant of proportionality. In either case, cooling or warming, if $T_m$ is a constant, it stands to reason that $k < 0$.

**SPREAD OF A DISEASE**   A contagious disease—for example, a flu virus—is spread throughout a community by people coming into contact with other people. Let $x(t)$ denote the number of people who have contracted the disease and $y(t)$ denote the number of people who have not yet been exposed. It seems reasonable to assume that the rate $dx/dt$ at which the disease spreads is proportional to the number of encounters, or *interactions,* between these two groups of people. If we assume that the number of interactions is jointly proportional to $x(t)$ and $y(t)$—that is, proportional to the product $xy$—then

$$\frac{dx}{dt} = kxy, \qquad (4)$$

where $k$ is the usual constant of proportionality. Suppose a small community has a fixed population of $n$ people. If one infected person is introduced into this community, then it could be argued that $x(t)$ and $y(t)$ are related by $x + y = n + 1$. Using this last equation to eliminate $y$ in (4) gives us the model

$$\frac{dx}{dt} = kx(n + 1 - x). \qquad (5)$$

An obvious initial condition accompanying equation (5) is $x(0) = 1$.

**CHEMICAL REACTIONS**   The disintegration of a radioactive substance, governed by the differential equation (1), is said to be a **first-order reaction.** In chemistry a few reactions follow this same empirical law: If the molecules of substance $A$ decompose into smaller molecules, it is a natural assumption that the rate at which this decomposition takes place is proportional to the amount of the first substance that has not undergone conversion; that is, if $X(t)$ is the amount of substance $A$ remaining at any time, then $dX/dt = kX$, where $k$ is a negative constant since $X$ is decreasing. An example of a first-order chemical reaction is the conversion of $t$-butyl chloride, $(CH_3)_3CCl$, into $t$-butyl alcohol, $(CH_3)_3COH$:

$$(CH_3)_3CCl + NaOH \rightarrow (CH_3)_3COH + NaCl.$$

Only the concentration of the $t$-butyl chloride controls the rate of reaction. But in the reaction

$$CH_3Cl + NaOH \rightarrow CH_3OH + NaCl$$

one molecule of sodium hydroxide, NaOH, is consumed for every molecule of methyl chloride, $CH_3Cl$, thus forming one molecule of methyl alcohol, $CH_3OH$, and one molecule of sodium chloride, NaCl. In this case the rate at which the reaction proceeds is proportional to the product of the remaining concentrations of $CH_3Cl$ and NaOH. To describe this second reaction in general, let us suppose *one* molecule of a substance $A$ combines with *one* molecule of a substance $B$ to form *one* molecule of a substance $C$. If $X$ denotes the amount of chemical $C$ formed at time $t$ and if $\alpha$ and $\beta$ are, in turn, the amounts of the two chemicals $A$ and $B$ at $t = 0$ (the initial amounts), then the instantaneous amounts of $A$ and $B$ not converted to chemical $C$ are $\alpha - X$ and $\beta - X$, respectively. Hence the rate of formation of $C$ is given by

$$\frac{dX}{dt} = k(\alpha - X)(\beta - X), \tag{6}$$

where $k$ is a constant of proportionality. A reaction whose model is equation (6) is said to be a **second-order reaction.**

**MIXTURES**    The mixing of two salt solutions of differing concentrations gives rise to a first-order differential equation for the amount of salt contained in the mixture. Let us suppose that a large mixing tank initially holds 300 gallons of brine (that is, water in which a certain number of pounds of salt has been dissolved). Another brine solution is pumped into the large tank at a rate of 3 gallons per minute; the concentration of the salt in this inflow is 2 pounds per gallon. When the solution in the tank is well stirred, it is pumped out at the same rate as the entering solution. See Figure 1.3.2. If $A(t)$ denotes the amount of salt (measured in pounds) in the tank at time $t$, then the rate at which $A(t)$ changes is a net rate:

$$\frac{dA}{dt} = \left(\begin{array}{c}input\ rate \\ of\ salt\end{array}\right) - \left(\begin{array}{c}output\ rate \\ of\ salt\end{array}\right) = R_{in} - R_{out}. \tag{7}$$

The input rate $R_{in}$ at which salt enters the tank is the product of the inflow concentration of salt and the inflow rate of fluid. Note that $R_{in}$ is measured in pounds per minute:

$$R_{in} = (2\ \text{lb/gal}) \cdot (3\ \text{gal/min}) = (6\ \text{lb/min}).$$

Now, since the solution is being pumped out of the tank at the same rate that it is pumped in, the number of gallons of brine in the tank at time $t$ is a constant 300 gallons. Hence the concentration of the salt in the tank as well as in the outflow is $c(t) = A(t)/300$ lb/gal, so the output rate $R_{out}$ of salt is

$$R_{out} = \left(\frac{A(t)}{300}\ \text{lb/gal}\right) \cdot (3\ \text{gal/min}) = \frac{A(t)}{100}\ \text{lb/min}.$$

The net rate (7) then becomes

$$\frac{dA}{dt} = 6 - \frac{A}{100} \qquad \text{or} \qquad \frac{dA}{dt} + \frac{1}{100}A = 6. \tag{8}$$

If $r_{in}$ and $r_{out}$ denote general input and output rates of the brine solutions,* then there are three possibilities: $r_{in} = r_{out}$, $r_{in} > r_{out}$, and $r_{in} < r_{out}$. In the analysis leading to (8) we have assumed that $r_{in} = r_{out}$. In the latter two cases the number of gallons of brine in the tank is either increasing ($r_{in} > r_{out}$) or decreasing ($r_{in} < r_{out}$) at the net rate $r_{in} - r_{out}$. See Problems 10–12 in Exercises 1.3.

---

*Don't confuse these symbols with $R_{in}$ and $R_{out}$, which are input and output rates of *salt*.

input rate of brine
3 gal/min

constant
300 gal

output rate of brine
3 gal/min

**FIGURE 1.3.2** Mixing tank

**FIGURE 1.3.3** Draining tank

**DRAINING A TANK** In hydrodynamics, **Torricelli's law** states that the speed $v$ of efflux of water though a sharp-edged hole at the bottom of a tank filled to a depth $h$ is the same as the speed that a body (in this case a drop of water) would acquire in falling freely from a height $h$—that is, $v = \sqrt{2gh}$, where $g$ is the acceleration due to gravity. This last expression comes from equating the kinetic energy $\frac{1}{2}mv^2$ with the potential energy $mgh$ and solving for $v$. Suppose a tank filled with water is allowed to drain through a hole under the influence of gravity. We would like to find the depth $h$ of water remaining in the tank at time $t$. Consider the tank shown in Figure 1.3.3. If the area of the hole is $A_h$ (in ft$^2$) and the speed of the water leaving the tank is $v = \sqrt{2gh}$ (in ft/s), then the volume of water leaving the tank per second is $A_h\sqrt{2gh}$ (in ft$^3$/s). Thus if $V(t)$ denotes the volume of water in the tank at time $t$, then

$$\frac{dV}{dt} = -A_h\sqrt{2gh}, \tag{9}$$

where the minus sign indicates that $V$ is decreasing. Note here that we are ignoring the possibility of friction at the hole that might cause a reduction of the rate of flow there. Now if the tank is such that the volume of water in it at time $t$ can be written $V(t) = A_w h$, where $A_w$ (in ft$^2$) is the *constant* area of the upper surface of the water (see Figure 1.3.3), then $dV/dt = A_w \, dh/dt$. Substituting this last expression into (9) gives us the desired differential equation for the height of the water at time $t$:

$$\frac{dh}{dt} = -\frac{A_h}{A_w}\sqrt{2gh}. \tag{10}$$

It is interesting to note that (10) remains valid even when $A_w$ is not constant. In this case we must express the upper surface area of the water as a function of $h$—that is, $A_w = A(h)$. See Problem 14 in Exercises 1.3.

**(a)** *LRC*-series circuit

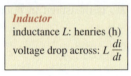

**Inductor**
inductance $L$: henries (h)
voltage drop across: $L\dfrac{di}{dt}$

**Resistor**
resistance $R$: ohms ($\Omega$)
voltage drop across: $iR$

**Capacitor**
capacitance $C$: farads (f)
voltage drop across: $\dfrac{1}{C}q$

**(b)**

**FIGURE 1.3.4** Symbols, units, and voltages. Current $i(t)$ and charge $q(t)$ are measured in amperes (A) and coulombs (C), respectively

**SERIES CIRCUITS** Consider the single-loop *LRC*-series circuit shown in Figure 1.3.4(a), containing an inductor, resistor, and capacitor. The current in a circuit after a switch is closed is denoted by $i(t)$; the charge on a capacitor at time $t$ is denoted by $q(t)$. The letters $L$, $R$, and $C$ are known as inductance, resistance, and capacitance, respectively, and are generally constants. Now according to **Kirchhoff's second law,** the impressed voltage $E(t)$ on a closed loop must equal the sum of the voltage drops in the loop. Figure 1.3.4(b) shows the symbols and the formulas for the respective voltage drops across an inductor, a capacitor, and a resistor. Since current $i(t)$ is related to charge $q(t)$ on the capacitor by $i = dq/dt$, adding the three voltages

$$\underset{\text{inductor}}{L\frac{di}{dt} = L\frac{d^2q}{dt^2},} \qquad \underset{\text{resistor}}{iR = R\frac{dq}{dt},} \qquad \text{and} \qquad \underset{\text{capacitor}}{\frac{1}{C}q}$$

and equating the sum to the impressed voltage yields a second-order differential equation

$$L\frac{d^2q}{dt^2} + R\frac{dq}{dt} + \frac{1}{C}q = E(t). \tag{11}$$

We will examine a differential equation analogous to (11) in great detail in Section 5.1.

**FALLING BODIES** To construct a mathematical model of the motion of a body moving in a force field, one often starts with the laws of motion formulated by the English mathematician **Isaac Newton** (1643–1727). Recall from elementary physics that **Newton's first law of motion** states that a body either will remain at rest or will continue to move with a constant velocity unless acted on by an external force. In each case this is equivalent to saying that when the sum of the forces

$F = \sum F_k$—that is, the *net* or resultant force—acting on the body is zero, then the acceleration $a$ of the body is zero. **Newton's second law of motion** indicates that when the net force acting on a body is not zero, then the net force is proportional to its acceleration $a$ or, more precisely, $F = ma$, where $m$ is the mass of the body.

Now suppose a rock is tossed upward from the roof of a building as illustrated in Figure 1.3.5. What is the position $s(t)$ of the rock relative to the ground at time $t$? The acceleration of the rock is the second derivative $d^2s/dt^2$. If we assume that the upward direction is positive and that no force acts on the rock other than the force of gravity, then Newton's second law gives

$$m\frac{d^2s}{dt^2} = -mg \quad \text{or} \quad \frac{d^2s}{dt^2} = -g. \tag{12}$$

In other words, the net force is simply the weight $F = F_1 = -W$ of the rock near the surface of the Earth. Recall that the magnitude of the weight is $W = mg$, where $m$ is the mass of the body and $g$ is the acceleration due to gravity. The minus sign in (12) is used because the weight of the rock is a force directed downward, which is opposite to the positive direction. If the height of the building is $s_0$ and the initial velocity of the rock is $v_0$, then $s$ is determined from the second-order initial-value problem

$$\frac{d^2s}{dt^2} = -g, \quad s(0) = s_0, \quad s'(0) = v_0. \tag{13}$$

Although we have not been stressing solutions of the equations we have constructed, note that (13) can be solved by integrating the constant $-g$ twice with respect to $t$. The initial conditions determine the two constants of integration. From elementary physics you might recognize the solution of (13) as the formula $s(t) = -\frac{1}{2}gt^2 + v_0t + s_0$.

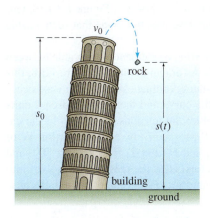

**FIGURE 1.3.5** Position of rock measured from ground level

**FALLING BODIES AND AIR RESISTANCE**   Before the famous experiment by the Italian mathematician and physicist **Galileo Galilei** (1564–1642) from the leaning tower of Pisa, it was generally believed that heavier objects in free fall, such as a cannonball, fell with a greater acceleration than lighter objects, such as a feather. Obviously, a cannonball and a feather when dropped simultaneously from the same height *do* fall at different rates, but it is not because a cannonball is heavier. The difference in rates is due to air resistance. The resistive force of air was ignored in the model given in (13). Under some circumstances a falling body of mass $m$, such as a feather with low density and irregular shape, encounters air resistance proportional to its instantaneous velocity $v$. If we take, in this circumstance, the positive direction to be oriented downward, then the net force acting on the mass is given by $F = F_1 + F_2 = mg - kv$, where the weight $F_1 = mg$ of the body is force acting in the positive direction and air resistance $F_2 = -kv$ is a force, called **viscous damping,** acting in the opposite or upward direction. See Figure 1.3.6. Now since $v$ is related to acceleration $a$ by $a = dv/dt$, Newton's second law becomes $F = ma = m\,dv/dt$. By equating the net force to this form of Newton's second law, we obtain a first-order differential equation for the velocity $v(t)$ of the body at time $t$,

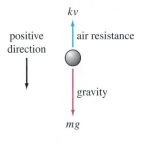

**FIGURE 1.3.6** Falling body of mass $m$

$$m\frac{dv}{dt} = mg - kv. \tag{14}$$

Here $k$ is a positive constant of proportionality. If $s(t)$ is the distance the body falls in time $t$ from its initial point of release, then $v = ds/dt$ and $a = dv/dt = d^2s/dt^2$. In terms of $s$, (14) is a second-order differential equation

$$m\frac{d^2s}{dt^2} = mg - k\frac{ds}{dt} \quad \text{or} \quad m\frac{d^2s}{dt^2} + k\frac{ds}{dt} = mg. \tag{15}$$

**(a)** suspension bridge cable

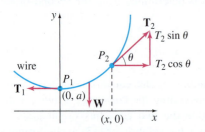

**(b)** telephone wires

**FIGURE 1.3.7** Cables suspended between vertical supports

**FIGURE 1.3.8** Element of cable

**SUSPENDED CABLES** Suppose a flexible cable, wire, or heavy rope is suspended between two vertical supports. Physical examples of this could be one of the two cables supporting the roadbed of a suspension bridge as shown in Figure 1.3.7(a) or a long telephone wire strung between two posts as shown in Figure 1.3.7(b). Our goal is to construct a mathematical model that describes the shape that such a cable assumes.

To begin, let's agree to examine only a portion or element of the cable between its lowest point $P_1$ and any arbitrary point $P_2$. As drawn in blue in Figure 1.3.8, this element of the cable is the curve in a rectangular coordinate system with $y$-axis chosen to pass through the lowest point $P_1$ on the curve and the $x$-axis chosen $a$ units below $P_1$. Three forces are acting on the cable: the tensions $\mathbf{T}_1$ and $\mathbf{T}_2$ in the cable that are tangent to the cable at $P_1$ and $P_2$, respectively, and the portion $\mathbf{W}$ of the total vertical load between the points $P_1$ and $P_2$. Let $T_1 = |\mathbf{T}_1|$, $T_2 = |\mathbf{T}_2|$, and $W = |\mathbf{W}|$ denote the magnitudes of these vectors. Now the tension $\mathbf{T}_2$ resolves into horizontal and vertical components (scalar quantities) $T_2 \cos \theta$ and $T_2 \sin \theta$. Because of static equilibrium we can write

$$T_1 = T_2 \cos \theta \qquad \text{and} \qquad W = T_2 \sin \theta.$$

By dividing the last equation by the first, we eliminate $T_2$ and get $\tan \theta = W/T_1$. But because $dy/dx = \tan \theta$, we arrive at

$$\frac{dy}{dx} = \frac{W}{T_1}. \tag{16}$$

This simple first-order differential equation serves as a model for both the shape of a flexible wire such as a telephone wire hanging under its own weight and the shape of the cables that support the roadbed of a suspension bridge. We will come back to equation (16) in Exercises 2.2 and Section 5.3.

**WHAT LIES AHEAD** Throughout this text you will see three different types of approaches to, or analyses of, differential equations. Over the centuries differential equations would often spring from the efforts of a scientist or engineer to describe some physical phenomenon or to translate an empirical or experimental law into mathematical terms. As a consequence, a scientist, engineer, or mathematician would often spend many years of his or her life trying to find the solutions of a DE. With a solution in hand, the study of its properties then followed. This quest for solutions is called by some the *analytical approach* to differential equations. Once they realized that explicit solutions are at best difficult to obtain and at worst impossible to obtain, mathematicians learned that a differential equation itself could be a font of valuable information. It is possible, in some instances, to glean directly from the differential equation answers to questions such as

*Does the DE actually have solutions? If a solution of the DE exists and satisfies an initial condition, is it the only such solution? What are some of the properties of the unknown solutions? What can we say about the geometry of the solution curves?*

Such an approach is *qualitative analysis*. Finally, if a differential equation cannot be solved by analytical methods, yet we can prove that a solution exists, the next logical query is

*Can we somehow approximate the values of an unknown solution?*

Here we enter the realm of *numerical analysis*. An affirmative answer to the last question stems from the fact that a differential equation can be used as a cornerstone for constructing very accurate approximation algorithms. In Chapter 2 we start with qualitative considerations of first-order ODEs, then examine analytical stratagems for solving some special first-order equations, and conclude with an introduction to an elementary numerical method. See Figure 1.3.9.

(a) analytical          (b) qualitative          (c) numerical

**FIGURE 1.3.9** Different approaches to the study of differential equations

## REMARKS

Most examples in this section describe a dynamical system—a system that changes or evolves with the flow of time $t$. Since the study of dynamical systems is a branch of mathematics currently in vogue, we shall occasionally relate the terminology of that field to the discussion at hand.

In more precise terms, a **dynamical system** consists of a set of time-dependent variables, called **state variables,** together with a rule that enables us to determine (without ambiguity) the state of the system (this may be a past, present, or future state) in terms of a state prescribed at some time $t_0$. Dynamical systems are classified as either discrete-time systems or continuous-time systems. In this course we shall be concerned only with continuous-time systems—systems in which *all* variables are defined over a continuous range of time. The rule, or mathematical model, in a continuous-time dynamical system is a differential equation or a system of differential equations. The **state of the system** at a time $t$ is the value of the state variables at that time; the specified state of the system at a time $t_0$ is simply the initial conditions that accompany the mathematical model. The solution of the initial-value problem is referred to as the **response of the system.** For example, in the case of radioactive decay, the rule is $dA/dt = kA$. Now if the quantity of a radioactive substance at some time $t_0$ is known, say $A(t_0) = A_0$, then by solving the rule we find that the response of the system for $t \geq t_0$ is $A(t) = A_0 e^{(t-t_0)}$ (see Section 3.1). The response $A(t)$ is the single state variable for this system. In the case of the rock tossed from the roof of a building, the response of the system—the solution of the differential equation $d^2s/dt^2 = -g$, subject to the initial state $s(0) = s_0$, $s'(0) = v_0$, is the function $s(t) = -\frac{1}{2}gt^2 + v_0 t + s_0$, $0 \leq t \leq T$, where $T$ represents the time when the rock hits the ground. The state variables are $s(t)$ and $s'(t)$, which are the vertical position of the rock above ground and its velocity at time $t$, respectively. The acceleration $s''(t)$ is *not* a state variable, since we have to know only any initial position and initial velocity at a time $t_0$ to uniquely determine the rock's position $s(t)$ and velocity $s'(t) = v(t)$ for any time in the interval $t_0 \leq t \leq T$. The acceleration $s''(t) = a(t)$ is, of course, given by the differential equation $s''(t) = -g$, $0 < t < T$.

One last point: Not every system studied in this text is a dynamical system. We shall also examine some static systems in which the model is a differential equation.

## EXERCISES **1.3**

*Answers to selected odd-numbered problems begin on page ANS-1.*

### Population Dynamics

1. Under the same assumptions that underlie the model in (1), determine a differential equation for the population $P(t)$ of a country when individuals are allowed to immigrate into the country at a constant rate $r > 0$. What is the differential equation for the population $P(t)$ of the country when individuals are allowed to emigrate from the country at a constant rate $r > 0$?

2. The population model given in (1) fails to take death into consideration; the growth rate equals the birth rate. In another model of a changing population of a community it is assumed that the rate at which the population changes is a *net* rate—that is, the difference between the rate of births and the rate of deaths in the community. Determine a model for the population $P(t)$ if both the birth rate and the death rate are proportional to the population present at time $t > 0$.

3. Using the concept of net rate introduced in Problem 2, determine a model for a population $P(t)$ if the birth rate is proportional to the population present at time $t$ but the death rate is proportional to the square of the population present at time $t$.

4. Modify the model in Problem 3 for net rate at which the population $P(t)$ of a certain kind of fish changes by also assuming that the fish are harvested at a constant rate $h > 0$.

### Newton's Law of Cooling/Warming

5. A cup of coffee cools according to Newton's law of cooling (3). Use data from the graph of the temperature $T(t)$ in Figure 1.3.10 to estimate the constants $T_m$, $T_0$, and $k$ in a model of the form of a first-order initial-value problem: $dT/dt = k(T - T_m)$, $T(0) = T_0$.

**FIGURE 1.3.10** Cooling curve in Problem 5

6. The ambient temperature $T_m$ in (3) could be a function of time $t$. Suppose that in an artificially controlled environment, $T_m(t)$ is periodic with a 24-hour period, as illustrated in Figure 1.3.11. Devise a mathematical model for the temperature $T(t)$ of a body within this environment.

**FIGURE 1.3.11** Ambient temperature in Problem 6

### Spread of a Disease/Technology

7. Suppose a student carrying a flu virus returns to an isolated college campus of 1000 students. Determine a differential equation for the number of people $x(t)$ who have contracted the flu if the rate at which the disease spreads is proportional to the number of interactions between the number of students who have the flu and the number of students who have not yet been exposed to it.

8. At a time denoted as $t = 0$ a technological innovation is introduced into a community that has a fixed population of $n$ people. Determine a differential equation for the number of people $x(t)$ who have adopted the innovation at time $t$ if it is assumed that the rate at which the innovations spread through the community is jointly proportional to the number of people who have adopted it and the number of people who have not adopted it.

### Mixtures

9. Suppose that a large mixing tank initially holds 300 gallons of water in which 50 pounds of salt have been dissolved. Pure water is pumped into the tank at a rate of 3 gal/min, and when the solution is well stirred, it is then pumped out at the same rate. Determine a differential equation for the amount of salt $A(t)$ in the tank at time $t > 0$. What is $A(0)$?

10. Suppose that a large mixing tank initially holds 300 gallons of water in which 50 pounds of salt have been dissolved. Another brine solution is pumped into the tank at a rate of 3 gal/min, and when the solution is well stirred, it is then pumped out at a *slower* rate of 2 gal/min. If the concentration of the solution entering is 2 lb/gal, determine a differential equation for the amount of salt $A(t)$ in the tank at time $t > 0$.

11. What is the differential equation in Problem 10, if the well-stirred solution is pumped out at a *faster* rate of 3.5 gal/min?

12. Generalize the model given in equation (8) of this section by assuming that the large tank initially contains $N_0$ number of gallons of brine, $r_{in}$ and $r_{out}$ are the input and output rates of the brine, respectively (measured in gallons per minute), $c_{in}$ is the concentration of the salt in the inflow, $c(t)$ the concentration of

the salt in the tank as well as in the outflow at time $t$ (measured in pounds of salt per gallon), and $A(t)$ is the amount of salt in the tank at time $t > 0$.

## Draining a Tank

**13.** Suppose water is leaking from a tank through a circular hole of area $A_h$ at its bottom. When water leaks through a hole, friction and contraction of the stream near the hole reduce the volume of water leaving the tank per second to $cA_h\sqrt{2gh}$, where $c$ $(0 < c < 1)$ is an empirical constant. Determine a differential equation for the height $h$ of water at time $t$ for the cubical tank shown in Figure 1.3.12. The radius of the hole is 2 in., and $g = 32$ ft/s$^2$.

**FIGURE 1.3.12** Cubical tank in Problem 13

**14.** The right-circular conical tank shown in Figure 1.3.13 loses water out of a circular hole at its bottom. Determine a differential equation for the height of the water $h$ at time $t > 0$. The radius of the hole is 2 in., $g = 32$ ft/s$^2$, and the friction/contraction factor introduced in Problem 13 is $c = 0.6$.

**FIGURE 1.3.13** Conical tank in Problem 14

## Series Circuits

**15.** A series circuit contains a resistor and an inductor as shown in Figure 1.3.14. Determine a differential equation for the current $i(t)$ if the resistance is $R$, the inductance is $L$, and the impressed voltage is $E(t)$.

**FIGURE 1.3.14** $LR$-series circuit in Problem 15

**16.** A series circuit contains a resistor and a capacitor as shown in Figure 1.3.15. Determine a differential equation for the charge $q(t)$ on the capacitor if the resistance is $R$, the capacitance is $C$, and the impressed voltage is $E(t)$.

**FIGURE 1.3.15** $RC$-series circuit in Problem 16

## Falling Bodies and Air Resistance

**17.** For high-speed motion through the air—such as the skydiver shown in Figure 1.3.16, falling before the parachute is opened—air resistance is closer to a power of the instantaneous velocity $v(t)$. Determine a differential equation for the velocity $v(t)$ of a falling body of mass $m$ if air resistance is proportional to the square of the instantaneous velocity. Assume the downward direction is positive.

**FIGURE 1.3.16** Air resistance proportional to square of velocity in Problem 17

## Newton's Second Law and Archimedes' Principle

**18.** A cylindrical barrel $s$ feet in diameter of weight $w$ lb is floating in water as shown in Figure 1.3.17(a). After an initial depression the barrel exhibits an up-and-down bobbing motion along a vertical line. Using Figure 1.3.17(b), determine a differential equation for the vertical displacement $y(t)$ if the origin is taken to be on the vertical axis at the surface of the water when the barrel is at rest. Use **Archimedes' principle:** Buoyancy, or upward force of the water on the barrel, is equal to the weight of the water displaced. Assume that the downward direction is positive, that the weight density of water is 62.4 lb/ft$^3$, and that there is no resistance between the barrel and the water.

**FIGURE 1.3.17** Bobbing motion of floating barrel in Problem 18

## Newton's Second Law and Hooke's Law

**19.** After a mass $m$ is attached to a spring, it stretches it $s$ units and then hangs at rest in the equilibrium position as shown in Figure 1.3.18(b). After the spring/mass system has been set in motion, let $x(t)$ denote the directed distance of the mass beyond the equilibrium position. As indicated in Figure 1.3.18(c), assume that the downward direction is positive, that the motion takes place in a vertical straight line through the center of gravity of the mass, and that the only forces acting on the system are the weight of the mass and the restoring force of the stretched spring. Use **Hooke's law:** The restoring force of a spring is proportional to its total elongation. Determine a differential equation for the displacement $x(t)$ at time $t > 0$.

unstretched spring

$x(t) < 0$

$-x = 0$

$x(t) > 0$

equilibrium position

**(a)** **(b)** **(c)**

**FIGURE 1.3.18** Spring/mass system in Problem 19

**20.** In Problem 19, what is a differential equation for the displacement $x(t)$ if the motion takes place in a medium that imparts a damping force on the spring/mass system that is proportional to the instantaneous velocity of the mass and acts in a direction opposite to that of motion?

## Newton's Second Law and Variable Mass

When the mass $m$ of a body is changing with time, Newton's second law of motion becomes

$$F = \frac{d}{dt}(mv), \tag{17}$$

where $F$ is the net force acting on the body and $mv$ is its momentum. Use (17) in Problems 21 and 22.

**21.** A small single-stage rocket is launched vertically as shown in Figure 1.3.19. Once launched, the rocket consumes its

James L. Davidson/Shutterstock.com

**FIGURE 1.3.19** Single-stage rocket in Problem 21

fuel, and so its total mass $m(t)$ varies with time $t > 0$. If it is assumed that the positive direction is upward, air resistance is proportional to the instantaneous velocity $v$ of the rocket, and $R$ is the upward thrust or force generated by the propulsion system, then construct a mathematical model for the velocity $v(t)$ of the rocket. [*Hint:* See (14) in Section 1.3.]

**22.** In Problem 21, the mass $m(t)$ is the sum of three different masses: $m(t) = m_p + m_v + m_f(t)$, where $m_p$ is the constant mass of the payload, $m_v$ is the constant mass of the vehicle, and $m_f(t)$ is the variable amount of fuel.

**(a)** Show that the rate at which the total mass $m(t)$ of the rocket changes is the same as the rate at which the mass $m_f(t)$ of the fuel changes.

**(b)** If the rocket consumes its fuel at a constant rate $\lambda$, find $m(t)$. Then rewrite the differential equation in Problem 21 in terms of $\lambda$ and the initial total mass $m(0) = m_0$.

**(c)** Under the assumption in part (b), show that the burnout time $t_b > 0$ of the rocket, or the time at which all the fuel is consumed, is $t_b = m_f(0)/\lambda$, where $m_f(0)$ is the initial mass of the fuel.

## Newton's Second Law and the Law of Universal Gravitation

**23.** By **Newton's universal law of gravitation** the free-fall acceleration $a$ of a body, such as the satellite shown in Figure 1.3.20, falling a great distance to the surface is *not* the constant $g$. Rather, the acceleration $a$ is inversely proportional to the square of the distance from the center of the Earth, $a = k/r^2$, where $k$ is the constant of proportionality. Use the fact that at the surface of the Earth $r = R$ and $a = g$ to determine $k$. If the positive direction is upward, use Newton's second law and his universal law of gravitation to find a differential equation for the distance $r$.

satellite of mass $m$

surface

$r$

$R$

Earth of mass $M$

**FIGURE 1.3.20** Satellite in Problem 23

**24.** Suppose a hole is drilled through the center of the Earth and a bowling ball of mass $m$ is dropped into the hole, as shown in Figure 1.3.21. Construct a mathematical model that describes the

surface

$m$

$r$

$R$

**FIGURE 1.3.21** Hole through Earth in Problem 24

motion of the ball. At time $t$ let $r$ denote the distance from the center of the Earth to the mass $m$, $M$ denote the mass of the Earth, $M_r$ denote the mass of that portion of the Earth within a sphere of radius $r$, and $\delta$ denote the constant density of the Earth.

### Additional Mathematical Models

**25. Learning Theory**   In the theory of learning, the rate at which a subject is memorized is assumed to be proportional to the amount that is left to be memorized. Suppose $M$ denotes the total amount of a subject to be memorized and $A(t)$ is the amount memorized in time $t > 0$. Determine a differential equation for the amount $A(t)$.

**26. Forgetfulness**   In Problem 25 assume that the rate at which material is *forgotten* is proportional to the amount memorized in time $t > 0$. Determine a differential equation for the amount $A(t)$ when forgetfulness is taken into account.

**27. Infusion of a Drug**   A drug is infused into a patient's bloodstream at a constant rate of $r$ grams per second. Simultaneously, the drug is removed at a rate proportional to the amount $x(t)$ of the drug present at time $t$. Determine a differential equation for the amount $x(t)$.

**28. Tractrix**   A motorboat starts at the origin and moves in the direction of the positive $x$-axis, pulling a waterskier along a curve $C$ called a **tractrix**. See Figure 1.3.22. The waterskier, initially located on the $y$-axis at the point $(0, a)$, is pulled by a rope of constant length $a$ that is kept taut throughout the motion. At time $t > 0$ the waterskier is at point $P(x, y)$. Assume that the rope is always tangent to $C$. Use the concept of slope to determine a differential equation for the path $C$ of motion.

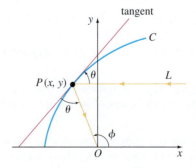

**FIGURE 1.3.22** Waterskier in Problem 28

**29. Reflecting Surface**   Assume that when the plane curve $C$ shown in Figure 1.3.23 is revolved about the $x$-axis, it generates

**FIGURE 1.3.23** Reflecting surface in Problem 29

a surface of revolution with the property that all light rays $L$ parallel to the $x$-axis striking the surface are reflected to a single point $O$ (the origin). Use the fact that the angle of incidence is equal to the angle of reflection to determine a differential equation that describes the shape of the curve $C$. Such a curve $C$ is important in applications ranging from construction of telescopes to satellite antennas, automobile headlights, and solar collectors. [*Hint*: Inspection of the figure shows that we can write $\phi = 2\theta$. Why? Now use an appropriate trigonometric identity.]

### Discussion Problems

**30.** Reread Problem 45 in Exercises 1.1 and then give an explicit solution $P(t)$ for equation (1). Find a one-parameter family of solutions of (1).

**31.** Reread the sentence following equation (3) and assume that $T_m$ is a positive constant. Discuss why we would expect $k < 0$ in (3) in both cases of cooling and warming. You might start by interpreting, say, $T(t) > T_m$ in a graphical manner.

**32.** Reread the discussion leading up to equation (8). If we assume that initially the tank holds, say, 50 lb of salt, it stands to reason that because salt is being added to the tank continuously for $t > 0$, $A(t)$ should be an increasing function. Discuss how you might determine from the DE, without actually solving it, the number of pounds of salt in the tank after a long period of time.

**33. Population Model**   The differential equation $\dfrac{dP}{dt} = (k \cos t)P$, where $k$ is a positive constant, is a model of human population $P(t)$ of a certain community. Discuss an interpretation for the solution of this equation. In other words, what kind of population do you think the differential equation describes?

**34. Rotating Fluid**   As shown in Figure 1.3.24(a), a right-circular cylinder partially filled with fluid is rotated with a constant angular velocity $\omega$ about a vertical $y$-axis through its center. The rotating fluid forms a surface of revolution $S$. To identify $S$, we first establish a coordinate system consisting of a vertical plane determined by the $y$-axis and an $x$-axis drawn perpendicular to the $y$-axis such that the point of intersection of the axes (the origin) is located at the lowest point on the surface $S$. We then seek a function $y = f(x)$ that represents the curve $C$ of intersection of the surface $S$ and the vertical coordinate plane. Let the point $P(x, y)$ denote the position of a particle of the rotating fluid of mass $m$ in the coordinate plane. See Figure 1.3.24(b).

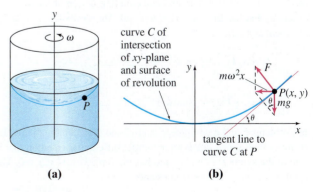

**(a)**          **(b)**

**FIGURE 1.3.24** Rotating fluid in Problem 34

(a) At $P$ there is a reaction force of magnitude $F$ due to the other particles of the fluid which is normal to the surface $S$. By Newton's second law the magnitude of the net force acting on the particle is $m\omega^2 x$. What is this force? Use Figure 1.3.24(b) to discuss the nature and origin of the equations

$$F \cos \theta = mg, \qquad F \sin \theta = m\omega^2 x.$$

(b) Use part (a) to find a first-order differential equation that defines the function $y = f(x)$.

35. **Falling Body**   In Problem 23, suppose $r = R + s$, where $s$ is the distance from the surface of the Earth to the falling body. What does the differential equation obtained in Problem 23 become when $s$ is very small in comparison to $R$? [*Hint*: Think binomial series for

$$(R + s)^{-2} = R^{-2}(1 + s/R)^{-2}.]$$

36. **Raindrops Keep Falling**   In meteorology the term *virga* refers to falling raindrops or ice particles that evaporate before they reach the ground. Assume that a typical raindrop is spherical. Starting at some time, which we can designate as $t = 0$, the raindrop of radius $r_0$ falls from rest from a cloud and begins to evaporate.

(a) If it is assumed that a raindrop evaporates in such a manner that its shape remains spherical, then it also makes sense to assume that the rate at which the raindrop evaporates—that is, the rate at which it loses mass—is proportional to its surface area. Show that this latter assumption implies that the rate at which the radius $r$ of the raindrop decreases is a constant. Find $r(t)$. [*Hint*: See Problem 55 in Exercises 1.1.]

(b) If the positive direction is downward, construct a mathematical model for the velocity $v$ of the falling raindrop at time $t > 0$. Ignore air resistance. [*Hint*: Use the form of Newton's second law of motion given in (17).]

37. **Let It Snow**   The "snowplow problem" is a classic and appears in many differential equations texts, but it was probably made famous by Ralph Palmer Agnew:

*One day it started snowing at a heavy and steady rate. A snowplow started out at noon, going 2 miles the first hour and 1 mile the second hour. What time did it start snowing?*

Find the textbook *Differential Equations,* Ralph Palmer Agnew, McGraw-Hill Book Co., and then discuss the construction and solution of the mathematical model.

38. **Population Dynamics**   Suppose that $dP/dt = 0.15 P(t)$ represents a mathematical model for the growth of a certain cell culture, where $P(t)$ is the size of the culture (measured in millions of cells) at time $t > 0$ (measured in hours). How fast is the culture growing at the time when the size of the culture reaches 2 million cells?

39. **Radioactive Decay**   Suppose that $dA/dt = -0.0004332 A(t)$ represents a mathematical model for the decay of radium-226, where $A(t)$ is the amount of radium (measured in grams) remaining at time $t > 0$ (measured in years). How much of the radium sample remains at the time when the sample is decaying at a rate of 0.002 grams per year?

40. Reread this section and classify each mathematical model as linear or nonlinear.

# Chapter 1 In Review

*Answers to selected odd-numbered problems begin on page ANS-1.*

In Problems 1 and 2 fill in the blank and then write this result as a linear first-order differential equation that is free of the symbol $c_1$ and has the form $dy/dx = f(x, y)$. The symbol $c_1$ represents a constant.

1. $\dfrac{d}{dx} c_1 e^{10x} = $ _____

2. $\dfrac{d}{dx} (5 + c_1 e^{-2x}) = $ _____

In Problems 3 and 4 fill in the blank and then write this result as a linear second-order differential equation that is free of the symbols $c_1$ and $c_2$ and has the form $F(y, y'') = 0$. The symbols $c_1$, $c_2$, and $k$ represent constants.

3. $\dfrac{d^2}{dx^2} (c_1 \cos kx + c_2 \sin kx) = $ _____

4. $\dfrac{d^2}{dx^2} (c_1 \cosh kx + c_2 \sinh kx) = $ _____

In Problems 5 and 6 compute $y'$ and $y''$ and then combine these derivatives with $y$ as a linear second-order differential equation that is free of the symbols $c_1$ and $c_2$ and has the form $F(y, y' \, y'') = 0$. The symbols $c_1$ and $c_2$ represent constants.

5. $y = c_1 e^x + c_2 x e^x$

6. $y = c_1 e^x \cos x + c_2 e^x \sin x$

In Problems 7–12 match each of the given differential equations with one or more of these solutions:

(a) $y = 0,$    (b) $y = 2,$    (c) $y = 2x,$    (d) $y = 2x^2.$

7. $xy' = 2y$

8. $y' = 2$

9. $y' = 2y - 4$

10. $xy' = y$

11. $y'' + 9y = 18$

12. $xy'' - y' = 0$

In Problems 13 and 14 determine by inspection at least one solution of the given differential equation.

13. $y'' = y'$

14. $y' = y(y - 3)$

In Problems 15 and 16 interpret each statement as a differential equation.

15. On the graph of $y = \phi(x)$ the slope of the tangent line at a point $P(x, y)$ is the square of the distance from $P(x, y)$ to the origin.

16. On the graph of $y = \phi(x)$ the rate at which the slope changes with respect to $x$ at a point $P(x, y)$ is the negative of the slope of the tangent line at $P(x, y)$.

**17. (a)** Give the domain of the function $y = x^{2/3}$.

**(b)** Give the largest interval $I$ of definition over which $y = x^{2/3}$ is a solution of the differential equation $3xy' - 2y = 0$.

**18. (a)** Verify that the one-parameter family $y^2 - 2y = x^2 - x + c$ is an implicit solution of the differential equation

$$(2y - 2)y' = 2x - 1.$$

**(b)** Find a member of the one-parameter family in part (a) that satisfies the initial condition $y(0) = 1$.

**(c)** Use your result in part (b) to find an explicit *function* $y = \phi(x)$ that satisfies $y(0) = 1$. Give the domain of the function $\phi$. Is $y = \phi(x)$ a *solution* of the initial-value problem? If so, give its interval $I$ of definition; if not, explain.

**19.** The function $y = x - 2/x$ is a solution of the DE $xy' + y = 2x$. Find $x_0$ and the largest interval $I$ for which $y(x)$ is a solution of the first-order IVP $xy' + y = 2x$, $y(x_0) = 1$.

**20.** Suppose that $y(x)$ denotes a solution of the first-order IVP $y' = x^2 + y^2$, $y(1) = -1$ and that $y(x)$ possesses at least a second derivative at $x = 1$. In some neighborhood of $x = 1$ use the DE to determine whether $y(x)$ is increasing or decreasing and whether the graph $y(x)$ is concave up or concave down.

**21.** A differential equation may possess more than one family of solutions.

**(a)** Plot different members of the families $y = \phi_1(x) = x^2 + c_1$ and $y = \phi_2(x) = -x^2 + c_2$.

**(b)** Verify that $y = \phi_1(x)$ and $y = \phi_2(x)$ are two solutions of the nonlinear first-order differential equation $(y')^2 = 4x^2$.

**(c)** Construct a piecewise-defined function that is a solution of the nonlinear DE in part (b) but is not a member of either family of solutions in part (a).

**22.** What is the slope of the tangent line to the graph of a solution of $y' = 6\sqrt{y} + 5x^3$ that passes through $(-1, 4)$?

In Problems 23–26 verify that the indicated function is an explicit solution of the given differential equation. Give an interval of definition $I$ for each solution.

**23.** $y'' + y = 2\cos x - 2\sin x; \quad y = x\sin x + x\cos x$

**24.** $y'' + y = \sec x; \quad y = x\sin x + (\cos x)\ln(\cos x)$

**25.** $x^2y'' + xy' + y = 0; \quad y = \sin(\ln x)$

**26.** $x^2y'' + xy' + y = \sec(\ln x);$
$y = \cos(\ln x)\ln(\cos(\ln x)) + (\ln x)\sin(\ln x)$

In Problems 27–30 use (12) of Section 1.1 to verify that the indicated function is a solution of the given differential equation. Assume an appropriate interval $I$ of definition of each solution.

**27.** $\dfrac{dy}{dx} + (\sin x)y = x; \quad y = e^{\cos x}\displaystyle\int_0^x te^{-\cos t}\,dt$

**28.** $\dfrac{dy}{dx} - 2xy = e^x; \quad y = e^{x^2}\displaystyle\int_0^x e^{t-t^2}\,dt$

**29.** $x^2y'' + (x^2 - x)y' + (1 - x)y = 0; \quad y = x\displaystyle\int_1^x \dfrac{e^{-t}}{t}\,dt$

**30.** $y'' + y = e^{x^2}; \quad y = \sin x\displaystyle\int_0^x e^{t^2}\cos t\,dt - \cos x\displaystyle\int_0^x e^{t^2}\sin t\,dt$

In Problems 31–34 verify that the indicated expression is an implicit solution of the given differential equation.

**31.** $x\dfrac{dy}{dx} + y = \dfrac{1}{y^2}; \quad x^3y^3 = x^3 + 1$

**32.** $\left(\dfrac{dy}{dx}\right)^2 + 1 = \dfrac{1}{y^2}; \quad (x - 5)^2 + y^2 = 1$

**33.** $y'' = 2y(y')^3; \quad y^3 + 3y = 1 - 3x$

**34.** $(1 - xy)y' = y^2; \quad y = e^{xy}$

In Problems 35–38, $y = c_1e^{3x} + c_2e^{-x} - 2x$ is a two-parameter family of the second-order DE $y'' - 2y' - 3y = 6x + 4$. Find a solution of the second-order IVP consisting of this differential equation and the given initial conditions.

**35.** $y(0) = 0, y'(0) = 0$ 　　**36.** $y(0) = 1, y'(0) = -3$

**37.** $y(1) = 4, y'(1) = -2$ 　　**38.** $y(-1) = 0, y'(-1) = 1$

**39.** The graph of a solution of a second-order initial-value problem $d^2y/dx^2 = f(x, y, y')$, $y(2) = y_0$, $y'(2) = y_1$, is given in Figure 1.R.1. Use the graph to estimate the values of $y_0$ and $y_1$.

**FIGURE 1.R.1** Graph for Problem 39

**40.** A tank in the form of a right-circular cylinder of radius 2 feet and height 10 feet is standing on end. If the tank is initially full of water and water leaks from a circular hole of radius $\frac{1}{2}$ inch at its bottom, determine a differential equation for the height $h$ of the water at time $t > 0$. Ignore friction and contraction of water at the hole.

# 2

# First-Order Differential Equations

Joggie Botma/Shutterstock.com

The history of mathematics is rife with stories of persons who devoted much of their lives to solving equations—algebraic equations at first and then eventually differential equations. In Sections 2.2–2.5 we will study some of the important analytical methods for solving first-order DEs. But before we start solving anything, you should be aware of two facts: A differential equation may have no solutions, and a differential equation may possess solutions yet there might not exist any analytical method for solving it. In Sections 2.1 and 2.6 we do not solve any DEs but show how to glean information about its solutions directly from the equation itself.

# Solution Curves Without a Solution

**INTRODUCTION**   Let us imagine for the moment that we have in front of us a first-order differential equation $dy/dx = f(x, y)$, and that we can neither find a solution nor invent a method for solving it analytically. This is not as bad a predicament as one might think, since the differential equation itself can be a fount of information about how its solutions "behave."

We begin our study of first-order differential equations with two ways of analyzing a DE qualitatively. Both of these ways enable us to determine, in an approximate sense, what a solution curve must look like without actually solving the equation.

## 2.1.1   DIRECTION FIELDS

**SOME FUNDAMENTAL QUESTIONS**   We saw in Section 1.2 that whenever $f(x, y)$ and $\partial f/\partial y$ satisfy certain continuity conditions, qualitative questions about existence and uniqueness of solutions can be answered. In this section we shall see that other qualitative questions about properties of solutions—How does a solution behave near a certain point? How does a solution behave as $x \rightarrow \infty$?—can often be answered when the function $f$ depends solely on the variable $y$. We begin, however, with a simple concept from calculus:

*A derivative $dy/dx$ of a differentiable function $y = y(x)$ gives slopes of tangent lines at points on its graph.*

**SLOPE**   Because a solution $y = y(x)$ of a first-order differential equation

$$\frac{dy}{dx} = f(x, y) \tag{1}$$

is necessarily a differentiable function on its interval $I$ of definition, it must also be continuous on $I$. Thus the corresponding solution curve on $I$ must have no breaks and must possess a tangent line at each point $(x, y(x))$. The function $f$ in the normal form (1) is called the **slope function** or **rate function.** The slope of the tangent line at $(x, y(x))$ on a solution curve is the value of the first derivative $dy/dx$ at this point, and we know from (1) that this is the value of the slope function $f(x, y(x))$. Now suppose that $(x, y)$ represents any point in a region of the $xy$-plane over which the function $f$ is defined. The value $f(x, y)$ that the function $f$ assigns to the point represents the slope of a line or, as we shall envision it, a line segment called a **lineal element.** For example, consider the equation $dy/dx = 0.2xy$, where $f(x, y) = 0.2xy$. At, say, the point $(2, 3)$ the slope of a lineal element is $f(2, 3) = 0.2(2)(3) = 1.2$. Figure 2.1.1(a) shows a line segment with slope 1.2 passing though $(2, 3)$. As shown in Figure 2.1.1(b), *if* a solution curve also passes through the point $(2, 3)$, it does so tangent to this line segment; in other words, the lineal element is a miniature tangent line at that point.

**DIRECTION FIELD**   If we systematically evaluate $f$ over a rectangular grid of points in the $xy$-plane and draw a line element at each point $(x, y)$ of the grid with slope $f(x, y)$, then the collection of all these line elements is called a **direction field** or a **slope field** of the differential equation $dy/dx = f(x, y)$. Visually, the direction field suggests the appearance or shape of a family of solution curves of the differential equation, and consequently, it may be possible to see at a glance certain qualitative aspects of the solutions—regions in the plane, for example, in which a solution exhibits an unusual behavior. A single solution curve that passes through

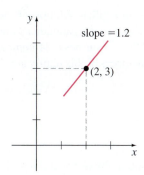

**(a)** lineal element at a point

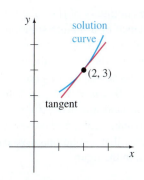

**(b)** lineal element is tangent to solution curve that passes through the point

**FIGURE 2.1.1**  A solution curve is tangent to lineal element at $(2, 3)$

**FIGURE 2.1.2** Solution curves following flow of a direction field

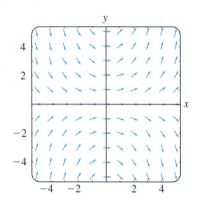

**(a)** direction field for $dy/dx = 0.2xy$

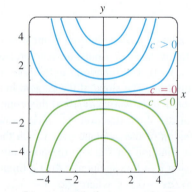

**(b)** some solution curves in the family $y = ce^{0.1x^2}$

**FIGURE 2.1.3** Direction field and solution curves in Example 1

a direction field must follow the flow pattern of the field; it is tangent to a lineal element when it intersects a point in the grid. Figure 2.1.2 shows a computer-generated direction field of the differential equation $dy/dx = \sin(x + y)$ over a region of the $xy$-plane. Note how the three solution curves shown in color follow the flow of the field.

### EXAMPLE 1 Direction Field

The direction field for the differential equation $dy/dx = 0.2xy$ shown in Figure 2.1.3(a) was obtained by using computer software in which a $5 \times 5$ grid of points $(mh, nh)$, $m$ and $n$ integers, was defined by letting $-5 \le m \le 5$, $-5 \le n \le 5$, and $h = 1$. Notice in Figure 2.1.3(a) that at any point along the $x$-axis $(y = 0)$ and the $y$-axis $(x = 0)$, the slopes are $f(x, 0) = 0$ and $f(0, y) = 0$, respectively, so the lineal elements are horizontal. Moreover, observe in the first quadrant that for a fixed value of $x$ the values of $f(x, y) = 0.2xy$ increase as $y$ increases; similarly, for a fixed $y$ the values of $f(x, y) = 0.2xy$ increase as $x$ increases. This means that as both $x$ and $y$ increase, the lineal elements almost become vertical and have positive slope ($f(x, y) = 0.2xy > 0$ for $x > 0$, $y > 0$). In the second quadrant, $|f(x, y)|$ increases as $|x|$ and $y$ increase, so the lineal elements again become almost vertical but this time have negative slope ($f(x, y) = 0.2xy < 0$ for $x < 0$, $y > 0$). Reading from left to right, imagine a solution curve that starts at a point in the second quadrant, moves steeply downward, becomes flat as it passes through the $y$-axis, and then, as it enters the first quadrant, moves steeply upward—in other words, its shape would be concave upward and similar to a horseshoe. From this it could be surmised that $y \to \infty$ as $x \to \pm\infty$. Now in the third and fourth quadrants, since $f(x, y) = 0.2xy > 0$ and $f(x, y) = 0.2xy < 0$, respectively, the situation is reversed: A solution curve increases and then decreases as we move from left to right.

We saw in (1) of Section 1.1 that $y = e^{0.1x^2}$ is an explicit solution of the differential equation $dy/dx = 0.2xy$; you should verify that a one-parameter family of solutions of the same equation is given by $y = ce^{0.1x^2}$. For purposes of comparison with Figure 2.1.3(a) some representative graphs of members of this family are shown in Figure 2.1.3(b). ∎

### EXAMPLE 2 Direction Field

Use a direction field to sketch an approximate solution curve for the initial-value problem $dy/dx = \sin y$, $y(0) = -\frac{3}{2}$.

**SOLUTION** Before proceeding, recall that from the continuity of $f(x, y) = \sin y$ and $\partial f/\partial y = \cos y$, Theorem 1.2.1 guarantees the existence of a unique solution curve passing through any specified point $(x_0, y_0)$ in the plane. Now we set our computer software again for a $5 \times 5$ rectangular region and specify (because of the initial condition) points in that region with vertical and horizontal separation of $\frac{1}{2}$ unit—that is, at points $(mh, nh)$, $h = \frac{1}{2}$, $m$ and $n$ integers such that $-10 \le m \le 10$, $-10 \le n \le 10$. The result is shown in Figure 2.1.4. Because the right-hand side of $dy/dx = \sin y$ is 0 at $y = 0$, and at $y = -\pi$, the lineal elements are horizontal at all points whose second coordinates are $y = 0$ or $y = -\pi$. It makes sense then that a solution curve passing through the initial point $\left(0, -\frac{3}{2}\right)$ has the shape shown in the figure. ∎

**INCREASING/DECREASING** Interpretation of the derivative $dy/dx$ as a function that gives slope plays the key role in the construction of a direction field. Another telling property of the first derivative will be used next, namely, if $dy/dx > 0$ (or $dy/dx < 0$) for all $x$ in an interval $I$, then a differentiable function $y = y(x)$ is increasing (or decreasing) on $I$.

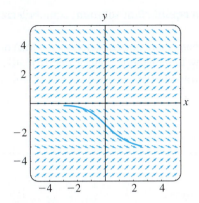

**FIGURE 2.1.4** Direction field in Example 2 on page 38

## REMARKS

Sketching a direction field by hand is straightforward but time consuming; it is probably one of those tasks about which an argument can be made for doing it once or twice in a lifetime, but it is overall most efficiently carried out by means of computer software. Before calculators, PCs, and software the **method of isoclines** was used to facilitate sketching a direction field by hand. For the DE $dy/dx = f(x, y)$, any member of the family of curves $f(x, y) = c$, $c$ a constant, is called an **isocline.** Lineal elements drawn through points on a specific isocline, say, $f(x, y) = c_1$ all have the same slope $c_1$. In Problem 15 in Exercises 2.1 you have your two opportunities to sketch a direction field by hand.

### 2.1.2 AUTONOMOUS FIRST-ORDER DEs

**AUTONOMOUS FIRST-ORDER DEs**    In Section 1.1 we divided the class of ordinary differential equations into two types: linear and nonlinear. We now consider briefly another kind of classification of ordinary differential equations, a classification that is of particular importance in the qualitative investigation of differential equations. An ordinary differential equation in which the independent variable does not appear explicitly is said to be **autonomous.** If the symbol $x$ denotes the independent variable, then an autonomous first-order differential equation can be written as $f(y, y') = 0$ or in normal form as

$$\frac{dy}{dx} = f(y). \tag{2}$$

We shall assume throughout that the function $f$ in (2) and its derivative $f'$ are continuous functions of $y$ on some interval $I$. The first-order equations

$$\overset{\displaystyle f(y)}{\underset{\textstyle \downarrow}{}}\qquad\qquad \overset{\displaystyle f(x, y)}{\underset{\textstyle \downarrow}{}}$$

$$\frac{dy}{dx} = 1 + y^2 \qquad \text{and} \qquad \frac{dy}{dx} = 0.2xy$$

are autonomous and nonautonomous, respectively.

Many differential equations encountered in applications or equations that are models of physical laws that do not change over time are autonomous. As we have already seen in Section 1.3, in an applied context, symbols other than $y$ and $x$ are routinely used to represent the dependent and independent variables. For example, if $t$ represents time then inspection of

$$\frac{dA}{dt} = kA, \qquad \frac{dx}{dt} = kx(n + 1 - x), \qquad \frac{dT}{dt} = k(T - T_m), \qquad \frac{dA}{dt} = 6 - \frac{1}{100}A,$$

where $k$, $n$, and $T_m$ are constants, shows that each equation is time independent. Indeed, *all* of the first-order differential equations introduced in Section 1.3 are time independent and so are autonomous.

**CRITICAL POINTS**    The zeros of the function $f$ in (2) are of special importance. We say that a real number $c$ is a **critical point** of the autonomous differential equation (2) if it is a zero of $f$—that is, $f(c) = 0$. A critical point is also called an **equilibrium point** or **stationary point.** Now observe that if we substitute the constant function $y(x) = c$ into (2), then both sides of the equation are zero. This means:

*If $c$ is a critical point of (2), then $y(x) = c$ is a constant solution of the autonomous differential equation.*

A constant solution $y(x) = c$ of (2) is called an **equilibrium solution;** equilibria are the *only* constant solutions of (2).

As was already mentioned, we can tell when a nonconstant solution $y = y(x)$ of (2) is increasing or decreasing by determining the algebraic sign of the derivative $dy/dx$; in the case of (2) we do this by identifying intervals on the $y$-axis over which the function $f(y)$ is positive or negative.

### EXAMPLE 3 An Autonomous DE

The differential equation

$$\frac{dP}{dt} = P(a - bP),$$

where $a$ and $b$ are positive constants, has the normal form $dP/dt = f(P)$, which is (2) with $t$ and $P$ playing the parts of $x$ and $y$, respectively, and hence is autonomous. From $f(P) = P(a - bP) = 0$ we see that $0$ and $a/b$ are critical points of the equation, so the equilibrium solutions are $P(t) = 0$ and $P(t) = a/b$. By putting the critical points on a vertical line, we divide the line into three intervals defined by $-\infty < P < 0$, $0 < P < a/b$, $a/b < P < \infty$. The arrows on the line shown in Figure 2.1.5 indicate the algebraic sign of $f(P) = P(a - bP)$ on these intervals and whether a nonconstant solution $P(t)$ is increasing or decreasing on an interval. The following table explains the figure.

**FIGURE 2.1.5** Phase portrait of the DE in Example 3

| Interval | Sign of $f(P)$ | $P(t)$ | Arrow |
|---|---|---|---|
| $(-\infty, 0)$ | minus | decreasing | points down |
| $(0, a/b)$ | plus | increasing | points up |
| $(a/b, \infty)$ | minus | decreasing | points down |

Figure 2.1.5 is called a **one-dimensional phase portrait,** or simply **phase portrait,** of the differential equation $dP/dt = P(a - bP)$. The vertical line is called a **phase line.**

**(a)** region $R$

**(b)** subregions $R_1, R_2,$ and $R_3$ of $R$

**FIGURE 2.1.6** Lines $y(x) = c_1$ and $y(x) = c_2$ partition $R$ into three horizontal subregions

**SOLUTION CURVES** Without solving an autonomous differential equation, we can usually say a great deal about its solution curves. Since the function $f$ in (2) is independent of the variable $x$, we may consider $f$ defined for $-\infty < x < \infty$ or for $0 \le x < \infty$. Also, since $f$ and its derivative $f'$ are continuous functions of $y$ on some interval $I$ of the $y$-axis, the fundamental results of Theorem 1.2.1 hold in some horizontal strip or region $R$ in the $xy$-plane corresponding to $I$, and so through any point $(x_0, y_0)$ in $R$ there passes only one solution curve of (2). See Figure 2.1.6(a). For the sake of discussion, let us suppose that (2) possesses exactly two critical points $c_1$ and $c_2$ and that $c_1 < c_2$. The graphs of the equilibrium solutions $y(x) = c_1$ and $y(x) = c_2$ are horizontal lines, and these lines partition the region $R$ into three subregions $R_1$, $R_2$, and $R_3$, as illustrated in Figure 2.1.6(b). Without proof here are some conclusions that we can draw about a nonconstant solution $y(x)$ of (2):

- If $(x_0, y_0)$ is in a subregion $R_i$, $i = 1, 2, 3$, and $y(x)$ is a solution whose graph passes through this point, then $y(x)$ remains in the subregion $R_i$ for all $x$. As illustrated in Figure 2.1.6(b), the solution $y(x)$ in $R_2$ is bounded below by $c_1$ and above by $c_2$, that is, $c_1 < y(x) < c_2$ for all $x$. The solution curve stays within $R_2$ for all $x$ because the graph of a nonconstant solution of (2) cannot cross the graph of either equilibrium solution $y(x) = c_1$ or $y(x) = c_2$. See Problem 33 in Exercises 2.1.

- By continuity of $f$ we must then have either $f(y) > 0$ or $f(y) < 0$ for all $x$ in a subregion $R_i$, $i = 1, 2, 3$. In other words, $f(y)$ cannot change signs in a subregion. See Problem 33 in Exercises 2.1.

- Since $dy/dx = f(y(x))$ is either positive or negative in a subregion $R_i$, $i = 1$, 2, 3, a solution $y(x)$ is strictly monotonic—that is, $y(x)$ is either increasing or decreasing in the subregion $R_i$. Therefore $y(x)$ cannot be oscillatory, nor can it have a relative extremum (maximum or minimum). See Problem 33 in Exercises 2.1.

- If $y(x)$ is *bounded above* by a critical point $c_1$ (as in subregion $R_1$ where $y(x) < c_1$ for all $x$), then the graph of $y(x)$ must approach the graph of the equilibrium solution $y(x) = c_1$ either as $x \to \infty$ or as $x \to -\infty$. If $y(x)$ is *bounded*—that is, bounded above and below by two consecutive critical points (as in subregion $R_2$ where $c_1 < y(x) < c_2$ for all $x$)—then the graph of $y(x)$ must approach the graphs of the equilibrium solutions $y(x) = c_1$ and $y(x) = c_2$, one as $x \to \infty$ and the other as $x \to -\infty$. If $y(x)$ is *bounded below* by a critical point (as in subregion $R_3$ where $c_2 < y(x)$ for all $x$), then the graph of $y(x)$ must approach the graph of the equilibrium solution $y(x) = c_2$ either as $x \to \infty$ or as $x \to -\infty$. See Problem 34 in Exercises 2.1.

With the foregoing facts in mind, let us reexamine the differential equation in Example 3.

### EXAMPLE 4 | Example 3 Revisited

The three intervals determined on the $P$-axis or phase line by the critical points 0 and $a/b$ now correspond in the $tP$-plane to three subregions defined by:

$$R_1: -\infty < P < 0, \qquad R_2: 0 < P < a/b, \qquad \text{and} \qquad R_3: a/b < P < \infty,$$

where $-\infty < t < \infty$. The phase portrait in Figure 2.1.7 tells us that $P(t)$ is decreasing in $R_1$, increasing in $R_2$, and decreasing in $R_3$. If $P(0) = P_0$ is an initial value, then in $R_1$, $R_2$, and $R_3$ we have, respectively, the following:

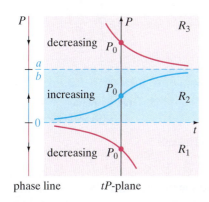

$(i)$   For $P_0 < 0$, $P(t)$ is bounded above. Since $P(t)$ is decreasing, $P(t)$ decreases without bound for increasing $t$, and so $P(t) \to 0$ as $t \to -\infty$. This means that the negative $t$-axis, the graph of the equilibrium solution $P(t) = 0$, is a horizontal asymptote for a solution curve.

$(ii)$  For $0 < P_0 < a/b$, $P(t)$ is bounded. Since $P(t)$ is increasing, $P(t) \to a/b$ as $t \to \infty$ and $P(t) \to 0$ as $t \to -\infty$. The graphs of the two equilibrium solutions, $P(t) = 0$ and $P(t) = a/b$, are horizontal lines that are horizontal asymptotes for any solution curve starting in this subregion.

$(iii)$ For $P_0 > a/b$, $P(t)$ is bounded below. Since $P(t)$ is decreasing, $P(t) \to a/b$ as $t \to \infty$. The graph of the equilibrium solution $P(t) = a/b$ is a horizontal asymptote for a solution curve.

**FIGURE 2.1.7** Phase portrait and solution curves in Example 4

In Figure 2.1.7 the phase line is the $P$-axis in the $tP$-plane. For clarity the original phase line from Figure 2.1.5 is reproduced to the left of the plane in which the subregions $R_1$, $R_2$, and $R_3$ are shaded. The graphs of the equilibrium solutions $P(t) = a/b$ and $P(t) = 0$ (the $t$-axis) are shown in the figure as blue dashed lines; the solid graphs represent typical graphs of $P(t)$ illustrating the three cases just discussed. ∎

In a subregion such as $R_1$ in Example 4, where $P(t)$ is decreasing and unbounded below, we must necessarily have $P(t) \to -\infty$. Do *not* interpret this last statement to mean $P(t) \to -\infty$ as $t \to \infty$; we could have $P(t) \to -\infty$ as $t \to T$, where $T > 0$ is a finite number that depends on the initial condition $P(t_0) = P_0$. Thinking in dynamic terms, $P(t)$ could "blow up" in finite time; thinking graphically, $P(t)$ could have a vertical asymptote at $t = T > 0$. A similar remark holds for the subregion $R_3$.

The differential equation $dy/dx = \sin y$ in Example 2 is autonomous and has an infinite number of critical points, since $\sin y = 0$ at $y = n\pi$, $n$ an integer. Moreover, we now know that because the solution $y(x)$ that passes through $\left(0, -\frac{3}{2}\right)$ is bounded above and below by two consecutive critical points ($-\pi < y(x) < 0$) and is decreasing ($\sin y < 0$ for $-\pi < y < 0$), the graph of $y(x)$ must approach the graphs of the equilibrium solutions as horizontal asymptotes: $y(x) \to -\pi$ as $x \to \infty$ and $y(x) \to 0$ as $x \to -\infty$.

**EXAMPLE 5** **Solution Curves of an Autonomous DE**

The autonomous equation $dy/dx = (y - 1)^2$ possesses the single critical point 1. From the phase portrait in Figure 2.1.8(a) we conclude that a solution $y(x)$ is an increasing function in the subregions defined by $-\infty < y < 1$ and $1 < y < \infty$, where $-\infty < x < \infty$. For an initial condition $y(0) = y_0 < 1$, a solution $y(x)$ is increasing and bounded above by 1, and so $y(x) \to 1$ as $x \to \infty$; for $y(0) = y_0 > 1$ a solution $y(x)$ is increasing and unbounded.

Now $y(x) = 1 - 1/(x + c)$ is a one-parameter family of solutions of the differential equation. (See Problem 4 in Exercises 2.2.) A given initial condition determines a value for $c$. For the initial conditions, say, $y(0) = -1 < 1$ and $y(0) = 2 > 1$, we find, in turn, that $y(x) = 1 - 1/\left(x + \frac{1}{2}\right)$, and $y(x) = 1 - 1/(x - 1)$. As shown in Figures 2.1.8(b) and 2.1.8(c), the graph of each of these rational functions possesses

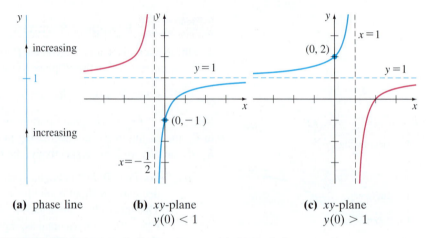

**(a)** phase line     **(b)** $xy$-plane     **(c)** $xy$-plane
                                     $y(0) < 1$             $y(0) > 1$

**FIGURE 2.1.8** Behavior of solutions near $y = 1$ in Example 5

a vertical asymptote. But bear in mind that the solutions of the IVPs

$$\frac{dy}{dx} = (y - 1)^2, \quad y(0) = -1 \quad \text{and} \quad \frac{dy}{dx} = (y - 1)^2, \quad y(0) = 2$$

are defined on special intervals. The two solutions are, respectively,

$$y(x) = 1 - \frac{1}{x + \frac{1}{2}}, \quad -\tfrac{1}{2} < x < \infty \quad \text{and} \quad y(x) = 1 - \frac{1}{x - 1}, \quad -\infty < x < 1.$$

The solution curves are the portions of the graphs in Figures 2.1.8(b) and 2.1.8(c) shown in blue. As predicted by the phase portrait, for the solution curve in Figure 2.1.8(b), $y(x) \to 1$ as $x \to \infty$; for the solution curve in Figure 2.1.8(c), $y(x) \to \infty$ as $x \to 1$ from the left. ■

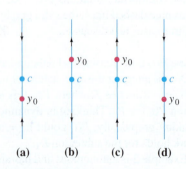

**FIGURE 2.1.9** Critical point $c$ is an attractor in (a), a repeller in (b), and semi-stable in (c) and (d).

**(a)**    **(b)**    **(c)**    **(d)**

**ATTRACTORS AND REPELLERS** Suppose that $y(x)$ is a nonconstant solution of the autonomous differential equation given in (1) and that $c$ is a critical point of the DE. There are basically three types of behavior that $y(x)$ can exhibit near $c$. In Figure 2.1.9 we have placed $c$ on four vertical phase lines. When both arrowheads on either side of the dot labeled $c$ point *toward* $c$, as in Figure 2.1.9(a), all solutions $y(x)$ of (1) that start from an initial point $(x_0, y_0)$ sufficiently near $c$ exhibit the asymptotic behavior $\lim_{x \to \infty} y(x) = c$. For this reason the critical point $c$ is said to be **asymptotically stable.** Using a physical analogy, a solution that starts near $c$ is like a charged particle that, over time, is drawn to a particle of opposite charge, and so $c$ is also referred to as an **attractor.** When both arrowheads on either side of the dot labeled $c$ point *away* from $c$, as in Figure 2.1.9(b), all solutions $y(x)$ of (1) that start from an initial point $(x_0, y_0)$ move away from $c$ as $x$ increases. In this case the critical point $c$ is said to be **unstable.** An unstable critical point is also called a **repeller,** for obvious reasons. The critical point $c$ illustrated in Figures 2.1.9(c) and 2.1.9(d) is neither an attractor nor a repeller. But since $c$ exhibits characteristics of both an attractor and a repeller—that is, a solution starting from an initial point $(x_0, y_0)$ sufficiently near $c$ is attracted to $c$ from one side and repelled from the other side—we say that the critical point $c$ is **semi-stable.** In Example 3 the critical point $a/b$ is asymptotically stable (an attractor) and the critical point 0 is unstable (a repeller). The critical point 1 in Example 5 is semi-stable.

**AUTONOMOUS DEs AND DIRECTION FIELDS**    If a first-order differential equation is autonomous, then we see from the right-hand side of its normal form $dy/dx = f(y)$ that slopes of lineal elements through points in the rectangular grid used to construct a direction field for the DE depend solely on the $y$-coordinate of the points. Put another way, lineal elements passing through points on any *horizontal* line must all have the same slope and therefore are parallel; slopes of lineal elements along any *vertical* line will, of course, vary. These facts are apparent from inspection of the horizontal yellow strip and vertical blue strip in Figure 2.1.10. The figure exhibits a direction field for the autonomous equation $dy/dx = 2(y - 1)$. The red lineal elements in Figure 2.1.10 have zero slope because they lie along the graph of the equilibrium solution $y = 1$.

slopes of lineal elements on a vertical line vary

slopes of lineal elements on a horizontal line are all the same

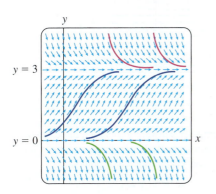

**FIGURE 2.1.10** Direction field for an autonomous DE

**TRANSLATION PROPERTY**    You may recall from precalculus mathematics that the graph of a function $y = f(x - k)$, where $k$ is a constant, is the graph of $y = f(x)$ rigidly translated or shifted horizontally along the $x$-axis by an amount $|k|$; the translation is to the right if $k > 0$ and to the left if $k < 0$. It turns out that under the conditions stipulated for (2), solution curves of an autonomous first-order DE are related by the concept of translation. To see this, let's consider the differential equation $dy/dx = y(3 - y)$, which is a special case of the autonomous equation considered in Examples 3 and 4. Because $y = 0$ and $y = 3$ are equilibrium solutions of the DE, their graphs divide the $xy$-plane into three subregions $R_1$, $R_2$, and $R_3$:

$$R_1: -\infty < y < 0, \quad R_2: 0 < y < 3, \quad \text{and} \quad R_3: 3 < y < \infty.$$

**FIGURE 2.1.11** Translated solution curves of an autonomous DE

In Figure 2.1.11 we have superimposed on a direction field of the DE six solutions curves. The figure illustrates that all solution curves of the same color, that is, solution curves lying within a particular subregion $R_i$, look alike. This is no coincidence but is a natural consequence of the fact that lineal elements passing through points on any horizontal line are parallel. That said, the following **translation property** of an autonomous DE should make sense:

*If $y(x)$ is a solution of an autonomous differential equation $dy/dx = f(y)$, then $y_1(x) = y(x - k)$, $k$ a constant, is also a solution.*

Thus, if $y(x)$ is a solution of the initial-value problem $dy/dx = f(y)$, $y(0) = y_0$, then $y_1(x) = y(x - x_0)$ is a solution of the IVP $dy/dx = f(y)$, $y(x_0) = y_0$. For example, it is easy to verify that $y(x) = e^x$, $-\infty < x < \infty$, is a solution of the IVP $dy/dx = y$, $y(0) = 1$ and so a solution $y_1(x)$ of, say, $dy/dx = y$, $y(5) = 1$ is $y(x) = e^x$ translated 5 units to the right:

$$y_1(x) = y(x - 5) = e^{x-5}, \quad -\infty < x < \infty.$$

# EXERCISES 2.1

*Answers to selected odd-numbered problems begin on page ANS-1.*

## 2.1.1 Direction Fields

In Problems 1–4 reproduce the given computer-generated direction field. Then sketch, by hand, an approximate solution curve that passes through each of the indicated points. Use different colored pencils for each solution curve.

**1.** $\dfrac{dy}{dx} = x^2 - y^2$

  **(a)** $y(-2) = 1$      **(b)** $y(3) = 0$

  **(c)** $y(0) = 2$      **(d)** $y(0) = 0$

**FIGURE 2.1.12** Direction field for Problem 1

**2.** $\dfrac{dy}{dx} = e^{-0.01xy^2}$

  **(a)** $y(-6) = 0$      **(b)** $y(0) = 1$

  **(c)** $y(0) = -4$      **(d)** $y(8) = -4$

**FIGURE 2.1.13** Direction field for Problem 2

**3.** $\dfrac{dy}{dx} = 1 - xy$

  **(a)** $y(0) = 0$      **(b)** $y(-1) = 0$

  **(c)** $y(2) = 2$      **(d)** $y(0) = -4$

**FIGURE 2.1.14** Direction field for Problem 3

**4.** $\dfrac{dy}{dx} = (\sin x) \cos y$

  **(a)** $y(0) = 1$      **(b)** $y(1) = 0$

  **(c)** $y(3) = 3$      **(d)** $y(0) = -\frac{5}{2}$

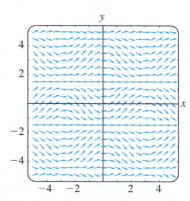

**FIGURE 2.1.15** Direction field for Problem 4

In Problems 5–12 use computer software to obtain a direction field for the given differential equation. By hand, sketch an approximate solution curve passing through each of the given points.

**5.** $y' = x$

  **(a)** $y(0) = 0$

  **(b)** $y(0) = -3$

**6.** $y' = x + y$

  **(a)** $y(-2) = 2$

  **(b)** $y(1) = -3$

**7.** $y\dfrac{dy}{dx} = -x$

  **(a)** $y(1) = 1$

  **(b)** $y(0) = 4$

**8.** $\dfrac{dy}{dx} = \dfrac{1}{y}$

  **(a)** $y(0) = 1$

  **(b)** $y(-2) = -1$

**9.** $\dfrac{dy}{dx} = 0.2x^2 + y$

  **(a)** $y(0) = \frac{1}{2}$

  **(b)** $y(2) = -1$

**10.** $\dfrac{dy}{dx} = xe^y$

  **(a)** $y(0) = -2$

  **(b)** $y(1) = 2.5$

**11.** $y' = y - \cos\dfrac{\pi}{2}x$

**(a)** $y(2) = 2$

**(b)** $y(-1) = 0$

**12.** $\dfrac{dy}{dx} = 1 - \dfrac{y}{x}$

**(a)** $y\left(-\tfrac{1}{2}\right) = 2$

**(b)** $y\left(\tfrac{3}{2}\right) = 0$

In Problems 13 and 14 the given figure represents the graph of $f(y)$ and $f(x)$, respectively. By hand, sketch a direction field over an appropriate grid for $dy/dx = f(y)$ (Problem 13) and then for $dy/dx = f(x)$ (Problem 14).

**13.**

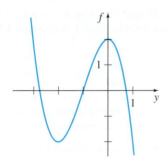

**FIGURE 2.1.16** Graph for Problem 13

**14.**

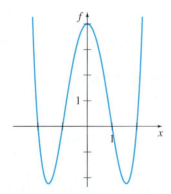

**FIGURE 2.1.17** Graph for Problem 14

**15.** In parts (a) and (b) sketch **isoclines** $f(x, y) = c$ (see the *Remarks* on page 39) for the given differential equation using the indicated values of $c$. Construct a direction field over a grid by carefully drawing lineal elements with the appropriate slope at chosen points on each isocline. In each case, use this rough direction field to sketch an approximate solution curve for the IVP consisting of the DE and the initial condition $y(0) = 1$.

**(a)** $dy/dx = x + y$; $c$ an integer satisfying $-5 \le c \le 5$

**(b)** $dy/dx = x^2 + y^2$; $c = \tfrac{1}{4}$, $c = 1$, $c = \tfrac{9}{4}$, $c = 4$

### Discussion Problems

**16. (a)** Consider the direction field of the differential equation $dy/dx = x(y - 4)^2 - 2$, but do not use technology to obtain it. Describe the slopes of the lineal elements on the lines $x = 0$, $y = 3$, $y = 4$, and $y = 5$.

**(b)** Consider the IVP $dy/dx = x(y - 4)^2 - 2$, $y(0) = y_0$, where $y_0 < 4$. Can a solution $y(x) \to \infty$ as $x \to \infty$? Based on the information in part (a), discuss.

**17.** For a first-order DE $dy/dx = f(x, y)$ a curve in the plane defined by $f(x, y) = 0$ is called a **nullcline** of the equation, since a lineal element at a point on the curve has zero slope. Use computer software to obtain a direction field over a rectangular grid of points for $dy/dx = x^2 - 2y$, and then superimpose the graph of the nullcline $y = \tfrac{1}{2}x^2$ over the direction field. Discuss the behavior of solution curves in regions of the plane defined by $y < \tfrac{1}{2}x^2$ and by $y > \tfrac{1}{2}x^2$. Sketch some approximate solution curves. Try to generalize your observations.

**18. (a)** Identify the nullclines (see Problem 17) in Problems 1, 3, and 4. With a colored pencil, circle any lineal elements in Figures 2.1.12, 2.1.14, and 2.1.15 that you think may be a lineal element at a point on a nullcline.

**(b)** What are the nullclines of an autonomous first-order DE?

### 2.1.2 Autonomous First-Order DEs

**19.** Consider the autonomous first-order differential equation $dy/dx = y - y^3$ and the initial condition $y(0) = y_0$. By hand, sketch the graph of a typical solution $y(x)$ when $y_0$ has the given values.

**(a)** $y_0 > 1$

**(b)** $0 < y_0 < 1$

**(c)** $-1 < y_0 < 0$

**(d)** $y_0 < -1$

**20.** Consider the autonomous first-order differential equation $dy/dx = y^2 - y^4$ and the initial condition $y(0) = y_0$. By hand, sketch the graph of a typical solution $y(x)$ when $y_0$ has the given values.

**(a)** $y_0 > 1$

**(b)** $0 < y_0 < 1$

**(c)** $-1 < y_0 < 0$

**(d)** $y_0 < -1$

In Problems 21–28 find the critical points and phase portrait of the given autonomous first-order differential equation. Classify each critical point as asymptotically stable, unstable, or semi-stable. By hand, sketch typical solution curves in the regions in the $xy$-plane determined by the graphs of the equilibrium solutions.

**21.** $\dfrac{dy}{dx} = y^2 - 3y$

**22.** $\dfrac{dy}{dx} = y^2 - y^3$

**23.** $\dfrac{dy}{dx} = (y - 2)^4$

**24.** $\dfrac{dy}{dx} = 10 + 3y - y^2$

**25.** $\dfrac{dy}{dx} = y^2(4 - y^2)$

**26.** $\dfrac{dy}{dx} = y(2 - y)(4 - y)$

**27.** $\dfrac{dy}{dx} = y \ln(y + 2)$

**28.** $\dfrac{dy}{dx} = \dfrac{ye^y - 9y}{e^y}$

In Problems 29 and 30 consider the autonomous differential equation $dy/dx = f(y)$, where the graph of $f$ is given. Use the graph to locate the critical points of each differential equation. Sketch a phase portrait of each differential equation. By hand, sketch typical solution curves in the subregions in the $xy$-plane determined by the graphs of the equilibrium solutions.

**29.**

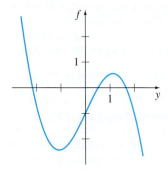

**FIGURE 2.1.18** Graph for Problem 29

**30.**

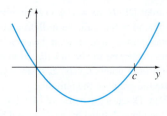

**FIGURE 2.1.19** Graph for Problem 30

## Discussion Problems

**31.** Consider the autonomous DE $dy/dx = (2/\pi)y - \sin y$. Determine the critical points of the equation. Discuss a way of obtaining a phase portrait of the equation. Classify the critical points as asymptotically stable, unstable, or semi-stable.

**32.** A critical point $c$ of an autonomous first-order DE is said to be **isolated** if there exists some open interval that contains $c$ but no other critical point. Can there exist an autonomous DE of the form given in (2) for which *every* critical point is nonisolated? Discuss; do not think profound thoughts.

**33.** Suppose that $y(x)$ is a nonconstant solution of the autonomous equation $dy/dx = f(y)$ and that $c$ is a critical point of the DE. Discuss: Why can't the graph of $y(x)$ cross the graph of the equilibrium solution $y = c$? Why can't $f(y)$ change signs in one of the subregions discussed on page 40? Why can't $y(x)$ be oscillatory or have a relative extremum (maximum or minimum)?

**34.** Suppose that $y(x)$ is a solution of the autonomous equation $dy/dx = f(y)$ and is bounded above and below by two consecutive critical points $c_1 < c_2$, as in subregion $R_2$ of Figure 2.1.6(b). If $f(y) > 0$ in the region, then $\lim_{x \to \infty} y(x) = c_2$. Discuss why there cannot exist a number $L < c_2$ such that $\lim_{x \to \infty} y(x) = L$. As part of your discussion, consider what happens to $y'(x)$ as $x \to \infty$.

**35.** Using the autonomous equation (2), discuss how it is possible to obtain information about the location of points of inflection of a solution curve.

**36.** Consider the autonomous DE $dy/dx = y^2 - y - 6$. Use your ideas from Problem 35 to find intervals on the $y$-axis for which solution curves are concave up and intervals for which solution

curves are concave down. Discuss why *each* solution curve of an initial-value problem of the form $dy/dx = y^2 - y - 6$, $y(0) = y_0$, where $-2 < y_0 < 3$, has a point of inflection with the same $y$-coordinate? What is that $y$-coordinate? Carefully sketch the solution curve for which $y(0) = -1$. Repeat for $y(2) = 2$.

**37.** Suppose the autonomous DE in (2) has no critical points. Discuss the behavior of the solutions.

## Mathematical Models

**38. Population Model** The differential equation in Example 3 is a well-known population model. Suppose the DE is changed to

$$\frac{dP}{dt} = P(aP - b),$$

where $a$ and $b$ are positive constants. Discuss what happens to the population $P$ as time $t$ increases.

**39. Population Model** Another population model is given by

$$\frac{dP}{dt} = kP - h,$$

where $h$ and $k$ are positive constants. For what initial values $P(0) = P_0$ does this model predict that the population will go extinct?

**40. Terminal Velocity** In Section 1.3 we saw that the autonomous differential equation

$$m\frac{dv}{dt} = mg - kv,$$

where $k$ is a positive constant and $g$ is the acceleration due to gravity, is a model for the velocity $v$ of a body of mass $m$ that is falling under the influence of gravity. Because the term $-kv$ represents air resistance, the velocity of a body falling from a great height does not increase without bound as time $t$ increases. Use a phase portrait of the differential equation to find the limiting, or terminal, velocity of the body. Explain your reasoning.

**41.** Suppose the model in Problem 40 is modified so that air resistance is proportional to $v^2$, that is,

$$m\frac{dv}{dt} = mg - kv^2.$$

See Problem 17 in Exercises 1.3. Use a phase portrait to find the terminal velocity of the body. Explain your reasoning.

**42. Chemical Reactions** When certain kinds of chemicals are combined, the rate at which the new compound is formed is modeled by the autonomous differential equation

$$\frac{dX}{dt} = k(\alpha - X)(\beta - X),$$

where $k > 0$ is a constant of proportionality and $\beta > \alpha > 0$. Here $X(t)$ denotes the number of grams of the new compound formed in time $t$.

**(a)** Use a phase portrait of the differential equation to predict the behavior of $X(t)$ as $t \to \infty$.

**(b)** Consider the case when $\alpha = \beta$. Use a phase portrait of the differential equation to predict the behavior of $X(t)$ as $t \to \infty$ when $X(0) < \alpha$. When $X(0) > \alpha$.

**(c)** Verify that an explicit solution of the DE in the case when $k = 1$ and $\alpha = \beta$ is $X(t) = \alpha - 1/(t + c)$. Find a solution that satisfies $X(0) = \alpha/2$. Then find a solution that satisfies $X(0) = 2\alpha$. Graph these two solutions. Does the behavior of the solutions as $t \to \infty$ agree with your answers to part (b)?

---

## 2.2 Separable Equations

**INTRODUCTION**    We begin our study of solution methods with the simplest of all differential equations: first-order equations with separable variables. Because the method discussed in this section and many other methods for solving differential equations involve integration, you are urged to refresh your memory on important formulas (such as $\int du/u$) and techniques (such as integration by parts) by consulting a calculus text.

**SOLUTION BY INTEGRATION**    Consider the first-order differential equation $dy/dx = f(x, y)$. When $f$ does not depend on the variable $y$, that is, $f(x, y) = g(x)$, the differential equation

$$\frac{dy}{dx} = g(x) \tag{1}$$

can be solved by integration. If $g(x)$ is a continuous function, then integrating both sides of (1) gives $y = \int g(x)\, dx = G(x) + c$, where $G(x)$ is an antiderivative (indefinite integral) of $g(x)$. For example, if $dy/dx = 1 + e^{2x}$, then its solution is $y = \int (1 + e^{2x})\, dx$ or $y = x + \frac{1}{2}e^{2x} + c$.

**A DEFINITION**    Equation (1), as well as its method of solution, is just a special case when the function $f$ in the normal form $dy/dx = f(x, y)$ can be factored into a function of $x$ times a function of $y$.

> **DEFINITION 2.2.1    Separable Equation**
>
> A first-order differential equation of the form
>
> $$\frac{dy}{dx} = g(x)h(y)$$
>
> is said to be **separable** or to have **separable variables.**

For example, the equations

$$\frac{dy}{dx} = y^2 x e^{3x+4y} \qquad \text{and} \qquad \frac{dy}{dx} = y + \sin x$$

are separable and nonseparable, respectively. In the first equation we can factor $f(x, y) = y^2 x e^{3x+4y}$ as

$$f(x, y) = y^2 x e^{3x+4y} = \underset{\underset{g(x)}{\downarrow}}{(x e^{3x})} \underset{\underset{h(y)}{\downarrow}}{(y^2 e^{4y})},$$

but in the second equation there is no way of expressing $y + \sin x$ as a product of a function of $x$ times a function of $y$.

Observe that by dividing by the function $h(y)$, we can write a separable equation $dy/dx = g(x)h(y)$ as

$$p(y) \frac{dy}{dx} = g(x), \tag{2}$$

where, for convenience, we have denoted $1/h(y)$ by $p(y)$. From this last form we can see immediately that (2) reduces to (1) when $h(y) = 1$.

Now if $y = \phi(x)$ represents a solution of (2), we must have $p(\phi(x))\phi'(x) = g(x)$, and therefore

$$\int p(\phi(x))\phi'(x)\, dx = \int g(x)\, dx. \tag{3}$$

But $dy = \phi'(x)\, dx$, and so (3) is the same as

$$\int p(y)\, dy = \int g(x)\, dx \qquad \text{or} \qquad H(y) = G(x) + c, \tag{4}$$

where $H(y)$ and $G(x)$ are antiderivatives of $p(y) = 1/h(y)$ and $g(x)$, respectively.

**METHOD OF SOLUTION**   Equation (4) indicates the procedure for solving separable equations. A one-parameter family of solutions, usually given implicitly, is obtained by integrating both sides of $p(y)\, dy = g(x)\, dx$.

**NOTE**   There is no need to use two constants in the integration of a separable equation, because if we write $H(y) + c_1 = G(x) + c_2$, then the difference $c_2 - c_1$ can be replaced by a single constant $c$, as in (4). In many instances throughout the chapters that follow, we will relabel constants in a manner convenient to a given equation. For example, multiples of constants or combinations of constants can sometimes be replaced by a single constant.

---

**EXAMPLE 1**   **Solving a Separable DE**

Solve $(1 + x)\, dy - y\, dx = 0$.

**SOLUTION**   Dividing by $(1 + x)y$, we can write $dy/y = dx/(1 + x)$, from which it follows that

$$\int \frac{dy}{y} = \int \frac{dx}{1 + x}$$

$$\ln|y| = \ln|1 + x| + c_1$$

$$|y| = e^{\ln|1+x|+c_1} = e^{\ln|1+x|} \cdot e^{c_1} \qquad \leftarrow \text{laws of exponents}$$

$$= |1 + x|\, e^{c_1}$$

$$y = \pm e^{c_1}(1 + x). \qquad \leftarrow \begin{cases} |1 + x| = 1 + x, & x \geq -1 \\ |1 + x| = -(1 + x), & x < -1 \end{cases}$$

Relabeling $\pm e^{c_1}$ as $c$ then gives $y = c(1 + x)$.   ■

In the solution of Example 1 because each integral results in a logarithm, a judicious choice for the constant of integration is $\ln|c|$ rather than $c$. Rewriting the second line of the solution as $\ln|y| = \ln|1 + x| + \ln|c|$ enables us to combine the terms on the right-hand side by the properties of logarithms. From $\ln|y| = \ln|c(1 + x)|$ we immediately get $y = c(1 + x)$. Even if the indefinite integrals are not *all* logarithms, it may still be advantageous to use $\ln|c|$. However, no firm rule can be given.

In Section 1.1 we saw that a solution curve may be only a segment or an arc of the graph of an implicit solution $G(x, y) = 0$.

### EXAMPLE 2    Solution Curve

Solve the initial-value problem $\dfrac{dy}{dx} = -\dfrac{x}{y}, \quad y(4) = -3$.

**SOLUTION**  Rewriting the equation as $y\,dy = -x\,dx$, we get

$$\int y\,dy = -\int x\,dx \quad \text{and} \quad \frac{y^2}{2} = -\frac{x^2}{2} + c_1.$$

We can write the result of the integration as $x^2 + y^2 = c^2$ by replacing the constant $2c_1$ by $c^2$. This solution of the differential equation represents a one-parameter family of concentric circles centered at the origin.

Now when $x = 4$, $y = -3$, so $16 + 9 = 25 = c^2$. Thus the initial-value problem determines the circle $x^2 + y^2 = 25$ with radius 5. Because of its simplicity we can solve this implicit solution for an explicit solution that satisfies the initial condition. We saw this solution as $y = \phi_2(x)$ or $y = -\sqrt{25 - x^2}, -5 < x < 5$ in Example 7 of Section 1.1. A solution curve is the graph of a differentiable function. In this case the solution curve is the lower semicircle, shown in dark blue in Figure 2.2.1 containing the point $(4, -3)$.

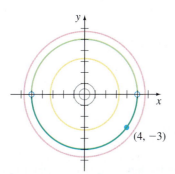

**FIGURE 2.2.1** Solution curve for the IVP in Example 2

**LOSING A SOLUTION**    Some care should be exercised in separating variables, since the variable divisors could be zero at a point. Specifically, if $r$ is a zero of the function $h(y)$, then substituting $y = r$ into $dy/dx = g(x)h(y)$ makes both sides zero; in other words, $y = r$ is a constant solution of the differential equation.

But after variables are separated, the left-hand side of $\dfrac{dy}{h(y)} = g(x)\,dx$ is undefined at $r$.

As a consequence, $y = r$ might not show up in the family of solutions that are obtained after integration and simplification. Recall that such a solution is called a singular solution.

### EXAMPLE 3    Losing a Solution

Solve $\dfrac{dy}{dx} = y^2 - 4$.

**SOLUTION**  We put the equation in the form

$$\frac{dy}{y^2 - 4} = dx \quad \text{or} \quad \left[\frac{\frac{1}{4}}{y - 2} - \frac{\frac{1}{4}}{y + 2}\right] dy = dx. \tag{5}$$

The second equation in (5) is the result of using partial fractions on the left-hand side of the first equation. Integrating and using the laws of logarithms gives

$$\frac{1}{4}\ln|y - 2| - \frac{1}{4}\ln|y + 2| = x + c_1$$

$$\ln\left|\frac{y - 2}{y + 2}\right| = 4x + c_2 \quad \text{or} \quad \frac{y - 2}{y + 2} = \pm e^{4x + c_2}.$$

Here we have replaced $4c_1$ by $c_2$. Finally, after replacing $\pm e^{c_2}$ by $c$ and solving the last equation for $y$, we get the one-parameter family of solutions

$$y = 2\frac{1 + ce^{4x}}{1 - ce^{4x}}.\tag{6}$$

Now if we factor the right-hand side of the differential equation as $dy/dx = (y - 2)(y + 2)$, we know from the discussion of critical points in Section 2.1 that $y = 2$ and $y = -2$ are two constant (equilibrium) solutions. The solution $y = 2$ is a member of the family of solutions defined by (6) corresponding to the value $c = 0$. However, $y = -2$ is a singular solution; it cannot be obtained from (6) for any choice of the parameter $c$. This latter solution was lost early on in the solution process. Inspection of (5) clearly indicates that we must preclude $y = \pm 2$ in these steps. ■

---

### EXAMPLE 4   An Initial-Value Problem

Solve $(e^{2y} - y)\cos x\,\dfrac{dy}{dx} = e^y \sin 2x, \quad y(0) = 0.$

**SOLUTION** Dividing the equation by $e^y \cos x$ gives

$$\frac{e^{2y} - y}{e^y}\,dy = \frac{\sin 2x}{\cos x}\,dx.$$

Before integrating, we use termwise division on the left-hand side and the trigonometric identity $\sin 2x = 2\sin x \cos x$ on the right-hand side. Then

$$\text{integration by parts} \rightarrow \quad \int (e^y - ye^{-y})\,dy = 2\int \sin x\,dx$$

yields

$$e^y + ye^{-y} + e^{-y} = -2\cos x + c.\tag{7}$$

The initial condition $y = 0$ when $x = 0$ implies $c = 4$. Thus a solution of the initial-value problem is

$$e^y + ye^{-y} + e^{-y} = 4 - 2\cos x.\tag{8}$$ ■

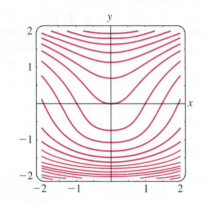

**FIGURE 2.2.2** Level curves of $G(x, y) = e^y + ye^{-y} + e^{-y} + 2\cos x$

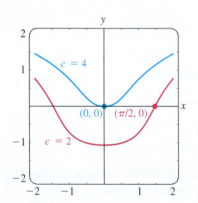

**FIGURE 2.2.3** Level curves $c = 2$ and $c = 4$

**USE OF COMPUTERS** The *Remarks* at the end of Section 1.1 mentioned that it may be difficult to use an implicit solution $G(x, y) = 0$ to find an explicit solution $y = \phi(x)$. Equation (8) shows that the task of solving for $y$ in terms of $x$ may present more problems than just the drudgery of symbol pushing—sometimes it simply cannot be done! Implicit solutions such as (8) are somewhat frustrating; neither the graph of the equation nor an interval over which a solution satisfying $y(0) = 0$ is defined is apparent. The problem of "seeing" what an implicit solution looks like can be overcome in some cases by means of technology. One way* of proceeding is to use the contour plot application of a computer algebra system (CAS). Recall from multivariate calculus that for a function of two variables $z = G(x, y)$ the *two-dimensional* curves defined by $G(x, y) = c$, where $c$ is constant, are called the *level curves* of the function. With the aid of a CAS, some of the level curves of the function $G(x, y) = e^y + ye^{-y} + e^{-y} + 2\cos x$ have been reproduced in Figure 2.2.2. The family of solutions defined by (7) is the level curves $G(x, y) = c$. Figure 2.2.3 illustrates the level curve $G(x, y) = 4$, which is the particular solution (8), in blue color. The other curve in Figure 2.2.3 is the level curve $G(x, y) = 2$, which is the member of the family $G(x, y) = c$ that satisfies $y(\pi/2) = 0$.

---

*In Section 2.6 we will discuss several other ways of proceeding that are based on the concept of a numerical solver.

If an initial condition leads to a particular solution by yielding a specific value of the parameter $c$ in a family of solutions for a first-order differential equation, there is a natural inclination for most students (and instructors) to relax and be content. However, a solution of an initial-value problem might not be unique. We saw in Example 4 of Section 1.2 that the initial-value problem

$$\frac{dy}{dx} = xy^{1/2}, \quad y(0) = 0 \tag{9}$$

has at least two solutions, $y = 0$ and $y = \frac{1}{16}x^4$. We are now in a position to solve the equation. Separating variables and integrating $y^{-1/2}\,dy = x\,dx$ gives $2y^{1/2} = \frac{1}{2}x^2 + c_1$. Solving for $y$ and replacing $\frac{1}{2}c_1$ by $c$ yields

$$y = \left(\frac{1}{4}x^2 + c\right)^2. \tag{10}$$

Each of the functions in the family given in (10) is a solution of the DE defined on the interval $(-\infty, \infty)$ provided we take $c \geq 0$. See Problem 52 in Exercises 2.2. Now when we substitute $x = 0$, $y = 0$ in (10) we see that $c = 0$. Therefore $y = \frac{1}{16}x^4$ is a solution of the IVP. The trivial solution $y = 0$ was lost by dividing by $y^{1/2}$. The initial-value problem (9) actually possesses many more solutions, since for any choice of the parameter $a \geq 0$ the piecewise-defined function

$$y = \begin{cases} 0, & x < a \\ \frac{1}{16}(x^2 - a^2)^2, & x \geq a \end{cases}$$

satisfies both the differential equation and the initial condition. See Figure 2.2.4.

**FIGURE 2.2.4** Piecewise-defined solutions of (9)

**AN INTEGRAL-DEFINED FUNCTION**  In (*ii*) of the *Remarks* at the end of Section 1.1 it was pointed out that a solution method for a certain kind of differential equation may lead to an integral-defined function. This is especially true for separable differential equations because integration *is* the method of solution. For example, if $g$ is continuous on some interval $I$ containing $x_0$ and $x$, then a solution of the simple initial-value problem $dy/dx = g(x)$, $y(x_0) = y_0$, defined on $I$, is given by

$$y(x) = y_0 + \int_{x_0}^{x} g(t)\,dt.$$

To see this, we have immediately from (12) of Section 1.1 that $dy/dx = g(x)$ and $y(x_0) = y_0$ because $\int_{x_0}^{x_0} g(t)\,dt = 0$. When $\int g(t)\,dt$ is nonelementary—that is, cannot be expressed in terms of elementary functions—the form $y(x) = y_0 + \int_{x_0}^{x} g(t)\,dt$ may be the best we can do in obtaining an explicit solution of an IVP. The next example illustrates this idea.

**EXAMPLE 5  An Initial-Value Problem**

Solve $\dfrac{dy}{dx} = e^{-x^2}$, $y(3) - 5$.

**SOLUTION**  The function $g(x) = e^{-x^2}$ is continuous on $(-\infty, \infty)$, but its antiderivative is not an elementary function. Using $t$ as dummy variable of integration, we can write

$$\int_{3}^{x} \frac{dy}{dt}\,dt = \int_{3}^{x} e^{-t^2}\,dt$$

$$y(t)\Big]_{3}^{x} = \int_{3}^{x} e^{-t^2}\,dt$$

$$y(x) - y(3) = \int_{3}^{x} e^{-t^2}\,dt$$

$$y(x) = y(3) + \int_{3}^{x} e^{-t^2}\,dt.$$

Using the initial condition $y(3) = 5$, we obtain the solution

$$y(x) = 5 + \int_3^x e^{-t^2}\, dt.$$

The procedure demonstrated in Example 5 works equally well on separable equations $dy/dx = g(x)f(y)$ where, say, $f(y)$ possesses an elementary antiderivative but $g(x)$ does not possess an elementary antiderivative. See Problems 29 and 30 in Exercises 2.2.

### REMARKS

In some of the preceding examples we saw that the constant in the one-parameter family of solutions for a first-order differential equation can be relabeled when convenient. Also, it can easily happen that two individuals solving the same equation correctly arrive at dissimilar expressions for their answers. For example, by separation of variables we can show that one-parameter families of solutions for the DE $(1 + y^2)\, dx + (1 + x^2)\, dy = 0$ are

$$\arctan x + \arctan y = c \qquad \text{or} \qquad \frac{x + y}{1 - xy} = c.$$

As you work your way through the next several sections, bear in mind that families of solutions may be equivalent in the sense that one family may be obtained from another by either relabeling the constant or applying algebra and trigonometry. See Problems 27 and 28 in Exercises 2.2.

## EXERCISES 2.2

Answers to selected odd-numbered problems begin on page ANS-1.

In Problems 1–22 solve the given differential equation by separation of variables.

**1.** $\dfrac{dy}{dx} = \sin 5x$

**2.** $\dfrac{dy}{dx} = (x + 1)^2$

**3.** $dx + e^{3x}\, dy = 0$

**4.** $dy - (y - 1)^2\, dx = 0$

**5.** $x\dfrac{dy}{dx} = 4y$

**6.** $\dfrac{dy}{dx} + 2xy^2 = 0$

**7.** $\dfrac{dy}{dx} = e^{3x+2y}$

**8.** $e^x y \dfrac{dy}{dx} = e^{-y} + e^{-2x-y}$

**9.** $y \ln x \dfrac{dx}{dy} = \left(\dfrac{y + 1}{x}\right)^2$

**10.** $\dfrac{dy}{dx} = \left(\dfrac{2y + 3}{4x + 5}\right)^2$

**11.** $\csc y\, dx + \sec^2 x\, dy = 0$

**12.** $\sin 3x\, dx + 2y \cos^3 3x\, dy = 0$

**13.** $(e^y + 1)^2 e^{-y}\, dx + (e^x + 1)^3 e^{-x}\, dy = 0$

**14.** $x(1 + y^2)^{1/2}\, dx = y(1 + x^2)^{1/2}\, dy$

**15.** $\dfrac{dS}{dr} = kS$

**16.** $\dfrac{dQ}{dt} = k(Q - 70)$

**17.** $\dfrac{dP}{dt} = P - P^2$

**18.** $\dfrac{dN}{dt} + N = Nte^{t+2}$

**19.** $\dfrac{dy}{dx} = \dfrac{xy + 3x - y - 3}{xy - 2x + 4y - 8}$

**20.** $\dfrac{dy}{dx} = \dfrac{xy + 2y - x - 2}{xy - 3y + x - 3}$

**21.** $\dfrac{dy}{dx} = x\sqrt{1 - y^2}$

**22.** $(e^x + e^{-x})\dfrac{dy}{dx} = y^2$

In Problems 23–28 find an explicit solution of the given initial-value problem.

**23.** $\dfrac{dx}{dt} = 4(x^2 + 1), \quad x(\pi/4) = 1$

**24.** $\dfrac{dy}{dx} = \dfrac{y^2 - 1}{x^2 - 1}, \quad y(2) = 2$

**25.** $x^2 \dfrac{dy}{dx} = y - xy, \quad y(-1) = -1$

**26.** $\dfrac{dy}{dt} + 2y = 1, \quad y(0) = \dfrac{5}{2}$

**27.** $\sqrt{1 - y^2}\, dx - \sqrt{1 - x^2}\, dy = 0, \quad y(0) = \dfrac{\sqrt{3}}{2}$

**28.** $(1 + x^4)\, dy + x(1 + 4y^2)\, dx = 0, \quad y(1) = 0$

In Problems 29 and 30 proceed as in Example 5 and find an explicit solution of the given initial-value problem.

**29.** $\dfrac{dy}{dx} = ye^{-x^2}, \quad y(4) = 1$

**30.** $\dfrac{dy}{dx} = y^2 \sin x^2, \quad y(-2) = \dfrac{1}{3}$

In Problems 31–34 find an explicit solution of the given initial-value problem. Determine the exact interval $I$ of definition by analytical methods. Use a graphing utility to plot the graph of the solution.

**31.** $\dfrac{dy}{dx} = \dfrac{2x + 1}{2y}$, $\quad y(-2) = -1$

**32.** $(2y - 2)\dfrac{dy}{dx} = 3x^2 + 4x + 2$, $\quad y(1) = -2$

**33.** $e^y\, dx - e^{-x}\, dy = 0$, $\quad y(0) = 0$

**34.** $\sin x\, dx + y\, dy = 0$, $\quad y(0) = 1$

**35. (a)** Find a solution of the initial-value problem consisting of the differential equation in Example 3 and each of the initial-conditions: $y(0) = 2$, $y(0) = -2$, and $y\left(\frac{1}{4}\right) = 1$.

**(b)** Find the solution of the differential equation in Example 3 when $\ln c_1$ is used as the constant of integration on the *left-hand* side in the solution and $4 \ln c_1$ is replaced by $\ln c$. Then solve the same initial-value problems in part (a).

**36.** Find a solution of $x\dfrac{dy}{dx} = y^2 - y$ that passes through the indicated points.

**(a)** $(0, 1)$    **(b)** $(0, 0)$    **(c)** $\left(\frac{1}{2}, \frac{1}{2}\right)$    **(d)** $\left(2, \frac{1}{4}\right)$

**37.** Find a singular solution of Problem 21. Of Problem 22.

**38.** Show that an implicit solution of

$$2x \sin^2 y\, dx - (x^2 + 10) \cos y\, dy = 0$$

is given by $\ln(x^2 + 10) + \csc y = c$. Find the constant solutions, if any, that were lost in the solution of the differential equation.

Often a radical change in the form of the solution of a differential equation corresponds to a very small change in either the initial condition or the equation itself. In Problems 39–42 find an explicit solution of the given initial-value problem. Use a graphing utility to plot the graph of each solution. Compare each solution curve in a neighborhood of $(0, 1)$.

**39.** $\dfrac{dy}{dx} = (y - 1)^2$, $\quad y(0) = 1$

**40.** $\dfrac{dy}{dx} = (y - 1)^2$, $\quad y(0) = 1.01$

**41.** $\dfrac{dy}{dx} = (y - 1)^2 + 0.01$, $\quad y(0) = 1$

**42.** $\dfrac{dy}{dx} = (y - 1)^2 - 0.01$, $\quad y(0) = 1$

**43.** Every autonomous first-order equation $dy/dx = f(y)$ is separable. Find explicit solutions $y_1(x)$, $y_2(x)$, $y_3(x)$, and $y_4(x)$ of the differential equation $dy/dx = y - y^3$ that satisfy, in turn, the initial conditions $y_1(0) = 2$, $y_2(0) = \frac{1}{2}$, $y_3(0) = -\frac{1}{2}$, and $y_4(0) = -2$. Use a graphing utility to plot the graphs of each solution. Compare these graphs with those predicted in Problem 19 of Exercises 2.1. Give the exact interval of definition for each solution.

**44. (a)** The autonomous first-order differential equation $dy/dx = 1/(y - 3)$ has no critical points. Nevertheless, place 3 on the phase line and obtain a phase portrait of the equation. Compute $d^2y/dx^2$ to determine where solution curves are concave up and where they are concave down

(see Problems 35 and 36 in Exercises 2.1). Use the phase portrait and concavity to sketch, by hand, some typical solution curves.

**(b)** Find explicit solutions $y_1(x)$, $y_2(x)$, $y_3(x)$, and $y_4(x)$ of the differential equation in part (a) that satisfy, in turn, the initial conditions $y_1(0) = 4$, $y_2(0) = 2$, $y_3(1) = 2$, and $y_4(-1) = 4$. Graph each solution and compare with your sketches in part (a). Give the exact interval of definition for each solution.

In Problems 45–50 use a technique of integration or a substitution to find an explicit solution of the given differential equation or initial-value problem.

**45.** $\dfrac{dy}{dx} = \dfrac{1}{1 + \sin x}$

**46.** $\dfrac{dy}{dx} = \dfrac{\sin\sqrt{x}}{\sqrt{y}}$

**47.** $\left(\sqrt{x} + x\right)\dfrac{dy}{dx} = \sqrt{y} + y$

**48.** $\dfrac{dy}{dx} = y^{2/3} - y$

**49.** $\dfrac{dy}{dx} = \dfrac{e^{\sqrt{x}}}{y}$, $\quad y(1) = 4$

**50.** $\dfrac{dy}{dx} = \dfrac{x \tan^{-1} x}{y}$, $\quad y(0) = 3$

## Discussion Problems

**51. (a)** Explain why the interval of definition of the explicit solution $y = \phi_2(x)$ of the initial-value problem in Example 2 is the *open* interval $(-5, 5)$.

**(b)** Can any solution of the differential equation cross the $x$-axis? Do you think that $x^2 + y^2 = 1$ is an implicit solution of the initial-value problem $dy/dx = -x/y$, $y(1) = 0$?

**52.** On page 51 we showed that a one-parameter family of solutions of the first-order differential equation $dy/dx = xy^{1/2}$ is $y = \left(\frac{1}{4}x^4 + c\right)^2$ for $c \geq 0$. Each solution in this family is defined on $(-\infty, \infty)$. The last statement is not true if we choose $c$ to be negative. For $c = -1$, explain why $y = \left(\frac{1}{4}x^4 - 1\right)^2$ *is not* a solution of the DE on the interval $(-\infty, \infty)$. Find an interval of definition $I$ on which $y = \left(\frac{1}{4}x^4 - 1\right)^2$ *is* a solution of the DE.

**53.** In Problems 43 and 44 we saw that every autonomous first-order differential equation $dy/dx = f(y)$ is separable. Does this fact help in the solution of the initial-value problem

$$\dfrac{dy}{dx} = \sqrt{1 + y^2}\, \sin^2 y, \quad y(0) = \tfrac{1}{2}?$$ Discuss. Sketch, by hand, a

plausible solution curve of the problem.

**54. (a)** Solve the two initial-value problems:

$$\dfrac{dy}{dx} = y, \quad y(0) = 1$$

and

$$\dfrac{dy}{dx} = y + \dfrac{y}{x \ln x}, \quad y(e) = 1.$$

**(b)** Show that there are more than 1.65 million digits in the $y$-coordinate of the point of intersection of the two solution curves in part (a).

**55.** Find a function whose square plus the square of its derivative is 1.

**56. (a)** The differential equation in Problem 27 is equivalent to the normal form

$$\frac{dy}{dx} = \sqrt{\frac{1 - y^2}{1 - x^2}}$$

in the square region in the $xy$-plane defined by $|x| < 1$, $|y| < 1$. But the quantity under the radical is nonnegative also in the regions defined by $|x| > 1$, $|y| > 1$. Sketch all regions in the $xy$-plane for which this differential equation possesses real solutions.

**(b)** Solve the DE in part (a) in the regions defined by $|x| > 1$, $|y| > 1$. Then find an implicit and an explicit solution of the differential equation subject to $y(2) = 2$.

## Mathematical Model

**57. Suspension Bridge** In (16) of Section 1.3 we saw that a mathematical model for the shape of a flexible cable strung between two vertical supports is

$$\frac{dy}{dx} = \frac{W}{T_1}, \tag{11}$$

where $W$ denotes the portion of the total vertical load between the points $P_1$ and $P_2$ shown in Figure 1.3.7. The DE (11) is separable under the following conditions that describe a suspension bridge.

Let us assume that the $x$- and $y$-axes are as shown in Figure 2.2.5—that is, the $x$-axis runs along the horizontal roadbed, and the $y$-axis passes through $(0, a)$, which is the lowest point on one cable over the span of the bridge, coinciding with the interval $[-L/2, L/2]$. In the case of a suspension bridge, the usual assumption is that the vertical load in (11) is only a uniform roadbed distributed along the horizontal axis. In other words, it is assumed that the weight of all cables is negligible in comparison to the weight of the roadbed and that the weight per unit length of the roadbed (say, pounds per horizontal foot) is a constant $\rho$. Use this information to set up and solve an appropriate initial-value problem from which the shape (a curve with equation $y = \phi(x)$) of each of the two cables in a suspension bridge is determined. Express your solution of the IVP in terms of the sag $h$ and span $L$. See Figure 2.2.5.

## Computer Lab Assignments

**58. (a)** Use a CAS and the concept of level curves to plot representative graphs of members of the family of solutions of the differential equation $\frac{dy}{dx} = -\frac{8x + 5}{3y^2 + 1}$. Experiment with different numbers of level curves as well as various rectangular regions defined by $a \le x \le b$, $c \le y \le d$.

**(b)** On separate coordinate axes plot the graphs of the particular solutions corresponding to the initial conditions: $y(0) = -1$; $y(0) = 2$; $y(-1) = 4$; $y(-1) = -3$.

**59. (a)** Find an implicit solution of the IVP

$$(2y + 2)\, dy - (4x^3 + 6x)\, dx = 0, \quad y(0) = -3.$$

**(b)** Use part (a) to find an explicit solution $y = \phi(x)$ of the IVP.

**(c)** Consider your answer to part (b) as a *function* only. Use a graphing utility or a CAS to graph this function, and then use the graph to estimate its domain.

**(d)** With the aid of a root-finding application of a CAS, determine the approximate largest interval $I$ of definition of the *solution* $y = \phi(x)$ in part (b). Use a graphing utility or a CAS to graph the solution curve for the IVP on this interval.

**60. (a)** Use a CAS and the concept of level curves to plot representative graphs of members of the family of solutions of the differential equation

$$\frac{dy}{dx} = \frac{x(1 - x)}{y(-2 + y)}.$$

Experiment with different numbers of level curves as well as various rectangular regions in the $xy$-plane until your result resembles Figure 2.2.6.

**(b)** On separate coordinate axes, plot the graph of the implicit solution corresponding to the initial condition $y(0) = \frac{3}{2}$. Use a colored pencil to mark off that segment of the graph that corresponds to the solution curve of a solution $\phi$ that satisfies the initial condition. With the aid of a root-finding application of a CAS, determine the approximate largest interval $I$ of definition of the solution $\phi$. [*Hint*: First find the points on the curve in part (a) where the tangent is vertical.]

**(c)** Repeat part (b) for the initial condition $y(0) = -2$.

**FIGURE 2.2.5** Shape of a cable in Problem 57

**FIGURE 2.2.6** Level curves in Problem 60

## 2.3 Linear Equations

**INTRODUCTION**   We continue our quest for solutions of first-order differential equations by next examining linear equations. Linear differential equations are an especially "friendly" family of differential equations, in that, given a linear equation, whether first order or a higher-order kin, there is always a good possibility that we can find some sort of solution of the equation that we can examine.

**A DEFINITION**   The form of a linear first-order DE was first given in (7) of Section 1.1. This form, in the case when $n = 1$ in (6) of that section, is reproduced here for convenience.

---

**DEFINITION 2.3.1   Linear Equation**

A first-order differential equation of the form

$$a_1(x)\frac{dy}{dx} + a_0(x)y = g(x), \tag{1}$$

is said to be a **linear equation** in the variable $y$.

---

**STANDARD FORM**   By dividing both sides of (1) by the lead coefficient $a_1(x)$, we obtain a more useful form, the **standard form,** of a linear equation:

$$\frac{dy}{dx} + P(x)y = f(x). \tag{2}$$

We seek a solution of (2) on an interval $I$ for which both coefficient functions $P$ and $f$ are continuous.

Before we examine a general procedure for solving equations of form (2) we note that in some instances (2) can be solved by separation of variables. For example, you should verify that the equations

$$\frac{dy}{dx} + 2xy = 0 \qquad \text{and} \qquad \frac{dy}{dx} = y + 5$$

We match each equation with (2). In the first equation $P(x) = 2x$, $f(x) = 0$ and in the second $P(x) = -1$, $f(x) = 5$.

are both linear and separable, but that the linear equation

$$\frac{dy}{dx} + y = x$$

is not separable.

**METHOD OF SOLUTION**   The method for solving (2) hinges on a remarkable fact that the *left-hand side* of the equation can be recast into the form of the exact derivative of a product by multiplying the both sides of (2) by a special function $\mu(x)$. It is relatively easy to find the function $\mu(x)$ because we want

$$\underbrace{\frac{d}{dx}[\mu(x)y]}_{\text{product}} = \underbrace{\mu\frac{dy}{dx} + \frac{d\mu}{dx}y}_{\text{product rule}} = \underbrace{\mu\frac{dy}{dx} + \mu Py}_{\substack{\text{left-hand side of}\\ \text{(2) multiplied by } \mu(x)}}.$$

these must be equal

The equality is true provided that

$$\frac{d\mu}{dx} = \mu P.$$

The last equation can be solved by separation of variables. Integrating

$$\frac{d\mu}{\mu} = P\,dx \quad \text{and solving} \quad \ln|\mu(x)| = \int P(x)\,dx + c_1$$

See Problem 55
in Exercises 2.3

gives $\mu(x) = c_2 e^{\int P(x)\,dx}$. Even though there are infinite choices of $\mu(x)$ (all constant multiples of $e^{\int P(x)\,dx}$), all produce the same desired result. Hence we can simplify life and choose $c_2 = 1$. The function

$$\mu(x) = e^{\int P(x)\,dx} \tag{3}$$

is called an **Integrating factor** for equation (2).

Here is what we have so far: We multiplied both sides of (2) by (3) and, by construction, the left-hand side is the derivative of a product of the integrating factor and $y$:

$$e^{\int P(x)\,dx}\frac{dy}{dx} + P(x)e^{\int P(x)\,dx}y = e^{\int P(x)\,dx}f(x)$$

$$\frac{d}{dx}\left[e^{\int P(x)\,dx}\,y\right] = e^{\int P(x)\,dx}f(x).$$

Finally, we discover why (3) is called an *integrating* factor. We can integrate both sides of the last equation,

$$e^{\int P(x)\,dx}\,y = \int e^{\int P(x)\,dx}f(x)\,dx + c$$

and solve for $y$. The result is a one-parameter family of solutions of (2):

$$y = e^{-\int P(x)\,dx}\int e^{\int P(x)\,dx}f(x)\,dx + ce^{-\int P(x)\,dx}. \tag{4}$$

We emphasize that you should **not memorize** formula (4). The following procedure should be worked through each time.

### Solving a Linear First-Order Equation

(*i*)    Remember to put a linear equation into the standard form (2).

(*ii*)   From the standard form of the equation identify $P(x)$ and then find the integrating factor $e^{\int P(x)\,dx}$. No constant need be used in evaluating the indefinite integral $\int P(x)\,dx$.

(*iii*)  Multiply the both sides of the standard form equation by the integrating factor. The left-hand side of the resulting equation is automatically the derivative of the product of the integrating factor $e^{\int P(x)\,dx}$ and $y$:

$$\frac{d}{dx}\left[e^{\int P(x)\,dx}y\right] = e^{\int P(x)\,dx}f(x). \tag{5}$$

(*iv*)   Integrate both sides of the last equation and solve for $y$.

---

**EXAMPLE 1**   Solving a Linear Equation

Solve $\dfrac{dy}{dx} - 3y = 0$.

**SOLUTION** This linear equation can be solved by separation of variables. Alternatively, since the differential equation is already in standard form (2), we identify

$P(x) = -3$, and so the integrating factor is $e^{\int (-3)\, dx} = e^{-3x}$. We then multiply the given equation by this factor and recognize that

$$e^{-3x}\frac{dy}{dx} - 3e^{-3x}y = e^{-3x} \cdot 0 \qquad \text{is the same as} \qquad \frac{d}{dx}[e^{-3x}y] = 0.$$

Integration of the last equation,

$$\int \frac{d}{dx}[e^{-3x}y]\, dx = \int 0\, dx$$

then yields $e^{-3x}y = c$ or $y = ce^{3x}$, $-\infty < x < \infty$. ∎

### EXAMPLE 2 | Solving a Linear Equation

Solve $\dfrac{dy}{dx} - 3y = 6$.

**SOLUTION** This linear equation, like the one in Example 1, is already in standard form with $P(x) = -3$. Thus the integrating factor is again $e^{-3x}$. This time multiplying the given equation by this factor gives

$$e^{-3x}\frac{dy}{dx} - 3e^{-3x}y = 6e^{-3x} \qquad \text{and so} \qquad \frac{d}{dx}[e^{-3x}y] = 6e^{-3x}.$$

Integrating the last equation,

$$\int \frac{d}{dx}[e^{-3x}y]\, dx = 6\int e^{-3x}\, dx \qquad \text{gives} \qquad e^{-3x}y = -6\left(\frac{e^{-3x}}{3}\right) + c,$$

or $y = -2 + ce^{3x}$, $-\infty < x < \infty$. ∎

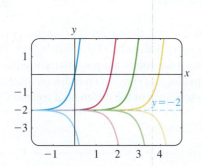

**FIGURE 2.3.1** Solution curves of DE in Example 2

When $a_1$, $a_0$, and $g$ in (1) are constants, the differential equation is autonomous. In Example 2 you can verify from the normal form $dy/dx = 3(y + 2)$ that $-2$ is a critical point and that it is unstable (a repeller). Thus a solution curve with an initial point either above or below the graph of the equilibrium solution $y = -2$ pushes away from this horizontal line as $x$ increases. Figure 2.3.1, obtained with the aid of a graphing utility, shows the graph of $y = -2$ along with some additional solution curves.

**GENERAL SOLUTION** Suppose again that the functions $P$ and $f$ in (2) are continuous on a common interval $I$. In the steps leading to (4) we showed that *if* (2) has a solution on $I$, then it must be of the form given in (4). Conversely, it is a straightforward exercise in differentiation to verify that any function of the form given in (4) is a solution of the differential equation (2) on $I$. In other words, (4) is a one-parameter family of solutions of equation (2) and *every solution of (2) defined on I is a member of this family*. Therefore we call (4) the **general solution** of the differential equation on the interval $I$. (See the *Remarks* at the end of Section 1.1.) Now by writing (2) in the normal form $y' = F(x, y)$, we can identify $F(x, y) = -P(x)y + f(x)$ and $\partial F/\partial y = -P(x)$. From the continuity of $P$ and $f$ on the interval $I$ we see that $F$ and $\partial F/\partial y$ are also continuous on $I$. With Theorem 1.2.1 as our justification, we conclude that there exists one and only one solution of the first-order initial-value problem

$$\frac{dy}{dx} + P(x)y = f(x), \quad y(x_0) = y_0 \qquad (6)$$

defined on *some* interval $I_0$ containing $x_0$. But when $x_0$ is in $I$, finding a solution of (6) is just a matter of finding an appropriate value of $c$ in (4)—that is, to each $x_0$ in $I$ there corresponds a distinct $c$. In other words, the interval $I_0$ of existence and uniqueness in Theorem 1.2.1 for the initial-value problem (6) is the entire interval $I$.

---

**EXAMPLE 3** General Solution

Solve $x\dfrac{dy}{dx} - 4y = x^6e^x$.

**SOLUTION** Dividing by $x$, the standard form of the given DE is

$$\frac{dy}{dx} - \frac{4}{x}y = x^5e^x. \tag{7}$$

From this form we identify $P(x) = -4/x$ and $f(x) = x^5e^x$ and further observe that $P$ and $f$ are continuous on $(0, \infty)$. Hence the integrating factor is

we can use ln $x$ instead of ln $|x|$ since $x > 0$
$$\downarrow$$
$$e^{-4\int dx/x} = e^{-4\ln x} = e^{\ln x^{-4}} = x^{-4}.$$

Here we have used the basic identity $b^{\log_b N} = N, N > 0$. Now we multiply (7) by $x^{-4}$ and rewrite

$$x^{-4}\frac{dy}{dx} - 4x^{-5}y = xe^x \qquad \text{as} \qquad \frac{d}{dx}[x^{-4}y] = xe^x.$$

It follows from integration by parts that the general solution defined on the interval $(0, \infty)$ is $x^{-4}y = xe^x - e^x + c$ or $y = x^5e^x - x^4e^x + cx^4$. ■

Except in the case in which the lead coefficient is 1, the recasting of equation (1) into the standard form (2) requires division by $a_1(x)$. Values of $x$ for which $a_1(x) = 0$ are called **singular points** of the equation. Singular points are potentially trouble-some. Specifically, in (2), if $P(x)$ (formed by dividing $a_0(x)$ by $a_1(x)$) is discontinuous at a point, the discontinuity may carry over to solutions of the differential equation.

In case you are wondering why the interval $(0, \infty)$ is important in Example 3, read this paragraph and the paragraph following Example 4.

---

**EXAMPLE 4** General Solution

Find the general solution of $(x^2 - 9)\dfrac{dy}{dx} + xy = 0$.

**SOLUTION** We write the differential equation in standard form

$$\frac{dy}{dx} + \frac{x}{x^2 - 9}y = 0 \tag{8}$$

and identify $P(x) = x/(x^2 - 9)$. Although $P$ is continuous on $(-\infty, -3)$, $(-3, 3)$, and $(3, \infty)$, we shall solve the equation on the first and third intervals. On these intervals the integrating factor is

$$e^{\int x\,dx/(x^2-9)} = e^{1/2\int 2x\,dx/(x^2-9)} = e^{1/2\ln|x^2-9|} = \sqrt{x^2 - 9}.$$

After multiplying the standard form (8) by this factor, we get

$$\frac{d}{dx}\left[\sqrt{x^2 - 9}\,y\right] = 0.$$

Integrating both sides of the last equation gives $\sqrt{x^2 - 9}\,y = c$. Thus on either $(-\infty, -3)$ or $(3, \infty)$ the general solution of the equation is $y = \dfrac{c}{\sqrt{x^2 - 9}}$. ■

Notice in Example 4 that $x = 3$ and $x = -3$ are singular points of the equation and that every function in the general solution $y = c/\sqrt{x^2 - 9}$ is discontinuous at these points. On the other hand, $x = 0$ is a singular point of the differential equation in Example 3, but the general solution $y = x^5 e^x - x^4 e^x + cx^4$ is noteworthy in that every function in this one-parameter family is continuous at $x = 0$ and is defined on the interval $(-\infty, \infty)$ and not just on $(0, \infty)$, as stated in the solution. However, the family $y = x^5 e^x - x^4 e^x + cx^4$ defined on $(-\infty, \infty)$ cannot be considered the general solution of the DE, since the singular point $x = 0$ still causes a problem. See Problems 50 and 51 in Exercises 2.3.

### EXAMPLE 5 An Initial-Value Problem

Solve $\dfrac{dy}{dx} + y = x$, $\quad y(0) = 4$.

**SOLUTION** The equation is in standard form, and $P(x) = 1$ and $f(x) = x$ are continuous on $(-\infty, \infty)$. The integrating factor is $e^{\int dx} = e^x$, so integrating

$$\frac{d}{dx}[e^x y] = xe^x$$

gives $e^x y = xe^x - e^x + c$. Solving this last equation for $y$ yields the general solution $y = x - 1 + ce^{-x}$. But from the initial condition we know that $y = 4$ when $x = 0$. Substituting these values into the general solution implies that $c = 5$. Hence the solution of the problem on the interval $(-\infty, \infty)$ is

$$y = x - 1 + 5e^{-x}. \tag{9} \blacksquare$$

Figure 2.3.2, obtained with the aid of a graphing utility, shows the graph of the solution (9) in dark blue along with the graphs of other members of the one-parameter family of solutions $y = x - 1 + ce^{-x}$. It is interesting to observe that as $x$ increases, the graphs of *all* members of this family are close to the graph of the solution $y = x - 1$. The last solution corresponds to $c = 0$ in the family and is shown in dark green in Figure 2.3.2. This asymptotic behavior of solutions is due to the fact that the contribution of $ce^{-x}$, $c \neq 0$, becomes negligible for increasing values of $x$. We say that $ce^{-x}$ is a **transient term**, since $e^{-x} \to 0$ as $x \to \infty$. While this behavior is not characteristic of all general solutions of linear equations (see Example 2), the notion of a transient is often important in applied problems.

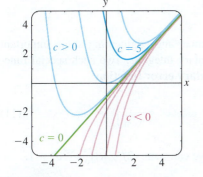

**FIGURE 2.3.2** Solution curves of DE in Example 5

**PIECEWISE-LINEAR DIFFERENTIAL EQUATION** In the construction of mathematical models (especially in the biological sciences and engineering) it can happen that one or more coefficients in a differential equation is a piecewise-defined function. In particular, when either $P(x)$ or $f(x)$ in (2) is a piecewise-defined function the equation is then referred to as a **piecewise-linear differential equation.** In the next example, $f(x)$ is piecewise continuous on the interval $[0, \infty)$ with a single jump discontinuity at $x = 1$. The basic idea is to solve the initial-value problem in two parts corresponding to the two intervals over which $f(x)$ is defined; each part consists of a linear equation solvable by the method of this section. As we will see, it is then possible to piece the two solutions together at $x = 1$ so that $y(x)$ is continuous on $[0, \infty)$. See Problems 37–42 in Exercises 2.3.

### EXAMPLE 6 An Initial-Value Problem

Solve $\dfrac{dy}{dx} + y = f(x)$, $\quad y(0) = 0$ where $\quad f(x) = \begin{cases} 1, & 0 \leq x \leq 1 \\ 0, & x > 1. \end{cases}$

**FIGURE 2.3.3** Discontinuous $f(x)$ in Example 6

**FIGURE 2.3.4** Graph of (10) in Example 6

**SOLUTION** The graph of the discontinuous function $f$ is shown in Figure 2.3.3. We solve the DE for $y(x)$ first on the interval $[0, 1]$ and then on the interval $(1, \infty)$. For $0 \leq x \leq 1$ we have

$$\frac{dy}{dx} + y = 1 \quad \text{or, equivalently,} \quad \frac{d}{dx}[e^x y] = e^x.$$

Integrating this last equation and solving for $y$ gives $y = 1 + c_1 e^{-x}$. Since $y(0) = 0$, we must have $c_1 = -1$, and therefore $y = 1 - e^{-x}$, $0 \leq x \leq 1$. Then for $x > 1$ the equation

$$\frac{dy}{dx} + y = 0$$

leads to $y = c_2 e^{-x}$. Hence we can write

$$y = \begin{cases} 1 - e^{-x}, & 0 \leq x \leq 1 \\ c_2 e^{-x}, & x > 1. \end{cases}$$

By appealing to the definition of continuity at a point, it is possible to determine $c_2$ so that the foregoing function is continuous at $x = 1$. The requirement that $\lim_{x \to 1^+} y(x) = y(1)$ implies that $c_2 e^{-1} = 1 - e^{-1}$ or $c_2 = e - 1$. As seen in Figure 2.3.4, the function

$$y = \begin{cases} 1 - e^{-x}, & 0 \leq x \leq 1 \\ (e - 1)e^{-x}, & x > 1 \end{cases} \tag{10}$$

is continuous on $(0, \infty)$. ■

It is worthwhile to think about (10) and Figure 2.3.4 a little bit; you are urged to read and answer Problem 53 in Exercises 2.3.

**ERROR FUNCTION** In mathematics, science, and engineering some important functions are defined in terms of nonelementary integrals. Two such special functions are the **error function** and **complementary error function:**

$$\text{erf}(x) = \frac{2}{\sqrt{\pi}} \int_0^x e^{-t^2} dt \quad \text{and} \quad \text{erfc}(x) = \frac{2}{\sqrt{\pi}} \int_x^\infty e^{-t^2} dt. \tag{11}$$

From the known result $\int_0^\infty e^{-t^2} dt = \sqrt{\pi}/2$* we can write $(2/\sqrt{\pi})\int_0^\infty e^{-t^2} dt = 1$.

Using the additive interval property of definite integrals $\int_0^\infty = \int_0^x + \int_x^\infty$ we can rewrite the last result in the alternative form

$$\frac{2}{\sqrt{\pi}} \int_0^\infty e^{-t^2} dt = \underbrace{\frac{2}{\sqrt{\pi}} \int_0^x e^{-t^2} dt}_{\text{erf}(x)} + \underbrace{\frac{2}{\sqrt{\pi}} \int_x^\infty e^{-t^2} dt}_{\text{erfc}(x)} = 1. \tag{12}$$

It is seen from (12) that the error function $\text{erf}(x)$ and complementary error function $\text{erfc}(x)$ are related by the identity

$$\text{erf}(x) + \text{erfc}(x) = 1.$$

Because of its importance in probability, statistics, and applied partial differential equations, the error function has been extensively tabulated. Note that $\text{erf}(0) = 0$ is one obvious function value. Numerical values of $\text{erf}(x)$ can also be found using a CAS such as *Mathematica*.

*This result is proved in Appendix A.

If we are solving an initial-value problem (6) and recognize that indefinite integration of the right-hand side of (5) would lead to a nonelementary integral, then as we saw in Example 5 of Section 2.2 it is convenient to use instead *definite* integration over the interval $[x_0, x]$. The last example illustrates that this procedure automatically incorporates the initial condition at $x_0$ into the solution of the DE; in other words, we do not have to solve for the constant $c$ in its general solution.

### EXAMPLE 7    The Error Function

Solve the initial-value problem $\dfrac{dy}{dx} - 2xy = 2, \quad y(0) = 1$.

**SOLUTION** The differential equation is already in standard form, and so we see that the integrating factor is $e^{\int(-2x\,dx)} = e^{-x^2}$. Multiplying both sides of the equation by this factor then gives $e^{-x^2}\dfrac{dy}{dx} - 2xe^{-x^2}y = 2e^{-x^2}$, which is the same as

$$\frac{d}{dx}\left[e^{-x^2}y\right] = 2e^{-x^2}. \tag{13}$$

Because indefinite integration of both sides of equation (13) leads to the nonelementary integral $\int e^{-x^2}\,dx$ we identify $x_0 = 0$ and use definite integration over the interval $[0, x]$:

$$\int_0^x \frac{d}{dt}\left(e^{-t^2}y(t)\right)dt = 2\int_0^x e^{-t^2}\,dt \quad \text{or} \quad e^{-x^2}y(x) - y(0) = 2\int_0^x e^{-t^2}\,dt.$$

Using the initial condition $y(0) = 1$ the last expression yields the solution

$$y = e^{x^2} + 2e^{x^2}\int_0^x e^{-t^2}\,dt. \tag{14}$$

Then by inserting the factor $\sqrt{\pi}/\sqrt{\pi}$ into this solution in the following manner—

$$y = e^{x^2} + 2e^{x^2}\int_0^x e^{-t^2}\,dt = e^{x^2}\left[1 + \sqrt{\pi}\left(\overbrace{\frac{2}{\sqrt{\pi}}\int_0^x e^{-t^2}\,dt}^{\text{erf}(x)}\right)\right]$$

we see from (11) that (14) can be rewritten in terms of the error function as

$$y = e^{x^2}\left[1 + \sqrt{\pi}\,\text{erf}(x)\right]. \tag{15}$$

The graph of solution (15), obtained with the aid of a CAS, is given in Figure 2.3.5.

See Problems 43–48 in Exercises 2.3.

**FIGURE 2.3.5** Graph of (15) in Example 7

**USE OF COMPUTERS** The computer algebra systems *Mathematica* and *Maple* are capable of producing implicit or explicit solutions for some kinds of differential equations using their *dsolve* commands.*

---

*Certain commands have the same spelling, but in *Mathematica* commands begin with a capital letter (**DSolve**), whereas in *Maple* the same command begins with a lower case letter (**dsolve**). When discussing such common syntax, we compromise and write, for example, *dsolve*. See the *Student Resource Manual* for the complete input commands used to solve a linear first-order DE.

## REMARKS

(*i*) Occasionally, a first-order differential equation is not linear in one variable but is linear in the other variable. For example, the differential equation

$$\frac{dy}{dx} = \frac{1}{x + y^2}$$

is not linear in the variable *y*. But its reciprocal

$$\frac{dx}{dy} = x + y^2 \quad \text{or} \quad \frac{dx}{dy} - x = y^2$$

is recognized as linear in the variable *x*. You should verify that the integrating factor $e^{\int(-1)\,dy} = e^{-y}$ and integration by parts yield the explicit solution $x = -y^2 - 2y - 2 + ce^y$ for the second equation. This expression is then an implicit solution of the first equation.

(*ii*) Mathematicians have adopted as their own certain words from engineering, which they found appropriately descriptive. The word *transient*, used earlier, is one of these terms. In future discussions the words *input* and *output* will occasionally pop up. The function *f* in (2) is called the **input** or **driving function;** a solution *y*(*x*) of the differential equation for a given input is called the **output** or **response.**

(*iii*) The term **special functions** mentioned in conjunction with the error function also applies to the **sine integral function** and the **Fresnel sine integral** introduced in Problems 47 and 48 in Exercises 2.3. "Special Functions" is actually a well-defined branch of mathematics. More special functions are studied in Section 6.4.

## EXERCISES **2.3**

*Answers to selected odd-numbered problems begin on page ANS-2.*

In Problems 1–24 find the general solution of the given differential equation. Give the largest interval *I* over which the general solution is defined. Determine whether there are any transient terms in the general solution.

**1.** $\dfrac{dy}{dx} = 5y$

**2.** $\dfrac{dy}{dx} + 2y = 0$

**3.** $\dfrac{dy}{dx} + y = e^{3x}$

**4.** $3\dfrac{dy}{dx} + 12y = 4$

**5.** $y' + 3x^2 y = x^2$

**6.** $y' + 2xy = x^3$

**7.** $x^2 y' + xy = 1$

**8.** $y' = 2y + x^2 + 5$

**9.** $x\dfrac{dy}{dx} - y = x^2 \sin x$

**10.** $x\dfrac{dy}{dx} + 2y = 3$

**11.** $x\dfrac{dy}{dx} + 4y = x^3 - x$

**12.** $(1 + x)\dfrac{dy}{dx} - xy = x + x^2$

**13.** $x^2 y' + x(x + 2)y = e^x$

**14.** $xy' + (1 + x)y = e^{-x}\sin 2x$

**15.** $y\,dx - 4(x + y^6)\,dy = 0$

**16.** $y\,dx = (ye^y - 2x)\,dy$

**17.** $\cos x\dfrac{dy}{dx} + (\sin x)y = 1$

**18.** $\cos^2 x\,\sin x\dfrac{dy}{dx} + (\cos^3 x)y = 1$

**19.** $(x + 1)\dfrac{dy}{dx} + (x + 2)y = 2xe^{-x}$

**20.** $(x + 2)^2\dfrac{dy}{dx} = 5 - 8y - 4xy$

**21.** $\dfrac{dr}{d\theta} + r\sec\theta = \cos\theta$

**22.** $\dfrac{dP}{dt} + 2tP = P + 4t - 2$

**23.** $x\dfrac{dy}{dx} + (3x + 1)y = e^{-3x}$

**24.** $(x^2 - 1)\dfrac{dy}{dx} + 2y = (x + 1)^2$

In Problems 25–36 solve the given initial-value problem. Give the largest interval $I$ over which the solution is defined.

**25.** $\dfrac{dy}{dx} = x + 5y, \quad y(0) = 3$

**26.** $\dfrac{dy}{dx} = 2x - 3y, \quad y(0) = \frac{1}{3}$

**27.** $xy' + y = e^x, \quad y(1) = 2$

**28.** $y\dfrac{dx}{dy} - x = 2y^2, \quad y(1) = 5$

**29.** $L\dfrac{di}{dt} + Ri = E, \quad i(0) = i_0, \quad L, R, E, i_0$ constants

**30.** $\dfrac{dT}{dt} = k(T - T_m), \quad T(0) = T_0, \quad k, T_m, T_0$ constants

**31.** $x\dfrac{dy}{dx} + y = 4x + 1, \quad y(1) = 8$

**32.** $y' + 4xy = x^3 e^{x^2}, \quad y(0) = -1$

**33.** $(x + 1)\dfrac{dy}{dx} + y = \ln x, \quad y(1) = 10$

**34.** $x(x + 1)\dfrac{dy}{dx} + xy = 1, \quad y(e) = 1$

**35.** $y' - (\sin x)y = 2\sin x, \quad y(\pi/2) = 1$

**36.** $y' + (\tan x)y = \cos^2 x, \quad y(0) = -1$

In Problems 37–40 proceed as in Example 6 to solve the given initial-value problem. Use a graphing utility to graph the continuous function $y(x)$.

**37.** $\dfrac{dy}{dx} + 2y = f(x), \quad y(0) = 0,$ where

$$f(x) = \begin{cases} 1, & 0 \le x \le 3 \\ 0, & x > 3 \end{cases}$$

**38.** $\dfrac{dy}{dx} + y = f(x), \quad y(0) = 1,$ where

$$f(x) = \begin{cases} 1, & 0 \le x \le 1 \\ -1, & x > 1 \end{cases}$$

**39.** $\dfrac{dy}{dx} + 2xy = f(x), \quad y(0) = 2,$ where

$$f(x) = \begin{cases} x, & 0 \le x < 1 \\ 0, & x \ge 1 \end{cases}$$

**40.** $(1 + x^2)\dfrac{dy}{dx} + 2xy = f(x), \quad y(0) = 0,$ where

$$f(x) = \begin{cases} x, & 0 \le x < 1 \\ -x, & x \ge 1 \end{cases}$$

In Problems 41 and 42 proceed as in Example 6 to solve the given initial-value problem. Use a graphing utility to graph the continuous function $y(x)$.

**41.** $\dfrac{dy}{dx} + P(x)y = 4x, \quad y(0) = 3,$ where

$$P(x) = \begin{cases} 2, & 0 \le x \le 1 \\ -\dfrac{2}{x}, & x > 1 \end{cases}$$

**42.** $\dfrac{dy}{dx} + P(x)y = 0, \quad y(0) = 4,$ where

$$P(x) = \begin{cases} 1, & 0 \le x \le 2 \\ 5, & x > 2 \end{cases}$$

In Problems 43 and 44 proceed as in Example 7 and express the solution of the given initial-value problem in terms of erf($x$) (Problem 43) and erfc($x$) (Problem 44).

**43.** $\dfrac{dy}{dx} - 2xy = 1, \quad y(1) = 1$

**44.** $\dfrac{dy}{dx} - 2xy = -1, \quad y(0) = \sqrt{\pi}/2$

In Problems 45 and 46 proceed as in Example 7 and express the solution of the given initial-value problem in terms of an integral-defined function.

**45.** $\dfrac{dy}{dx} + e^x y = 1, \quad y(0) = 1$

**46.** $x^2\dfrac{dy}{dx} - y = x^3, \quad y(1) = 0$

**47.** The **sine integral function** is defined as

$$\text{Si}(x) = \int_0^x \frac{\sin t}{t}\,dt,$$

where the integrand is defined to be 1 at $x = 0$. See Appendix A. Express the solution of the initial-value problem

$$x^3\frac{dy}{dx} + 2x^2 y = 10\sin x, \quad y(1) = 0$$

in terms of Si($x$).

**48.** The **Fresnel sine integral** function is defined as

$$S(x) = \int_0^x \sin\left(\frac{\pi}{2}t^2\right)dt.$$

See Appendix A. Express the solution of the initial-value problem

$$\frac{dy}{dx} - (\sin x^2)y = 0, \quad y(0) = 5$$

in terms of S($x$).

## Discussion Problems

**49.** Reread the discussion following Example 2. Construct a linear first-order differential equation for which all nonconstant solutions approach the horizontal asymptote $y = 4$ as $x \rightarrow \infty$.

**50.** Reread Example 3 and then discuss, with reference to Theorem 1.2.1, the existence and uniqueness of a solution of the initial-value problem consisting of $xy' - 4y = x^6 e^x$ and the given initial condition.

**(a)** $y(0) = 0$

**(b)** $y(0) = y_0, y_0 > 0$

**(c)** $y(x_0) = y_0, x_0 > 0, y_0 > 0$

**51.** Reread Example 4 and then find the general solution of the differential equation on the interval $(-3, 3)$.

**52.** Reread the discussion following Example 5. Construct a linear first-order differential equation for which all solutions are asymptotic to the line $y = 3x - 5$ as $x \to \infty$.

**53.** Reread Example 6 and then discuss why it is technically incorrect to say that the function in (10) is a "solution" of the IVP on the interval $[0, \infty)$.

**54. (a)** Construct a linear first-order differential equation of the form $xy' + 3y = g(x)$ for which $y = x^3 + c/x^3$ is its general solution. Give an interval $I$ of definition of this solution.

**(b)** Give an initial condition $y(x_0) = y_0$ for the DE found in part (a) so that the solution of the IVP is $y = x^3 - 1/x^3$. Repeat if the solution is $y = x^3 + 2/x^3$. Give an interval $I$ of definition of each of these solutions. Graph the solution curves. Is there an initial-value problem whose solution is defined on $(-\infty, \infty)$?

**(c)** Is each IVP found in part (b) unique? That is, can there be more than one IVP for which, say, $y = x^3 - 1/x^3$, $x$ in some interval $I$, is the solution?

**55.** In determining the integrating factor (3), we did not use a constant of integration in the evaluation of $\int P(x)\,dx$. Explain why using $\int P(x)\,dx + c_1$ has no effect on the solution of (2).

**56.** Suppose $P(x)$ is continuous on some interval $I$ and $a$ is a number in $I$. What can be said about the solution of the initial-value problem $y' + P(x)y = 0$, $y(a) = 0$?

**Mathematical Models**

**57. Radioactive Decay Series** The following system of differential equations is encountered in the study of the decay of a special type of radioactive series of elements:

$$\frac{dx}{dt} = -\lambda_1 x$$

$$\frac{dy}{dt} = \lambda_1 x - \lambda_2 y,$$

where $\lambda_1$ and $\lambda_2$ are constants. Discuss how to solve this system subject to $x(0) = x_0$, $y(0) = y_0$. Carry out your ideas.

**58. Heart Pacemaker** A heart pacemaker consists of a switch, a battery of constant voltage $E_0$, a capacitor with constant capacitance $C$, and the heart as a resistor with constant resistance $R$. When the switch is closed, the capacitor charges; when the switch is open, the capacitor discharges, sending an electrical stimulus to the heart. During the time the heart is being stimulated, the voltage $E$ across the heart satisfies the linear differential equation

$$\frac{dE}{dt} = -\frac{1}{RC}E.$$

Solve the DE, subject to $E(4) = E_0$.

**Computer Lab Assignments**

**59. (a)** Use a CAS to graph the solution curve of the initial-value problem in Problem 44 on the interval $(-\infty, \infty)$.

**(b)** Use tables or a CAS to value the value $y(2)$.

**60. (a)** Use a CAS to graph the solution curve of the initial-value problem in Problem 47 on the interval $[0, \infty)$.

**(b)** Use a CAS to find the value of the absolute maximum of the solution $y(x)$ on the interval.

**61. (a)** Use a CAS to graph the solution curve of the initial-value problem in Problem 48 on the interval $(-\infty, \infty)$.

**(b)** It is known that Fresnel sine integral $S(x) \to \frac{1}{2}$ as $x \to \infty$ and $S(x) \to -\frac{1}{2}$ as $x \to -\infty$. What does the solution $y(x)$ approach as $x \to \infty$? As $x \to -\infty$?

**(c)** Use a CAS to find the values of the absolute maximum and the absolute minimum of the solution $y(x)$ on the interval.

## 2.4 Exact Equations

**INTRODUCTION** Although the simple first-order differential equation

$$y\,dx + x\,dy = 0$$

is both separable and linear, we can solve the equation by an alternative manner by recognizing that the expression on the left-hand side of the equality is the differential of the function $f(x, y) = xy$; that is,

$$d(xy) = y\,dx + x\,dy.$$

In this section we examine first-order equations written in differential form $M(x, y)\,dx + N(x, y)\,dy = 0$. By applying a simple test to the coefficients $M$ and $N$, we can determine whether $M(x, y)\,dx + N(x, y)\,dy$ is the differential of a function $f(x, y)$. If the answer is yes, we can construct $f$ by partial integration.

**DIFFERENTIAL OF A FUNCTION OF TWO VARIABLES**  If $z = f(x, y)$ is a function of two variables with continuous first partial derivatives in a region $R$ of the $xy$-plane, then recall from calculus that its **differential** is defined to be

$$dz = \frac{\partial f}{\partial x} dx + \frac{\partial f}{\partial y} dy. \tag{1}$$

In the special case when $f(x, y) = c$, where $c$ is a constant, then (1) gives

$$\frac{\partial f}{\partial x} dx + \frac{\partial f}{\partial y} dy = 0. \tag{2}$$

In other words, given a one-parameter family of functions $f(x, y) = c$, we can generate a first-order differential equation by computing the differential of both sides of the equality. For example, if $x^2 - 5xy + y^3 = c$, then (2) gives the first-order DE

$$(2x - 5y)\, dx + (-5x + 3y^2)\, dy = 0. \tag{3}$$

**A DEFINITION**  Of course, not every first-order DE written in differential form $M(x, y)\, dx + N(x, y)\, dy = 0$ corresponds to a differential of $f(x, y) = c$. So for our purposes it is more important to turn the foregoing example around; namely, if we are given a first-order DE such as (3), is there some way we can recognize that the differential expression $(2x - 5y)\, dx + (-5x + 3y^2)\, dy$ is the differential $d(x^2 - 5xy + y^3)$? If there is, then an implicit solution of (3) is $x^2 - 5xy + y^3 = c$. We answer this question after the next definition.

---

**DEFINITION** 2.4.1 **Exact Equation**

A differential expression $M(x, y)\, dx + N(x, y)\, dy$ is an **exact differential** in a region $R$ of the $xy$-plane if it corresponds to the differential of some function $f(x, y)$ defined in $R$. A first-order differential equation of the form

$$M(x, y)\, dx + N(x, y)\, dy = 0$$

is said to be an **exact equation** if the expression on the left-hand side is an exact differential.

---

For example, $x^2 y^3\, dx + x^3 y^2\, dy = 0$ is an exact equation, because its left-hand side is an exact differential:

$$d\left(\tfrac{1}{3} x^3 y^3\right) = x^2 y^3\, dx + x^3 y^2\, dy.$$

Notice that if we make the identifications $M(x, y) = x^2 y^3$ and $N(x, y) = x^3 y^2$, then $\partial M / \partial y = 3x^2 y^2 = \partial N / \partial x$. Theorem 2.4.1, given next, shows that the equality of the partial derivatives $\partial M / \partial y$ and $\partial N / \partial x$ is no coincidence.

---

**THEOREM** 2.4.1 **Criterion for an Exact Differential**

Let $M(x, y)$ and $N(x, y)$ be continuous and have continuous first partial derivatives in a rectangular region $R$ defined by $a < x < b$, $c < y < d$. Then a necessary and sufficient condition that $M(x, y)\, dx + N(x, y)\, dy$ be an exact differential is

$$\frac{\partial M}{\partial y} = \frac{\partial N}{\partial x}. \tag{4}$$

**PROOF OF THE NECESSITY** For simplicity let us assume that $M(x, y)$ and $N(x, y)$ have continuous first partial derivatives for all $(x, y)$. Now if the expression $M(x, y)\, dx + N(x, y)\, dy$ is exact, there exists some function $f$ such that for all $x$ in $R$,

$$M(x, y)\, dx + N(x, y)\, dy = \frac{\partial f}{\partial x}\, dx + \frac{\partial f}{\partial y}\, dy.$$

Therefore

$$M(x, y) = \frac{\partial f}{\partial x}, \qquad N(x, y) = \frac{\partial f}{\partial y},$$

and

$$\frac{\partial M}{\partial y} = \frac{\partial}{\partial y}\left(\frac{\partial f}{\partial x}\right) = \frac{\partial^2 f}{\partial y\, \partial x} = \frac{\partial}{\partial x}\left(\frac{\partial f}{\partial y}\right) = \frac{\partial N}{\partial x}.$$

The equality of the mixed partials is a consequence of the continuity of the first partial derivatives of $M(x, y)$ and $N(x, y)$. ∎

The sufficiency part of Theorem 2.4.1 consists of showing that there exists a function $f$ for which $\partial f/\partial x = M(x, y)$ and $\partial f/\partial y = N(x, y)$ whenever (4) holds. The construction of the function $f$ actually reflects a basic procedure for solving exact equations.

**METHOD OF SOLUTION** Given an equation in the differential form $M(x, y)\, dx + N(x, y)\, dy = 0$, determine whether the equality in (4) holds. If it does, then there exists a function $f$ for which

$$\frac{\partial f}{\partial x} = M(x, y).$$

We can find $f$ by integrating $M(x, y)$ with respect to $x$ while holding $y$ constant:

$$f(x, y) = \int M(x, y)\, dx + g(y), \tag{5}$$

where the arbitrary function $g(y)$ is the "constant" of integration. Now differentiate (5) with respect to $y$ and assume that $\partial f/\partial y = N(x, y)$:

$$\frac{\partial f}{\partial y} = \frac{\partial}{\partial y}\int M(x, y)\, dx + g'(y) = N(x, y).$$

This gives

$$g'(y) = N(x, y) - \frac{\partial}{\partial y}\int M(x, y)\, dx. \tag{6}$$

Finally, integrate (6) with respect to $y$ and substitute the result in (5). The implicit solution of the equation is $f(x, y) = c$.

Some observations are in order. First, it is important to realize that the expression $N(x, y) - (\partial/\partial y)\int M(x, y)\, dx$ in (6) is independent of $x$, because

$$\frac{\partial}{\partial x}\left[N(x, y) - \frac{\partial}{\partial y}\int M(x, y)\, dx\right] = \frac{\partial N}{\partial x} - \frac{\partial}{\partial y}\left(\frac{\partial}{\partial x}\int M(x, y)\, dx\right) = \frac{\partial N}{\partial x} - \frac{\partial M}{\partial y} = 0.$$

Second, we could just as well start the foregoing procedure with the assumption that $\partial f/\partial y = N(x, y)$. After integrating $N$ with respect to $y$ and then differentiating that result, we would find the analogues of (5) and (6) to be, respectively,

$$f(x, y) = \int N(x, y)\, dy + h(x) \qquad \text{and} \qquad h'(x) = M(x, y) - \frac{\partial}{\partial x}\int N(x, y)\, dy.$$

In either case *none of these formulas should be memorized.*

**EXAMPLE 1**  Solving an Exact DE

Solve $2xy\,dx + (x^2 - 1)\,dy = 0$.

**SOLUTION**  With $M(x, y) = 2xy$ and $N(x, y) = x^2 - 1$ we have

$$\frac{\partial M}{\partial y} = 2x = \frac{\partial N}{\partial x}.$$

Thus the equation is exact, and so by Theorem 2.4.1 there exists a function $f(x, y)$ such that

$$\frac{\partial f}{\partial x} = 2xy \qquad \text{and} \qquad \frac{\partial f}{\partial y} = x^2 - 1.$$

From the first of these equations we obtain, after integrating,

$$f(x, y) = x^2 y + g(y).$$

Taking the partial derivative of the last expression with respect to $y$ and setting the result equal to $N(x, y)$ gives

$$\frac{\partial f}{\partial y} = x^2 + g'(y) = x^2 - 1. \qquad \leftarrow N(x, y)$$

It follows that $g'(y) = -1$ and $g(y) = -y$. Hence $f(x, y) = x^2 y - y$, so the solution of the equation in implicit form is $x^2 y - y = c$. The explicit form of the solution is easily seen to be $y = c/(x^2 - 1)$ and is defined on any interval not containing either $x = 1$ or $x = -1$. ∎

**NOTE**  The solution of the DE in Example 1 is *not* $f(x, y) = x^2 y - y$. Rather, it is $f(x, y) = c$; if a constant is used in the integration of $g'(y)$, we can then write the solution as $f(x, y) = 0$.

**EXAMPLE 2**  Solving an Exact DE

Solve $(e^{2y} - y\cos xy)\,dx + (2xe^{2y} - x\cos xy + 2y)\,dy = 0$.

**SOLUTION**  The equation is exact because

$$\frac{\partial M}{\partial y} = 2e^{2y} + xy\sin xy - \cos xy = \frac{\partial N}{\partial x}.$$

Hence a function $f(x, y)$ exists for which

$$M(x, y) = \frac{\partial f}{\partial x} \qquad \text{and} \qquad N(x, y) = \frac{\partial f}{\partial y}.$$

Now, for variety, we shall start with the assumption that $\partial f/\partial y = N(x, y)$; that is,

$$\frac{\partial f}{\partial y} = 2xe^{2y} - x\cos xy + 2y$$

$$f(x, y) = 2x\int e^{2y}\,dy - x\int \cos xy\,dy + 2\int y\,dy + h(x).$$

Remember, the reason $x$ can come out in front of the symbol $\int$ is that in the integration with respect to $y$, $x$ is treated as an ordinary constant. It follows that

$$f(x, y) = xe^{2y} - \sin xy + y^2 + h(x)$$

$$\frac{\partial f}{\partial x} = e^{2y} - y\cos xy + h'(x) = e^{2y} - y\cos xy, \qquad \leftarrow M(x, y)$$

and so $h'(x) = 0$ or $h(x) = c$. Hence a family of solutions is

$$xe^{2y} - \sin xy + y^2 + c = 0.$$

## EXAMPLE 3   An Initial-Value Problem

Solve $\dfrac{dy}{dx} = \dfrac{xy^2 - \cos x \sin x}{y(1 - x^2)}, \quad y(0) = 2.$

**SOLUTION**   By writing the differential equation in the form

$$(\cos x \sin x - xy^2)\,dx + y(1 - x^2)\,dy = 0,$$

we recognize that the equation is exact because

$$\frac{\partial M}{\partial y} = -2xy = \frac{\partial N}{\partial x}.$$

Now
$$\frac{\partial f}{\partial y} = y(1 - x^2)$$

$$f(x, y) = \frac{y^2}{2}(1 - x^2) + h(x)$$

$$\frac{\partial f}{\partial x} = -xy^2 + h'(x) = \cos x \sin x - xy^2.$$

The last equation implies that $h'(x) = \cos x \sin x$. Integrating gives

$$h(x) = -\int (\cos x)(-\sin x\,dx) = -\frac{1}{2}\cos^2 x.$$

Thus     $\dfrac{y^2}{2}(1 - x^2) - \dfrac{1}{2}\cos^2 x = c_1$     or     $y^2(1 - x^2) - \cos^2 x = c$,     (7)

where $2c_1$ has been replaced by $c$. The initial condition $y = 2$ when $x = 0$ demands that $4(1) - \cos^2(0) = c$, and so $c = 3$. An implicit solution of the problem is then $y^2(1 - x^2) - \cos^2 x = 3$.

The solution curve of the IVP is the curve drawn in blue in Figure 2.4.1; it is part of an interesting family of curves. The graphs of the members of the one-parameter family of solutions given in (7) can be obtained in several ways, two of which are using software to graph level curves (as discussed in Section 2.2) and using a graphing utility to carefully graph the explicit functions obtained for various values of $c$ by solving $y^2 = (c + \cos^2 x)/(1 - x^2)$ for $y$.

**FIGURE 2.4.1**   Solution curves of DE in Example 3

**INTEGRATING FACTORS**   Recall from Section 2.3 that the left-hand side of the linear equation $y' + P(x)y = f(x)$ can be transformed into a derivative when we multiply the equation by an integrating factor. The same basic idea sometimes works for a nonexact differential equation $M(x, y)\,dx + N(x, y)\,dy = 0$. That is, it is sometimes possible to find an **integrating factor** $\mu(x, y)$ so that after multiplying, the left-hand side of

$$\mu(x, y)M(x, y)\,dx + \mu(x, y)N(x, y)\,dy = 0 \tag{8}$$

is an exact differential. In an attempt to find $\mu$, we turn to the criterion (4) for exactness. Equation (8) is exact if and only if $(\mu M)_y = (\mu N)_x$, where the subscripts denote partial derivatives. By the Product Rule of differentiation the last equation is the same as $\mu M_y + \mu_y M = \mu N_x + \mu_x N$ or

$$\mu_x N - \mu_y M = (M_y - N_x)\mu. \tag{9}$$

Although $M$, $N$, $M_y$, and $N_x$ are known functions of $x$ and $y$, the difficulty here in determining the unknown $\mu(x, y)$ from (9) is that we must solve a partial differential equation. Since we are not prepared to do that, we make a simplifying assumption. Suppose $\mu$ is a function of one variable; for example, say that $\mu$ depends only on $x$. In this case, $\mu_x = d\mu/dx$ and $\mu_y = 0$, so (9) can be written as

$$\frac{d\mu}{dx} = \frac{M_y - N_x}{N}\mu. \tag{10}$$

We are still at an impasse if the quotient $(M_y - N_x)/N$ depends on both $x$ and $y$. However, if after all obvious algebraic simplifications are made, the quotient $(M_y - N_x)/N$ turns out to depend solely on the variable $x$, then (10) is a first-order ordinary differential equation. We can finally determine $\mu$ because (10) is *separable* as well as *linear*. It follows from either Section 2.2 or Section 2.3 that $\mu(x) = e^{\int((M_y - N_x)/N)dx}$. In like manner, it follows from (9) that if $\mu$ depends only on the variable $y$, then

$$\frac{d\mu}{dy} = \frac{N_x - M_y}{M}\mu. \tag{11}$$

In this case, if $(N_x - M_y)/M$ is a function of $y$ only, then we can solve (11) for $\mu$.

We summarize the results for the differential equation

$$M(x, y)\, dx + N(x, y)\, dy = 0. \tag{12}$$

- If $(M_y - N_x)/N$ is a function of $x$ alone, then an integrating factor for (12) is

$$\mu(x) = e^{\int \frac{M_y - N_x}{N}dx}. \tag{13}$$

- If $(N_x - M_y)/M$ is a function of $y$ alone, then an integrating factor for (12) is

$$\mu(y) = e^{\int \frac{N_x - M_y}{M}dy}. \tag{14}$$

### EXAMPLE 4 A Nonexact DE Made Exact

The nonlinear first-order differential equation

$$xy\, dx + (2x^2 + 3y^2 - 20)\, dy = 0$$

is not exact. With the identifications $M = xy$, $N = 2x^2 + 3y^2 - 20$, we find the partial derivatives $M_y = x$ and $N_x = 4x$. The first quotient from (13) gets us nowhere, since

$$\frac{M_y - N_x}{N} = \frac{x - 4x}{2x^2 + 3y^2 - 20} = \frac{-3x}{2x^2 + 3y^2 - 20}$$

depends on $x$ and $y$. However, (14) yields a quotient that depends only on $y$:

$$\frac{N_x - M_y}{M} = \frac{4x - x}{xy} = \frac{3x}{xy} = \frac{3}{y}.$$

The integrating factor is then $e^{\int 3\, dy/y} = e^{3\ln y} = e^{\ln y^3} = y^3$. After we multiply the given DE by $\mu(y) = y^3$, the resulting equation is

$$xy^4\, dx + (2x^2y^3 + 3y^5 - 20y^3)\, dy = 0.$$

You should verify that the last equation is now exact as well as show, using the method of this section, that a family of solutions is $\frac{1}{2}x^2y^4 + \frac{1}{2}y^6 - 5y^4 = c$. ∎

### REMARKS

(*i*) When testing an equation for exactness, make sure it is of the precise form $M(x, y)\,dx + N(x, y)\,dy = 0$. Sometimes a differential equation is written $G(x, y)\,dx = H(x, y)\,dy$. In this case, first rewrite it as $G(x, y)\,dx - H(x, y)\,dy = 0$ and then identify $M(x, y) = G(x, y)$ and $N(x, y) = -H(x, y)$ before using (4).

(*ii*) In some texts on differential equations the study of exact equations precedes that of linear DEs. Then the method for finding integrating factors just discussed can be used to derive an integrating factor for $y' + P(x)y = f(x)$. By rewriting the last equation in the differential form $(P(x)y - f(x))\,dx + dy = 0$, we see that

$$\frac{M_y - N_x}{N} = P(x).$$

From (13) we arrive at the already familiar integrating factor $e^{\int P(x)dx}$ used in Section 2.3.

## EXERCISES 2.4

Answers to selected odd-numbered problems begin on page ANS-2.

In Problems 1–20 determine whether the given differential equation is exact. If it is exact, solve it.

**1.** $(2x - 1)\,dx + (3y + 7)\,dy = 0$

**2.** $(2x + y)\,dx - (x + 6y)\,dy = 0$

**3.** $(5x + 4y)\,dx + (4x - 8y^3)\,dy = 0$

**4.** $(\sin y - y \sin x)\,dx + (\cos x + x \cos y - y)\,dy = 0$

**5.** $(2xy^2 - 3)\,dx + (2x^2y + 4)\,dy = 0$

**6.** $\left(2y - \dfrac{1}{x} + \cos 3x\right)\dfrac{dy}{dx} + \dfrac{y}{x^2} - 4x^3 + 3y \sin 3x = 0$

**7.** $(x^2 - y^2)\,dx + (x^2 - 2xy)\,dy = 0$

**8.** $\left(1 + \ln x + \dfrac{y}{x}\right)dx = (1 - \ln x)\,dy$

**9.** $(x - y^3 + y^2 \sin x)\,dx = (3xy^2 + 2y \cos x)\,dy$

**10.** $(x^3 + y^3)\,dx + 3xy^2\,dy = 0$

**11.** $(y \ln y - e^{-xy})\,dx + \left(\dfrac{1}{y} + x \ln y\right)dy = 0$

**12.** $(3x^2y + e^y)\,dx + (x^3 + xe^y - 2y)\,dy = 0$

**13.** $x\dfrac{dy}{dx} = 2xe^x - y + 6x^2$

**14.** $\left(1 - \dfrac{3}{y} + x\right)\dfrac{dy}{dx} + y = \dfrac{3}{x} - 1$

**15.** $\left(x^2y^3 - \dfrac{1}{1 + 9x^2}\right)\dfrac{dx}{dy} + x^3y^2 = 0$

**16.** $(5y - 2x)y' - 2y = 0$

**17.** $(\tan x - \sin x \sin y)\,dx + \cos x \cos y\,dy = 0$

**18.** $(2y \sin x \cos x - y + 2y^2 e^{xy^2})\,dx$
$= (x - \sin^2 x - 4xye^{xy^2})\,dy$

**19.** $(4t^3y - 15t^2 - y)\,dt + (t^4 + 3y^2 - t)\,dy = 0$

**20.** $\left(\dfrac{1}{t} + \dfrac{1}{t^2} - \dfrac{y}{t^2 + y^2}\right)dt + \left(ye^y + \dfrac{t}{t^2 + y^2}\right)dy = 0$

In Problems 21–26 solve the given initial-value problem.

**21.** $(x + y)^2\,dx + (2xy + x^2 - 1)\,dy = 0, \quad y(1) = 1$

**22.** $(e^x + y)\,dx + (2 + x + ye^y)\,dy = 0, \quad y(0) = 1$

**23.** $(4y + 2t - 5)\,dt + (6y + 4t - 1)\,dy = 0, \quad y(-1) = 2$

**24.** $\left(\dfrac{3y^2 - t^2}{y^5}\right)\dfrac{dy}{dt} + \dfrac{t}{2y^4} = 0, \quad y(1) = 1$

**25.** $(y^2 \cos x - 3x^2y - 2x)\,dx$
$+ (2y \sin x - x^3 + \ln y)\,dy = 0, \quad y(0) = e$

**26.** $\left(\dfrac{1}{1 + y^2} + \cos x - 2xy\right)\dfrac{dy}{dx} = y(y + \sin x), \quad y(0) = 1$

In Problems 27 and 28 find the value of $k$ so that the given differential equation is exact.

**27.** $(y^3 + kxy^4 - 2x)\,dx + (3xy^2 + 20x^2y^3)\,dy = 0$

**28.** $(6xy^3 + \cos y)\,dx + (2kx^2y^2 - x \sin y)\,dy = 0$

In Problems 29 and 30 verify that the given differential equation is not exact. Multiply the given differential equation by the indicated integrating factor $\mu(x, y)$ and verify that the new equation is exact. Solve.

**29.** $(-xy \sin x + 2y \cos x)\,dx + 2x \cos x\,dy = 0;$
$\mu(x, y) = xy$

**30.** $(x^2 + 2xy - y^2) dx + (y^2 + 2xy - x^2) dy = 0;$
$\mu(x, y) = (x + y)^{-2}$

In Problems 31–36 solve the given differential equation by finding, as in Example 4, an appropriate integrating factor.

**31.** $(2y^2 + 3x) dx + 2xy \, dy = 0$

**32.** $y(x + y + 1) dx + (x + 2y) dy = 0$

**33.** $6xy \, dx + (4y + 9x^2) dy = 0$

**34.** $\cos x \, dx + \left(1 + \dfrac{2}{y}\right) \sin x \, dy = 0$

**35.** $(10 - 6y + e^{-3x}) dx - 2 \, dy = 0$

**36.** $(y^2 + xy^3) dx + (5y^2 - xy + y^3 \sin y) dy = 0$

In Problems 37 and 38 solve the given initial-value problem by finding, as in Example 4, an appropriate integrating factor.

**37.** $x \, dx + (x^2 y + 4y) dy = 0, \quad y(4) = 0$

**38.** $(x^2 + y^2 - 5) dx = (y + xy) dy, \quad y(0) = 1$

**39. (a)** Show that a one-parameter family of solutions of the equation

$$(4xy + 3x^2) dx + (2y + 2x^2) dy = 0$$

is $x^3 + 2x^2 y + y^2 = c.$

**(b)** Show that the initial conditions $y(0) = -2$ and $y(1) = 1$ determine the same implicit solution.

**(c)** Find explicit solutions $y_1(x)$ and $y_2(x)$ of the differential equation in part (a) such that $y_1(0) = -2$ and $y_2(1) = 1$. Use a graphing utility to graph $y_1(x)$ and $y_2(x)$.

## Discussion Problems

**40.** Consider the concept of an integrating factor used in Problems 29–38. Are the two equations $M \, dx + N \, dy = 0$ and $\mu M \, dx + \mu N \, dy = 0$ necessarily equivalent in the sense that a solution of one is also a solution of the other? Discuss.

**41.** Reread Example 3 and then discuss why we can conclude that the interval of definition of the explicit solution of the IVP (the blue curve in Figure 2.4.1) is $(-1, 1)$.

**42.** Discuss how the functions $M(x, y)$ and $N(x, y)$ can be found so that each differential equation is exact. Carry out your ideas.

**(a)** $M(x, y) dx + \left(xe^{xy} + 2xy + \dfrac{1}{x}\right) dy = 0$

**(b)** $\left(x^{-1/2} y^{1/2} + \dfrac{x}{x^2 + y}\right) dx + N(x, y) dy = 0$

**43.** Differential equations are sometimes solved by having a clever idea. Here is a little exercise in cleverness: Although the differential equation $(x - \sqrt{x^2 + y^2}) dx + y \, dy = 0$ is not exact, show how the rearrangement $(x \, dx + y \, dy)/\sqrt{x^2 + y^2} = dx$ and the observation $\frac{1}{2} d(x^2 + y^2) = x \, dx + y \, dy$ can lead to a solution.

**44.** True or False: Every separable first-order equation $dy/dx = g(x)h(y)$ is exact.

## Mathematical Model

**45. Falling Chain**  A portion of a uniform chain of length 8 ft is loosely coiled around a peg at the edge of a high horizontal platform, and the remaining portion of the chain hangs at rest over the edge of the platform. See Figure 2.4.2. Suppose that the length of the overhanging chain is 3 ft, that the chain weighs 2 lb/ft, and that the positive direction is downward. Starting at $t = 0$ seconds, the weight of the overhanging portion causes the chain on the table to uncoil smoothly and to fall to the floor. If $x(t)$ denotes the length of the chain overhanging the table at time $t > 0$, then $v = dx/dt$ is its velocity. When all resistive forces are ignored, it can be shown that a mathematical model relating $v$ to $x$ is given by

$$xv \dfrac{dv}{dx} + v^2 = 32x.$$

**(a)** Rewrite this model in differential form. Proceed as in Problems 31–36 and solve the DE for $v$ in terms of $x$ by finding an appropriate integrating factor. Find an explicit solution $v(x)$.

**(b)** Determine the velocity with which the chain leaves the platform.

**FIGURE 2.4.2**  Uncoiling chain in Problem 45

## Computer Lab Assignments

**46. Streamlines**

**(a)** The solution of the differential equation

$$\dfrac{2xy}{(x^2 + y^2)^2} dx + \left[1 + \dfrac{y^2 - x^2}{(x^2 + y^2)^2}\right] dy = 0$$

is a family of curves that can be interpreted as streamlines of a fluid flow around a circular object whose boundary is described by the equation $x^2 + y^2 = 1$. Solve this DE and note the solution $f(x, y) = c$ for $c = 0$.

**(b)** Use a CAS to plot the streamlines for $c = 0, \pm 0.2, \pm 0.4, \pm 0.6,$ and $\pm 0.8$ in three different ways. First, use the *contourplot* of a CAS. Second, solve for $x$ in terms of the variable $y$. Plot the resulting two functions of $y$ for the given values of $c$, and then combine the graphs. Third, use the CAS to solve a cubic equation for $y$ in terms of $x$.

# 2.5  Solutions by Substitutions

**INTRODUCTION**   We usually solve a differential equation by recognizing it as a certain kind of equation (say, separable, linear, exact, and so on) and then carrying out a procedure consisting of *equation-specific mathematical steps* (say, separating variables and integrating) that yields a solution of the equation. But it is not uncommon to be stumped by a differential equation because it does not fall into one of the classes of equations that we know how to solve. The substitution procedures that are discussed in this section may be helpful in this situation.

**SUBSTITUTIONS**   Often the first step in solving a differential equation consists of transforming it into another differential equation by means of a **substitution**. For example, suppose we wish to transform the first-order differential equation $dy/dx = f(x, y)$ by the substitution $y = g(x, u)$, where $u$ is regarded as a function of the variable $x$. If $g$ possesses first-partial derivatives, then the Chain Rule

$$\frac{dy}{dx} = \frac{\partial g}{\partial x}\frac{dx}{dx} + \frac{\partial g}{\partial u}\frac{du}{dx} \qquad \text{gives} \qquad \frac{dy}{dx} = g_x(x, u) + g_u(x, u)\frac{du}{dx}.$$

If we replace $dy/dx$ by the foregoing derivative and replace $y$ in $f(x, y)$ by $g(x, u)$, then the DE $dy/dx = f(x, y)$ becomes $g_x(x, u) + g_u(x, u)\dfrac{du}{dx} = f(x, g(x, u))$, which, solved for $du/dx$, has the form $\dfrac{du}{dx} = F(x, u)$. If we can determine a solution $u = \phi(x)$ of this last equation, then a solution of the original differential equation is $y = g(x, \phi(x))$.

In the discussion that follows we examine three different kinds of first-order differential equations that are solvable by means of a substitution.

**HOMOGENEOUS EQUATIONS**   If a function $f$ possesses the property $f(tx, ty) = t^\alpha f(x, y)$ for some real number $\alpha$, then $f$ is said to be a **homogeneous function** of degree $\alpha$. For example, $f(x, y) = x^3 + y^3$ is a homogeneous function of degree 3, since

$$f(tx, ty) = (tx)^3 + (ty)^3 = t^3(x^3 + y^3) = t^3 f(x, y),$$

whereas $f(x, y) = x^3 + y^3 + 1$ is not homogeneous. A first-order DE in differential form

$$M(x, y)\, dx + N(x, y)\, dy = 0 \tag{1}$$

is said to be **homogeneous** if both coefficient functions $M$ and $N$ are homogeneous functions of the *same* degree. In other words, (1) is homogeneous if

$$M(tx, ty) = t^\alpha M(x, y) \qquad \text{and} \qquad N(tx, ty) = t^\alpha N(x, y).$$

In addition, if $M$ and $N$ are homogeneous functions of degree $\alpha$, we can also write

$$M(x, y) = x^\alpha M(1, u) \qquad \text{and} \qquad N(x, y) = x^\alpha N(1, u), \quad \text{where } u = y/x, \tag{2}$$

and

$$M(x, y) = y^\alpha M(v, 1) \qquad \text{and} \qquad N(x, y) = y^\alpha N(v, 1), \quad \text{where } v = x/y. \tag{3}$$

See Problem 31 in Exercises 2.5. Properties (2) and (3) suggest the substitutions that can be used to solve a homogeneous differential equation. Specifically, *either* of the substitutions $y = ux$ or $x = vy$, where $u$ and $v$ are new dependent variables, will reduce a homogeneous equation to a *separable* first-order differential equation. To show this, observe that as a consequence of (2) a homogeneous equation $M(x, y) \, dx + N(x, y) \, dy = 0$ can be rewritten as

$$x^\alpha M(1, u) \, dx + x^\alpha N(1, u) \, dy = 0 \qquad \text{or} \qquad M(1, u) \, dx + N(1, u) \, dy = 0,$$

where $u = y/x$ or $y = ux$. By substituting the differential $dy = u \, dx + x \, du$ into the last equation and gathering terms, we obtain a separable DE in the variables $u$ and $x$:

$$M(1, u) \, dx + N(1, u)[u \, dx + x \, du] = 0$$

$$[M(1, u) + uN(1, u)] \, dx + xN(1, u) \, du = 0$$

or

$$\frac{dx}{x} + \frac{N(1, u) \, du}{M(1, u) + uN(1, u)} = 0.$$

At this point we offer the same advice as in the preceding sections: Do not memorize anything here (especially the last formula); rather, *work through the procedure each time.* The proof that the substitutions $x = vy$ and $dx = v \, dy + y \, dv$ also lead to a separable equation follows in an analogous manner from (3).

---

### EXAMPLE 1    Solving a Homogeneous DE

Solve $(x^2 + y^2) \, dx + (x^2 - xy) \, dy = 0$.

**SOLUTION** Inspection of $M(x, y) = x^2 + y^2$ and $N(x, y) = x^2 - xy$ shows that these coefficients are homogeneous functions of degree 2. If we let $y = ux$, then $dy = u \, dx + x \, du$, so after substituting, the given equation becomes

$$(x^2 + u^2 x^2) \, dx + (x^2 - ux^2)[u \, dx + x \, du] = 0$$

$$x^2(1 + u) \, dx + x^3(1 - u) \, du = 0$$

$$\frac{1 - u}{1 + u} \, du + \frac{dx}{x} = 0$$

$$\left[ -1 + \frac{2}{1 + u} \right] du + \frac{dx}{x} = 0. \qquad \leftarrow \text{long division}$$

After integration the last line gives

$$-u + 2 \ln|1 + u| + \ln|x| = \ln|c|$$

$$-\frac{y}{x} + 2 \ln\left|1 + \frac{y}{x}\right| + \ln|x| = \ln|c|. \qquad \leftarrow \text{resubstituting } u = y/x$$

Using the properties of logarithms, we can write the preceding solution as

$$\ln\left|\frac{(x + y)^2}{cx}\right| = \frac{y}{x} \qquad \text{or} \qquad (x + y)^2 = cxe^{y/x}. \qquad \blacksquare$$

Although either of the indicated substitutions can be used for every homogeneous differential equation, in practice we try $x = vy$ whenever the function $M(x, y)$ is simpler than $N(x, y)$. Also it could happen that after using one substitution, we may encounter integrals that are difficult or impossible to evaluate in closed form; switching substitutions may result in an easier problem.

**BERNOULLI'S EQUATION**    The differential equation

$$\frac{dy}{dx} + P(x)y = f(x)y^n, \tag{4}$$

where $n$ is any real number, is called **Bernoulli's equation.** Note that for $n = 0$ and $n = 1$, equation (4) is linear. For $n \neq 0$ and $n \neq 1$ the substitution $u = y^{1-n}$ reduces any equation of form (4) to a linear equation.

**EXAMPLE 2** Solving a Bernoulli DE

Solve $x\dfrac{dy}{dx} + y = x^2 y^2$.

**SOLUTION** We begin by rewriting the equation in the form given in (4) by dividing by $x$:

$$\frac{dy}{dx} + \frac{1}{x}y = xy^2.$$

With $n = 2$ we have $u = y^{-1}$ or $y = u^{-1}$. We then substitute

$$\frac{dy}{dx} = \frac{dy}{du}\frac{du}{dx} = -u^{-2}\frac{du}{dx} \qquad \leftarrow \text{Chain Rule}$$

into the given equation and simplify. The result is

$$\frac{du}{dx} - \frac{1}{x}u = -x.$$

The integrating factor for this linear equation on, say, $(0, \infty)$ is

$$e^{-\int dx/x} = e^{-\ln x} = e^{\ln x^{-1}} = x^{-1}.$$

Integrating 
$$\frac{d}{dx}[x^{-1}u] = -1$$

gives $x^{-1}u = -x + c$ or $u = -x^2 + cx$. Since $u = y^{-1}$, we have $y = 1/u$, so a solution of the given equation is $y = 1/(-x^2 + cx)$. ∎

Note that we have not obtained the general solution of the original nonlinear differential equation in Example 2, since $y = 0$ is a singular solution of the equation.

**REDUCTION TO SEPARATION OF VARIABLES** A differential equation of the form

$$\frac{dy}{dx} = f(Ax + By + C) \tag{5}$$

can always be reduced to an equation with separable variables by means of the substitution $u = Ax + By + C$, $B \neq 0$. Example 3 illustrates the technique.

**EXAMPLE 3** An Initial-Value Problem

Solve $\dfrac{dy}{dx} = (-2x + y)^2 - 7$, $y(0) = 0$.

**SOLUTION** If we let $u = -2x + y$, then $du/dx = -2 + dy/dx$, so the differential equation is transformed into

$$\frac{du}{dx} + 2 = u^2 - 7 \qquad \text{or} \qquad \frac{du}{dx} = u^2 - 9.$$

The last equation is separable. Using partial fractions

$$\frac{du}{(u - 3)(u + 3)} = dx \qquad \text{or} \qquad \frac{1}{6}\left[\frac{1}{u - 3} - \frac{1}{u + 3}\right]du = dx$$

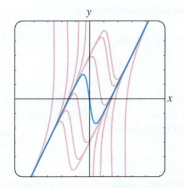

**FIGURE 2.5.1** Solutions of DE in Example 3

and then integrating yields

$$\frac{1}{6}\ln\left|\frac{u-3}{u+3}\right| = x + c_1 \qquad \text{or} \qquad \frac{u-3}{u+3} = e^{6x+6c_1} = ce^{6x}. \quad \leftarrow \text{replace } e^{6c_1} \text{ by } c$$

Solving the last equation for $u$ and then resubstituting gives the solution

$$u = \frac{3(1 + ce^{6x})}{1 - ce^{6x}} \qquad \text{or} \qquad y = 2x + \frac{3(1 + ce^{6x})}{1 - ce^{6x}}. \qquad (6)$$

Finally, applying the initial condition $y(0) = 0$ to the last equation in (6) gives $c = -1$. Figure 2.5.1, obtained with the aid of a graphing utility, shows the graph of the particular solution $y = 2x + \dfrac{3(1 - e^{6x})}{1 + e^{6x}}$ in blue, along with the graphs of some other members of the family of solutions (6). ■

## EXERCISES 2.5

Answers to selected odd-numbered problems begin on page ANS-2.

*Each DE in Problems 1–14 is homogeneous.*

In Problems 1–10 solve the given differential equation by using an appropriate substitution.

**1.** $(x - y)\,dx + x\,dy = 0$

**2.** $(x + y)\,dx + x\,dy = 0$

**3.** $x\,dx + (y - 2x)\,dy = 0$

**4.** $y\,dx = 2(x + y)\,dy$

**5.** $(y^2 + yx)\,dx - x^2\,dy = 0$

**6.** $(y^2 + yx)\,dx + x^2\,dy = 0$

**7.** $\dfrac{dy}{dx} = \dfrac{y - x}{y + x}$

**8.** $\dfrac{dy}{dx} = \dfrac{x + 3y}{3x + y}$

**9.** $-y\,dx + \left(x + \sqrt{xy}\right)dy = 0$

**10.** $x\dfrac{dy}{dx} = y + \sqrt{x^2 - y^2}, \quad x > 0$

In Problems 11–14 solve the given initial-value problem.

**11.** $xy^2\dfrac{dy}{dx} = y^3 - x^3, \quad y(1) = 2$

**12.** $(x^2 + 2y^2)\dfrac{dx}{dy} = xy, \quad y(-1) = 1$

**13.** $(x + ye^{y/x})\,dx - xe^{y/x}\,dy = 0, \quad y(1) = 0$

**14.** $y\,dx + x(\ln x - \ln y - 1)\,dy = 0, \quad y(1) = e$

*Each DE in Problems 15–22 is a Bernoulli equation.*

In Problems 15–20 solve the given differential equation by using an appropriate substitution.

**15.** $x\dfrac{dy}{dx} + y = \dfrac{1}{y^2}$

**16.** $\dfrac{dy}{dx} - y = e^x y^2$

**17.** $\dfrac{dy}{dx} = y(xy^3 - 1)$

**18.** $x\dfrac{dy}{dx} - (1 + x)y = xy^2$

**19.** $t^2\dfrac{dy}{dt} + y^2 = ty$

**20.** $3(1 + t^2)\dfrac{dy}{dt} = 2ty(y^3 - 1)$

In Problems 21 and 22 solve the given initial-value problem.

**21.** $x^2\dfrac{dy}{dx} - 2xy = 3y^4, \quad y(1) = \frac{1}{2}$

**22.** $y^{1/2}\dfrac{dy}{dx} + y^{3/2} = 1, \quad y(0) = 4$

*Each DE in Problems 23–30 is of the form given in (5).*

In Problems 23–28 solve the given differential equation by using an appropriate substitution.

**23.** $\dfrac{dy}{dx} = (x + y + 1)^2$

**24.** $\dfrac{dy}{dx} = \dfrac{1 - x - y}{x + y}$

**25.** $\dfrac{dy}{dx} = \tan^2(x + y)$

**26.** $\dfrac{dy}{dx} = \sin(x + y)$

**27.** $\dfrac{dy}{dx} = 2 + \sqrt{y - 2x + 3}$

**28.** $\dfrac{dy}{dx} = 1 + e^{y - x + 5}$

In Problems 29 and 30 solve the given initial-value problem.

**29.** $\dfrac{dy}{dx} = \cos(x + y), \quad y(0) = \pi/4$

**30.** $\dfrac{dy}{dx} = \dfrac{3x + 2y}{3x + 2y + 2}, \quad y(-1) = -1$

### Discussion Problems

**31.** Explain why it is always possible to express any homogeneous differential equation $M(x, y)\,dx + N(x, y)\,dy = 0$ in the form

$$\frac{dy}{dx} = F\left(\frac{y}{x}\right).$$

You might start by proving that

$$M(x, y) = x^\alpha M(1, y/x) \qquad \text{and} \qquad N(x, y) = x^\alpha N(1, y/x).$$

**32.** Put the homogeneous differential equation

$$(5x^2 - 2y^2)\,dx - xy\,dy = 0$$

into the form given in Problem 31.

**33. (a)** Determine two singular solutions of the DE in Problem 10.

**(b)** If the initial condition $y(5) = 0$ is as prescribed in Problem 10, then what is the largest interval $I$ over which the solution is defined? Use a graphing utility to graph the solution curve for the IVP.

**34.** In Example 3 the solution $y(x)$ becomes unbounded as $x \to \pm\infty$. Nevertheless, $y(x)$ is asymptotic to a curve as $x \to -\infty$ and to a different curve as $x \to \infty$. What are the equations of these curves?

**35.** The differential equation $dy/dx = P(x) + Q(x)y + R(x)y^2$ is known as **Riccati's equation.**

**(a)** A Riccati equation can be solved by a succession of two substitutions *provided* that we know a particular solution $y_1$ of the equation. Show that the substitution $y = y_1 + u$ reduces Riccati's equation to a Bernoulli equation (4) with $n = 2$. The Bernoulli equation can then be reduced to a linear equation by the substitution $w = u^{-1}$.

**(b)** Find a one-parameter family of solutions for the differential equation

$$\frac{dy}{dx} = -\frac{4}{x^2} - \frac{1}{x}y + y^2$$

where $y_1 = 2/x$ is a known solution of the equation.

**36.** Determine an appropriate substitution to solve

$$xy' = y \ln(xy).$$

## Mathematical Models

**37. Falling Chain**   In Problem 45 in Exercises 2.4 we saw that a mathematical model for the velocity $v$ of a chain slipping off the edge of a high horizontal platform is

$$xv\frac{dv}{dx} + v^2 = 32x.$$

In that problem you were asked to solve the DE by converting it into an exact equation using an integrating factor. This time solve the DE using the fact that it is a Bernoulli equation.

**38. Population Growth**   In the study of population dynamics one of the most famous models for a growing but bounded population is the **logistic equation**

$$\frac{dP}{dt} = P(a - bP),$$

where $a$ and $b$ are positive constants. Although we will come back to this equation and solve it by an alternative method in Section 3.2, solve the DE this first time using the fact that it is a Bernoulli equation.

---

## 2.6   A Numerical Method

**INTRODUCTION**   A differential equation can be a source of information. We started this chapter by observing that we could garner *qualitative* information from a first-order DE about its solutions even before we attempted to solve the equation. Then in Sections 2.2–2.5 we examined first-order DEs *analytically*—that is, we developed some procedures for obtaining explicit and implicit solutions. But a differential equation can possess a solution and yet it may not be possible to obtain it analytically. So to round out the picture of the different types of analyses of differential equations, we conclude this chapter with a method by which we can "solve" the differential equation *numerically*; this means that the DE is used as the foundation on which we construct an algorithm for approximating the unknown solution.

In this section we are going to develop only the simplest of numerical methods—a method that utilizes the idea that a tangent line can be used to approximate the values of a function in a small neighborhood of the point of tangency. A more extensive treatment of numerical methods for ordinary differential equations is given in Chapter 9.

**USING THE TANGENT LINE**   Let us assume that the first-order initial-value problem

$$y' = f(x, y), \quad y(x_0) = y_0 \tag{1}$$

possesses a solution. One way of approximating this solution is to use tangent lines. For example, let $y(x)$ denote the unknown solution of the first-order initial-value problem $y' = 0.1\sqrt{y} + 0.4x^2$, $y(2) = 4$. The nonlinear differential equation in this IVP cannot be solved directly by any of the methods considered in

Sections 2.2, 2.4, and 2.5; nevertheless, we can still find approximate numerical values of the unknown $y(x)$. Specifically, suppose we wish to know the value of $y(2.5)$. The IVP has a solution, and as the flow of the direction field of the DE in Figure 2.6.1(a) suggests, a solution curve must have a shape similar to the curve shown in blue.

The direction field in Figure 2.6.1(a) was generated with lineal elements passing through points in a grid with integer coordinates. As the solution curve passes through the initial point $(2, 4)$, the lineal element at this point is a tangent line with slope given by $f(2, 4) = 0.1\sqrt{4} + 0.4(2)^2 = 1.8$. As is apparent in Figure 2.6.1(a) and the "zoom in" in Figure 2.6.1(b), when $x$ is close to 2, the points on the solution curve are close to the points on the tangent line (the lineal element). Using the point $(2, 4)$, the slope $f(2, 4) = 1.8$, and the point-slope form of a line, we find that an equation of the tangent line is $y = L(x)$, where $L(x) = 1.8x + 0.4$. This last equation, called a **linearization** of $y(x)$ at $x = 2$, can be used to approximate values of $y(x)$ within a small neighborhood of $x = 2$. If $y_1 = L(x_1)$ denotes the $y$-coordinate on the tangent line and $y(x_1)$ is the $y$-coordinate on the solution curve corresponding to an $x$-coordinate $x_1$ that is close to $x = 2$, then $y(x_1) \approx y_1$. If we choose, say, $x_1 = 2.1$, then $y_1 = L(2.1) = 1.8(2.1) + 0.4 = 4.18$, so $y(2.1) \approx 4.18$.

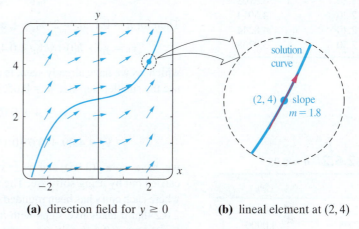

(a) direction field for $y \geq 0$      (b) lineal element at $(2, 4)$

**FIGURE 2.6.1** Magnification of a neighborhood about the point $(2, 4)$

**FIGURE 2.6.2** Approximating $y(x_1)$ using a tangent line

**EULER'S METHOD** To generalize the procedure just illustrated, we use the linearization of the unknown solution $y(x)$ of (1) at $x = x_0$:

$$L(x) = y_0 + f(x_0, y_0)(x - x_0). \tag{2}$$

The graph of this linearization is a straight line tangent to the graph of $y = y(x)$ at the point $(x_0, y_0)$. We now let $h$ be a positive increment of the $x$-axis, as shown in Figure 2.6.2. Then by replacing $x$ by $x_1 = x_0 + h$ in (2), we get

$$L(x_1) = y_0 + f(x_0, y_0)(x_0 + h - x_0) \quad \text{or} \quad y_1 = y_0 + hf(x_1, y_1),$$

where $y_1 = L(x_1)$. The point $(x_1, y_1)$ on the tangent line is an approximation to the point $(x_1, y(x_1))$ on the solution curve. Of course, the accuracy of the approximation $L(x_1) \approx y(x_1)$ or $y_1 \approx y(x_1)$ depends heavily on the size of the increment $h$. Usually, we must choose this **step size** to be "reasonably small." We now repeat the process using a second "tangent line" at $(x_1, y_1)$.* By identifying the new starting

*This is not an actual tangent line, since $(x_1, y_1)$ lies on the first tangent and not on the solution curve.

point as $(x_1, y_1)$ with $(x_0, y_0)$ in the above discussion, we obtain an approximation $y_2 \approx y(x_2)$ corresponding to two steps of length $h$ from $x_0$, that is, $x_2 = x_1 + h = x_0 + 2h$, and

$$y(x_2) = y(x_0 + 2h) = y(x_1 + h) \approx y_2 = y_1 + hf(x_1, y_1).$$

Continuing in this manner, we see that $y_1, y_2, y_3, \ldots$, can be defined recursively by the general formula

$$y_{n+1} = y_n + hf(x_n, y_n), \tag{3}$$

where $x_n = x_0 + nh$, $n = 0, 1, 2, \ldots$. This procedure of using successive "tangent lines" is called **Euler's method.**

### EXAMPLE 1 | Euler's Method

Consider the initial-value problem $y' = 0.1\sqrt{y} + 0.4x^2$, $y(2) = 4$. Use Euler's method to obtain an approximation of $y(2.5)$ using first $h = 0.1$ and then $h = 0.05$.

**SOLUTION** With the identification $f(x, y) = 0.1\sqrt{y} + 0.4x^2$, (3) becomes

$$y_{n+1} = y_n + h\left(0.1\sqrt{y_n} + 0.4x_n^2\right).$$

Then for $h = 0.1$, $x_0 = 2$, $y_0 = 4$, and $n = 0$ we find

$$y_1 = y_0 + h\left(0.1\sqrt{y_0} + 0.4x_0^2\right) = 4 + 0.1\left(0.1\sqrt{4} + 0.4(2)^2\right) = 4.18,$$

which, as we have already seen, is an estimate to the value of $y(2.1)$. However, if we use the smaller step size $h = 0.05$, it takes two steps to reach $x = 2.1$. From

$$y_1 = 4 + 0.05\left(0.1\sqrt{4} + 0.4(2)^2\right) = 4.09$$

$$y_2 = 4.09 + 0.05\left(0.1\sqrt{4.09} + 0.4(2.05)^2\right) = 4.18416187$$

we have $y_1 \approx y(2.05)$ and $y_2 \approx y(2.1)$. The remainder of the calculations were carried out by using software. The results are summarized in Tables 2.6.1 and 2.6.2, where each entry has been rounded to four decimal places. We see in Tables 2.6.1 and 2.6.2 that it takes five steps with $h = 0.1$ and 10 steps with $h = 0.05$, respectively, to get to $x = 2.5$. Intuitively, we would expect that $y_{10} = 5.0997$ corresponding to $h = 0.05$ is the better approximation of $y(2.5)$ than the value $y_5 = 5.0768$ corresponding to $h = 0.1$.

In Example 2 we apply Euler's method to a differential equation for which we have already found a solution. We do this to compare the values of the approximations $y_n$ at each step with the true or actual values of the solution $y(x_n)$ of the initial-value problem.

### EXAMPLE 2 | Comparison of Approximate and Actual Values

Consider the initial-value problem $y' = 0.2xy$, $y(1) = 1$. Use Euler's method to obtain an approximation of $y(1.5)$ using first $h = 0.1$ and then $h = 0.05$.

**SOLUTION** With the identification $f(x, y) = 0.2xy$, (3) becomes

$$y_{n+1} = y_n + h(0.2x_n y_n)$$

where $x_0 = 1$ and $y_0 = 1$. Again with the aid of computer software we obtain the values in Tables 2.6.3 and 2.6.4 on page 79.

In Example 1 the true or actual values were calculated from the known solution $y = e^{0.1(x^2-1)}$. (Verify.) The **absolute error** is defined to be

$$|actual\ value - approximation|.$$

**TABLE 2.6.1**   $h = 0.1$

| $x_n$ | $y_n$ |
|-------|-------|
| 2.00 | 4.0000 |
| 2.10 | 4.1800 |
| 2.20 | 4.3768 |
| 2.30 | 4.5914 |
| 2.40 | 4.8244 |
| 2.50 | 5.0768 |

**TABLE 2.6.2**   $h = 0.05$

| $x_n$ | $y_n$ |
|-------|-------|
| 2.00 | 4.0000 |
| 2.05 | 4.0900 |
| 2.10 | 4.1842 |
| 2.15 | 4.2826 |
| 2.20 | 4.3854 |
| 2.25 | 4.4927 |
| 2.30 | 4.6045 |
| 2.35 | 4.7210 |
| 2.40 | 4.8423 |
| 2.45 | 4.9686 |
| 2.50 | 5.0997 |

**TABLE 2.6.3**  $h = 0.1$

| $x_n$ | $y_n$ | Actual value | Abs. error | % Rel. error |
|------|--------|--------------|------------|--------------|
| 1.00 | 1.0000 | 1.0000 | 0.0000 | 0.00 |
| 1.10 | 1.0200 | 1.0212 | 0.0012 | 0.12 |
| 1.20 | 1.0424 | 1.0450 | 0.0025 | 0.24 |
| 1.30 | 1.0675 | 1.0714 | 0.0040 | 0.37 |
| 1.40 | 1.0952 | 1.1008 | 0.0055 | 0.50 |
| 1.50 | 1.1259 | 1.1331 | 0.0073 | 0.64 |

**TABLE 2.6.4**  $h = 0.05$

| $x_n$ | $y_n$ | Actual value | Abs. error | % Rel. error |
|------|--------|--------------|------------|--------------|
| 1.00 | 1.0000 | 1.0000 | 0.0000 | 0.00 |
| 1.05 | 1.0100 | 1.0103 | 0.0003 | 0.03 |
| 1.10 | 1.0206 | 1.0212 | 0.0006 | 0.06 |
| 1.15 | 1.0318 | 1.0328 | 0.0009 | 0.09 |
| 1.20 | 1.0437 | 1.0450 | 0.0013 | 0.12 |
| 1.25 | 1.0562 | 1.0579 | 0.0016 | 0.16 |
| 1.30 | 1.0694 | 1.0714 | 0.0020 | 0.19 |
| 1.35 | 1.0833 | 1.0857 | 0.0024 | 0.22 |
| 1.40 | 1.0980 | 1.1008 | 0.0028 | 0.25 |
| 1.45 | 1.1133 | 1.1166 | 0.0032 | 0.29 |
| 1.50 | 1.1295 | 1.1331 | 0.0037 | 0.32 |

The **relative error** and **percentage relative error** are, in turn,

$$\frac{absolute\ error}{|actual\ value|} \quad \text{and} \quad \frac{absolute\ error}{|actual\ value|} \times 100.$$

It is apparent from Tables 2.6.3 and 2.6.4 that the accuracy of the approximations improves as the step size $h$ decreases. Also, we see that even though the percentage relative error is growing with each step, it does not appear to be that bad. But you should not be deceived by one example. If we simply change the coefficient of the right side of the DE in Example 2 from 0.2 to 2, then at $x_n = 1.5$ the percentage relative errors increase dramatically. See Problem 4 in Exercises 2.6.

**A CAVEAT**    Euler's method is just one of many different ways in which a solution of a differential equation can be approximated. Although attractive for its simplicity, *Euler's method is seldom used in serious calculations.* It was introduced here simply to give you a first taste of numerical methods. We will go into greater detail in discussing numerical methods that give significantly greater accuracy, notably the **fourth order Runge-Kutta method,** referred to as the **RK4 method,** in Chapter 9.

**NUMERICAL SOLVERS**    Regardless of whether we can actually find an explicit or implicit solution, if a solution of a differential equation exists, it represents a smooth curve in the Cartesian plane. The basic idea behind *any* numerical method for first-order ordinary differential equations is to somehow approximate the $y$-values of a solution for preselected values of $x$. We start at a specified initial point $(x_0, y_0)$ on a solution curve and proceed to calculate in a step-by-step fashion a sequence of points $(x_1, y_1), (x_2, y_2), \ldots, (x_n, y_n)$ whose $y$-coordinates $y_i$ approximate the $y$-coordinates $y(x_i)$ of points $(x_1, y(x_1)), (x_2, y(x_2)), \ldots, (x_n, y(x_n))$ that lie on the graph of the usually unknown solution $y(x)$. By taking the $x$-coordinates close together (that is, for small values of $h$) and by joining the points $(x_1, y_1), (x_2, y_2), \ldots, (x_n, y_n)$ with short line segments, we obtain a polygonal curve whose qualitative characteristics we hope are close to those of an actual solution curve. Drawing curves is something that is well suited to a computer. A computer program written to either implement a numerical method or render a visual representation of an approximate solution curve fitting the numerical data produced by this method is referred to as a **numerical solver.** Many different numerical solvers are commercially available, either embedded in a larger software package, such as a computer algebra system, or provided as a stand-alone package. Some software packages simply plot the generated numerical approximations, whereas others generate hard numerical data as well as the corresponding approximate or

**FIGURE 2.6.3** Comparison of the Runge-Kutta (RK4) and Euler methods

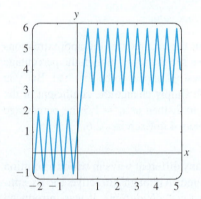

**FIGURE 2.6.4** A not-very helpful numerical solution curve

**numerical solution curves.** By way of illustration of the connect-the-dots nature of the graphs produced by a numerical solver, the two colored polygonal graphs in Figure 2.6.3 are the numerical solution curves for the initial-value problem $y' = 0.2xy$, $y(0) = 1$ on the interval $[0, 4]$ obtained from Euler's method and the RK4 method using the step size $h = 1$. The blue smooth curve is the graph of the exact solution $y = e^{0.1x^2}$ of the IVP. Notice in Figure 2.6.3 that, even with the ridiculously large step size of $h = 1$, the RK4 method produces the more believable "solution curve." The numerical solution curve obtained from the RK4 method is indistinguishable from the actual solution curve on the interval $[0, 4]$ when a more typical step size of $h = 0.1$ is used.

**USING A NUMERICAL SOLVER** Knowledge of the various numerical methods is not necessary in order to use a numerical solver. A solver usually requires that the differential equation be expressed in normal form $dy/dx = f(x, y)$. Numerical solvers that generate only curves usually require that you supply $f(x, y)$ and the initial data $x_0$ and $y_0$ and specify the desired numerical method. If the idea is to approximate the numerical value of $y(a)$, then a solver may additionally require that you state a value for $h$ or, equivalently, give the number of steps that you want to take to get from $x = x_0$ to $x = a$. For example, if we wanted to approximate $y(4)$ for the IVP illustrated in Figure 2.6.3, then, starting at $x = 0$ it would take four steps to reach $x = 4$ with a step size of $h = 1$; 40 steps is equivalent to a step size of $h = 0.1$. Although we will not delve here into the many problems that one can encounter when attempting to approximate mathematical quantities, you should at least be aware of the fact that a numerical solver may break down near certain points or give an incomplete or misleading picture when applied to some first-order differential equations in the normal form. Figure 2.6.4 illustrates the graph obtained by applying Euler's method to a certain first-order initial-value problem $dy/dx = f(x, y)$, $y(0) = 1$. Equivalent results were obtained using three different commercial numerical solvers, yet the graph is hardly a plausible solution curve. (Why?) There are several avenues of recourse when a numerical solver has difficulties; three of the more obvious are decrease the step size, use another numerical method, and try a different numerical solver.

# EXERCISES 2.6

*Answers to selected odd-numbered problems begin on page ANS-2.*

In Problems 1 and 2 use Euler's method to obtain a four-decimal approximation of the indicated value. Carry out the recursion of (3) by hand, first using $h = 0.1$ and then using $h = 0.05$.

**1.** $y' = 2x - 3y + 1$, $y(1) = 5$; $y(1.2)$

**2.** $y' = x + y^2$, $y(0) = 0$; $y(0.2)$

In Problems 3 and 4 use Euler's method to obtain a four-decimal approximation of the indicated value. First use $h = 0.1$ and then use $h = 0.05$. Find an explicit solution for each initial-value problem and then construct tables similar to Tables 2.6.3 and 2.6.4.

**3.** $y' = y$, $y(0) = 1$; $y(1.0)$

**4.** $y' = 2xy$, $y(1) = 1$; $y(1.5)$

In Problems 5–10 use a numerical solver and Euler's method to obtain a four-decimal approximation of the indicated value. First use $h = 0.1$ and then use $h = 0.05$.

**5.** $y' = e^{-y}$, $y(0) = 0$; $y(0.5)$

**6.** $y' = x^2 + y^2$, $y(0) = 1$; $y(0.5)$

**7.** $y' = (x - y)^2$, $y(0) = 0.5$; $y(0.5)$

**8.** $y' = xy + \sqrt{y}$, $y(0) = 1$; $y(0.5)$

**9.** $y' = xy^2 - \dfrac{y}{x}$, $y(1) = 1$; $y(1.5)$

**10.** $y' = y - y^2$, $y(0) = 0.5$; $y(0.5)$

In Problems 11 and 12 use a numerical solver to obtain a numerical solution curve for the given initial-value problem. First use Euler's method and then the RK4 method. Use $h = 0.25$ in each case. Superimpose both solution curves on the same coordinate axes. If possible, use a different color for each curve. Repeat, using $h = 0.1$ and $h = 0.05$.

**11.** $y' = 2(\cos x)y$, $y(0) = 1$

**12.** $y' = y(10 - 2y)$, $y(0) = 1$

### Discussion Problems

**13.** Use a numerical solver and Euler's method to approximate $y(1.0)$, where $y(x)$ is the solution to $y' = 2xy^2$, $y(0) = 1$. First use $h = 0.1$ and then use $h = 0.05$. Repeat, using the RK4 method. Discuss what might cause the approximations to $y(1.0)$ to differ so greatly.

### Computer Lab Assignments

**14. (a)** Use a numerical solver and the RK4 method to graph the solution of the initial-value problem $y' = -2xy + 1$, $y(0) = 0$.

**(b)** Solve the initial-value problem by one of the analytic procedures developed earlier in this chapter.

**(c)** Use the analytic solution $y(x)$ found in part (b) and a CAS to find the coordinates of all relative extrema.

## Chapter 2 In Review

*Answers to selected odd-numbered problems begin on page ANS-3.*

Answer Problems 1–12 without referring back to the text. Fill in the blanks or answer true or false.

**1.** The linear DE, $y' - ky = A$, where $k$ and $A$ are constants, is autonomous. The critical point _____ of the equation is a(n) _____ (attractor or repeller) for $k > 0$ and a(n) _____ (attractor or repeller) for $k < 0$.

**2.** The initial-value problem $x\dfrac{dy}{dx} - 4y = 0$, $y(0) = k$, has an infinite number of solutions for $k = $ _____ and no solution for $k = $ _____.

**3.** The linear DE, $y' + k_1 y = k_2$, where $k_1$ and $k_2$ are nonzero constants, always possesses a constant solution. _____

**4.** The linear DE, $a_1(x)y' + a_0(x)y = 0$ is also separable. _____

**5.** An example of a nonlinear third-order differential equation in normal form is _____.

**6.** The first-order DE, $\dfrac{dr}{d\theta} = r\theta + r + \theta + 1$ is not separable. _____

**7.** Every autonomous DE $dy/dx = f(y)$ is separable. _____

**8.** By inspection, two solutions of the differential equation $y' + |y| = 2$ are _____.

**9.** If $y' = e^x y$, then $y = $ _____.

**10.** If a differentiable function $y(x)$ satisfies $y' = |x|$, $y(-1) = 2$, then $y(x) = $ _____.

**11.** $y = e^{\cos x}\displaystyle\int_0^x te^{-\cos t}\,dt$ is a solution of the linear first-order differential equation _____.

**12.** An example of an autonomous linear first-order DE with a single critical point $-3$ is _____, whereas an autonomous nonlinear first-order DE with a single critical point $-3$ is _____.

In Problems 13 and 14 construct an autonomous first-order differential equation $dy/dx = f(y)$ whose phase portrait is consistent with the given figure.

**13.**

**FIGURE 2.R.1** Graph for Problem 13

**14.**

**FIGURE 2.R.2** Graph for Problem 14

**15.** The number 0 is a critical point of the autonomous differential equation $dx/dt = x^n$, where $n$ is a positive integer. For what values of $n$ is 0 asymptotically stable? Semi-stable? Unstable? Repeat for the differential equation $dx/dt = -x^n$.

**16.** Consider the differential equation $dP/dt = f(P)$, where

$$f(P) = -0.5P^3 - 1.7P + 3.4.$$

The function $f(P)$ has one real zero, as shown in Figure 2.R.3. Without attempting to solve the differential equation, estimate the value of $\lim_{t\to\infty} P(t)$.

**FIGURE 2.R.3** Graph for Problem 16

**17.** Figure 2.R.4 is a portion of a direction field of a differential equation $dy/dx = f(x, y)$. By hand, sketch two different solution curves—one that is tangent to the lineal element shown in black and one that is tangent to the lineal element shown in red.

**FIGURE 2.R.4** Portion of a direction field for Problem 17

**18.** Classify each differential equation as separable, exact, linear, homogeneous, or Bernoulli. Some equations may be more than one kind. Do not solve.

**(a)** $\dfrac{dy}{dx} = \dfrac{x - y}{x}$

**(b)** $\dfrac{dy}{dx} = \dfrac{1}{y - x}$

**(c)** $(x + 1)\dfrac{dy}{dx} = -y + 10$

**(d)** $\dfrac{dy}{dx} = \dfrac{1}{x(x - y)}$

**(e)** $\dfrac{dy}{dx} = \dfrac{y^2 + y}{x^2 + x}$

**(f)** $\dfrac{dy}{dx} = 5y + y^2$

**(g)** $y\,dx = (y - xy^2)\,dy$

**(h)** $x\dfrac{dy}{dx} = ye^{x/y} - x$

**(i)** $xy\,y' + y^2 = 2x$

**(j)** $2xy\,y' + y^2 = 2x^2$

**(k)** $y\,dx + x\,dy = 0$

**(l)** $\left(x^2 + \dfrac{2y}{x}\right)dx = (3 - \ln x^2)\,dy$

**(m)** $\dfrac{dy}{dx} = \dfrac{x}{y} + \dfrac{y}{x} + 1$

**(n)** $\dfrac{y}{x^2}\dfrac{dy}{dx} + e^{2x^3 + y^2} = 0$

In Problems 19–26 solve the given differential equation.

**19.** $(y^2 + 1)\,dx = y\sec^2 x\,dy$

**20.** $y(\ln x - \ln y)\,dx = (x \ln x - x \ln y - y)\,dy$

**21.** $(6x + 1)y^2\dfrac{dy}{dx} + 3x^2 + 2y^3 = 0$

**22.** $\dfrac{dx}{dy} = -\dfrac{4y^2 + 6xy}{3y^2 + 2x}$

**23.** $t\dfrac{dQ}{dt} + Q = t^4 \ln t$

**24.** $(2x + y + 1)y' = 1$

**25.** $(x^2 + 4)\,dy = (2x - 8xy)\,dx$

**26.** $(2r^2 \cos\theta\sin\theta + r\cos\theta)\,d\theta$
$\quad + (4r + \sin\theta - 2r\cos^2\theta)\,dr = 0$

In Problems 27–30 express the solution of the given initial-value problem in terms of an integral-defined function.

**27.** $2\dfrac{dy}{dx} + (4\cos x)y = x, \quad y(0) = 1$

**28.** $\dfrac{dy}{dx} - 4xy = \sin x^2, \quad y(0) = 7$

**29.** $x\dfrac{dy}{dx} + 2y = xe^{x^2}, \quad y(1) = 3$

**30.** $x\dfrac{dy}{dx} + (\sin x)y = 0, \quad y(0) = 10$

In Problems 31 and 32 solve the given initial-value problem.

**31.** $\dfrac{dy}{dx} + y = f(x), \quad y(0) = 5$, where

$$f(x) = \begin{cases} e^{-x}, & 0 \le x < 1 \\ 0, & x \ge 1 \end{cases}$$

**32.** $\dfrac{dy}{dx} + P(x)y = e^x, \quad y(0) = -1$, where

$$P(x) = \begin{cases} 1, & 0 \le x < 1 \\ -1, & x \ge 1 \end{cases}$$

In Problems 33 and 34 solve the given initial-value problem and give the largest interval $I$ on which the solution is defined.

**33.** $\sin x\dfrac{dy}{dx} + (\cos x)y = 0, \quad y(7\pi/6) = -2$

**34.** $\dfrac{dy}{dt} + 2(t + 1)y^2 = 0, \quad y(0) = -\tfrac{1}{8}$

**35. (a)** Without solving, explain why the initial-value problem

$$\dfrac{dy}{dx} = \sqrt{y}, \quad y(x_0) = y_0$$

has no solution for $y_0 < 0$.

**(b)** Solve the initial-value problem in part (a) for $y_0 > 0$ and find the largest interval $I$ on which the solution is defined.

**36. (a)** Find an implicit solution of the initial-value problem

$$\dfrac{dy}{dx} = \dfrac{y^2 - x^2}{xy}, \quad y(1) = -\sqrt{2}.$$

**(b)** Find an explicit solution of the problem in part (a) and give the largest interval $I$ over which the solution is defined. A graphing utility may be helpful here.

**37.** Graphs of some members of a family of solutions for a first-order differential equation $dy/dx = f(x, y)$ are shown in Figure 2.R.5. The graphs of two implicit solutions, one that passes through the point $(1, -1)$ and one that passes through $(-1, 3)$, are shown in blue. Reproduce the figure on a piece of paper. With colored pencils trace out the solution curves for the solutions $y = y_1(x)$ and $y = y_2(x)$ defined by the implicit solutions such that

**FIGURE 2.R.5** Graph for Problem 37

$y_1(1) = -1$ and $y_2(-1) = 3$, respectively. Estimate the intervals on which the solutions $y = y_1(x)$ and $y = y_2(x)$ are defined.

**38.** Use Euler's method with step size $h = 0.1$ to approximate $y(1.2)$, where $y(x)$ is a solution of the initial-value problem $y' = 1 + x\sqrt{y}$, $y(1) = 9$.

In Problems 39 and 40 each figure represents a portion of a direction field of an autonomous first-order differential equation $dy/dx = f(y)$. Reproduce the figure on a separate piece of paper and then complete the direction field over the grid. The points of the grid are $(mh, nh)$, where $h = \frac{1}{2}$, $m$ and $n$ integers, $-7 \leq m \leq 7$, $-7 \leq n \leq 7$. In each direction field, sketch by hand an approximate solution curve that passes through each of the solid points shown in red. Discuss: Does it appear that the DE possesses critical points in the interval $-3.5 \leq y \leq 3.5$? If so, classify the critical points as asymptotically stable, unstable, or semi-stable.

**40.**

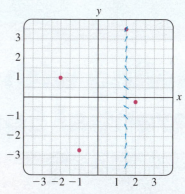

**FIGURE 2.R.7** Portion of a direction field for Problem 40

**39.**

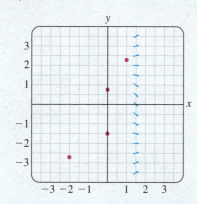

**FIGURE 2.R.6** Portion of a direction field for Problem 39

# 3

# Modeling with First-Order Differential Equations

Fotos593/Shutterstock.com

In Section 1.3 we saw how a first-order differential equation could be used as a mathematical model in the study of population growth, radioactive decay, cooling of bodies, mixtures, chemical reactions, fluid draining from a tank, velocity of a falling body, and current in a series circuit. Using the methods discussed in Chapter 2, we are now able to solve some of the linear DEs in Section 3.1 and nonlinear DEs in Section 3.2 that frequently appear in applications.

## 3.1 Linear Models

**INTRODUCTION**    In this section we solve some of the linear first-order models that were introduced in Section 1.3.

**GROWTH AND DECAY**    The initial-value problem

$$\frac{dx}{dt} = kx, \quad x(t_0) = x_0, \tag{1}$$

where $k$ is a constant of proportionality, serves as a model for diverse phenomena involving either growth or decay. We saw in Section 1.3 that in biological applications the rate of growth of certain populations (bacteria, small animals) over short periods of time is proportional to the population present at time $t$. Knowing the population at some arbitrary initial time $t_0$, we can then use the solution of (1) to predict the population in the future—that is, at times $t > t_0$. The constant of proportionality $k$ in (1) can be determined from the solution of the initial-value problem, using a subsequent measurement of $x$ at a time $t_1 > t_0$. In physics and chemistry (1) is seen in the form of a *first-order reaction*—that is, a reaction whose rate, or velocity, $dx/dt$ is directly proportional to the amount $x$ of a substance that is unconverted or remaining at time $t$. The decomposition, or decay, of U-238 (uranium) by radioactivity into Th-234 (thorium) is a first-order reaction.

### EXAMPLE 1    Bacterial Growth

A culture initially has $P_0$ number of bacteria. At $t = 1$ h the number of bacteria is measured to be $\frac{3}{2}P_0$. If the rate of growth is proportional to the number of bacteria $P(t)$ present at time $t$, determine the time necessary for the number of bacteria to triple.

**SOLUTION**    We first solve the differential equation in (1), with the symbol $x$ replaced by $P$. With $t_0 = 0$ the initial condition is $P(0) = P_0$. We then use the empirical observation that $P(1) = \frac{3}{2}P_0$ to determine the constant of proportionality $k$.

Notice that the differential equation $dP/dt = kP$ is both separable and linear. When it is put in the standard form of a linear first-order DE,

$$\frac{dP}{dt} - kP = 0,$$

we can see by inspection that the integrating factor is $e^{-kt}$. Multiplying both sides of the equation by this term and integrating gives, in turn,

$$\frac{d}{dt}[e^{-kt}P] = 0 \quad \text{and} \quad e^{-kt}P = c.$$

Therefore $P(t) = ce^{kt}$. At $t = 0$ it follows that $P_0 = ce^0 = c$, so $P(t) = P_0 e^{kt}$. At $t = 1$ we have $\frac{3}{2}P_0 = P_0 e^k$ or $e^k = \frac{3}{2}$. From the last equation we get $k = \ln \frac{3}{2} = 0.4055$, so $P(t) = P_0 e^{0.4055t}$. To find the time at which the number of bacteria has tripled, we solve $3P_0 = P_0 e^{0.4055t}$ for $t$. It follows that $0.4055t = \ln 3$, or

$$t = \frac{\ln 3}{0.4055} \approx 2.71 \text{ h}.$$

See Figure 3.1.1.

**FIGURE 3.1.1** Time in which population triples in Example 1

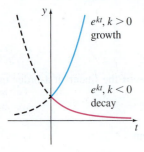

**FIGURE 3.1.2** Growth ($k > 0$) and decay ($k < 0$)

Notice in Example 1 that the actual number $P_0$ of bacteria present at time $t = 0$ played no part in determining the time required for the number in the culture to triple. The time necessary for an initial population of, say, 100 or 1,000,000 bacteria to triple is still approximately 2.71 hours.

As shown in Figure 3.1.2, the exponential function $e^{kt}$ increases as $t$ increases for $k > 0$ and decreases as $t$ increases for $k < 0$. Thus problems describing growth (whether of populations, bacteria, or even capital) are characterized by a positive value of $k$, whereas problems involving decay (as in radioactive disintegration) yield a negative $k$ value. Accordingly, we say that $k$ is either a **growth constant** ($k > 0$) or a **decay constant** ($k < 0$).

**HALF-LIFE**   In physics the **half-life** is a measure of the stability of a radioactive substance. The half-life is simply the time it takes for one-half of the atoms in an initial amount $A_0$ to disintegrate, or transmute, into the atoms of another element. The longer the half-life of a substance, the more stable it is. For example, the half-life of highly radioactive radium, Ra-226, is about 1700 years. In 1700 years one-half of a given quantity of Ra-226 is transmuted into radon, Rn-222. The most commonly occurring uranium isotope, U-238, has a half-life of approximately 4,500,000,000 years. In about 4.5 billion years, one-half of a quantity of U-238 is transmuted into lead, Pb-206.

**EXAMPLE 2**   Half-Life of Plutonium

A breeder reactor converts relatively stable uranium-238 into the isotope plutonium-239. After 15 years it is determined that 0.043% of the initial amount $A_0$ of plutonium has disintegrated. Find the half-life of this isotope if the rate of disintegration is proportional to the amount remaining.

**SOLUTION**   Let $A(t)$ denote the amount of plutonium remaining at time $t$. As in Example 1 the solution of the initial-value problem

$$\frac{dA}{dt} = kA, \quad A(0) = A_0$$

is $A(t) = A_0 e^{kt}$. If 0.043% of the atoms of $A_0$ have disintegrated, then 99.957% of the substance remains. To find the decay constant $k$, we use $0.99957A_0 = A(15)$—that is, $0.99957A_0 = A_0 e^{15k}$. Solving for $k$ then gives $k = \frac{1}{15} \ln 0.99957 = -0.00002867$. Hence $A(t) = A_0 e^{-0.00002867t}$. Now the half-life is the corresponding value of time at which $A(t) = \frac{1}{2}A_0$. Solving for $t$ gives $\frac{1}{2}A_0 = A_0 e^{-0.00002867t}$, or $\frac{1}{2} = e^{-0.00002867t}$. The last equation yields

$$t = \frac{\ln 2}{0.00002867} \approx 24{,}180 \text{ yr.} \qquad \blacksquare$$

**CARBON DATING**   About 1950, a team of scientists at the University of Chicago led by the chemist **Willard Libby** devised a method using a radioactive isotope of carbon as a means of determining the approximate ages of carbonaceous fossilized matter. See Figure 3.1.3. The theory of **carbon dating** is based on the fact that the radioisotope carbon-14 is produced in the atmosphere by the action of cosmic radiation on nitrogen-14. The ratio of the amount of C-14 to the stable C-12 in the atmosphere appears to be a constant, and as a consequence the proportionate amount of the isotope present in all living organisms is the same as that in the atmosphere. When a living organism dies, the absorption of C-14, by breathing, eating, or photosynthesis, ceases. By comparing the proportionate amount of C-14, say, in a fossil with the constant amount ratio found in the atmosphere, it is possible to obtain a reasonable estimation of its age. The method is based on the knowledge of the

**FIGURE 3.1.3** Willard Libby (1908–1980)

**FIGURE 3.1.4** A page of the Gnostic Gospel of Judas

half-life of C-14. Libby's calculated value of the half-life of C-14 was approximately 5600 years, and is called the **Libby half-life.** Today the commonly accepted value for the half-life of C-14 is the **Cambridge half-life** that is close to 5730 years. For his work, Libby was awarded the Nobel Prize for chemistry in 1960. Libby's method has been used to date wooden furniture found in Egyptian tombs, the woven flax wrappings of the Dead Sea Scrolls, a recently discovered copy of the Gnostic Gospel of Judas written on papyrus, and the cloth of the enigmatic Shroud of Turin. See Figure 3.1.4 and Problem 12 in Exercises 3.1.

**EXAMPLE 3    Age of a Fossil**

A fossilized bone is found to contain 0.1% of its original amount of C-14. Determine the age of the fossil.

**SOLUTION**  As in Example 2 the starting point is $A(t) = A_0 e^{kt}$. To determine the value of the decay constant $k$ we use the fact that $\frac{1}{2} A_0 = A(5730)$ or $\frac{1}{2} A_0 = A_0 e^{5730k}$. The last equation implies $5730k = \ln \frac{1}{2} = -\ln 2$ and so we get $k = -(\ln 2)/5730 = -0.00012097$. Therefore $A(t) = A_0 e^{-0.00012097t}$. With $A(t) = 0.001A_0$ we have $0.001A_0 = A_0 e^{-0.00012097t}$ and $-0.00012097t = \ln(0.001) = -\ln 1000$. Thus

$$t = \frac{\ln 1000}{0.00012097} \approx 57{,}100 \text{ years.}$$

The age found in Example 3 is really at the border of accuracy of this method. The usual carbon-14 technique is limited to about 10 half-lives of the isotope, or roughly 60,000 years. One fundamental reason for this limitation is the relatively short half-life of C-14. There are other problems as well: The chemical analysis needed to obtain an accurate measurement of the remaining C-14 becomes somewhat formidable around the point $0.001A_0$. Moreover, this analysis requires the destruction of a rather large sample of the specimen. If this measurement is accomplished indirectly, based on the actual radioactivity of the specimen, then it is very difficult to distinguish between the radiation from the specimen and the normal background radiation. But recently the use of a particle accelerator has enabled scientists to separate the C-14 from the stable C-12 directly. When the precise value of the ratio of C-14 to C-12 is computed, the accuracy of the dating method can be extended to 70,000–100,000 years. For these reasons and the fact that the C-14 dating is restricted to organic materials, this method is used mainly by archeologists. On the other hand, geologists who are interested in questions about the age of rocks or the age of the Earth use **radiometric-dating** techniques. Radiometric dating, invented by the physicist chemist **Ernest Rutherford** (1871–1937) around 1905, is based on the radioactive decay of a naturally occurring radioactive isotope with a very long half-life and a comparison between a measured quantity of this decaying isotope and one of its decay products using known decay rates. Radiometric methods using potassium-argon, rubidium-strontium, or uranium-lead can give ages of certain kinds of rocks of several billion years. See Problems 5 and 6 in Exercises 3.3 for a brief discussion of the **potassium-argon method of dating.**

 The size and location of the sample caused major difficulties when a team of scientists were invited to use carbon-14 dating on the Shroud of Turin in 1988.

 The half-life of uranium-238 is about 4.47 billion years.

**NEWTON'S LAW OF COOLING/WARMING**    In equation (3) of Section 1.3 we saw that the mathematical formulation of Newton's empirical law of cooling/warming of an object is given by the linear first-order differential equation

$$\frac{dT}{dt} = k(T - T_m), \tag{2}$$

where $k$ is a constant of proportionality, $T(t)$ is the temperature of the object for $t > 0$, and $T_m$ is the ambient temperature—that is, the temperature of the medium around the object. In Example 4 we assume that $T_m$ is constant.

## EXAMPLE 4   Cooling of a Cake

When a cake is removed from an oven, its temperature is measured at 300° F. Three minutes later its temperature is 200° F. How long will it take for the cake to cool off to a room temperature of 70° F?

**SOLUTION**   In (2) we make the identification $T_m = 70$. We must then solve the initial-value problem

$$\frac{dT}{dt} = k(T - 70), \quad T(0) = 300 \tag{3}$$

and determine the value of $k$ so that $T(3) = 200$.

Equation (3) is both linear and separable. If we separate variables,

$$\frac{dT}{T - 70} = k\,dt,$$

yields $\ln|T - 70| = kt + c_1$, and so $T = 70 + c_2 e^{kt}$. When $t = 0$, $T = 300$, so $300 = 70 + c_2$ gives $c_2 = 230$; therefore $T = 70 + 230e^{kt}$. Finally, the measurement $T(3) = 200$ leads to $e^{3k} = \frac{13}{23}$, or $k = \frac{1}{3}\ln\frac{13}{23} = -0.19018$. Thus

$$T(t) = 70 + 230e^{-0.19018t}. \tag{4}$$

We note that (4) furnishes no finite solution to $T(t) = 70$, since $\lim_{t\to\infty}T(t) = 70$. Yet we intuitively expect the cake to reach room temperature after a reasonably long period of time. How long is "long"? Of course, we should not be disturbed by the fact that the model (3) does not quite live up to our physical intuition. Parts (a) and (b) of Figure 3.1.5 clearly show that the cake will be approximately at room temperature in about one-half hour.  ∎

The ambient temperature in (2) need not be a constant but could be a function $T_m(t)$ of time $t$. See Problem 18 in Exercises 3.1.

**MIXTURES**   The mixing of two fluids sometimes gives rise to a linear first-order differential equation. When we discussed the mixing of two brine solutions in Section 1.3, we assumed that the rate $A'(t)$ at which the amount of salt in the mixing tank changes was a net rate:

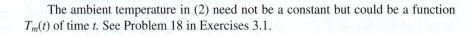

$$\frac{dA}{dt} = (\text{input rate of salt}) - (\text{output rate of salt}) = R_{in} - R_{out}. \tag{5}$$

In Example 5 we solve equation (8) on page 25 of Section 1.3.

## EXAMPLE 5   Mixture of Two Salt Solutions

Recall that the large tank considered in Section 1.3 held 300 gallons of a brine solution. Salt was entering and leaving the tank; a brine solution was being pumped into the tank at the rate of 3 gal/min; it mixed with the solution there, and then the mixture was pumped out at the rate of 3 gal/min. The concentration of the salt in the inflow, or solution entering, was 2 lb/gal, so salt was entering the tank at the rate $R_{in} = (2\text{ lb/gal}) \cdot (3\text{ gal/min}) = 6\text{ lb/min}$ and leaving the tank at the rate $R_{out} = (A/300\text{ lb/gal}) \cdot (3\text{ gal/min}) = A/100\text{ lb/min}$. From this data and (5) we get equation (8) of Section 1.3. Let us pose the question: If 50 pounds of salt were dissolved initially in the 300 gallons, how much salt is in the tank after a long time?

T

300

150    T = 70

15    30    t

**(a)**

| $T(t)$ | $t$ (min) |
|--------|-----------|
| 75°    | 20.1      |
| 74°    | 21.3      |
| 73°    | 22.8      |
| 72°    | 24.9      |
| 71°    | 28.6      |
| 70.5°  | 32.3      |

**(b)**

**FIGURE 3.1.5**  Temperature of cooling cake in Example 4

| $t$ (min) | $A$ (lb) |
|-----------|----------|
| 50  | 266.41 |
| 100 | 397.67 |
| 150 | 477.27 |
| 200 | 525.57 |
| 300 | 572.62 |
| 400 | 589.93 |

**(b)**

**FIGURE 3.1.6** Pounds of salt in the tank in Example 5

**SOLUTION** To find the amount of salt $A(t)$ in the tank at time $t$, we solve the initial-value problem

$$\frac{dA}{dt} + \frac{1}{100}A = 6, \quad A(0) = 50.$$

Note here that the side condition is the initial amount of salt $A(0) = 50$ in the tank and *not* the initial amount of liquid in the tank. Now since the integrating factor of the linear differential equation is $e^{t/100}$, we can write the equation as

$$\frac{d}{dt}[e^{t/100}A] = 6e^{t/100}.$$

Integrating the last equation and solving for $A$ gives the general solution $A(t) = 600 + ce^{-t/100}$. When $t = 0$, $A = 50$, so we find that $c = -550$. Thus the amount of salt in the tank at time $t$ is given by

$$A(t) = 600 - 550e^{-t/100}. \tag{6}$$

The solution (6) was used to construct the table in Figure 3.1.6(b). Also, it can be seen from (6) and Figure 3.1.6(a) that $A(t) \to 600$ as $t \to \infty$. Of course, this is what we would intuitively expect; over a long time the number of pounds of salt in the solution must be (300 gal)(2 lb/gal) = 600 lb.  ■

In Example 5 we assumed that the rate at which the solution was pumped in was the same as the rate at which the solution was pumped out. However, this need not be the case; the mixed brine solution could be pumped out at a rate $r_{out}$ that is faster or slower than the rate $r_{in}$ at which the other brine solution is pumped in. The next example illustrates the case when the mixture is pumped out at rate that is *slower* than the rate at which the brine solution is being pumped into the tank.

**EXAMPLE 6**  **Example 5 Revisited**

If the well-stirred solution in Example 5 is pumped out at a slower rate of, say, $r_{out} = 2$ gal/min, then liquid will accumulate in the tank at the rate of

$$r_{in} - r_{out} = (3 - 2) \text{ gal/min} = 1 \text{ gal/min}.$$

After $t$ minutes,

$$(1 \text{ gal/min}) \cdot (t \text{ min}) = t \text{ gal}$$

will accumulate, so the tank will contain $300 + t$ gallons of brine. The concentration of the outflow is then $c(t) = A/(300 + t)$ lb/gal, and the output rate of salt is $R_{out} = c(t) \cdot r_{out}$, or

$$R_{out} = \left(\frac{A}{300 + t} \text{ lb/gal}\right) \cdot (2 \text{ gal/min}) = \frac{2A}{300 + t} \text{ lb/min}.$$

Hence equation (5) becomes

$$\frac{dA}{dt} = 6 - \frac{2A}{300 + t} \quad \text{or} \quad \frac{dA}{dt} + \frac{2}{300 + t}A = 6.$$

The integrating factor for the last equation is

$$e^{\int 2\,dt/(300+t)} = e^{2\ln(300+t)} = e^{\ln(300+t)^2} = (300 + t)^2$$

and so after multiplying by the factor the equation is cast into the form

$$\frac{d}{dt}[(300 + t)^2 A] = 6(300 + t)^2.$$

**FIGURE 3.1.7** Graph of $A(t)$ in Example 6

**FIGURE 3.1.8** *LR*-series circuit

**FIGURE 3.1.9** *RC*-series circuit

Integrating the last equation gives $(300 + t)^2 A = 2(300 + t)^3 + c$. By applying the initial condition $A(0) = 50$ and solving for $A$ yields the solution $A(t) = 600 + 2t - (4.95 \times 10^7)(300 + t)^{-2}$. As Figure 3.1.7 shows, not unexpectedly, salt builds up in the tank over time, that is, $A \to \infty$ as $t \to \infty$. ∎

**SERIES CIRCUITS**    For a series circuit containing only a resistor and an inductor, Kirchhoff's second law states that the sum of the voltage drop across the inductor ($L(di/dt)$) and the voltage drop across the resistor ($iR$) is the same as the impressed voltage ($E(t)$) on the circuit. See Figure 3.1.8.

Thus we obtain the linear differential equation for the current $i(t)$,

$$L\frac{di}{dt} + Ri = E(t), \tag{7}$$

where $L$ and $R$ are known as the inductance and the resistance, respectively. The current $i(t)$ is also called the **response** of the system.

The voltage drop across a capacitor with capacitance $C$ is given by $q(t)/C$, where $q$ is the charge on the capacitor. Hence, for the series circuit shown in Figure 3.1.9, Kirchhoff's second law gives

$$Ri + \frac{1}{C}q = E(t). \tag{8}$$

But current $i$ and charge $q$ are related by $i = dq/dt$, so (8) becomes the linear differential equation

$$R\frac{dq}{dt} + \frac{1}{C}q = E(t). \tag{9}$$

**EXAMPLE 7**    *LR*-Series Circuit

A 12-volt battery is connected to a series circuit in which the inductance is $\frac{1}{2}$ henry and the resistance is 10 ohms. Determine the current $i$ if the initial current is zero.

**SOLUTION**    From (7) we see that we must solve the equation

$$\frac{1}{2}\frac{di}{dt} + 10i = 12,$$

subject to $i(0) = 0$. First, we multiply the differential equation by 2 and read off the integrating factor $e^{20t}$. We then obtain

$$\frac{d}{dt}[e^{20t}i] = 24e^{20t}.$$

Integrating each side of the last equation and solving for $i$ gives $i(t) = \frac{6}{5} + ce^{-20t}$. Now $i(0) = 0$ implies that $0 = \frac{6}{5} + c$ or $c = -\frac{6}{5}$. Therefore the response is $i(t) = \frac{6}{5} - \frac{6}{5}e^{-20t}$. ∎

From (4) of Section 2.3 we can write a general solution of (7):

$$i(t) = \frac{e^{-(R/L)t}}{L}\int e^{(R/L)t}E(t)\,dt + ce^{-(R/L)t}. \tag{10}$$

In particular, when $E(t) = E_0$ is a constant, (10) becomes

$$i(t) = \frac{E_0}{R} + ce^{-(R/L)t}. \tag{11}$$

Note that as $t \to \infty$, the second term in equation (11) approaches zero. Such a term is usually called a **transient term;** any remaining terms are called the **steady-state** part of the solution. In this case $E_0/R$ is also called the **steady-state current;** for large values of time it appears that the current in the circuit is simply governed by Ohm's law ($E = iR$).

### REMARKS

The solution $P(t) = P_0 e^{0.4055t}$ of the initial-value problem in Example 1 described the population of a colony of bacteria at any time $t \geq 0$. Of course, $P(t)$ is a continuous function that takes on *all* real numbers in the interval $P_0 \leq P < \infty$. But since we are talking about a population, common sense dictates that $P$ can take on only positive integer values. Moreover, we would not expect the population to grow continuously—that is, every second, every microsecond, and so on—as predicted by our solution; there may be intervals of time $[t_1, t_2]$ over which there is no growth at all. Perhaps, then, the graph shown in Figure 3.1.10(a) is a more realistic description of $P$ than is the graph of an exponential function. Using a continuous function to describe a discrete phenomenon is often more a matter of convenience than of accuracy. However, for some purposes we may be satisfied if our model describes the system fairly closely when viewed macroscopically in time, as in Figures 3.1.10(b) and 3.1.10(c), rather than microscopically, as in Figure 3.1.10(a).

**FIGURE 3.1.10** Population growth is a discrete process

---

### Growth and Decay

1. The population of a community is known to increase at a rate proportional to the number of people present at time $t$. If an initial population $P_0$ has doubled in 5 years, how long will it take to triple? To quadruple?

2. Suppose it is known that the population of the community in Problem 1 is 10,000 after 3 years. What was the initial population $P_0$? What will be the population in 10 years? How fast is the population growing at $t = 10$?

3. The population of a town grows at a rate proportional to the population present at time $t$. The initial population of 500 increases by 15% in 10 years. What will be the population in 30 years? How fast is the population growing at $t = 30$?

4. The population of bacteria in a culture grows at a rate proportional to the number of bacteria present at time $t$. After 3 hours it is observed that 400 bacteria are present. After 10 hours 2000 bacteria are present. What was the initial number of bacteria?

5. The radioactive isotope of lead, Pb-209, decays at a rate proportional to the amount present at time $t$ and has a half-life of 3.3 hours. If 1 gram of this isotope is present initially, how long will it take for 90% of the lead to decay?

6. Initially 100 milligrams of a radioactive substance was present. After 6 hours the mass had decreased by 3%. If the rate of decay is proportional to the amount of the substance present at time $t$, find the amount remaining after 24 hours.

7. Determine the half-life of the radioactive substance described in Problem 6.

8. **(a)** Consider the initial-value problem $dA/dt = kA$, $A(0) = A_0$ as the model for the decay of a radioactive substance. Show that, in general, the half-life $T$ of the substance is $T = -(\ln 2)/k$.

   **(b)** Show that the solution of the initial-value problem in part (a) can be written $A(t) = A_0 2^{-t/T}$.

   **(c)** If a radioactive substance has the half-life $T$ given in part (a), how long will it take an initial amount $A_0$ of the substance to decay to $\frac{1}{8}A_0$?

9. When a vertical beam of light passes through a transparent medium, the rate at which its intensity $I$ decreases is proportional to $I(t)$, where $t$ represents the thickness of the medium (in feet). In clear seawater, the intensity 3 feet below the surface is 25% of the initial intensity $I_0$ of the incident beam. What is the intensity of the beam 15 feet below the surface?

10. When interest is compounded continuously, the amount of money increases at a rate proportional to the amount $S$ present at time $t$, that is, $dS/dt = rS$, where $r$ is the annual rate of interest.

    **(a)** Find the amount of money accrued at the end of 5 years when $5000 is deposited in a savings account drawing $5\frac{3}{4}\%$ annual interest compounded continuously.

    **(b)** In how many years will the initial sum deposited have doubled?

    **(c)** Use a calculator to compare the amount obtained in part (a) with the amount $S = 5000\left(1 + \frac{1}{4}(0.0575)\right)^{5(4)}$ that is accrued when interest is compounded quarterly.

## Carbon Dating

11. Archaeologists used pieces of burned wood, or charcoal, found at the site to date prehistoric paintings and drawings on walls and ceilings of a cave in Lascaux, France. See Figure 3.1.11. Use the information on page 87 to determine the approximate age of a piece of burned wood, if it was found that 85.5% of the C-14 found in living trees of the same type had decayed.

**FIGURE 3.1.11** Cave wall painting in Problem 11

12. The Shroud of Turin, which shows the negative image of the body of a man who appears to have been crucified, is believed by many to be the burial shroud of Jesus of Nazareth. See Figure 3.1.12. In 1988 the Vatican granted permission to have the shroud carbon-dated. Three independent scientific laboratories analyzed the cloth and concluded that the shroud was approximately 660 years old, an age consistent with its historical appearance. Using this age, determine what percentage of the original amount of C-14 remained in the cloth as of 1988.

**FIGURE 3.1.12** Shroud image in Problem 12

## Newton's Law of Cooling/Warming

13. A thermometer is removed from a room where the temperature is 70° F and is taken outside, where the air temperature is 10° F. After one-half minute the thermometer reads 50° F. What is the reading of the thermometer at $t = 1$ min? How long will it take for the thermometer to reach 15° F?

14. A thermometer is taken from an inside room to the outside, where the air temperature is 5° F. After 1 minute the thermometer reads 55° F, and after 5 minutes it reads 30° F. What is the initial temperature of the inside room?

15. A small metal bar, whose initial temperature was 20° C, is dropped into a large container of boiling water. How long will it take the bar to reach 90° C if it is known that its temperature increases 2° in 1 second? How long will it take the bar to reach 98° C?

16. Two large containers $A$ and $B$ of the same size are filled with different fluids. The fluids in containers $A$ and $B$ are maintained at 0° C and 100° C, respectively. A small metal bar, whose initial temperature is 100° C, is lowered into container $A$. After 1 minute the temperature of the bar is 90° C. After 2 minutes the bar is removed and instantly transferred to the other container. After 1 minute in container $B$ the temperature of the bar rises 10°. How long, measured from the start of the entire process, will it take the bar to reach 99.9° C?

17. A thermometer reading 70° F is placed in an oven preheated to a constant temperature. Through a glass window in the oven door, an observer records that the thermometer reads 110° F after $\frac{1}{2}$ minute and 145° F after 1 minute. How hot is the oven?

18. At $t = 0$ a sealed test tube containing a chemical is immersed in a liquid bath. The initial temperature of the chemical in the test tube is 80° F. The liquid bath has a controlled

temperature (measured in degrees Fahrenheit) given by $T_m(t) = 100 - 40e^{-0.1t}$, $t \geq 0$, where $t$ is measured in minutes.

(a) Assume that $k = -0.1$ in (2). Before solving the IVP, describe in words what you expect the temperature $T(t)$ of the chemical to be like in the short term. In the long term.

(b) Solve the initial-value problem. Use a graphing utility to plot the graph of $T(t)$ on time intervals of various lengths. Do the graphs agree with your predictions in part (a)?

19. A dead body was found within a closed room of a house where the temperature was a constant 70° F. At the time of discovery the core temperature of the body was determined to be 85° F. One hour later a second measurement showed that the core temperature of the body was 80° F. Assume that the time of death corresponds to $t = 0$ and that the core temperature at that time was 98.6° F. Determine how many hours elapsed before the body was found. [*Hint*: Let $t_1 > 0$ denote the time that the body was discovered.]

20. The rate at which a body cools also depends on its exposed surface area $S$. If $S$ is a constant, then a modification of (2) is

$$\frac{dT}{dt} = kS(T - T_m),$$

where $k < 0$ and $T_m$ is a constant. Suppose that two cups $A$ and $B$ are filled with coffee at the same time. Initially, the temperature of the coffee is 150° F. The exposed surface area of the coffee in cup $B$ is twice the surface area of the coffee in cup $A$. After 30 min the temperature of the coffee in cup $A$ is 100° F. If $T_m = 70°$ F, then what is the temperature of the coffee in cup $B$ after 30 min?

## Mixtures

21. A tank contains 200 liters of fluid in which 30 grams of salt is dissolved. Brine containing 1 gram of salt per liter is then pumped into the tank at a rate of 4 L/min; the well-mixed solution is pumped out at the same rate. Find the number $A(t)$ of grams of salt in the tank at time $t$.

22. Solve Problem 21 assuming that pure water is pumped into the tank.

23. A large tank is filled to capacity with 500 gallons of pure water. Brine containing 2 pounds of salt per gallon is pumped into the tank at a rate of 5 gal/min. The well-mixed solution is pumped out at the same rate. Find the number $A(t)$ of pounds of salt in the tank at time $t$.

24. In Problem 23, what is the concentration $c(t)$ of the salt in the tank at time $t$? At $t = 5$ min? What is the concentration of the salt in the tank after a long time, that is, as $t \to \infty$? At what time is the concentration of the salt in the tank equal to one-half this limiting value?

25. Solve Problem 23 under the assumption that the solution is pumped out at a faster rate of 10 gal/min. When is the tank empty?

26. Determine the amount of salt in the tank at time $t$ in Example 5 if the concentration of salt in the inflow is variable and given by $c_{in}(t) = 2 + \sin(t/4)$ lb/gal. Without actually graphing, conjecture what the solution curve of the IVP should look like. Then use a graphing utility to plot the graph of the solution on the interval [0, 300]. Repeat for the interval [0, 600] and compare your graph with that in Figure 3.1.6(a).

27. A large tank is partially filled with 100 gallons of fluid in which 10 pounds of salt is dissolved. Brine containing $\frac{1}{2}$ pound of salt

per gallon is pumped into the tank at a rate of 6 gal/min. The well-mixed solution is then pumped out at a slower rate of 4 gal/min. Find the number of pounds of salt in the tank after 30 minutes.

28. In Example 5 the size of the tank containing the salt mixture was not given. Suppose, as in the discussion following Example 5, that the rate at which brine is pumped into the tank is 3 gal/min but that the well-stirred solution is pumped out at a rate of 2 gal/min. It stands to reason that since brine is accumulating in the tank at the rate of 1 gal/min, any finite tank must eventually overflow. Now suppose that the tank has an open top and has a total capacity of 400 gallons.

(a) When will the tank overflow?

(b) What will be the number of pounds of salt in the tank at the instant it overflows?

(c) Assume that although the tank is overflowing, brine solution continues to be pumped in at a rate of 3 gal/min and the well-stirred solution continues to be pumped out at a rate of 2 gal/min. Devise a method for determining the number of pounds of salt in the tank at $t = 150$ minutes.

(d) Determine the number of pounds of salt in the tank as $t \to \infty$. Does your answer agree with your intuition?

(e) Use a graphing utility to plot the graph of $A(t)$ on the interval [0, 500].

## Series Circuits

29. A 30-volt electromotive force is applied to an $LR$-series circuit in which the inductance is 0.1 henry and the resistance is 50 ohms. Find the current $i(t)$ if $i(0) = 0$. Determine the current as $t \to \infty$.

30. Solve equation (7) under the assumption that $E(t) = E_0 \sin \omega t$ and $i(0) = i_0$.

31. A 100-volt electromotive force is applied to an $RC$-series circuit in which the resistance is 200 ohms and the capacitance is $10^{-4}$ farad. Find the charge $q(t)$ on the capacitor if $q(0) = 0$. Find the current $i(t)$.

32. A 200-volt electromotive force is applied to an $RC$-series circuit in which the resistance is 1000 ohms and the capacitance is $5 \times 10^{-6}$ farad. Find the charge $q(t)$ on the capacitor if $i(0) = 0.4$. Determine the charge and current at $t = 0.005$ s. Determine the charge as $t \to \infty$.

33. An electromotive force

$$E(t) = \begin{cases} 120, & 0 \leq t \leq 20 \\ 0, & t > 20 \end{cases}$$

is applied to an $LR$-series circuit in which the inductance is 20 henries and the resistance is 2 ohms. Find the current $i(t)$ if $i(0) = 0$.

34. An $LR$-series circuit has a variable inductor with the inductance defined by

$$L(t) = \begin{cases} 1 - \frac{1}{10}t, & 0 \leq t < 10 \\ 0, & t > 10. \end{cases}$$

Find the current $i(t)$ if the resistance is 0.2 ohm, the impressed voltage is $E(t) = 4$, and $i(0) = 0$. Graph $i(t)$.

## Additional Linear Models

**35. Air Resistance**   In (14) of Section 1.3 we saw that a differential equation describing the velocity $v$ of a falling mass subject to air resistance proportional to the instantaneous velocity is

$$m\frac{dv}{dt} = mg - kv,$$

where $k > 0$ is a constant of proportionality. The positive direction is downward.

**(a)** Solve the equation subject to the initial condition $v(0) = v_0$.

**(b)** Use the solution in part (a) to determine the limiting, or terminal, velocity of the mass. We saw how to determine the terminal velocity without solving the DE in Problem 40 in Exercises 2.1.

**(c)** If the distance $s$, measured from the point where the mass was released above ground, is related to velocity $v$ by $ds/dt = v(t)$, find an explicit expression for $s(t)$ if $s(0) = 0$.

**36. How High?—No Air Resistance**   Suppose a small cannonball weighing 16 pounds is shot vertically upward, as shown in Figure 3.1.13, with an initial velocity $v_0 = 300$ ft/s. The answer to the question "How high does the cannonball go?" depends on whether we take air resistance into account.

**(a)** Suppose air resistance is ignored. If the positive direction is upward, then a model for the state of the cannonball is given by $d^2s/dt^2 = -g$ (equation (12) of Section 1.3). Since $ds/dt = v(t)$ the last differential equation is the same as $dv/dt = -g$, where we take $g = 32$ ft/s². Find the velocity $v(t)$ of the cannonball at time $t$.

**(b)** Use the result obtained in part (a) to determine the height $s(t)$ of the cannonball measured from ground level. Find the maximum height attained by the cannonball.

**FIGURE 3.1.13** Find the maximum height of the cannonball in Problem 36

**37. How High?—Linear Air Resistance**   Repeat Problem 36, but this time assume that air resistance is proportional to instantaneous velocity. It stands to reason that the maximum height attained by the cannonball must be *less* than that in part (b) of Problem 36. Show this by supposing that the constant of proportionality is $k = 0.0025$. [*Hint:* Slightly modify the differential equation in Problem 35.]

**38. Skydiving**   A skydiver weighs 125 pounds, and her parachute and equipment combined weigh another 35 pounds. After exiting from a plane at an altitude of 15,000 feet, she waits 15 seconds and opens her parachute. Assume that the constant of proportionality in the model in Problem 35 has the value $k = 0.5$ during free fall and $k = 10$ after the parachute is opened. Assume that her initial velocity on leaving the plane is zero. What is her velocity and how far has she traveled 20 seconds after leaving the plane? See Figure 3.1.14. How does her velocity at 20 seconds compare with her terminal velocity? How long does it take her to reach the ground? [*Hint:* Think in terms of two distinct IVPs.]

**FIGURE 3.1.14** Find the time to reach the ground in Problem 38

**39. Rocket Motion**   Suppose a small single-stage rocket of total mass $m(t)$ is launched vertically, the positive direction is upward, the air resistance is linear, and the rocket consumes its fuel at a constant rate. In Problem 22 of Exercises 1.3 you were asked to use Newton's second law of motion in the form given in (17) of that exercise set to show that a mathematical model for the velocity $v(t)$ of the rocket is given by

$$\frac{dv}{dt} + \frac{k - \lambda}{m_0 - \lambda t}v = -g + \frac{R}{m_0 - \lambda t},$$

where $k$ is the air resistance constant of proportionality, $\lambda$ is the constant rate at which fuel is consumed, $R$ is the thrust of the rocket, $m(t) = m_0 - \lambda t$, $m_0$ is the total mass of the rocket at $t = 0$, and $g$ is the acceleration due to gravity.

**(a)** Find the velocity $v(t)$ of the rocket if $m_0 = 200$ kg, $R = 2000$ N, $\lambda = 1$ kg/s, $g = 9.8$ m/s², $k = 3$ kg/s, and $v(0) = 0$.

**(b)** Use $ds/dt = v$ and the result in part (a) to find the height $s(t)$ of the rocket at time $t$.

**40. Rocket Motion—Continued**   In Problem 39 suppose of the rocket's initial mass $m_0$ that 50 kg is the mass of the fuel.

**(a)** What is the burnout time $t_b$, or the time at which all the fuel is consumed?

**(b)** What is the velocity of the rocket at burnout?

**(c)** What is the height of the rocket at burnout?

**(d)** Why would you expect the rocket to attain an altitude higher than the number in part (b)?

**(e)** After burnout what is a mathematical model for the velocity of the rocket?

**41. Evaporating Raindrop** As a raindrop falls, it evaporates while retaining its spherical shape. If we make the further assumptions that the rate at which the raindrop evaporates is proportional to its surface area and that air resistance is negligible, then a model for the velocity $v(t)$ of the raindrop is

$$\frac{dv}{dt} + \frac{3(k/\rho)}{(k/\rho)t + r_0} v = g.$$

Here $\rho$ is the density of water, $r_0$ is the radius of the raindrop at $t = 0$, $k < 0$ is the constant of proportionality, and the downward direction is taken to be the positive direction.

(a) Solve for $v(t)$ if the raindrop falls from rest.

(b) Reread Problem 36 of Exercises 1.3 and then show that the radius of the raindrop at time $t$ is $r(t) = (k/\rho)t + r_0$.

(c) If $r_0 = 0.01$ ft and $r = 0.007$ ft 10 seconds after the raindrop falls from a cloud, determine the time at which the raindrop has evaporated completely.

**42. Fluctuating Population** The differential equation $dP/dt = (k \cos t)P$, where $k$ is a positive constant, is a mathematical model for a population $P(t)$ that undergoes yearly seasonal fluctuations. Solve the equation subject to $P(0) = P_0$. Use a graphing utility to graph the solution for different choices of $P_0$.

**43. Population Model** In one model of the changing population $P(t)$ of a community, it is assumed that

$$\frac{dP}{dt} = \frac{dB}{dt} - \frac{dD}{dt},$$

where $dB/dt$ and $dD/dt$ are the birth and death rates, respectively.

(a) Solve for $P(t)$ if $dB/dt = k_1P$ and $dD/dt = k_2P$.

(b) Analyze the cases $k_1 > k_2$, $k_1 = k_2$, and $k_1 < k_2$.

**44. Constant-Harvest Model** A model that describes the population of a fishery in which harvesting takes place at a constant rate is given by

$$\frac{dP}{dt} = kP - h,$$

where $k$ and $h$ are positive constants.

(a) Solve the DE subject to $P(0) = P_0$.

(b) Describe the behavior of the population $P(t)$ for increasing time in the three cases $P_0 > h/k$, $P_0 = h/k$, and $0 < P_0 < h/k$.

(c) Use the results from part (b) to determine whether the fish population will ever go extinct in finite time, that is, whether there exists a time $T > 0$ such that $P(T) = 0$. If the population goes extinct, then find $T$.

**45. Drug Dissemination** A mathematical model for the rate at which a drug disseminates into the bloodstream is given by

$$\frac{dx}{dt} = r - kx,$$

where $r$ and $k$ are positive constants. The function $x(t)$ describes the concentration of the drug in the bloodstream at time $t$.

(a) Since the DE is autonomous, use the phase portrait concept of Section 2.1 to find the limiting value of $x(t)$ as $t \to \infty$.

(b) Solve the DE subject to $x(0) = 0$. Sketch the graph of $x(t)$ and verify your prediction in part (a). At what time is the concentration one-half this limiting value?

**46. Memorization** When forgetfulness is taken into account, the rate of memorization of a subject is given by

$$\frac{dA}{dt} = k_1(M - A) - k_2A,$$

where $k_1 > 0$, $k_2 > 0$, $A(t)$ is the amount memorized in time $t$, $M$ is the total amount to be memorized, and $M - A$ is the amount remaining to be memorized.

(a) Since the DE is autonomous, use the phase portrait concept of Section 2.1 to find the limiting value of $A(t)$ as $t \to \infty$. Interpret the result.

(b) Solve the DE subject to $A(0) = 0$. Sketch the graph of $A(t)$ and verify your prediction in part (a).

**47. Heart Pacemaker** A heart pacemaker, shown in Figure 3.1.15, consists of a switch, a battery, a capacitor, and the heart as a resistor. When the switch $S$ is at $P$, the capacitor charges; when $S$ is at $Q$, the capacitor discharges, sending an electrical stimulus to the heart. In Problem 58 in Exercises 2.3 we saw that during this time the electrical stimulus is being applied to the heart, the voltage $E$ across the heart satisfies the linear DE

$$\frac{dE}{dt} = -\frac{1}{RC} E.$$

(a) Let us assume that over the time interval of length $t_1$, $0 < t < t_1$, the switch $S$ is at position $P$ shown in Figure 3.1.15 and the capacitor is being charged. When the switch is moved to position $Q$ at time $t_1$ the capacitor discharges, sending an impulse to the heart over the time interval of length $t_2$: $t_1 \leq t < t_1 + t_2$. Thus over the initial charging/discharging interval $0 < t < t_1 + t_2$ the voltage to the heart is actually modeled by the piecewise-linear differential equation

$$\frac{dE}{dt} = \begin{cases} 0, & 0 \leq t < t_1 \\ -\dfrac{1}{RC}E, & t_1 \leq t < t_1 + t_2. \end{cases}$$

By moving $S$ between $P$ and $Q$, the charging and discharging over time intervals of lengths $t_1$ and $t_2$ is

**FIGURE 3.1.15** Model of a pacemaker in Problem 47

repeated indefinitely. Suppose $t_1 = 4$ s, $t_2 = 2$ s, $E_0 = 12$ V, and $E(0) = 0$, $E(4) = 12$, $E(6) = 0$, $E(10) = 12$, $E(12) = 0$, and so on. Solve for $E(t)$ for $0 \le t \le 24$.

**(b)** Suppose for the sake of illustration that $R = C = 1$. Use a graphing utility to graph the solution for the IVP in part (a) for $0 \le t \le 24$.

**48. Sliding Box** **(a)** A box of mass $m$ slides down an inclined plane that makes an angle $\theta$ with the horizontal as shown in Figure 3.1.16. Find a differential equation for the velocity $v(t)$ of the box at time $t$ in each of the following three cases:

    **(*i*)**  No sliding friction and no air resistance

    **(*ii*)**  With sliding friction and no air resistance

    **(*iii*)** With sliding friction and air resistance

In cases (*ii*) and (*iii*), use the fact that the force of friction opposing the motion of the box is $\mu N$, where $\mu$ is the coefficient of sliding friction and $N$ is the normal component of the weight of the box. In case (*iii*) assume that air resistance is proportional to the instantaneous velocity.

**(b)** In part (a), suppose that the box weighs 96 pounds, that the angle of inclination of the plane is $\theta = 30°$, that the coefficient of sliding friction is $\mu = \sqrt{3}/4$, and that the additional retarding force due to air resistance is numerically equal to $\frac{1}{4}v$. Solve the differential equation in each of the three cases, assuming that the box starts from rest from the highest point 50 ft above ground.

friction

motion

$W = mg$    50 ft

$\theta$

**FIGURE 3.1.16** Box sliding down inclined plane in Problem 48

**49. Sliding Box—Continued** **(a)** In Problem 48 let $s(t)$ be the distance measured down the inclined plane from the highest point. Use $ds/dt = v(t)$ and the solution for each of the three cases in part (b) of Problem 48 to find the time that it takes the box to slide completely down the inclined plane. A root-finding application of a CAS may be useful here.

**(b)** In the case in which there is friction ($\mu \ne 0$) but no air resistance, explain why the box will not slide down the plane starting from *rest* from the highest point above ground when the inclination angle $\theta$ satisfies $\tan \theta \le \mu$.

**(c)** The box *will* slide downward on the plane when $\tan \theta \le \mu$ if it is given an initial velocity $v(0) = v_0 > 0$. Suppose that $\mu = \sqrt{3}/4$ and $\theta = 23°$. Verify that $\tan \theta \le \mu$. How far will the box slide down the plane if $v_0 = 1$ ft/s?

**(d)** Using the values $\mu = \sqrt{3}/4$ and $\theta = 23°$, approximate the smallest initial velocity $v_0$ that can be given to the box so that, starting at the highest point 50 ft above ground, it will slide completely down the inclined plane. Then find the corresponding time it takes to slide down the plane.

**50. What Goes Up . . .** **(a)** It is well known that the model in which air resistance is ignored, part (a) of Problem 36, predicts that the time $t_a$ it takes the cannonball to attain its maximum height is the same as the time $t_d$ it takes the cannonball to fall from the maximum height to the ground. Moreover, the magnitude of the impact velocity $v_i$ will be the same as the initial velocity $v_0$ of the cannonball. Verify both of these results.

**(b)** Then, using the model in Problem 37 that takes air resistance into account, compare the value of $t_a$ with $t_d$ and the value of the magnitude of $v_i$ with $v_0$. A root-finding application of a CAS (or graphic calculator) may be useful here.

## 3.2 Nonlinear Models

**INTRODUCTION** We finish our study of single first-order differential equations with an examination of some nonlinear models.

**POPULATION DYNAMICS** If $P(t)$ denotes the size of a population at time $t$, the model for exponential growth begins with the assumption that $dP/dt = kP$ for some $k > 0$. In this model, the **relative,** or **specific, growth rate** defined by

$$\frac{dP/dt}{P} \tag{1}$$

is a constant $k$. True cases of exponential growth over long periods of time are hard to find because the limited resources of the environment will at some time exert restrictions on the growth of a population. Thus for other models, (1) can be expected to decrease as the population $P$ increases in size.

The assumption that the rate at which a population grows (or decreases) is dependent only on the number $P$ present and not on any time-dependent mechanisms such as seasonal phenomena (see Problem 33 in Exercises 1.3) can be stated as

$$\frac{dP/dt}{P} = f(P) \quad \text{or} \quad \frac{dP}{dt} = Pf(P). \tag{2}$$

The differential equation in (2), which is widely assumed in models of animal populations, is called the **density-dependent hypothesis.**

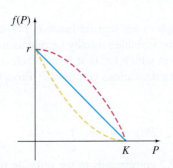

**FIGURE 3.2.1** Simplest assumption for $f(P)$ is a straight line (blue color)

**LOGISTIC EQUATION** Suppose an environment is capable of sustaining no more than a fixed number $K$ of individuals in its population. The quantity $K$ is called the **carrying capacity** of the environment. Hence for the function $f$ in (2) we have $f(K) = 0$, and we simply let $f(0) = r$. Figure 3.2.1 shows three functions $f$ that satisfy these two conditions. The simplest assumption that we can make is that $f(P)$ is linear—that is, $f(P) = c_1 P + c_2$. If we use the conditions $f(0) = r$ and $f(K) = 0$, we find, in turn, $c_2 = r$ and $c_1 = -r/K$, and so $f$ takes on the form $f(P) = r - (r/K)P$. Equation (2) becomes

$$\frac{dP}{dt} = P\left(r - \frac{r}{K}P\right). \tag{3}$$

With constants relabeled, the nonlinear equation (3) is the same as

$$\frac{dP}{dt} = P(a - bP). \tag{4}$$

Around 1840 the Belgian mathematician-biologist **P. F. Verhulst** (1804–1849) was concerned with mathematical models for predicting the human populations of various countries. One of the equations he studied was (4), where $a > 0$ and $b > 0$. Equation (4) came to be known as the **logistic equation,** and its solution is called the **logistic function.** The graph of a logistic function is called a **logistic curve.**

The linear differential equation $dP/dt = kP$ does not provide a very accurate model for population when the population itself is very large. Overcrowded conditions, with the resulting detrimental effects on the environment such as pollution and excessive and competitive demands for food and fuel, can have an inhibiting effect on population growth. As we shall now see, the solution of (4) is bounded as $t \to \infty$. If we rewrite (4) as $dP/dt = aP - bP^2$, the nonlinear term $-bP^2$, $b > 0$, can be interpreted as an "inhibition" or "competition" term. Also, in most applications the positive constant $a$ is much larger than the constant $b$.

Logistic curves have proved to be quite accurate in predicting the growth patterns, in a limited space, of certain types of bacteria, protozoa, water fleas (*Daphnia*), and fruit flies (*Drosophila*).

**SOLUTION OF THE LOGISTIC EQUATION** One method of solving (4) is separation of variables. Decomposing the left side of $dP/P(a - bP) = dt$ into partial fractions and integrating gives

$$\left(\frac{1/a}{P} + \frac{b/a}{a - bP}\right)dP = dt$$

$$\frac{1}{a}\ln|P| - \frac{1}{a}\ln|a - bP| = t + c$$

$$\ln\left|\frac{P}{a - bP}\right| = at + ac$$

$$\frac{P}{a - bP} = c_1 e^{at}.$$

It follows from the last equation that

$$P(t) = \frac{ac_1e^{at}}{1 + bc_1e^{at}} = \frac{ac_1}{bc_1 + e^{-at}}.$$

If $P(0) = P_0$, $P_0 \neq a/b$, we find $c_1 = P_0/(a - bP_0)$, and so after substituting and simplifying, the solution becomes

$$P(t) = \frac{aP_0}{bP_0 + (a - bP_0)e^{-at}}. \tag{5}$$

**GRAPHS OF $P(t)$**   The basic shape of the graph of the logistic function $P(t)$ can be obtained without too much effort. Although the variable $t$ usually represents time and we are seldom concerned with applications in which $t < 0$, it is nonetheless of some interest to include this interval in displaying the various graphs of $P$. From (5) we see that

$$P(t) \to \frac{aP_0}{bP_0} = \frac{a}{b} \quad \text{as} \quad t \to \infty \quad \text{and} \quad P(t) \to 0 \quad \text{as} \quad t \to -\infty.$$

The dashed line $P = a/2b$ shown in Figure 3.2.2 corresponds to the ordinate of a point of inflection of the logistic curve. To show this, we differentiate (4) by the Product Rule:

$$\frac{d^2P}{dt^2} = P\left(-b\frac{dP}{dt}\right) + (a - bP)\frac{dP}{dt} = \frac{dP}{dt}(a - 2bP)$$

$$= P(a - bP)(a - 2bP)$$

$$= 2b^2P\left(P - \frac{a}{b}\right)\left(P - \frac{a}{2b}\right).$$

From calculus recall that the points where $d^2P/dt^2 = 0$ are possible points of inflection, but $P = 0$ and $P = a/b$ can obviously be ruled out. Hence $P = a/2b$ is the only possible ordinate value at which the concavity of the graph can change. For $0 < P < a/2b$ it follows that $P'' > 0$, and $a/2b < P < a/b$ implies that $P'' < 0$. Thus, as we read from left to right, the graph changes from concave up to concave down at the point corresponding to $P = a/2b$. When the initial value satisfies $0 < P_0 < a/2b$, the graph of $P(t)$ assumes the shape of an S, as we see in Figure 3.2.2(b). For $a/2b < P_0 < a/b$ the graph is still S-shaped, but the point of inflection occurs at a negative value of $t$, as shown in Figure 3.2.2(c).

**FIGURE 3.2.2** Logistic curves for different initial conditions

We have already seen equation (4) in (5) of Section 1.3 in the form $dx/dt = kx(n + 1 - x)$, $k > 0$. This differential equation provides a reasonable model for describing the spread of an epidemic brought about initially by introducing an infected individual into a static population. The solution $x(t)$ represents the number of individuals infected with the disease at time $t$.

EXAMPLE 1    Logistic Growth

Suppose a student carrying a flu virus returns to an isolated college campus of 1000 students. If it is assumed that the rate at which the virus spreads is proportional not only to the number $x$ of infected students but also to the number of students not infected, determine the number of infected students after 6 days if it is further observed that after 4 days $x(4) = 50$.

**SOLUTION**  Assuming that no one leaves the campus throughout the duration of the disease, we must solve the initial-value problem

$$\frac{dx}{dt} = kx(1000 - x), \quad x(0) = 1.$$

By making the identification $a = 1000k$ and $b = k$, we have immediately from (5) that

$$x(t) = \frac{1000k}{k + 999ke^{-1000kt}} = \frac{1000}{1 + 999e^{-1000kt}}.$$

Now, using the information $x(4) = 50$, we determine $k$ from

$$50 = \frac{1000}{1 + 999e^{-4000k}}.$$

We find $-1000k = \frac{1}{4} \ln \frac{19}{999} = -0.9906$. Thus

$$x(t) = \frac{1000}{1 + 999e^{-0.9906t}}.$$

**(a)**

| $t$ (days) | $x$ (number infected) |
|---|---|
| 4 | 50 (observed) |
| 5 | 124 |
| 6 | 276 |
| 7 | 507 |
| 8 | 735 |
| 9 | 882 |
| 10 | 953 |

**(b)**

**FIGURE 3.2.3**  Number of infected students in Example 1

Finally, $\qquad x(6) = \dfrac{1000}{1 + 999e^{-5.9436}} = 276$ students.

Additional calculated values of $x(t)$ are given in the table in Figure 3.2.3(b). Note that the number of infected students $x(t)$ approaches 1000 as $t$ increases. ∎

**MODIFICATIONS OF THE LOGISTIC EQUATION**    There are many variations of the logistic equation. For example, the differential equations

$$\frac{dP}{dt} = P(a - bP) - h \quad \text{and} \quad \frac{dP}{dt} = P(a - bP) + h \qquad (6)$$

could serve, in turn, as models for the population in a fishery where fish are **harvested** or are **restocked** at rate $h$. When $h > 0$ is a constant, the DEs in (6) can be readily analyzed qualitatively or solved analytically by separation of variables. The equations in (6) could also serve as models of the human population decreased by **emigration** or increased by immigration, respectively. The rate $h$ in (6) could be a function of time $t$ or could be population dependent; for example, harvesting might be done periodically over time or might be done at a rate proportional to the population $P$ at time $t$. In the latter instance, the model would look like $P' = P(a - bP) - cP$, $c > 0$. The human population of a community might change because of **immigration** in such a manner that the contribution due to immigration was large when the population $P$ of the community was itself small but small when $P$ was large; a reasonable model for the population of the community would then be $P' = P(a - bP) + ce^{-kP}$, $c > 0$, $k > 0$. See Problem 24 in Exercises 3.2. Another equation of the form given in (2),

$$\frac{dP}{dt} = P(a - b \ln P), \qquad (7)$$

is a modification of the logistic equation known as the **Gompertz differential equation** named after the English mathematician **Benjamin Gompertz** (1779–1865). This DE is sometimes used as a model in the study of the growth or decline of populations, the growth of solid tumors, and certain kinds of actuarial predictions. See Problem 8 in Exercises 3.2.

**CHEMICAL REACTIONS**   Suppose that $a$ grams of chemical $A$ are combined with $b$ grams of chemical $B$. If there are $M$ parts of $A$ and $N$ parts of $B$ formed in the compound and $X(t)$ is the number of grams of chemical $C$ formed, then the number of grams of chemical $A$ and the number of grams of chemical $B$ remaining at time $t$ are, respectively,

$$a - \frac{M}{M+N}X \quad \text{and} \quad b - \frac{N}{M+N}X.$$

The law of mass action states that when no temperature change is involved, the rate at which the two substances react is proportional to the product of the amounts of $A$ and $B$ that are untransformed (remaining) at time $t$:

$$\frac{dX}{dt} \propto \left(a - \frac{M}{M+N}X\right)\left(b - \frac{N}{M+N}X\right). \tag{8}$$

If we factor out $M/(M+N)$ from the first factor and $N/(M+N)$ from the second and introduce a constant of proportionality $k > 0$, (8) has the form

$$\frac{dX}{dt} = k(\alpha - X)(\beta - X), \tag{9}$$

where $\alpha = a(M+N)/M$ and $\beta = b(M+N)/N$. Recall from (6) of Section 1.3 that a chemical reaction governed by the nonlinear differential equation (9) is said to be a **second-order reaction.**

## EXAMPLE 2  Second-Order Chemical Reaction

A compound $C$ is formed when two chemicals $A$ and $B$ are combined. The resulting reaction between the two chemicals is such that for each gram of $A$, 4 grams of $B$ is used. It is observed that 30 grams of the compound $C$ is formed in 10 minutes. Determine the amount of $C$ at time $t$ if the rate of the reaction is proportional to the amounts of $A$ and $B$ remaining and if initially there are 50 grams of $A$ and 32 grams of $B$. How much of the compound $C$ is present at 15 minutes? Interpret the solution as $t \rightarrow \infty$.

**SOLUTION**   Let $X(t)$ denote the number of grams of the compound $C$ present at time $t$. Clearly, $X(0) = 0$ g and $X(10) = 30$ g.

If, for example, 2 grams of compound $C$ is present, we must have used, say, $a$ grams of $A$ and $b$ grams of $B$, so $a + b = 2$ and $b = 4a$. Thus we must use $a = \frac{2}{5} = 2\left(\frac{1}{5}\right)$ g of chemical $A$ and $b = \frac{8}{5} = 2\left(\frac{4}{5}\right)$ g of $B$. In general, for $X$ grams of $C$ we must use

$$\frac{1}{5}X \text{ grams of } A \quad \text{and} \quad \frac{4}{5}X \text{ grams of } B.$$

The amounts of $A$ and $B$ remaining at time $t$ are then

$$50 - \frac{1}{5}X \quad \text{and} \quad 32 - \frac{4}{5}X,$$

respectively.

Now we know that the rate at which compound $C$ is formed satisfies

$$\frac{dX}{dt} \propto \left(50 - \frac{1}{5}X\right)\left(32 - \frac{4}{5}X\right).$$

To simplify the subsequent algebra, we factor $\frac{1}{5}$ from the first term and $\frac{4}{5}$ from the second and then introduce the constant of proportionality:

$$\frac{dX}{dt} = k(250 - X)(40 - X).$$

By separation of variables and partial fractions we can write

$$-\frac{\frac{1}{210}}{250 - X}\,dX + \frac{\frac{1}{210}}{40 - X}\,dX = k\,dt.$$

Integrating gives

$$\ln\frac{250 - X}{40 - X} = 210kt + c_1 \qquad \text{or} \qquad \frac{250 - X}{40 - X} = c_2 e^{210kt}. \tag{10}$$

When $t = 0$, $X = 0$, so it follows at this point that $c_2 = \frac{25}{4}$. Using $X = 30$ g at $t = 10$, we find $210k = \frac{1}{10}\ln\frac{88}{25} = 0.1258$. With this information we solve the last equation in (10) for $X$:

$$X(t) = 1000\frac{1 - e^{-0.1258t}}{25 - 4e^{-0.1258t}}. \tag{11}$$

From (11) we find $X(15) = 34.78$ grams. The behavior of $X$ as a function of time is displayed in Figure 3.2.4. It is clear from the accompanying table and (11) that $X \to 40$ as $t \to \infty$. This means that 40 grams of compound $C$ is formed, leaving

$$50 - \frac{1}{5}(40) = 42 \text{ g of } A \qquad \text{and} \qquad 32 - \frac{4}{5}(40) = 0 \text{ g of } B. \qquad \blacksquare$$

(a)

| $t$ (min) | $X$ (g) |
|-----------|---------|
| 10 | 30 (measured) |
| 15 | 34.78 |
| 20 | 37.25 |
| 25 | 38.54 |
| 30 | 39.22 |
| 35 | 39.59 |

(b)

**FIGURE 3.2.4** Number of grams of compound $C$ in Example 2

### REMARKS

The indefinite integral $\int du/(a^2 - u^2)$ can be evaluated in terms of logarithms, the inverse hyperbolic tangent, or the inverse hyperbolic cotangent. For example, of the two results

$$\int \frac{du}{a^2 - u^2} = \frac{1}{a}\tanh^{-1}\frac{u}{a} + c, \quad |u| < a \tag{12}$$

$$\int \frac{du}{a^2 - u^2} = \frac{1}{2a}\ln\left|\frac{a + u}{a - u}\right| + c, \quad |u| \neq a, \tag{13}$$

(12) may be convenient in Problems 15 and 26 in Exercises 3.2, whereas (13) may be preferable in Problem 27.

## EXERCISES 3.2

Answers to selected odd-numbered problems begin on page ANS-3.

### Logistic Equation

1. The number $N(t)$ of supermarkets throughout the country that are using a computerized checkout system is described by the initial-value problem

$$\frac{dN}{dt} = N(1 - 0.0005N), \quad N(0) = 1.$$

(a) Use the phase portrait concept of Section 2.1 to predict how many supermarkets are expected to adopt the new procedure over a long period of time. By hand, sketch a solution curve of the given initial-value problem.

(b) Solve the initial-value problem and then use a graphing utility to verify the solution curve in part (a). How many companies are expected to adopt the new technology when $t = 10$?

2. The number $N(t)$ of people in a community who are exposed to a particular advertisement is governed by the logistic equation. Initially, $N(0) = 500$, and it is observed that $N(1) = 1000$. Solve

for $N(t)$ if it is predicted that the limiting number of people in the community who will see the advertisement is 50,000.

**3.** A model for the population $P(t)$ in a suburb of a large city is given by the initial-value problem

$$\frac{dP}{dt} = P(10^{-1} - 10^{-7}P), \quad P(0) = 5000,$$

where $t$ is measured in months. What is the limiting value of the population? At what time will the population be equal to one-half of this limiting value?

**4. (a)** Census data for the United States between 1790 and 1950 are given in Table 3.2.1. Construct a logistic population model using the data from 1790, 1850, and 1910.

**(b)** Construct a table comparing actual census population with the population predicted by the model in part (a). Compute the error and the percentage error for each entry pair.

**TABLE 3.2.1**

| Year | Population (in millions) |
|------|-------------------------|
| 1790 | 3.929 |
| 1800 | 5.308 |
| 1810 | 7.240 |
| 1820 | 9.638 |
| 1830 | 12.866 |
| 1840 | 17.069 |
| 1850 | 23.192 |
| 1860 | 31.433 |
| 1870 | 38.558 |
| 1880 | 50.156 |
| 1890 | 62.948 |
| 1900 | 75.996 |
| 1910 | 91.972 |
| 1920 | 105.711 |
| 1930 | 122.775 |
| 1940 | 131.669 |
| 1950 | 150.697 |

## Modifications of the Logistic Model

**5. (a)** If a constant number $h$ of fish are harvested from a fishery per unit time, then a model for the population $P(t)$ of the fishery at time $t$ is given by

$$\frac{dP}{dt} = P(a - bP) - h, \quad P(0) = P_0,$$

where $a$, $b$, $h$, and $P_0$ are positive constants. Suppose $a = 5$, $b = 1$, and $h = 4$. Since the DE is autonomous, use the phase portrait concept of Section 2.1 to sketch representative solution curves corresponding to the cases $P_0 > 4$, $1 < P_0 < 4$, and $0 < P_0 < 1$. Determine the long-term behavior of the population in each case.

**(b)** Solve the IVP in part (a). Verify the results of your phase portrait in part (a) by using a graphing utility to plot the

graph of $P(t)$ with an initial condition taken from each of the three intervals given.

**(c)** Use the information in parts (a) and (b) to determine whether the fishery population becomes extinct in finite time. If so, find that time.

**6.** Investigate the harvesting model in Problem 5 both qualitatively and analytically in the case $a = 5$, $b = 1$, $h = \frac{25}{4}$. Determine whether the population becomes extinct in finite time. If so, find that time.

**7.** Repeat Problem 6 in the case $a = 5$, $b = 1$, $h = 7$.

**8. (a)** Suppose $a = b = 1$ in the Gompertz differential equation (7). Since the DE is autonomous, use the phase portrait concept of Section 2.1 to sketch representative solution curves corresponding to the cases $P_0 > e$ and $0 < P_0 < e$.

**(b)** Suppose $a = 1$, $b = -1$ in (7). Use a new phase portrait to sketch representative solution curves corresponding to the cases $P_0 > e^{-1}$ and $0 < P_0 < e^{-1}$.

**(c)** Find an explicit solution of (7) subject to $P(0) = P_0$.

## Chemical Reactions

**9.** Two chemicals $A$ and $B$ are combined to form a chemical $C$. The rate, or velocity, of the reaction is proportional to the product of the instantaneous amounts of $A$ and $B$ not converted to chemical $C$. Initially, there are 40 grams of $A$ and 50 grams of $B$, and for each gram of $B$, 2 grams of $A$ is used. It is observed that 10 grams of $C$ is formed in 5 minutes. How much is formed in 20 minutes? What is the limiting amount of $C$ after a long time? How much of chemicals $A$ and $B$ remains after a long time?

**10.** Solve Problem 9 if 100 grams of chemical $A$ is present initially. At what time is chemical $C$ half-formed?

## Additional Nonlinear Models

**11. Leaking Cylindrical Tank** A tank in the form of a right-circular cylinder standing on end is leaking water through a circular hole in its bottom. As we saw in (10) of Section 1.3, when friction and contraction of water at the hole are ignored, the height $h$ of water in the tank is described by

$$\frac{dh}{dt} = -\frac{A_h}{A_w}\sqrt{2gh},$$

where $A_w$ and $A_h$ are the cross-sectional areas of the water and the hole, respectively.

**(a)** Solve the DE if the initial height of the water is $H$. By hand, sketch the graph of $h(t)$ and give its interval $I$ of definition in terms of the symbols $A_w$, $A_h$, and $H$. Use $g = 32$ ft/s$^2$.

**(b)** Suppose the tank is 10 feet high and has radius 2 feet and the circular hole has radius $\frac{1}{2}$ inch. If the tank is initially full, how long will it take to empty?

**12. Leaking Cylindrical Tank—Continued**   When friction and contraction of the water at the hole are taken into account, the model in Problem 11 becomes

$$\frac{dh}{dt} = -c\,\frac{A_h}{A_w}\sqrt{2gh},$$

where $0 < c < 1$. How long will it take the tank in Problem 11(b) to empty if $c = 0.6$? See Problem 13 in Exercises 1.3.

**13. Leaking Conical Tank**   A tank in the form of a right-circular cone standing on end, vertex down, is leaking water through a circular hole in its bottom.

   **(a)** Suppose the tank is 20 feet high and has radius 8 feet and the circular hole has radius 2 inches. In Problem 14 in Exercises 1.3 you were asked to show that the differential equation governing the height $h$ of water leaking from a tank is

$$\frac{dh}{dt} = -\frac{5}{6h^{3/2}}.$$

   In this model, friction and contraction of the water at the hole were taken into account with $c = 0.6$, and $g$ was taken to be 32 ft/s². See Figure 1.3.12. If the tank is initially full, how long will it take the tank to empty?

   **(b)** Suppose the tank has a vertex angle of 60° and the circular hole has radius 2 inches. Determine the differential equation governing the height $h$ of water. Use $c = 0.6$ and $g = 32$ ft/s². If the height of the water is initially 9 feet, how long will it take the tank to empty?

**14. Inverted Conical Tank**   Suppose that the conical tank in Problem 13(a) is inverted, as shown in Figure 3.2.5, and that water leaks out a circular hole of radius 2 inches in the center of its circular base. Is the time it takes to empty a full tank the same as for the tank with vertex down in Problem 13? Take the friction/contraction coefficient to be $c = 0.6$ and $g = 32$ ft/s².

**FIGURE 3.2.5**  Inverted conical tank in Problem 14

**15. Air Resistance**   A differential equation for the velocity $v$ of a falling mass $m$ subjected to air resistance proportional to the square of the instantaneous velocity is

$$m\,\frac{dv}{dt} = mg - kv^2,$$

where $k > 0$ is a constant of proportionality. The positive direction is downward.

   **(a)** Solve the equation subject to the initial condition $v(0) = v_0$.

   **(b)** Use the solution in part (a) to determine the limiting, or terminal, velocity of the mass. We saw how to determine the terminal velocity without solving the DE in Problem 41 in Exercises 2.1.

   **(c)** If the distance $s$, measured from the point where the mass was released above ground, is related to velocity $v$ by $ds/dt = v(t)$, find an explicit expression for $s(t)$ if $s(0) = 0$.

**16. How High?—Nonlinear Air Resistance**   Consider the 16-pound cannonball shot vertically upward in Problems 36 and 37 in Exercises 3.1 with an initial velocity $v_0 = 300$ ft/s. Determine the maximum height attained by the cannonball if air resistance is assumed to be proportional to the square of the instantaneous velocity. Assume that the positive direction is upward and take $k = 0.0003$. [*Hint*: Slightly modify the DE in Problem 15.]

**17. That Sinking Feeling**   **(a)** Determine a differential equation for the velocity $v(t)$ of a mass $m$ sinking in water that imparts a resistance proportional to the square of the instantaneous velocity and also exerts an upward buoyant force whose magnitude is given by Archimedes' principle. See Problem 18 in Exercises 1.3. Assume that the positive direction is downward.

   **(b)** Solve the differential equation in part (a).

   **(c)** Determine the limiting, or terminal, velocity of the sinking mass.

**18. Solar Collector**   The differential equation

$$\frac{dy}{dx} = \frac{-x + \sqrt{x^2 + y^2}}{y}$$

describes the shape of a plane curve $C$ that will reflect all incoming light beams to the same point and could be a model for the mirror of a reflecting telescope, a satellite antenna, or a solar collector. See Problem 29 in Exercises 1.3. There are several ways of solving this DE.

   **(a)** Verify that the differential equation is homogeneous (see Section 2.5). Show that the substitution $y = ux$ yields

$$\frac{u\,du}{\sqrt{1 + u^2}\left(1 - \sqrt{1 + u^2}\right)} = \frac{dx}{x}.$$

   Use a CAS (or another judicious substitution) to integrate the left-hand side of the equation. Show that the curve $C$ must be a parabola with focus at the origin and is symmetric with respect to the $x$-axis.

   **(b)** Show that the first differential equation can also be solved by means of the substitution $u = x^2 + y^2$.

**19. Tsunami**   **(a)** A simple model for the shape of a tsunami is given by

$$\frac{dW}{dx} = W\sqrt{4 - 2W},$$

where $W(x) > 0$ is the height of the wave expressed as a function of its position relative to a point offshore. By inspection, find all constant solutions of the DE.

**(b)** Solve the differential equation in part (a). A CAS may be useful for integration.

**(c)** Use a graphing utility to obtain the graphs of all solutions that satisfy the initial condition $W(0) = 2$.

**20. Evaporation** An outdoor decorative pond in the shape of a hemispherical tank is to be filled with water pumped into the tank through an inlet in its bottom. Suppose that the radius of the tank is $R = 10$ ft, that water is pumped in at a rate of $\pi$ ft³/min, and that the tank is initially empty. See Figure 3.2.6. As the tank fills, it loses water through evaporation. Assume that the rate of evaporation is proportional to the area $A$ of the surface of the water and that the constant of proportionality is $k = 0.01$.

**(a)** The rate of change $dV/dt$ of the volume of the water at time $t$ is a net rate. Use this net rate to determine a differential equation for the height $h$ of the water at time $t$. The volume of the water shown in the figure is $V = \pi R h^2 - \frac{1}{3}\pi h^3$, where $R = 10$. Express the area of the surface of the water $A = \pi r^2$ in terms of $h$.

**(b)** Solve the differential equation in part (a). Graph the solution.

**(c)** If there were no evaporation, how long would it take the tank to fill?

**(d)** With evaporation, what is the depth of the water at the time found in part (c)? Will the tank ever be filled? Prove your assertion.

Output: water evaporates
at rate proportional
to area $A$ of surface

Input: water pumped in
at rate $\pi$ ft³/min

**(a)** hemispherical tank       **(b)** cross-section of tank

**FIGURE 3.2.6** Decorative pond in Problem 20

**21. Doomsday Equation** Consider the differential equation

$$\frac{dP}{dt} = kP^{1+c},$$

where $k > 0$ and $c \geq 0$. In Section 3.1 we saw that in the case $c = 0$ the linear differential equation $dP/dt = kP$ is a mathematical model of a population $P(t)$ that exhibits unbounded growth over the infinite time interval $[0, \infty)$, that is, $P(t) \to \infty$ as $t \to \infty$. See Example 1 on page 85.

**(a)** Suppose for $c = 0.01$ that the nonlinear differential equation

$$\frac{dP}{dt} = kP^{1.01}, \quad k > 0,$$

is a mathematical model for a population of small animals, where time $t$ is measured in months. Solve the differential

equation subject to the initial condition $P(0) = 10$ and the fact that the animal population has doubled in 5 months.

**(b)** The differential equation in part (a) is called a **doomsday equation** because the population $P(t)$ exhibits unbounded growth over a finite time interval $(0, T)$, that is, there is some time $T$ such that $P(t) \to \infty$ as $t \to T^-$. Find $T$.

**(c)** From part (a), what is $P(50)$? $P(100)$?

**22. Doomsday or Extinction** Suppose the population model (4) is modified to be

$$\frac{dP}{dt} = P(bP - a).$$

**(a)** If $a > 0, b > 0$ show by means of a phase portrait (see page 40) that, depending on the initial condition $P(0) = P_0$, the mathematical model could include a doomsday scenario $(P(t) \to \infty)$ or an extinction scenario $(P(t) \to 0)$.

**(b)** Solve the initial-value problem

$$\frac{dP}{dt} = P(0.0005P - 0.1), \quad P(0) = 300.$$

Show that this model predicts a doomsday for the population in a finite time $T$.

**(c)** Solve the differential equation in part (b) subject to the initial condition $P(0) = 100$. Show that this model predicts extinction for the population as $t \to \infty$.

## Project Problems

**23. Regression Line** Read the documentation for your CAS on *scatter plots* (or *scatter diagrams*) and *least-squares linear fit*. The straight line that best fits a set of data points is called a **regression line** or a **least squares line.** Your task is to construct a logistic model for the population of the United States, defining $f(P)$ in (2) as an equation of a regression line based on the population data in the table in Problem 4. One way of doing this is to approximate the left-hand side $\frac{1}{P}\frac{dP}{dt}$ of the first equation in (2), using the forward difference quotient in place of $dP/dt$:

$$Q(t) = \frac{1}{P(t)}\frac{P(t + h) - P(t)}{h}.$$

**(a)** Make a table of the values $t$, $P(t)$, and $Q(t)$ using $t = 0, 10, 20, \ldots, 160$ and $h = 10$. For example, the first line of the table should contain $t = 0$, $P(0)$, and $Q(0)$. With $P(0) = 3.929$ and $P(10) = 5.308$,

$$Q(0) = \frac{1}{P(0)}\frac{P(10) - P(0)}{10} = 0.035.$$

Note that $Q(160)$ depends on the 1960 census population $P(170)$. Look up this value.

**(b)** Use a CAS to obtain a scatter plot of the data $(P(t), Q(t))$ computed in part (a). Also use a CAS to find an equation of the regression line and to superimpose its graph on the scatter plot.

**(c)** Construct a logistic model $dP/dt = Pf(P)$, where $f(P)$ is the equation of the regression line found in part (b).

**(d)** Solve the model in part (c) using the initial condition $P(0) = 3.929$.

**(e)** Use a CAS to obtain another scatter plot, this time of the ordered pairs $(t, P(t))$ from your table in part (a). Use your CAS to superimpose the graph of the solution in part (d) on the scatter plot.

**(f)** Look up the U.S. census data for 1970, 1980, and 1990. What population does the logistic model in part (c) predict for these years? What does the model predict for the U.S. population $P(t)$ as $t \rightarrow \infty$?

**24. Immigration Model**    **(a)** In Examples 3 and 4 of Section 2.1 we saw that any solution $P(t)$ of (4) possesses the asymptotic behavior $P(t) \rightarrow a/b$ as $t \rightarrow \infty$ for $P_0 > a/b$ and for $0 < P_0 < a/b$; as a consequence the equilibrium solution $P = a/b$ is called an attractor. Use a root-finding application of a CAS (or a graphic calculator) to approximate the equilibrium solution of the immigration model

$$\frac{dP}{dt} = P(1 - P) + 0.3e^{-P}.$$

**(b)** Use a graphing utility to graph the function $F(P) = P(1 - P) + 0.3e^{-P}$. Explain how this graph can be used to determine whether the number found in part (a) is an attractor.

**(c)** Use a numerical solver to compare the solution curves for the IVPs

$$\frac{dP}{dt} = P(1 - P), \quad P(0) = P_0$$

for $P_0 = 0.2$ and $P_0 = 1.2$ with the solution curves for the IVPs

$$\frac{dP}{dt} = P(1 - P) + 0.3e^{-P}, \quad P(0) = P_0$$

for $P_0 = 0.2$ and $P_0 = 1.2$. Superimpose all curves on the same coordinate axes but, if possible, use a different color for the curves of the second initial-value problem. Over a long period of time, what percentage increase does the immigration model predict in the population compared to the logistic model?

**25. What Goes Up . . .**    In Problem 16 let $t_a$ be the time it takes the cannonball to attain its maximum height and let $t_d$ be the time it takes the cannonball to fall from the maximum height to the ground. Compare the value of $t_a$ with the value of $t_d$ and compare the magnitude of the impact velocity $v_i$ with the initial velocity $v_0$. See Problem 50 in Exercises 3.1. A root-finding application of a CAS might be useful here. [*Hint*: Use the model in Problem 15 when the cannonball is falling.]

**26. Skydiving**    A skydiver is equipped with a stopwatch and an altimeter. As shown in Figure 3.2.7, he opens his parachute 25 seconds after exiting a plane flying at an altitude of 20,000 feet and observes that his altitude is 14,800 feet. Assume that air resistance is proportional to the square of the instantaneous

velocity, his initial velocity on leaving the plane is zero, and $g = 32$ ft/s$^2$.

**(a)** Find the distance $s(t)$, measured from the plane, the skydiver has traveled during freefall in time $t$. [*Hint*: The constant of proportionality $k$ in the model given in Problem 15 is not specified. Use the expression for terminal velocity $v_t$ obtained in part (b) of Problem 15 to eliminate $k$ from the IVP. Then eventually solve for $v_t$.]

**(b)** How far does the skydiver fall and what is his velocity at $t = 15$ s?

**FIGURE 3.2.7** Skydiver in Problem 26

**27. Hitting Bottom**    A helicopter hovers 500 feet above a large open tank full of liquid (not water). A dense compact object weighing 160 pounds is dropped (released from rest) from the helicopter into the liquid. Assume that air resistance is proportional to instantaneous velocity $v$ while the object is in the air and that viscous damping is proportional to $v^2$ after the object has entered the liquid. For air take $k = \frac{1}{4}$, and for the liquid take $k = 0.1$. Assume that the positive direction is downward. If the tank is 75 feet high, determine the time and the impact velocity when the object hits the bottom of the tank. [*Hint*: Think in terms of two distinct IVPs. If you use (13), be careful in removing the absolute value sign. You might compare the velocity when the object hits the liquid—the initial velocity for the second problem—with the terminal velocity $v_t$ of the object falling through the liquid.]

**28. Old Man River . . .**    In Figure 3.2.8(a) suppose that the $y$-axis and the dashed vertical line $x = 1$ represent, respectively, the straight west and east beaches of a river that is 1 mile wide. The river flows northward with a velocity $\mathbf{v}_r$, where $|\mathbf{v}_r| = v_r$ mi/h is a constant. A man enters the current at the point $(1, 0)$ on the east shore and swims in a direction and rate relative to the river given by the vector $\mathbf{v}_s$, where the speed $|\mathbf{v}_s| = v_s$ mi/h is a constant. The man wants to reach the west beach exactly at $(0, 0)$ and so swims in such a manner that keeps his velocity vector $\mathbf{v}_s$ always directed toward the point $(0, 0)$. Use Figure 3.2.8(b) as an aid in showing that a mathematical model for the path of the swimmer in the river is

$$\frac{dy}{dx} = \frac{v_s y - v_r\sqrt{x^2 + y^2}}{v_s x}.$$

[*Hint*: The velocity $\mathbf{v}$ of the swimmer along the path or curve shown in Figure 3.2.8 is the resultant $\mathbf{v} = \mathbf{v}_s + \mathbf{v}_r$. Resolve $\mathbf{v}_s$ and $\mathbf{v}_r$ into components in the $x$- and $y$-directions. If $x = x(t)$, $y = y(t)$ are parametric equations of the swimmer's path, then $\mathbf{v} = (dx/dt, dy/dt)$.]

**(a)**

**(b)**

**FIGURE 3.2.8** Path of swimmer in Problem 28

**29. (a)** Solve the DE in Problem 28 subject to $y(1) = 0$. For convenience let $k = v_r/v_s$.

**(b)** Determine the values of $v_s$ for which the swimmer will reach the point $(0, 0)$ by examining $\lim_{x \to 0^+} y(x)$ in the cases $k = 1, k > 1$, and $0 < k < 1$.

**30. Old Man River Keeps Moving . . .** Suppose the man in Problem 28 again enters the current at $(1, 0)$ but this time decides to swim so that his velocity vector $\mathbf{v}_s$ is always directed toward the west beach. Assume that the speed $|\mathbf{v}_s| = v_s$ mi/h is a constant. Show that a mathematical model for the path of the swimmer in the river is now

$$\frac{dy}{dx} = -\frac{v_r}{v_s}.$$

**31.** The current speed $v_r$ of a straight river such as that in Problem 28 is usually not a constant. Rather, an approximation to the current speed (measured in miles per hour) could be a function such as $v_r(x) = 30x(1 - x), 0 \le x \le 1$, whose values are small at the shores (in this case, $v_r(0) = 0$ and $v_r(1) = 0$) and largest in the middle of the river. Solve the DE in Problem 30 subject to $y(1) = 0$, where $v_s = 2$ mi/h and $v_r(x)$ is as given. When the swimmer makes it across the river, how far will he have to walk along the beach to reach the point $(0, 0)$?

**32. Raindrops Keep Falling . . .** When a bottle of liquid refreshment was opened recently, the following factoid was found inside the bottle cap:

*The average velocity of a falling raindrop is 7 miles/hour.*

A quick search of the Internet found that meteorologist Jeff Haby offers the additional information that an "average" spherical raindrop has a radius of 0.04 in. and an approximate volume of 0.000000155 ft³. Use this data and, if need be, dig up other data and make other reasonable assumptions to determine whether "*average velocity of . . . 7 mi/h*" is consistent with the models in Problems 35 and 36 in Exercises 3.1 and Problem 15 in this exercise set. Also see Problem 36 in Exercises 1.3.

**33. Time Drips By**    The **clepsydra**, or water clock, was a device that the ancient Egyptians, Greeks, Romans, and Chinese used to measure the passage of time by observing the change in the height of water that was permitted to flow out of a small hole in the bottom of a container or tank.

**(a)** Suppose a tank is made of glass and has the shape of a right-circular cylinder of radius 1 ft. Assume that $h(0) = 2$ ft corresponds to water filled to the top of the tank, a hole in the bottom is circular with radius $\frac{1}{32}$ in., $g = 32$ ft/s², and $c = 0.6$. Use the differential equation in Problem 12 to find the height $h(t)$ of the water.

**(b)** For the tank in part (a), how far up from its bottom should a mark be made on its side, as shown in Figure 3.2.9, that corresponds to the passage of one hour? Next determine where to place the marks corresponding to the passage of 2 hr, 3 hr, . . . , 12 hr. Explain why these marks are not evenly spaced.

**FIGURE 3.2.9** Clepsydra in Problem 33

**34. (a)** Suppose that a glass tank has the shape of a cone with circular cross section as shown in Figure 3.2.10. As in part (a) of Problem 33, assume that $h(0) = 2$ ft corresponds to water filled to the top of the tank, a hole in the bottom is circular with radius $\frac{1}{32}$ in., $g = 32$ ft/s², and $c = 0.6$. Use the differential equation in Problem 12 to find the height $h(t)$ of the water.

**(b)** Can this water clock measure 12 time intervals of length equal to 1 hour? Explain using sound mathematics.

**FIGURE 3.2.10** Clepsydra in Problem 34

**35.** Suppose that $r = f(h)$ defines the shape of a water clock for which the time marks are equally spaced. Use the differential equation in Problem 12 to find $f(h)$ and sketch a typical graph of $h$ as a function of $r$. Assume that the cross-sectional area $A_h$ of the hole is constant. [*Hint*: In this situation $dh/dt = -a$, where $a > 0$ is a constant.]

# 3.3 Modeling with Systems of First-Order DEs

**INTRODUCTION**   This section is similar to Section 1.3 in that we are just going to discuss certain mathematical models, but instead of a single differential equation the models will be systems of first-order differential equations. Although some of the models will be based on topics that we explored in the preceding two sections, we are not going to develop any general methods for solving these systems. There are reasons for this: First, we do not possess the necessary mathematical tools for solving systems at this point. Second, some of the systems that we discuss—notably the systems of *nonlinear* first-order DEs—simply cannot be solved analytically. We shall examine solution methods for systems of *linear* DEs in Chapters 4, 7, and 8.

**LINEAR/NONLINEAR SYSTEMS**   We have seen that a single differential equation can serve as a mathematical model for a single population in an environment. But if there are, say, two interacting and perhaps competing species living in the same environment (for example, rabbits and foxes), then a model for their populations $x(t)$ and $y(t)$ might be a system of two first-order differential equations such as

$$\frac{dx}{dt} = g_1(t, x, y)$$

$$\frac{dy}{dt} = g_2(t, x, y). \tag{1}$$

When $g_1$ and $g_2$ are linear in the variables $x$ and $y$—that is, $g_1$ and $g_2$ have the forms

$$g_1(t, x, y) = c_1 x + c_2 y + f_1(t) \qquad \text{and} \qquad g_2(t, x, y) = c_3 x + c_4 y + f_2(t),$$

where the coefficients $c_i$ could depend on $t$—then (1) is said to be a **linear system.** A system of differential equations that is not linear is said to be **nonlinear.**

**RADIOACTIVE SERIES**   In the discussion of radioactive decay in Sections 1.3 and 3.1 we assumed that the rate of decay was proportional to the number $A(t)$ of nuclei of the substance present at time $t$. When a substance decays by radioactivity, it usually doesn't just transmute in one step into a stable substance; rather, the first substance decays into another radioactive substance, which in turn decays into a third substance, and so on. This process, called a **radioactive decay series,** continues until a stable element is reached. For example, the uranium decay series is U-238 $\rightarrow$ Th-234 $\rightarrow \cdots \rightarrow$ Pb-206, where Pb-206 is a stable isotope of lead. The half-lives of the various elements in a radioactive series can range from billions of years ($4.5 \times 10^9$ years for U-238) to a fraction of a second. Suppose a radioactive series is described schematically by $X \overset{-\lambda_1}{\rightarrow} Y \overset{-\lambda_2}{\rightarrow} Z$, where $k_1 = -\lambda_1 < 0$ and $k_2 = -\lambda_2 < 0$ are the decay constants for substances $X$ and $Y$, respectively, and $Z$ is a stable element. Suppose, too, that $x(t)$, $y(t)$, and $z(t)$ denote amounts of substances $X$, $Y$, and $Z$, respectively, remaining at time $t$. The decay of element $X$ is described by

$$\frac{dx}{dt} = -\lambda_1 x,$$

whereas the rate at which the second element $Y$ decays is the net rate

$$\frac{dy}{dt} = \lambda_1 x - \lambda_2 y,$$

since $Y$ is *gaining* atoms from the decay of $X$ and at the same time *losing* atoms because of its own decay. Since $Z$ is a stable element, it is simply gaining atoms from the decay of element $Y$:

$$\frac{dz}{dt} = \lambda_2 y.$$

In other words, a model of the radioactive decay series for three elements is the linear system of three first-order differential equations

$$\frac{dx}{dt} = -\lambda_1 x$$

$$\frac{dy}{dt} = \lambda_1 x - \lambda_2 y \tag{2}$$

$$\frac{dz}{dt} = \lambda_2 y.$$

**MIXTURES**   Consider the two tanks shown in Figure 3.3.1. Let us suppose for the sake of discussion that tank $A$ contains 50 gallons of water in which 25 pounds of salt is dissolved. Suppose tank $B$ contains 50 gallons of pure water. Liquid is pumped into and out of the tanks as indicated in the figure; the mixture exchanged between the two tanks and the liquid pumped out of tank $B$ are assumed to be well stirred. We wish to construct a mathematical model that describes the number of pounds $x_1(t)$ and $x_2(t)$ of salt in tanks $A$ and $B$, respectively, at time $t$.

pure water
3 gal/min

mixture
1 gal/min

$A$   $B$

mixture
4 gal/min

mixture
3 gal/min

**FIGURE 3.3.1**   Connected mixing tanks

By an analysis similar to that on page 25 in Section 1.3 and Example 5 of Section 3.1 we see that the net rate of change of $x_1(t)$ for tank $A$ is

$$\frac{dx_1}{dt} = \overbrace{(3 \text{ gal/min}) \cdot (0 \text{ lb/gal}) + (1 \text{ gal/min}) \cdot \left(\frac{x_2}{50} \text{ lb/gal}\right)}^{\substack{\text{input rate} \\ \text{of salt}}} - \overbrace{(4 \text{ gal/min}) \cdot \left(\frac{x_1}{50} \text{ lb/gal}\right)}^{\substack{\text{output rate} \\ \text{of salt}}}$$

$$= -\frac{2}{25} x_1 + \frac{1}{50} x_2.$$

Similarly, for tank $B$ the net rate of change of $x_2(t)$ is

$$\frac{dx_2}{dt} = 4 \cdot \frac{x_1}{50} - 3 \cdot \frac{x_2}{50} - 1 \cdot \frac{x_2}{50}$$

$$= \frac{2}{25} x_1 - \frac{2}{25} x_2.$$

Thus we obtain the linear system

$$\frac{dx_1}{dt} = -\frac{2}{25} x_1 + \frac{1}{50} x_2$$

$$\frac{dx_2}{dt} = \frac{2}{25} x_1 - \frac{2}{25} x_2. \tag{3}$$

Observe that the foregoing system is accompanied by the initial conditions $x_1(0) = 25$, $x_2(0) = 0$.

**A PREDATOR-PREY MODEL**   Suppose that two different species of animals interact within the same environment or ecosystem, and suppose further that the first species eats only vegetation and the second eats only the first species. In other words, one species is a predator, and the other is a prey. For example, wolves hunt grass-eating caribou, sharks devour little fish, and the snowy owl pursues an arctic rodent called the lemming. For the sake of discussion, let us imagine that the predators are foxes and the prey are rabbits.

Let $x(t)$ and $y(t)$ denote the fox and rabbit populations, respectively, at time $t$. If there were no rabbits, then one might expect that the foxes, lacking an adequate food supply, would decline in number according to

$$\frac{dx}{dt} = -ax, \quad a > 0. \tag{4}$$

When rabbits are present in the environment, however, it seems reasonable that the number of encounters or interactions between these two species per unit time is jointly proportional to their populations $x$ and $y$—that is, proportional to the product $xy$. Thus when rabbits are present, there is a supply of food, so foxes are added to the system at a rate $bxy$, $b > 0$. Adding this last rate to (4) gives a model for the fox population:

$$\frac{dx}{dt} = -ax + bxy. \tag{5}$$

On the other hand, if there were no foxes, then the rabbits would, with an added assumption of unlimited food supply, grow at a rate that is proportional to the number of rabbits present at time $t$:

$$\frac{dy}{dt} = dy, \quad d > 0. \tag{6}$$

But when foxes are present, a model for the rabbit population is (6) decreased by $cxy$, $c > 0$—that is, decreased by the rate at which the rabbits are eaten during their encounters with the foxes:

$$\frac{dy}{dt} = dy - cxy. \tag{7}$$

Equations (5) and (7) constitute a system of nonlinear differential equations

$$\begin{aligned}
\frac{dx}{dt} &= -ax + bxy = x(-a + by) \\
\frac{dy}{dt} &= dy - cxy = y(d - cx),
\end{aligned} \tag{8}$$

where $a$, $b$, $c$, and $d$ are positive constants. This famous system of equations is known as the **Lotka-Volterra predator-prey model.**

Except for two constant solutions, $x(t) = 0$, $y(t) = 0$ and $x(t) = d/c$, $y(t) = a/b$, the nonlinear system (8) cannot be solved in terms of elementary functions. However, we can analyze such systems quantitatively and qualitatively. See Chapter 9, "Numerical Solutions of Ordinary Differential Equations," and Chapter 10, "Systems of Nonlinear First-Order Differential Equations."*

### EXAMPLE 1   Predator-Prey Model

Suppose

$$\frac{dx}{dt} = -0.16x + 0.08xy$$

$$\frac{dy}{dt} = 4.5y - 0.9xy$$

---

*Chapters 10–15 are in the expanded version of this text, *Differential Equations with Boundary-Value Problems.*

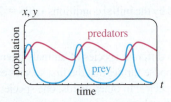

**FIGURE 3.3.2** Populations of predators (red) and prey (blue) in Example 1

represents a predator-prey model. Because we are dealing with populations, we have $x(t) \geq 0$, $y(t) \geq 0$. Figure 3.3.2, obtained with the aid of a numerical solver, shows typical population curves of the predators and prey for this model superimposed on the same coordinate axes. The initial conditions used were $x(0) = 4$, $y(0) = 4$. The curve in red represents the population $x(t)$ of the predators (foxes), and the blue curve is the population $y(t)$ of the prey (rabbits). Observe that the model seems to predict that both populations $x(t)$ and $y(t)$ are periodic in time. This makes intuitive sense because as the number of prey decreases, the predator population eventually decreases because of a diminished food supply; but attendant to a decrease in the number of predators is an increase in the number of prey; this in turn gives rise to an increased number of predators, which ultimately brings about another decrease in the number of prey. ■

**COMPETITION MODELS**   Now suppose two different species of animals occupy the same ecosystem, not as predator and prey but rather as competitors for the same resources (such as food and living space) in the system. In the absence of the other, let us assume that the rate at which each population grows is given by

$$\frac{dx}{dt} = ax \quad \text{and} \quad \frac{dy}{dt} = cy, \tag{9}$$

respectively.

Since the two species compete, another assumption might be that each of these rates is diminished simply by the influence, or existence, of the other population. Thus a model for the two populations is given by the linear system

$$\frac{dx}{dt} = ax - by$$
$$\frac{dy}{dt} = cy - dx, \tag{10}$$

where $a$, $b$, $c$, and $d$ are positive constants.

On the other hand, we might assume, as we did in (5), that each growth rate in (9) should be reduced by a rate proportional to the number of interactions between the two species:

$$\frac{dx}{dt} = ax - bxy$$
$$\frac{dy}{dt} = cy - dxy. \tag{11}$$

Inspection shows that this nonlinear system is similar to the Lotka-Volterra predator-prey model. Finally, it might be more realistic to replace the rates in (9), which indicate that the population of each species in isolation grows exponentially, with rates indicating that each population grows logistically (that is, over a long time the population is bounded):

$$\frac{dx}{dt} = a_1 x - b_1 x^2 \quad \text{and} \quad \frac{dy}{dt} = a_2 y - b_2 y^2. \tag{12}$$

When these new rates are decreased by rates proportional to the number of interactions, we obtain another nonlinear model:

$$\frac{dx}{dt} = a_1 x - b_1 x^2 - c_1 xy = x(a_1 - b_1 x - c_1 y)$$
$$\frac{dy}{dt} = a_2 y - b_2 y^2 - c_2 xy = y(a_2 - b_2 y - c_2 x), \tag{13}$$

where all coefficients are positive. The linear system (10) and the nonlinear systems (11) and (13) are, of course, called **competition models.**

**NETWORKS**    An electrical network having more than one loop also gives rise to simultaneous differential equations. As shown in Figure 3.3.3, the current $i_1(t)$ splits in the directions shown at point $B_1$, called a *branch point* of the network. By **Kirchhoff's first law** we can write

$$i_1(t) = i_2(t) + i_3(t). \tag{14}$$

We can also apply **Kirchhoff's second law** to each loop. For loop $A_1B_1B_2A_2A_1$, summing the voltage drops across each part of the loop gives

$$E(t) = i_1R_1 + L_1\frac{di_2}{dt} + i_2R_2. \tag{15}$$

**FIGURE 3.3.3** Network whose model is given in (17)

Similarly, for loop $A_1B_1C_1C_2B_2A_2A_1$ we find

$$E(t) = i_1R_1 + L_2\frac{di_3}{dt}. \tag{16}$$

Using (14) to eliminate $i_1$ in (15) and (16) yields two linear first-order equations for the currents $i_2(t)$ and $i_3(t)$:

$$L_1\frac{di_2}{dt} + (R_1 + R_2)i_2 + R_1i_3 = E(t)$$
$$L_2\frac{di_3}{dt} + \qquad R_1i_2 + R_1i_3 = E(t). \tag{17}$$

We leave it as an exercise (see Problem 16 in Exercises 3.3) to show that the system of differential equations describing the currents $i_1(t)$ and $i_2(t)$ in the network containing a resistor, an inductor, and a capacitor shown in Figure 3.3.4 is

$$L\frac{di_1}{dt} + Ri_2 \quad = E(t)$$
$$RC\frac{di_2}{dt} + i_2 - i_1 = 0. \tag{18}$$

**FIGURE 3.3.4** Network whose model is given in (18)

## EXERCISES 3.3

*Answers to selected odd-numbered problems begin on page ANS-4.*

### Radioactive Series

1. We have not discussed methods by which systems of first-order differential equations can be solved. Nevertheless, systems such as (2) can be solved with no knowledge other than how to solve a single linear first-order equation. Find a solution of (2) subject to the initial conditions $x(0) = x_0$, $y(0) = 0$, $z(0) = 0$.

2. In Problem 1 suppose that time is measured in days, that the decay constants are $k_1 = -0.138629$ and $k_2 = -0.004951$, and that $x_0 = 20$. Use a graphing utility to obtain the graphs of the solutions $x(t)$, $y(t)$, and $z(t)$ on the same set of coordinate axes. Use the graphs to approximate the half-lives of substances $X$ and $Y$.

3. Use the graphs in Problem 2 to approximate the times when the amounts $x(t)$ and $y(t)$ are the same, the times when the amounts $x(t)$ and $z(t)$ are the same, and the times when the amounts $y(t)$ and $z(t)$ are the same. Why does the time that is determined when the amounts $y(t)$ and $z(t)$ are the same make intuitive sense?

4. Construct a mathematical model for a radioactive series of four elements $W$, $X$, $Y$, and $Z$, where $Z$ is a stable element.

5. **Potassium-40 Decay**    The chemical element potassium is a soft metal that can be found extensively throughout the Earth's crust and oceans. Although potassium occurs naturally in the form of three isotopes, only the isotope potassium-40 (K-40) is radioactive. This isotope is also unusual in that it decays by two different nuclear reactions. Over time, by emitting beta particles a great percentage of an initial amount $K_0$ of K-40 decays into the stable isotope calcium-40 (Ca-40), whereas by electron capture a smaller percentage of $K_0$ decays into the stable isotope argon-40 (Ar-40). Because the rates at which the amounts $C(t)$ of Ca-40 and $A(t)$ of Ar-40 increase are proportional to the amount $K(t)$ of potassium present, and the rate at which $K(t)$ decays is also proportional to $K(t)$, we obtain the system of linear first-order equations

$$\frac{dC}{dt} = \lambda_1K$$
$$\frac{dA}{dt} = \lambda_2K$$
$$\frac{dK}{dt} = -(\lambda_1 + \lambda_2)K,$$

where $\lambda_1$ and $\lambda_2$ are positive constants of proportionality. By proceeding as in Problem 1 we can solve the foregoing mathematical model.

**(a)** From the last equation in the given system of differential equations find $K(t)$ if $K(0) = K_0$. Then use $K(t)$ to find $C(t)$ and $A(t)$ from the first and second equations. Assume that $C(0) = 0$ and $A(0) = 0$.

**(b)** It is known that $\lambda_1 = 4.7526 \times 10^{-10}$ and $\lambda_2 = 0.5874 \times 10^{-10}$. Find the half-life of K-40.

**(c)** Use $C(t)$ and $A(t)$ found in part (a) to determine the percentage of an initial amount $K_0$ of K-40 that decays into Ca-40 and the percentage that decays into Ar-40 over a very long period of time.

**6. Potassium-Argon Dating** The knowledge of how K-40 decays can be used to determine the age of very old igneous rocks. See Figure 3.3.5.

**(a)** Use the solutions obtained in part (a) of Problem 5 to find the ratio $A(t)/K(t)$.

**(b)** Use $A(t)/K(t)$ found in part (a) to show that

$$t = \frac{1}{\lambda_1 + \lambda_2} \ln\left[1 + \frac{\lambda_1 + \lambda_2}{\lambda_2}\frac{A(t)}{K(t)}\right].$$

**(c)** Suppose it is determined that each gram of an igneous rock sample contains $8.5 \times 10^{-7}$ grams of Ar-40 and $5.4 \times 10^{-6}$ grams of K-40. Use the result in part (b) to find the approximate age of the rock.

**FIGURE 3.3.5** Igneous rocks are formed through solidification of volcanic lava

## Mixtures

**7.** Consider two tanks $A$ and $B$, with liquid being pumped in and out at the same rates, as described by the system of equations (3). What is the system of differential equations if, instead of pure water, a brine solution containing 2 pounds of salt per gallon is pumped into tank $A$?

**8.** Use the information given in Figure 3.3.6 to construct a mathematical model for the number of pounds of salt $x_1(t)$, $x_2(t)$, and $x_3(t)$ at time $t$ in tanks $A$, $B$, and $C$, respectively.

**FIGURE 3.3.6** Mixing tanks in Problem 8

**9.** Two very large tanks $A$ and $B$ are each partially filled with 100 gallons of brine. Initially, 100 pounds of salt is dissolved in the solution in tank $A$ and 50 pounds of salt is dissolved in the solution in tank $B$. The system is closed in that the well-stirred liquid is pumped only between the tanks, as shown in Figure 3.3.7.

**(a)** Use the information given in the figure to construct a mathematical model for the number of pounds of salt $x_1(t)$ and $x_2(t)$ at time $t$ in tanks $A$ and $B$, respectively.

**(b)** Find a relationship between the variables $x_1(t)$ and $x_2(t)$ that holds at time $t$. Explain why this relationship makes intuitive sense. Use this relationship to help find the amount of salt in tank $B$ at $t = 30$ min.

**FIGURE 3.3.7** Mixing tanks in Problem 9

**10.** Three large tanks contain brine, as shown in Figure 3.3.8. Use the information in the figure to construct a mathematical model for the number of pounds of salt $x_1(t)$, $x_2(t)$, and $x_3(t)$ at time $t$ in tanks $A$, $B$, and $C$, respectively. Without solving the system, predict limiting values of $x_1(t)$, $x_2(t)$, and $x_3(t)$ as $t \to \infty$.

**FIGURE 3.3.8** Mixing tanks in Problem 10

### Predator-Prey Models

**11.** Consider the Lotka-Volterra predator-prey model defined by

$$\frac{dx}{dt} = -0.1x + 0.02xy$$

$$\frac{dy}{dt} = 0.2y - 0.025xy,$$

where the populations $x(t)$ (predators) and $y(t)$ (prey) are measured in thousands. Suppose $x(0) = 6$ and $y(0) = 6$. Use a numerical solver to graph $x(t)$ and $y(t)$. Use the graphs to approximate the time $t > 0$ when the two populations are first equal. Use the graphs to approximate the period of each population.

### Competition Models

**12.** Consider the competition model defined by

$$\frac{dx}{dt} = x(2 - 0.4x - 0.3y)$$

$$\frac{dy}{dt} = y(1 - 0.1y - 0.3x),$$

where the populations $x(t)$ and $y(t)$ are measured in thousands and $t$ in years. Use a numerical solver to analyze the populations over a long period of time for each of the following cases:

**(a)** $x(0) = 1.5$, $y(0) = 3.5$

**(b)** $x(0) = 1$, $y(0) = 1$

**(c)** $x(0) = 2$, $y(0) = 7$

**(d)** $x(0) = 4.5$, $y(0) = 0.5$

**13.** Consider the competition model defined by

$$\frac{dx}{dt} = x(1 - 0.1x - 0.05y)$$

$$\frac{dy}{dt} = y(1.7 - 0.1y - 0.15x),$$

where the populations $x(t)$ and $y(t)$ are measured in thousands and $t$ in years. Use a numerical solver to analyze the populations over a long period of time for each of the following cases:

**(a)** $x(0) = 1$, $y(0) = 1$

**(b)** $x(0) = 4$, $y(0) = 10$

**(c)** $x(0) = 9$, $y(0) = 4$

**(d)** $x(0) = 5.5$, $y(0) = 3.5$

### Networks

**14.** Show that a system of differential equations that describes the currents $i_2(t)$ and $i_3(t)$ in the electrical network shown in Figure 3.3.9 is

$$L\frac{di_2}{dt} + L\frac{di_3}{dt} + R_1i_2 = E(t)$$

$$-R_1\frac{di_2}{dt} + R_2\frac{di_3}{dt} + \frac{1}{C}i_3 = 0.$$

**FIGURE 3.3.9** Network in Problem 14

**15.** Determine a system of first-order differential equations that describes the currents $i_2(t)$ and $i_3(t)$ in the electrical network shown in Figure 3.3.10.

**FIGURE 3.3.10** Network in Problem 15

**16.** Show that the linear system given in (18) describes the currents $i_1(t)$ and $i_2(t)$ in the network shown in Figure 3.3.4. [*Hint:* $dq/dt = i_3$.]

### Additional Nonlinear Models

**17. SIR Model** A communicable disease is spread throughout a small community, with a fixed population of $n$ people, by contact between infected individuals and people who are susceptible to the disease. Suppose that everyone is initially susceptible to the disease and that no one leaves the community while the epidemic is spreading. At time $t$, let $s(t)$, $i(t)$, and $r(t)$ denote, in turn, the number of people in the community (measured in hundreds) who are *susceptible* to the disease but not yet infected with it, the number of people who are *infected* with the disease, and the number of people who have *recovered* from the disease. Explain why the system of differential equations

$$\frac{ds}{dt} = -k_1si$$

$$\frac{di}{dt} = -k_2i + k_1si$$

$$\frac{dr}{dt} = k_2i,$$

where $k_1$ (called the *infection rate*) and $k_2$ (called the *removal rate*) are positive constants, is a reasonable mathematical model, commonly called a **SIR model,** for the spread of the epidemic throughout the community. Give plausible initial conditions associated with this system of equations.

**18. (a)** In Problem 17, explain why it is sufficient to analyze only

$$\frac{ds}{dt} = -k_1si$$

$$\frac{di}{dt} = -k_2i + k_1si.$$

**(b)** Suppose $k_1 = 0.2$, $k_2 = 0.7$, and $n = 10$. Choose various values of $i(0) = i_0$, $0 < i_0 < 10$. Use a numerical solver to determine what the model predicts about the epidemic in the two cases $s_0 > k_2/k_1$ and $s_0 \le k_2/k_1$. In the case of an epidemic, estimate the number of people who are eventually infected.

19. **Concentration of a Nutrient** Suppose compartments $A$ and $B$ shown in Figure 3.3.11 are filled with fluids and are separated by a permeable membrane. The figure is a compartmental representation of the exterior and interior of a cell. Suppose, too, that a nutrient necessary for cell growth passes through the membrane. A model for the concentrations $x(t)$ and $y(t)$ of the nutrient in compartments $A$ and $B$, respectively, at time $t$ is given by the linear system of differential equations

$$\frac{dx}{dt} = \frac{\kappa}{V_A}(y - x)$$

$$\frac{dy}{dt} = \frac{\kappa}{V_B}(x - y),$$

where $V_A$ and $V_B$ are the volumes of the compartments, and $\kappa > 0$ is a permeability factor. Let $x(0) = x_0$ and $y(0) = y_0$ denote the initial concentrations of the nutrient. Solely on the basis of the equations in the system and the assumption $x_0 > y_0 > 0$, sketch, on the same set of coordinate axes, possible solution curves of the system. Explain your reasoning. Discuss the behavior of the solutions over a long period of time.

20. The system in Problem 19, like the system in (2), can be solved with no advanced knowledge. Solve for $x(t)$ and $y(t)$ and compare their graphs with your sketches in Problem 19. Determine the limiting values of $x(t)$ and $y(t)$ as $t \to \infty$. Explain why the answer to the last question makes intuitive sense.

21. **Mixtures** Solely on the basis of the physical description of the mixture problem on page 108 and in Figure 3.3.1, discuss the nature of the functions $x_1(t)$ and $x_2(t)$. What is the behavior of each function over a long period of time? Sketch possible graphs of $x_1(t)$ and $x_2(t)$. Check your conjectures by using a numerical solver to obtain numerical solution curves of (3) subject to the initial conditions $x_1(0) = 25$, $x_2(0) = 0$.

22. **Newton's Law of Cooling/Warming** As shown in Figure 3.3.12, a small metal bar is placed inside container $A$, and container $A$ then is placed within a much larger container $B$. As the metal bar cools, the ambient temperature $T_A(t)$ of the medium within container $A$ changes according to Newton's law of cooling. As container $A$ cools, the temperature of the medium inside container $B$ does not change significantly and can be considered to be a constant $T_B$. Construct a mathematical model for the temperatures $T(t)$ and $T_A(t)$, where $T(t)$ is the temperature of the metal bar inside container $A$. As in Problems 1, 5, and 20, this model can be solved by using prior knowledge. Find a solution of the system subject to the initial conditions $T(0) = T_0$, $T_A(0) = T_1$.

FIGURE 3.3.11 Nutrient flow through a membrane in Problem 19

FIGURE 3.3.12 Container within a container in Problem 22

# Chapter 3 In Review

*Answers to selected odd-numbered problems begin on page ANS-4.*

Answer Problems 1 and 2 without referring back to the text. Fill in the blank or answer true or false.

1. If $P(t) = P_0 e^{0.15t}$ gives the population in an environment at time $t$, then a differential equation satisfied by $P(t)$ is _____.

2. If the rate of decay of a radioactive substance is proportional to the amount $A(t)$ remaining at time $t$, then the half-life of the substance is necessarily $T = -(\ln 2)/k$. The rate of decay of the substance at time $t = T$ is one-half the rate of decay at $t = 0$. _____

3. In March 1976 the world population reached 4 billion. At that time, a popular news magazine predicted that with an average yearly growth rate of 1.8%, the world population would be 8 billion in 45 years. How does this value compare with the value predicted by the model that assumes that the rate of increase in population is proportional to the population present at time $t$?

4. Air containing 0.06% carbon dioxide is pumped into a room whose volume is 8000 ft$^3$. The air is pumped in at a rate of 2000 ft$^3$/min, and the circulated air is then pumped out at the same rate. If there is an initial concentration of 0.2% carbon dioxide in the room, determine the subsequent amount in the room at time $t$. What is the concentration of carbon dioxide at 10 minutes? What is the steady-state, or equilibrium, concentration of carbon dioxide?

**5. Ötzi the Iceman** In September of 1991 two German tourists found the well-preserved body of a man partially frozen in a glacier in the Ötztal Alps on the border between Austria and Italy. See Figure 3.R.1. Through the carbon-dating technique it was found that the body of Ötzi the iceman—as he came to be called—contained 53% as much C-14 as found in a living person. Assume that the iceman was carbon dated in 1991. Use the method illustrated in Example 3 of Section 3.1 to find the approximate date of his death.

**FIGURE 3.R.1** Ötzi the iceman in Problem 5

**6.** In the treatment of cancer of the thyroid, the radioactive liquid Iodine-131 is often used. Suppose that after one day in storage, analysis shows that an initial amount $A_0$ of iodine-131 in a sample has decreased by 8.3%.

  **(a)** Find the amount of iodine-131 remaining in the sample after 8 days.

  **(b)** Explain the significance of the result in part (a).

**7.** Solve the differential equation

$$\frac{dy}{dx} = -\frac{y}{\sqrt{a^2 - y^2}}$$

of the tractrix. See Problem 28 in Exercises 1.3. Assume that the initial point on the $y$-axis in $(0, 10)$ and that the length of the rope is $x = 10$ ft.

**8.** Suppose a cell is suspended in a solution containing a solute of constant concentration $C_s$. Suppose further that the cell has constant volume $V$ and that the area of its permeable membrane is the constant $A$. By **Fick's law** the rate of change of its mass $m$ is directly proportional to the area $A$ and the difference $C_s - C(t)$, where $C(t)$ is the concentration of the solute inside the cell at time $t$. Find $C(t)$ if $m = V \cdot C(t)$ and $C(0) = C_0$. See Figure 3.R.2.

**FIGURE 3.R.2** Cell in Problem 8

**9.** Suppose that as a body cools, the temperature of the surrounding medium increases because it completely absorbs the heat being lost by the body. Let $T(t)$ and $T_m(t)$ be the temperatures of the body and the medium at time $t$, respectively.

If the initial temperature of the body is $T_1$ and the initial temperature of the medium is $T_2$, then it can be shown in this case that Newton's law of cooling is $dT/dt = k(T - T_m)$, $k < 0$, where $T_m = T_2 + B(T_1 - T)$, $B > 0$ is a constant.

  **(a)** The foregoing DE is autonomous. Use the phase portrait concept of Section 2.1 to determine the limiting value of the temperature $T(t)$ as $t \to \infty$. What is the limiting value of $T_m(t)$ as $t \to \infty$?

  **(b)** Verify your answers in part (a) by actually solving the differential equation.

  **(c)** Discuss a physical interpretation of your answers in part (a).

**10.** According to **Stefan's law of radiation** the absolute temperature $T$ of a body cooling in a medium at constant absolute temperature $T_m$ is given by

$$\frac{dT}{dt} = k(T^4 - T_m^4),$$

where $k$ is a constant. Stefan's law can be used over a greater temperature range than Newton's law of cooling.

  **(a)** Solve the differential equation.

  **(b)** Show that when $T - T_m$ is small in comparison to $T_m$ then Newton's law of cooling approximates Stefan's law. [*Hint:* Think binomial series of the right-hand side of the DE.]

**11.** Suppose an $RC$-series circuit has a variable resistor. If the resistance at time $t$ is defined by $R(t) = k_1 + k_2 t$, where $k_1$ and $k_2$ are known positive constants, then the differential equation in (9) of Section 3.1 becomes

$$(k_1 + k_2 t)\frac{dq}{dt} + \frac{1}{C}q = E(t),$$

where $C$ is a constant. If $E(t) = E_0$ and $q(0) = q_0$, where $E_0$ and $q_0$ are constants, then show that

$$q(t) = E_0 C + (q_0 - E_0 C)\left(\frac{k_1}{k_1 + k_2 t}\right)^{1/Ck_2}.$$

**12.** A classical problem in the calculus of variations is to find the shape of a curve $\mathscr{C}$ such that a bead, under the influence of gravity, will slide from point $A(0, 0)$ to point $B(x_1, y_1)$ in the least time. See Figure 3.R.3. It can be shown that a nonlinear differential for the shape $y(x)$ of the path is $y[1 + (y')^2] = k$, where $k$ is a constant. First solve for $dx$ in terms of $y$ and $dy$, and then use the substitution $y = k \sin^2\theta$ to obtain a parametric form of the solution. The curve $\mathscr{C}$ turns out to be a **cycloid**.

**FIGURE 3.R.3** Sliding bead in Problem 12

**13.** A model for the populations of two interacting species of animals is

$$\frac{dx}{dt} = k_1 x(\alpha - x)$$

$$\frac{dy}{dt} = k_2 xy.$$

Solve for $x$ and $y$ in terms of $t$.

**14.** Initially, two large tanks $A$ and $B$ each hold 100 gallons of brine. The well-stirred liquid is pumped between the tanks as shown in Figure 3.R.4. Use the information given in the figure to construct a mathematical model for the number of pounds of salt $x_1(t)$ and $x_2(t)$ at time $t$ in tanks $A$ and $B$, respectively.

**FIGURE 3.R.4** Mixing tanks in Problem 14

When all the curves in a family $G(x, y, c_1) = 0$ intersect orthogonally all the curves in another family $H(x, y, c_2) = 0$, the families are said to be **orthogonal trajectories** of each other. See Figure 3.R.5. If $dy/dx = f(x, y)$ is the differential equation of one family, then the differential equation for the orthogonal trajectories of this family is $dy/dx = -1/f(x, y)$. In Problems 15–18 find the differential equation of the given family by computing $dy/dx$ and eliminating $c_1$ from this equation. Then find the orthogonal trajectories of the family. Use a graphing utility to graph both families on the same set of coordinate axes.

**FIGURE 3.R.5** Orthogonal families

**15.** $y = c_1 x$

**16.** $x^2 - 2y^2 = c_1$

**17.** $y = c_1 e^x$

**18.** $y = \dfrac{1}{x + c_1}$

**19.** With the identifications $a = r$, $b = r/K$, and $a/b = K$, Figures 2.1.7 and 3.2.2 show that the logistic population model, (3) of Section 3.2, predicts that for an initial population $P_0$, $0 < P_0 < K$, regardless of how small $P_0$ is, the population increases over time but does not surpass the carrying capacity $K$. Also, for $P_0 > K$ the same model predicts that a population cannot sustain itself over time, so it decreases but yet never falls below the carrying capacity $K$ of the ecosystem. The American ecologist **Warder Clyde Allee** (1885–1955) showed that by depleting certain fisheries beyond a certain level, the fish population never recovers. How would you modify the differential equation (3) to describe a population $P$ that has these same two characteristics of (3) but additionally has a **threshold level** $A$, $0 < A < K$, below which the population cannot sustain itself and approaches extinction over time. [*Hint*: Construct a phase portrait of what you want and then form a differential equation.]

**20. Sawing Wood** A long *uniform* piece of wood (cross sections are the same) is cut through perpendicular to its length by a vertical saw blade. See Figure 3.R.6. If the friction between the sides of the saw blade and the wood through which the blade passes is ignored, then it can be assumed that the rate at which the saw blade moves through the piece of wood is

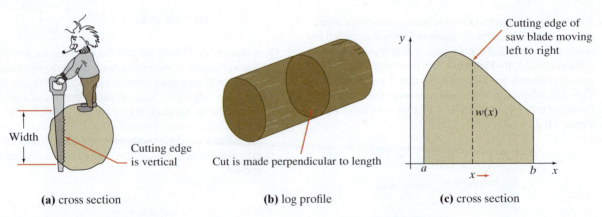

**(a)** cross section    **(b)** log profile    **(c)** cross section

**FIGURE 3.R.6** Sawing a log in Problem 20

inversely proportional to the width of the wood in contact with its cutting edge. As the blade advances through the wood (moving, say, left to right) the width of a cross section changes as a nonnegative continuous function $w$. If a cross section of the wood is described as a region in the $xy$-plane defined over an interval $[a, b]$ then, as shown in Figure 3.R.6(c), the position $x$ of the saw blade is a function of time $t$ and the vertical cut made by the blade can be represented by a vertical line segment. The length of this vertical line is the width $w(x)$ of the wood at that point. Thus the position $x(t)$ of the saw blade and the rate $dx/dt$ at which it moves to the right are related to $w(x)$ by

$$w(x)\frac{dx}{dt} = k, \quad x(0) = a.$$

Here $k$ represents the number of square units of the material removed by the saw blade per unit time.

(a) Suppose the saw is computerized and can be programmed so that $k = 1$. Find an *implicit* solution of the foregoing initial-value problem when the piece of wood is a circular log. Assume a cross section is a circle of radius 2 centered at $(0, 0)$. [*Hint*: To save time see formula 41 in the table of integrals given on the right inside page of the front cover.]

(b) Solve the implicit solution obtained in part (b) for time $t$ as a function of $x$. Graph the function $t(x)$. With the aid of the graph, approximate the time that it takes the saw to cut through this piece of wood. Then find the exact value of this time.

21. Solve the initial-value problem in Problem 20 when a cross section of a uniform piece of wood is the triangular region given in Figure 3.R.7. Assume again that $k = 1$. How long does it take to cut through this piece of wood?

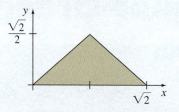

**FIGURE 3.R.7** Triangular cross section in Problem 21

22. **Chemical Kinetics**    Suppose a gas consists of molecules of type A. When the gas is heated a second substance B is formed by molecular collision. Let $A(t)$ and $B(t)$ denote, in turn, the number of molecules of types A and B present at time $t \geq 0$. A mathematical model for the rate at which the number of molecules of type A decreases is

$$\frac{dA}{dt} = -kA^2, \quad k > 0.$$

(a) Determine $A(t)$ if $A(0) = A_0$.

(b) Determine the number of molecules of substance B present at time $t$ if it is assumed that $A(t) + B(t) = A_0$.

(c) By hand, sketch rough graphs of $A(t)$ and $B(t)$ for $t \geq 0$.

# 4

# Higher-Order Differential Equations

Bill Ingalls/NASA

W e turn now to the solution of ordinary differential equations of order two or higher. In the first seven sections of this chapter we examine the underlying theory and solution methods for certain kinds of linear DEs. The chapter concludes with a brief examination of solution methods for nonlinear higher-order equations.

# 4.1 Preliminary Theory—Linear Equations

Reread (vi) of the *Remarks* in Section 1.1 and page 57 of Section 2.3.

**INTRODUCTION** In Chapter 2 we were able to solve a few first-order differential equations by recognizing them as separable, linear, exact, or having homogeneous coefficients. Even though the solutions obtained were in the form of a one-parameter family, this family, with one exception, did not represent the *general solution* of the differential equation. Recall that a **general solution** is a family of solutions defined on some interval $I$ that contains *all* solutions of the DE that are defined on $I$. Only in the case of *linear* first-order differential equations were we able to obtain general solutions by paying close attention to certain continuity conditions imposed on the coefficients in the equation. Because our primary goal in this chapter is to find general solutions of linear higher-order DEs, we first need to examine some of the basic theory of linear equations.

## 4.1.1 INITIAL-VALUE AND BOUNDARY-VALUE PROBLEMS

**INITIAL-VALUE PROBLEM** In Section 1.2 we defined an initial-value problem for a general $n$th-order differential equation. For a linear differential equation an **$n$th-order initial-value problem (IVP)** is

*Solve:*
$$a_n(x)\frac{d^n y}{dx^n} + a_{n-1}(x)\frac{d^{n-1} y}{dx^{n-1}} + \cdots + a_1(x)\frac{dy}{dx} + a_0(x)y = g(x)$$

*Subject to:* $y(x_0) = y_0, \quad y'(x_0) = y_1, \ldots, y^{(n-1)}(x_0) = y_{n-1}.$

(1)

Recall that for a problem such as this one we seek a function defined on some interval $I$, containing $x_0$, that satisfies the differential equation and the $n$ initial conditions specified at $x_0$: $y(x_0) = y_0, y'(x_0) = y_1, \ldots, y^{(n-1)}(x_0) = y_{n-1}$. We have already seen that in the case of a second-order initial-value problem a solution curve must pass through the point $(x_0, y_0)$ and have slope $y_1$ at this point.

**EXISTENCE AND UNIQUENESS** In Section 1.2 we stated a theorem that gave conditions under which the existence and uniqueness of a solution of a first-order initial-value problem were guaranteed. The theorem that follows gives sufficient conditions for the existence of a unique solution of the problem in (1).

---

**THEOREM 4.1.1** Existence of a Unique Solution

Let $a_n(x), a_{n-1}(x), \ldots, a_1(x), a_0(x)$ and $g(x)$ be continuous on an interval $I$ and let $a_n(x) \neq 0$ for every $x$ in this interval. If $x = x_0$ is any point in this interval, then a solution $y(x)$ of the initial-value problem (1) exists on the interval and is unique.

---

**EXAMPLE 1** Unique Solution of an IVP

The initial-value problem

$$3y''' + 5y'' - y' + 7y = 0, \quad y(1) = 0, \quad y'(1) = 0, \quad y''(1) = 0$$

possesses the trivial solution $y = 0$. Because the third-order equation is linear with constant coefficients, it follows that all the conditions of Theorem 4.1.1 are fulfilled. Hence $y = 0$ is the *only* solution on any interval containing $x = 1$. ∎

EXAMPLE 2    Unique Solution of an IVP

You should verify that the function $y = 3e^{2x} + e^{-2x} - 3x$ is a solution of the initial-value problem

$$y'' - 4y = 12x, \quad y(0) = 4, \quad y'(0) = 1.$$

Now the differential equation is linear, the coefficients as well as $g(x) = 12x$ are continuous, and $a_2(x) = 1 \neq 0$ on any interval $I$ containing $x = 0$. We conclude from Theorem 4.1.1 that the given function is the unique solution on $I$.    ∎

The requirements in Theorem 4.1.1 that $a_i(x)$, $i = 0, 1, 2, \ldots, n$ be continuous and $a_n(x) \neq 0$ for every $x$ in $I$ are both important. Specifically, if $a_n(x) = 0$ for some $x$ in the interval, then the solution of a linear initial-value problem may not be unique or even exist. For example, you should verify that the function $y = cx^2 + x + 3$ is a solution of the initial-value problem

$$x^2 y'' - 2xy' + 2y = 6, \quad y(0) = 3, \quad y'(0) = 1$$

on the interval $(-\infty, \infty)$ for any choice of the parameter $c$. In other words, there is no unique solution of the problem. Although most of the conditions of Theorem 4.1.1 are satisfied, the obvious difficulties are that $a_2(x) = x^2$ is zero at $x = 0$ and that the initial conditions are also imposed at $x = 0$.

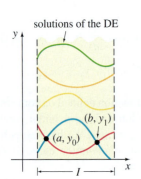

solutions of the DE

$(b, y_1)$

$(a, y_0)$

$I$

**FIGURE 4.1.1** Solution curves of a BVP that pass through two points

**BOUNDARY-VALUE PROBLEM**    Another type of problem consists of solving a linear differential equation of order two or greater in which the dependent variable $y$ or its derivatives are specified at *different points*. A problem such as

**Solve:**    $a_2(x) \dfrac{d^2y}{dx^2} + a_1(x) \dfrac{dy}{dx} + a_0(x)y = g(x)$

**Subject to:**    $y(a) = y_0, \quad y(b) = y_1$

is called a **boundary-value problem (BVP).** The prescribed values $y(a) = y_0$ and $y(b) = y_1$ are called **boundary conditions (BC).** A solution of the foregoing problem is a function satisfying the differential equation on some interval $I$, containing $a$ and $b$, whose graph passes through the two points $(a, y_0)$ and $(b, y_1)$. See Figure 4.1.1.

For a second-order differential equation other pairs of boundary conditions could be

$$y'(a) = y_0, \qquad y(b) = y_1$$
$$y(a) = y_0, \qquad y'(b) = y_1$$
$$y'(a) = y_0, \qquad y'(b) = y_1,$$

where $y_0$ and $y_1$ denote arbitrary constants. These three pairs of conditions are just special cases of the general boundary conditions

$$\alpha_1 y(a) + \beta_1 y'(a) = \gamma_1$$
$$\alpha_2 y(b) + \beta_2 y'(b) = \gamma_2.$$

The next example shows that even when the conditions of Theorem 4.1.1 are fulfilled, a boundary-value problem may have several solutions (as suggested in Figure 4.1.1), a unique solution, or no solution at all.

EXAMPLE 3    A BVP Can Have Many, One, or No Solutions

In Example 9 of Section 1.1 we saw that the two-parameter family of solutions of the differential equation $x'' + 16x = 0$ is

$$x = c_1 \cos 4t + c_2 \sin 4t. \tag{2}$$

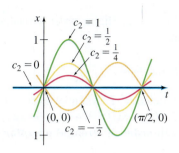

**FIGURE 4.1.2** Solution curves for BVP in part (a) of Example 3

**(a)** Suppose we now wish to determine the solution of the equation that further satisfies the boundary conditions $x(0) = 0$, $x(\pi/2) = 0$. Observe that the first condition $0 = c_1 \cos 0 + c_2 \sin 0$ implies that $c_1 = 0$, so $x = c_2 \sin 4t$. But when $t = \pi/2$, $0 = c_2 \sin 2\pi$ is satisfied for any choice of $c_2$, since $\sin 2\pi = 0$. Hence the boundary-value problem

$$x'' + 16x = 0, \quad x(0) = 0, \quad x\left(\frac{\pi}{2}\right) = 0 \tag{3}$$

has infinitely many solutions. Figure 4.1.2 shows the graphs of some of the members of the one-parameter family $x = c_2 \sin 4t$ that pass through the two points $(0, 0)$ and $(\pi/2, 0)$.

**(b)** If the boundary-value problem in (3) is changed to

$$x'' + 16x = 0, \quad x(0) = 0, \quad x\left(\frac{\pi}{8}\right) = 0, \tag{4}$$

then $x(0) = 0$ still requires $c_1 = 0$ in the solution (2). But applying $x(\pi/8) = 0$ to $x = c_2 \sin 4t$ demands that $0 = c_2 \sin(\pi/2) = c_2 \cdot 1$. Hence $x = 0$ is a solution of this new boundary-value problem. Indeed, it can be proved that $x = 0$ is the *only* solution of (4).

**(c)** Finally, if we change the problem to

$$x'' + 16x = 0, \quad x(0) = 0, \quad x\left(\frac{\pi}{2}\right) = 1, \tag{5}$$

we find again from $x(0) = 0$ that $c_1 = 0$, but applying $x(\pi/2) = 1$ to $x = c_2 \sin 4t$ leads to the contradiction $1 = c_2 \sin 2\pi = c_2 \cdot 0 = 0$. Hence the boundary-value problem (5) has no solution. ∎

### 4.1.2 HOMOGENEOUS EQUATIONS

A linear *n*th-order differential equation of the form

$$a_n(x)\frac{d^n y}{dx^n} + a_{n-1}(x)\frac{d^{n-1}y}{dx^{n-1}} + \cdots + a_1(x)\frac{dy}{dx} + a_0(x)y = 0 \tag{6}$$

is said to be **homogeneous,** whereas an equation

$$a_n(x)\frac{d^n y}{dx^n} + a_{n-1}(x)\frac{d^{n-1}y}{dx^{n-1}} + \cdots + a_1(x)\frac{dy}{dx} + a_0(x)y = g(x), \tag{7}$$

with $g(x)$ not identically zero, is said to be **nonhomogeneous.** For example, $2y'' + 3y' - 5y = 0$ is a homogeneous linear second-order differential equation, whereas $x^3 y''' + 6y' + 10y - e^x$ is a nonhomogeneous linear third-order differential equation. The word *homogeneous* in this context does not refer to coefficients that are homogeneous functions, as in Section 2.5.

We shall see that to solve a nonhomogeneous linear equation (7), we must first be able to solve the **associated homogeneous equation** (6).

To avoid needless repetition throughout the remainder of this text, we shall, as a matter of course, make the following important assumptions when stating definitions and theorems about linear equations (1). On some common interval $I$,

Please remember these two assumptions.

- the coefficient functions $a_i(x)$, $i = 0, 1, 2, \ldots, n$ and $g(x)$ are continuous;
- $a_n(x) \neq 0$ for every $x$ in the interval.

**DIFFERENTIAL OPERATORS** In calculus differentiation is often denoted by the capital letter $D$—that is, $dy/dx = Dy$. The symbol $D$ is called a **differential operator** because it transforms a differentiable function into another function. For example,

$D(\cos 4x) = -4 \sin 4x$ and $D(5x^3 - 6x^2) = 15x^2 - 12x$. Higher-order derivatives can be expressed in terms of $D$ in a natural manner:

$$\frac{d}{dx}\left(\frac{dy}{dx}\right) = \frac{d^2y}{dx^2} = D(Dy) = D^2y \quad \text{and, in general,} \quad \frac{d^ny}{dx^n} = D^ny,$$

where $y$ represents a sufficiently differentiable function. Polynomial expressions involving $D$, such as $D + 3$, $D^2 + 3D - 4$, and $5x^3D^3 - 6x^2D^2 + 4xD + 9$, are also differential operators. In general, we define an **$n$th-order differential operator** or **polynomial operator** to be

$$L = a_n(x)D^n + a_{n-1}(x)D^{n-1} + \cdots + a_1(x)D + a_0(x). \tag{8}$$

As a consequence of two basic properties of differentiation, $D(cf(x)) = cDf(x)$, $c$ is a constant, and $D\{f(x) + g(x)\} = Df(x) + Dg(x)$, the differential operator $L$ possesses a linearity property; that is, $L$ operating on a linear combination of two differentiable functions is the same as the linear combination of $L$ operating on the individual functions. In symbols this means that

$$L\{\alpha f(x) + \beta g(x)\} = \alpha L(f(x)) + \beta L(g(x)), \tag{9}$$

where $\alpha$ and $\beta$ are constants. Because of (9) we say that the $n$th-order differential operator $L$ is a **linear operator.**

**DIFFERENTIAL EQUATIONS**   Any linear differential equation can be expressed in terms of the $D$ notation. For example, the differential equation $y'' + 5y' + 6y = 5x - 3$ can be written as $D^2y + 5Dy + 6y = 5x - 3$ or $(D^2 + 5D + 6)y = 5x - 3$. Using (8), we can write the linear $n$th-order differential equations (6) and (7) compactly as

$$L(y) = 0 \quad \text{and} \quad L(y) = g(x),$$

respectively.

**SUPERPOSITION PRINCIPLE**   In the next theorem we see that the sum, or **superposition,** of two or more solutions of a homogeneous linear differential equation is also a solution.

---

**THEOREM 4.1.2   Superposition Principle—Homogeneous Equations**

Let $y_1, y_2, \ldots, y_k$ be solutions of the homogeneous $n$th-order differential equation (6) on an interval $I$. Then the linear combination

$$y = c_1y_1(x) + c_2y_2(x) + \cdots + c_ky_k(x),$$

where the $c_i$, $i = 1, 2, \ldots, k$ are arbitrary constants, is also a solution on the interval.

---

**PROOF**   We prove the case $k = 2$. Let $L$ be the differential operator defined in (8), and let $y_1(x)$ and $y_2(x)$ be solutions of the homogeneous equation $L(y) = 0$. If we define $y = c_1y_1(x) + c_2y_2(x)$, then by linearity of $L$ we have

$$L(y) = L\{c_1y_1(x) + c_2y_2(x)\} = c_1 L(y_1) + c_2 L(y_2) = c_1 \cdot 0 + c_2 \cdot 0 = 0. \quad \blacksquare$$

---

**COROLLARIES TO THEOREM 4.1.2**

**(A)** A constant multiple $y = c_1y_1(x)$ of a solution $y_1(x)$ of a homogeneous linear differential equation is also a solution.

**(B)** A homogeneous linear differential equation always possesses the trivial solution $y = 0$.

**EXAMPLE 4**  **Superposition—Homogeneous DE**

The functions $y_1 = x^2$ and $y_2 = x^2 \ln x$ are both solutions of the homogeneous linear equation $x^3 y''' - 2xy' + 4y = 0$ on the interval $(0, \infty)$. By the superposition principle the linear combination

$$y = c_1 x^2 + c_2 x^2 \ln x$$

is also a solution of the equation on the interval. ∎

The function $y = e^{7x}$ is a solution of $y'' - 9y' + 14y = 0$. Because the differential equation is linear and homogeneous, the constant multiple $y = ce^{7x}$ is also a solution. For various values of $c$ we see that $y = 9e^{7x}$, $y = 0$, $y = -\sqrt{5}e^{7x}$, ... are all solutions of the equation.

**LINEAR DEPENDENCE AND LINEAR INDEPENDENCE**  The next two concepts are basic to the study of linear differential equations.

> **DEFINITION 4.1.1**  Linear Dependence/Independence
>
> A set of functions $f_1(x), f_2(x), \ldots, f_n(x)$ is said to be **linearly dependent** on an interval $I$ if there exist constants $c_1, c_2, \ldots, c_n$, not all zero, such that
>
> $$c_1 f_1(x) + c_2 f_2(x) + \cdots + c_n f_n(x) = 0$$
>
> for every $x$ in the interval. If the set of functions is not linearly dependent on the interval, it is said to be **linearly independent.**

In other words, a set of functions is linearly independent on an interval $I$ if the only constants for which

$$c_1 f_1(x) + c_2 f_2(x) + \cdots + c_n f_n(x) = 0$$

for every $x$ in the interval are $c_1 = c_2 = \cdots = c_n = 0$.

It is easy to understand these definitions for a set consisting of two functions $f_1(x)$ and $f_2(x)$. If the set of functions is linearly dependent on an interval, then there exist constants $c_1$ and $c_2$ that are not both zero such that for every $x$ in the interval, $c_1 f_1(x) + c_2 f_2(x) = 0$. Therefore if we assume that $c_1 \neq 0$, it follows that $f_1(x) = (-c_2/c_1)f_2(x)$; that is, *if a set of two functions is linearly dependent, then one function is simply a constant multiple of the other.* Conversely, if $f_1(x) = c_2 f_2(x)$ for some constant $c_2$, then $(-1) \cdot f_1(x) + c_2 f_2(x) = 0$ for every $x$ in the interval. Hence the set of functions is linearly dependent because at least one of the constants (namely, $c_1 = -1$) is not zero. We conclude that *a set of two functions $f_1(x)$ and $f_2(x)$ is linearly independent when neither function is a constant multiple of the other* on the interval. For example, the set of functions $f_1(x) = \sin 2x$, $f_2(x) = \sin x \cos x$ is linearly dependent on $(-\infty, \infty)$ because $f_1(x)$ is a constant multiple of $f_2(x)$. Recall from the double-angle formula for the sine that $\sin 2x = 2 \sin x \cos x$. On the other hand, the set of functions $f_1(x) = x$, $f_2(x) = |x|$ is linearly independent on $(-\infty, \infty)$. Inspection of Figure 4.1.3 should convince you that neither function is a constant multiple of the other on the interval.

It follows from the preceding discussion that the quotient $f_2(x)/f_1(x)$ is not a constant on an interval on which the set $f_1(x), f_2(x)$ is linearly independent. This little fact will be used in the next section.

**(a)**

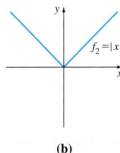

**(b)**

**FIGURE 4.1.3** Set consisting of $f_1$ and $f_2$ is linearly independent on $(-\infty, \infty)$

**EXAMPLE 5**  **Linearly Dependent Set of Functions**

The set of functions $f_1(x) = \cos^2 x$, $f_2(x) = \sin^2 x$, $f_3(x) = \sec^2 x$, $f_4(x) = \tan^2 x$ is linearly dependent on the interval $(-\pi/2, \pi/2)$ because

$$c_1 \cos^2 x + c_2 \sin^2 x + c_3 \sec^2 x + c_4 \tan^2 x = 0$$

for every real number $x$ in the interval when $c_1 = c_2 = 1$, $c_3 = -1$, $c_4 = 1$. We used here $\cos^2 x + \sin^2 x = 1$ and $1 + \tan^2 x = \sec^2 x$. ■

A set of $n$ functions $f_1(x), f_2(x), \ldots, f_n(x)$ is linearly dependent on an interval $I$ if at least one of the functions can be expressed as a linear combination of the remaining functions. For example, three functions $f_1(x)$, $f_2(x)$, and $f_3(x)$ are linearly dependent on $I$ if at least one of these functions is a linear combination of the other two, say,

$$f_3(x) = c_1 f_1(x) + c_2 f_2(x)$$

for all $x$ in $I$. A set of $n$ functions is linearly independent on $I$ if no one function is a linear combination of the other functions.

---

### EXAMPLE 6   Linearly Dependent Set of Functions

The set of functions $f_1(x) = \sqrt{x} + 5$, $f_2(x) = \sqrt{x} + 5x$, $f_3(x) = x - 1$, $f_4(x) = x^2$ is linearly dependent on the interval $(0, \infty)$ because $f_2$ can be written as a linear combination of $f_1$, $f_3$, and $f_4$. Observe that

$$f_2(x) = 1 \cdot f_1(x) + 5 \cdot f_3(x) + 0 \cdot f_4(x)$$

for every $x$ in the interval $(0, \infty)$. ■

**SOLUTIONS OF DIFFERENTIAL EQUATIONS** We are primarily interested in linearly independent functions or, more to the point, linearly independent *solutions* of a linear differential equation. Although we could always appeal directly to Definition 4.1.1, it turns out that the question of whether the set of $n$ solutions $y_1, y_2, \ldots, y_n$ of a homogeneous linear $n$th-order differential equation (6) is linearly independent can be settled somewhat mechanically by using a determinant.

---

**DEFINITION 4.1.2** Wronskian

Suppose each of the functions $f_1(x), f_2(x), \ldots, f_n(x)$ possesses at least $n - 1$ derivatives. The determinant

$$W(f_1, f_2, \ldots, f_n) = \begin{vmatrix} f_1 & f_2 & \cdots & f_n \\ f_1' & f_2' & \cdots & f_n' \\ \vdots & \vdots & & \vdots \\ f_1^{(n-1)} & f_2^{(n-1)} & \cdots & f_n^{(n-1)} \end{vmatrix},$$

where the primes denote derivatives, is called the **Wronskian** of the functions.

---

The Wronskian determinant is named after the Polish philosopher and mathematician **Jósef Maria Hoëné-Wronski** (1778–1853).

---

**THEOREM 4.1.3** Criterion for Linearly Independent Solutions

Let $y_1, y_2, \ldots, y_n$ be $n$ solutions of the homogeneous linear $n$th-order differential equation (6) on an interval $I$. Then the set of solutions is **linearly independent** on $I$ if and only if $W(y_1, y_2, \ldots, y_n) \neq 0$ for every $x$ in the interval.

It follows from Theorem 4.1.3 that when $y_1, y_2, \ldots, y_n$ are $n$ solutions of (6) on an interval $I$, the Wronskian $W(y_1, y_2, \ldots, y_n)$ is either identically zero or never zero on the interval. Thus, if we can show that $W(y_1, y_2, \ldots, y_n) \neq 0$ for *some* $x_0$ in $I$, then the solutions $y_1, y_2, \ldots, y_n$ are linearly independent on $I$. For example, the functions

$$y_1(x) = \frac{\cos(2\ln x)}{x^3}, \quad y_2(x) = \frac{\sin(2\ln x)}{x^3}$$

and solutions of the differential equation

$$x^2 y'' + 7xy' + 13y = 0$$

on the interval $(0, \infty)$. Note that the coefficient functions $a_2(x) = x^2$, $a_1(x) = 7x$, and $a_0(x) = 13$ are continuous on $(0, \infty)$ and that $a_2(x) \neq 0$ for every value of $x$ in the interval. The Wronskian is

$$W(y_1(x), y_2(x)) = \begin{vmatrix} \dfrac{\cos(2\ln x)}{x^3} & \dfrac{\sin(2\ln x)}{x^3} \\ \dfrac{-2x^2 \sin(2\ln x) - 3x^2 \cos(2\ln x)}{x^6} & \dfrac{2x^2 \cos(2\ln x) - 3x^2 \sin(2\ln x)}{x^6} \end{vmatrix}.$$

Rather than expanding this unwieldy determinant, we choose $x = 1$ in the interval $(0, \infty)$ and find

$$W(y_1(1), y_2(1)) = \begin{vmatrix} 1 & 0 \\ -3 & 2 \end{vmatrix} = 2.$$

The fact that $W(y_1(1), y_2(1)) = 2 \neq 0$ is sufficient to conclude that $y_1(x)$ and $y_2(x)$ are linearly independent on $(0, \infty)$.

A set of $n$ linearly independent solutions of a homogeneous linear $n$th-order differential equation is given a special name.

### DEFINITION 4.1.3 Fundamental Set of Solutions

Any set $y_1, y_2, \ldots, y_n$ of $n$ linearly independent solutions of the homogeneous linear $n$th-order differential equation (6) on an interval $I$ is said to be a **fundamental set of solutions** on the interval.

The basic question of whether a fundamental set of solutions exists for a linear equation is answered in the next theorem.

### THEOREM 4.1.4 Existence of a Fundamental Set

There exists a fundamental set of solutions for the homogeneous linear $n$th-order differential equation (6) on an interval $I$.

Analogous to the fact that any vector in three dimensions can be expressed as a linear combination of the *linearly independent* vectors **i, j, k,** any solution of an $n$th-order homogeneous linear differential equation on an interval $I$ can be expressed as a linear combination of $n$ linearly independent solutions on $I$. In other words, $n$ linearly independent solutions $y_1, y_2, \ldots, y_n$ are the basic building blocks for the general solution of the equation.

---

**THEOREM 4.1.5** General Solution—Homogeneous Equations

Let $y_1, y_2, \ldots, y_n$ be a fundamental set of solutions of the homogeneous linear $n$th-order differential equation (6) on an interval $I$. Then the **general solution** of the equation on the interval is

$$y = c_1 y_1(x) + c_2 y_2(x) + \cdots + c_n y_n(x),$$

where $c_i, i = 1, 2, \ldots, n$ are arbitrary constants.

---

Theorem 4.1.5 states that if $Y(x)$ is any solution of (6) on the interval, then constants $C_1, C_2, \ldots, C_n$ can always be found so that

$$Y(x) = C_1 y_1(x) + C_2 y_2(x) + \cdots + C_n y_n(x).$$

We will prove the case when $n = 2$.

**PROOF** Let $Y$ be a solution and let $y_1$ and $y_2$ be linearly independent solutions of $a_2 y'' + a_1 y' + a_0 y = 0$ on an interval $I$. Suppose that $x = t$ is a point in $I$ for which $W(y_1(t), y_2(t)) \neq 0$. Suppose also that $Y(t) = k_1$ and $Y'(t) = k_2$. If we now examine the equations

$$C_1 y_1(t) + C_2 y_2(t) = k_1$$

$$C_1 y_1'(t) + C_2 y_2'(t) = k_2,$$

it follows that we can determine $C_1$ and $C_2$ uniquely, provided that the determinant of the coefficients satisfies

$$\begin{vmatrix} y_1(t) & y_2(t) \\ y_1'(t) & y_2'(t) \end{vmatrix} \neq 0.$$

But this determinant is simply the Wronskian evaluated at $x = t$, and by assumption, $W \neq 0$. If we define $G(x) = C_1 y_1(x) + C_2 y_2(x)$, we observe that $G(x)$ satisfies the differential equation since it is a superposition of two known solutions; $G(x)$ satisfies the initial conditions

$$G(t) = C_1 y_1(t) + C_2 y_2(t) = k_1 \quad \text{and} \quad G'(t) = C_1 y_1'(t) + C_2 y_2'(t) = k_2;$$

and $Y(x)$ satisfies the *same* linear equation and the *same* initial conditions. Because the solution of this linear initial-value problem is unique (Theorem 4.1.1), we have $Y(x) = G(x)$ or $Y(x) = C_1 y_1(x) + C_2 y_2(x)$. ∎

---

**EXAMPLE 7** General Solution of a Homogeneous DE

The functions $y_1 = e^{3x}$ and $y_2 = e^{-3x}$ are both solutions of the homogeneous linear equation $y'' - 9y = 0$ on the interval $(-\infty, \infty)$. By inspection the solutions are linearly independent on the $x$-axis. This fact can be corroborated by observing that the Wronskian

$$W(e^{3x}, e^{-3x}) = \begin{vmatrix} e^{3x} & e^{-3x} \\ 3e^{3x} & -3e^{-3x} \end{vmatrix} = -6 \neq 0$$

for every $x$. We conclude that $y_1$ and $y_2$ form a fundamental set of solutions, and consequently, $y = c_1 e^{3x} + c_2 e^{-3x}$ is the general solution of the equation on the interval. ∎

EXAMPLE 8  A Solution Obtained from a General Solution

The function $y = 4\sinh 3x - 5e^{-3x}$ is a solution of the differential equation in Example 7. (Verify this.) In view of Theorem 4.1.5 we must be able to obtain this solution from the general solution $y = c_1e^{3x} + c_2e^{-3x}$. Observe that if we choose $c_1 = 2$ and $c_2 = -7$, then $y = 2e^{3x} - 7e^{-3x}$ can be rewritten as

$$y = 2e^{3x} - 2e^{-3x} - 5e^{-3x} = 4\left(\frac{e^{3x} - e^{-3x}}{2}\right) - 5e^{-3x}.$$

The last expression is recognized as $y = 4\sinh 3x - 5e^{-3x}$.  ∎

EXAMPLE 9  General Solution of a Homogeneous DE

The functions $y_1 = e^x$, $y_2 = e^{2x}$, and $y_3 = e^{3x}$ satisfy the third-order equation $y''' - 6y'' + 11y' - 6y = 0$. Since

$$W(e^x, e^{2x}, e^{3x}) = \begin{vmatrix} e^x & e^{2x} & e^{3x} \\ e^x & 2e^{2x} & 3e^{3x} \\ e^x & 4e^{2x} & 9e^{3x} \end{vmatrix} = 2e^{6x} \neq 0$$

for every real value of $x$, the functions $y_1$, $y_2$, and $y_3$ form a fundamental set of solutions on $(-\infty, \infty)$. We conclude that $y = c_1e^x + c_2e^{2x} + c_3e^{3x}$ is the general solution of the differential equation on the interval.  ∎

### 4.1.3  NONHOMOGENEOUS EQUATIONS

Any function $y_p$, free of arbitrary parameters, that satisfies (7) is said to be a **particular solution** of the equation. For example, it is a straightforward task to show that the constant function $y_p = 3$ is a particular solution of the nonhomogeneous equation $y'' + 9y = 27$.

Now if $y_1, y_2, \ldots, y_k$ are solutions of (6) on an interval $I$ and $y_p$ is any particular solution of (7) on $I$, then the linear combination

$$y = c_1y_1(x) + c_2y_2(x) + \cdots + c_ky_k(x) + y_p(x) \tag{10}$$

is also a solution of the nonhomogeneous equation (7). If you think about it, this makes sense, because the linear combination $c_1y_1(x) + c_2y_2(x) + \cdots + c_ky_k(x)$ is transformed into 0 by the operator $L = a_nD^n + a_{n-1}D^{n-1} + \cdots + a_1D + a_0$, whereas $y_p$ is transformed into $g(x)$. If we use $k = n$ linearly independent solutions of the $n$th-order equation (6), then the expression in (10) becomes the general solution of (7).

THEOREM 4.1.6  General Solution—Nonhomogeneous Equations

Let $y_p$ be any particular solution of the nonhomogeneous linear $n$th-order differential equation (7) on an interval $I$, and let $y_1, y_2, \ldots, y_n$ be a fundamental set of solutions of the associated homogeneous differential equation (6) on $I$. Then the **general solution** of the equation on the interval is

$$y = c_1y_1(x) + c_2y_2(x) + \cdots + c_ny_n(x) + y_p(x),$$

where the $c_i$, $i = 1, 2, \ldots, n$ are arbitrary constants.

**PROOF**  Let $L$ be the differential operator defined in (8) and let $Y(x)$ and $y_p(x)$ be particular solutions of the nonhomogeneous equation $L(y) = g(x)$. If we define $u(x) = Y(x) - y_p(x)$, then by linearity of $L$ we have

$$L(u) = L\{Y(x) - y_p(x)\} = L(Y(x)) - L(y_p(x)) = g(x) - g(x) = 0.$$

This shows that $u(x)$ is a solution of the homogeneous equation $L(y) = 0$. Hence by Theorem 4.1.5, $u(x) = c_1y_1(x) + c_2y_2(x) + \cdots + c_ny_n(x)$, and so

$$Y(x) - y_p(x) = c_1y_1(x) + c_2y_2(x) + \cdots + c_ny_n(x)$$

or $$Y(x) = c_1y_1(x) + c_2y_2(x) + \cdots + c_ny_n(x) + y_p(x).$$ ■

**COMPLEMENTARY FUNCTION**   We see in Theorem 4.1.6 that the general solution of a nonhomogeneous linear equation consists of the sum of two functions:

$$y = c_1y_1(x) + c_2y_2(x) + \cdots + c_ny_n(x) + y_p(x) = y_c(x) + y_p(x).$$

The linear combination $y_c(x) = c_1y_1(x) + c_2y_2(x) + \cdots + c_ny_n(x)$, which is the general solution of (6), is called the **complementary function** for equation (7). In other words, to solve a nonhomogeneous linear differential equation, we first solve the associated homogeneous equation and then find any particular solution of the nonhomogeneous equation. The general solution of the nonhomogeneous equation is then

$$y = \textit{complementary function} + \textit{any particular solution}$$
$$= y_c + y_p.$$

**EXAMPLE 10**   General Solution of a Nonhomogeneous DE

By substitution the function $y_p = -\frac{11}{12} - \frac{1}{2}x$ is readily shown to be a particular solution of the nonhomogeneous equation

$$y''' - 6y'' + 11y' - 6y = 3x. \tag{11}$$

To write the general solution of (11), we must also be able to solve the associated homogeneous equation

$$y''' - 6y'' + 11y' - 6y = 0.$$

But in Example 9 we saw that the general solution of this latter equation on the interval $(-\infty, \infty)$ was $y_c = c_1e^x + c_2e^{2x} + c_3e^{3x}$. Hence the general solution of (11) on the interval is

$$y = y_c + y_p = c_1e^x + c_2e^{2x} + c_3e^{3x} - \frac{11}{12} - \frac{1}{2}x.$$ ■

**ANOTHER SUPERPOSITION PRINCIPLE**   The last theorem of this discussion will be useful in Section 4.4 when we consider a method for finding particular solutions of nonhomogeneous equations.

**THEOREM 4.1.7** Superposition Principle—Nonhomogeneous Equations

Let $y_{p_1}, y_{p_2}, \ldots, y_{p_k}$ be $k$ particular solutions of the nonhomogeneous linear $n$th-order differential equation (7) on an interval $I$ corresponding, in turn, to $k$ distinct functions $g_1, g_2, \ldots, g_k$. That is, suppose $y_{p_i}$ denotes a particular solution of the corresponding differential equation

$$a_n(x)y^{(n)} + a_{n-1}(x)y^{(n-1)} + \cdots + a_1(x)y' + a_0(x)y = g_i(x), \tag{12}$$

where $i = 1, 2, \ldots, k$. Then

$$y_p(x) = y_{p_1}(x) + y_{p_2}(x) + \cdots + y_{p_k}(x) \tag{13}$$

is a particular solution of

$$a_n(x)y^{(n)} + a_{n-1}(x)y^{(n-1)} + \cdots + a_1(x)y' + a_0(x)y$$

$$= g_1(x) + g_2(x) + \cdots + g_k(x). \tag{14}$$

**PROOF**   We prove the case $k = 2$. Let $L$ be the differential operator defined in (8) and let $y_{p_1}(x)$ and $y_{p_2}(x)$ be particular solutions of the nonhomogeneous equations $L(y) = g_1(x)$ and $L(y) = g_2(x)$, respectively. If we define $y_p(x) = y_{p_1}(x) + y_{p_2}(x)$, we want to show that $y_p$ is a particular solution of $L(y) = g_1(x) + g_2(x)$. The result follows again by the linearity of the operator $L$:

$$L(y_p) = L\{y_{p_1}(x) + y_{p_2}(x)\} = L(y_{p_1}(x)) + L(y_{p_2}(x)) = g_1(x) + g_2(x). \quad \blacksquare$$

---

**EXAMPLE 11**   **Superposition—Nonhomogeneous DE**

---

You should verify that

$y_{p_1} = -4x^2$   is a particular solution of   $y'' - 3y' + 4y = -16x^2 + 24x - 8$,

$y_{p_2} = e^{2x}$   is a particular solution of   $y'' - 3y' + 4y = 2e^{2x}$,

$y_{p_3} = xe^x$   is a particular solution of   $y'' - 3y' + 4y = 2xe^x - e^x$.

It follows from (13) of Theorem 4.1.7 that the superposition of $y_{p_1}$, $y_{p_2}$, and $y_{p_3}$,

$$y = y_{p_1} + y_{p_2} + y_{p_3} = -4x^2 + e^{2x} + xe^x,$$

is a solution of

$$y'' - 3y' + 4y = \underbrace{-16x^2 + 24x - 8}_{g_1(x)} + \underbrace{2e^{2x}}_{g_2(x)} + \underbrace{2xe^x - e^x}_{g_3(x)}. \quad \blacksquare$$

**NOTE**   If the $y_{p_i}$ are particular solutions of (12) for $i = 1, 2, \ldots, k$, then the linear combination

$$y_p = c_1 y_{p_1} + c_2 y_{p_2} + \cdots + c_k y_{p_k},$$

where the $c_i$ are constants, is also a particular solution of (14) when the right-hand member of the equation is the linear combination

$$c_1 g_1(x) + c_2 g_2(x) + \cdots + c_k g_k(x).$$

Before we actually start solving homogeneous and nonhomogeneous linear differential equations, we need one additional bit of theory, which is presented in the next section.

---

**REMARKS**

This remark is a continuation of the brief discussion of dynamical systems given at the end of Section 1.3.

A dynamical system whose rule or mathematical model is a linear $n$th-order differential equation

$$a_n(t)y^{(n)} + a_{n-1}(t)y^{(n-1)} + \cdots + a_1(t)y' + a_0(t)y = g(t)$$

is said to be an $n$th-order **linear system.** The $n$ time-dependent functions $y(t), y'(t), \ldots, y^{(n-1)}(t)$ are the **state variables** of the system. Recall that their values at some time $t$ give the **state of the system.** The function $g$ is called the **input function** or **forcing function.** A solution $y(t)$ of the differential equation is said to be the **output** or **response** of the system. Under the conditions stated

*(continued)*

in Theorem 4.1.1, the output or response $y(t)$ is uniquely determined by the input and the state of the system prescribed at a time $t_0$—that is, by the initial conditions $y(t_0), y'(t_0), \ldots, y^{(n-1)}(t_0)$.

For a dynamical system to be a linear system, it is necessary that the superposition principle (Theorem 4.1.7) holds in the system; that is, the response of the system to a superposition of inputs is a superposition of outputs. We have already examined some simple linear systems in Section 3.1 (linear first-order equations); in Section 5.1 we examine linear systems in which the mathematical models are second-order differential equations.

## EXERCISES 4.1

*Answers to selected odd-numbered problems begin on page ANS-4.*

### 4.1.1 Initial-Value and Boundary-Value Problems

In Problems 1–4 the given family of functions is the general solution of the differential equation on the indicated interval. Find a member of the family that is a solution of the initial-value problem.

**1.** $y = c_1 e^x + c_2 e^{-x}$, $(-\infty, \infty)$;

$y'' - y = 0$, $y(0) = 0$, $y'(0) = 1$

**2.** $y = c_1 e^{4x} + c_2 e^{-x}$, $(-\infty, \infty)$;

$y'' - 3y' - 4y = 0$, $y(0) = 1$, $y'(0) = 2$

**3.** $y = c_1 x + c_2 x \ln x$, $(0, \infty)$;

$x^2 y'' - xy' + y = 0$, $y(1) = 3$, $y'(1) = -1$

**4.** $y = c_1 + c_2 \cos x + c_3 \sin x$, $(-\infty, \infty)$;

$y''' + y' = 0$, $y(\pi) = 0$, $y'(\pi) = 2$, $y''(\pi) = -1$

**5.** Given that $y = c_1 + c_2 x^2$ is a two-parameter family of solutions of $xy'' - y' = 0$ on the interval $(-\infty, \infty)$, show that constants $c_1$ and $c_2$ cannot be found so that a member of the family satisfies the initial conditions $y(0) = 0$, $y'(0) = 1$. Explain why this does not violate Theorem 4.1.1.

**6.** Find two members of the family of solutions in Problem 5 that satisfy the initial conditions $y(0) = 0$, $y'(0) = 0$.

**7.** Given that $x(t) = c_1 \cos \omega t + c_2 \sin \omega t$ is the general solution of $x'' + \omega^2 x = 0$ on the interval $(-\infty, \infty)$, show that a solution satisfying the initial conditions $x(0) = x_0$, $x'(0) = x_1$ is given by

$$x(t) = x_0 \cos \omega t + \frac{x_1}{\omega} \sin \omega t.$$

**8.** Use the general solution of $x'' + \omega^2 x = 0$ given in Problem 7 to show that a solution satisfying the initial conditions $x(t_0) = x_0$, $x'(t_0) = x_1$ is the solution given in Problem 7 shifted by an amount $t_0$:

$$x(t) = x_0 \cos \omega(t - t_0) + \frac{x_1}{\omega} \sin \omega(t - t_0).$$

In Problems 9 and 10 find an interval centered about $x = 0$ for which the given initial-value problem has a unique solution.

**9.** $(x - 2)y'' + 3y = x$, $y(0) = 0$, $y'(0) = 1$

**10.** $y'' + (\tan x)y = e^x$, $y(0) = 1$, $y'(0) = 0$

**11. (a)** Use the family in Problem 1 to find a solution of $y'' - y = 0$ that satisfies the boundary conditions $y(0) = 0$, $y(1) = 1$.

**(b)** The DE in part (a) has the alternative general solution $y = c_3 \cosh x + c_4 \sinh x$ on $(-\infty, \infty)$. Use this family to find a solution that satisfies the boundary conditions in part (a).

**(c)** Show that the solutions in parts (a) and (b) are equivalent

**12.** Use the family in Problem 5 to find a solution of $xy'' - y' = 0$ that satisfies the boundary conditions $y(0) = 1$, $y'(1) = 6$.

In Problems 13 and 14 the given two-parameter family is a solution of the indicated differential equation on the interval $(-\infty, \infty)$. Determine whether a member of the family can be found that satisfies the boundary conditions.

**13.** $y = c_1 e^x \cos x + c_2 e^x \sin x$; $y'' - 2y' + 2y = 0$

**(a)** $y(0) = 1$, $y'(\pi) = 0$

**(b)** $y(0) = 1$, $y(\pi) = -1$

**(c)** $y(0) = 1$, $y(\pi/2) = 1$

**(d)** $y(0) = 0$, $y(\pi) = 0$.

**14.** $y = c_1 x^2 + c_2 x^4 + 3$; $x^2 y'' - 5xy' + 8y = 24$

**(a)** $y(-1) = 0$, $y(1) = 4$

**(b)** $y(0) = 1$, $y(1) = 2$

**(c)** $y(0) = 3$, $y(1) = 0$

**(d)** $y(1) = 3$, $y(2) = 15$

### 4.1.2 Homogeneous Equations

In Problems 15–22 determine whether the given set of functions is linearly independent on the interval $(-\infty, \infty)$.

**15.** $f_1(x) = x$, $f_2(x) = x^2$, $f_3(x) = 4x - 3x^2$

**16.** $f_1(x) = 0$, $f_2(x) = x$, $f_3(x) = e^x$

**17.** $f_1(x) = 5$, $f_2(x) = \cos^2 x$, $f_3(x) = \sin^2 x$

**18.** $f_1(x) = \cos 2x$, $f_2(x) = 1$, $f_3(x) = \cos^2 x$

**19.** $f_1(x) = x$, $f_2(x) = x - 1$, $f_3(x) = x + 3$

**20.** $f_1(x) = 2 + x$, $f_2(x) = 2 + |x|$

**21.** $f_1(x) = 1 + x$, $f_2(x) = x$, $f_3(x) = x^2$

**22.** $f_1(x) = e^x$, $f_2(x) = e^{-x}$, $f_3(x) = \sinh x$

In Problems 23–30 verify that the given functions form a fundamental set of solutions of the differential equation on the indicated interval. Form the general solution.

**23.** $y'' - y' - 12y = 0$; $e^{-3x}, e^{4x}, (-\infty, \infty)$

**24.** $y'' - 4y = 0$; $\cosh 2x, \sinh 2x, (-\infty, \infty)$

**25.** $y'' - 2y' + 5y = 0$; $e^x \cos 2x, e^x \sin 2x, (-\infty, \infty)$

**26.** $4y'' - 4y' + y = 0$; $e^{x/2}, xe^{x/2}, (-\infty, \infty)$

**27.** $x^2 y'' - 6xy' + 12y = 0$; $x^3, x^4, (0, \infty)$

**28.** $x^2 y'' + xy' + y = 0$; $\cos(\ln x), \sin(\ln x), (0, \infty)$

**29.** $x^3 y''' + 6x^2 y'' + 4xy' - 4y = 0$; $x, x^{-2}, x^{-2} \ln x, (0, \infty)$

**30.** $y^{(4)} + y'' = 0$; $1, x, \cos x, \sin x, (-\infty, \infty)$

### 4.1.3 Nonhomogeneous Equations

In Problems 31–34 verify that the given two-parameter family of functions is the general solution of the nonhomogeneous differential equation on the indicated interval.

**31.** $y'' - 7y' + 10y = 24e^x$;
$y = c_1 e^{2x} + c_2 e^{5x} + 6e^x, (-\infty, \infty)$

**32.** $y'' + y = \sec x$;
$y = c_1 \cos x + c_2 \sin x + x \sin x + (\cos x)\ln(\cos x), (-\pi/2, \pi/2)$

**33.** $y'' - 4y' + 4y = 2e^{2x} + 4x - 12$;
$y = c_1 e^{2x} + c_2 xe^{2x} + x^2 e^{2x} + x - 2, (-\infty, \infty)$

**34.** $2x^2 y'' + 5xy' + y = x^2 - x$;
$y = c_1 x^{-1/2} + c_2 x^{-1} + \frac{1}{15}x^2 - \frac{1}{6}x, (0, \infty)$

**35. (a)** Verify that $y_{p_1} = 3e^{2x}$ and $y_{p_2} = x^2 + 3x$ are, respectively, particular solutions of

$$y'' - 6y' + 5y = -9e^{2x}$$

and $\quad y'' - 6y' + 5y = 5x^2 + 3x - 16$.

**(b)** Use part (a) to find particular solutions of

$$y'' - 6y' + 5y = 5x^2 + 3x - 16 - 9e^{2x}$$

and $\quad y'' - 6y' + 5y = -10x^2 - 6x + 32 + e^{2x}$.

**36. (a)** By inspection find a particular solution of

$$y'' + 2y = 10.$$

**(b)** By inspection find a particular solution of

$$y'' + 2y = -4x.$$

**(c)** Find a particular solution of $y'' + 2y = -4x + 10$.

**(d)** Find a particular solution of $y'' + 2y = 8x + 5$.

## Discussion Problems

**37.** Let $n = 1, 2, 3, \ldots$. Discuss how the observations $D^n x^{n-1} = 0$ and $D^n x^n = n!$ can be used to find the general solutions of the given differential equations.

**(a)** $y'' = 0$    **(b)** $y''' = 0$    **(c)** $y^{(4)} = 0$

**(d)** $y'' = 2$    **(e)** $y''' = 6$    **(f)** $y^{(4)} = 24$

**38.** Suppose that $y_1 = e^x$ and $y_2 = e^{-x}$ are two solutions of a homogeneous linear differential equation. Explain why $y_3 = \cosh x$ and $y_4 = \sinh x$ are also solutions of the equation.

**39. (a)** Verify that $y_1 = x^3$ and $y_2 = |x|^3$ are linearly independent solutions of the differential equation $x^2 y'' - 4xy' + 6y = 0$ on the interval $(-\infty, \infty)$.

**(b)** For the functions $y_1$ and $y_2$ in part (a), show that $W(y_1, y_2) = 0$ for every real number $x$. Does this result violate Theorem 4.1.3? Explain.

**(c)** Verify that $Y_1 = x^3$ and $Y_2 = x^2$ are also linearly independent solutions of the differential equation in part (a) on the interval $(-\infty, \infty)$.

**(d)** Besides the functions $y_1$, $y_2$, $Y_1$, and $Y_2$ in parts (a) and (c), find a solution of the differential equation that satisfies $y(0) = 0, y'(0) = 0$.

**(e)** By the superposition principle, Theorem 4.1.2, both linear combinations $y = c_1 y_1 + c_2 y_2$ and $Y = c_1 Y_1 + c_2 Y_2$ are solutions of the differential equation. Discuss whether one, both, or neither of the linear combinations is a general solution of the differential equation on the interval $(-\infty, \infty)$.

**40.** Is the set of functions $f_1(x) = e^{x+2}, f_2(x) = e^{x-3}$ linearly dependent or linearly independent on $(-\infty, \infty)$? Discuss.

**41.** Suppose $y_1, y_2, \ldots, y_k$ are $k$ linearly independent solutions on $(-\infty, \infty)$ of a homogeneous linear $n$th-order differential equation with constant coefficients. By Theorem 4.1.2 it follows that $y_{k+1} = 0$ is also a solution of the differential equation. Is the set of solutions $y_1, y_2, \ldots, y_k, y_{k+1}$ linearly dependent or linearly independent on $(-\infty, \infty)$? Discuss.

**42.** Suppose that $y_1, y_2, \ldots, y_k$ are $k$ nontrivial solutions of a homogeneous linear $n$th-order differential equation with constant coefficients and that $k = n + 1$. Is the set of solutions $y_1, y_2, \ldots, y_k$ linearly dependent or linearly independent on $(-\infty, \infty)$? Discuss.

# 4.2 Reduction of Order

**INTRODUCTION** In the preceding section we saw that the general solution of a homogeneous linear second-order differential equation

$$a_2(x)y'' + a_1(x)y' + a_0(x)y = 0 \qquad (1)$$

is a linear combination $y = c_1 y_1 + c_2 y_2$, where $y_1$ and $y_2$ are solutions that constitute a linearly independent set on some interval $I$. Beginning in the next section, we examine a method for determining these solutions when the coefficients of the differential equation in (1) are constants. This method, which is a straightforward exercise in algebra, breaks down in a few cases and yields only a single solution $y_1$ of the DE. It turns out that we can construct a second solution $y_2$ of a homogeneous equation (1) (even when the coefficients in (1) are variable) provided that we know a nontrivial solution $y_1$ of the DE. The basic idea described in this section is that *equation* (1) *can be reduced to a linear first-order DE by means of a substitution* involving the known solution $y_1$. A second solution $y_2$ of (1) is apparent after this first-order differential equation is solved.

**REDUCTION OF ORDER** Suppose that $y_1$ denotes a nontrivial solution of (1) and that $y_1$ is defined on an interval $I$. We seek a second solution $y_2$ so that the set consisting of $y_1$ and $y_2$ is linearly independent on $I$. Recall from Section 4.1 that if $y_1$ and $y_2$ are linearly independent, then their quotient $y_2/y_1$ is nonconstant on $I$—that is, $y_2(x)/y_1(x) = u(x)$ or $y_2(x) = u(x)y_1(x)$. The function $u(x)$ can be found by substituting $y_2(x) = u(x)y_1(x)$ into the given differential equation. This method is called **reduction of order** because we must solve a linear first-order differential equation to find $u$.

---

**EXAMPLE 1    A Second Solution by Reduction of Order**

Given that $y_1 = e^x$ is a solution of $y'' - y = 0$ on the interval $(-\infty, \infty)$, use reduction of order to find a second solution $y_2$.

**SOLUTION** If $y = u(x)y_1(x) = u(x)e^x$, then the Product Rule gives

$$y' = ue^x + e^x u', \quad y'' = ue^x + 2e^x u' + e^x u'',$$

and so
$$y'' - y = e^x(u'' + 2u') = 0.$$

Since $e^x \neq 0$, the last equation requires $u'' + 2u' = 0$. If we make the substitution $w = u'$, this linear second-order equation in $u$ becomes $w' + 2w = 0$, which is a linear first-order equation in $w$. Using the integrating factor $e^{2x}$, we can write $\frac{d}{dx}[e^{2x}w] = 0$. After integrating, we get $w = c_1 e^{-2x}$ or $u' = c_1 e^{-2x}$. Integrating again then yields $u = -\frac{1}{2}c_1 e^{-2x} + c_2$. Thus

$$y = u(x)e^x = -\frac{c_1}{2}e^{-x} + c_2 e^x. \qquad (2)$$

By choosing $c_2 = 0$ and $c_1 = -2$, we obtain the desired second solution, $y_2 = e^{-x}$. Because $W(e^x, e^{-x}) \neq 0$ for every $x$, the solutions are linearly independent on $(-\infty, \infty)$. ■

Since we have shown that $y_1 = e^x$ and $y_2 = e^{-x}$ are linearly independent solutions of a linear second-order equation, the expression in (2) is actually the general solution of $y'' - y = 0$ on $(-\infty, \infty)$.

**GENERAL CASE** Suppose we divide by $a_2(x)$ to put equation (1) in the **standard form**

$$y'' + P(x)y' + Q(x)y = 0, \qquad (3)$$

where $P(x)$ and $Q(x)$ are continuous on some interval $I$. Let us suppose further that $y_1(x)$ is a known solution of (3) on $I$ and that $y_1(x) \neq 0$ for every $x$ in the interval. If we define $y = u(x)y_1(x)$, it follows that

$$y' = uy_1' + y_1u', \quad y'' = uy_1'' + 2y_1'u' + y_1u''$$

$$y'' + Py' + Qy = u[\underbrace{y_1'' + Py_1' + Qy_1}_{\text{zero}}] + y_1u'' + (2y_1' + Py_1)u' = 0.$$

This implies that we must have

$$y_1u'' + (2y_1' + Py_1)u' = 0 \qquad \text{or} \qquad y_1w' + (2y_1' + Py_1)w = 0, \qquad (4)$$

where we have let $w = u'$. Observe that the last equation in (4) is both linear and separable. Separating variables and integrating, we obtain

$$\frac{dw}{w} + 2\frac{y_1'}{y_1}\,dx + P\,dx = 0$$

$$\ln|wy_1^2| = -\int P\,dx + c \qquad \text{or} \qquad wy_1^2 = c_1 e^{-\int P\,dx}.$$

We solve the last equation for $w$, use $w = u'$, and integrate again:

$$u = c_1 \int \frac{e^{-\int P\,dx}}{y_1^2}\,dx + c_2.$$

By choosing $c_1 = 1$ and $c_2 = 0$, we find from $y = u(x)y_1(x)$ that a second solution of equation (3) is

$$y_2 = y_1(x) \int \frac{e^{-\int P(x)\,dx}}{y_1^2(x)}\,dx. \qquad (5)$$

It makes a good review of differentiation to verify that the function $y_2(x)$ defined in (5) satisfies equation (3) and that $y_1$ and $y_2$ are linearly independent on any interval on which $y_1(x)$ is not zero.

---

**EXAMPLE 2**   **A Second Solution by Formula (5)**

The function $y_1 = x^2$ is a solution of $x^2y'' - 3xy' + 4y = 0$. Find the general solution of the differential equation on the interval $(0, \infty)$.

**SOLUTION** From the standard form of the equation,

$$y'' - \frac{3}{x}y' + \frac{4}{x^2}y = 0,$$

we find from (5)    $y_2 = x^2 \int \frac{e^{3\int dx/x}}{x^4}\,dx \qquad \leftarrow e^{3\int dx/x} = e^{\ln x^3} = x^3$

$$= x^2 \int \frac{dx}{x} = x^2 \ln x.$$

The general solution on the interval $(0, \infty)$ is given by $y = c_1y_1 + c_2y_2$; that is, $y = c_1x^2 + c_2x^2 \ln x$. $\blacksquare$

**REMARKS**

(*i*) The derivation and use of formula (5) have been illustrated here because this formula appears again in the next section and in Sections 4.7 and 6.3. We use (5) simply to save time in obtaining a desired result. Your instructor will tell you whether you should memorize (5) or whether you should know the first principles of reduction of order.

(*ii*) Reduction of order can be used to find the general solution of a nonhomogeneous equation $a_2(x)y'' + a_1(x)y' + a_0(x)y = g(x)$ whenever a solution $y_1$ of the associated homogeneous equation is known. See Problems 17–20 in Exercises 4.2.

(*iii*) The integral in (5) may be nonelementary. In this case we simply write the second solution in terms of an integral-defined function:

$$y_2(x) = y_1(x) \int_{x_0}^{x} \frac{e^{-\int P(t)\,dt}}{y_1^2(t)}\, dt,$$

where we assume that the integrand is continuous on the interval $[x_0, x]$. See Problems 21 and 22 in Exercises 4.2.

---

# EXERCISES 4.2

*Answers to selected odd-numbered problems begin on page ANS-4.*

In Problems 1–16 the indicated function $y_1(x)$ is a solution of the given differential equation. Use reduction of order or formula (5), as instructed, to find a second solution $y_2(x)$.

**1.** $y'' - 4y' + 4y = 0$;   $y_1 = e^{2x}$

**2.** $y'' + 2y' + y = 0$;   $y_1 = xe^{-x}$

**3.** $y'' + 16y = 0$;   $y_1 = \cos 4x$

**4.** $y'' + 9y = 0$;   $y_1 = \sin 3x$

**5.** $y'' - y = 0$;   $y_1 = \cosh x$

**6.** $y'' - 25y = 0$;   $y_1 = e^{5x}$

**7.** $9y'' - 12y' + 4y = 0$;   $y_1 = e^{2x/3}$

**8.** $6y'' + y' - y = 0$;   $y_1 = e^{x/3}$

**9.** $x^2y'' - 7xy' + 16y = 0$;   $y_1 = x^4$

**10.** $x^2y'' + 2xy' - 6y = 0$;   $y_1 = x^2$

**11.** $xy'' + y' = 0$;   $y_1 = \ln x$

**12.** $4x^2y'' + y = 0$;   $y_1 = x^{1/2}\ln x$

**13.** $x^2y'' - xy' + 2y = 0$;   $y_1 = x\sin(\ln x)$

**14.** $x^2y'' - 3xy' + 5y = 0$;   $y_1 = x^2\cos(\ln x)$

**15.** $(1 - 2x - x^2)y'' + 2(1 + x)y' - 2y = 0$;   $y_1 = x + 1$

**16.** $(1 - x^2)y'' + 2xy' = 0$;   $y_1 = 1$

In Problems 17–20 the indicated function $y_1(x)$ is a solution of the associated homogeneous equation. Use the method of reduction of order to find a second solution $y_2(x)$ of the homogeneous equation and a particular solution $y_p(x)$ of the given nonhomogeneous equation.

**17.** $y'' - 4y = 2$;   $y_1 = e^{-2x}$

**18.** $y'' + y' = 1$;   $y_1 = 1$

**19.** $y'' - 3y' + 2y = 5e^{3x}$;   $y_1 = e^x$

**20.** $y'' - 4y' + 3y = x$;   $y_1 = e^x$

In Problems 21 and 22 the indicated function $y_1(x)$ is a solution of the given differential equation. Use formula (5) to find a second solution $y_2(x)$ expressed in terms of an integral-defined function. See (*iii*) in the *Remarks*.

**21.** $x^2y'' + (x^2 - x)y' + (1 - x)y = 0$;   $y_1 = x$

**22.** $2xy'' - (2x + 1)y' + y = 0$;   $y_1 = e^x$

## Discussion Problems

**23.** **(a)** Give a convincing demonstration that the second-order equation $ay'' + by' + cy = 0$, $a$, $b$, and $c$ constants, always possesses at least one solution of the form $y_1 = e^{m_1 x}$, $m_1$ a constant.

**(b)** Explain why the differential equation in part (a) must then have a second solution either of the form $y_2 = e^{m_2 x}$ or of the form $y_2 = xe^{m_1 x}$, $m_1$ and $m_2$ constants.

**(c)** Reexamine Problems 1–8. Can you explain why the statements in parts (a) and (b) above are not contradicted by the answers to Problems 3–5?

**24.** Verify that $y_1(x) = x$ is a solution of $xy'' - xy' + y = 0$. Use reduction of order to find a second solution $y_2(x)$ in the form of an infinite series. Conjecture an interval of definition for $y_2(x)$.

**Computer Lab Assignments**

**25. (a)** Verify that $y_1(x) = e^x$ is a solution of
$$xy'' - (x + 10)y' + 10y = 0.$$

**(b)** Use (5) to find a second solution $y_2(x)$. Use a CAS to carry out the required integration.

**(c)** Explain, using Corollary (A) of Theorem 4.1.2, why the second solution can be written compactly as
$$y_2(x) = \sum_{n=0}^{10} \frac{1}{n!} x^n.$$

---

## 4.3 Homogeneous Linear Equations with Constant Coefficients

**INTRODUCTION** As a means of motivating the discussion in this section, let us return to first-order differential equations—more specifically, to *homogeneous* linear equations $ay' + by = 0$, where the coefficients $a \neq 0$ and $b$ are constants. This type of equation can be solved either by separation of variables or with the aid of an integrating factor, but there is another solution method, one that uses only algebra. Before illustrating this alternative method, we make one observation: Solving $ay' + by = 0$ for $y'$ yields $y' = ky$, where $k$ is a constant. This observation reveals the nature of the unknown solution $y$; the only nontrivial elementary function whose derivative is a constant multiple of itself is an exponential function $e^{mx}$. Now the new solution method: If we substitute $y = e^{mx}$ and $y' = me^{mx}$ into $ay' + by = 0$, we get

$$ame^{mx} + be^{mx} = 0 \quad \text{or} \quad e^{mx}(am + b) = 0.$$

Since $e^{mx}$ is never zero for real values of $x$, the last equation is satisfied only when $m$ is a solution or root of the first-degree polynomial equation $am + b = 0$. For this single value of $m$, $y = e^{mx}$ is a solution of the DE. To illustrate, consider the constant-coefficient equation $2y' + 5y = 0$. It is not necessary to go through the differentiation and substitution of $y = e^{mx}$ into the DE; we merely have to form the equation $2m + 5 = 0$ and solve it for $m$. From $m = -\frac{5}{2}$ we conclude that $y = e^{-5x/2}$ is a solution of $2y' + 5y = 0$, and its general solution on the interval $(-\infty, \infty)$ is $y = c_1 e^{-5x/2}$.

In this section we will see that the foregoing procedure can produce exponential solutions for homogeneous linear higher-order DEs,

$$a_n y^{(n)} + a_{n-1} y^{(n-1)} + \cdots + a_2 y'' + a_1 y' + a_0 y = 0, \tag{1}$$

where the coefficients $a_i$, $i = 0, 1, \ldots, n$ are real constants and $a_n \neq 0$.

**AUXILIARY EQUATION** We begin by considering the special case of the second-order equation

$$ay'' + by' + cy = 0, \tag{2}$$

where $a$, $b$, and $c$ are constants. If we try to find a solution of the form $y = e^{mx}$, then after substitution of $y' = me^{mx}$ and $y'' = m^2 e^{mx}$, equation (2) becomes

$$am^2 e^{mx} + bm e^{mx} + c e^{mx} = 0 \quad \text{or} \quad e^{mx}(am^2 + bm + c) = 0.$$

As in the introduction we argue that because $e^{mx} \neq 0$ for all $x$, it is apparent that the only way $y = e^{mx}$ can satisfy the differential equation (2) is when $m$ is chosen as a root of the quadratic equation

$$am^2 + bm + c = 0. \tag{3}$$

This last equation is called the **auxiliary equation** of the differential equation (2). Since the two roots of (3) are $m_1 = (-b + \sqrt{b^2 - 4ac})/2a$ and $m_2 = (-b - \sqrt{b^2 - 4ac})/2a$, there will be three forms of the general solution of (2) corresponding to the three cases:

- $m_1$ and $m_2$ real and distinct ($b^2 - 4ac > 0$),
- $m_1$ and $m_2$ real and equal ($b^2 - 4ac = 0$), and
- $m_1$ and $m_2$ conjugate complex numbers ($b^2 - 4ac < 0$).

We discuss each of these cases in turn.

**CASE I: DISTINCT REAL ROOTS**   Under the assumption that the auxiliary equation (3) has two unequal real roots $m_1$ and $m_2$, we find two solutions, $y_1 = e^{m_1 x}$ and $y_2 = e^{m_2 x}$. We see that these functions are linearly independent on $(-\infty, \infty)$ and hence form a fundamental set. It follows that the general solution of (2) on this interval is

$$y = c_1 e^{m_1 x} + c_2 e^{m_2 x}. \tag{4}$$

**CASE II: REPEATED REAL ROOTS**   When $m_1 = m_2$, we necessarily obtain only one exponential solution, $y_1 = e^{m_1 x}$. From the quadratic formula we find that $m_1 = -b/2a$ since the only way to have $m_1 = m_2$ is to have $b^2 - 4ac = 0$. It follows from (5) in Section 4.2 that a second solution of the equation is

$$y_2 = e^{m_1 x} \int \frac{e^{2m_1 x}}{e^{2m_1 x}} \, dx = e^{m_1 x} \int dx = x e^{m_1 x}. \tag{5}$$

In (5) we have used the fact that $-b/a = 2m_1$. The general solution is then

$$y = c_1 e^{m_1 x} + c_2 x e^{m_1 x}. \tag{6}$$

**CASE III: CONJUGATE COMPLEX ROOTS**   If $m_1$ and $m_2$ are complex, then we can write $m_1 = \alpha + i\beta$ and $m_2 = \alpha - i\beta$, where $\alpha$ and $\beta > 0$ are real and $i^2 = -1$. Formally, there is no difference between this case and Case I, and hence

$$y = C_1 e^{(\alpha + i\beta)x} + C_2 e^{(\alpha - i\beta)x}.$$

However, in practice we prefer to work with real functions instead of complex exponentials. To this end we use **Euler's formula:**

$$e^{i\theta} = \cos\theta + i\sin\theta,$$

where $\theta$ is any real number.* It follows from this formula that

$$e^{i\beta x} = \cos\beta x + i\sin\beta x \quad \text{and} \quad e^{-i\beta x} = \cos\beta x - i\sin\beta x, \tag{7}$$

where we have used $\cos(-\beta x) = \cos\beta x$ and $\sin(-\beta x) = -\sin\beta x$. Note that by first adding and then subtracting the two equations in (7), we obtain, respectively,

$$e^{i\beta x} + e^{-i\beta x} = 2\cos\beta x \quad \text{and} \quad e^{i\beta x} - e^{-i\beta x} = 2i\sin\beta x.$$

Since $y = C_1 e^{(\alpha + i\beta)x} + C_2 e^{(\alpha - i\beta)x}$ is a solution of (2) for any choice of the constants $C_1$ and $C_2$, the choices $C_1 = C_2 = 1$ and $C_1 = 1, C_2 = -1$ give, in turn, two solutions:

$$y_1 = e^{(\alpha + i\beta)x} + e^{(\alpha - i\beta)x} \quad \text{and} \quad y_2 = e^{(\alpha + i\beta)x} - e^{(\alpha - i\beta)x}.$$

But

$$y_1 = e^{\alpha x}(e^{i\beta x} + e^{-i\beta x}) = 2e^{\alpha x}\cos\beta x$$

and

$$y_2 = e^{\alpha x}(e^{i\beta x} - e^{-i\beta x}) = 2ie^{\alpha x}\sin\beta x.$$

---

*A formal derivation of Euler's formula can be obtained from the Maclaurin series $e^x = \sum_{n=0}^{\infty} \frac{x^n}{n!}$ by substituting $x = i\theta$, using $i^2 = -1, i^3 = -i, \ldots$, and then separating the series into real and imaginary parts. The plausibility thus established, we can adopt $\cos\theta + i\sin\theta$ as the *definition* of $e^{i\theta}$.

Hence from Corollary (A) of Theorem 4.1.2 the last two results show that $e^{\alpha x} \cos \beta x$ and $e^{\alpha x} \sin \beta x$ are *real* solutions of (2). Moreover, these solutions form a fundamental set on $(-\infty, \infty)$. Consequently, the general solution is

$$y = c_1 e^{\alpha x} \cos \beta x + c_2 e^{\alpha x} \sin \beta x = e^{\alpha x}(c_1 \cos \beta x + c_2 \sin \beta x). \qquad (8)$$

### EXAMPLE 1  Second-Order DEs

Solve the following differential equations.

**(a)** $2y'' - 5y' - 3y = 0$     **(b)** $y'' - 10y' + 25y = 0$     **(c)** $y'' + 4y' + 7y = 0$

**SOLUTION** We give the auxiliary equations, the roots, and the corresponding general solutions.

**(a)** $2m^2 - 5m - 3 = (2m + 1)(m - 3) = 0, \quad m_1 = -\frac{1}{2}, m_2 = 3$
From (4), $y = c_1 e^{-x/2} + c_2 e^{3x}$.

**(b)** $m^2 - 10m + 25 = (m - 5)^2 = 0, \quad m_1 = m_2 = 5$
From (6), $y = c_1 e^{5x} + c_2 x e^{5x}$.

**(c)** $m^2 + 4m + 7 = 0, \quad m_1 = -2 + \sqrt{3}i, \quad m_2 = -2 - \sqrt{3}i$
From (8) with $\alpha = -2, \beta = \sqrt{3}, y = e^{-2x}\left(c_1 \cos \sqrt{3}x + c_2 \sin \sqrt{3}x\right)$. ■

### EXAMPLE 2  An Initial-Value Problem

Solve $4y'' + 4y' + 17y = 0, y(0) = -1, y'(0) = 2$.

**SOLUTION** By the quadratic formula we find that the roots of the auxiliary equation $4m^2 + 4m + 17 = 0$ are $m_1 = -\frac{1}{2} + 2i$ and $m_2 = -\frac{1}{2} - 2i$. Thus from (8) we have $y = e^{-x/2}(c_1 \cos 2x + c_2 \sin 2x)$. Applying the condition $y(0) = -1$, we see from $e^0(c_1 \cos 0 + c_2 \sin 0) = -1$ that $c_1 = -1$. Differentiating $y = e^{-x/2}(-\cos 2x + c_2 \sin 2x)$ and then using $y'(0) = 2$ gives $2c_2 + \frac{1}{2} = 2$ or $c_2 = \frac{3}{4}$. Hence the solution of the IVP is $y = e^{-x/2}(-\cos 2x + \frac{3}{4}\sin 2x)$. In Figure 4.3.1 we see that the solution is oscillatory, but $y \to 0$ as $x \to \infty$. ■

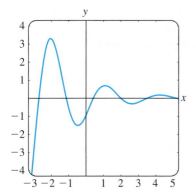

**FIGURE 4.3.1** Solution curve of IVP in Example 2

**TWO EQUATIONS WORTH KNOWING** The two differential equations

$$y'' + k^2 y = 0 \qquad \text{and} \qquad y'' - k^2 y = 0,$$

where $k$ is real, are important in applied mathematics. For $y'' + k^2 y = 0$ the auxiliary equation $m^2 + k^2 = 0$ has imaginary roots $m_1 = ki$ and $m_2 = -ki$. With $\alpha = 0$ and $\beta = k$ in (8) the general solution of the DE is seen to be

$$y = c_1 \cos kx + c_2 \sin kx. \qquad (9)$$

On the other hand, the auxiliary equation $m^2 - k^2 = 0$ for $y'' - k^2 y = 0$ has distinct real roots $m_1 = k$ and $m_2 = -k$, and so by (4) the general solution of the DE is

$$y = c_1 e^{kx} + c_2 e^{-kx}. \qquad (10)$$

Notice that if we choose $c_1 = c_2 = \frac{1}{2}$ and $c_1 = \frac{1}{2}, c_2 = -\frac{1}{2}$ in (10), we get the particular solutions $y = \frac{1}{2}(e^{kx} + e^{-kx}) = \cosh kx$ and $y = \frac{1}{2}(e^{kx} - e^{-kx}) = \sinh kx$. Since $\cosh kx$ and $\sinh kx$ are linearly independent on any interval of the $x$-axis, an alternative form for the general solution of $y'' - k^2 y = 0$ is

$$y = c_1 \cosh kx + c_2 \sinh kx. \qquad (11)$$

See Problems 41 and 42 in Exercises 4.3.

**HIGHER-ORDER EQUATIONS** In general, to solve an $n$th-order differential equation (1), where the $a_i$, $i = 0, 1, \ldots, n$ are real constants, we must solve an $n$th-degree polynomial equation

$$a_n m^n + a_{n-1} m^{n-1} + \cdots + a_2 m^2 + a_1 m + a_0 = 0. \qquad (12)$$

If all the roots of (12) are real and distinct, then the general solution of (1) is

$$y = c_1 e^{m_1 x} + c_2 e^{m_2 x} + \cdots + c_n e^{m_n x}.$$

It is somewhat harder to summarize the analogues of Cases II and III because the roots of an auxiliary equation of degree greater than two can occur in many combinations. For example, a fifth-degree equation could have five distinct real roots, or three distinct real and two complex roots, or one real and four complex roots, or five real but equal roots, or five real roots but two of them equal, and so on. When $m_1$ is a root of multiplicity $k$ of an $n$th-degree auxiliary equation (that is, $k$ roots are equal to $m_1$), it can be shown that the linearly independent solutions are

$$e^{m_1 x}, \quad x e^{m_1 x}, \quad x^2 e^{m_1 x}, \ldots, \quad x^{k-1} e^{m_1 x}$$

and the general solution must contain the linear combination

$$c_1 e^{m_1 x} + c_2 x e^{m_1 x} + c_3 x^2 e^{m_1 x} + \cdots + c_k x^{k-1} e^{m_1 x}.$$

Finally, it should be remembered that when the coefficients are real, complex roots of an auxiliary equation always appear in conjugate pairs. Thus, for example, a cubic polynomial equation can have at most two complex roots.

---

**EXAMPLE 3** **Third-Order DE**

Solve $y''' + 3y'' - 4y = 0$.

**SOLUTION** It should be apparent from inspection of $m^3 + 3m^2 - 4 = 0$ that one root is $m_1 = 1$, so $m - 1$ is a factor of $m^3 + 3m^2 - 4$. By division we find

$$m^3 + 3m^2 - 4 = (m - 1)(m^2 + 4m + 4) = (m - 1)(m + 2)^2,$$

so the other roots are $m_2 = m_3 = -2$. Thus the general solution of the DE is $y = c_1 e^x + c_2 e^{-2x} + c_3 x e^{-2x}$. ∎

---

**EXAMPLE 4** **Fourth-Order DE**

Solve $\dfrac{d^4 y}{dx^4} + 2\dfrac{d^2 y}{dx^2} + y = 0$.

**SOLUTION** The auxiliary equation $m^4 + 2m^2 + 1 = (m^2 + 1)^2 = 0$ has roots $m_1 = m_3 = i$ and $m_2 = m_4 = -i$. Thus from Case II the solution is

$$y = C_1 e^{ix} + C_2 e^{-ix} + C_3 x e^{ix} + C_4 x e^{-ix}.$$

By Euler's formula the grouping $C_1 e^{ix} + C_2 e^{-ix}$ can be rewritten as

$$c_1 \cos x + c_2 \sin x$$

after a relabeling of constants. Similarly, $x(C_3 e^{ix} + C_4 e^{-ix})$ can be expressed as $x(c_3 \cos x + c_4 \sin x)$. Hence the general solution is

$$y = c_1 \cos x + c_2 \sin x + c_3 x \cos x + c_4 x \sin x.$$ ∎

Example 4 illustrates a special case when the auxiliary equation has repeated complex roots. In general, if $m_1 = \alpha + i\beta$, $\beta > 0$ is a complex root of multiplicity $k$

of an auxiliary equation with real coefficients, then its conjugate $m_2 = \alpha - i\beta$ is also a root of multiplicity $k$. From the $2k$ complex-valued solutions

$$e^{(\alpha+i\beta)x}, \quad xe^{(\alpha+i\beta)x}, \quad x^2e^{(\alpha+i\beta)x}, \quad \ldots, \quad x^{k-1}e^{(\alpha+i\beta)x},$$

$$e^{(\alpha-i\beta)x}, \quad xe^{(\alpha-i\beta)x}, \quad x^2e^{(\alpha-i\beta)x}, \quad \ldots, \quad x^{k-1}e^{(\alpha-i\beta)x},$$

we conclude, with the aid of Euler's formula, that the general solution of the corresponding differential equation must then contain a linear combination of the $2k$ real linearly independent solutions

$$e^{\alpha x}\cos\beta x, \quad xe^{\alpha x}\cos\beta x, \quad x^2e^{\alpha x}\cos\beta x, \quad \ldots, \quad x^{k-1}e^{\alpha x}\cos\beta x,$$

$$e^{\alpha x}\sin\beta x, \quad xe^{\alpha x}\sin\beta x, \quad x^2e^{\alpha x}\sin\beta x, \quad \ldots, \quad x^{k-1}e^{\alpha x}\sin\beta x.$$

In Example 4 we identify $k = 2$, $\alpha = 0$, and $\beta = 1$.

**RATIONAL ROOTS**　Of course the most difficult aspect of solving constant-coefficient differential equations is finding roots of auxiliary equations of degree greater than two. Something we can try is to test the auxiliary equation for rational roots. Recall from precalculus mathematics, if $m_1 = p/q$ is a rational root (expressed in lowest terms) of a polynomial equation $a_n m^n + \cdots + a_1 m + a_0 = 0$ with integer coefficients, then the integer $p$ is a factor of the constant term $a_0$ and the integer $q$ is a factor of the leading coefficient $a_n$.

---

**EXAMPLE 5**　**Finding Rational Roots**

Solve $3y''' + 5y'' + 10y' - 4y = 0$.

**SOLUTION**　To solve the equation we must solve the cubic polynomial auxiliary equation $3m^3 + 5m^2 + 10m - 4 = 0$. With the identifications $a_0 = -4$ and $a_3 = 3$ then the integer factors of $a_0$ and $a_3$ are, respectively, $p$: $\pm 1$, $\pm 2$, $\pm 4$ and $q$: $\pm 1$, $\pm 3$. So the possible rational roots of the cubic equation are

$$\frac{p}{q}: \pm 1, \ \pm 2, \ \pm 4, \ \pm\frac{1}{3}, \ \pm\frac{2}{3}, \ \pm\frac{4}{3}.$$

Each of these numbers can then be tested—say, by synthetic division. In this way we discover both the root $m_1 = \frac{1}{3}$ and the factorization

$$3m^3 + 5m^2 + 10m - 4 = \left(m - \tfrac{1}{3}\right)(3m^2 + 6m + 12).$$

The quadratic formula applied to $3m^2 + 6m + 12 = 0$ then yields the remaining two roots $m_2 = -1 - \sqrt{3}i$ and $m_3 = -1 + \sqrt{3}i$. Therefore the general solution of the given differential equation is $y = c_1 e^{x/3} + e^{-x}(c_2 \cos\sqrt{3}x + c_3 \sin\sqrt{3}x)$.　∎

**USE OF COMPUTERS**　Finding roots or approximation of roots of auxiliary equations is a routine problem with an appropriate calculator or computer software. Polynomial equations (in one variable) of degree less than five can be solved by means of algebraic formulas using the *solve* commands in *Mathematica* and *Maple*. For auxiliary equations of degree five or greater it might be necessary to resort to numerical commands such as **NSolve** and **FindRoot** in *Mathematica*. Because of their capability of solving polynomial equations, it is not surprising that these computer algebra systems are also able, by means of their *dsolve* commands, to provide explicit solutions of homogeneous linear constant-coefficient differential equations.

There is more on this in the *SRM*.

In the classic text *Differential Equations* by Ralph Palmer Agnew* (used by the author as a student) the following statement is made:

*It is not reasonable to expect students in this course to have computing skill and equipment necessary for efficient solving of equations such as*

$$4.317 \frac{d^4y}{dx^4} + 2.179 \frac{d^3y}{dx^3} + 1.416 \frac{d^2y}{dx^2} + 1.295 \frac{dy}{dx} + 3.169y = 0. \quad (13)$$

Although it is debatable whether computing skills have improved in the intervening years, it is a certainty that technology has. If one has access to a computer algebra system, equation (13) could now be considered reasonable. After simplification and some relabeling of output, *Mathematica* yields the (approximate) general solution

$$y = c_1 e^{-0.728852x} \cos(0.618605x) + c_2 e^{-0.728852x} \sin(0.618605x)$$

$$+ c_3 e^{0.476478x} \cos(0.759081x) + c_4 e^{0.476478x} \sin(0.759081x).$$

Finally, if we are faced with an initial-value problem consisting of, say, a fourth-order equation, then to fit the general solution of the DE to the four initial conditions, we must solve four linear equations in four unknowns (the $c_1, c_2, c_3, c_4$ in the general solution). Using a CAS to solve the system can save lots of time. See Problems 69 and 70 in Exercises 4.3 and Problem 45 in Chapter 4 in Review.

---

*McGraw-Hill, New York, 1960.

In Problems 1–14 find the general solution of the given second-order differential equation.

**1.** $4y'' + y' = 0$

**2.** $y'' - 36y = 0$

**3.** $y'' - y' - 6y = 0$

**4.** $y'' - 3y' + 2y = 0$

**5.** $y'' + 8y' + 16y = 0$

**6.** $y'' - 10y' + 25y = 0$

**7.** $12y'' - 5y' - 2y = 0$

**8.** $y'' + 4y' - y = 0$

**9.** $y'' + 9y = 0$

**10.** $3y'' + y = 0$

**11.** $y'' - 4y' + 5y = 0$

**12.** $2y'' + 2y' + y = 0$

**13.** $3y'' + 2y' + y = 0$

**14.** $2y'' - 3y' + 4y = 0$

In Problems 15–28 find the general solution of the given higher-order differential equation.

**15.** $y''' - 4y'' - 5y' = 0$

**16.** $y''' - y = 0$

**17.** $y''' - 5y'' + 3y' + 9y = 0$

**18.** $y''' + 3y'' - 4y' - 12y = 0$

**19.** $\dfrac{d^3u}{dt^3} + \dfrac{d^2u}{dt^2} - 2u = 0$

**20.** $\dfrac{d^3x}{dt^3} - \dfrac{d^2x}{dt^2} - 4x = 0$

**21.** $y''' + 3y'' + 3y' + y = 0$

**22.** $y''' - 6y'' + 12y' - 8y = 0$

**23.** $y^{(4)} + y''' + y'' = 0$

**24.** $y^{(4)} - 2y'' + y = 0$

**25.** $16 \dfrac{d^4y}{dx^4} + 24 \dfrac{d^2y}{dx^2} + 9y = 0$

**26.** $\dfrac{d^4y}{dx^4} - 7 \dfrac{d^2y}{dx^2} - 18y = 0$

**27.** $\dfrac{d^5u}{dr^5} + 5 \dfrac{d^4u}{dr^4} - 2 \dfrac{d^3u}{dr^3} - 10 \dfrac{d^2u}{dr^2} + \dfrac{du}{dr} + 5u = 0$

**28.** $2 \dfrac{d^5x}{ds^5} - 7 \dfrac{d^4x}{ds^4} + 12 \dfrac{d^3x}{ds^3} + 8 \dfrac{d^2x}{ds^2} = 0$

In Problems 29–36 solve the given initial-value problem.

**29.** $y'' + 16y = 0, \quad y(0) = 2, y'(0) = -2$

**30.** $\dfrac{d^2y}{d\theta^2} + y = 0, \quad y(\pi/3) = 0, y'(\pi/3) = 2$

**31.** $\dfrac{d^2y}{dt^2} - 4 \dfrac{dy}{dt} - 5y = 0, \quad y(1) = 0, y'(1) = 2$

**32.** $4y'' - 4y' - 3y = 0, \quad y(0) = 1, y'(0) = 5$

**33.** $y'' + y' + 2y = 0$, $y(0) = y'(0) = 0$

**34.** $y'' - 2y' + y = 0$, $y(0) = 5, y'(0) = 10$

**35.** $y''' + 12y'' + 36y' = 0$, $y(0) = 0, y'(0) = 1, y''(0) = -7$

**36.** $y''' + 2y'' - 5y' - 6y = 0$, $y(0) = y'(0) = 0, y''(0) = 1$

In Problems 37–40 solve the given boundary-value problem.

**37.** $y'' - 10y' + 25y = 0$, $y(0) = 1, y(1) = 0$

**38.** $y'' + 4y = 0$, $y(0) = 0, y(\pi) = 0$

**39.** $y'' + y = 0$, $y'(0) = 0, y'(\pi/2) = 0$

**40.** $y'' - 2y' + 2y = 0$, $y(0) = 1, y(\pi) = 1$

In Problems 41 and 42 solve the given problem first using the form of the general solution given in (10). Solve again, this time using the form given in (11).

**41.** $y'' - 3y = 0$, $y(0) = 1, y'(0) = 5$

**42.** $y'' - y = 0$, $y(0) = 1, y'(1) = 0$

In Problems 43–48 each figure represents the graph of a particular solution of one of the following differential equations:

(a) $y'' - 3y' - 4y = 0$    (b) $y'' + 4y = 0$

(c) $y'' + 2y' + y = 0$    (d) $y'' + y = 0$

(e) $y'' + 2y' + 2y = 0$    (f) $y'' - 3y' + 2y = 0$

Match a solution curve with one of the differential equations. Explain your reasoning.

**43.**

**FIGURE 4.3.2** Graph for Problem 43

**44.**

**FIGURE 4.3.3** Graph for Problem 44

**45.**

**FIGURE 4.3.4** Graph for Problem 45

**46.**

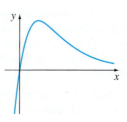

**FIGURE 4.3.5** Graph for Problem 46

**47.**

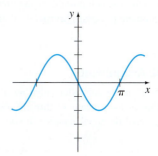

**FIGURE 4.3.6** Graph for Problem 47

**48.**

**FIGURE 4.3.7** Graph for Problem 48

In Problems 49–58 find a homogeneous linear differential equation with constant coefficients whose general solution is given.

**49.** $y = c_1 e^x + c_2 e^{5x}$

**50.** $y = c_1 e^{-4x} + c_2 e^{-3x}$

**51.** $y = c_1 + c_2 e^{2x}$

**52.** $y = c_1 e^{10x} + c_2 x e^{10x}$

**53.** $y = c_1 \cos 3x + c_2 \sin 3x$

**54.** $y = c_1 \cosh 7x + c_2 \sinh 7x$

**55.** $y = c_1 e^{-x} \cos x + c_2 e^{-x} \sin x$

**56.** $y = c_1 + c_2 e^{2x} \cos 5x + c_3 e^{2x} \sin 5x$

**57.** $y = c_1 + c_2 x + c_3 e^{8x}$

**58.** $y = c_1 \cos x + c_2 \sin x + c_3 \cos 2x + c_4 \sin 2x$

## Discussion Problems

**59.** Two roots of a cubic auxiliary equation with real coefficients are $m_1 = -\frac{1}{2}$ and $m_2 = 3 + i$. What is the corresponding homogeneous linear differential equation? Discuss: Is your answer unique?

**60.** Find the general solution of $2y''' + 7y'' + 4y' - 4y = 0$ if $m_1 = \frac{1}{2}$ is one root of its auxiliary equation.

**61.** Find the general solution of $y''' + 6y'' + y' - 34y = 0$ if it is known that $y_1 = e^{-4x} \cos x$ is one solution.

**62.** To solve $y^{(4)} + y = 0$, we must find the roots of $m^4 + 1 = 0$. This is a trivial problem using a CAS but can also be done by hand working with complex numbers. Observe that $m^4 + 1 = (m^2 + 1)^2 - 2m^2$. How does this help? Solve the differential equation.

**63.** Verify that $y = \sinh x - 2 \cos (x + \pi/6)$ is a particular solution of $y^{(4)} - y = 0$. Reconcile this particular solution with the general solution of the DE.

**64.** Consider the boundary-value problem $y'' + \lambda y = 0$, $y(0) = 0$, $y(\pi/2) = 0$. Discuss: Is it possible to determine real values of $\lambda$ so that the problem possesses **(a)** trivial solutions? **(b)** nontrivial solutions?

### Computer Lab Assignments

In Problems 65–68 use a computer either as an aid in solving the auxiliary equation or as a means of directly obtaining the general solution of the given differential equation. If you use a CAS to obtain the general solution, simplify the output and, if necessary, write the solution in terms of real functions.

**65.** $y''' - 6y'' + 2y' + y = 0$

**66.** $6.11y''' + 8.59y'' + 7.93y' + 0.778y = 0$

**67.** $3.15y^{(4)} - 5.34y'' + 6.33y' - 2.03y = 0$

**68.** $y^{(4)} + 2y'' - y' + 2y = 0$

In Problems 69 and 70 use a CAS as an aid in solving the auxiliary equation. Form the general solution of the differential equation. Then use a CAS as an aid in solving the system of equations for the coefficients $c_i$, $i = 1, 2, 3, 4$ that results when the initial conditions are applied to the general solution.

**69.** $2y^{(4)} + 3y''' - 16y'' + 15y' - 4y = 0$,
$y(0) = -2$, $y'(0) = 6$, $y''(0) = 3$, $y'''(0) = \frac{1}{2}$

**70.** $y^{(4)} - 3y''' + 3y'' - y' = 0$,
$y(0) = y'(0) = 0$, $y''(0) = y'''(0) = 1$

---

## 4.4 Undetermined Coefficients—Superposition Approach*

**INTRODUCTION**  To solve a nonhomogeneous linear differential equation

$$a_n y^{(n)} + a_{n-1} y^{(n-1)} + \cdots + a_1 y' + a_0 y = g(x), \tag{1}$$

we must do two things:

- find the complementary function $y_c$ and
- find *any* particular solution $y_p$ of the nonhomogeneous equation (1).

Then, as was discussed in Section 4.1, the general solution of (1) is $y = y_c + y_p$. The complementary function $y_c$ is the general solution of the associated homogeneous DE of (1), that is,

$$a_n y^{(n)} + a_{n-1} y^{(n-1)} + \cdots + a_1 y' + a_0 y = 0.$$

In Section 4.3 we saw how to solve these kinds of equations when the coefficients were constants. Our goal in the present section is to develop a method for obtaining particular solutions.

**METHOD OF UNDETERMINED COEFFICIENTS**  The first of two ways we shall consider for obtaining a particular solution $y_p$ for a nonhomogeneous linear DE is called the **method of undetermined coefficients.** The underlying idea behind this method is a conjecture about the form of $y_p$, an educated guess really, that is motivated by the kinds of functions that make up the input function $g(x)$. The general method is limited to linear DEs such as (1) where

- the coefficients $a_i$, $i = 0, 1, \ldots, n$ are constants and
- $g(x)$ is a constant $k$, a polynomial function, an exponential function $e^{\alpha x}$, a sine or cosine function $\sin \beta x$ or $\cos \beta x$, or finite sums and products of these functions.

---

*Note to the Instructor:** In this section the method of undetermined coefficients is developed from the viewpoint of the superposition principle for nonhomogeneous equations (Theorem 4.1.7). In Section 4.5 an entirely different approach will be presented, one utilizing the concept of differential annihilator operators. Take your pick.

**NOTE** Strictly speaking, $g(x) = k$ (constant) is a polynomial function. Since a constant function is probably not the first thing that comes to mind when you think of polynomial functions, for emphasis we shall continue to use the redundancy "constant functions, polynomials, . . . ."

The following functions are some examples of the types of inputs $g(x)$ that are appropriate for this discussion:

$$g(x) = 10, \quad g(x) = x^2 - 5x, \quad g(x) = 15x - 6 + 8e^{-x},$$
$$g(x) = \sin 3x - 5x \cos 2x, \quad g(x) = xe^x \sin x + (3x^2 - 1)e^{-4x}.$$

That is, $g(x)$ is a linear combination of functions of the type

$$P(x) = a_n x^n + a_{n-1} x^{n-1} + \cdots + a_1 x + a_0, \quad P(x) e^{\alpha x}, \quad P(x) e^{\alpha x} \sin \beta x, \quad \text{and} \quad P(x) e^{\alpha x} \cos \beta x,$$

where $n$ is a nonnegative integer and $\alpha$ and $\beta$ are real numbers. The method of undetermined coefficients is not applicable to equations of form (1) when

$$g(x) = \ln x, \quad g(x) = \frac{1}{x}, \quad g(x) = \tan x, \quad g(x) = \sin^{-1} x,$$

and so on. Differential equations in which the input $g(x)$ is a function of this last kind will be considered in Section 4.6.

The set of functions that consists of constants, polynomials, exponentials $e^{\alpha x}$, sines, and cosines has the remarkable property that derivatives of their sums and products are again sums and products of constants, polynomials, exponentials $e^{\alpha x}$, sines, and cosines. Because the linear combination of derivatives $a_n y_p^{(n)} + a_{n-1} y_p^{(n-1)} + \cdots + a_1 y_p' + a_0 y_p$ must be identical to $g(x)$, it seems reasonable to assume that $y_p$ *has the same form as $g(x)$.*

The next two examples illustrate the basic method.

---

**EXAMPLE 1**    **General Solution Using Undetermined Coefficients**

Solve $y'' + 4y' - 2y = 2x^2 - 3x + 6.$         (2)

**SOLUTION Step 1.** We first solve the associated homogeneous equation $y'' + 4y' - 2y = 0$. From the quadratic formula we find that the roots of the auxiliary equation $m^2 + 4m - 2 = 0$ are $m_1 = -2 - \sqrt{6}$ and $m_2 = -2 + \sqrt{6}$. Hence the complementary function is

$$y_c = c_1 e^{-(2+\sqrt{6})x} + c_2 e^{(-2+\sqrt{6})x}.$$

**Step 2.** Now, because the function $g(x)$ is a quadratic polynomial, let us assume a particular solution that is also in the form of a quadratic polynomial:

$$y_p = Ax^2 + Bx + C.$$

We seek to determine *specific* coefficients $A$, $B$, and $C$ for which $y_p$ is a solution of (2). Substituting $y_p$ and the derivatives

$$y_p' = 2Ax + B \quad \text{and} \quad y_p'' = 2A$$

into the given differential equation (2), we get

$$y_p'' + 4y_p' - 2y_p = 2A + 8Ax + 4B - 2Ax^2 - 2Bx - 2C = 2x^2 - 3x + 6.$$

Because the last equation is supposed to be an identity, the coefficients of like powers of $x$ must be equal:

That is,        $-2A = 2$,        $8A - 2B = -3$,        $2A + 4B - 2C = 6$.

Solving this system of equations leads to the values $A = -1$, $B = -\frac{5}{2}$, and $C = -9$. Thus a particular solution is

$$y_p = -x^2 - \frac{5}{2}x - 9.$$

**Step 3.**   The general solution of the given equation is

$$y = y_c + y_p = c_1 e^{-(2+\sqrt{6})x} + c_2 e^{(-2+\sqrt{6})x} - x^2 - \frac{5}{2}x - 9. \qquad \blacksquare$$

## EXAMPLE 2   Particular Solution Using Undetermined Coefficients

Find a particular solution of $y'' - y' + y = 2 \sin 3x$.

**SOLUTION**   A natural first guess for a particular solution would be $A \sin 3x$. But because successive differentiations of $\sin 3x$ produce $\sin 3x$ *and* $\cos 3x$, we are prompted instead to assume a particular solution that includes both of these terms:

$$y_p = A \cos 3x + B \sin 3x.$$

Differentiating $y_p$ and substituting the results into the differential equation gives, after regrouping,

$$y_p'' - y_p' + y_p = (-8A - 3B) \cos 3x + (3A - 8B) \sin 3x = 2 \sin 3x$$

or

From the resulting system of equations,

$$-8A - 3B = 0, \qquad 3A - 8B = 2,$$

we get $A = \frac{6}{73}$ and $B = -\frac{16}{73}$. A particular solution of the equation is

$$y_p = \frac{6}{73} \cos 3x - \frac{16}{73} \sin 3x. \qquad \blacksquare$$

As we mentioned, the form that we assume for the particular solution $y_p$ is an educated guess; it is not a blind guess. This educated guess must take into consideration not only the types of functions that make up $g(x)$ but also, as we shall see in Example 4, the functions that make up the complementary function $y_c$.

### EXAMPLE 3   Forming $y_p$ by Superposition

Solve $y'' - 2y' - 3y = 4x - 5 + 6xe^{2x}$. $\qquad\qquad$ (3)

**SOLUTION Step 1.**  First, the solution of the associated homogeneous equation $y'' - 2y' - 3y = 0$ is found to be $y_c = c_1 e^{-x} + c_2 e^{3x}$.

**Step 2.**  Next, the presence of $4x - 5$ in $g(x)$ suggests that the particular solution includes a linear polynomial. Furthermore, because the derivative of the product $xe^{2x}$ produces $2xe^{2x}$ and $e^{2x}$, we also assume that the particular solution includes both $xe^{2x}$ and $e^{2x}$. In other words, $g$ is the sum of two basic kinds of functions:

$$g(x) = g_1(x) + g_2(x) = polynomial + exponentials.$$

Correspondingly, the superposition principle for nonhomogeneous equations (Theorem 4.1.7) suggests that we seek a particular solution

$$y_p = y_{p_1} + y_{p_2},$$

where $y_{p_1} = Ax + B$ and $y_{p_2} = Cxe^{2x} + Ee^{2x}$. Substituting

$$y_p = Ax + B + Cxe^{2x} + Ee^{2x}$$

into the given equation (3) and grouping like terms gives

$$y_p'' - 2y_p' - 3y_p = -3Ax - 2A - 3B - 3Cxe^{2x} + (2C - 3E)e^{2x} = 4x - 5 + 6xe^{2x}. \quad (4)$$

From this identity we obtain the four equations

$$-3A = 4, \qquad -2A - 3B = -5, \qquad -3C = 6, \qquad 2C - 3E = 0.$$

The last equation in this system results from the interpretation that the coefficient of $e^{2x}$ in the right member of (4) is zero. Solving, we find $A = -\frac{4}{3}$, $B = \frac{23}{9}$, $C = -2$, and $E = -\frac{4}{3}$. Consequently,

$$y_p = -\frac{4}{3}x + \frac{23}{9} - 2xe^{2x} - \frac{4}{3}e^{2x}.$$

**Step 3.**  The general solution of the equation is

$$y = c_1 e^{-x} + c_2 e^{3x} - \frac{4}{3}x + \frac{23}{9} - \left(2x + \frac{4}{3}\right)e^{2x}. \qquad\blacksquare$$

In light of the superposition principle (Theorem 4.1.7) we can also approach Example 3 from the viewpoint of solving two simpler problems. You should verify that substituting

$$y_{p_1} = Ax + B \qquad \text{into} \qquad y'' - 2y' - 3y = 4x - 5$$

and

$$y_{p_2} = Cxe^{2x} + Ee^{2x} \qquad \text{into} \qquad y'' - 2y' - 3y = 6xe^{2x}$$

yields, in turn, $y_{p_1} = -\frac{4}{3}x + \frac{23}{9}$ and $y_{p_2} = -\left(2x + \frac{4}{3}\right)e^{2x}$. A particular solution of (3) is then $y_p = y_{p_1} + y_{p_2}$.

The next example illustrates that sometimes the "obvious" assumption for the form of $y_p$ is not a correct assumption.

### EXAMPLE 4   A Glitch in the Method

Find a particular solution of $y'' - 5y' + 4y = 8e^x$.

**SOLUTION**   Differentiation of $e^x$ produces no new functions. Therefore proceeding as we did in the earlier examples, we can reasonably assume a particular solution of the form $y_p = Ae^x$. But substitution of this expression into the differential equation yields the contradictory statement $0 = 8e^x$, so we have clearly made the wrong guess for $y_p$.

The difficulty here is apparent on examining the complementary function $y_c = c_1 e^x + c_2 e^{4x}$. Observe that our assumption $Ae^x$ is already present in $y_c$. This means that $e^x$ is a solution of the associated homogeneous differential equation, and a constant multiple $Ae^x$ when substituted into the differential equation necessarily produces zero.

What then should be the form of $y_p$? Inspired by Case II of Section 4.3, let's see whether we can find a particular solution of the form

$$y_p = Axe^x.$$

Substituting $y_p' = Axe^x + Ae^x$ and $y_p'' = Axe^x + 2Ae^x$ into the differential equation and simplifying gives

$$y_p'' - 5y_p' + 4y_p = -3Ae^x = 8e^x.$$

From the last equality we see that the value of $A$ is now determined as $A = -\frac{8}{3}$. Therefore a particular solution of the given equation is $y_p = -\frac{8}{3}xe^x$. ■

The difference in the procedures used in Examples 1–3 and in Example 4 suggests that we consider two cases. The first case reflects the situation in Examples 1–3.

**CASE I**   No function in the assumed particular solution is a solution of the associated homogeneous differential equation.

In Table 4.4.1 we illustrate some specific examples of $g(x)$ in (1) along with the corresponding form of the particular solution. We are, of course, taking for granted that no function in the assumed particular solution $y_p$ is duplicated by a function in the complementary function $y_c$.

**TABLE 4.4.1**   **Trial Particular Solutions**

| $g(x)$ | Form of $y_p$ |
|---|---|
| **1.** 1 (any constant) | $A$ |
| **2.** $5x + 7$ | $Ax + B$ |
| **3.** $3x^2 - 2$ | $Ax^2 + Bx + C$ |
| **4.** $x^3 - x + 1$ | $Ax^3 + Bx^2 + Cx + E$ |
| **5.** $\sin 4x$ | $A \cos 4x + B \sin 4x$ |
| **6.** $\cos 4x$ | $A \cos 4x + B \sin 4x$ |
| **7.** $e^{5x}$ | $Ae^{5x}$ |
| **8.** $(9x - 2)e^{5x}$ | $(Ax + B)e^{5x}$ |
| **9.** $x^2 e^{5x}$ | $(Ax^2 + Bx + C)e^{5x}$ |
| **10.** $e^{3x} \sin 4x$ | $Ae^{3x} \cos 4x + Be^{3x} \sin 4x$ |
| **11.** $5x^2 \sin 4x$ | $(Ax^2 + Bx + C) \cos 4x + (Ex^2 + Fx + G) \sin 4x$ |
| **12.** $xe^{3x} \cos 4x$ | $(Ax + B)e^{3x} \cos 4x + (Cx + E)e^{3x} \sin 4x$ |

EXAMPLE 5   Forms of Particular Solutions — Case I

Determine the form of a particular solution of

**(a)** $y'' - 8y' + 25y = 5x^3 e^{-x} - 7e^{-x}$          **(b)** $y'' + 4y = x \cos x$

**SOLUTION (a)** We can write $g(x) = (5x^3 - 7)e^{-x}$. Using entry 9 in Table 4.4.1 as a model, we assume a particular solution of the form

$$y_p = (Ax^3 + Bx^2 + Cx + E)e^{-x}.$$

Note that there is no duplication between any of the terms in $y_p$ and the terms in the complementary function $y_c = e^{4x}(c_1 \cos 3x + c_2 \sin 3x)$.

**(b)** The function $g(x) = x \cos x$ is similar to entry 11 in Table 4.4.1 except, of course, that we use a linear rather than a quadratic polynomial and $\cos x$ and $\sin x$ instead of $\cos 4x$ and $\sin 4x$ in the form of $y_p$:

$$y_p = (Ax + B) \cos x + (Cx + E) \sin x.$$

Again observe that there is no duplication of terms between $y_p$ and $y_c = c_1 \cos 2x + c_2 \sin 2x$.  ∎

If $g(x)$ consists of a sum of, say, $m$ terms of the kind listed in the table, then (as in Example 3) the assumption for a particular solution $y_p$ consists of the sum of the trial forms $y_{p_1}, y_{p_2}, \ldots, y_{p_m}$ corresponding to these terms:

$$y_p = y_{p_1} + y_{p_2} + \cdots + y_{p_m}.$$

The foregoing sentence can be put another way.

> *Form Rule for Case I*   *The form of $y_p$ is a linear combination of all linearly independent functions that are generated by repeated differentiations of $g(x)$.*

EXAMPLE 6   Finding $y_p$ by Superposition—Case I

Determine the form of a particular solution of

$$y'' - 9y' + 14y = 3x^2 - 5 \sin 2x + 8xe^{6x}.$$

**SOLUTION** The right-hand side of the equation $g(x) = 3x^2 - 5 \sin 2x + 8xe^{6x}$ consists of three different types of functions: $x^2$, $\sin 2x$, and $xe^{6x}$. The derivatives of these functions yield, in turn, the additional functions $x$, 1; $\cos 2x$; and $e^{6x}$. Therefore:

corresponding to $x^2$ we assume        $y_{p_1} = Ax^2 + Bx + C$,

corresponding to $\sin 2x$ we assume        $y_{p_2} = E \cos 2x + F \sin 2x$,

corresponding to $xe^{6x}$ we assume        $y_{p_3} = Gxe^{6x} + He^{6x}$.

The assumption for a particular solution of the given nonhomogeneous differential equation is then

$$y_p = y_{p_1} + y_{p_2} + y_{p_3} = Ax^2 + Bx + C + E \cos 2x + F \sin 2x + (Gx + H)e^{6x}.$$

Note that none of the seven terms in this assumption for $y_p$ duplicates a term in the complementary function $y_c = c_1 e^{2x} + c_2 e^{7x}$.  ∎

**CASE II**   A function in the assumed particular solution is also a solution of the associated homogeneous differential equation.

The next example is similar to Example 4.

**EXAMPLE 7**   **Particular Solution—Case II**

Find a particular solution of $y'' - 2y' + y = e^x$.

**SOLUTION**  The complementary function is $y_c = c_1 e^x + c_2 x e^x$. As in Example 4, the assumption $y_p = A e^x$ will fail, since it is apparent from $y_c$ that $e^x$ is a solution of the associated homogeneous equation $y'' - 2y' + y = 0$. Moreover, we will not be able to find a particular solution of the form $y_p = A x e^x$, since the term $x e^x$ is also duplicated in $y_c$. We next try

$$y_p = A x^2 e^x.$$

Substituting into the given differential equation yields $2A e^x = e^x$, so $A = \frac{1}{2}$. Thus a particular solution is $y_p = \frac{1}{2} x^2 e^x$.  ■

Suppose again that $g(x)$ consists of $m$ terms of the kind given in Table 4.4.1, and suppose further that the usual assumption for a particular solution is

$$y_p = y_{p_1} + y_{p_2} + \cdots + y_{p_m},$$

where the $y_{p_i}$, $i = 1, 2, \ldots, m$ are the trial particular solution forms corresponding to these terms. Under the circumstances described in Case II, we can make up the following general rule.

> *__Multiplication Rule for Case II__   If any $y_{p_i}$ contains terms that duplicate terms in $y_{c}$, then that $y_{p_i}$ must be multiplied by $x^n$, where n is the smallest positive integer that eliminates that duplication.*

**EXAMPLE 8**   **An Initial-Value Problem**

Solve $y'' + y = 4x + 10 \sin x$, $y(\pi) = 0$, $y'(\pi) = 2$.

**SOLUTION**  The solution of the associated homogeneous equation $y'' + y = 0$ is $y_c = c_1 \cos x + c_2 \sin x$. Because $g(x) = 4x + 10 \sin x$ is the sum of a linear polynomial and a sine function, our normal assumption for $y_p$, from entries 2 and 5 of Table 4.4.1, would be the sum of $y_{p_1} = Ax + B$ and $y_{p_2} = C \cos x + E \sin x$:

$$y_p = Ax + B + C \cos x + E \sin x. \tag{5}$$

But there is an obvious duplication of the terms $\cos x$ and $\sin x$ in this assumed form and two terms in the complementary function. This duplication can be eliminated by simply multiplying $y_{p_2}$ by $x$. Instead of (5) we now use

$$y_p = Ax + B + Cx \cos x + Ex \sin x. \tag{6}$$

Differentiating this expression and substituting the results into the differential equation gives

$$y_p'' + y_p = Ax + B - 2C \sin x + 2E \cos x = 4x + 10 \sin x,$$

and so $A = 4$, $B = 0$, $-2C = 10$, and $2E = 0$. The solutions of the system are immediate: $A = 4$, $B = 0$, $C = -5$, and $E = 0$. Therefore from (6) we obtain $y_p = 4x - 5x \cos x$. The general solution of the given equation is

$$y = y_c + y_p = c_1 \cos x + c_2 \sin x + 4x - 5x \cos x.$$

We now apply the prescribed initial conditions to the general solution of the equation. First, $y(\pi) = c_1 \cos \pi + c_2 \sin \pi + 4\pi - 5\pi \cos \pi = 0$ yields $c_1 = 9\pi$, since $\cos \pi = -1$ and $\sin \pi = 0$. Next, from the derivative

$$y' = -9\pi \sin x + c_2 \cos x + 4 + 5x \sin x - 5 \cos x$$

and $\quad y'(\pi) = -9\pi \sin \pi + c_2 \cos \pi + 4 + 5\pi \sin \pi - 5 \cos \pi = 2$

we find $c_2 = 7$. The solution of the initial-value is then

$$y = 9\pi \cos x + 7 \sin x + 4x - 5x \cos x.$$

■

**EXAMPLE 9** | **Using the Multiplication Rule**

Solve $y'' - 6y' + 9y = 6x^2 + 2 - 12e^{3x}$.

**SOLUTION** The complementary function is $y_c = c_1 e^{3x} + c_2 x e^{3x}$. And so, based on entries 3 and 7 of Table 4.4.1, the usual assumption for a particular solution would be

$$y_p = \underbrace{Ax^2 + Bx + C}_{y_{p_1}} + \underbrace{Ee^{3x}}_{y_{p_2}}.$$

Inspection of these functions shows that the one term in $y_{p_2}$ is duplicated in $y_c$. If we multiply $y_{p_2}$ by $x$, we note that the term $xe^{3x}$ is still part of $y_c$. But multiplying $y_{p_2}$ by $x^2$ eliminates all duplications. Thus the operative form of a particular solution is

$$y_p = Ax^2 + Bx + C + Ex^2 e^{3x}.$$

Differentiating this last form, substituting into the differential equation, and collecting like terms gives

$$y_p'' - 6y_p' + 9y_p = 9Ax^2 + (-12A + 9B)x + 2A - 6B + 9C + 2Ee^{3x} = 6x^2 + 2 - 12e^{3x}.$$

It follows from this identity that $A = \frac{2}{3}$, $B = \frac{8}{9}$, $C = \frac{2}{3}$, and $E = -6$. Hence the general solution $y = y_c + y_p$ is $y = c_1 e^{3x} + c_2 x e^{3x} + \frac{2}{3}x^2 + \frac{8}{9}x + \frac{2}{3} - 6x^2 e^{3x}$.

■

**EXAMPLE 10** | **Third-Order DE—Case I**

Solve $y''' + y'' = e^x \cos x$.

**SOLUTION** From the characteristic equation $m^3 + m^2 = 0$ we find $m_1 = m_2 = 0$ and $m_3 = -1$. Hence the complementary function of the equation is $y_c = c_1 + c_2 x + c_3 e^{-x}$. With $g(x) = e^x \cos x$, we see from entry 10 of Table 4.4.1 that we should assume that

$$y_p = Ae^x \cos x + Be^x \sin x.$$

Because there are no functions in $y_p$ that duplicate functions in the complementary solution, we proceed in the usual manner. From

$$y_p''' + y_p'' = (-2A + 4B)e^x \cos x + (-4A - 2B)e^x \sin x = e^x \cos x$$

we get $-2A + 4B = 1$ and $-4A - 2B = 0$. This system gives $A = -\frac{1}{10}$ and $B = \frac{1}{5}$, so a particular solution is $y_p = -\frac{1}{10} e^x \cos x + \frac{1}{5} e^x \sin x$. The general solution of the equation is

$$y = y_c + y_p = c_1 + c_2 x + c_3 e^{-x} - \frac{1}{10} e^x \cos x + \frac{1}{5} e^x \sin x.$$

■

**EXAMPLE 11** | **Fourth-Order DE—Case II**

Determine the form of a particular solution of $y^{(4)} + y''' = 1 - x^2 e^{-x}$.

**SOLUTION** Comparing $y_c = c_1 + c_2 x + c_3 x^2 + c_4 e^{-x}$ with our normal assumption for a particular solution

$$y_p = \underbrace{A}_{y_{p_1}} + \underbrace{Bx^2 e^{-x} + Cxe^{-x} + Ee^{-x}}_{y_{p_2}},$$

we see that the duplications between $y_c$ and $y_p$ are eliminated when $y_{p_1}$ is multiplied by $x^3$ and $y_{p_2}$ is multiplied by $x$. Thus the correct assumption for a particular solution is $y_p = Ax^3 + Bx^3 e^{-x} + Cx^2 e^{-x} + Exe^{-x}$. ■

## REMARKS

(*i*) In Problems 27–36 in Exercises 4.4 you are asked to solve initial-value problems, and in Problems 37–40 you are asked to solve boundary-value problems. As illustrated in Example 8, be sure to apply the initial conditions or the boundary conditions to the general solution $y = y_c + y_p$. Students often make the mistake of applying these conditions only to the complementary function $y_c$ because it is that part of the solution that contains the constants $c_1, c_2, \ldots, c_n$.

(*ii*) From the "Form Rule for Case I" on page 147 of this section you see why the method of undetermined coefficients is not well suited to nonhomogeneous linear DEs when the input function $g(x)$ is something other than one of the four basic types highlighted in color on page 143. For example, if $P(x)$ is a polynomial, then continued differentiation of $P(x)e^{\alpha x} \sin \beta x$ will generate an independent set containing only a *finite* number of functions—all of the same type, namely, a polynomial times $e^{\alpha x} \sin \beta x$ or a polynomial times $e^{\alpha x} \cos \beta x$. On the other hand, repeated differentiation of input functions such as $g(x) = \ln x$ or $g(x) = \tan^{-1} x$ generates an independent set containing an *infinite* number of functions:

$$\text{derivatives of } \ln x: \quad \frac{1}{x}, \frac{-1}{x^2}, \frac{2}{x^3}, \ldots,$$

$$\text{derivatives of } \tan^{-1} x: \quad \frac{1}{1+x^2}, \frac{-2x}{(1+x^2)^2}, \frac{-2+6x^2}{(1+x^2)^3}, \ldots.$$

## EXERCISES 4.4

Answers to selected odd-numbered problems begin on page ANS-5.

In Problems 1–26 solve the given differential equation by undetermined coefficients.

**1.** $y'' + 3y' + 2y = 6$

**2.** $4y'' + 9y = 15$

**3.** $y'' - 10y' + 25y = 30x + 3$

**4.** $y'' + y' - 6y = 2x$

**5.** $\frac{1}{4}y'' + y' + y = x^2 - 2x$

**6.** $y'' - 8y' + 20y = 100x^2 - 26xe^x$

**7.** $y'' + 3y = -48x^2 e^{3x}$

**8.** $4y'' - 4y' - 3y = \cos 2x$

**9.** $y'' - y' = -3$

**10.** $y'' + 2y' = 2x + 5 - e^{-2x}$

**11.** $y'' - y' + \frac{1}{4}y = 3 + e^{x/2}$

**12.** $y'' - 16y = 2e^{4x}$

**13.** $y'' + 4y = 3 \sin 2x$

**14.** $y'' - 4y = (x^2 - 3) \sin 2x$

**15.** $y'' + y = 2x \sin x$

**16.** $y'' - 5y' = 2x^3 - 4x^2 - x + 6$

**17.** $y'' - 2y' + 5y = e^x \cos 2x$

**18.** $y'' - 2y' + 2y = e^{2x}(\cos x - 3\sin x)$

**19.** $y'' + 2y' + y = \sin x + 3\cos 2x$

**20.** $y'' + 2y' - 24y = 16 - (x + 2)e^{4x}$

**21.** $y''' - 6y'' = 3 - \cos x$

**22.** $y''' - 2y'' - 4y' + 8y = 6xe^{2x}$

**23.** $y''' - 3y'' + 3y' - y = x - 4e^x$

**24.** $y''' - y'' - 4y' + 4y = 5 - e^x + e^{2x}$

**25.** $y^{(4)} + 2y'' + y = (x - 1)^2$

**26.** $y^{(4)} - y'' = 4x + 2xe^{-x}$

In Problems 27–36 solve the given initial-value problem.

**27.** $y'' + 4y = -2$, $y(\pi/8) = \frac{1}{2}$, $y'(\pi/8) = 2$

**28.** $2y'' + 3y' - 2y = 14x^2 - 4x - 11$, $y(0) = 0$, $y'(0) = 0$

**29.** $5y'' + y' = -6x$, $y(0) = 0$, $y'(0) = -10$

**30.** $y'' + 4y' + 4y = (3 + x)e^{-2x}$, $y(0) = 2$, $y'(0) = 5$

**31.** $y'' + 4y' + 5y = 35e^{-4x}$, $y(0) = -3$, $y'(0) = 1$

**32.** $y'' - y = \cosh x$, $y(0) = 2$, $y'(0) = 12$

**33.** $\dfrac{d^2x}{dt^2} + \omega^2 x = F_0 \sin \omega t$, $x(0) = 0$, $x'(0) = 0$

**34.** $\dfrac{d^2x}{dt^2} + \omega^2 x = F_0 \cos \gamma t$, $x(0) = 0$, $x'(0) = 0$

**35.** $y''' - 2y'' + y' = 2 - 24e^x + 40e^{5x}$, $y(0) = \frac{1}{2}$, $y'(0) = \frac{5}{2}$, $y''(0) = -\frac{9}{2}$

**36.** $y''' + 8y = 2x - 5 + 8e^{-2x}$, $y(0) = -5$, $y'(0) = 3$, $y''(0) = -4$

In Problems 37–40 solve the given boundary-value problem.

**37.** $y'' + y = x^2 + 1$, $y(0) = 5$, $y(1) = 0$

**38.** $y'' - 2y' + 2y = 2x - 2$, $y(0) = 0$, $y(\pi) = \pi$

**39.** $y'' + 3y = 6x$, $y(0) = 0$, $y(1) + y'(1) = 0$

**40.** $y'' + 3y = 6x$, $y(0) + y'(0) = 0$, $y(1) = 0$

In Problems 41 and 42 solve the given initial-value problem in which the input function $g(x)$ is discontinuous. [*Hint*: Solve each problem on two intervals, and then find a solution so that $y$ and $y'$ are continuous at $x = \pi/2$ (Problem 41) and at $x = \pi$ (Problem 42).]

**41.** $y'' + 4y = g(x)$, $y(0) = 1$, $y'(0) = 2$, where

$$g(x) = \begin{cases} \sin x, & 0 \le x \le \pi/2 \\ 0, & x > \pi/2 \end{cases}$$

**42.** $y'' - 2y' + 10y = g(x)$, $y(0) = 0$, $y'(0) = 0$, where

$$g(x) = \begin{cases} 20, & 0 \le x \le \pi \\ 0, & x > \pi \end{cases}$$

## Discussion Problems

**43.** Consider the differential equation $ay'' + by' + cy = e^{kx}$, where $a$, $b$, $c$, and $k$ are constants. The auxiliary equation of the associated homogeneous equation is $am^2 + bm + c = 0$.

**(a)** If $k$ is not a root of the auxiliary equation, show that we can find a particular solution of the form $y_p = Ae^{kx}$, where $A = 1/(ak^2 + bk + c)$.

**(b)** If $k$ is a root of the auxiliary equation of multiplicity one, show that we can find a particular solution of the form $y_p = Axe^{kx}$, where $A = 1/(2ak + b)$. Explain how we know that $k \neq -b/(2a)$.

**(c)** If $k$ is a root of the auxiliary equation of multiplicity two, show that we can find a particular solution of the form $y = Ax^2 e^{kx}$, where $A = 1/(2a)$.

**44.** Discuss how the method of this section can be used to find a particular solution of $y'' + y = \sin x \cos 2x$. Carry out your idea.

In Problems 45–48 without solving, match a solution curve of $y'' + y = f(x)$ shown in the figure with one of the following functions:

(*i*) $f(x) = 1$,  (*ii*) $f(x) = e^{-x}$,

(*iii*) $f(x) = e^x$,  (*iv*) $f(x) = \sin 2x$,

(*v*) $f(x) = e^x \sin x$,  (*vi*) $f(x) = \sin x$.

Briefly discuss your reasoning.

**45.**

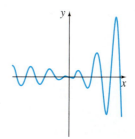

**FIGURE 4.4.1** Graph for Problem 45

**46.**

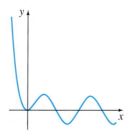

**FIGURE 4.4.2** Graph for Problem 46

**47.**

**FIGURE 4.4.3** Graph for Problem 47

**48.**

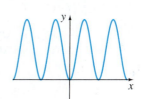

**FIGURE 4.4.4** Graph for Problem 48

### Computer Lab Assignments

In Problems 49 and 50 find a particular solution of the given differential equation. Use a CAS as an aid in carrying out differentiations, simplifications, and algebra.

**49.** $y'' - 4y' + 8y = (2x^2 - 3x)e^{2x} \cos 2x$
$$+ (10x^2 - x - 1)e^{2x} \sin 2x$$

**50.** $y^{(4)} + 2y'' + y = 2 \cos x - 3x \sin x$

---

## 4.5 Undetermined Coefficients—Annihilator Approach

**INTRODUCTION** We saw in Section 4.1 that an $n$th-order differential equation can be written

$$a_n D^n y + a_{n-1} D^{n-1} y + \cdots + a_1 Dy + a_0 y = g(x), \tag{1}$$

where $D^k y = d^k y / dx^k$, $k = 0, 1, \ldots, n$. When it suits our purpose, (1) is also written as $L(y) = g(x)$, where $L$ denotes the linear $n$th-order differential, or polynomial, operator

$$a_n D^n + a_{n-1} D^{n-1} + \cdots + a_1 D + a_0. \tag{2}$$

Not only is the operator notation a helpful shorthand, but also on a very practical level the application of differential operators enables us to justify the somewhat mind-numbing rules for determining the form of particular solution $y_p$ that were presented in the preceding section. In this section there are no special rules; the form of $y_p$ follows almost automatically once we have found an appropriate linear differential operator that *annihilates* $g(x)$ in (1). Before investigating how this is done, we need to examine two concepts.

**FACTORING OPERATORS** When the coefficients $a_i$, $i = 0, 1, \ldots, n$ are real constants, a linear differential operator (1) can be factored whenever the characteristic polynomial $a_n m^n + a_{n-1} m^{n-1} + \cdots + a_1 m + a_0$ factors. In other words, if $r_1$ is a root of the auxiliary equation

$$a_n m^n + a_{n-1} m^{n-1} + \cdots + a_1 m + a_0 = 0,$$

then $L = (D - r_1) P(D)$, where the polynomial expression $P(D)$ is a linear differential operator of order $n - 1$. For example, if we treat $D$ as an algebraic quantity, then the operator $D^2 + 5D + 6$ can be factored as $(D + 2)(D + 3)$ or as $(D + 3)(D + 2)$. Thus if a function $y = f(x)$ possesses a second derivative, then

$$(D^2 + 5D + 6)y = (D + 2)(D + 3)y = (D + 3)(D + 2)y.$$

This illustrates a general property:

*Factors of a linear differential operator with constant coefficients commute.*

A differential equation such as $y'' + 4y' + 4y = 0$ can be written as

$$(D^2 + 4D + 4)y = 0 \quad \text{or} \quad (D + 2)(D + 2)y = 0 \quad \text{or} \quad (D + 2)^2 y = 0.$$

**ANNIHILATOR OPERATOR** If $L$ is a linear differential operator with constant coefficients and $f$ is a sufficiently differentiable function such that

$$L(f(x)) = 0,$$

then $L$ is said to be an **annihilator** of the function. For example, a constant function $y = k$ is annihilated by $D$, since $Dk = 0$. The function $y = x$ is annihilated by the differential operator $D^2$ since the first and second derivatives of $x$ are 1 and 0, respectively. Similarly, $D^3 x^2 = 0$, and so on.

> The differential operator $D^n$ annihilates each of the functions
> $$1, \quad x, \quad x^2, \quad \ldots, \quad x^{n-1}. \tag{3}$$

As an immediate consequence of (3) and the fact that differentiation can be done term by term, a polynomial

$$c_0 + c_1 x + c_2 x^2 + \cdots + c_{n-1} x^{n-1} \tag{4}$$

can be annihilated by finding an operator that annihilates the highest power of $x$.

The functions that are annihilated by a linear $n$th-order differential operator $L$ are simply those functions that can be obtained from the general solution of the homogeneous differential equation $L(y) = 0$.

> The differential operator $(D - \alpha)^n$ annihilates each of the functions
> $$e^{\alpha x}, \quad x e^{\alpha x}, \quad x^2 e^{\alpha x}, \quad \ldots, \quad x^{n-1} e^{\alpha x}. \tag{5}$$

To see this, note that the auxiliary equation of the homogeneous equation $(D - \alpha)^n y = 0$ is $(m - \alpha)^n = 0$. Since $\alpha$ is a root of multiplicity $n$, the general solution is

$$y = c_1 e^{\alpha x} + c_2 x e^{\alpha x} + \cdots + c_n x^{n-1} e^{\alpha x}. \tag{6}$$

### EXAMPLE 1  Annihilator Operators

Find a differential operator that annihilates the given function.

**(a)** $1 - 5x^2 + 8x^3$  **(b)** $e^{-3x}$  **(c)** $4e^{2x} - 10xe^{2x}$

**SOLUTION**  **(a)** From (3) we know that $D^4 x^3 = 0$, so it follows from (4) that

$$D^4(1 - 5x^2 + 8x^3) = 0.$$

**(b)** From (5), with $\alpha = -3$ and $n = 1$, we see that

$$(D + 3)e^{-3x} = 0.$$

**(c)** From (5) and (6), with $\alpha = 2$ and $n = 2$, we have

$$(D - 2)^2(4e^{2x} - 10xe^{2x}) = 0. \qquad \blacksquare$$

When $\alpha$ and $\beta$, $\beta > 0$ are real numbers, the quadratic formula reveals that $[m^2 - 2\alpha m + (\alpha^2 + \beta^2)]^n = 0$ has complex roots $\alpha + i\beta$, $\alpha - i\beta$, both of multiplicity $n$. From the discussion at the end of Section 4.3 we have the next result.

> The differential operator $[D^2 - 2\alpha D + (\alpha^2 + \beta^2)]^n$ annihilates each of the functions
> $$e^{\alpha x} \cos \beta x, \quad x e^{\alpha x} \cos \beta x, \quad x^2 e^{\alpha x} \cos \beta x, \quad \ldots, \quad x^{n-1} e^{\alpha x} \cos \beta x,$$
> $$e^{\alpha x} \sin \beta x, \quad x e^{\alpha x} \sin \beta x, \quad x^2 e^{\alpha x} \sin \beta x, \quad \ldots, \quad x^{n-1} e^{\alpha x} \sin \beta x. \tag{7}$$

### EXAMPLE 2 Annihilator Operator

Find a differential operator that annihilates $5e^{-x} \cos 2x - 9e^{-x} \sin 2x$.

**SOLUTION** Inspection of the functions $e^{-x} \cos 2x$ and $e^{-x} \sin 2x$ shows that $\alpha = -1$ and $\beta = 2$. Hence from (7) we conclude that $D^2 + 2D + 5$ will annihilate each function. Since $D^2 + 2D + 5$ is a linear operator, it will annihilate *any* linear combination of these functions such as $5e^{-x} \cos 2x - 9e^{-x} \sin 2x$. ∎

When $\alpha = 0$ and $n = 1$, a special case of (7) is

$$(D^2 + \beta^2) \begin{cases} \cos \beta x \\ \sin \beta x \end{cases} = 0. \tag{8}$$

For example, $D^2 + 16$ will annihilate any linear combination of $\sin 4x$ and $\cos 4x$.

We are often interested in annihilating the sum of two or more functions. As we have just seen in Examples 1 and 2, if $L$ is a linear differential operator such that $L(y_1) = 0$ and $L(y_2) = 0$, then $L$ will annihilate the linear combination $c_1 y_1(x) + c_2 y_2(x)$. This is a direct consequence of Theorem 4.1.2. Let us now suppose that $L_1$ and $L_2$ are linear differential operators with constant coefficients such that $L_1$ annihilates $y_1(x)$ and $L_2$ annihilates $y_2(x)$, but $L_1(y_2) \neq 0$ and $L_2(y_1) \neq 0$. Then the *product* of differential operators $L_1 L_2$ annihilates the sum $c_1 y_1(x) + c_2 y_2(x)$. We can easily demonstrate this, using linearity and the fact that $L_1 L_2 = L_2 L_1$:

$$
\begin{aligned}
L_1 L_2(y_1 + y_2) &= L_1 L_2(y_1) + L_1 L_2(y_2) \\
&= L_2 L_1(y_1) + L_1 L_2(y_2) \\
&= L_2 [\underbrace{L_1(y_1)}_{\text{zero}}] + L_1 [\underbrace{L_2(y_2)}_{\text{zero}}] = 0.
\end{aligned}
$$

For example, we know from (3) that $D^2$ annihilates $7 - x$ and from (8) that $D^2 + 16$ annihilates $\sin 4x$. Therefore the product of operators $D^2(D^2 + 16)$ will annihilate the linear combination $7 - x + 6 \sin 4x$.

**NOTE** The differential operator that annihilates a function is not unique. We saw in part (b) of Example 1 that $D + 3$ will annihilate $e^{-3x}$, but so will differential operators of higher order as long as $D + 3$ is one of the factors of the operator. For example, $(D + 3)(D + 1)$, $(D + 3)^2$, and $D^3(D + 3)$ all annihilate $e^{-3x}$. (Verify this.) As a matter of course, when we seek a differential annihilator for a function $y = f(x)$, we want the operator of *lowest possible order* that does the job.

**UNDETERMINED COEFFICIENTS** This brings us to the point of the preceding discussion. Suppose that $L(y) = g(x)$ is a linear differential equation with constant coefficients and that the input $g(x)$ consists of finite sums and products of the functions listed in (3), (5), and (7)—that is, $g(x)$ is a linear combination of functions of the form

$$k \text{ (constant)}, \quad x^m, \quad x^m e^{\alpha x}, \quad x^m e^{\alpha x} \cos \beta x, \quad \text{and} \quad x^m e^{\alpha x} \sin \beta x,$$

where $m$ is a nonnegative integer and $\alpha$ and $\beta$ are real numbers. We now know that such a function $g(x)$ can be annihilated by a differential operator $L_1$ of lowest order, consisting of a product of the operators $D^n$, $(D - \alpha)^n$, and $(D^2 - 2\alpha D + \alpha^2 + \beta^2)^n$. Applying $L_1$ to both sides of the equation $L(y) = g(x)$ yields $L_1 L(y) = L_1(g(x)) = 0$. By solving the *homogeneous higher-order* equation $L_1 L(y) = 0$, we can discover the *form* of a particular solution $y_p$ for the original *nonhomogeneous* equation $L(y) = g(x)$. We then substitute this assumed form into $L(y) = g(x)$ to find an explicit particular solution. This procedure for determining

$y_p$, called the **method of undetermined coefficients,** is illustrated in the next several examples.

Before proceeding, recall that the general solution of a nonhomogeneous linear differential equation $L(y) = g(x)$ is $y = y_c + y_p$, where $y_c$ is the complementary function—that is, the general solution of the associated homogeneous equation $L(y) = 0$. The general solution of each equation $L(y) = g(x)$ is defined on the interval $(-\infty, \infty)$.

---

**EXAMPLE 3   General Solution Using Undetermined Coefficients**

Solve $y'' + 3y' + 2y = 4x^2$.  (9)

**SOLUTION Step 1.**   First, we solve the homogeneous equation $y'' + 3y' + 2y = 0$. Then, from the auxiliary equation $m^2 + 3m + 2 = (m + 1)(m + 2) = 0$ we find $m_1 = -1$ and $m_2 = -2$, and so the complementary function is

$$y_c = c_1 e^{-x} + c_2 e^{-2x}.$$

**Step 2.**   Now, since $4x^2$ is annihilated by the differential operator $D^3$, we see that $D^3(D^2 + 3D + 2)y = 4D^3 x^2$ is the same as

$$D^3(D^2 + 3D + 2)y = 0.  \tag{10}$$

The auxiliary equation of the fifth-order equation in (10),

$$m^3(m^2 + 3m + 2) = 0 \quad \text{or} \quad m^3(m + 1)(m + 2) = 0,$$

has roots $m_1 = m_2 = m_3 = 0$, $m_4 = -1$, and $m_5 = -2$. Thus its general solution must be

$$y = c_1 + c_2 x + c_3 x^2 + \boxed{c_4 e^{-x} + c_5 e^{-2x}}.  \tag{11}$$

The terms in the shaded box in (11) constitute the complementary function of the original equation (9). We can then argue that a particular solution $y_p$ of (9) should also satisfy equation (10). This means that the terms remaining in (11) must be the basic form of $y_p$:

$$y_p = A + Bx + Cx^2,  \tag{12}$$

where, for convenience, we have replaced $c_1$, $c_2$, and $c_3$ by $A$, $B$, and $C$, respectively. For (12) to be a particular solution of (9), it is necessary to find *specific* coefficients $A$, $B$, and $C$. Differentiating (12), we have

$$y_p' = B + 2Cx, \qquad y_p'' = 2C,$$

and substitution into (9) then gives

$$y_p'' + 3y_p' + 2y_p = 2C + 3B + 6Cx + 2A + 2Bx + 2Cx^2 = 4x^2.$$

Because the last equation is supposed to be an identity, the coefficients of like powers of $x$ must be equal:

That is    $2C = 4, \qquad 2B + 6C = 0, \qquad 2A + 3B + 2C = 0.$  (13)

Solving the equations in (13) gives $A = 7$, $B = -6$, and $C = 2$. Thus $y_p = 7 - 6x + 2x^2$.

**Step 3.**   The general solution of the equation in (9) is $y = y_c + y_p$ or

$$y = c_1 e^{-x} + c_2 e^{-2x} + 7 - 6x + 2x^2.$$  ■

---

**EXAMPLE 4**   **General Solution Using Undetermined Coefficients**

---

Solve $y'' - 3y' = 8e^{3x} + 4 \sin x$.   (14)

**SOLUTION  Step 1.**   The auxiliary equation for the associated homogeneous equation $y'' - 3y' = 0$ is $m^2 - 3m = m(m - 3) = 0$, so $y_c = c_1 + c_2 e^{3x}$.

**Step 2.**   Now, since $(D - 3)e^{3x} = 0$ and $(D^2 + 1) \sin x = 0$, we apply the differential operator $(D - 3)(D^2 + 1)$ to both sides of (14):

$$(D - 3)(D^2 + 1)(D^2 - 3D)y = 0.$$   (15)

The auxiliary equation of (15) is

$$(m - 3)(m^2 + 1)(m^2 - 3m) = 0 \qquad \text{or} \qquad m(m - 3)^2(m^2 + 1) = 0.$$

Thus    $y = \boxed{c_1 + c_2 e^{3x}} + c_3 x e^{3x} + c_4 \cos x + c_5 \sin x.$

After excluding the linear combination of terms in the box that corresponds to $y_c$, we arrive at the form of $y_p$:

$$y_p = Ax e^{3x} + B \cos x + C \sin x.$$

Substituting $y_p$ in (14) and simplifying yield

$$y_p'' - 3y_p' = 3Ae^{3x} + (-B - 3C) \cos x + (3B - C) \sin x = 8e^{3x} + 4 \sin x.$$

Equating coefficients gives $3A = 8$, $-B - 3C = 0$, and $3B - C = 4$. We find $A = \frac{8}{3}$, $B = \frac{6}{5}$, and $C = -\frac{2}{5}$, and consequently,

$$y_p = \frac{8}{3} x e^{3x} + \frac{6}{5} \cos x - \frac{2}{5} \sin x.$$

**Step 3.**   The general solution of (14) is then

$$y = c_1 + c_2 e^{3x} + \frac{8}{3} x e^{3x} + \frac{6}{5} \cos x - \frac{2}{5} \sin x.$$  ■

---

**EXAMPLE 5**   **General Solution Using Undetermined Coefficients**

---

Solve $y'' + y = x \cos x - \cos x$.   (16)

**SOLUTION**  The complementary function is $y_c = c_1 \cos x + c_2 \sin x$. Now by comparing $\cos x$ and $x \cos x$ with the functions in the first row of (7), we see that $\alpha = 0$ and $n = 1$, and so $(D^2 + 1)^2$ is an annihilator for the right-hand member of the equation in (16). Applying this operator to the differential equation gives

$$(D^2 + 1)^2 (D^2 + 1)y = 0 \qquad \text{or} \qquad (D^2 + 1)^3 y = 0.$$

Since $i$ and $-i$ are both complex roots of multiplicity 3 of the auxiliary equation of the last differential equation, we conclude that

$$y = \boxed{c_1 \cos x + c_2 \sin x} + c_3 x \cos x + c_4 x \sin x + c_5 x^2 \cos x + c_6 x^2 \sin x.$$

We substitute

$$y_p = Ax \cos x + Bx \sin x + Cx^2 \cos x + Ex^2 \sin x$$

into (16) and simplify:

$$y_p'' + y_p = 4Ex \cos x - 4Cx \sin x + (2B + 2C) \cos x + (-2A + 2E) \sin x$$
$$= x \cos x - \cos x.$$

Equating coefficients gives the equations $4E = 1$, $-4C = 0$, $2B + 2C = -1$, and $-2A + 2E = 0$, from which we find $A = \frac{1}{4}$, $B = -\frac{1}{2}$, $C = 0$, and $E = \frac{1}{4}$. Hence the general solution of (16) is

$$y = c_1 \cos x + c_2 \sin x + \frac{1}{4}x \cos x - \frac{1}{2}x \sin x + \frac{1}{4}x^2 \sin x. \qquad \blacksquare$$

### EXAMPLE 6   Form of a Particular Solution

Determine the form of a particular solution for

$$y'' - 2y' + y = 10e^{-2x} \cos x. \qquad (17)$$

**SOLUTION** The complementary function for the given equation is $y_c = c_1 e^x + c_2 x e^x$. Now from (7), with $\alpha = -2$, $\beta = 1$, and $n = 1$, we know that

$$(D^2 + 4D + 5)e^{-2x} \cos x = 0.$$

Applying the operator $D^2 + 4D + 5$ to (17) gives

$$(D^2 + 4D + 5)(D^2 - 2D + 1)y = 0. \qquad (18)$$

Since the roots of the auxiliary equation of (18) are $-2 - i$, $-2 + i$, 1, and 1, we see from

$$y = \boxed{c_1 e^x + c_2 x e^x} + c_3 e^{-2x} \cos x + c_4 e^{-2x} \sin x$$

that a particular solution of (17) can be found with the form

$$y_p = Ae^{-2x} \cos x + Be^{-2x} \sin x. \qquad \blacksquare$$

### EXAMPLE 7   Form of a Particular Solution

Determine the form of a particular solution for

$$y''' - 4y'' + 4y' = 5x^2 - 6x + 4x^2 e^{2x} + 3e^{5x}. \qquad (19)$$

**SOLUTION** Observe that

$$D^3(5x^2 - 6x) = 0, \qquad (D - 2)^3 x^2 e^{2x} = 0, \qquad \text{and} \qquad (D - 5)e^{5x} = 0.$$

Therefore $D^3(D - 2)^3(D - 5)$ applied to (19) gives

$$D^3(D - 2)^3(D - 5)(D^3 - 4D^2 + 4D)y = 0$$

or

$$D^4(D - 2)^5(D - 5)y = 0.$$

The roots of the auxiliary equation for the last differential equation are easily seen to be 0, 0, 0, 0, 2, 2, 2, 2, 2, and 5. Hence

$$y = \boxed{c_1} + c_2 x + c_3 x^2 + c_4 x^3 + \boxed{c_5 e^{2x} + c_6 x e^{2x}} + c_7 x^2 e^{2x} + c_8 x^3 e^{2x} + c_9 x^4 e^{2x} + c_{10} e^{5x}. \qquad (20)$$

Because the linear combination $c_1 + c_5e^{2x} + c_6xe^{2x}$ corresponds to the complementary function of (19), the remaining terms in (20) give the form of a particular solution of the differential equation:

$$y_p = Ax + Bx^2 + Cx^3 + Ex^2e^{2x} + Fx^3e^{2x} + Gx^4e^{2x} + He^{5x}. \qquad ■$$

**SUMMARY OF THE METHOD**   For your convenience the method of undetermined coefficients is summarized as follows.

### Undetermined Coefficients—Annihilator Approach

The differential equation $L(y) = g(x)$ has constant coefficients, and the function $g(x)$ consists of finite sums and products of constants, polynomials, exponential functions $e^{\alpha x}$, sines, and cosines.

*(i)* Find the complementary solution $y_c$ for the homogeneous equation $L(y) = 0$.

*(ii)* Operate on both sides of the nonhomogeneous equation $L(y) = g(x)$ with a differential operator $L_1$ that annihilates the function $g(x)$.

*(iii)* Find the general solution of the higher-order homogeneous differential equation $L_1L(y) = 0$.

*(iv)* Delete from the solution in step *(iii)* all those terms that are duplicated in the complementary solution $y_c$ found in step *(i)*. Form a linear combination $y_p$ of the terms that remain. This is the form of a particular solution of $L(y) = g(x)$.

*(v)* Substitute $y_p$ found in step *(iv)* into $L(y) = g(x)$. Match coefficients of the various functions on each side of the equality, and solve the resulting system of equations for the unknown coefficients in $y_p$.

*(vi)* With the particular solution found in step *(v)*, form the general solution $y = y_c + y_p$ of the given differential equation.

### REMARKS

The method of undetermined coefficients is not applicable to linear differential equations with variable coefficients nor is it applicable to linear equations with constant coefficients when $g(x)$ is a function such as

$$g(x) = \ln x, \qquad g(x) = \frac{1}{x}, \qquad g(x) = \tan x, \qquad g(x) = \sin^{-1} x,$$

and so on. Differential equations in which the input $g(x)$ is a function of this last kind will be considered in the next section.

## EXERCISES 4.5

*Answers to selected odd-numbered problems begin on page ANS-5.*

In Problems 1–10 write the given differential equation in the form $L(y) = g(x)$, where $L$ is a linear differential operator with constant coefficients. If possible, factor $L$.

**1.** $9y'' - 4y = \sin x$

**2.** $y'' - 5y = x^2 - 2x$

**3.** $y'' - 4y' - 12y = x - 6$

**4.** $2y'' - 3y' - 2y = 1$

**5.** $y''' + 10y'' + 25y' = e^x$

**6.** $y''' + 4y' = e^x \cos 2x$

**7.** $y''' + 2y'' - 13y' + 10y = xe^{-x}$

**8.** $y''' + 4y'' + 3y' = x^2 \cos x - 3x$

**9.** $y^{(4)} + 8y' = 4$

**10.** $y^{(4)} - 8y'' + 16y = (x^3 - 2x)e^{4x}$

In Problems 11–14 verify that the given differential operator annihilates the indicated functions.

**11.** $D^4$; $y = 10x^3 - 2x$

**12.** $2D - 1$; $y = 4e^{x/2}$

**13.** $(D - 2)(D + 5)$; $y = e^{2x} + 3e^{-5x}$

**14.** $D^2 + 64$; $y = 2 \cos 8x - 5 \sin 8x$

In Problems 15–26 find a linear differential operator that annihilates the given function.

**15.** $1 + 6x - 2x^3$

**16.** $x^3(1 - 5x)$

**17.** $1 + 7e^{2x}$

**18.** $x + 3xe^{6x}$

**19.** $\cos 2x$

**20.** $1 + \sin x$

**21.** $13x + 9x^2 - \sin 4x$

**22.** $8x - \sin x + 10 \cos 5x$

**23.** $e^{-x} + 2xe^x - x^2e^x$

**24.** $(2 - e^x)^2$

**25.** $3 + e^x \cos 2x$

**26.** $e^{-x} \sin x - e^{2x} \cos x$

In Problems 27–34 find linearly independent functions that are annihilated by the given differential operator.

**27.** $D^5$

**28.** $D^2 + 4D$

**29.** $(D - 6)(2D + 3)$

**30.** $D^2 - 9D - 36$

**31.** $D^2 + 5$

**32.** $D^2 - 6D + 10$

**33.** $D^3 - 10D^2 + 25D$

**34.** $D^2(D - 5)(D - 7)$

In Problems 35–64 solve the given differential equation by undetermined coefficients.

**35.** $y'' - 9y = 54$

**36.** $2y'' - 7y' + 5y = -29$

**37.** $y'' + y' = 3$

**38.** $y''' + 2y'' + y' = 10$

**39.** $y'' + 4y' + 4y = 2x + 6$

**40.** $y'' + 3y' = 4x - 5$

**41.** $y''' + y'' = 8x^2$

**42.** $y'' - 2y' + y = x^3 + 4x$

**43.** $y'' - y' - 12y = e^{4x}$

**44.** $y'' + 2y' + 2y = 5e^{6x}$

**45.** $y'' - 2y' - 3y = 4e^x - 9$

**46.** $y'' + 6y' + 8y = 3e^{-2x} + 2x$

**47.** $y'' + 25y = 6 \sin x$

**48.** $y'' + 4y = 4 \cos x + 3 \sin x - 8$

**49.** $y'' + 6y' + 9y = -xe^{4x}$

**50.** $y'' + 3y' - 10y = x(e^x + 1)$

**51.** $y'' - y = x^2e^x + 5$

**52.** $y'' + 2y' + y = x^2e^{-x}$

**53.** $y'' - 2y' + 5y = e^x \sin x$

**54.** $y'' + y' + \dfrac{1}{4}y = e^x(\sin 3x - \cos 3x)$

**55.** $y'' + 25y = 20 \sin 5x$

**56.** $y'' + y = 4 \cos x - \sin x$

**57.** $y'' + y' + y = x \sin x$

**58.** $y'' + 4y = \cos^2 x$

**59.** $y''' + 8y'' = -6x^2 + 9x + 2$

**60.** $y''' - y'' + y' - y = xe^x - e^{-x} + 7$

**61.** $y''' - 3y'' + 3y' - y = e^x - x + 16$

**62.** $2y''' - 3y'' - 3y' + 2y = (e^x + e^{-x})^2$

**63.** $y^{(4)} - 2y''' + y'' = e^x + 1$

**64.** $y^{(4)} - 4y'' = 5x^2 - e^{2x}$

In Problems 65–72 solve the given initial-value problem.

**65.** $y'' - 64y = 16$, $y(0) = 1, y'(0) = 0$

**66.** $y'' + y' = x$, $y(0) = 1, y'(0) = 0$

**67.** $y'' - 5y' = x - 2$, $y(0) = 0, y'(0) = 2$

**68.** $y'' + 5y' - 6y = 10e^{2x}$, $y(0) = 1, y'(0) = 1$

**69.** $y'' + y = 8 \cos 2x - 4 \sin x$, $y(\pi/2) = -1, y'(\pi/2) = 0$

**70.** $y''' - 2y'' + y' = xe^x + 5$, $y(0) = 2, y'(0) = 2$, $y''(0) = -1$

**71.** $y'' - 4y' + 8y = x^3$, $y(0) = 2, y'(0) = 4$

**72.** $y^{(4)} - y''' = x + e^x$, $y(0) = 0, y'(0) = 0, y''(0) = 0$, $y'''(0) = 0$

### Discussion Problems

**73.** Suppose $L$ is a linear differential operator that factors but has variable coefficients. Do the factors of $L$ commute? Defend your answer.

---

## 4.6 VARIATION OF PARAMETERS

**INTRODUCTION** We pointed out in the discussions in Sections 4.4 and 4.5 that the method of undetermined coefficients has two inherent weaknesses that limit its wider application to linear equations: The DE must have constant coefficients and the input function $g(x)$ must be of the type listed in Table 4.4.1. In this section we examine a method for determining a particular solution $y_p$ of a nonhomogeneous linear DE that has, in theory, no such restrictions on it. This method, due to the eminent astronomer and mathematician **Joseph Louis Lagrange** (1736–1813), is known as **variation of parameters.**

Before examining this powerful method for higher-order equations we revisit the solution of linear first-order differential equations that have been put into standard form. The discussion under the first heading in this section is optional and is intended to motivate the main discussion of this section that starts under the second heading. If pressed for time this motivational material could be assigned for reading.

### LINEAR FIRST-ORDER DEs REVISITED

In Section 2.3 we saw that the general solution of a linear first-order differential equation $a_1(x)y' + a_0(x)y = g(x)$ can be found by first rewriting it in the standard form

$$\frac{dy}{dx} + P(x)y = f(x) \tag{1}$$

and assuming that $P(x)$ and $f(x)$ are continuous on a common interval $I$. Using the integrating factor method, the general solution of (1) on the interval $I$, was found to be

See (4) of Section 2.3.

$$y = c_1 e^{-\int P(x)\,dx} + e^{-\int P(x)\,dx} \int e^{\int P(x)\,dx} f(x)\,dx.$$

The foregoing solution has the same form as that given in Theorem 4.1.6, namely, $y = y_c + y_p$. In this case $y_c = c_1 e^{-\int P(x)\,dx}$ is a solution of the associated homogeneous equation

$$\frac{dy}{dx} + P(x)y = 0 \tag{2}$$

and

$$y_p = e^{-\int P(x)\,dx} \int e^{\int P(x)\,dx} f(x)\,dx \tag{3}$$

is a particular solution of the nonhomogeneous equation (1). As a means of motivating a method for solving nonhomogeneous linear equations of higher-order we propose to rederive the particular solution (3) by a method known as **variation of parameters.**

The basic procedure is that used in Section 4.2.

Suppose that $y_1$ is a known solution of the homogeneous equation (2), that is,

$$\frac{dy_1}{dx} + P(x)y_1 = 0. \tag{4}$$

It is easily shown that $y_1 = e^{-\int P(x)\,dx}$ is a solution of (4) and because the equation is linear, $c_1 y_1(x)$ is its general solution. Variation of parameters consists of finding a particular solution of (1) of the form $y_p = u_1(x)y_1(x)$. In other words, we have replaced the *parameter* $c_1$ by a *function* $u_1$.

Substituting $y_p = u_1 y_1$ into (1) and using the Product Rule gives

$$\frac{d}{dx}[u_1 y_1] + P(x)u_1 y_1 = f(x)$$

$$u_1 \frac{dy_1}{dx} + y_1 \frac{du_1}{dx} + P(x)u_1 y_1 = f(x)$$

0 because of (4)

$$u_1 \left[ \frac{dy_1}{dx} + P(x)y_1 \right] + y_1 \frac{du_1}{dx} = f(x)$$

so

$$y_1 \frac{du_1}{dx} = f(x).$$

By separating variables and integrating, we find $u_1$:

$$du_1 = \frac{f(x)}{y_1(x)}\,dx \quad \text{yields} \quad u_1 = \int \frac{f(x)}{y_1(x)}\,dx.$$

Hence the sought-after particular solution is

$$y_p = u_1 y_1 = y_1 \int \frac{f(x)}{y_1(x)}\,dx.$$

From the fact that $y_1 = e^{-\int P(x)\,dx}$ we see the last result is identical to (3).

**LINEAR SECOND-ORDER DEs**   Next we consider the case of a linear second-order equation

$$a_2(x)y'' + a_1(x)y' + a_0(x)y = g(x), \tag{5}$$

although, as we shall see, variation of parameters extends to higher-order equations. The method again begins by putting (5) into the standard form

$$y'' + P(x)y' + Q(x)y = f(x) \tag{6}$$

by dividing by the leading coefficient $a_2(x)$. In (6) we suppose that coefficient functions $P(x)$, $Q(x)$, and $f(x)$ are continuous on some common interval $I$. As we have already seen in Section 4.3, there is no difficulty in obtaining the complementary solution $y_c = c_1 y_1(x) + c_2 y_2(x)$, the general solution of the associated homogeneous equation of (6), when the coefficients are constants. Analogous to the preceding discussion, we now ask: Can the parameters $c_1$ and $c_2$ in $y_c$ can be replaced with functions $u_1$ and $u_2$, or "variable parameters," so that

$$y_p(x) = u_1(x)y_1(x) + u_2(x)y_2(x) \tag{7}$$

is a particular solution of (6)? To answer this question we substitute (7) into (6). Using the Product Rule to differentiate $y_p$ twice, we get

$$y_p' = u_1 y_1' + y_1 u_1' + u_2 y_2' + y_2 u_2'$$

$$y_p'' = u_1 y_1'' + y_1' u_1' + y_1 u_1'' + u_1' y_1' + u_2 y_2'' + y_2' u_2' + y_2 u_2'' + u_2' y_2'.$$

Substituting (7) and the foregoing derivatives into (6) and grouping terms yields

$$y_p'' + P(x)y_p' + Q(x)y_p = u_1[\overbrace{y_1'' + Py_1' + Qy_1}^{\text{zero}}] + u_2[\overbrace{y_2'' + Py_2' + Qy_2}^{\text{zero}}] + y_1 u_1'' + u_1' y_1'$$

$$+ y_2 u_2'' + u_2' y_2' + P[y_1 u_1' + y_2 u_2'] + y_1' u_1' + y_2' u_2'$$

$$= \frac{d}{dx}[y_1 u_1'] + \frac{d}{dx}[y_2 u_2'] + P[y_1 u_1' + y_2 u_2'] + y_1' u_1' + y_2' u_2'$$

$$= \frac{d}{dx}[y_1 u_1' + y_2 u_2'] + P[y_1 u_1' + y_2 u_2'] + y_1' u_1' + y_2' u_2' = f(x). \tag{8}$$

Because we seek to determine two unknown functions $u_1$ and $u_2$, reason dictates that we need two equations. We can obtain these equations by making the further assumption that the functions $u_1$ and $u_2$ satisfy $y_1 u_1' + y_2 u_2' = 0$. This assumption does not come out of the blue but is prompted by the first two terms in (8), since if we demand that $y_1 u_1' + y_2 u_2' = 0$, then (8) reduces to $y_1' u_1' + y_2' u_2' = f(x)$. We now have our desired two equations, albeit two equations for determining the derivatives $u_1'$ and $u_2'$. By Cramer's Rule, the solution of the system

$$y_1 u_1' + y_2 u_2' = 0$$

$$y_1' u_1' + y_2' u_2' = f(x)$$

can be expressed in terms of determinants:

$$u_1' = \frac{W_1}{W} = -\frac{y_2 f(x)}{W} \quad \text{and} \quad u_2' = \frac{W_2}{W} = \frac{y_1 f(x)}{W}, \tag{9}$$

where $\quad W = \begin{vmatrix} y_1 & y_2 \\ y_1' & y_2' \end{vmatrix}, \quad W_1 = \begin{vmatrix} 0 & y_2 \\ f(x) & y_2' \end{vmatrix}, \quad W_2 = \begin{vmatrix} y_1 & 0 \\ y_1' & f(x) \end{vmatrix}. \tag{10}$

The functions $u_1$ and $u_2$ are found by integrating the results in (9). The determinant $W$ is recognized as the Wronskian of $y_1$ and $y_2$. By linear independence of $y_1$ and $y_2$ on $I$, we know that $W(y_1(x), y_2(x)) \neq 0$ for every $x$ in the interval.

**SUMMARY OF THE METHOD**   Usually, it is not a good idea to memorize formulas in lieu of understanding a procedure. However, the foregoing procedure is too long and complicated to use each time we wish to solve a differential equation. In this case it is more efficient to simply use the formulas in (9). Thus to solve $a_2 y'' + a_1 y' + a_0 y = g(x)$, first find the complementary function $y_c = c_1 y_1 + c_2 y_2$ and then compute the Wronskian $W(y_1(x), y_2(x))$. By dividing by $a_2$, we put the equation into the standard form $y'' + Py' + Qy = f(x)$ to determine $f(x)$. We find $u_1$ and $u_2$ by integrating $u_1' = W_1/W$ and $u_2' = W_2/W$, where $W_1$ and $W_2$ are defined as in (10). A particular solution is $y_p = u_1 y_1 + u_2 y_2$. The general solution of the equation is then $y = y_c + y_p$.

**EXAMPLE 1**   **General Solution Using Variation of Parameters**

Solve $y'' - 4y' + 4y = (x + 1)e^{2x}$.

**SOLUTION** From the auxiliary equation $m^2 - 4m + 4 = (m - 2)^2 = 0$ we have $y_c = c_1 e^{2x} + c_2 x e^{2x}$. With the identifications $y_1 = e^{2x}$ and $y_2 = x e^{2x}$, we next compute the Wronskian:

$$W(e^{2x}, x e^{2x}) = \begin{vmatrix} e^{2x} & x e^{2x} \\ 2e^{2x} & 2x e^{2x} + e^{2x} \end{vmatrix} = e^{4x}.$$

Since the given differential equation is already in form (6) (that is, the coefficient of $y''$ is 1), we identify $f(x) = (x + 1)e^{2x}$. From (10) we obtain

$$W_1 = \begin{vmatrix} 0 & x e^{2x} \\ (x + 1)e^{2x} & 2x e^{2x} + e^{2x} \end{vmatrix} = -(x + 1)x e^{4x}, \quad W_2 = \begin{vmatrix} e^{2x} & 0 \\ 2e^{2x} & (x + 1)e^{2x} \end{vmatrix} = (x + 1)e^{4x},$$

and so from (9)

$$u_1' = -\frac{(x + 1)x e^{4x}}{e^{4x}} = -x^2 - x, \quad u_2' = \frac{(x + 1)e^{4x}}{e^{4x}} = x + 1.$$

Integrating the foregoing derivatives gives

$$u_1 = -\frac{1}{3}x^3 - \frac{1}{2}x^2 \quad \text{and} \quad u_2 = \frac{1}{2}x^2 + x.$$

Hence from (7) we have

$$y_p = \left(-\frac{1}{3}x^3 - \frac{1}{2}x^2\right)e^{2x} + \left(\frac{1}{2}x^2 + x\right)x e^{2x} = \frac{1}{6}x^3 e^{2x} + \frac{1}{2}x^2 e^{2x}$$

and $\quad y = y_c + y_p = c_1 e^{2x} + c_2 x e^{2x} + \frac{1}{6}x^3 e^{2x} + \frac{1}{2}x^2 e^{2x}.$ ∎

EXAMPLE 2   **General Solution Using Variation of Parameters**

Solve $4y'' + 36y = \csc 3x$.

**SOLUTION**  We first put the equation in the standard form (6) by dividing by 4:

$$y'' + 9y = \frac{1}{4} \csc 3x.$$

Because the roots of the auxiliary equation $m^2 + 9 = 0$ are $m_1 = 3i$ and $m_2 = -3i$, the complementary function is $y_c = c_1 \cos 3x + c_2 \sin 3x$. Using $y_1 = \cos 3x$, $y_2 = \sin 3x$, and $f(x) = \frac{1}{4} \csc 3x$, we obtain

$$W(\cos 3x, \sin 3x) = \begin{vmatrix} \cos 3x & \sin 3x \\ -3 \sin 3x & 3 \cos 3x \end{vmatrix} = 3,$$

$$W_1 = \begin{vmatrix} 0 & \sin 3x \\ \frac{1}{4} \csc 3x & 3 \cos 3x \end{vmatrix} = -\frac{1}{4}, \qquad W_2 = \begin{vmatrix} \cos 3x & 0 \\ -3 \sin 3x & \frac{1}{4} \csc 3x \end{vmatrix} = \frac{1}{4} \frac{\cos 3x}{\sin 3x}.$$

Integrating     $u'_1 = \dfrac{W_1}{W} = -\dfrac{1}{12}$     and     $u'_2 = \dfrac{W_2}{W} = \dfrac{1}{12} \dfrac{\cos 3x}{\sin 3x}$

gives

$$u_1 = -\frac{1}{12}x \qquad \text{and} \qquad u_2 = \frac{1}{36} \ln |\sin 3x|.$$

Thus from (7) a particular solution is

$$y_p = -\frac{1}{12} x \cos 3x + \frac{1}{36} (\sin 3x) \ln|\sin 3x|.$$

The general solution of the equation is

$$y = y_c + y_p = c_1 \cos 3x + c_2 \sin 3x - \frac{1}{12} x \cos 3x + \frac{1}{36} (\sin 3x) \ln|\sin 3x|. \quad (11) \quad \blacksquare$$

Equation (11) represents the general solution of the differential equation on, say, the interval $(0, \pi/6)$.

**CONSTANTS OF INTEGRATION**   When computing the indefinite integrals of $u'_1$ and $u'_2$, we need not introduce any constants. This is because

$$y = y_c + y_p = c_1 y_1 + c_2 y_2 + (u_1 + a_1) y_1 + (u_2 + b_1) y_2$$

$$= (c_1 + a_1) y_1 + (c_2 + b_1) y_2 + u_1 y_1 + u_2 y_2$$

$$= C_1 y_1 + C_2 y_2 + u_1 y_1 + u_2 y_2.$$

**INTEGRAL-DEFINED FUNCTIONS**   We have seen several times in the preceding sections and chapters that when a solution method involves integration we may encounter nonelementary integrals. As the next example shows, sometimes the best we can do in constructing a particular solution (7) of a linear second-order differential equation is to use the integral-defined functions

$$u_1(x) = -\int_{x_0}^{x} \frac{y_2(t) f(t)}{W(t)} \, dt \qquad \text{and} \qquad u_2(x) = \int_{x_0}^{x} \frac{y_1(t) f(t)}{W(t)} \, dt.$$

Here we assume that the integrand is continuous on the interval $[x_0, x]$.
    See Problems 23–26 in Exercises 4.6.

**EXAMPLE 3**    **General Solution Using Variation of Parameters**

Solve $y'' - y = \dfrac{1}{x}$.

**SOLUTION**    The auxiliary equation $m^2 - 1 = 0$ yields $m_1 = -1$ and $m_2 = 1$. Therefore $y_c = c_1 e^x + c_2 e^{-x}$. Now $W(e^x, e^{-x}) = -2$, and

$$u_1' = -\frac{e^{-x}(1/x)}{-2}, \quad u_2' = \frac{e^{-x}(1/x)}{-2}$$

$$u_1 = \frac{1}{2}\int_{x_0}^x \frac{e^t}{t}\,dt, \quad u_2 = -\frac{1}{2}\int_{x_0}^x \frac{e^t}{t}\,dt.$$

Since the foregoing integrals are nonelementary, we are forced to write

$$y_p = \frac{1}{2}e^x \int_{x_0}^x \frac{e^{-t}}{t}\,dt - \frac{1}{2}e^{-x}\int_{x_0}^x \frac{e^t}{t}\,dt,$$

and so $\quad y = y_c + y_p = c_1 e^x + c_2 e^{-x} + \dfrac{1}{2}e^x \displaystyle\int_{x_0}^x \frac{e^{-t}}{t}\,dt - \dfrac{1}{2}e^{-x}\displaystyle\int_{x_0}^x \frac{e^t}{t}\,dt. \quad (12)$ ∎

In Example 3 we can integrate on any interval $[x_0, x]$ that does not contain the origin. We will solve the equation in Example 3 by an alternative method in Section 4.8.

**HIGHER-ORDER EQUATIONS**    The method that we have just examined for non-homogeneous second-order differential equations can be generalized to linear $n$th-order equations that have been put into the standard form

$$y^{(n)} + P_{n-1}(x)y^{(n-1)} + \cdots + P_1(x)y' + P_0(x)y = f(x). \quad (13)$$

If $y_c = c_1 y_1 + c_2 y_2 + \cdots + c_n y_n$ is the complementary function for (13), then a particular solution is

$$y_p(x) = u_1(x)y_1(x) + u_2(x)y_2(x) + \cdots + u_n(x)y_n(x),$$

where the $u_k'$, $k = 1, 2, \ldots, n$ are determined by the $n$ equations

$$
\begin{aligned}
y_1 u_1' + \quad y_2 u_2' + \cdots + \quad y_n u_n' &= 0 \\
y_1' u_1' + \quad y_2' u_2' + \cdots + \quad y_n' u_n' &= 0 \\
\vdots \qquad\qquad\qquad\qquad\qquad \vdots & \\
y_1^{(n-1)} u_1' + y_2^{(n-1)} u_2' + \cdots + y_n^{(n-1)} u_n' &= f(x).
\end{aligned}
\quad (14)
$$

The first $n - 1$ equations in this system, like $y_1 u_1' + y_2 u_2' = 0$ in (8), are assumptions that are made to simplify the resulting equation after $y_p = u_1(x)y_1(x) + \cdots + u_n(x)y_n(x)$ is substituted in (13). In this case Cramer's Rule gives

$$u_k' = \frac{W_k}{W}, \quad k = 1, 2, \ldots, n,$$

where $W$ is the Wronskian of $y_1, y_2, \ldots, y_n$ and $W_k$ is the determinant obtained by replacing the $k$th column of the Wronskian by the column consisting of the right-hand side of (14)—that is, the column consisting of $(0, 0, \ldots, f(x))$. When $n = 2$, we get (9). When $n = 3$, the particular solution is $y_p = u_1 y_1 + u_2 y_2 + u_3 y_3$, where $y_1$, $y_2$, and $y_3$ constitute a linearly independent set of solutions of the associated homogeneous DE and $u_1, u_2, u_3$ are determined from

$$u_1' = \frac{W_1}{W}, \quad u_2' = \frac{W_2}{W}, \quad u_3' = \frac{W_3}{W}, \quad (15)$$

$$
W = \begin{vmatrix} y_1 & y_2 & y_3 \\ y_1' & y_2' & y_3' \\ y_1'' & y_2'' & y_3'' \end{vmatrix}, \quad
W_1 = \begin{vmatrix} 0 & y_2 & y_3 \\ 0 & y_2' & y_3' \\ f(x) & y_2'' & y_3'' \end{vmatrix}, \quad
W_2 = \begin{vmatrix} y_1 & 0 & y_3 \\ y_1' & 0 & y_3' \\ y_1'' & f(x) & y_3'' \end{vmatrix}, \quad
W_3 = \begin{vmatrix} y_1 & y_2 & 0 \\ y_1' & y_2' & 0 \\ y_1'' & y_2'' & f(x) \end{vmatrix}.
$$

See Problems 29–32 in Exercises 4.6.

## REMARKS

(*i*) Variation of parameters has a distinct advantage over the method of undetermined coefficients in that it will *always* yield a particular solution $y_p$ provided that the associated homogeneous equation can be solved. The present method is not limited to a function $f(x)$ that is a combination of the four types listed on page 143. As we shall see in the next section, variation of parameters, unlike undetermined coefficients, is applicable to linear DEs with variable coefficients.

(*ii*) In the problems that follow, do not hesitate to simplify the form of $y_p$. Depending on how the antiderivatives of $u_1'$ and $u_2'$ are found, you might not obtain the same $y_p$ as given in the answer section. For example, in Problem 3 in Exercises 4.6 both $y_p = \frac{1}{2}\sin x - \frac{1}{2}x\cos x$ and $y_p = \frac{1}{4}\sin x - \frac{1}{2}x\cos x$ are valid answers. In either case the general solution $y = y_c + y_p$ simplifies to $y = c_1\cos x + c_2\sin x - \frac{1}{2}x\cos x$. Why?

## EXERCISES 4.6

Answers to selected odd-numbered problems begin on page ANS-5.

In Problems 1–18 solve each differential equation by variation of parameters.

**1.** $y'' + y = \sec x$

**2.** $y'' + y = \tan x$

**3.** $y'' + y = \sin x$

**4.** $y'' + y = \sec\theta\tan\theta$

**5.** $y'' + y = \cos^2 x$

**6.** $y'' + y = \sec^2 x$

**7.** $y'' - y = \cosh x$

**8.** $y'' - y = \sinh 2x$

**9.** $y'' - 9y = \dfrac{9x}{e^{3x}}$

**10.** $4y'' - y = e^{x/2} + 3$

**11.** $y'' + 3y' + 2y = \dfrac{1}{1 + e^x}$

**12.** $y'' - 2y' + y = \dfrac{e^x}{1 + x^2}$

**13.** $y'' + 3y' + 2y = \sin e^x$

**14.** $y'' - 2y' + y = e^t\arctan t$

**15.** $y'' + 2y' + y = e^{-t}\ln t$

**16.** $2y'' + y' = 6x$

**17.** $3y'' - 6y' + 6y = e^x\sec x$

**18.** $4y'' - 4y' + y = e^{x/2}\sqrt{1 - x^2}$

In Problems 19–22 solve each differential equation by variation of parameters, subject to the initial conditions $y(0) = 1$, $y'(0) = 0$.

**19.** $4y'' - y = xe^{x/2}$

**20.** $2y'' + y' - y = x + 1$

**21.** $y'' + 2y' - 8y = 2e^{-2x} - e^{-x}$

**22.** $y'' - 4y' + 4y = (12x^2 - 6x)e^{2x}$

In Problems 23–26 proceed as in Example 3 and solve each differential equation by variation of parameters.

**23.** $y'' + y = e^{x^2}$

**24.** $y'' - 4y = \dfrac{e^{2x}}{x}$

**25.** $y'' + y' - 2y = \ln x$

**26.** $2y'' + 2y' + y = 4\sqrt{x}$

In Problems 27 and 28 the indicated functions are known linearly independent solutions of the associated homogeneous differential equation on $(0, \infty)$. Find the general solution of the given nonhomogeneous equation.

**27.** $x^2 y'' + xy' + \left(x^2 - \frac{1}{4}\right)y = x^{3/2}$;

$\quad y_1 = x^{-1/2}\cos x$, $y_2 = x^{-1/2}\sin x$

**28.** $x^2 y'' + xy' + y = \sec(\ln x)$;

$\quad y_1 = \cos(\ln x)$, $y_2 = \sin(\ln x)$

In Problems 29–32 solve the given third-order differential equation by variation of parameters.

**29.** $y''' + y' = \tan x$

**30.** $y''' + 4y' = \sec 2x$

**31.** $y''' - 2y'' - y' + 2y = e^{4x}$

**32.** $y''' - 3y'' + 2y' = \dfrac{e^{2x}}{1 + e^x}$

### Discussion Problems

In Problems 33 and 34 discuss how the methods of undetermined coefficients and variation of parameters can be combined to solve the given differential equation. Carry out your ideas.

**33.** $3y'' - 6y' + 30y = 15\sin x + e^x\tan 3x$

**34.** $y'' - 2y' + y = 4x^2 - 3 + x^{-1}e^x$

**35.** What are the intervals of definition of the general solutions in Problems 1, 7, 15, and 18? Discuss why the interval of definition of the general solution in Problem 28 is *not* $(0, \infty)$.

**36.** Find the general solution of $x^4 y'' + x^3 y' - 4x^2 y = 1$ given that $y_1 = x^2$ is a solution of the associated homogeneous equation.

# 4.7 Cauchy-Euler Equations

**INTRODUCTION** The same relative ease with which we were able to find explicit solutions of higher-order linear differential equations with constant coefficients in the preceding sections does not, in general, carry over to linear equations with variable coefficients. We shall see in Chapter 6 that when a linear DE has variable coefficients, the best that we can *usually* expect is to find a solution in the form of an infinite series. However, the type of differential equation that we consider in this section is an exception to this rule; it is a linear equation with variable coefficients whose general solution can always be expressed in terms of powers of $x$, sines, cosines, and logarithmic functions. Moreover, its method of solution is quite similar to that for constant-coefficient equations in that an auxiliary equation must be solved.

**CAUCHY-EULER EQUATION**   A linear differential equation of the form

$$a_n x^n \frac{d^n y}{dx^n} + a_{n-1} x^{n-1} \frac{d^{n-1} y}{dx^{n-1}} + \cdots + a_1 x \frac{dy}{dx} + a_0 y = g(x), \tag{1}$$

where the coefficients $a_n, a_{n-1}, \ldots, a_0$ are constants, is known as a **Cauchy-Euler equation.** The differential equation is named in honor of two of the most prolific mathematicians of all time, **Augustin-Louis Cauchy** (French, 1789–1857) and **Leonhard Euler** (Swiss, 1707–1783). The observable characteristic of this type of equation is that the degree $k = n, n - 1, \ldots, 1, 0$ of the monomial coefficients $x^k$ matches the order $k$ of differentiation $d^k y / dx^k$:

$$\overset{\text{same}}{\underset{\downarrow}{a_n x^n}} \; \overset{\text{same}}{\underset{\downarrow}{\frac{d^n y}{dx^n}}} + \overset{}{\underset{}{a_{n-1} x^{n-1}}} \; \frac{d^{n-1} y}{dx^{n-1}} + \cdots.$$

As in Section 4.3, we start the discussion with a detailed examination of the forms of the general solutions of the homogeneous second-order equation

$$ax^2 \frac{d^2 y}{dx^2} + bx \frac{dy}{dx} + cy = 0. \tag{2}$$

The solution of higher-order equations follows analogously. Also, we can solve the nonhomogeneous equation $ax^2 y'' + bxy' + cy = g(x)$ by variation of parameters, once we have determined the complementary function $y_c$.

**NOTE**   The lead coefficient $a_n x^n$ of any Cauchy-Euler equation (1) is zero at $x = 0$. Hence in order to guarantee that the fundamental results of Theorem 4.1.1 are applicable to a Cauchy-Euler equation we will focus our attention on finding general solutions defined on the interval $(0, \infty)$. The interval $(-\infty, 0)$ can also be used.

**METHOD OF SOLUTION**   We try a solution of the form $y = x^m$, where $m$ is to be determined. Analogous to what happened when we substituted $e^{mx}$ into a linear equation with constant coefficients, when we substitute $x^m$, each term of a Cauchy-Euler equation becomes a polynomial in $m$ times $x^m$, since

$$a_k x^k \frac{d^k y}{dx^k} = a_k x^k m(m-1)(m-2) \cdots (m-k+1)x^{m-k} = a_k m(m-1)(m-2) \cdots (m-k+1)x^m.$$

For example, when we substitute $y = x^m$, the second-order equation (2) becomes

$$ax^2 \frac{d^2 y}{dx^2} + bx \frac{dy}{dx} + cy = am(m-1)x^m + bmx^m + cx^m = (am(m-1) + bm + c)x^m = 0.$$

Thus $y = x^m$ is a solution of the differential equation whenever $m$ is a solution of the **auxiliary equation**

$$am(m-1) + bm + c = 0 \quad \text{or} \quad am^2 + (b-a)m + c = 0. \tag{3}$$

There are three different cases to be considered, depending on whether the roots of this quadratic equation are real and distinct, real and equal, or complex. In the last case the roots appear as a conjugate pair.

**CASE I: DISTINCT REAL ROOTS**   Let $m_1$ and $m_2$ denote the real roots of (3) such that $m_1 \neq m_2$. Then $y_1 = x^{m_1}$ and $y_2 = x^{m_2}$ form a fundamental set of solutions. Hence the general solution of (2) is

$$y = c_1 x^{m_1} + c_2 x^{m_2}. \tag{4}$$

---

**EXAMPLE 1**   **Distinct Roots**

Solve $x^2 \dfrac{d^2 y}{dx^2} - 2x \dfrac{dy}{dx} - 4y = 0$.

**SOLUTION** Rather than just memorizing equation (3), it is preferable to assume $y = x^m$ as the solution a few times to understand the origin and the difference between this new form of the auxiliary equation and that obtained in Section 4.3. Differentiate twice,

$$\frac{dy}{dx} = mx^{m-1}, \qquad \frac{d^2 y}{dx^2} = m(m-1)x^{m-2},$$

and substitute back into the differential equation:

$$x^2 \frac{d^2 y}{dx^2} - 2x \frac{dy}{dx} - 4y = x^2 \cdot m(m-1)x^{m-2} - 2x \cdot mx^{m-1} - 4x^m$$

$$= x^m(m(m-1) - 2m - 4) = x^m(m^2 - 3m - 4) = 0$$

if $m^2 - 3m - 4 = 0$. Now $(m+1)(m-4) = 0$ implies $m_1 = -1$, $m_2 = 4$, so the general solution is $y = c_1 x^{-1} + c_2 x^4$.   ■

**CASE II: REPEATED REAL ROOTS**   If the roots of (3) are repeated (that is, $m_1 = m_2$), then we obtain only one solution—namely, $y = x^{m_1}$. When the roots of the quadratic equation $am^2 + (b-a)m + c = 0$ are equal, the discriminant of the coefficients is necessarily zero. It follows from the quadratic formula that the root must be $m_1 = -(b-a)/2a$.

Now we can construct a second solution $y_2$, using (5) of Section 4.2. We first write the Cauchy-Euler equation in the standard form

$$\frac{d^2 y}{dx^2} + \frac{b}{ax} \frac{dy}{dx} + \frac{c}{ax^2} y = 0$$

and make the identifications $P(x) = b/ax$ and $\int (b/ax)\, dx = (b/a) \ln x$. Thus

$$y_2 = x^{m_1} \int \frac{e^{-(b/a)\ln x}}{x^{2m_1}}\, dx$$

$$= x^{m_1} \int x^{-b/a} \cdot x^{-2m_1}\, dx \qquad \leftarrow e^{-(b/a)\ln x} = e^{\ln x^{-b/a}} = x^{-b/a}$$

$$= x^{m_1} \int x^{-b/a} \cdot x^{(b-a)/a}\, dx \qquad \leftarrow -2m_1 = (b-a)/a$$

$$= x^{m_1} \int \frac{dx}{x} = x^{m_1} \ln x.$$

The general solution of (2) is then

$$y = c_1 x^{m_1} + c_2 x^{m_1} \ln x. \tag{5}$$

### EXAMPLE 2   Repeated Roots

Solve $4x^2 \dfrac{d^2 y}{dx^2} + 8x \dfrac{dy}{dx} + y = 0$.

**SOLUTION**  The substitution $y = x^m$ yields

$$4x^2 \frac{d^2 y}{dx^2} + 8x \frac{dy}{dx} + y = x^m(4m(m-1) + 8m + 1) = x^m(4m^2 + 4m + 1) = 0$$

when $4m^2 + 4m + 1 = 0$ or $(2m + 1)^2 = 0$. Since $m_1 = -\frac{1}{2}$, it follows from (5) that the general solution is $y = c_1 x^{-1/2} + c_2 x^{-1/2} \ln x$. ■

For higher-order equations, if $m_1$ is a root of multiplicity $k$, then it can be shown that

$$x^{m_1}, \quad x^{m_1} \ln x, \quad x^{m_1} (\ln x)^2, \ldots, \quad x^{m_1} (\ln x)^{k-1}$$

are $k$ linearly independent solutions. Correspondingly, the general solution of the differential equation must then contain a linear combination of these $k$ solutions.

**CASE III: CONJUGATE COMPLEX ROOTS**   If the roots of (3) are the conjugate pair $m_1 = \alpha + i\beta$, $m_2 = \alpha - i\beta$, where $\alpha$ and $\beta > 0$ are real, then a solution is

$$y = C_1 x^{\alpha + i\beta} + C_2 x^{\alpha - i\beta}.$$

But when the roots of the auxiliary equation are complex, as in the case of equations with constant coefficients, we wish to write the solution in terms of real functions only. We note the identity

$$x^{i\beta} = (e^{\ln x})^{i\beta} = e^{i\beta \ln x},$$

which, by Euler's formula, is the same as

$$x^{i\beta} = \cos(\beta \ln x) + i \sin(\beta \ln x).$$

Similarly, $\qquad\qquad x^{-i\beta} = \cos(\beta \ln x) - i \sin(\beta \ln x).$

Adding and subtracting the last two results yields

$$x^{i\beta} + x^{-i\beta} = 2\cos(\beta \ln x) \qquad \text{and} \qquad x^{i\beta} - x^{-i\beta} = 2i \sin(\beta \ln x),$$

respectively. From the fact that $y = C_1 x^{\alpha + i\beta} + C_2 x^{\alpha - i\beta}$ is a solution for any values of the constants, we see, in turn, for $C_1 = C_2 = 1$ and $C_1 = 1$, $C_2 = -1$ that

$$y_1 = x^\alpha (x^{i\beta} + x^{-i\beta}) \qquad \text{and} \qquad y_2 = x^\alpha (x^{i\beta} - x^{-i\beta})$$

or $\qquad\qquad y_1 = 2x^\alpha \cos(\beta \ln x) \qquad \text{and} \qquad y_2 = 2ix^\alpha \sin(\beta \ln x)$

are also solutions. Since $W(x^\alpha \cos(\beta \ln x), x^\alpha \sin(\beta \ln x)) = \beta x^{2\alpha - 1} \neq 0$, $\beta > 0$ on the interval $(0, \infty)$, we conclude that

$$y_1 = x^\alpha \cos(\beta \ln x) \qquad \text{and} \qquad y_2 = x^\alpha \sin(\beta \ln x)$$

constitute a fundamental set of real solutions of the differential equation. Hence the general solution of (2) is

$$y = x^\alpha [c_1 \cos(\beta \ln x) + c_2 \sin(\beta \ln x)]. \tag{6}$$

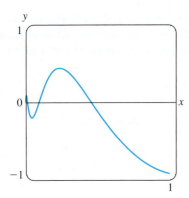

**(a)** solution for $0 < x \leq 1$

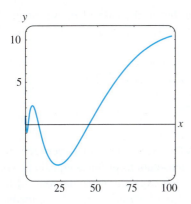

**(b)** solution for $0 < x \leq 100$

**FIGURE 4.7.1** Solution curve of IVP in Example 3

**EXAMPLE 3** An Initial-Value Problem

Solve $4x^2 y'' + 17y = 0$, $y(1) = -1$, $y'(1) = -\frac{1}{2}$.

**SOLUTION** The $y'$ term is missing in the given Cauchy-Euler equation; nevertheless, the substitution $y = x^m$ yields

$$4x^2 y'' + 17y = x^m(4m(m-1) + 17) = x^m(4m^2 - 4m + 17) = 0$$

when $4m^2 - 4m + 17 = 0$. From the quadratic formula we find that the roots are $m_1 = \frac{1}{2} + 2i$ and $m_2 = \frac{1}{2} - 2i$. With the identifications $\alpha = \frac{1}{2}$ and $\beta = 2$ we see from (6) that the general solution of the differential equation is

$$y = x^{1/2}[c_1 \cos(2 \ln x) + c_2 \sin(2 \ln x)].$$

By applying the initial conditions $y(1) = -1$, $y'(1) = -\frac{1}{2}$ to the foregoing solution and using $\ln 1 = 0$, we then find, in turn, that $c_1 = -1$ and $c_2 = 0$. Hence the solution of the initial-value problem is $y = -x^{1/2} \cos(2 \ln x)$. The graph of this function, obtained with the aid of computer software, is given in Figure 4.7.1. The particular solution is seen to be oscillatory and unbounded as $x \to \infty$.

The next example illustrates the solution of a third-order Cauchy-Euler equation.

**EXAMPLE 4** Third-Order Equation

Solve $x^3 \dfrac{d^3 y}{dx^3} + 5x^2 \dfrac{d^2 y}{dx^2} + 7x \dfrac{dy}{dx} + 8y = 0$.

**SOLUTION** The first three derivatives of $y = x^m$ are

$$\frac{dy}{dx} = mx^{m-1}, \qquad \frac{d^2 y}{dx^2} = m(m-1)x^{m-2}, \qquad \frac{d^3 y}{dx^3} = m(m-1)(m-2)x^{m-3},$$

so the given differential equation becomes

$$x^3 \frac{d^3 y}{dx^3} + 5x^2 \frac{d^2 y}{dx^2} + 7x \frac{dy}{dx} + 8y = x^3 m(m-1)(m-2)x^{m-3} + 5x^2 m(m-1)x^{m-2} + 7xmx^{m-1} + 8x^m$$

$$= x^m(m(m-1)(m-2) + 5m(m-1) + 7m + 8)$$

$$= x^m(m^3 + 2m^2 + 4m + 8) = x^m(m+2)(m^2+4) = 0.$$

In this case we see that $y = x^m$ will be a solution of the differential equation for $m_1 = -2$, $m_2 = 2i$, and $m_3 = -2i$. Hence the general solution is $y = c_1 x^{-2} + c_2 \cos(2 \ln x) + c_3 \sin(2 \ln x)$.

**NONHOMOGENEOUS EQUATIONS** The method of undetermined coefficients described in Sections 4.4 and 4.5 does not carry over, *in general*, to nonhomogeneous linear differential equations with variable coefficients. Consequently, in our next example the method of variation of parameters is employed.

**EXAMPLE 5** Variation of Parameters

Solve $x^2 y'' - 3xy' + 3y = 2x^4 e^x$.

**SOLUTION** Since the equation is nonhomogeneous, we first solve the associated homogeneous equation. From the auxiliary equation $(m - 1)(m - 3) = 0$ we find $y_c = c_1 x + c_2 x^3$. Now before using variation of parameters to find a particular solution $y_p = u_1 y_1 + u_2 y_2$, recall that the formulas $u_1' = W_1/W$ and $u_2' = W_2/W$, where $W_1$, $W_2$, and $W$ are the determinants defined on page 162, were derived under the assumption that the differential equation has been put into the standard form $y'' + P(x)y' + Q(x)y = f(x)$. Therefore we divide the given equation by $x^2$, and from

$$y'' - \frac{3}{x}y' + \frac{3}{x^2}y = 2x^2 e^x$$

we make the identification $f(x) = 2x^2 e^x$. Now with $y_1 = x$, $y_2 = x^3$, and

$$W = \begin{vmatrix} x & x^3 \\ 1 & 3x^2 \end{vmatrix} = 2x^3, \quad W_1 = \begin{vmatrix} 0 & x^3 \\ 2x^2 e^x & 3x^2 \end{vmatrix} = -2x^5 e^x, \quad W_2 = \begin{vmatrix} x & 0 \\ 1 & 2x^2 e^x \end{vmatrix} = 2x^3 e^x,$$

we find

$$u_1' = -\frac{2x^5 e^x}{2x^3} = -x^2 e^x \quad \text{and} \quad u_2' = \frac{2x^3 e^x}{2x^3} = e^x.$$

The integral of the last function is immediate, but in the case of $u_1'$ we integrate by parts twice. The results are $u_1 = -x^2 e^x + 2x e^x - 2e^x$ and $u_2 = e^x$. Hence $y_p = u_1 y_1 + u_2 y_2$ is

$$y_p = (-x^2 e^x + 2x e^x - 2e^x)x + e^x x^3 = 2x^2 e^x - 2x e^x.$$

Finally, $y = y_c + y_p = c_1 x + c_2 x^3 + 2x^2 e^x - 2x e^x$.    ∎

**REDUCTION TO CONSTANT COEFFICIENTS**   The similarities between the forms of solutions of Cauchy-Euler equations and solutions of linear equations with constant coefficients are not just a coincidence. For example, when the roots of the auxiliary equations for $ay'' + by' + cy = 0$ and $ax^2 y'' + bxy' + cy = 0$ are distinct and real, the respective general solutions are

$$y = c_1 e^{m_1 x} + c_2 e^{m_2 x} \quad \text{and} \quad y = c_1 x^{m_1} + c_2 x^{m_2}, \quad x > 0. \tag{7}$$

In view of the identity $e^{\ln x} = x$, $x > 0$, the second solution given in (7) can be expressed in the same form as the first solution:

$$y = c_1 e^{m_1 \ln x} + c_2 e^{m_2 \ln x} = c_1 e^{m_1 t} + c_2 e^{m_2 t},$$

where $t = \ln x$. This last result illustrates the fact that any Cauchy-Euler equation can *always* be rewritten as a linear differential equation with constant coefficients by means of the substitution $x = e^t$. The idea is to solve the new differential equation in terms of the variable $t$, using the methods of the previous sections, and, once the general solution is obtained, resubstitute $t = \ln x$. This method, illustrated in the last example, requires the use of the Chain Rule of differentiation.

---

**EXAMPLE 6**   **Changing to Constant Coefficients**

Solve $x^2 y'' - xy' + y = \ln x$.

**SOLUTION**   With the substitution $x = e^t$ or $t = \ln x$, it follows that

$$\frac{dy}{dx} = \frac{dy}{dt}\frac{dt}{dx} = \frac{1}{x}\frac{dy}{dt} \qquad \leftarrow \text{Chain Rule}$$

$$\frac{d^2 y}{dx^2} = \frac{1}{x}\frac{d}{dx}\left(\frac{dy}{dt}\right) + \frac{dy}{dt}\left(-\frac{1}{x^2}\right) \qquad \leftarrow \text{Product Rule and Chain Rule}$$

$$= \frac{1}{x}\left(\frac{d^2 y}{dt^2}\frac{1}{x}\right) + \frac{dy}{dt}\left(-\frac{1}{x^2}\right) = \frac{1}{x^2}\left(\frac{d^2 y}{dt^2} - \frac{dy}{dt}\right).$$

Substituting in the given differential equation and simplifying yields

$$\frac{d^2y}{dt^2} - 2\frac{dy}{dt} + y = t.$$

Since this last equation has constant coefficients, its auxiliary equation is $m^2 - 2m + 1 = 0$, or $(m - 1)^2 = 0$. Thus we obtain $y_c = c_1e^t + c_2te^t$.

By undetermined coefficients we try a particular solution of the form $y_p = A + Bt$. This assumption leads to $-2B + A + Bt = t$, so $A = 2$ and $B = 1$. Using $y = y_c + y_p$, we get

$$y = c_1e^t + c_2te^t + 2 + t.$$

By resubstituting $e^t = x$ and $t = \ln x$ we see that the general solution of the original differential equation on the interval $(0, \infty)$ is $y = c_1x + c_2x \ln x + 2 + \ln x$.  ∎

**SOLUTIONS FOR $x < 0$**    In the preceding discussion we have solved Cauchy-Euler equations for $x > 0$. One way of solving a Cauchy-Euler equation for $x < 0$ is to change the independent variable by means of the substitution $t = -x$ (which implies $t > 0$) and using the Chain Rule:

$$\frac{dy}{dx} = \frac{dy}{dt}\frac{dt}{dx} = -\frac{dy}{dt} \quad \text{and} \quad \frac{d^2y}{dx^2} = \frac{d}{dt}\left(-\frac{dy}{dt}\right)\frac{dt}{dx} = \frac{d^2y}{dt^2}.$$

See Problems 37 and 38 in Exercises 4.7.

**A DIFFERENT FORM**    A second-order equation of the form

$$a(x - x_0)^2\frac{d^2y}{dx^2} + b(x - x_0)\frac{dy}{dx} + cy = 0 \tag{8}$$

is also a Cauchy-Euler equation. Observe that (8) reduces to (2) when $x_0 = 0$.

We can solve (8) as we did (2), namely, seeking solutions of $y = (x - x_0)^m$ and using

$$\frac{dy}{dx} = m(x - x_0)^{m-1} \quad \text{and} \quad \frac{d^2y}{dx^2} = m(m - 1)(x - x_0)^{m-2}.$$

Alternatively, we can reduce (8) to the familiar form (2) by means of the change of independent variable $t = x - x_0$, solving the reduced equation, and resubstituting. See Problems 39–42 in Exercises 4.7.

## EXERCISES **4.7**

*Answers to selected odd numbered problems begin on page ANS 5.*

In Problems 1–18 solve the given differential equation.

**1.** $x^2y'' - 2y = 0$

**2.** $4x^2y'' + y = 0$

**3.** $xy'' + y' = 0$

**4.** $xy'' - 3y' = 0$

**5.** $x^2y'' + xy' + 4y = 0$

**6.** $x^2y'' + 5xy' + 3y = 0$

**7.** $x^2y'' - 3xy' - 2y = 0$

**8.** $x^2y'' + 3xy' - 4y = 0$

**9.** $25x^2y'' + 25xy' + y = 0$

**10.** $4x^2y'' + 4xy' - y = 0$

**11.** $x^2y'' + 5xy' + 4y = 0$

**12.** $x^2y'' + 8xy' + 6y = 0$

**13.** $3x^2y'' + 6xy' + y = 0$

**14.** $x^2y'' - 7xy' + 41y = 0$

**15.** $x^3y''' - 6y = 0$

**16.** $x^3y''' + xy' - y = 0$

**17.** $xy^{(4)} + 6y''' = 0$

**18.** $x^4y^{(4)} + 6x^3y''' + 9x^2y'' + 3xy' + y = 0$

In Problems 19–24 solve the given differential equation by variation of parameters.

**19.** $xy'' - 4y' = x^4$

**20.** $2x^2y'' + 5xy' + y = x^2 - x$

**21.** $x^2y'' - xy' + y = 2x$

**22.** $x^2y'' - 2xy' + 2y = x^4e^x$

**23.** $x^2y'' + xy' - y = \ln x$

**24.** $x^2y'' + xy' - y = \dfrac{1}{x + 1}$

In Problems 25–30 solve the given initial-value problem. Use a graphing utility to graph the solution curve.

**25.** $x^2y'' + 3xy' = 0$,   $y(1) = 0, y'(1) = 4$

**26.** $x^2y'' - 5xy' + 8y = 0$,   $y(2) = 32, y'(2) = 0$

**27.** $x^2y'' + xy' + y = 0$,   $y(1) = 1, y'(1) = 2$

**28.** $x^2y'' - 3xy' + 4y = 0$,   $y(1) = 5, y'(1) = 3$

**29.** $xy'' + y' = x$,   $y(1) = 1, y'(1) = -\frac{1}{2}$

**30.** $x^2y'' - 5xy' + 8y = 8x^6$,   $y\left(\frac{1}{2}\right) = 0, y'\left(\frac{1}{2}\right) = 0$

In Problems 31–36 use the substitution $x = e^t$ to transform the given Cauchy-Euler equation to a differential equation with constant coefficients. Solve the original equation by solving the new equation using the procedures in Sections 4.3–4.5.

**31.** $x^2y'' + 9xy' - 20y = 0$

**32.** $x^2y'' - 9xy' + 25y = 0$

**33.** $x^2y'' + 10xy' + 8y = x^2$

**34.** $x^2y'' - 4xy' + 6y = \ln x^2$

**35.** $x^2y'' - 3xy' + 13y = 4 + 3x$

**36.** $x^3y''' - 3x^2y'' + 6xy' - 6y = 3 + \ln x^3$

In Problems 37 and 38 use the substitution $t = -x$ to solve the given initial-value problem on the interval $(-\infty, 0)$.

**37.** $4x^2y'' + y = 0$,   $y(-1) = 2, y'(-1) = 4$

**38.** $x^2y'' - 4xy' + 6y = 0$,   $y(-2) = 8, y'(-2) = 0$

In Problems 39 and 40 use $y = (x - x_0)^m$ to solve the given differential equation.

**39.** $(x + 3)^2y'' - 8(x + 3)y' + 14y = 0$

**40.** $(x - 1)^2y'' - (x - 1)y' + 5y = 0$

In Problems 41 and 42 use the substitution $t = x - x_0$ to solve the given differential equation.

**41.** $(x + 2)^2y'' + (x + 2)y' + y = 0$

**42.** $(x - 4)^2y'' - 5(x - 4)y' + 9y = 0$

## Discussion Problems

**43.** Give the largest interval over which the general solution of Problem 42 is defined.

**44.** Can a Cauchy-Euler differential equation of lowest order with real coefficients be found if it is known that 2 and $1 - i$ are roots of its auxiliary equation? Carry out your ideas.

**45.** The initial-conditions $y(0) = y_0$, $y'(0) = y_1$ apply to each of the following differential equations:

$$x^2y'' = 0,$$
$$x^2y'' - 2xy' + 2y = 0,$$
$$x^2y'' - 4xy' + 6y = 0.$$

For what values of $y_0$ and $y_1$ does each initial-value problem have a solution?

**46.** What are the $x$-intercepts of the solution curve shown in Figure 4.7.1? How many $x$-intercepts are there for $0 < x < \frac{1}{2}$?

## Mathematical Model

**47. Bending of a Circular Plate**   In the analysis of the bending of a uniformly loaded circular plate, the equation $w(r)$ of the deflection curve of the plate can be shown to satisfy the differential equation

$$\frac{d^3w}{dr^3} + \frac{1}{r}\frac{d^2w}{dr^2} - \frac{1}{r^2}\frac{dw}{dr} = \frac{q}{2D}r, \tag{9}$$

where $q$ and $D$ are constants. Here $r$ is the radial distance from a point on the circular plate to its center.

**(a)** Use the method of this section along with variation of parameters as given in (15) of Section 4.6 to find the general solution of equation (9).

**(b)** Find a solution of (9) that satisfies the boundary conditions

$$w'(0) = 0, \quad w(a) = 0, \quad w'(a) = 0,$$

where $a > 0$ is the radius of the plate. [*Hint*: The condition $w'(0) = 0$ is correct. Use this condition to determine one of the constants in the general solution found in part (a).]

**48.** In the engineering textbook where equation (9) was found, the author states that the differential equation is readily solved by integration. True, but a big hint is required.

**(a)** Verify that equation (9) can be written in the alternative form

$$\frac{d}{dr}\left[\frac{1}{r}\frac{d}{dr}\left(r\frac{dw}{dr}\right)\right] = \frac{q}{2D}r. \tag{10}$$

**(b)** Solve equation (10) using only integration with respect to $r$. Show your result is equivalent to the solution obtained in part (a) of Problem 47.

## Computer Lab Assignments

In Problems 49–52 solve the given differential equation by using a CAS to find the (approximate) roots of the auxiliary equation.

**49.** $2x^3y''' - 10.98x^2y'' + 8.5xy' + 1.3y = 0$

**50.** $x^3y''' + 4x^2y'' + 5xy' - 9y = 0$

**51.** $x^4y^{(4)} + 6x^3y''' + 3x^2y'' - 3xy' + 4y = 0$

**52.** $x^4y^{(4)} - 6x^3y''' + 33x^2y'' - 105xy' + 169y = 0$

**53.** Solve $x^3y''' - x^2y'' - 2xy' + 6y = x^2$ by variation of parameters. Use a CAS as an aid in computing roots of the auxiliary equation and the determinants given in (15) of Section 4.6.

## 4.8   Green's Functions

**INTRODUCTION**   We will see in Chapter 5 that the linear second-order differential equation

$$a_2(x)\frac{d^2y}{dx^2} + a_1(x)\frac{dy}{dx} + a_0(x)y = g(x) \tag{1}$$

plays an important role in many applications. In the mathematical analysis of physical systems it is often desirable to express the **response** or **output** $y(x)$ of (1) subject to either initial conditions or boundary conditions directly in terms of the **forcing function** or **input** $g(x)$. In this manner the response of the system can quickly be analyzed for different forcing functions.

To see how this is done, we start by examining solutions of initial-value problems in which the DE (1) has been put into the standard form

$$y'' + P(x)y' + Q(x)y = f(x) \tag{2}$$

by dividing the equation by the lead coefficient $a_2(x)$. We also assume throughout this section that the coefficient functions $P(x)$, $Q(x)$, and $f(x)$ are continuous on some common interval $I$.

### 4.8.1   INITIAL-VALUE PROBLEMS

**THREE INITIAL-VALUE PROBLEMS**   We will see as the discussion unfolds that the solution $y(x)$ of the second order initial-value problem

$$y'' + P(x)y' + Q(x)y = f(x), \quad y(x_0) = y_0, \ y'(x_0) = y_1 \tag{3}$$

can be expressed as the superposition of two solutions:

$$y(x) = y_h(x) + y_p(x), \tag{4}$$

where $y_h(x)$ is the solution of the associated homogeneous DE with nonhomogeneous initial conditions

$$y'' + P(x)y' + Q(x)y = 0, \quad y(x_0) = y_0, \ y'(x_0) = y_1 \tag{5}$$

 Here at least one of the numbers $y_0$ or $y_1$ is assumed to be nonzero. If both $y_0$ and $y_1$ are 0, then the solution of the IVP is $y = 0$.

and $y_p(x)$ is the solution of the nonhomogeneous DE with homogeneous (that is, zero) initial conditions

$$y'' + P(x)y' + Q(x)y = f(x), \quad y(x_0) = 0, \ y'(x_0) = 0. \tag{6}$$

In the case where the coefficients $P$ and $Q$ are constants the solution of the IVP (5) presents no difficulties: We use the method of Section 4.3 to find the general solution of the homogeneous DE and then use the given initial conditions to determine the two constants in that solution. So we will focus on the solution of the IVP (6). Because of the zero initial conditions, the solution of (6) could describe a physical system that is initially at rest and so is sometimes called a **rest solution.**

**GREEN'S FUNCTION**   If $y_1(x)$ and $y_2(x)$ form a fundamental set of solutions on the interval $I$ of the associated homogeneous form of (2), then a particular solution of the nonhomogeneous equation (2) on the interval $I$ can be found by variation of parameters. Recall from (3) of Section 4.6, the form of this solution is

$$y_p(x) = u_1(x)y_1(x) + u_2(x)y_2(x). \tag{7}$$

The variable coefficients $u_1(x)$ and $u_2(x)$ in (7) are defined by (9) of Section 4.6:

$$u_1'(x) = -\frac{y_2(x)f(x)}{W}, \quad u_2'(x) = \frac{y_1(x)f(x)}{W}. \tag{8}$$

The linear independence of $y_1(x)$ and $y_2(x)$ on the interval $I$ guarantees that the Wronskian $W = W(y_1(x), y_2(x)) \neq 0$ for all $x$ in $I$. If $x$ and $x_0$ are numbers in $I$, then integrating the derivatives $u_1'(x)$ and $u_2'(x)$ in (8) on the interval $[x_0, x]$ and substituting the results into (7) give

Because $y_1(x)$ and $y_2(x)$ are constant with respect to the integration on $t$, we can move these functions inside the definite intergrals.

$$
\begin{aligned}
y_p(x) &= y_1(x) \int_{x_0}^{x} \frac{-y_2(t)f(t)}{W(t)}\, dt + y_2(x) \int_{x_0}^{x} \frac{y_1(t)f(t)}{W(t)}\, dt \\
&= \int_{x_0}^{x} \frac{-y_1(x)y_2(t)}{W(t)} f(t)\, dt \;+\; \int_{x_0}^{x} \frac{y_1(t)y_2(x)}{W(t)} f(t)\, dt,
\end{aligned}
\tag{9}
$$

where
$$W(t) = W(y_1(t), y_2(t)) = \begin{vmatrix} y_1(t) & y_2(t) \\ y_1'(t) & y_2'(t) \end{vmatrix}$$

From the properties of the definite integral, the two integrals in the second line of (9) can be rewritten as a single integral

$$y_p(x) = \int_{x_0}^{x} G(x, t) f(t)\, dt. \tag{10}$$

The function $G(x, t)$ in (10),

$$G(x, t) = \frac{y_1(t)y_2(x) - y_1(x)y_2(t)}{W(t)} \tag{11}$$

is called the **Green's function** for the differential equation (2).

Important. Read this paragraph a second time.

Observe that a Green's function (11) depends only on the fundamental solutions $y_1(x)$ and $y_2(x)$ of the associated homogeneous differential equation for (2) and *not* on the forcing function $f(x)$. Therefore all linear second-order differential equations (2) with the same left-hand side but with different forcing functions have the same Green's function. So an alternative title for (11) is the **Green's function for the second-order differential operator** $L = D^2 + P(x)D + Q(x)$.

### EXAMPLE 1   Particular Solution

Use (10) and (11) to find a particular solution of $y'' - y = f(x)$.

**SOLUTION** The solutions of the associated homogeneous equation $y'' - y = 0$ are $y_1 = e^x$, $y_2 = e^{-x}$, and $W(y_1(t), y_2(t)) = -2$. It follows from (11) that the Green's function is

$$G(x, t) = \frac{e^t e^{-x} - e^x e^{-t}}{-2} = \frac{e^{x-t} - e^{-(x-t)}}{2} = \sinh(x - t). \tag{12}$$

Thus from (10), a particular solution of the DE is

$$y_p(x) = \int_{x_0}^{x} \sinh(x - t) f(t)\, dt. \tag{13} \quad\blacksquare$$

### EXAMPLE 2   General Solutions

Find the general solution of following nonhomogeneous differential equations.

**(a)** $y'' - y = 1/x$                 **(b)** $y'' - y = e^{2x}$

**SOLUTION** From Example 1, both DEs possess the same complementary function $y_c = c_1e^{-x} + c_2e^{x}$. Moreover, as pointed out in the paragraph preceding Example 1, the Green's function for both differential equations is (12).

**(a)** With the identifications $f(x) = 1/x$ and $f(t) = 1/t$ we see from (13) that a particular solution of $y'' - y = 1/x$ is $y_p(x) = \displaystyle\int_{x_0}^{x} \frac{\sinh(x - t)}{t} \, dt$. Thus the general solution $y = y_c + y_p$ of the given DE on any interval $[x_0, x]$ not containing the origin is

$$y = c_1e^{x} + c_2e^{-x} + \int_{x_0}^{x} \frac{\sinh(x - t)}{t} \, dt. \tag{14}$$

You should compare this solution with that found in Example 3 of Section 4.6.

**(b)** With $f(x) = e^{2x}$ in (13), a particular solution of $y'' - y = e^{2x}$ is $y_p(x) = \int_{x_0}^{x} \sinh(x - t) \, e^{2t} \, dt$. The general solution $y = y_c + y_p$ is then

$$y = c_1e^{x} + c_2e^{-x} + \int_{x_0}^{x} \sinh(x - t) \, e^{2t} \, dt. \tag{15} \blacksquare$$

Now consider the special initial-value problem (6) with homogeneous initial conditions. One way of solving the problem when $f(x) \neq 0$ has already been illustrated in Sections 4.4 and 4.6, that is, apply the initial conditions $y(x_0) = 0$, $y'(x_0) = 0$ to the general solution of the nonhomogeneous DE. But there is no actual need to do this because we already have a solution of the IVP at hand; it is the function defined in (10).

> **THEOREM 4.8.1**  Solution of the IVP (6)
>
> The function $y_p(x)$ defined in (10) is the solution of the initial-value problem (6).

**PROOF** By construction we know that $y_p(x)$ in (10) satisfies the nonhomogeneous DE. Next, because a definite integral has the property $\int_a^a = 0$ we have

$$y_p(x_0) = \int_{x_0}^{x_0} G(x_0, t) f(t) \, dt = 0.$$

Finally, to show that $y_p'(x_0) = 0$ we utilize the Leibniz formula* for the derivative of an integral:

$$y_p'(x) = \overbrace{G(x, x)f(x)}^{0 \text{ from } (11)} + \int_{x_0}^{x} \frac{y_1(t)y_2'(x) - y_1'(x)y_2(t)}{W(t)} f(t) \, dt.$$

Hence,    $$y_p'(x_0) = \int_{x_0}^{x_0} \frac{y_1(t)y_2'(x_0) - y_1'(x_0)y_2(t)}{W(t)} f(t) \, dt = 0. \qquad \blacksquare$$

**EXAMPLE 3**  **Example 2 Revisited**

Solve the initial-value problems

**(a)** $y'' - y = 1/x$,  $y(1) = 0$, $y'(1) = 0$      **(b)** $y'' - y = e^{2x}$,  $y(0) = 0$, $y'(0) = 0$

*See Appendix A.

**SOLUTION** **(a)** With $x_0 = 1$ and $f(t) = 1/t$, it follows from (14) of Example 2 and Theorem 4.8.1 that the solution of the initial-value problem is

$$y_p(x) = \int_1^x \frac{\sinh(x-t)}{t}\, dt,$$

where $[1, x]$, $x > 0$.

**(b)** Identifying $x_0 = 0$ and $f(t) = e^{2t}$, we see from (15) that the solution of the IVP is

$$y_p(x) = \int_0^x \sinh(x-t)\, e^{2t}\, dt. \qquad (16) \quad \blacksquare$$

In part (b) of Example 3, we can carry out the integration in (16), but bear in mind that $x$ is held constant throughout the integration with respect to $t$:

$$y_p(x) = \int_0^x \sinh(x-t)\, e^{2t}\, dt = \int_0^x \frac{e^{x-t} - e^{-(x-t)}}{2}\, e^{2t}\, dt$$

$$= \tfrac{1}{2} e^x \int_0^x e^t\, dt - \tfrac{1}{2} e^{-x} \int_0^x e^{3t}\, dt$$

$$= \tfrac{1}{3} e^{2x} - \tfrac{1}{2} e^x + \tfrac{1}{6} e^{-x}.$$

---

### EXAMPLE 4   Using (10) and (11)

Solve the initial-value problem

$$y'' + 4y = x, \quad y(0) = 0,\ y'(0) = 0.$$

**SOLUTION** We begin by constructing the Green's function for the given differential equation.

    The two linearly independent solutions of $y'' + 4y = 0$ are $y_1(x) = \cos 2x$ and $y_2(x) = \sin 2x$. From (11), with $W(\cos 2t, \sin 2t) = 2$, we find

$$G(x, t) = \frac{\cos 2t \sin 2x - \cos 2x \sin 2t}{2} = \tfrac{1}{2} \sin 2(x-t).$$

*Here we have used the trigonometric identity*
$\sin(2x - 2t) =$
$\sin 2x \cos 2t - \cos 2x \sin 2t$

With the further identifications $x_0 = 0$ and $f(t) = t$ in (10) we see that a solution of the given initial-value problem is

$$y_p(x) = \tfrac{1}{2} \int_0^x t \sin 2(x-t)\, dt.$$

If we wish to evaluate the integral, we first write

$$y_p(x) = \tfrac{1}{2} \sin 2x \int_0^x t \cos 2t\, dt - \tfrac{1}{2} \cos 2x \int_0^x t \sin 2t\, dt$$

and then use integration by parts:

$$y_p(x) = \tfrac{1}{2} \sin 2x \left[ \tfrac{1}{2} t \sin 2t + \tfrac{1}{4} \cos 2t \right]_0^x - \tfrac{1}{2} \cos 2x \left[ -\tfrac{1}{2} t \cos 2t + \tfrac{1}{4} \sin 2t \right]_0^x$$

or

$$y_p(x) = \tfrac{1}{4} x - \tfrac{1}{8} \sin 2x. \qquad \blacksquare$$

**INITIAL-VALUE PROBLEMS—CONTINUED**   Finally, we are now in a position to make use of Theorem 4.8.1 to find the solution of the initial-value problem posed in (3). It is simply the function already given in (4).

> ### THEOREM 4.8.2 Solution of the IVP (3)
>
> If $y_h(x)$ is the solution of the initial-value problem (5) and $y_p(x)$ is the solution (10) of the initial-value problem (6) on the interval $I$, then
>
> $$y(x) = y_h(x) + y_p(x) \qquad (17)$$
>
> is the solution of the initial-value problem (3).

**PROOF** Because $y_h(x)$ is a linear combination of the fundamental solutions, it follows from (10) of Section 4.1 that $y = y_h + y_p$ is a solution of the nonhomogeneous DE. Moreover, since $y_h$ satisfies the initial conditions in (5) and $y_p$ satisfies the initial conditions in (6), we have,

$$y(x_0) = y_h(x_0) + y_p(x_0) = y_0 + 0 = y_0$$

$$y'(x_0) = y'_h(x_0) + y'_p(x_0) = y_1 + 0 = y_1. \qquad \blacksquare$$

Keeping in mind the absence of a forcing function in (5) and the presence of such a term in (6), we see from (17) that the response $y(x)$ of a physical system described by the initial-value problem (3) can be separated into two different responses:

$$y(x) = \underbrace{y_h(x)}_{\substack{\text{response of system} \\ \text{due to initial conditions} \\ y(x_0) = y_0,\ y(x_0) = y_1}} + \underbrace{y_p(x)}_{\substack{\text{response of system} \\ \text{due to the forcing} \\ \text{function } f}} \qquad (18)$$

If you wish to peek ahead, the following initial-value problem represents a pure resonance situation for a driven spring/mass system. See page 207.

### EXAMPLE 5 Using Theorem 4.8.2

Solve the initial-value problem

$$y'' + 4y = \sin 2x, \quad y(0) = 1, \ y'(0) = -2.$$

**SOLUTION** We solve two initial-value problems.

First, we solve $y'' + 4y = 0$, $y(0) = 1$, $y'(0) = -2$. By applying the initial conditions to the general solution $y(x) = c_1 \cos 2x + c_2 \sin 2x$ of the homogeneous DE, we find that $c_1 = 1$ and $c_2 = -1$. Therefore, $y_h(x) = \cos 2x - \sin 2x$.

Next we solve $y'' + 4y = \sin 2x$, $y(0) = 0$, $y'(0) = 0$. Since the left-hand side of the differential equation is the same as the DE in Example 4, the Green's function is the same, namely, $G(x, t) = \frac{1}{2}\sin 2(x - t)$. With $f(t) = \sin 2t$ we see from (10) that the solution of this second problem is $y_p(x) = \frac{1}{2}\int_0^x \sin 2(x - t)\sin 2t\, dt$.

Finally, in view of (17) in Theorem 4.8.2, the solution of the original IVP is

$$y(x) = y_h(x) + y_p(x) = \cos 2x - \sin 2x + \frac{1}{2}\int_0^x \sin 2(x - t)\sin 2t\, dt, \qquad (19) \quad \blacksquare$$

If desired, we can integrate the definite integral in (19) by using the trigonometric identity

$$\sin A \sin B = \tfrac{1}{2}[\cos(A - B) - \cos(A + B)]$$

with $A = 2(x - t)$ and $B = 2t$:

$$y_p(x) = \frac{1}{2}\int_0^x \sin 2(x - t)\sin 2t\, dt$$

$$= \frac{1}{4}\int_0^x [\cos(2x - 4t) - \cos 2x]\, dt \qquad (20)$$

$$= \frac{1}{4}\Big[-\tfrac{1}{4}\sin(2x - 4t) - t\cos 2x\Big]_0^x$$

$$= \tfrac{1}{8}\sin 2x - \tfrac{1}{4}x\cos 2x.$$

Hence, the solution (19) can be rewritten as

$$y(x) = y_h(x) + y_p(x) = \cos 2x - \sin 2x + \left(\tfrac{1}{8} \sin 2x - \tfrac{1}{4} x \cos 2x\right),$$

or                                    $y(x) = \cos 2x - \tfrac{7}{8} \sin 2x - \tfrac{1}{4} x \cos 2x.$          (21)

Note that the physical significance indicated in (18) is lost in (21) after combining like terms in the two parts of the solution $y(x) = y_h(x) + y_p(x)$.

The beauty of the solution given in (19) is that we can immediately write down the response of a system if the initial conditions remain the same, but the forcing function is changed. For example, if the problem in Example 5 is changed to

$$y'' + 4y = x, \quad y(0) = 1, y'(0) = -2,$$

we simply replace $\sin 2t$ in the integral in (19) by $t$ and the solution is then

$$y(x) = y_h(x) + y_p(x)$$

$$= \cos 2x - \sin 2x + \tfrac{1}{2} \int_0^x t \sin 2(x - t) \, dt \leftarrow \text{see Example 4}$$

$$= \tfrac{1}{4} x + \cos 2x - \tfrac{9}{8} \sin 2x.$$

Because the forcing function $f$ is isolated in the particular solution $y_p(x) = \int_{x_0}^x G(x, t) f(t) \, dt$, the solution in (17) is useful when $f$ is piecewise defined. The next example illustrates this idea.

### EXAMPLE 6   An Initial-Value Problem

Solve the initial-value problem

$$y'' + 4y = f(x), \quad y(0) = 1, y'(0) = -2,$$

where the forcing function $f$ is piecewise defined:

$$f(x) = \begin{cases} 0, & x < 0 \\ \sin 2x, & 0 \le x \le 2\pi \\ 0, & x > 2\pi. \end{cases}$$

**SOLUTION**  From (19), with $\sin 2t$ replaced by $f(t)$, we can write

$$y(x) = \cos 2x - \sin 2x + \tfrac{1}{2} \int_0^x \sin 2(x - t) f(t) \, dt.$$

Because $f$ is defined in three pieces, we consider three cases in the evaluation of the definite integral. For $x < 0$,

$$y_p(x) = \tfrac{1}{2} \int_0^x \sin 2(x - t) \, 0 \, dt = 0,$$

for $0 \le x \le 2$,

$$y_p(x) = \tfrac{1}{2} \int_0^x \sin 2(x - t) \sin 2t \, dt \leftarrow \text{using the integration in (20)}$$

$$= \tfrac{1}{8} \sin 2x - \tfrac{1}{4} x \cos 2x,$$

and finally for $x > 2\pi$, we can use the integration following Example 5:

$$
\begin{aligned}
y_p(x) &= \tfrac{1}{2}\int_0^{2\pi} \sin 2(x-t)\sin 2t\, dt + \tfrac{1}{2}\int_{2\pi}^x \sin 2(x-t)\, 0\, dt \\
&= \tfrac{1}{2}\pi\int_0^{2\pi} \sin 2(x-t)\sin 2t\, dt \\
&= \tfrac{1}{4}\Big[-\tfrac{1}{4}\sin(2x-4t) - t\cos 2x\Big]_0^{2\pi} \quad \leftarrow \text{using the integration in (20)} \\
&= -\tfrac{1}{16}\sin(2x-8\pi) - \tfrac{1}{2}\pi\cos 2x + \tfrac{1}{16}\sin 2x \quad \leftarrow \sin(2x-8\pi) = \sin 2x \\
&= -\tfrac{1}{2}\pi\cos 2x.
\end{aligned}
$$

Hence $y_p(x)$ is

$$
y_p(x) = \begin{cases}
0, & x < 0 \\
\tfrac{1}{8}\sin 2x - \tfrac{1}{4}x\cos 2x, & 0 \le x \le 2\pi \\
-\tfrac{1}{2}\pi\cos 2x, & x > 2\pi.
\end{cases}
$$

and so

$$
y(x) = y_h(x) + y_p(x) = \cos 2x - \sin 2x + y_p(x).
$$

Putting all the pieces together we get

$$
y(x) = \begin{cases}
\cos 2x - \sin 2x, & x < 0 \\
\left(1 - \tfrac{1}{4}x\right)\cos 2x - \tfrac{7}{8}\sin 2x, & 0 \le x \le 2\pi \\
\left(1 - \tfrac{1}{2}\pi\right)\cos 2x - \sin 2x, & x > 2\pi.
\end{cases}
$$

The three parts of $y(x)$ are shown in different colors in Figure 4.8.1. ◼

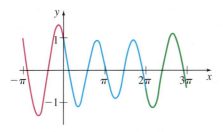

**FIGURE 4.8.1** Graph of $y(x)$ in Example 6

We next examine how a boundary value problem (BVP) can be solved using a different kind of Green's function.

### 4.8.2 BOUNDARY-VALUE PROBLEMS

In contrast to a second-order IVP, in which $y(x)$ and $y'(x)$ are specified at the same point, a BVP for a second-order DE involves conditions on $y(x)$ and $y'(x)$ that are specified at two different points $x = a$ and $x = b$. Conditions such as

$$y(a) = 0, \quad y(b) = 0; \qquad y(a) = 0, \quad y'(b) = 0; \qquad y'(a) = 0, \quad y'(b) = 0$$

are just special cases of the more general homogeneous boundary conditions:

$$A_1 y(a) + B_1 y'(a) = 0 \tag{22}$$

$$A_2 y(b) + B_2 y'(b) = 0, \tag{23}$$

where $A_1$, $A_2$, $B_1$, and $B_2$ are constants. Specifically, our goal is to find an integral solution $y_p(x)$ that is analogous to (10) for nonhomogeneous boundary value problems of the form

$$y'' + P(x)y' + Q(x)y = f(x),$$

$$A_1 y(a) + B_1 y'(a) = 0 \tag{24}$$

$$A_2 y(b) + B_2 y'(b) = 0.$$

In addition to the usual assumptions that $P(x)$, $Q(x)$, and $f(x)$ are continuous on $[a, b]$, we assume that the homogeneous problem

$$y'' + P(x)y' + Q(x)y = 0,$$

$$A_1 y(a) + B_1 y'(a) = 0$$

$$A_2 y(b) + B_2 y'(b) = 0,$$

possesses only the trivial solution $y = 0$. This latter assumption is sufficient to guarantee that a unique solution of (24) exists and is given by an integral $y_p(x) = \int_a^b G(x, t) f(t)\, dt$, where $G(x, t)$ is a Green's function.

The starting point in the construction of $G(x, t)$ is again the variation of parameters formulas (7) and (8).

**ANOTHER GREEN'S FUNCTION**   Suppose $y_1(x)$ and $y_2(x)$ are linearly independent solutions on $[a, b]$ of the associated homogeneous form of the DE in (24) and that $x$ is a number in the interval $[a, b]$. Unlike the construction of (9) where we started by integrating the derivatives in (8) over the same interval, we now integrate the first equation in (8) on $[b, x]$ and the second equation in (8) on $[a, x]$:

$$u_1(x) = -\int_b^x \frac{y_2(t)f(t)}{W(t)}\, dt \quad \text{and} \quad u_2(x) = \int_a^x \frac{y_1(t)f(t)}{W(t)}\, dt. \tag{25}$$

The reason for integrating $u_1'(x)$ and $u_2'(x)$ over different intervals will become clear shortly. From (25), a particular solution $y_p(x) = u_1(x)y_1(x) + u_2(x)y_2(x)$ of the DE is

$$\overbrace{\rule{3cm}{0pt}}^{\substack{\text{here we used the minus}\\ \text{sign in (25) to reverse}\\ \text{the limits of integration}}}$$

$$y_p(x) = y_1(x)\int_x^b \frac{y_2(t)f(t)}{W(t)}\, dt + y_2(x)\int_a^x \frac{y_1(t)f(t)}{W(t)}\, dt$$

or $\qquad y_p(x) = \int_a^x \frac{y_2(x)y_1(t)}{W(t)} f(t)\, dt + \int_x^b \frac{y_1(x)y_2(t)}{W(t)} f(t)\, dt. \tag{26}$

The right-hand side of (26) can be written compactly as a single integral

$$y_p(x) = \int_a^b G(x, t) f(t)\, dt, \tag{27}$$

where the function $G(x, t)$ is

$$G(x, t) = \begin{cases} \dfrac{y_1(t)y_2(x)}{W(t)}, & a \le t \le x \\[2mm] \dfrac{y_1(x)y_2(t)}{W(t)}, & x \le t \le b. \end{cases} \tag{28}$$

The piecewise-defined function (28) is called a **Green's function** for the boundary-value problem (24). It can be proved that $G(x, t)$ is a continuous function of $x$ on the interval $[a, b]$.

Now if the solutions $y_1(x)$ and $y_2(x)$ used in the construction of $G(x, t)$ in (28) are chosen in such a manner that at $x = a$, $y_1(x)$ satisfies $A_1 y_1(a) + B_1 y_1'(a) = 0$, and at $x = b$, $y_2(x)$ satisfies $A_2 y_2(b) + B_2 y_2'(b) = 0$, then, wondrously, $y_p(x)$ defined in (27) satisfies both homogeneous boundary conditions in (24).

To see this we will need

The second line in (30) results from the fact that

$y_1(x)u_1'(x) + y_2(x)u_2'(x) = 0$.

See the discussion in Section 4.6 following (4).

and

$$y_p(x) = u_1(x)y_1(x) + u_2(x)y_2(x) \tag{29}$$

$$y_p'(x) = u_1(x)y_1'(x) + y_1(x)u_1'(x) + u_2(x)y_2'(x) + y_2(x)u_2'(x)$$
$$= u_1(x)y_1'(x) + u_2(x)y_2'(x). \tag{30}$$

Before proceeding, observe in (25) that $u_1(b) = 0$ and $u_2(a) = 0$. In view of the second of these two properties we can show that $y_p(x)$ satisfies (22) whenever $y_1(x)$ satisfies the same boundary condition. From (29) and (30) we have

$$A_1 y_p(a) + B_1 y_p'(a) = A_1[u_1(a)y_1(a) + \overset{0}{\overbrace{u_2(a)}}y_2(a)] + B_1[u_1(a)y_1'(a) + \overset{0}{\overbrace{u_2(a)}}y_2'(a)]$$
$$= u_1(a)\underbrace{[A_1 y_1(a) + B_1 y_1'(a)]}_{\text{0 from (22)}} = 0.$$

Likewise, $u_1(b) = 0$ implies that whenever $y_2(x)$ satisfies (23) so does $y_p(x)$:

$$A_2 y_p(b) + B_2 y_p'(b) = A_2[\overset{0}{\overbrace{u_1(b)}}y_1(b) + u_2(b)y_2(b)] + B_2[\overset{0}{\overbrace{u_1(b)}}y_1'(b) + u_2(b)y_2'(b)]$$
$$= u_2(b)\underbrace{[A_2 y_2(b) + B_2 y_2'(b)]}_{\text{0 from (22)}} = 0.$$

The next theorem summarizes these results.

---

**THEOREM 4.8.3** **Solution of the BVP (24)**

Let $y_1(x)$ and $y_2(x)$ be linearly independent solutions of

$$y'' + P(x)y' + Q(x)y = 0$$

on $[a,\ b]$, and suppose $y_1(x)$ and $y_2(x)$ satisfy (22) and (23), respectively. Then the function $y_p(x)$ defined in (27) is a solution of the boundary-value problem (24).

---

**EXAMPLE 7** **Using Theorem 4.8.3**

Solve the boundary-value problem

$$y'' + 4y = 3, \quad y'(0) = 0, \quad y(\pi/2) = 0.$$

The boundary condition $y'(0) = 0$ is a special case of (22) with $a = 0$, $A_1 = 0$, and $B_1 = 1$. The boundary condition $y(\pi/2) = 0$ is a special case of (23) with $b = \pi/2$, $A_2 = 1$, $B_2 = 0$.

**SOLUTION** The solutions of the associated homogeneous equation $y'' + 4y = 0$ are $y_1(x) = \cos 2x$ and $y_2(x) = \sin 2x$ and $y_1(x)$ satisfies $y'(0) = 0$ whereas $y_2(x)$ satisfies $y(\pi/2) = 0$. The Wronskian is $W(y_1, y_2) = 2$, and so from (28) we see that the Green's function for the boundary-value problem is

$$G(x, t) = \begin{cases} \frac{1}{2}\cos 2t \sin 2x, & 0 \le t \le x \\ \frac{1}{2}\cos 2x \sin 2t, & x \le t \le \pi/2. \end{cases}$$

It follows from Theorem 4.8.3 that a solution of the BVP is (27) with the identifications $a = 0$, $b = \pi/2$, and $f(t) = 3$.

$$y_p(x) = 3\int_0^{\pi/2} G(x, t)\, dt$$

$$= 3 \cdot \frac{1}{2}\sin 2x \int_0^x \cos 2t\, dt + 3 \cdot \frac{1}{2}\cos 2x \int_x^{\pi/2} \sin 2t\, dt,$$

or, after evaluating the definite integrals, $y_p(x) = \frac{3}{4} + \frac{3}{4}\cos 2x$. ∎

Don't infer from the preceding example that the demand that $y_1(x)$ satisfy (22) and $y_2(x)$ satisfy (23) uniquely determines these functions. As we see in the last example, there is a certain arbitrariness in the selection of these functions.

### EXAMPLE 8    A Boundary-Value Problem

Solve the boundary-value problem

$$x^2 y'' - 3xy' + 3y = 24x^5, \quad y(1) = 0, \quad y(2) = 0.$$

**SOLUTION** The differential equation is recognized as a Cauchy-Euler DE. From the auxiliary equation $m(m - 1) - 3m + 3 = (m - 1)(m - 3) = 0$ the general solution of the associated homogeneous equation is $y = c_1 x + c_2 x^3$. Applying $y(1) = 0$ to this solution implies $c_1 + c_2 = 0$ or $c_1 = -c_2$. By choosing $c_2 = -1$ we get $c_1 = 1$ and $y_1 = x - x^3$. On the other hand, $y(2) = 0$ applied to the general solution shows $2c_1 + 8c_2 = 0$ or $c_1 = -4c_2$. The choice $c_2 = -1$ now gives $c_1 = 4$ and so $y_2(x) = 4x - x^3$. The Wronskian of these two functions is

$$W(y_1(x), y_2(x)) = \begin{vmatrix} x - x^3 & 4x - x^3 \\ 1 - 3x^2 & 4 - 3x^2 \end{vmatrix} = 6x^3.$$

Hence the Green's function for the boundary-value problem is

$$G(x, t) = \begin{cases} \dfrac{(t - t^3)(4x - x^3)}{6t^3}, & 1 \le t \le x \\[2ex] \dfrac{(x - x^3)(4t - t^3)}{6t^3}, & x \le t \le 2. \end{cases}$$

In order to identify the correct forcing function $f$ we must write the DE in standard form:

$$y'' - \frac{3}{x} y' + \frac{3}{x^2} y = 24x^3.$$

From this equation we see that $f(t) = 24t^3$ and so $y_p(x)$ in (27) becomes

$$y_p(x) = 24 \int_1^2 G(x, t)\, t^3 dt$$

$$= 4(4x - x^3) \int_1^x (t - t^3)\, dt + 4(x - x^3) \int_x^2 (4t - t^3)\, dt.$$

Straightforward definite integration and algebraic simplification yield the solution
$y_p(x) = 3x^5 - 15x^3 + 12x.$ ∎

 Verify $y_p(x)$ that satisfies the differential equation and the two boundary conditions.

---

### REMARKS

We have barely scratched the surface of the elegant, albeit complicated, theory of Green's functions. Green's functions can also be constructed for linear second-order partial differential equations, but we leave coverage of the latter topic to an advanced course.

---

## EXERCISES 4.8

Answers to selected odd-numbered problems begin on page ANS-6.

### 4.8.1 Initial-Value Problems

In Problems 1–6 proceed as in Example 1 to find a particular solution $y_p(x)$ of the given differential equation in the integral form (10).

**1.** $y'' - 16y = f(x)$

**2.** $y'' + 3y' - 10y = f(x)$

**3.** $y'' + 2y' + y = f(x)$

**4.** $4y'' - 4y' + y = f(x)$

**5.** $y'' + 9y = f(x)$

**6.** $y'' - 2y' + 2y = f(x)$

In Problems 7–12 proceed as in Example 2 to find the general solution of the given differential equation. Use the results obtained in Problems 1–6. Do not evaluate the integral that defines $y_p(x)$.

**7.** $y'' - 16y = xe^{-2x}$

**8.** $y'' + 3y' - 10y = x^2$

**9.** $y'' + 2y' + y = e^{-x}$

**10.** $4y'' - 4y' + y = \arctan x$

**11.** $y'' + 9y = x + \sin x$

**12.** $y'' - 2y' + 2y = \cos^2 x$

In Problems 13–18 proceed as in Example 3 to find a solution of the given initial-value problem. Evaluate the integral that defines $y_p(x)$.

**13.** $y'' - 4y = e^{2x}, y(0) = 0, y'(0) = 0$

**14.** $y'' - y' = 1, y(0) = 0, y'(0) = 0$

**15.** $y'' - 10y' + 25y = e^{5x}, y(0) = 0, y'(0) = 0$

**16.** $y'' + 6y' + 9y = x, y(0) = 0, y'(0) = 0$

**17.** $y'' + y = \csc x \cot x, y(\pi/2) = 0, y'(\pi/2) = 0$

**18.** $y'' + y = \sec^2 x, y(\pi) = 0, y'(\pi) = 0$

In Problems 19–30 proceed as in Example 5 to find a solution of the given initial-value problem.

**19.** $y'' - 4y = e^{2x}, y(0) = 1, y'(0) = -4$

**20.** $y'' - y' = 1, y(0) = 10, y'(0) = 1$

**21.** $y'' - 10y' + 25y = e^{5x}, y(0) = -1, y'(0) = 1$

**22.** $y'' + 6y' + 9y = x, y(0) = 1, y'(0) = -3$

**23.** $y'' + y = \csc x \cot x, y(\pi/2) = -\pi/2, y'(\pi/2) = -1$

**24.** $y'' + y = \sec^2 x, y(\pi) = \frac{1}{2}, y'(\pi) = -1$

**25.** $y'' + 3y' + 2y = \sin e^x, y(0) = -1, y'(0) = 0$

**26.** $y'' + 3y' + 2y = \dfrac{1}{1 + e^x}, y(0) = 0, y'(0) = 1$

**27.** $x^2 y'' - 2xy' + 2y = x, y(1) = 2, y'(1) = -1$

**28.** $x^2 y'' - 2xy' + 2y = x \ln x, y(1) = 1, y'(1) = 0$

**29.** $x^2 y'' - 6y = \ln x, y(1) = 1, y'(1) = 3$

**30.** $x^2 y'' - xy' + y = x^2, y(1) = 4, y'(1) = 3$

In Problems 31–34 proceed as in Example 6 to find a solution of the initial-value problem with the given piecewise-defined forcing function.

**31.** $y'' - y = f(x), y(0) = 8, y'(0) = 2,$

where $f(x) = \begin{cases} -1, x < 0 \\ 1, x \geq 0 \end{cases}$

**32.** $y'' - y = f(x), y(0) = 3, y'(0) = 2,$

where $f(x) = \begin{cases} 0, x < 0 \\ x, x \geq 0 \end{cases}$

**33.** $y'' + y = f(x), y(0) = 1, y'(0) = -1,$

where $f(x) = \begin{cases} 0, & x < 0 \\ 10, & 0 \leq x \leq 3\pi \\ 0, & x > 3\pi \end{cases}$

**34.** $y'' + y = f(x), y(0) = 0, y'(0) = 1,$

where $f(x) = \begin{cases} 0, & x < 0 \\ \cos x, & 0 \leq x \leq 4\pi \\ 0, & x > 4\pi \end{cases}$

### 4.8.2 Boundary-Value Problems

In Problems 35 and 36, **(a)** use (27) and (28) to find a solution of the boundary-value problem. **(b)** Verify that the function $y_p(x)$ satisfies the differential equations and both boundary-conditions.

**35.** $y'' = f(x), y(0) = 0, y(1) = 0$

**36.** $y'' = f(x), y(0) = 0, y(1) + y'(1) = 0$

**37.** In Problem 35 find a solution of the BVP when $f(x) = 1$.

**38.** In Problem 36 find a solution of the BVP when $f(x) = x$.

In Problems 39–44 proceed as in Examples 7 and 8 to find a solution of the given boundary-value problem.

**39.** $y'' + y = 1, y(0) = 0, y(1) = 0$

**40.** $y'' + 9y = 1, y(0) = 0, y'(\pi) = 0$

**41.** $y'' - 2y' + 2y = e^x, y(0) = 0, y(\pi/2) = 0$

**42.** $y'' - y' = e^{2x}, y(0) = 0, y(1) = 0$

**43.** $x^2 y'' + xy' = 1, y(e^{-1}) = 0, y(1) = 0$

**44.** $x^2 y'' - 4xy' + 6y = x^4, y(1) - y'(1) = 0, y(3) = 0$

## 4.9 Solving Systems of Linear DEs by Elimination

**INTRODUCTION**   Simultaneous ordinary differential equations involve two or more equations that contain derivatives of two or more dependent variables—the unknown functions—with respect to a single independent variable. The method of **systematic elimination** for solving systems of differential equations with constant coefficients is based on the algebraic principle of elimination of variables. We shall see that the analogue of *multiplying* an algebraic equation by a constant is *operating* on an ODE with some combination of derivatives.

**SYSTEMATIC ELIMINATION**   The elimination of an unknown in a system of linear differential equations is expedited by rewriting each equation in the system in differential operator notation. Recall from Section 4.1 that a single linear equation

$$a_n y^{(n)} + a_{n-1} y^{(n-1)} + \cdots + a_1 y' + a_0 y = g(t),$$

where the $a_i$, $i = 0, 1, \ldots, n$ are constants, can be written as

$$(a_n D^n + a_{n-1} D^{(n-1)} + \cdots + a_1 D + a_0)y = g(t).$$

If the $n$th-order differential operator $a_n D^n + a_{n-1} D^{(n-1)} + \cdots + a_1 D + a_0$ factors into differential operators of lower order, then the factors commute. Now, for example, to rewrite the system

$$x'' + 2x' + y'' = x + 3y + \sin t$$
$$x' + y' = -4x + 2y + e^{-t}$$

in terms of the operator $D$, we first bring all terms involving the dependent variables to one side and group the same variables:

$$x'' + 2x' - x + y'' - 3y = \sin t \qquad \text{is the same as} \qquad (D^2 + 2D - 1)x + (D^2 - 3)y = \sin t$$
$$x' + 4x + y' - 2y = e^{-t} \qquad\qquad\qquad (D + 4)x + (D - 2)y = e^{-t}.$$

**SOLUTION OF A SYSTEM**  A **solution** of a system of differential equations is a set of sufficiently differentiable functions $x = \phi_1(t)$, $y = \phi_2(t)$, $z = \phi_3(t)$, and so on that satisfies each equation in the system on some common interval $I$.

**METHOD OF SOLUTION**  Consider the simple system of linear first-order equations

$$\frac{dx}{dt} = 3y \qquad \text{or, equivalently,} \qquad Dx - 3y = 0 \qquad (1)$$
$$\frac{dy}{dt} = 2x \qquad\qquad\qquad\qquad 2x - Dy = 0.$$

Operating on the first equation in (1) by $D$ while multiplying the second by $-3$ and then adding eliminates $y$ from the system and gives $D^2 x - 6x = 0$. Since the roots of the auxiliary equation of the last DE are $m_1 = \sqrt{6}$ and $m_2 = -\sqrt{6}$, we obtain

$$x(t) = c_1 e^{-\sqrt{6}t} + c_2 e^{\sqrt{6}t}. \qquad (2)$$

Multiplying the first equation in (1) by 2 while operating on the second by $D$ and then subtracting gives the differential equation for $y$, $D^2 y - 6y = 0$. It follows immediately that

$$y(t) = c_3 e^{-\sqrt{6}t} + c_4 e^{\sqrt{6}t}. \qquad (3)$$

Now (2) and (3) do not satisfy the system (1) for every choice of $c_1$, $c_2$, $c_3$, and $c_4$ because the system itself puts a constraint on the number of parameters in a solution that can be chosen arbitrarily. To see this, observe that substituting $x(t)$ and $y(t)$ into the first equation of the original system (1) gives, after simplification,

$$\left(-\sqrt{6}c_1 - 3c_3\right)e^{-\sqrt{6}t} + \left(\sqrt{6}c_2 - 3c_4\right)e^{\sqrt{6}t} = 0.$$

Since the latter expression is to be zero for all values of $t$, we must have $-\sqrt{6}c_1 - 3c_3 = 0$ and $\sqrt{6}c_2 - 3c_4 = 0$. These two equations enable us to write $c_3$ as a multiple of $c_1$ and $c_4$ as a multiple of $c_2$:

$$c_3 = -\frac{\sqrt{6}}{3}c_1 \qquad \text{and} \qquad c_4 = \frac{\sqrt{6}}{3}c_2. \qquad (4)$$

Hence we conclude that a solution of the system must be

$$x(t) = c_1 e^{-\sqrt{6}t} + c_2 e^{\sqrt{6}t}, \qquad y(t) = -\frac{\sqrt{6}}{3} c_1 e^{-\sqrt{6}t} + \frac{\sqrt{6}}{3} c_2 e^{\sqrt{6}t}.$$

You are urged to substitute (2) and (3) into the second equation of (1) and verify that the same relationship (4) holds between the constants.

---

### EXAMPLE 1   Solution by Elimination

Solve
$$Dx + (D + 2)y = 0$$
$$(D - 3)x - \qquad 2y = 0. \tag{5}$$

**SOLUTION** Operating on the first equation by $D - 3$ and on the second by $D$ and then subtracting eliminates $x$ from the system. It follows that the differential equation for $y$ is

$$[(D - 3)(D + 2) + 2D]y = 0 \qquad \text{or} \qquad (D^2 + D - 6)y = 0.$$

Since the characteristic equation of this last differential equation is $m^2 + m - 6 = (m - 2)(m + 3) = 0$, we obtain the solution

$$y(t) = c_1 e^{2t} + c_2 e^{-3t}. \tag{6}$$

Eliminating $y$ in a similar manner yields $(D^2 + D - 6)x = 0$, from which we find

$$x(t) = c_3 e^{2t} + c_4 e^{-3t}. \tag{7}$$

As we noted in the foregoing discussion, a solution of (5) does not contain four independent constants. Substituting (6) and (7) into the first equation of (5) gives

$$(4c_1 + 2c_3)e^{2t} + (-c_2 - 3c_4)e^{-3t} = 0.$$

From $4c_1 + 2c_3 = 0$ and $-c_2 - 3c_4 = 0$ we get $c_3 = -2c_1$ and $c_4 = -\frac{1}{3}c_2$. Accordingly, a solution of the system is

$$x(t) = -2c_1 e^{2t} - \frac{1}{3}c_2 e^{-3t}, \qquad y(t) = c_1 e^{2t} + c_2 e^{-3t}. \qquad \blacksquare$$

Because we could just as easily solve for $c_3$ and $c_4$ in terms of $c_1$ and $c_2$, the solution in Example 1 can be written in the alternative form

$$x(t) = c_3 e^{2t} + c_4 e^{-3t}, \qquad y(t) = -\frac{1}{2}c_3 e^{2t} - 3c_4 e^{-3t}.$$

It sometimes pays to keep one's eyes open when solving systems. Had we solved for $x$ first in Example 1, then $y$ could be found, along with the relationship between the constants, using the last equation in the system (5). You should verify that substituting $x(t)$ into $y = \frac{1}{2}(Dx - 3x)$ yields $y = -\frac{1}{2}c_3 e^{2t} - 3c_4 e^{-3t}$. Also note in the initial discussion that the relationship given in (4) and the solution $y(t)$ of (1) could also have been obtained by using $x(t)$ in (2) and the first equation of (1) in the form

*This might save you some time.* ⊙

$$y = \frac{1}{3}Dx = -\frac{1}{3}\sqrt{6}c_1 e^{-\sqrt{6}t} + \frac{1}{3}\sqrt{6}c_2 e^{\sqrt{6}t}.$$

---

### EXAMPLE 2   Solution by Elimination

Solve
$$x' - 4x + y'' = t^2$$
$$x' + x + y' = 0. \tag{8}$$

**SOLUTION** First we write the system in differential operator notation:

$$(D - 4)x + D^2 y = t^2$$
$$(D + 1)x + Dy = 0. \tag{9}$$

Then, by eliminating $x$, we obtain

$$[(D + 1)D^2 - (D - 4)D]y = (D + 1)t^2 - (D - 4)0$$

or

$$(D^3 + 4D)y = t^2 + 2t.$$

Since the roots of the auxiliary equation $m(m^2 + 4) = 0$ are $m_1 = 0$, $m_2 = 2i$, and $m_3 = -2i$, the complementary function is $y_c = c_1 + c_2 \cos 2t + c_3 \sin 2t$. To determine the particular solution $y_p$, we use undetermined coefficients by assuming that $y_p = At^3 + Bt^2 + Ct$. Therefore $y_p' = 3At^2 + 2Bt + C$, $y_p'' = 6At + 2B$, $y_p''' = 6A$,

$$y_p''' + 4y_p' = 12At^2 + 8Bt + 6A + 4C = t^2 + 2t.$$

The last equality implies that $12A = 1$, $8B = 2$, and $6A + 4C = 0$; hence $A = \frac{1}{12}$, $B = \frac{1}{4}$, and $C = -\frac{1}{8}$. Thus

$$y = y_c + y_p = c_1 + c_2 \cos 2t + c_3 \sin 2t + \frac{1}{12}t^3 + \frac{1}{4}t^2 - \frac{1}{8}t. \tag{10}$$

Eliminating $y$ from the system (9) leads to

$$[(D - 4) - D(D + 1)]x = t^2 \qquad \text{or} \qquad (D^2 + 4)x = -t^2.$$

It should be obvious that $x_c = c_4 \cos 2t + c_5 \sin 2t$ and that undetermined coefficients can be applied to obtain a particular solution of the form $x_p = At^2 + Bt + C$. In this case the usual differentiations and algebra yield $x_p = -\frac{1}{4}t^2 + \frac{1}{8}$, and so

$$x = x_c + x_p = c_4 \cos 2t + c_5 \sin 2t - \frac{1}{4}t^2 + \frac{1}{8}. \tag{11}$$

Now $c_4$ and $c_5$ can be expressed in terms of $c_2$ and $c_3$ by substituting (10) and (11) into either equation of (8). By using the second equation, we find, after combining terms,

$$(c_5 - 2c_4 - 2c_2) \sin 2t + (2c_5 + c_4 + 2c_3) \cos 2t = 0,$$

so $c_5 - 2c_4 - 2c_2 = 0$ and $2c_5 + c_4 + 2c_3 = 0$. Solving for $c_4$ and $c_5$ in terms of $c_2$ and $c_3$ gives $c_4 = -\frac{1}{5}(4c_2 + 2c_3)$ and $c_5 = \frac{1}{5}(2c_2 - 4c_3)$. Finally, a solution of (8) is found to be

$$x(t) = -\frac{1}{5}(4c_2 + 2c_3) \cos 2t + \frac{1}{5}(2c_2 - 4c_3) \sin 2t - \frac{1}{4}t^2 + \frac{1}{8},$$

$$y(t) = c_1 + c_2 \cos 2t + c_3 \sin 2t + \frac{1}{12}t^3 + \frac{1}{4}t^2 - \frac{1}{8}t. \qquad ■$$

---

**EXAMPLE 3**  **A Mixture Problem Revisited**

In (3) of Section 3.3 we saw that the system of linear first-order differential equations

$$\frac{dx_1}{dt} = -\frac{2}{25}x_1 + \frac{1}{50}x_2$$

$$\frac{dx_2}{dt} = \frac{2}{25}x_1 - \frac{2}{25}x_2$$

is a model for the number of pounds of salt $x_1(t)$ and $x_2(t)$ in brine mixtures in tanks $A$ and $B$, respectively, shown in Figure 3.3.1. At that time we were not able to solve the system. But now, in terms of differential operators, the foregoing system can be written as

$$\left(D + \frac{2}{25}\right)x_1 - \frac{1}{50}x_2 = 0$$

$$-\frac{2}{25}x_1 + \left(D + \frac{2}{25}\right)x_2 = 0.$$

Operating on the first equation by $D + \frac{2}{25}$, multiplying the second equation by $\frac{1}{50}$, adding, and then simplifying gives $(625D^2 + 100D + 3)x_1 = 0$. From the auxiliary equation

$$625m^2 + 100m + 3 = (25m + 1)(25m + 3) = 0$$

we see immediately that $x_1(t) = c_1e^{-t/25} + c_2e^{-3t/25}$. We can now obtain $x_2(t)$ by using the first DE of the system in the form $x_2 = 50\left(D + \frac{2}{25}\right)x_1$. In this manner we find the solution of the system to be

$$x_1(t) = c_1e^{-t/25} + c_2e^{-3t/25}, \qquad x_2(t) = 2c_1e^{-t/25} - 2c_2e^{-3t/25}.$$

In the original discussion on pages 108–109 we assumed that the initial conditions were $x_1(0) = 25$ and $x_2(0) = 0$. Applying these conditions to the solution yields $c_1 + c_2 = 25$ and $2c_1 - 2c_2 = 0$. Solving these equations simultaneously gives $c_1 = c_2 = \frac{25}{2}$. Finally, a solution of the initial-value problem is

$$x_1(t) = \frac{25}{2}e^{-t/25} + \frac{25}{2}e^{-3t/25}, \qquad x_2(t) = 25e^{-t/25} - 25e^{-3t/25}.$$

The graphs of both of these equations are given in Figure 4.9.1. Consistent with the fact that pure water is being pumped into tank $A$ we see in the figure that $x_1(t) \to 0$ and $x_2(t) \to 0$ as $t \to \infty$.

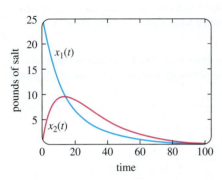

**FIGURE 4.9.1** Pounds of salt in tanks $A$ and $B$ in Example 3

## EXERCISES 4.9

Answers to selected odd-numbered problems begin on page ANS-6.

In Problems 1–20 solve the given system of differential equations by systematic elimination.

**1.** $\dfrac{dx}{dt} = 2x - y$

$\dfrac{dy}{dt} = x$

**2.** $\dfrac{dx}{dt} = 4x + 7y$

$\dfrac{dy}{dt} = x - 2y$

**3.** $\dfrac{dx}{dt} = -y + t$

$\dfrac{dy}{dt} = x - t$

**4.** $\dfrac{dx}{dt} - 4y = 1$

$\dfrac{dy}{dt} + x = 2$

**5.** $(D^2 + 5)x - \qquad 2y = 0$

$-2x + (D^2 + 2)y = 0$

**6.** $(D + 1)x + (D - 1)y = 2$

$3x + (D + 2)y = -1$

**7.** $\dfrac{d^2x}{dt^2} = 4y + e^t$

$\dfrac{d^2y}{dt^2} = 4x - e^t$

**8.** $\dfrac{d^2x}{dt^2} + \dfrac{dy}{dt} = -5x$

$\dfrac{dx}{dt} + \dfrac{dy}{dt} = -x + 4y$

**9.** $Dx + \qquad D^2y = e^{3t}$

$(D + 1)x + (D - 1)y = 4e^{3t}$

**10.** $D^2x - \qquad Dy = t$

$(D + 3)x + (D + 3)y = 2$

**11.** $(D^2 - 1)x - \quad y = 0$

$(D - 1)x + Dy = 0$

**12.** $(2D^2 - D - 1)x - (2D + 1)y = 1$

$(D - 1)x + \qquad Dy = -1$

**13.** $2\dfrac{dx}{dt} - 5x + \dfrac{dy}{dt} = e^t$

$\dfrac{dx}{dt} - \quad x + \dfrac{dy}{dt} = 5e^t$

**14.** $\dfrac{dx}{dt} + \dfrac{dy}{dt} \qquad = e^t$

$-\dfrac{d^2x}{dt^2} + \dfrac{dx}{dt} + x + y = 0$

**15.** $(D - 1)x + (D^2 + 1)y = 1$

$(D^2 - 1)x + \quad (D + 1)y = 2$

**16.** $D^2x - 2(D^2 + D)y = \sin t$

$x + \qquad Dy = 0$

**17.** $Dx = y$

$Dy = z$

$Dz = x$

**18.** $Dx + \qquad z = e^t$

$(D - 1)x + Dy + Dz = 0$

$x + 2y + Dz = e^t$

**19.** $\dfrac{dx}{dt} = 6y$

$\dfrac{dy}{dt} = x + z$

$\dfrac{dz}{dt} = x + y$

**20.** $\dfrac{dx}{dt} = -x + z$

$\dfrac{dy}{dt} = -y + z$

$\dfrac{dz}{dt} = -x + y$

In Problems 21 and 22 solve the given initial-value problem.

**21.** $\dfrac{dx}{dt} = -5x - y$

$\dfrac{dy}{dt} = 4x - y$

$x(1) = 0, y(1) = 1$

**22.** $\dfrac{dx}{dt} = y - 1$

$\dfrac{dy}{dt} = -3x + 2y$

$x(0) = 0, y(0) = 0$

## Mathematical Models

**23. Projectile Motion**   A projectile shot from a gun has weight $w = mg$ and velocity $\mathbf{v}$ tangent to its path of motion. Ignoring air resistance and all other forces acting on the projectile except its weight, determine a system of differential equations that describes its path of motion. See Figure 4.9.2. Solve the system. [*Hint*: Use Newton's second law of motion in the $x$ and $y$ directions.]

**FIGURE 4.9.2**  Path of projectile in Problem 23

**24. Projectile Motion with Air Resistance**   Determine a system of differential equations that describes the path of motion in Problem 23 if air resistance is a retarding force $\mathbf{k}$ (of magnitude $k$) acting tangent to the path of the projectile but opposite to its motion. See Figure 4.9.3. Solve the system. [*Hint*: $\mathbf{k}$ is a multiple of velocity, say, $\beta\mathbf{v}$.]

**FIGURE 4.9.3**  Forces in Problem 24

## Discussion Problems

**25.** Examine and discuss the following system:

$$Dx \qquad - 2Dy = t^2$$
$$(D + 1)x - 2(D + 1)y = 1.$$

## Computer Lab Assignments

**26.** Reexamine Figure 4.9.1 in Example 3. Then use a root-finding application to determine when tank $B$ contains more salt than tank $A$.

**27. (a)** Reread Problem 10 of Exercises 3.3. In that problem you were asked to show that the system of differential equations

$$\frac{dx_1}{dt} = -\frac{1}{50}x_1$$
$$\frac{dx_2}{dt} = \frac{1}{50}x_1 - \frac{2}{75}x_2$$
$$\frac{dx_3}{dt} = \frac{2}{75}x_2 - \frac{1}{25}x_3$$

is a model for the amounts of salt in the connected mixing tanks $A$, $B$, and $C$ shown in Figure 3.3.8. Solve the system subject to $x_1(0) = 15$, $x_2(0) = 10$, $x_3(0) = 5$.

**(b)** Use a CAS to graph $x_1(t)$, $x_2(t)$, and $x_3(t)$ in the same coordinate plane (as in Figure 4.9.1) on the interval [0, 200].

**(c)** Because only pure water is pumped into Tank $A$, it stands to reason that the salt will eventually be flushed out of all three tanks. Use a root-finding application of a CAS to determine the time when the amount of salt in each tank is less than or equal to 0.5 pound. When will the amounts of salt $x_1(t)$, $x_2(t)$, and $x_3(t)$ be simultaneously less than or equal to 0.5 pound?

---

## 4.10  Nonlinear Differential Equations

**INTRODUCTION**   The difficulties that surround higher-order *nonlinear* differential equations and the few methods that yield analytic solutions are examined next. Two of the solution methods considered in this section employ a change of variable to reduce a nonlinear second-order DE to a first-order DE. In that sense these methods are analogous to the material in Section 4.2.

**SOME DIFFERENCES**   There are several significant differences between linear and nonlinear differential equations. We saw in Section 4.1 that homogeneous linear equations of order two or higher have the property that a linear combination of solutions is also a solution (Theorem 4.1.2). Nonlinear equations do not possess this property of superposability. See Problems 1, 2, and 22 in Exercises 4.10. We can find general solutions of linear first-order DEs and higher-order equations with constant coefficients. Even when we can solve a nonlinear first-order differential equation in the form of a one-parameter family, this family does not, as a rule, represent a general solution. Stated another way, nonlinear first-order DEs can possess singular solutions, whereas linear equations cannot. But the major difference between linear and nonlinear equations of order two or higher lies in the realm of solvability. Given a linear equation, there is a chance that we can find some form of a solution that we can look at—an explicit solution or perhaps a solution in the form of an infinite series (see Chapter 6). On the other hand, nonlinear higher-order differential equations virtually defy solution by analytical methods. Although this might sound disheartening, there are still things that can be done. As was pointed out at the end of Section 1.3, we can always analyze a nonlinear DE qualitatively and numerically.

Let us make it clear at the outset that nonlinear higher-order differential equations are important—dare we say even more important than linear equations?—because as we fine-tune the mathematical model of, say, a physical system, we also increase the likelihood that this higher-resolution model will be nonlinear.

We begin by illustrating an analytical method that *occasionally* enables us to find explicit/implicit solutions of special kinds of nonlinear second-order differential equations.

**REDUCTION OF ORDER**    Nonlinear second-order differential equations $F(x, y', y'') = 0$, where the dependent variable $y$ is missing, and $F(y, y', y'') = 0$, where the independent variable $x$ is missing, can sometimes be solved by using first-order methods. Each equation can be reduced to a first-order equation by means of the substitution $u = y'$.

**DEPENDENT VARIABLE MISSING**    The next example illustrates the substitution technique for an equation of the form $F(x, y', y'') = 0$. If $u = y'$, then the differential equation becomes $F(x, u, u') = 0$. If we can solve this last equation for $u$, we can find $y$ by integration. Note that since we are solving a second-order equation, its solution will contain two arbitrary constants.

---

**EXAMPLE 1**    **Dependent Variable $y$ Is Missing**

Solve $y'' = 2x(y')^2$.

**SOLUTION**   If we let $u = y'$, then $du/dx = y''$. After substituting, the second-order equation reduces to a first-order equation with separable variables; the independent variable is $x$ and the dependent variable is $u$:

$$\frac{du}{dx} = 2xu^2 \quad \text{or} \quad \frac{du}{u^2} = 2x\,dx$$

$$\int u^{-2}\,du = \int 2x\,dx$$

$$-u^{-1} = x^2 + c_1^2.$$

The constant of integration is written as $c_1^2$ for convenience. The reason should be obvious in the next few steps. Because $u^{-1} = 1/y'$, it follows that

$$\frac{dy}{dx} = -\frac{1}{x^2 + c_1^2},$$

and so      $$y = \int \frac{dx}{x^2 + c_1^2} \quad \text{or} \quad y - -\frac{1}{c_1}\tan^{-1}\frac{x}{c_1} + c_2.$$   ■

**INDEPENDENT VARIABLE MISSING**    Next we show how to solve an equation that has the form $F(y, y', y'') = 0$. Once more we let $u = y'$, but because the independent variable $x$ is missing, we use this substitution to transform the differential equation into one in which the independent variable is $y$ and the dependent variable is $u$. To this end we use the Chain Rule to compute the second derivative of $y$:

$$y'' = \frac{du}{dx} = \frac{du}{dy}\frac{dy}{dx} = u\frac{du}{dy}.$$

In this case the first-order equation that we must now solve is

$$F\!\left(y, u, u\frac{du}{dy}\right) = 0.$$

### EXAMPLE 2 Independent Variable x Is Missing

Solve $yy'' = (y')^2$.

**SOLUTION** With the aid of $u = y'$, the Chain Rule shown above, and separation of variables, the given differential equation becomes

$$y\left(u\frac{du}{dy}\right) = u^2 \qquad \text{or} \qquad \frac{du}{u} = \frac{dy}{y}.$$

Integrating the last equation then yields $\ln|u| = \ln|y| + c_1$, which, in turn, gives $u = c_2 y$, where the constant $\pm e^{c_1}$ has been relabeled as $c_2$. We now resubstitute $u = dy/dx$, separate variables once again, integrate, and relabel constants a second time:

$$\int \frac{dy}{y} = c_2 \int dx \qquad \text{or} \qquad \ln|y| = c_2 x + c_3 \qquad \text{or} \qquad y = c_4 e^{c_2 x}. \quad \blacksquare$$

**USE OF TAYLOR SERIES** In some instances a solution of a nonlinear initial-value problem, in which the initial conditions are specified at $x_0$, can be approximated by a Taylor series centered at $x_0$.

### EXAMPLE 3 Taylor Series Solution of an IVP

Let us assume that a solution of the initial-value problem

$$y'' = x + y - y^2, \quad y(0) = -1, \quad y'(0) = 1 \tag{1}$$

exists. If we further assume that the solution $y(x)$ of the problem is analytic at 0, then $y(x)$ possesses a Taylor series expansion centered at 0:

$$y(x) = y(0) + \frac{y'(0)}{1!}x + \frac{y''(0)}{2!}x^2 + \frac{y'''(0)}{3!}x^3 + \frac{y^{(4)}(0)}{4!}x^4 + \frac{y^{(5)}(0)}{5!}x^5 + \cdots. \tag{2}$$

Note that the values of the first and second terms in the series (2) are known since those values are the specified initial conditions $y(0) = -1$, $y'(0) = 1$. Moreover, the differential equation itself defines the value of the second derivative at 0: $y''(0) = 0 + y(0) - y(0)^2 = 0 + (-1) - (-1)^2 = -2$. We can then find expressions for the higher derivatives $y'''$, $y^{(4)}$, ... by calculating the successive derivatives of the differential equation:

$$y'''(x) = \frac{d}{dx}(x + y - y^2) = 1 + y' - 2yy' \tag{3}$$

$$y^{(4)}(x) = \frac{d}{dx}(1 + y' - 2yy') = y'' - 2yy'' - 2(y')^2 \tag{4}$$

$$y^{(5)}(x) = \frac{d}{dx}(y'' - 2yy'' - 2(y')^2) = y''' - 2yy''' - 6y'y'' \tag{5}$$

and so on. Now using $y(0) = -1$ and $y'(0) = 1$, we find from (3) that $y'''(0) = 4$. From the values $y(0) = -1$, $y'(0) = 1$, and $y''(0) = -2$ we find $y^{(4)}(0) = -8$ from (4). With the additional information that $y'''(0) = 4$, we then see from (5) that $y^{(5)}(0) = 24$. Hence from (2) the first six terms of a series solution of the initial-value problem (1) are

$$y(x) = -1 + x - x^2 + \frac{2}{3}x^3 - \frac{1}{3}x^4 + \frac{1}{5}x^5 + \cdots. \quad \blacksquare$$

**USE OF A NUMERICAL SOLVER** Numerical methods, such as Euler's method or the Runge-Kutta method, are developed solely for first-order differential equations and then are extended to systems of first-order equations. To analyze an $n$th-order initial-value problem numerically, we express the $n$th-order ODE as a system of $n$

first-order equations. In brief, here is how it is done for a second-order initial-value problem: First, solve for $y''$—that is, put the DE into normal form $y'' = f(x, y, y')$—and then let $y' = u$. For example, if we substitute $y' = u$ in

$$\frac{d^2y}{dx^2} = f(x, y, y'), \quad y(x_0) = y_0, \quad y'(x_0) = u_0, \tag{6}$$

then $y'' = u'$ and $y'(x_0) = u(x_0)$, so the initial-value problem (6) becomes

$$\textbf{\textit{Solve}}: \quad \begin{cases} y' = u \\ u' = f(x, y, u) \end{cases}$$

$$\textbf{\textit{Subject to}}: \quad y(x_0) = y_0, \quad u(x_0) = u_0.$$

However, it should be noted that a commercial numerical solver *might not* require* that you supply the system.

---

### EXAMPLE 4   Graphical Analysis of Example 3

Following the foregoing procedure, we find that the second-order initial-value problem in Example 3 is equivalent to

$$\frac{dy}{dx} = u$$

$$\frac{du}{dx} = x + y - y^2$$

with initial conditions $y(0) = -1, u(0) = 1$. With the aid of a numerical solver we get the solution curve shown in blue in Figure 4.10.1. For comparison the graph of the fifth-degree Taylor polynomial $T_5(x) = -1 + x - x^2 + \frac{2}{3}x^3 - \frac{1}{3}x^4 + \frac{1}{5}x^5$ is shown in red. Although we do not know the interval of convergence of the Taylor series obtained in Example 3, the closeness of the two curves in a neighborhood of the origin suggests that the power series may converge on the interval $(-1, 1)$. ■

**QUALITATIVE QUESTIONS**   The blue numerical solution curve in Figure 4.10.1 raises some questions of a qualitative nature: Is the solution of the original initial-value problem oscillatory as $x \to \infty$? The graph generated by a numerical solver on the larger interval shown in Figure 4.10.2 would seem to *suggest* that the answer is yes. But this single example—or even an assortment of examples—does not answer the basic question as to whether *all* solutions of the differential equation $y'' = x + y - y^2$ are oscillatory in nature. Also, what is happening to the solution curve in Figure 4.10.2 when $x$ is near $-1$? What is the behavior of solutions of the differential equation as $x \to -\infty$? Are solutions bounded as $x \to \infty$? Questions such as these are not easily answered, in general, for nonlinear second-order differential equations. But certain kinds of second-order equations lend themselves to a systematic qualitative analysis, and these, like their first-order relatives encountered in Section 2.1, are the kind that have no explicit dependence on the independent variable. Second-order ODEs of the form

$$F(y, y', y'') = 0 \quad \text{or} \quad \frac{d^2y}{dx^2} = f(y, y'),$$

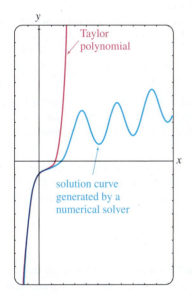

**FIGURE 4.10.1** Comparison of two approximate solutions in Example 3

**FIGURE 4.10.2** Numerical solution curve for the IVP in (1)

equations free of the independent variable $x$, are called **autonomous**. The differential equation in Example 2 is autonomous, and because of the presence of the $x$ term on its right-hand side, the equation in Example 3 is nonautonomous. For an in-depth treatment of the topic of stability of autonomous second-order differential equations and autonomous systems of differential equations, refer to Chapter 10 in *Differential Equations with Boundary-Value Problems*.

---

*Some numerical solvers require only that a second-order differential equation be expressed in normal form $y'' = f(x, y, y')$. The translation of the single equation into a system of two equations is then built into the computer program, since the first equation of the system is always $y' = u$ and the second equation is $u' = f(x, y, u)$.

# EXERCISES **4.10**

*Answers to selected odd-numbered problems begin on page ANS-6.*

In Problems 1 and 2 verify that $y_1$ and $y_2$ are solutions of the given differential equation but that $y = c_1 y_1 + c_2 y_2$ is, in general, not a solution.

**1.** $(y'')^2 = y^2$;   $y_1 = e^x, y_2 = \cos x$

**2.** $yy'' = \frac{1}{2}(y')^2$;   $y_1 = 1, y_2 = x^2$

In Problems 3–6 the dependent variable $y$ is missing in the given differential equation. Proceed as in Example 1 and solve the equation by using the substitution $u = y'$.

**3.** $y'' + (y')^2 + 1 = 0$      **4.** $y'' = 1 + (y')^2$

**5.** $x^2 y'' + (y')^2 = 0$      **6.** $e^{-x}y'' = (y')^2$

In Problems 7–10 the independent variable $x$ is missing in the given differential equation. Proceed as in Example 2 and solve the equation by using the substitution $u = y'$.

**7.** $yy'' + (y')^2 + 1 = 0$      **8.** $(y + 1)y'' = (y')^2$

**9.** $y'' + 2y(y')^3 = 0$      **10.** $y^2 y'' = y'$

In Problems 11 and 12 solve the given initial-value problem.

**11.** $2y'y'' = 1, y(0) = 2, y'(0) = 1$

**12.** $y'' + x(y')^2 = 0, y(1) = 4, y'(1) = 2$

**13.** Consider the initial-value problem

$$y'' + yy' = 0, \quad y(0) = 1, \quad y'(0) = -1.$$

(a) Use the DE and a numerical solver to graph the solution curve.

(b) Find an explicit solution of the IVP. Use a graphing utility to graph this solution.

(c) Find an interval of definition for the solution in part (b).

**14.** Find two solutions of the initial-value problem

$$(y'')^2 + (y')^2 = 1, \quad y\left(\frac{\pi}{2}\right) = \frac{1}{2}, \quad y'\left(\frac{\pi}{2}\right) = \frac{\sqrt{3}}{2}.$$

Use a numerical solver to graph the solution curves.

In Problems 15 and 16 show that the substitution $u = y'$ leads to a Bernoulli equation. Solve this equation (see Section 2.5).

**15.** $xy'' = y' + (y')^3$      **16.** $xy'' = y' + x(y')^2$

In Problems 17–20 proceed as in Example 3 and obtain the first six nonzero terms of a Taylor series solution, centered at 0, of the given initial-value problem. Use a numerical solver and a graphing utility to compare the solution curve with the graph of the Taylor polynomial.

**17.** $y'' = x + y^2$,   $y(0) = 1, y'(0) = 1$

**18.** $y'' + y^2 = 1$,   $y(0) = 2, y'(0) = 3$

**19.** $y'' = x^2 + y^2 - 2y'$,   $y(0) = 1, y'(0) = 1$

**20.** $y'' = e^y$,   $y(0) = 0, y'(0) = -1$

**21.** In calculus the curvature of a curve that is defined by a function $y = f(x)$ is defined as

$$\kappa = \frac{y''}{[1 + (y')^2]^{3/2}}.$$

Find $y = f(x)$ for which $\kappa = 1$. [*Hint:* For simplicity, ignore constants of integration.]

## Discussion Problems

**22.** In Problem 1 we saw that $\cos x$ and $e^x$ were solutions of the nonlinear equation $(y'')^2 - y^2 = 0$. Verify that $\sin x$ and $e^{-x}$ are also solutions. Without attempting to solve the differential equation, discuss how these explicit solutions can be found by using knowledge about linear equations. Without attempting to verify, discuss why the linear combinations $y = c_1 e^x + c_2 e^{-x} + c_3 \cos x + c_4 \sin x$ and $y = c_2 e^{-x} + c_4 \sin x$ are not, in general, solutions, but the two special linear combinations $y = c_1 e^x + c_2 e^{-x}$ and $y = c_3 \cos x + c_4 \sin x$ must satisfy the differential equation.

**23.** Discuss how the method of reduction of order considered in this section can be applied to the third-order differential equation $y''' = \sqrt{1 + (y'')^2}$. Carry out your ideas and solve the equation.

**24.** Discuss how to find an alternative two-parameter family of solutions for the nonlinear differential equation $y'' = 2x(y')^2$ in Example 1. [*Hint:* Suppose that $-c_1^2$ is used as the constant of integration instead of $+c_1^2$.]

## Mathematical Models

**25. Motion in a Force Field**   A mathematical model for the position $x(t)$ of a body moving rectilinearly on the $x$-axis in an inverse-square force field is given by

$$\frac{d^2 x}{dt^2} = -\frac{k^2}{x^2}.$$

Suppose that at $t = 0$ the body starts from rest from the position $x = x_0, x_0 > 0$. Show that the velocity of the body at time $t$ is given by $v^2 = 2k^2(1/x - 1/x_0)$. Use the last expression and a CAS to carry out the integration to express time $t$ in terms of $x$.

**26.** A mathematical model for the position $x(t)$ of a moving object is

$$\frac{d^2 x}{dt^2} + \sin x = 0.$$

Use a numerical solver to graphically investigate the solutions of the equation subject to $x(0) = 0, x'(0) = x_1, x_1 \geq 0$. Discuss the motion of the object for $t \geq 0$ and for various choices of $x_1$. Investigate the equation

$$\frac{d^2 x}{dt^2} + \frac{dx}{dt} + \sin x = 0$$

in the same manner. Give a possible physical interpretation of the $dx/dt$ term.

## Chapter 4 In Review

Answers to selected odd-numbered problems begin on page ANS-7.

Answer Problems 1–10 without referring back to the text. Fill in the blank or answer true or false.

**1.** The only solution of the initial-value problem
$y'' + x^2 y = 0$, $y(0) = 0$, $y'(0) = 0$ is _____.

**2.** For the method of undetermined coefficients, the assumed form of the particular solution $y_p$ for $y'' - y = 1 + e^x$ is _____.

**3.** A constant multiple of a solution of a linear differential equation is also a solution. _____

**4.** If the set consisting of two functions $f_1$ and $f_2$ is linearly independent on an interval $I$, then the Wronskian $W(f_1, f_2) \neq 0$ for all $x$ in $I$. _____

**5.** If $y = \sin 5x$ is a solution of a homogeneous linear second-order differential with constant coefficients, then the general solution of the DE is _____.

**6.** If $y = 1 - x + 6x^2 + 3e^x$ is a solution of a homogeneous fourth-order linear differential equation with constant coefficients, then the roots of the auxiliary equation are _____.

**7.** If $y = c_1 x^2 + c_2 x^2 \ln x$, $x > 0$, is the general solution of a homogeneous second-order Cauchy-Euler equation, then the DE is _____.

**8.** $y_p = Ax^2$ is particular solution of $y''' + y'' = 1$ for $A = $ _____.

**9.** If $y_{p_1} = x$ is a particular solution of $y'' + y = x$ and $y_{p_2} = x^2 - 2$ is a particular solution of $y'' + y = x^2$, then a particular solution of $y'' + y = x^2 + x$ is _____.

**10.** If $y_1 = e^x$ and $y_2 = e^{-x}$ are solutions of homogeneous linear differential equation, then necessarily $y = -5e^{-x} + 10e^x$ is also a solution of the DE. _____

**11.** Give an interval over which the set of two functions $f_1(x) = x^2$ and $f_2(x) = x|x|$ is linearly independent. Then give an interval over which the set consisting of $f_1$ and $f_2$ is linearly dependent.

**12.** Without the aid of the Wronskian, determine whether the given set of functions is linearly independent or linearly dependent on the indicated interval.

(a) $f_1(x) = \ln x$, $f_2(x) = \ln x^2$, $(0, \infty)$

(b) $f_1(x) = x^n$, $f_2(x) = x^{n+1}$, $n = 1, 2, \ldots$, $(-\infty, \infty)$

(c) $f_1(x) = x$, $f_2(x) = x + 1$, $(-\infty, \infty)$

(d) $f_1(x) = \cos\left(x + \dfrac{\pi}{2}\right)$, $f_2(x) = \sin x$, $(-\infty, \infty)$

(e) $f_1(x) = 0$, $f_2(x) = x$, $(-5, 5)$

(f) $f_1(x) = 2$, $f_2(x) = 2x$, $(-\infty, \infty)$

(g) $f_1(x) = x^2$, $f_2(x) = 1 - x^2$, $f_3(x) = 2 + x^2$, $(-\infty, \infty)$

(h) $f_1(x) = xe^{x+1}$, $f_2(x) = (4x - 5)e^x$, $f_3(x) = xe^x$, $(-\infty, \infty)$

**13.** Suppose $m_1 = 3$, $m_2 = -5$, and $m_3 = 1$ are roots of multiplicity one, two, and three, respectively, of an auxiliary equation. Write down the general solution of the corresponding homogeneous linear DE if it is

(a) an equation with constant coefficients,

(b) a Cauchy-Euler equation.

**14.** Consider the differential equation $ay'' + by' + cy = g(x)$, where $a$, $b$, and $c$ are constants. Choose the input functions $g(x)$ for which the method of undetermined coefficients is applicable and the input functions for which the method of variation of parameters is applicable.

(a) $g(x) = e^x \ln x$      (b) $g(x) = x^3 \cos x$

(c) $g(x) = \dfrac{\sin x}{e^x}$      (d) $g(x) = 2x^{-2}e^x$

(e) $g(x) = \sin^2 x$      (f) $g(x) = \dfrac{e^x}{\sin x}$

In Problems 15 and 16 find a homogeneous second-order Cauchy-Euler equation with real coefficients if the given numbers are roots of its auxiliary equation.

**15.** $m_1 = 4$, $m_2 = -1$      **16.** $m_1 = i$

In Problems 17–32 use the procedures developed in this chapter to find the general solution of each differential equation.

**17.** $y'' - 2y' - 2y = 0$

**18.** $2y'' + 2y' + 3y = 0$

**19.** $y''' + 10y'' + 25y' = 0$

**20.** $2y''' + 9y'' + 12y' + 5y = 0$

**21.** $3y''' + 10y'' + 15y' + 4y = 0$

**22.** $2y^{(4)} + 3y''' + 2y'' + 6y' - 4y = 0$

**23.** $y'' - 3y' + 5y = 4x^3 - 2x$

**24.** $y'' - 2y' + y = x^2 e^x$

**25.** $y''' - 5y'' + 6y' = 8 + 2\sin x$

**26.** $y''' - y'' = 6$

**27.** $y'' - 2y' + 2y = e^x \tan x$

**28.** $y'' - y = \dfrac{2e^x}{e^x + e^{-x}}$

**29.** $6x^2 y'' + 5xy' - y = 0$

**30.** $2x^3 y''' + 19x^2 y'' + 39xy' + 9y = 0$

**31.** $x^2 y'' - 4xy' + 6y = 2x^4 + x^2$

**32.** $x^2 y'' - xy' + y = x^3$

In Problems 33 and 34 write down the form of the general solution $y = y_c + y_p$ of the given differential equation in the two cases $\omega \neq \alpha$ and $\omega = \alpha$. Do not determine the coefficients in $y_p$.

**33.** $y'' + \omega^2 y = \sin \alpha x$      **34.** $y'' - \omega^2 y = e^{\alpha x}$

**35.** Given that $y = \sin x$ is a solution of

$$y^{(4)} + 2y''' + 11y'' + 2y' + 10y = 0,$$

find the general solution of the DE *without the aid of a calculator or a computer.*

**36.** Find a linear second-order differential equation with constant coefficients for which $y_1 = 1$ and $y_2 = e^{-x}$ are solutions of the associated homogeneous equation and $y_p = \frac{1}{2}x^2 - x$ is a particular solution of the nonhomogeneous equation.

**37. (a)** Write the general solution of the fourth-order DE $y^{(4)} - 2y'' + y = 0$ entirely in terms of hyperbolic functions.

**(b)** Write down the form of a particular solution of $y^{(4)} - 2y'' + y = \sinh x$.

**38.** Consider the differential equation

$$x^2y'' - (x^2 + 2x)y' + (x + 2)y = x^3.$$

Verify that $y_1 = x$ is one solution of the associated homogeneous equation. Then show that the method of reduction of order discussed in Section 4.2 leads to a second solution $y_2$ of the homogeneous equation as well as a particular solution $y_p$ of the nonhomogeneous equation. Form the general solution of the DE on the interval $(0, \infty)$.

In Problems 39–44 solve the given differential equation subject to the indicated conditions.

**39.** $y'' - 2y' + 2y = 0, \quad y(\pi/2) = 0, y(\pi) = -1$

**40.** $y'' + 2y' + y = 0, \quad y(-1) = 0, y'(0) = 0$

**41.** $y'' - y = x + \sin x, \quad y(0) = 2, y'(0) = 3$

**42.** $y'' + y = \sec^3 x, \quad y(0) = 1, y'(0) = \frac{1}{2}$

**43.** $y'y'' = 4x, \quad y(1) = 5, y'(1) = 2$

**44.** $2y'' = 3y^2, \quad y(0) = 1, y'(0) = 1$

**45. (a)** Use a CAS as an aid in finding the roots of the auxiliary equation for

$$12y^{(4)} + 64y''' + 59y'' - 23y' - 12y = 0.$$

Give the general solution of the equation.

**(b)** Solve the DE in part (a) subject to the initial conditions $y(0) = -1, y'(0) = 2, y''(0) = 5, y'''(0) = 0$. Use a CAS as an aid in solving the resulting systems of four equations in four unknowns.

**46.** Find a member of the family of solutions of $xy'' + y' + \sqrt{x} = 0$ whose graph is tangent to the $x$-axis at $x = 1$. Use a graphing utility to graph the solution curve.

In Problems 47–50 use systematic elimination to solve the given system.

**47.** $\dfrac{dx}{dt} + \dfrac{dy}{dt} = 2x + 2y + 1$

$\dfrac{dx}{dt} + 2\dfrac{dy}{dt} = y + 3$

**48.** $\dfrac{dx}{dt} = 2x + \quad y + \quad t - 2$

$\dfrac{dy}{dt} = 3x + 4y - 4t$

**49.** $(D - 2)x \qquad - y = -e^t$

$\qquad -3x + (D - 4)y = -7e^t$

**50.** $(D + 2)x + (D + 1)y = \sin 2t$

$\qquad 5x + (D + 3)y = \cos 2t$

# 5

# Modeling with Higher-Order Differential Equations

Brian A Jackson/Shutterstock.com

We have seen that a single differential equation can serve as a mathematical model for diverse physical systems. Forms of the linear second-order equation

$$a\frac{d^2y}{dt^2} + b\frac{dy}{dt} + cy = g(t),$$

appear in the analysis of problems in many different areas of science and engineering. In Section 5.1 we see that except for terminology and physical interpretations of the four terms in this differential equation the mathematics of, say, an electrical series circuit is identical to that of a vibrating spring/mass system.

# 5.1 Linear Models: Initial-Value Problems

**INTRODUCTION**    In this section we are going to consider several linear dynamical systems in which each mathematical model is a second-order differential equation with constant coefficients along with initial conditions specified at a time that we shall take to be $t = 0$:

$$a\frac{d^2y}{dt^2} + b\frac{dy}{dt} + cy = g(t), \quad y(0) = y_0, \quad y'(0) = y_1.$$

The function $g$ is variously called the **driving function, forcing function,** or **input** of the system. A solution $y(t)$ of the differential equation on an interval $I$ containing $t = 0$ that satisfies the initial conditions is called the **response** or **output** of the system.

## 5.1.1 SPRING/MASS SYSTEMS: FREE UNDAMPED MOTION

**HOOKE'S LAW**    Suppose that a flexible spring is suspended vertically from a rigid support and then a mass $m$ is attached to its free end. The amount of stretch, or elongation, of the spring will of course depend on the mass; masses with different weights stretch the spring by differing amounts. By Hooke's law the spring itself exerts a restoring force $F$ opposite to the direction of elongation and proportional to the amount of elongation $s$. Simply stated, $F = -ks$, where $k > 0$ is a constant of proportionality called the **spring constant.** The spring is essentially characterized by the number $k$. Using $|F| = k|s|$, we see that if a mass weighing 10 pounds stretches a spring $\frac{1}{2}$ foot, then $10 = k\left(\frac{1}{2}\right)$ implies $k = 20$ lb/ft. Necessarily then, a mass weighing, say, 8 pounds stretches the same spring only $\frac{2}{5}$ foot.

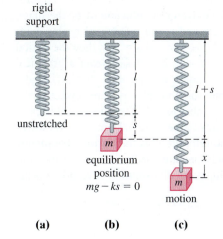

rigid support

unstretched

$l$

$l$

$l + s$

$s$

equilibrium position
$mg - ks = 0$

$m$

$x$

$m$

motion

**(a)**     **(b)**     **(c)**

**FIGURE 5.1.1**  Spring/mass system

$x < 0$

$x = 0$

$x > 0$

$m$

**FIGURE 5.1.2**  Direction below the equilibrium position is positive

**NEWTON'S SECOND LAW**    When a mass $m$ is attached to the lower end of a spring of negligible mass, it stretches the spring by an amount $s$ and attains an equilibrium position (or rest position) at which its weight $W$ is balanced by the restoring force $ks$ of the spring. Recall that weight is defined by $W = mg$, where $m$ is measured in slugs, kilograms, or grams and $g$ is the acceleration due to gravity ($32$ ft/s², $9.8$ m/s², or $980$ cm/s²). As indicated in Figure 5.1.1 the condition of equilibrium is $mg = ks$ or $mg - ks = 0$. Now suppose the mass on the spring is set in motion by giving it an initial displacement (an elongation or a compression) and an initial velocity. Let us assume that the motion takes place in a vertical line, that the displacements $x(t)$ of the mass are measured along this line such that $x = 0$ corresponds to the equilibrium position, and that displacements measured *below* the equilibrium position are *positive*. See Figure 5.1.2. To construct a mathematical model that describes this dynamic case, we employ Newton's second law of motion: the net or resultant force on a moving body of mass $m$ is given by $\Sigma F_k = ma$, where $a = d^2x/dt^2$ is its acceleration. If we further assume that the mass vibrates free of all other external forces—**free motion**—then Newton's second law gives

$$m\frac{d^2x}{dt^2} = \underbrace{-k(x + s) + mg}_{\text{net force}} = -kx + \underbrace{mg - ks}_{\text{zero}} = -kx. \tag{1}$$

The first term $F_1 = -k(x + s)$ on the right-hand side of equation (1) is the restoring force of the spring; the negative sign indicates that this force acts opposite to the direction of motion. The second term $F_2 = mg$ is the weight of the mass which always acts in the downward or positive direction.

**DE OF FREE UNDAMPED MOTION** By dividing (1) by the mass $m$, we obtain the second-order differential equation $d^2x/dt^2 + (k/m)x = 0$, or

$$\frac{d^2x}{dt^2} + \omega^2 x = 0, \tag{2}$$

where $\omega^2 = k/m$. Equation (2) is said to describe **simple harmonic motion** or **free undamped motion.** Two obvious initial conditions associated with (2) are $x(0) = x_0$ and $x'(0) = x_1$, the initial displacement and initial velocity of the mass, respectively. For example, if $x_0 > 0$, $x_1 < 0$, the mass starts from a point *below* the equilibrium position with an imparted *upward* velocity. When $x'(0) = 0$, the mass is said to be released from rest. For example, if $x_0 < 0$, $x_1 = 0$, the mass is released from *rest* from a point $|x_0|$ units *above* the equilibrium position.

**EQUATION OF MOTION** To solve equation (2), we note that the solutions of its auxiliary equation $m^2 + \omega^2 = 0$ are the complex numbers $m_1 = \omega i$, $m_2 = -\omega i$. Thus from (8) of Section 4.3 we find the general solution of (2) to be

$$x(t) = c_1 \cos \omega t + c_2 \sin \omega t. \tag{3}$$

The **period** of motion described by (3) is $T = 2\pi/\omega$. The number $T$ represents the time (measured in seconds) it takes the mass to execute one cycle of motion. A cycle is one complete oscillation of the mass, that is, the mass $m$ moving from, say, the lowest point below the equilibrium position to the point highest above the equilibrium position and then back to the lowest point. From a graphical viewpoint $T = 2\pi/\omega$ seconds is the length of the time interval between two successive maxima (or minima) of $x(t)$. Keep in mind that a maximum of $x(t)$ is a positive displacement corresponding to the mass attaining its greatest distance below the equilibrium position, whereas a minimum of $x(t)$ is negative displacement corresponding to the mass attaining its greatest height above the equilibrium position. We refer to either case as an **extreme displacement** of the mass. The **frequency** of motion is $f = 1/T = \omega/2\pi$ and is the number of cycles completed each second. For example, if $x(t) = 2 \cos 3\pi t - 4 \sin 3\pi t$, then the period is $T = 2\pi/3\pi = 2/3$ s, and the frequency is $f = 3/2$ cycles/s. From a graphical viewpoint the graph of $x(t)$ repeats every $\frac{2}{3}$ second, that is, $x\left(t + \frac{2}{3}\right) = x(t)$, and $\frac{3}{2}$ cycles of the graph are completed each second (or, equivalently, three cycles of the graph are completed every 2 seconds). The number $\omega = \sqrt{k/m}$ (measured in radians per second) is called the **circular frequency** of the system. Depending on which text you read, both $f = \omega/2\pi$ and $\omega$ are also referred to as the **natural frequency** of the system. Finally, when the initial conditions are used to determine the constants $c_1$ and $c_2$ in (3), we say that the resulting particular solution or response is the **equation of motion.**

**EXAMPLE 1** Free Undamped Motion

A mass weighing 2 pounds stretches a spring 6 inches. At $t = 0$ the mass is released from a point 8 inches below the equilibrium position with an upward velocity of $\frac{4}{3}$ ft/s. Determine the equation of motion.

**SOLUTION** Because we are using the engineering system of units, the measurements given in terms of inches must be converted into feet: 6 in. $= \frac{1}{2}$ ft; 8 in. $= \frac{2}{3}$ ft. In addition, we must convert the units of weight given in pounds into units of mass. From $m = W/g$ we have $m = \frac{2}{32} = \frac{1}{16}$ slug. Also, from Hooke's law, $2 = k\left(\frac{1}{2}\right)$ implies that the spring constant is $k = 4$ lb/ft. Hence (1) gives

$$\frac{1}{16}\frac{d^2x}{dt^2} = -4x \qquad \text{or} \qquad \frac{d^2x}{dt^2} + 64x = 0.$$

The initial displacement and initial velocity are $x(0) = \frac{2}{3}$, $x'(0) = -\frac{4}{3}$, where the negative sign in the last condition is a consequence of the fact that the mass is given an initial velocity in the negative, or upward, direction.

Now $\omega^2 = 64$ or $\omega = 8$, so the general solution of the differential equation is

$$x(t) = c_1 \cos 8t + c_2 \sin 8t. \tag{4}$$

Applying the initial conditions to $x(t)$ and $x'(t)$ gives $c_1 = \frac{2}{3}$ and $c_2 = -\frac{1}{6}$. Thus the equation of motion is

$$x(t) = \frac{2}{3} \cos 8t - \frac{1}{6} \sin 8t. \tag{5} \quad \blacksquare$$

**ALTERNATIVE FORMS OF $x(t)$**  When $c_1 \neq 0$ and $c_2 \neq 0$, the actual **amplitude** $A$ of free vibrations is not obvious from inspection of equation (3). For example, although the mass in Example 1 is initially displaced $\frac{2}{3}$ foot beyond the equilibrium position, the amplitude of vibrations is a number larger than $\frac{2}{3}$. Hence it is often convenient to convert a solution of form (3) to the simpler form

$$x(t) = A \sin(\omega t + \phi), \tag{6}$$

where $A = \sqrt{c_1^2 + c_2^2}$ and $\phi$ is a **phase angle** defined by

$$\left. \begin{array}{l} \sin \phi = \dfrac{c_1}{A} \\[2mm] \cos \phi = \dfrac{c_2}{A} \end{array} \right\} \quad \tan \phi = \dfrac{c_1}{c_2}. \tag{7}$$

To verify this, we expand (6) by the addition formula for the sine function:

$$A \sin \omega t \cos \phi + A \cos \omega t \sin \phi = (A \sin \phi) \cos \omega t + (A \cos \phi) \sin \omega t. \tag{8}$$

It follows from Figure 5.1.3 that if $\phi$ is defined by

$$\sin \phi = \frac{c_1}{\sqrt{c_1^2 + c_2^2}} = \frac{c_1}{A}, \qquad \cos \phi = \frac{c_2}{\sqrt{c_1^2 + c_2^2}} = \frac{c_2}{A},$$

then (8) becomes

$$A \frac{c_1}{A} \cos \omega t + A \frac{c_2}{A} \sin \omega t = c_1 \cos \omega t + c_2 \sin \omega t = x(t).$$

**FIGURE 5.1.3** A relationship between $c_1 > 0$, $c_2 > 0$ and phase angle $\phi$

**EXAMPLE 2**  Alternative Form of Solution (5)

In view of the foregoing discussion we can write solution (5) in the alternative form $x(t) = A \sin(8t + \phi)$. Computation of the amplitude is straightforward,

$$A = \sqrt{\left(\frac{2}{3}\right)^2 + \left(-\frac{1}{6}\right)^2} = \sqrt{\frac{17}{36}} \approx 0.69 \text{ ft,}$$

but some care should be exercised in computing the phase angle $\phi$ defined by (7). With $c_1 = \frac{2}{3}$ and $c_2 = -\frac{1}{6}$ we find $\tan \phi = -4$, and a calculator then gives $\tan^{-1}(-4) = -1.326$ rad. This is *not* the phase angle, since $\tan^{-1}(-4)$ is located in the *fourth quadrant* and therefore contradicts the fact that $\sin \phi > 0$ and $\cos \phi < 0$ because $c_1 > 0$ and $c_2 < 0$. Hence we must take $\phi$ to be the *second-quadrant* angle $\phi = \pi + (-1.326) = 1.816$ rad. Thus (5) is the same as

$$x(t) = \frac{\sqrt{17}}{6} \sin(8t + 1.816). \tag{9}$$

The period of this function is $T = 2\pi/8 = \pi/4$ s. $\quad \blacksquare$

You should be aware that some instructors in science and engineering prefer that (3) be expressed as a shifted cosine function

$$x(t) = A \cos(\omega t - \phi), \qquad (6')$$

where $A = \sqrt{c_1^2 + c_2^2}$. In this case the radian measured angle $\phi$ is defined in a slightly different manner than in (7):

$$\left.\begin{array}{r} \sin \phi = \dfrac{c_2}{A} \\[2mm] \cos \phi = \dfrac{c_1}{A} \end{array}\right\} \ \tan \phi = \dfrac{c_2}{c_1}. \qquad (7')$$

For example, in Example 2 with $c_1 = \frac{2}{3}$ and $c_2 = -\frac{1}{6}$, $(7')$ indicates that $\tan \phi = -\frac{1}{4}$. Because $\sin \phi < 0$ and $\cos \phi > 0$ the angle $\phi$ lies in the fourth quadrant and so rounded to three decimal places $\phi = \tan^{-1}(-\frac{1}{4}) = -0.245$ rad. From $(6')$ we obtain a second alternative form of solution (5):

$$x(t) = \frac{\sqrt{17}}{6} \cos(8t - (-0.245)) \quad \text{or} \quad x(t) = \frac{\sqrt{17}}{6} \cos(8t + 0.245).$$

**GRAPHICAL INTERPRETATION**   Figure 5.1.4(a) illustrates the mass in Example 2 going through approximately two complete cycles of motion. Reading left to right, the first five positions (marked with black dots) correspond to the initial position of the mass below the equilibrium position $\left(x = \frac{2}{3}\right)$, the mass passing through the equilibrium position for the first time heading upward ($x = 0$), the mass at its extreme

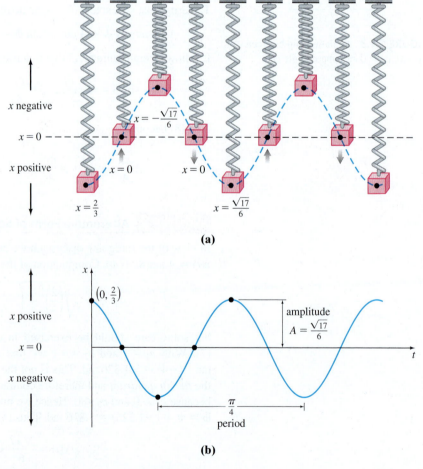

**(a)**

**(b)**

**FIGURE 5.1.4**  Simple harmonic motion

displacement above the equilibrium position ($x = -\sqrt{17}/6$), the mass at the equilibrium position for the second time heading downward ($x = 0$), and the mass at its extreme displacement below the equilibrium position ($x = \sqrt{17}/6$). The black dots on the graph of (9), given in Figure 5.1.4(b), also agree with the five positions just given. Note, however, that in Figure 5.1.4(b) the positive direction in the $tx$-plane is the usual upward direction and so is opposite to the positive direction indicated in Figure 5.1.4(a). Hence the solid blue graph representing the motion of the mass in Figure 5.1.4(b) is the reflection through the $t$-axis of the blue dashed curve in Figure 5.1.4(a).

Form (6) is very useful because it is easy to find values of time for which the graph of $x(t)$ crosses the positive $t$-axis (the line $x = 0$). We observe that $\sin(\omega t + \phi) = 0$ when $\omega t + \phi = n\pi$, where $n$ is a nonnegative integer.

**DOUBLE SPRING SYSTEMS** Suppose two *parallel* springs, with constants $k_1$ and $k_2$, are attached to a common rigid support and then to a single mass $m$ as shown in Figure 5.1.5. If the mass is displaced from its equilibrium position, the displacement $x$ is the same for both springs and so the net restoring force of the spring in (1) is simply $-k_1x - k_2x = -(k_1 + k_2)x$. We say that

$$k_{\text{eff}} = k_1 + k_2$$

is the **effective spring constant** of the system.

rigid
support

$k_1$      $k_2$

$m$

**FIGURE 5.1.5** Parallel springs

On the other hand, suppose two springs supporting a single mass $m$ are in *series*, that is, the springs are attached end to end as shown in Figure 5.1.6. In this case, a displacement $x$ of the mass from its equilibrium consists of the sum $x = x_1 + x_2$, where $x_1$ and $x_2$ are the displacements of the respective springs. But the restoring force is the same for both springs, so if $k_{\text{eff}}$ is the **effective spring constant** of the system we have

$$-k_{\text{eff}}(x_1 + x_2) = -k_1x_1 = -k_2x_2.$$

From $k_1x_1 = k_2x_2$ we see $x_1 = (k_2/k_1)x_2$ and so $-k_{\text{eff}}(x_1 + x_2) = -k_2x_2$ is the same as

$$k_{\text{eff}}\left(\frac{k_2}{k_1}x_2 + x_2\right) = k_2x_2.$$

Solving the last equation for $k_{\text{eff}}$ yields

$$k_{\text{eff}} = \frac{k_1k_2}{k_1 + k_2}.$$

So in either of the above cases, the differential equation of motion is (1) with $k$ replaced by $k_{\text{eff}}$. See Problems 13–18 in Exercises 5.1.

rigid
support

$k_1$

$k_2$

$m$

**FIGURE 5.1.6** Springs in series

**SYSTEMS WITH VARIABLE SPRING CONSTANTS** In the model discussed above we assumed an ideal world—a world in which the physical characteristics of the spring do not change over time. In the nonideal world, however, it seems reasonable to expect that when a spring/mass system is in motion for a long period, the spring will weaken; in other words, the "spring constant" will vary—or, more specifically, decay—with time. In one model for the **aging spring** the spring constant $k$ in (1) is replaced by the decreasing function $K(t) = ke^{-\alpha t}$, $k > 0$, $\alpha > 0$. The linear differential equation $mx'' + ke^{-\alpha t}x = 0$ cannot be solved by the methods that were considered in Chapter 4. Nevertheless, we can obtain two linearly independent solutions using the methods in Chapter 6. See Problem 19 in Exercises 5.1 and Example 5 in Section 6.4.

When a spring/mass system is subjected to an environment in which the temperature is rapidly decreasing, it might make sense to replace the constant $k$ with $K(t) = kt$, $k > 0$, a function that increases with time. The resulting model, $mx'' + ktx = 0$, is a form of **Airy's differential equation.** Like the equation for an aging spring, Airy's equation can be solved by the methods of Chapter 6. See Problem 20 in Exercises 5.1.

**(a)**

**(b)**

**FIGURE 5.1.7**  Damping devices

### 5.1.2   SPRING/MASS SYSTEMS: FREE DAMPED MOTION

The concept of free harmonic motion is somewhat unrealistic, since the motion described by equation (1) assumes that there are no retarding forces acting on the moving mass. Unless the mass is suspended in a perfect vacuum, there will be at least a resisting force due to the surrounding medium. As Figure 5.1.7 shows, the mass could be suspended in a viscous medium or connected to a dashpot damping device.

**DE OF FREE DAMPED MOTION**   In the study of mechanics, damping forces acting on a body are considered to be proportional to a power of the instantaneous velocity. In particular, we shall assume throughout the subsequent discussion that this force is given by a constant multiple of $dx/dt$. When no other external forces are impressed on the system, it follows from Newton's second law that

$$m\frac{d^2x}{dt^2} = -kx - \beta\frac{dx}{dt}, \tag{10}$$

where $\beta$ is a positive *damping constant* and the negative sign is a consequence of the fact that the damping force acts in a direction opposite to the motion.

Dividing (10) by the mass $m$, we find that the differential equation of **free damped motion** is

$$\frac{d^2x}{dt^2} + \frac{\beta}{m}\frac{dx}{dt} + \frac{k}{m}x = 0$$

or        $$\frac{d^2x}{dt^2} + 2\lambda\frac{dx}{dt} + \omega^2x = 0, \tag{11}$$

where        $$2\lambda = \frac{\beta}{m}, \qquad \omega^2 = \frac{k}{m}. \tag{12}$$

The symbol $2\lambda$ is used only for algebraic convenience because the auxiliary equation is $m^2 + 2\lambda m + \omega^2 = 0$, and the corresponding roots are then

$$m_1 = -\lambda + \sqrt{\lambda^2 - \omega^2}, \qquad m_2 = -\lambda - \sqrt{\lambda^2 - \omega^2}.$$

We can now distinguish three possible cases depending on the algebraic sign of $\lambda^2 - \omega^2$. Since each solution contains the *damping factor* $e^{-\lambda t}$, $\lambda > 0$, the displacements of the mass become negligible as time $t$ increases.

**FIGURE 5.1.8**  Motion of an overdamped system

**CASE I: $\lambda^2 - \omega^2 > 0$**   In this situation the system is said to be **overdamped** because the damping coefficient $\beta$ is large when compared to the spring constant $k$. The corresponding solution of (11) is $x(t) = c_1 e^{m_1 t} + c_2 e^{m_2 t}$ or

$$x(t) = e^{-\lambda t}\left(c_1 e^{\sqrt{\lambda^2 - \omega^2}\,t} + c_2 e^{-\sqrt{\lambda^2 - \omega^2}\,t}\right). \tag{13}$$

This equation represents a smooth and nonoscillatory motion. Figure 5.1.8 shows two possible graphs of $x(t)$.

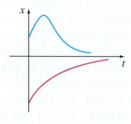

**FIGURE 5.1.9**  Motion of a critically damped system

**CASE II: $\lambda^2 - \omega^2 = 0$**   The system is said to be **critically damped** because any slight decrease in the damping force would result in oscillatory motion. The general solution of (11) is $x(t) = c_1 e^{m_1 t} + c_2 t e^{m_1 t}$ or

$$x(t) = e^{-\lambda t}(c_1 + c_2 t). \tag{14}$$

Some graphs of typical motion are given in Figure 5.1.9. Notice that the motion is quite similar to that of an overdamped system. It is also apparent from (14) that the mass can pass through the equilibrium position at most one time.

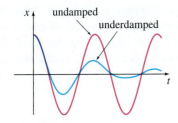

**FIGURE 5.1.10** Motion of an underdamped system

**CASE III:** $\lambda^2 - \omega^2 < 0$    In this case the system is said to be **underdamped,** since the damping coefficient is small in comparison to the spring constant. The roots $m_1$ and $m_2$ are now complex:

$$m_1 = -\lambda + \sqrt{\omega^2 - \lambda^2}\,i, \qquad m_2 = -\lambda - \sqrt{\omega^2 - \lambda^2}\,i.$$

Thus the general solution of equation (11) is

$$x(t) = e^{-\lambda t}\left(c_1 \cos \sqrt{\omega^2 - \lambda^2}\,t + c_2 \sin \sqrt{\omega^2 - \lambda^2}\,t\right). \tag{15}$$

As indicated in Figure 5.1.10, the motion described by (15) is oscillatory; but because of the coefficient $e^{-\lambda t}$, the amplitudes of vibration $\to 0$ as $t \to \infty$.

**EXAMPLE 3**    **Overdamped Motion**

It is readily verified that the solution of the initial-value problem

$$\frac{d^2x}{dt^2} + 5\frac{dx}{dt} + 4x = 0, \quad x(0) = 1, \quad x'(0) = 1$$

is

$$x(t) = \frac{5}{3}\,e^{-t} - \frac{2}{3}\,e^{-4t}. \tag{16}$$

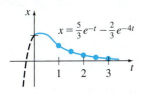

(a)

| $t$ | $x(t)$ |
|-----|--------|
| 1   | 0.601  |
| 1.5 | 0.370  |
| 2   | 0.225  |
| 2.5 | 0.137  |
| 3   | 0.083  |

(b)

**FIGURE 5.1.11** Overdamped system in Example 3

The problem can be interpreted as representing the overdamped motion of a mass on a spring. The mass is initially released from a position 1 unit *below* the equilibrium position with a *downward* velocity of 1 ft/s.

To graph $x(t)$, we find the value of $t$ for which the function has an extremum—that is, the value of time for which the first derivative (velocity) is zero. Differentiating (16) gives $x'(t) = -\frac{5}{3}e^{-t} + \frac{8}{3}e^{-4t}$, so $x'(t) = 0$ implies that $e^{3t} = \frac{8}{5}$ or $t = \frac{1}{3}\ln\frac{8}{5} = 0.157$. It follows from the first derivative test, as well as our physical intuition, that $x(0.157) = 1.069$ ft is actually a maximum. In other words, the mass attains an extreme displacement of 1.069 feet below the equilibrium position.

We should also check to see whether the graph crosses the $t$-axis—that is, whether the mass passes through the equilibrium position. This cannot happen in this instance because the equation $x(t) = 0$, or $e^{3t} = \frac{2}{5}$, has the physically irrelevant solution $t = \frac{1}{3}\ln\frac{2}{5} = -0.305$.

The graph of $x(t)$, along with some other pertinent data, is given in Figure 5.1.11.

**EXAMPLE 4**    **Critically Damped Motion**

A mass weighing 8 pounds stretches a spring 2 feet. Assuming that a damping force numerically equal to 2 times the instantaneous velocity acts on the system, determine the equation of motion if the mass is initially released from the equilibrium position with an upward velocity of 3 ft/s.

**SOLUTION** From Hooke's law we see that $8 = k(2)$ gives $k = 4$ lb/ft and that $W = mg$ gives $m = \frac{8}{32} = \frac{1}{4}$ slug. The differential equation of motion is then

$$\frac{1}{4}\frac{d^2x}{dt^2} = -4x - 2\frac{dx}{dt} \qquad \text{or} \qquad \frac{d^2x}{dt^2} + 8\frac{dx}{dt} + 16x = 0. \tag{17}$$

The auxiliary equation for (17) is $m^2 + 8m + 16 = (m + 4)^2 = 0$, so $m_1 = m_2 = -4$. Hence the system is critically damped, and

$$x(t) = c_1 e^{-4t} + c_2 t e^{-4t}. \tag{18}$$

Applying the initial conditions $x(0) = 0$ and $x'(0) = -3$, we find, in turn, that $c_1 = 0$ and $c_2 = -3$. Thus the equation of motion is

$$x(t) = -3t e^{-4t}. \tag{19}$$

**FIGURE 5.1.12** Critically damped system in Example 4

To graph $x(t)$, we proceed as in Example 3. From $x'(t) = -3e^{-4t}(1 - 4t)$ we see that $x'(t) = 0$ when $t = \frac{1}{4}$. The corresponding extreme displacement is $x(\frac{1}{4}) = -3(\frac{1}{4})e^{-1} = -0.276$ ft. As shown in Figure 5.1.12, we interpret this value to mean that the mass reaches a maximum height of 0.276 foot above the equilibrium position. ■

**EXAMPLE 5** Underdamped Motion

A mass weighing 16 pounds is attached to a 5-foot-long spring. At equilibrium the spring measures 8.2 feet. If the mass is initially released from rest at a point 2 feet above the equilibrium position, find the displacements $x(t)$ if it is further known that the surrounding medium offers a resistance numerically equal to the instantaneous velocity.

**SOLUTION** The elongation of the spring after the mass is attached is $8.2 - 5 = 3.2$ ft, so it follows from Hooke's law that $16 = k(3.2)$ or $k = 5$ lb/ft. In addition, $m = \frac{16}{32} = \frac{1}{2}$ slug, so the differential equation is given by

$$\frac{1}{2}\frac{d^2x}{dt^2} = -5x - \frac{dx}{dt} \quad \text{or} \quad \frac{d^2x}{dt^2} + 2\frac{dx}{dt} + 10x = 0. \qquad (20)$$

Proceeding, we find that the roots of $m^2 + 2m + 10 = 0$ are $m_1 = -1 + 3i$ and $m_2 = -1 - 3i$, which then implies that the system is underdamped, and

$$x(t) = e^{-t}(c_1 \cos 3t + c_2 \sin 3t). \qquad (21)$$

Finally, the initial conditions $x(0) = -2$ and $x'(0) = 0$ yield $c_1 = -2$ and $c_2 = -\frac{2}{3}$, so the equation of motion is

$$x(t) = e^{-t}\left(-2 \cos 3t - \frac{2}{3} \sin 3t\right). \qquad (22) ■$$

**ALTERNATIVE FORM OF $x(t)$** In a manner identical to the procedure used on page 199, we can write any solution

$$x(t) = e^{-\lambda t}\left(c_1 \cos \sqrt{\omega^2 - \lambda^2}\, t + c_2 \sin \sqrt{\omega^2 - \lambda^2}\, t\right)$$

in the alternative form

$$x(t) = Ae^{-\lambda t} \sin\left(\sqrt{\omega^2 - \lambda^2}\, t + \phi\right), \qquad (23)$$

where $A = \sqrt{c_1^2 + c_2^2}$ and the phase angle $\phi$ is determined from the equations

$$\sin \phi = \frac{c_1}{A}, \qquad \cos \phi = \frac{c_2}{A}, \qquad \tan \phi = \frac{c_1}{c_2}.$$

The coefficient $Ae^{-\lambda t}$ is sometimes called the **damped amplitude** of vibrations. Because (23) is not a periodic function, the number $2\pi/\sqrt{\omega^2 - \lambda^2}$ is called the **quasi period** and $\sqrt{\omega^2 - \lambda^2}/2\pi$ is the **quasi frequency.** The quasi period is the time interval between two successive maxima of $x(t)$. You should verify, for the equation of motion in Example 5, that $A = 2\sqrt{10}/3$ and $\phi = 4.391$. Therefore an equivalent form of (22) is

$$x(t) = \frac{2\sqrt{10}}{3}e^{-t} \sin(3t + 4.391).$$

### 5.1.3 SPRING/MASS SYSTEMS: DRIVEN MOTION

**DE OF DRIVEN MOTION WITH DAMPING** Suppose we now take into consideration an external force $f(t)$ acting on a vibrating mass on a spring. For example,

**FIGURE 5.1.13** Oscillatory vertical motion of the support

$f(t)$ could represent a driving force causing an oscillatory vertical motion of the support of the spring. See Figure 5.1.13. The inclusion of $f(t)$ in the formulation of Newton's second law gives the differential equation of **driven** or **forced motion:**

$$m \frac{d^2x}{dt^2} = -kx - \beta \frac{dx}{dt} + f(t). \tag{24}$$

Dividing (24) by $m$ gives

$$\frac{d^2x}{dt^2} + 2\lambda \frac{dx}{dt} + \omega^2 x = F(t), \tag{25}$$

where $F(t) = f(t)/m$ and, as in the preceding section, $2\lambda = \beta/m$, $\omega^2 = k/m$. To solve the latter nonhomogeneous equation, we can use either the method of undetermined coefficients or variation of parameters.

---

**EXAMPLE 6** | **Interpretation of an Initial-Value Problem**

Interpret and solve the initial-value problem

$$\frac{1}{5} \frac{d^2x}{dt^2} + 1.2 \frac{dx}{dt} + 2x = 5 \cos 4t, \quad x(0) = \frac{1}{2}, \quad x'(0) = 0. \tag{26}$$

**SOLUTION** We can interpret the problem to represent a vibrational system consisting of a mass ($m = \frac{1}{5}$ slug or kilogram) attached to a spring ($k = 2$ lb/ft or N/m). The mass is initially released from rest $\frac{1}{2}$ unit (foot or meter) below the equilibrium position. The motion is damped ($\beta = 1.2$) and is being driven by an external periodic ($T = \pi/2$ s) force beginning at $t = 0$. Intuitively, we would expect that even with damping, the system would remain in motion until such time as the forcing function was "turned off," in which case the amplitudes would diminish. However, as the problem is given, $f(t) = 5 \cos 4t$ will remain "on" forever.

We first multiply the differential equation in (26) by 5 and solve

$$\frac{dx^2}{dt^2} + 6 \frac{dx}{dt} + 10x = 0$$

by the usual methods. Because $m_1 = -3 + i$, $m_2 = -3 - i$, it follows that $x_c(t) = e^{-3t}(c_1 \cos t + c_2 \sin t)$. Using the method of undetermined coefficients, we assume a particular solution of the form $x_p(t) = A \cos 4t + B \sin 4t$. Differentiating $x_p(t)$ and substituting into the DE gives

$$x_p'' + 6x_p' + 10x_p = (-6A + 24B) \cos 4t + (-24A - 6B) \sin 4t = 25 \cos 4t.$$

The resulting system of equations

$$-6A + 24B = 25, \quad -24A - 6B = 0$$

yields $A = -\frac{25}{102}$ and $B = \frac{50}{51}$. It follows that

$$x(t) = e^{-3t}(c_1 \cos t + c_2 \sin t) - \frac{25}{102} \cos 4t + \frac{50}{51} \sin 4t. \tag{27}$$

When we set $t = 0$ in the above equation, we obtain $c_1 = \frac{38}{51}$. By differentiating the expression and then setting $t = 0$, we also find that $c_2 = -\frac{86}{51}$. Therefore the equation of motion is

$$x(t) = e^{-3t}\left(\frac{38}{51} \cos t - \frac{86}{51} \sin t\right) - \frac{25}{102} \cos 4t + \frac{50}{51} \sin 4t. \tag{28} \quad \blacksquare$$

**(a)**

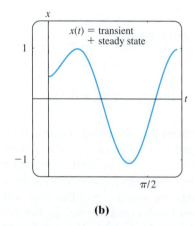

**(b)**

**FIGURE 5.1.14** Graph of solution in (28) of Example 6

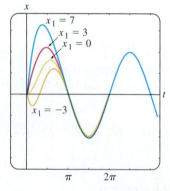

**FIGURE 5.1.15** Graph of solution in Example 7 for various initial velocities $x_1$

**TRANSIENT AND STEADY-STATE TERMS** When $F$ is a periodic function, such as $F(t) = F_0 \sin \gamma t$ or $F(t) = F_0 \cos \gamma t$, the general solution of (25) for $\lambda > 0$ is the sum of a nonperiodic function $x_c(t)$ and a periodic function $x_p(t)$. Moreover, $x_c(t)$ dies off as time increases—that is, $\lim_{t \to \infty} x_c(t) = 0$. Thus for large values of time, the displacements of the mass are closely approximated by the particular solution $x_p(t)$. The complementary function $x_c(t)$ is said to be a **transient term** or **transient solution,** and the function $x_p(t)$, the part of the solution that remains after an interval of time, is called a **steady-state term** or **steady-state solution.** Note therefore that the effect of the initial conditions on a spring/mass system driven by $F$ is transient. In the particular solution (28), $e^{-3t} \left( \frac{38}{51} \cos t - \frac{86}{51} \sin t \right)$ is a transient term, and $x_p(t) = -\frac{25}{102} \cos 4t + \frac{50}{51} \sin 4t$ is a steady-state term. The graphs of these two terms and the solution (28) are given in Figures 5.1.14(a) and 5.1.14(b), respectively.

---

**EXAMPLE 7** Transient/Steady-State Solutions

The solution of the initial-value problem

$$\frac{d^2x}{dt^2} + 2\frac{dx}{dt} + 2x = 4 \cos t + 2 \sin t, \quad x(0) = 0, \quad x'(0) = x_1,$$

where $x_1$ is constant, is given by

$$x(t) = \underbrace{(x_1 - 2)\, e^{-t} \sin t}_{\text{transient}} + \underbrace{2 \sin t}_{\text{steady-state}}.$$

Solution curves for selected values of the initial velocity $x_1$ are shown in Figure 5.1.15. The graphs show that the influence of the transient term is negligible for about $t > 3\pi/2$. ■

**DE OF DRIVEN MOTION WITHOUT DAMPING** With a periodic impressed force and no damping force, there is no transient term in the solution of a problem. Also, we shall see that a periodic impressed force with a frequency near or the same as the frequency of free undamped vibrations can cause a severe problem in any oscillatory mechanical system.

---

**EXAMPLE 8** Undamped Forced Motion

Solve the initial-value problem

$$\frac{d^2x}{dt^2} + \omega^2 x = F_0 \sin \gamma t, \quad x(0) = 0, \quad x'(0) = 0, \tag{29}$$

where $F_0$ is a constant and $\gamma \neq \omega$.

**SOLUTION** The complementary function is $x_c(t) = c_1 \cos \omega t + c_2 \sin \omega t$. To obtain a particular solution, we assume $x_p(t) = A \cos \gamma t + B \sin \gamma t$ so that

$$x_p'' + \omega^2 x_p = A(\omega^2 - \gamma^2) \cos \gamma t + B(\omega^2 - \gamma^2) \sin \gamma t = F_0 \sin \gamma t.$$

Equating coefficients immediately gives $A = 0$ and $B = F_0/(\omega^2 - \gamma^2)$. Therefore

$$x_p(t) = \frac{F_0}{\omega^2 - \gamma^2} \sin \gamma t.$$

Applying the given initial conditions to the general solution

$$x(t) = c_1 \cos \omega t + c_2 \sin \omega t + \frac{F_0}{\omega^2 - \gamma^2} \sin \gamma t$$

yields $c_1 = 0$ and $c_2 = -\gamma F_0/\omega(\omega^2 - \gamma^2)$. Thus the solution is

$$x(t) = \frac{F_0}{\omega(\omega^2 - \gamma^2)}(-\gamma \sin \omega t + \omega \sin \gamma t), \qquad \gamma \neq \omega. \qquad (30)$$ ■

**PURE RESONANCE**    Although equation (30) is not defined for $\gamma = \omega$, it is interesting to observe that its limiting value as $\gamma \to \omega$ can be obtained by applying L'Hôpital's Rule. This limiting process is analogous to "tuning in" the frequency of the driving force ($\gamma/2\pi$) to the frequency of free vibrations ($\omega/2\pi$). Intuitively, we expect that over a length of time we should be able to substantially increase the amplitudes of vibration. For $\gamma = \omega$ we define the solution to be

$$x(t) = \lim_{\gamma \to \omega} F_0 \frac{-\gamma \sin \omega t + \omega \sin \gamma t}{\omega(\omega^2 - \gamma^2)} = F_0 \lim_{\gamma \to \omega} \frac{\dfrac{d}{d\gamma}(-\gamma \sin \omega t + \omega \sin \gamma t)}{\dfrac{d}{d\gamma}(\omega^3 - \omega\gamma^2)}$$

$$= F_0 \lim_{\gamma \to \omega} \frac{-\sin \omega t + \omega t \cos \gamma t}{-2\omega\gamma} \qquad (31)$$

$$= F_0 \frac{-\sin \omega t + \omega t \cos \omega t}{-2\omega^2}$$

$$= \frac{F_0}{2\omega^2} \sin \omega t - \frac{F_0}{2\omega} t \cos \omega t.$$

As suspected, when $t \to \infty$, the displacements become large; in fact, $|x(t_n)| \to \infty$ when $t_n = n\pi/\omega$, $n = 1, 2, \ldots$. The phenomenon that we have just described is known as **pure resonance.** The graph given in Figure 5.1.16 shows typical motion in this case.

In conclusion it should be noted that there is no actual need to use a limiting process on (30) to obtain the solution for $\gamma = \omega$. Alternatively, the last equation in (31) can be obtained by solving the initial-value problem

$$\frac{d^2x}{dt^2} + \omega^2 x = F_0 \sin \omega t, \quad x(0) = 0, \quad x'(0) = 0$$

directly by the methods of undetermined coefficients or variation of parameter.

If the displacements of a spring/mass system were actually described by a function such as (31), the system would necessarily fail. Large oscillations of the mass would eventually force the spring beyond its elastic limit. One might argue too that the resonating model presented in Figure 5.1.16 is completely unrealistic because it ignores the retarding effects of ever-present damping forces. Although it is true that pure resonance cannot occur when the smallest amount of damping is taken into consideration, large and equally destructive amplitudes of vibration (although bounded as $t \to \infty$) can occur. See Problem 47 in Exercises 5.1.

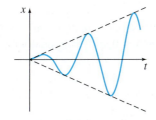

**FIGURE 5.1.16** Pure resonance

### 5.1.4 SERIES CIRCUIT ANALOGUE

*LRC*-SERIES CIRCUITS    As was mentioned in the introduction to this chapter, many different physical systems can be described by a linear second-order differential equation similar to the differential equation of forced motion with damping:

$$m\frac{d^2x}{dt^2} + \beta\frac{dx}{dt} + kx = f(t). \qquad (32)$$

If $i(t)$ denotes current in the ***LRC*-series electrical circuit** shown in Figure 5.1.17, then the voltage drops across the inductor, resistor, and capacitor are as shown in Figure 1.3.4. By Kirchhoff's second law the sum of these voltages equals the voltage $E(t)$ impressed on the circuit; that is,

$$L\frac{di}{dt} + Ri + \frac{1}{C}q = E(t). \qquad (33)$$

**FIGURE 5.1.17** *LRC*-series circuit

But the charge $q(t)$ on the capacitor is related to the current $i(t)$ by $i = dq/dt$, so (33) becomes the linear second-order differential equation

$$L\frac{d^2q}{dt^2} + R\frac{dq}{dt} + \frac{1}{C}q = E(t). \tag{34}$$

The nomenclature used in the analysis of circuits is similar to that used to describe spring/mass systems.

If $E(t) = 0$, the **electrical vibrations** of the circuit are said to be **free.** Because the auxiliary equation for (34) is $Lm^2 + Rm + 1/C = 0$, there will be three forms of the solution with $R \neq 0$, depending on the value of the discriminant $R^2 - 4L/C$. We say that the circuit is

$$\text{**overdamped** if} \qquad R^2 - 4L/C > 0,$$

$$\text{**critically damped** if} \quad R^2 - 4L/C = 0,$$

and $\qquad \text{**underdamped** if} \qquad R^2 - 4L/C < 0.$

In each of these three cases the general solution of (34) contains the factor $e^{-Rt/2L}$, so $q(t) \to 0$ as $t \to \infty$. In the underdamped case when $q(0) = q_0$, the charge on the capacitor oscillates as it decays; in other words, the capacitor is charging and discharging as $t \to \infty$. When $E(t) = 0$ and $R = 0$, the circuit is said to be undamped, and the electrical vibrations do not approach zero as $t$ increases without bound; the response of the circuit is **simple harmonic.**

### EXAMPLE 9   Underdamped Series Circuit

Find the charge $q(t)$ on the capacitor in an $LRC$-series circuit when $L = 0.25$ henry (h), $R = 10$ ohms ($\Omega$), $C = 0.001$ farad (f), $E(t) = 0$, $q(0) = q_0$ coulombs (C), and $i(0) = 0$.

**SOLUTION** Since $1/C = 1000$, equation (34) becomes

$$\frac{1}{4}q'' + 10q' + 1000q = 0 \qquad \text{or} \qquad q'' + 40q' + 4000q = 0.$$

Solving this homogeneous equation in the usual manner, we find that the circuit is underdamped and $q(t) = e^{-20t}(c_1 \cos 60t + c_2 \sin 60t)$. Applying the initial conditions, we find $c_1 = q_0$ and $c_2 = \frac{1}{3}q_0$. Thus

$$q(t) = q_0 e^{-20t}\left(\cos 60t + \frac{1}{3}\sin 60t\right).$$

Using (23), we can write the foregoing solution as

$$q(t) = \frac{q_0\sqrt{10}}{3}e^{-20t}\sin(60t + 1.249). \qquad \blacksquare$$

When there is an impressed voltage $E(t)$ on the circuit, the electrical vibrations are said to be **forced.** In the case when $R \neq 0$, the complementary function $q_c(t)$ of (34) is called a **transient solution.** If $E(t)$ is periodic or a constant, then the particular solution $q_p(t)$ of (34) is a **steady-state solution.**

### EXAMPLE 10   Steady-State Current

Find the steady-state solution $q_p(t)$ and the **steady-state current** in an $LRC$-series circuit when the impressed voltage is $E(t) = E_0 \sin \gamma t$.

**SOLUTION** The steady-state solution $q_p(t)$ is a particular solution of the differential equation

$$L\frac{d^2q}{dt^2} + R\frac{dq}{dt} + \frac{1}{C}q = E_0 \sin \gamma t.$$

Using the method of undetermined coefficients, we assume a particular solution of the form $q_p(t) = A \sin \gamma t + B \cos \gamma t$. Substituting this expression into the differential equation, simplifying, and equating coefficients gives

$$A = \frac{E_0\left(L\gamma - \dfrac{1}{C\gamma}\right)}{-\gamma\left(L^2\gamma^2 - \dfrac{2L}{C} + \dfrac{1}{C^2\gamma^2} + R^2\right)}, \qquad B = \frac{E_0 R}{-\gamma\left(L^2\gamma^2 - \dfrac{2L}{C} + \dfrac{1}{C^2\gamma^2} + R^2\right)}.$$

It is convenient to express $A$ and $B$ in terms of some new symbols.

If $\qquad X = L\gamma - \dfrac{1}{C\gamma}, \qquad$ then $\qquad X^2 = L^2\gamma^2 - \dfrac{2L}{C} + \dfrac{1}{C^2\gamma^2}.$

If $\qquad Z = \sqrt{X^2 + R^2}, \qquad$ then $\qquad Z^2 = L^2\gamma^2 - \dfrac{2L}{C} + \dfrac{1}{C^2\gamma^2} + R^2.$

Therefore $A = E_0 X/(-\gamma Z^2)$ and $B = E_0 R/(-\gamma Z^2)$, so the steady-state charge is

$$q_p(t) = -\frac{E_0 X}{\gamma Z^2} \sin \gamma t - \frac{E_0 R}{\gamma Z^2} \cos \gamma t.$$

Now the steady-state current is given by $i_p(t) = q_p'(t)$:

$$i_p(t) = \frac{E_0}{Z}\left(\frac{R}{Z} \sin \gamma t - \frac{X}{Z} \cos \gamma t\right). \qquad (35) \quad \blacksquare$$

The quantities $X = L\gamma - 1/C\gamma$ and $Z = \sqrt{X^2 + R^2}$ defined in Example 10 are called the **reactance** and **impedance,** respectively, of the circuit. Both the reactance and the impedance are measured in ohms.

---

## EXERCISES 5.1

Answers to selected odd-numbered problems begin on page ANS-7.

### 5.1.1 Spring/Mass Systems: Free Undamped Motion

1. A mass weighing 4 pounds is attached to a spring whose spring constant is 16 lb/ft. What is the period of simple harmonic motion?

2. A 20-kilogram mass is attached to a spring. If the frequency of simple harmonic motion is $2/\pi$ cycles/s, what is the spring constant $k$? What is the frequency of simple harmonic motion if the original mass is replaced with an 80-kilogram mass?

3. A mass weighing 24 pounds, attached to the end of a spring, stretches it 4 inches. Initially, the mass is released from rest from a point 3 inches above the equilibrium position. Find the equation of motion.

4. Determine the equation of motion if the mass in Problem 3 is initially released from the equilibrium position with a downward velocity of 2 ft/s.

5. A mass weighing 20 pounds stretches a spring 6 inches. The mass is initially released from rest from a point 6 inches below the equilibrium position.

   (a) Find the position of the mass at the times $t = \pi/12$, $\pi/8$, $\pi/6$, $\pi/4$, and $9\pi/32$ s.

   (b) What is the velocity of the mass when $t = 3\pi/16$ s? In which direction is the mass heading at this instant?

   (c) At what times does the mass pass through the equilibrium position?

6. A force of 400 newtons stretches a spring 2 meters. A mass of 50 kilograms is attached to the end of the spring and is initially released from the equilibrium position with an upward velocity of 10 m/s. Find the equation of motion.

7. Another spring whose constant is 20 N/m is suspended from the same rigid support but parallel to the spring/mass system in Problem 6. A mass of 20 kilograms is attached to the second spring, and both masses are initially released from the equilibrium position with an upward velocity of 10 m/s.

   (a) Which mass exhibits the greater amplitude of motion?

   (b) Which mass is moving faster at $t = \pi/4$ s? At $\pi/2$ s?

   (c) At what times are the two masses in the same position? Where are the masses at these times? In which directions are the masses moving?

8. A mass weighing 32 pounds stretches a spring 2 feet. Determine the amplitude and period of motion if the mass is initially released from a point 1 foot above the equilibrium position with an upward velocity of 2 ft/s. How many complete cycles will the mass have completed at the end of $4\pi$ seconds?

**9.** A mass weighing 8 pounds is attached to a spring. When set in motion, the spring/mass system exhibits simple harmonic motion.

   **(a)** Determine the equation of motion if the spring constant is 1 lb/ft and the mass is initially released from a point 6 inches below the equilibrium position with a downward velocity of $\frac{3}{2}$ ft/s.

   **(b)** Express the equation of motion in the form given in (6).

   **(c)** Express the equation of motion in the form given in (6').

**10.** A mass weighing 10 pounds stretches a spring $\frac{1}{4}$ foot. This mass is removed and replaced with a mass of 1.6 slugs, which is initially released from a point $\frac{1}{3}$ foot above the equilibrium position with a downward velocity of $\frac{5}{4}$ ft/s.

   **(a)** Express the equation of motion in the form given in (6).

   **(b)** Express the equation of motion in the form given in (6')

   **(c)** Use one of the solutions obtained in parts (a) and (b) to determine the times the mass attains a displacement below the equilibrium position numerically equal to $\frac{1}{2}$ the amplitude of motion.

**11.** A mass weighing 64 pounds stretches a spring 0.32 foot. The mass is initially released from a point 8 inches above the equilibrium position with a downward velocity of 5 ft/s.

   **(a)** Find the equation of motion.

   **(b)** What are the amplitude and period of motion?

   **(c)** How many complete cycles will the mass have completed at the end of $3\pi$ seconds?

   **(d)** At what time does the mass pass through the equilibrium position heading downward for the second time?

   **(e)** At what times does the mass attain its extreme displacements on either side of the equilibrium position?

   **(f)** What is the position of the mass at $t = 3$ s?

   **(g)** What is the instantaneous velocity at $t = 3$ s?

   **(h)** What is the acceleration at $t = 3$ s?

   **(i)** What is the instantaneous velocity at the times when the mass passes through the equilibrium position?

   **(j)** At what times is the mass 5 inches below the equilibrium position?

   **(k)** At what times is the mass 5 inches below the equilibrium position heading in the upward direction?

**12.** A mass of 1 slug is suspended from a spring whose spring constant is 9 lb/ft. The mass is initially released from a point 1 foot above the equilibrium position with an upward velocity of $\sqrt{3}$ ft/s. Find the times at which the mass is heading downward at a velocity of 3 ft/s.

**13.** A mass weighing 20 pounds stretches a spring 6 inches and another spring 2 inches. The two springs are then attached in parallel to a common rigid support in the manner shown in Figure 5.1.5. Determine the effective spring constant of the double-spring system. Find the equation of motion if the mass is initially released from the equilibrium position with a downward velocity of 2 ft/s.

**14.** A certain mass stretches one spring $\frac{1}{3}$ foot and another spring $\frac{1}{2}$ foot. The two springs are then attached in parallel to a common rigid support in the manner shown in Figure 5.1.5. The first mass is set aside, and a mass weighing 8 pounds is attached to the double-spring arrangement, and the system is set in motion. If the period of motion is $\pi/15$ second, determine how much the first mass weighs.

**15.** Solve Problem 13 again, but this time assume that the springs are in series as shown in Figure 5.1.6.

**16.** Solve Problem 14 again, but this time assume that the springs are in series as shown in Figure 5.1.6.

**17.** Find the effective spring constant of the parallel-spring system shown in Figure 5.1.5 when both springs have the spring constant $k$. Give a physical interpretation of this result.

**18.** Find the effective spring constant of the series-spring system shown in Figure 5.1.6 when both springs have the spring constant $k$. Give a physical interpretation of this result.

**19.** A model of a spring/mass system is $4x'' + e^{-0.1t}x = 0$. By inspection of the differential equation only, discuss the behavior of the system over a long period of time.

**20.** A model of a spring/mass system is $4x'' + tx = 0$. By inspection of the differential equation only, discuss the behavior of the system over a long period of time.

## 5.1.2 Spring/Mass Systems: Free Damped Motion

In Problems 21–24 the given figure represents the graph of an equation of motion for a damped spring/mass system. Use the graph to determine

   **(a)** whether the initial displacement is above or below the equilibrium position and

   **(b)** whether the mass is initially released from rest, heading downward, or heading upward.

**21.**

**FIGURE 5.1.18** Graph for Problem 21

**22.**

**FIGURE 5.1.19** Graph for Problem 22

**23.**

**FIGURE 5.1.20** Graph for Problem 23

**24.**

**FIGURE 5.1.21** Graph for Problem 24

**25.** A mass weighing 4 pounds is attached to a spring whose constant is 2 lb/ft. The medium offers a damping force that is numerically equal to the instantaneous velocity. The mass is initially released from a point 1 foot above the equilibrium position with a downward velocity of 8 ft/s. Determine the time at which the mass passes through the equilibrium position. Find the time at which the mass attains its extreme displacement from the equilibrium position. What is the position of the mass at this instant?

**26.** A 4-foot spring measures 8 feet long after a mass weighing 8 pounds is attached to it. The medium through which the mass moves offers a damping force numerically equal to $\sqrt{2}$ times the instantaneous velocity. Find the equation of motion if the mass is initially released from the equilibrium position with a downward velocity of 5 ft/s. Find the time at which the mass attains its extreme displacement from the equilibrium position. What is the position of the mass at this instant?

**27.** A 1-kilogram mass is attached to a spring whose constant is 16 N/m, and the entire system is then submerged in a liquid that imparts a damping force numerically equal to 10 times the instantaneous velocity. Determine the equations of motion if

**(a)** the mass is initially released from rest from a point 1 meter below the equilibrium position, and then

**(b)** the mass is initially released from a point 1 meter below the equilibrium position with an upward velocity of 12 m/s.

**28.** In parts (a) and (b) of Problem 27 determine whether the mass passes through the equilibrium position. In each case find the time at which the mass attains its extreme displacement from the equilibrium position. What is the position of the mass at this instant?

**29.** A force of 2 pounds stretches a spring 1 foot. A mass weighing 3.2 pounds is attached to the spring, and the system is then immersed in a medium that offers a damping force that is numerically equal to 0.4 times the instantaneous velocity.

**(a)** Find the equation of motion if the mass is initially released from rest from a point 1 foot above the equilibrium position.

**(b)** Express the equation of motion in the form given in (23).

**(c)** Find the first time at which the mass passes through the equilibrium position heading upward.

**30.** After a mass weighing 10 pounds is attached to a 5-foot spring, the spring measures 7 feet. This mass is removed and replaced with another mass that weighs 8 pounds. The entire system is placed in a medium that offers a damping force that is numerically equal to the instantaneous velocity.

**(a)** Find the equation of motion if the mass is initially released from a point $\frac{1}{2}$ foot below the equilibrium position with a downward velocity of 1 ft/s.

**(b)** Express the equation of motion in the form given in (23).

**(c)** Find the times at which the mass passes through the equilibrium position heading downward.

**(d)** Graph the equation of motion.

**31.** A mass weighing 10 pounds stretches a spring 2 feet. The mass is attached to a dashpot device that offers a damping force numerically equal to $\beta$ ($\beta > 0$) times the instantaneous velocity. Determine the values of the damping constant $\beta$ so that the subsequent motion is **(a)** overdamped, **(b)** critically damped, and **(c)** underdamped.

**32.** A mass weighing 24 pounds stretches a spring 4 feet. The subsequent motion takes place in medium that offers a damping force numerically equal to $\beta$ ($\beta > 0$) times the instantaneous velocity. If the mass is initially released from the equilibrium position with an upward velocity of 2 ft/s, show that when $\beta > 3\sqrt{2}$ the equation of motion is

$$x(t) = \frac{-3}{\sqrt{\beta^2 - 18}} e^{-2\beta t/3} \sinh \frac{2}{3} \sqrt{\beta^2 - 18}\, t.$$

### 5.1.3 Spring/Mass Systems: Driven Motion

**33.** A mass weighing 16 pounds stretches a spring $\frac{8}{3}$ feet. The mass is initially released from rest from a point 2 feet below the equilibrium position, and the subsequent motion takes place in a medium that offers a damping force that is numerically equal to $\frac{1}{2}$ the instantaneous velocity. Find the equation of motion if the mass is driven by an external force equal to $f(t) = 10 \cos 3t$.

**34.** A mass of 1 slug is attached to a spring whose constant is 5 lb/ft. Initially, the mass is released 1 foot below the equilibrium position with a downward velocity of 5 ft/s, and the subsequent motion takes place in a medium that offers a damping force that is numerically equal to 2 times the instantaneous velocity.

**(a)** Find the equation of motion if the mass is driven by an external force equal to $f(t) = 12 \cos 2t + 3 \sin 2t$.

**(b)** Graph the transient and steady-state solutions on the same coordinate axes.

**(c)** Graph the equation of motion.

**35.** A mass of 1 slug, when attached to a spring, stretches it 2 feet and then comes to rest in the equilibrium position. Starting at $t = 0$, an external force equal to $f(t) = 8 \sin 4t$ is applied to the system. Find the equation of motion if the surrounding medium offers a damping force that is numerically equal to 8 times the instantaneous velocity.

**36.** In Problem 35 determine the equation of motion if the external force is $f(t) = e^{-t} \sin 4t$. Analyze the displacements for $t \to \infty$.

**37.** When a mass of 2 kilograms is attached to a spring whose constant is 32 N/m, it comes to rest in the equilibrium position. Starting at $t = 0$, a force equal to $f(t) = 68e^{-2t} \cos 4t$ is applied to the system. Find the equation of motion in the absence of damping.

**38.** In Problem 37 write the equation of motion in the form $x(t) = A\sin(\omega t + \phi) + Be^{-2t}\sin(4t + \theta)$. What is the amplitude of vibrations after a very long time?

**39.** A mass $m$ is attached to the end of a spring whose constant is $k$. After the mass reaches equilibrium, its support begins to oscillate vertically about a horizontal line $L$ according to a formula $h(t)$. The value of $h$ represents the distance in feet measured from $L$. See Figure 5.1.22.

**(a)** Determine the differential equation of motion if the entire system moves through a medium offering a damping force that is numerically equal to $\beta(dx/dt)$.

**(b)** Solve the differential equation in part (a) if the spring is stretched 4 feet by a mass weighing 16 pounds and $\beta = 2$, $h(t) = 5 \cos t$, $x(0) = x'(0) = 0$.

**FIGURE 5.1.22** Oscillating support in Problem 39

**40.** A mass of 100 grams is attached to a spring whose constant is 1600 dynes/cm. After the mass reaches equilibrium, its support oscillates according to the formula $h(t) = \sin 8t$, where $h$ represents displacement from its original position. See Problem 39 and Figure 5.1.22.

**(a)** In the absence of damping, determine the equation of motion if the mass starts from rest from the equilibrium position.

**(b)** At what times does the mass pass through the equilibrium position?

**(c)** At what times does the mass attain its extreme displacements?

**(d)** What are the maximum and minimum displacements?

**(e)** Graph the equation of motion.

In Problems 41 and 42 solve the given initial-value problem.

**41.** $\dfrac{d^2x}{dt^2} + 4x = -5 \sin 2t + 3 \cos 2t,$

$x(0) = -1, \quad x'(0) = 1$

**42.** $\dfrac{d^2x}{dt^2} + 9x = 5 \sin 3t, \quad x(0) = 2, \quad x'(0) = 0$

**43. (a)** Show that the solution of the initial-value problem

$$\frac{d^2x}{dt^2} + \omega^2 x = F_0 \cos \gamma t, \quad x(0) = 0, \quad x'(0) = 0$$

is        $x(t) = \dfrac{F_0}{\omega^2 - \gamma^2} (\cos \gamma t - \cos \omega t).$

**(b)** Evaluate $\displaystyle\lim_{\gamma \to \omega} \frac{F_0}{\omega^2 - \gamma^2} (\cos \gamma t - \cos \omega t).$

**44.** Compare the result obtained in part (b) of Problem 43 with the solution obtained using variation of parameters when the external force is $F_0 \cos \omega t$.

**45. (a)** Show that $x(t)$ given in part (a) of Problem 43 can be written in the form

$$x(t) = \frac{-2F_0}{\omega^2 - \gamma^2} \sin \frac{1}{2}(\gamma - \omega)t \sin \frac{1}{2}(\gamma + \omega)t.$$

**(b)** If we define $\varepsilon = \frac{1}{2}(\gamma - \omega)$, show that when $\varepsilon$ is small an approximate solution is

$$x(t) = \frac{F_0}{2\varepsilon\gamma} \sin \varepsilon t \sin \gamma t.$$

When $\varepsilon$ is small, the frequency $\gamma/2\pi$ of the impressed force is close to the frequency $\omega/2\pi$ of free vibrations. When this occurs, the motion is as indicated in Figure 5.1.23. Oscillations of this kind are called **beats** and are due to the fact that the frequency of $\sin \varepsilon t$ is quite small in comparison to the frequency of $\sin \gamma t$. The dashed curves, or envelope of the graph of $x(t)$, are obtained from the graphs of $\pm(F_0/2\varepsilon\gamma) \sin \varepsilon t$. Use a graphing utility with various values of $F_0$, $\varepsilon$, and $\gamma$ to verify the graph in Figure 5.1.23.

**FIGURE 5.1.23** Beats phenomenon in Problem 45

## Computer Lab Assignments

**46.** Can there be beats when a damping force is added to the model in part (a) of Problem 43? Defend your position with graphs obtained either from the explicit solution of the problem

$$\frac{d^2x}{dt^2} + 2\lambda \frac{dx}{dt} + \omega^2 x = F_0 \cos \gamma t, \quad x(0) = 0, \quad x'(0) = 0$$

or from solution curves obtained using a numerical solver.

**47. (a)** Show that the general solution of

$$\frac{d^2x}{dt^2} + 2\lambda \frac{dx}{dt} + \omega^2 x = F_0 \sin \gamma t$$

is

$$x(t) = Ae^{-\lambda t} \sin(\sqrt{\omega^2 - \lambda^2}\, t + \phi)$$

$$+ \frac{F_0}{\sqrt{(\omega^2 - \gamma^2)^2 + 4\lambda^2\gamma^2}} \sin(\gamma t + \theta),$$

where $A = \sqrt{c_1^2 + c_2^2}$ and the phase angles $\phi$ and $\theta$ are, respectively, defined by $\sin \phi = c_1/A$, $\cos \phi = c_2/A$ and

$$\sin \theta = \frac{-2\lambda\gamma}{\sqrt{(\omega^2 - \gamma^2)^2 + 4\lambda^2\gamma^2}},$$

$$\cos \theta = \frac{\omega^2 - \gamma^2}{\sqrt{(\omega^2 - \gamma^2)^2 + 4\lambda^2\gamma^2}}.$$

**(b)** The solution in part (a) has the form $x(t) = x_c(t) + x_p(t)$. Inspection shows that $x_c(t)$ is transient, and hence for large values of time, the solution is approximated by $x_p(t) = g(\gamma) \sin(\gamma t + \theta)$, where

$$g(\gamma) = \frac{F_0}{\sqrt{(\omega^2 - \gamma^2)^2 + 4\lambda^2\gamma^2}}.$$

Although the amplitude $g(\gamma)$ of $x_p(t)$ is bounded as $t \to \infty$, show that the maximum oscillations will occur at the value $\gamma_1 = \sqrt{\omega^2 - 2\lambda^2}$. What is the maximum value of $g$? The number $\sqrt{\omega^2 - 2\lambda^2}/2\pi$ is said to be the **resonance frequency** of the system.

**(c)** When $F_0 = 2$, $m = 1$, and $k = 4$, $g$ becomes

$$g(\gamma) = \frac{2}{\sqrt{(4 - \gamma^2)^2 + \beta^2\gamma^2}}.$$

Construct a table of the values of $\gamma_1$ and $g(\gamma_1)$ corresponding to the damping coefficients $\beta = 2$, $\beta = 1$, $\beta = \frac{3}{4}$, $\beta = \frac{1}{2}$, and $\beta = \frac{1}{4}$. Use a graphing utility to obtain the graphs of $g$ corresponding to these damping coefficients. Use the same coordinate axes. This family of graphs is called the **resonance curve** or **frequency response curve** of the system. What is $\gamma_1$ approaching as $\beta \to 0$? What is happening to the resonance curve as $\beta \to 0$?

**48.** Consider a driven undamped spring/mass system described by the initial-value problem

$$\frac{d^2x}{dt^2} + \omega^2 x = F_0 \sin^n \gamma t, \quad x(0) = 0, \quad x'(0) = 0.$$

**(a)** For $n = 2$, discuss why there is a single frequency $\gamma_1/2\pi$ at which the system is in pure resonance.

**(b)** For $n = 3$, discuss why there are two frequencies $\gamma_1/2\pi$ and $\gamma_2/2\pi$ at which the system is in pure resonance.

**(c)** Suppose $\omega = 1$ and $F_0 = 1$. Use a numerical solver to obtain the graph of the solution of the initial-value problem for $n = 2$ and $\gamma = \gamma_1$ in part (a). Obtain the graph of the solution of the initial-value problem for $n = 3$ corresponding, in turn, to $\gamma = \gamma_1$ and $\gamma = \gamma_2$ in part (b).

### 5.1.4 Series Circuit Analogue

**49.** Find the charge on the capacitor in an *LRC*-series circuit at $t = 0.01$ s when $L = 0.05$ h, $R = 2$ $\Omega$, $C = 0.01$ f, $E(t) = 0$ V, $q(0) = 5$ C, and $i(0) = 0$ A. Determine the first time at which the charge on the capacitor is equal to zero.

**50.** Find the charge on the capacitor in an *LRC*-series circuit when $L = \frac{1}{4}$ h, $R = 20$ $\Omega$, $C = \frac{1}{300}$ f, $E(t) = 0$ V, $q(0) = 4$ C, and $i(0) = 0$ A. Is the charge on the capacitor ever equal to zero?

In Problems 51 and 52 find the charge on the capacitor and the current in the given *LRC*-series circuit. Find the maximum charge on the capacitor.

**51.** $L = \frac{5}{3}$ h, $R = 10$ $\Omega$, $C = \frac{1}{30}$ f, $E(t) = 300$ V, $q(0) = 0$ C, $i(0) = 0$ A

**52.** $L = 1$ h, $R = 100$ $\Omega$, $C = 0.0004$ f, $E(t) = 30$ V, $q(0) = 0$ C, $i(0) = 2$ A

**53.** Find the steady-state charge and the steady-state current in an *LRC*-series circuit when $L = 1$ h, $R = 2$ $\Omega$, $C = 0.25$ f, and $E(t) = 50 \cos t$ V.

**54.** Show that the amplitude of the steady-state current in the *LRC*-series circuit in Example 10 is given by $E_0/Z$, where $Z$ is the impedance of the circuit.

**55.** Use Problem 54 to show that the steady-state current in an *LRC*-series circuit when $L = \frac{1}{2}$ h, $R = 20$ $\Omega$, $C = 0.001$ f, and $E(t) = 100 \sin 60t$ V, is given by $i_p(t) = 4.160 \sin(60t - 0.588)$.

**56.** Find the steady-state current in an *LRC*-series circuit when $L = \frac{1}{2}$ h, $R = 20$ $\Omega$, $C = 0.001$ f, and $E(t) = 100 \sin 60t + 200 \cos 40t$ V.

**57.** Find the charge on the capacitor in an *LRC*-series circuit when $L = \frac{1}{2}$ h, $R = 10$ $\Omega$, $C = 0.01$ f, $E(t) = 150$ V, $q(0) = 1$ C, and $i(0) = 0$ A. What is the charge on the capacitor after a long time?

**58.** Show that if $L$, $R$, $C$, and $E_0$ are constant, then the amplitude of the steady-state current in Example 10 is a maximum when $\gamma = 1/\sqrt{LC}$. What is the maximum amplitude?

**59.** Show that if $L$, $R$, $E_0$, and $\gamma$ are constant, then the amplitude of the steady-state current in Example 10 is a maximum when the capacitance is $C = 1/L\gamma^2$.

**60.** Find the charge on the capacitor and the current in an *LC*-series circuit when $L = 0.1$ h, $C = 0.1$ f, $E(t) = 100 \sin \gamma t$ V, $q(0) = 0$ C, and $i(0) = 0$ A.

**61.** Find the charge on the capacitor and the current in an *LC*-series circuit when $E(t) = E_0 \cos \gamma t$ V, $q(0) = q_0$ C, and $i(0) = i_0$ A.

**62.** In Problem 61 find the current when the circuit is in resonance.

---

## 5.2 Linear Models: Boundary-Value Problems

**INTRODUCTION** The preceding section was devoted to systems in which a second-order mathematical model was accompanied by initial conditions—that is, side conditions that are specified on the unknown function and its first derivative at a single point. But often the mathematical description of a physical system demands that we solve a linear differential equation subject to boundary conditions—that is, conditions specified on the unknown function, or on one of its derivatives, or even on a linear combination of the unknown function and one of its derivatives at two (or more) different points.

**DEFLECTION OF A BEAM**    Many structures are constructed by using girders or beams, and these beams deflect or distort under their own weight or under the influence of some external force. As we shall now see, this deflection $y(x)$ is governed by a relatively simple linear fourth-order differential equation.

To begin, let us assume that a beam of length $L$ is homogeneous and has uniform cross sections along its length. In the absence of any load on the beam (including its weight), a curve joining the centroids of all its cross sections is a straight line called the **axis of symmetry.** See Figure 5.2.1(a). If a load is applied to the beam in a vertical plane containing the axis of symmetry, the beam, as shown in Figure 5.2.1(b), undergoes a distortion, and the curve connecting the centroids of all cross sections is called the **deflection curve** or **elastic curve.** The deflection curve approximates the shape of the beam. Now suppose that the $x$-axis coincides with the axis of symmetry and that the deflection $y(x)$, measured from this axis, is positive if downward. In the theory of elasticity it is shown that the bending moment $M(x)$ at a point $x$ along the beam is related to the load per unit length $w(x)$ by the equation

$$\frac{d^2 M}{dx^2} = w(x). \tag{1}$$

In addition, the bending moment $M(x)$ is proportional to the curvature $\kappa$ of the elastic curve

$$M(x) = EI\kappa, \tag{2}$$

where $E$ and $I$ are constants; $E$ is Young's modulus of elasticity of the material of the beam, and $I$ is the moment of inertia of a cross section of the beam (about an axis known as the neutral axis). The product $EI$ is called the **flexural rigidity** of the beam.

Now, from calculus, curvature is given by $\kappa = y''/[1 + (y')^2]^{3/2}$. When the deflection $y(x)$ is small, the slope $y' \approx 0$, and so $[1 + (y')^2]^{3/2} \approx 1$. If we let $\kappa \approx y''$, equation (2) becomes $M = EI\, y''$. The second derivative of this last expression is

$$\frac{d^2 M}{dx^2} = EI \frac{d^2}{dx^2} y'' = EI \frac{d^4 y}{dx^4}. \tag{3}$$

Using the given result in (1) to replace $d^2 M/dx^2$ in (3), we see that the deflection $y(x)$ satisfies the fourth-order differential equation

$$EI \frac{d^4 y}{dx^4} = w(x). \tag{4}$$

Boundary conditions associated with equation (4) depend on how the ends of the beam are supported. A cantilever beam is **embedded** or **clamped** at one end and **free** at the other. A diving board, an outstretched arm, an airplane wing, and a balcony are common examples of such beams, but even trees, flagpoles, skyscrapers, and the George Washington Monument can act as cantilever beams because they are embedded at one end and are subject to the bending force of the wind. For a cantilever beam the deflection $y(x)$ must satisfy the following two conditions at the embedded end $x = 0$:

- $y(0) = 0$ because there is no deflection, and
- $y'(0) = 0$ because the deflection curve is tangent to the $x$-axis (in other words, the slope of the deflection curve is zero at this point).

At $x = L$ the free-end conditions are

- $y''(L) = 0$ because the bending moment is zero, and
- $y'''(L) = 0$ because the shear force is zero.

axis of symmetry

**(a)**

deflection curve

**(b)**

**FIGURE 5.2.1** Deflection of a homogeneous beam

$x = 0$        $x = L$

**(a)** embedded at both ends

$x = 0$        $x = L$

**(b)** cantilever beam: embedded at the left end, free at the right end

$x = 0$        $x = L$

**(c)** simply supported at both ends

**FIGURE 5.2.2** Beams with various end conditions

**TABLE 5.2.1**

| Ends of the Beam | Boundary Conditions | |
|---|---|---|
| embedded | $y = 0,$ | $y' = 0$ |
| free | $y'' = 0,$ | $y''' = 0$ |
| simply supported or hinged | $y = 0,$ | $y'' = 0$ |

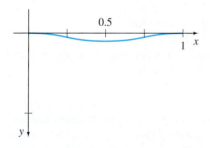

**FIGURE 5.2.3** Deflection curve for BVP in Example 1

The function $F(x) = dM/dx = EI\, d^3y/dx^3$ is called the shear force. If an end of a beam is **simply supported** or **hinged** (also called **pin supported** and **fulcrum supported**) then we must have $y = 0$ and $y'' = 0$ at that end. Table 5.2.1 summarizes the boundary conditions that are associated with (4). See Figure 5.2.2.

**EXAMPLE 1**   An Embedded Beam

A beam of length $L$ is embedded at both ends. Find the deflection of the beam if a constant load $w_0$ is uniformly distributed along its length—that is, $w(x) = w_0, 0 < x < L$.

**SOLUTION** From (4) we see that the deflection $y(x)$ satisfies

$$EI \frac{d^4y}{dx^4} = w_0.$$

Because the beam is embedded at both its left end ($x = 0$) and its right end ($x = L$), there is no vertical deflection and the line of deflection is horizontal at these points. Thus the boundary conditions are

$$y(0) = 0, \qquad y'(0) = 0, \qquad y(L) = 0, \qquad y'(L) = 0.$$

We can solve the nonhomogeneous differential equation in the usual manner (find $y_c$ by observing that $m = 0$ is root of multiplicity four of the auxiliary equation $m^4 = 0$ and then find a particular solution $y_p$ by undetermined coefficients), or we can simply integrate the equation $d^4y/dx^4 = w_0/EI$ four times in succession. Either way, we find the general solution of the equation $y = y_c + y_p$ to be

$$y(x) = c_1 + c_2x + c_3x^2 + c_4x^3 + \frac{w_0}{24EI}x^4.$$

Now the conditions $y(0) = 0$ and $y'(0) = 0$ give, in turn, $c_1 = 0$ and $c_2 = 0$, whereas the remaining conditions $y(L) = 0$ and $y'(L) = 0$ applied to $y(x) = c_3x^2 + c_4x^3 + \frac{w_0}{24EI}x^4$ yield the simultaneous equations

$$c_3L^2 + c_4L^3 + \frac{w_0}{24EI}L^4 = 0$$

$$2c_3L + 3c_4L^2 + \frac{w_0}{6EI}L^3 = 0.$$

Solving this system gives $c_3 = w_0L^2/24EI$ and $c_4 = -w_0L/12EI$. Thus the deflection is

$$y(x) = \frac{w_0L^2}{24EI}x^2 - \frac{w_0L}{12EI}x^3 + \frac{w_0}{24EI}x^4$$

or $y(x) = \dfrac{w_0}{24EI}x^2(x - L)^2$. By choosing $w_0 = 24EI$, and $L = 1$, we obtain the deflection curve in Figure 5.2.3. ∎

**EIGENVALUES AND EIGENFUNCTIONS**   Many applied problems demand that we solve a two-point boundary-value problem (BVP) involving a linear differential equation that contains a parameter $\lambda$. We seek the values of $\lambda$ for which the boundary-value problem has *nontrivial*, that is, *nonzero*, solutions.

**EXAMPLE 2**   Nontrivial Solutions of a BVP

Solve the boundary-value problem

$$y'' + \lambda y = 0, \quad y(0) = 0, \quad y(L) = 0.$$

**SOLUTION** We shall consider three cases: $\lambda = 0$, $\lambda < 0$, and $\lambda > 0$.

**Case I:** For $\lambda = 0$ the solution of $y'' = 0$ is $y = c_1 x + c_2$. The conditions $y(0) = 0$ and $y(L) = 0$ applied to this solution imply, in turn, $c_2 = 0$ and $c_1 = 0$. Hence for $\lambda = 0$ the only solution of the boundary-value problem is the trivial solution $y = 0$.

**Case II:** For $\lambda < 0$ it is convenient to write $\lambda = -\alpha^2$, where $\alpha$ denotes a positive number. With this notation the roots of the auxiliary equation $m^2 - \alpha^2 = 0$ are $m_1 = \alpha$ and $m_2 = -\alpha$. Since the interval on which we are working is finite, we choose to write the general solution of $y'' - \alpha^2 y = 0$ as $y = c_1 \cosh \alpha x + c_2 \sinh \alpha x$. Now $y(0)$ is

$$y(0) = c_1 \cosh 0 + c_2 \sinh 0 = c_1 \cdot 1 + c_2 \cdot 0 = c_1,$$

and so $y(0) = 0$ implies that $c_1 = 0$. Thus $y = c_2 \sinh \alpha x$. The second condition, $y(L) = 0$, demands that $c_2 \sinh \alpha L = 0$. For $\alpha \neq 0$, $\sinh \alpha L \neq 0$; consequently, we are forced to choose $c_2 = 0$. Again the only solution of the BVP is the trivial solution $y = 0$.

**Case III:** For $\lambda > 0$ we write $\lambda = \alpha^2$, where $\alpha$ is a positive number. Because the auxiliary equation $m^2 + \alpha^2 = 0$ has complex roots $m_1 = i\alpha$ and $m_2 = -i\alpha$, the general solution of $y'' + \alpha^2 y = 0$ is $y = c_1 \cos \alpha x + c_2 \sin \alpha x$. As before, $y(0) = 0$ yields $c_1 = 0$, and so $y = c_2 \sin \alpha x$. Now the last condition $y(L) = 0$, or

$$c_2 \sin \alpha L = 0,$$

is satisfied by choosing $c_2 = 0$. But this means that $y = 0$. If we require $c_2 \neq 0$, then $\sin \alpha L = 0$ is satisfied whenever $\alpha L$ is an integer multiple of $\pi$.

$$\alpha L = n\pi \quad \text{or} \quad \alpha = \frac{n\pi}{L} \quad \text{or} \quad \lambda_n = \alpha_n^2 = \left(\frac{n\pi}{L}\right)^2, \quad n = 1, 2, 3, \ldots.$$

Therefore for any real nonzero $c_2$, $y_n(x) = c_2 \sin(n\pi x/L)$ is a solution of the problem for each positive integer $n$. Because the differential equation is homogeneous, any constant multiple of a solution is also a solution, so we may, if desired, simply take $c_2 = 1$. In other words, for each number in the sequence

$$\lambda_1 = \frac{\pi^2}{L^2}, \qquad \lambda_2 = \frac{4\pi^2}{L^2}, \qquad \lambda_3 = \frac{9\pi^2}{L^2}, \ldots,$$

the corresponding function in the sequence

$$y_1 = \sin\frac{\pi}{L}x, \qquad y_2 = \sin\frac{2\pi}{L}x, \qquad y_3 = \sin\frac{3\pi}{L}x, \ldots$$

is a nontrivial solution of the problem $y'' + \lambda_n y = 0$, $y(0) = 0$, $y(L) = 0$ for $n = 1, 2, 3, \ldots$, respectively. ■

The numbers $\lambda_n = n^2\pi^2/L^2$, $n = 1, 2, 3, \ldots$ for which the boundary-value problem in Example 2 possesses nontrivial solutions are known as **eigenvalues**. The nontrivial solutions that depend on these values of $\lambda_n$, $y_n(x) = c_2 \sin(n\pi x/L)$ or simply $y_n(x) = \sin(n\pi x/L)$, are called **eigenfunctions**. The graphs of the eigenfunctions for $n = 1, 2, 3, 4, 5$ are shown in Figure 5.2.4. Note that each graph passes through the two points $(0, 0)$ and $(0, L)$.

### EXAMPLE 3    Example 2 Revisited

It follows from Example 2 and the preceding disucussion that the boundary-value problem

$$y'' + 5y = 0, \quad y(0) = 0, \quad y(L) = 0$$

possesses only the trivial solution $y = 0$ because 5 is *not* an eigenvalue. ■

> Note that we use hyperbolic functions here. Reread "Two Equations Worth Knowing" on page 137.

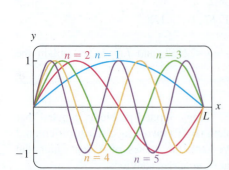

**FIGURE 5.2.4** Graphs of eigenfunctions $y_n = \sin(n\pi x/L)$, for $n = 1, 2, 3, 4, 5$

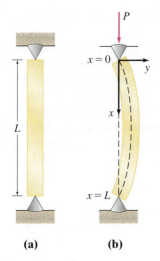

**(a)**       **(b)**

**FIGURE 5.2.5** Elastic column buckling under a compressive force

**BUCKLING OF A THIN VERTICAL COLUMN**    In the eighteenth century Leonhard Euler was one of the first mathematicians to study an eigenvalue problem in analyzing how a thin elastic column buckles under a compressive axial force.

Consider a long, slender vertical column of uniform cross section and length $L$. Let $y(x)$ denote the deflection of the column when a constant vertical compressive force, or load, $P$ is applied to its top, as shown in Figure 5.2.5. By comparing bending moments at any point along the column, we obtain

$$EI \frac{d^2y}{dx^2} = -Py \quad \text{or} \quad EI \frac{d^2y}{dx^2} + Py = 0, \tag{5}$$

where $E$ is Young's modulus of elasticity and $I$ is the moment of inertia of a cross section about a vertical line through its centroid.

### EXAMPLE 4   The Euler Load

Find the deflection of a thin vertical homogeneous column of length $L$ subjected to a constant axial load $P$ if the column is simply supported or hinged at both ends.

**SOLUTION**   The boundary-value problem to be solved is

$$EI \frac{d^2y}{dx^2} + Py = 0, \quad y(0) = 0, \quad y(L) = 0.$$

First note that $y = 0$ is a perfectly good solution of this problem. This solution has a simple intuitive interpretation: If the load $P$ is not great enough, there is no deflection. The question then is this: For what values of $P$ will the column bend? In mathematical terms: For what values of $P$ does the given boundary-value problem possess nontrivial solutions?

By writing $\lambda = P/EI$, we see that

$$y'' + \lambda y = 0, \quad y(0) = 0, \quad y(L) = 0$$

is identical to the problem in Example 2. From Case III of that discussion we see that the deflections are $y_n(x) = c_2 \sin(n\pi x/L)$ corresponding to the eigenvalues $\lambda_n = P_n/EI = n^2\pi^2/L^2$, $n = 1, 2, 3, \ldots$. Physically, this means that the column will buckle or deflect only when the compressive force is one of the values $P_n = n^2\pi^2 EI/L^2$, $n = 1, 2, 3, \ldots$. These different forces are called **critical loads.** The deflection corresponding to the smallest critical load $P_1 = \pi^2 EI/L^2$, called the **Euler load,** is $y_1(x) = c_2 \sin(\pi x/L)$ and is known as the **first buckling mode.**   ■

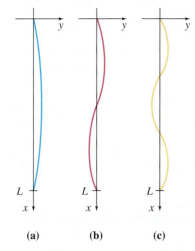

**(a)**     **(b)**     **(c)**

**FIGURE 5.2.6** Deflection curves corresponding to compressive forces $P_1, P_2, P_3$

The deflection curves in Example 4 corresponding to $n = 1$, $n = 2$, and $n = 3$ are shown in Figure 5.2.6. Note that if the original column has some sort of physical restraint put on it at $x = L/2$, then the smallest critical load will be $P_2 = 4\pi^2 EI/L^2$, and the deflection curve will be as shown in Figure 5.2.6(b). If restraints are put on the column at $x = L/3$ and at $x = 2L/3$, then the column will not buckle until the critical load $P_3 = 9\pi^2 EI/L^2$ is applied, and the deflection curve will be as shown in Figure 5.2.6(c). See Problem 25 in Exercises 5.2.

**ROTATING STRING**   The simple linear second-order differential equation

$$y'' + \lambda y = 0 \tag{6}$$

occurs again and again as a mathematical model. In Section 5.1 we saw (6) in the forms $d^2x/dt^2 + (k/m)x = 0$ and $d^2q/dt^2 + (1/LC)q = 0$ as models for, respectively, the simple harmonic motion of a spring/mass system and the simple harmonic response of a series circuit. It is apparent when the model for the deflection of a thin column in (5) is written as $d^2y/dx^2 + (P/EI)y = 0$ that it is the same as (6). We encounter the basic equation (6) one more time in this section: as a model that defines the deflection curve or the shape $y(x)$ assumed by a rotating string. The physical

**(a)**

**(b)**

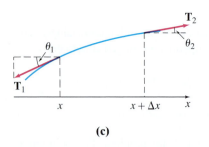

**(c)**

**FIGURE 5.2.7** Rotating string and forces acting on it

situation is analogous to when two people hold a jump rope and twirl it in a synchronous manner. See Figures 5.2.7(a) and 5.2.7(b).

Suppose a string of length $L$ with constant linear density $\rho$ (mass per unit length) is stretched along the $x$-axis and fixed at $x = 0$ and $x = L$. Suppose the string is then rotated about that axis at a constant angular speed $\omega$. Consider a portion of the string on the interval $[x, x + \Delta x]$, where $\Delta x$ is small. If the magnitude $T$ of the tension $\mathbf{T}$, acting tangential to the string, is constant along the string, then the desired differential equation can be obtained by equating two different formulations of the net force acting on the string on the interval $[x, x + \Delta x]$. First, we see from Figure 5.2.7(c) that the net vertical force is

$$F = T \sin \theta_2 - T \sin \theta_1. \tag{7}$$

When angles $\theta_1$ and $\theta_2$ (measured in radians) are small, we have $\sin \theta_2 \approx \tan \theta_2$ and $\sin \theta_1 \approx \tan \theta_1$. Moreover, since $\tan \theta_2$ and $\tan \theta_1$ are, in turn, slopes of the lines containing the vectors $\mathbf{T}_2$ and $\mathbf{T}_1$, we can also write

$$\tan \theta_2 = y'(x + \Delta x) \qquad \text{and} \qquad \tan \theta_1 = y'(x).$$

Thus (7) becomes

$$F \approx T[y'(x + \Delta x) - y'(x)]. \tag{8}$$

Second, we can obtain a different form of this same net force using Newton's second law, $F = ma$. Here the mass of the string on the interval is $m = \rho \, \Delta x$; the centripetal acceleration of a body rotating with angular speed $\omega$ in a circle of radius $r$ is $a = r\omega^2$. With $\Delta x$ small we take $r = y$. Thus the net vertical force is also approximated by

$$F \approx -(\rho \, \Delta x)y\omega^2, \tag{9}$$

where the minus sign comes from the fact that the acceleration points in the direction opposite to the positive $y$-direction. Now by equating (8) and (9), we have

$$T[y'(x + \Delta x) - y'(x)] = -(\rho \Delta x)y\omega^2 \qquad \text{or} \qquad T\,\frac{\overset{\text{difference quotient}}{\overbrace{y'(x + \Delta x) - y'(x)}}}{\Delta x} + \rho\omega^2 y = 0. \tag{10}$$

For $\Delta x$ close to zero the difference quotient in (10) is approximately the second derivative $d^2y/dx^2$. Finally, we arrive at the model

$$T\,\frac{d^2y}{dx^2} + \rho\omega^2 y = 0. \tag{11}$$

Since the string is anchored at its ends $x = 0$ and $x = L$, we expect that the solution $y(x)$ of equation (11) should also satisfy the boundary conditions $y(0) = 0$ and $y(L) = 0$.

---

### REMARKS

(*i*) Eigenvalues are not always easily found as they were in Example 2. You may have to approximate roots of equations such as $\cos x \cosh x = 1$ or $\tan x = -x$. See Problems 32 and 38 in Exercises 5.2.

(*ii*) Boundary conditions applied to a general solution of a linear differential equation can lead to a homogeneous algebraic system of linear equations in which the unknowns are the coefficients $c_i$ in the general solution. A homogeneous algebraic system of linear equations is always consistent because it possesses at least a trivial solution. But a homogeneous system of $n$ linear equations in $n$ unknowns has a nontrivial solution if and only if the determinant of the coefficients equals zero. You might need to use this last fact in Problems 21, 22, and 32 in Exercises 5.2.

### Deflection of a Beam

In Problems 1–5 solve equation (4) subject to the appropriate boundary conditions. The beam is of length $L$, and $w_0$ is a constant.

**1. (a)** The beam is embedded at its left end and free at its right end, and $w(x) = w_0, 0 < x < L$.

  **(b)** Use a graphing utility to graph the deflection curve when $w_0 = 24EI$ and $L = 1$.

**2. (a)** The beam is simply supported at both ends, and $w(x) = w_0, 0 < x < L$.

  **(b)** Use a graphing utility to graph the deflection curve when $w_0 = 24EI$ and $L = 1$.

**3. (a)** The beam is embedded at its left end and simply supported at its right end, and $w(x) = w_0, 0 < x < L$.

  **(b)** Use a graphing utility to graph the deflection curve when $w_0 = 48EI$ and $L = 1$.

**4. (a)** The beam is embedded at its left end and simply supported at its right end, and $w(x) = w_0 \sin(\pi x/L), 0 < x < L$.

  **(b)** Use a graphing utility to graph the deflection curve when $w_0 = 2\pi^3 EI$ and $L = 1$.

  **(c)** Use a root-finding application of a CAS (or a graphic calculator) to approximate the point in the graph in part (b) at which the maximum deflection occurs. What is the maximum deflection?

**5. (a)** The beam is simply supported at both ends, and $w(x) = w_0 x, 0 < x < L$.

  **(b)** Use a graphing utility to graph the deflection curve when $w_0 = 36EI$ and $L = 1$.

  **(c)** Use a root-finding application of a CAS (or a graphic calculator) to approximate the point in the graph in part (b) at which the maximum deflection occurs. What is the maximum deflection?

**6. (a)** Find the maximum deflection of the cantilever beam in Problem 1.

  **(b)** How does the maximum deflection of a beam that is half as long compare with the value in part (a)?

  **(c)** Find the maximum deflection of the simply supported beam in Problem 2.

  **(d)** How does the maximum deflection of the simply supported beam in part (c) compare with the value of maximum deflection of the embedded beam in Example 1?

**7.** A cantilever beam of length $L$ is embedded at its right end, and a horizontal tensile force of $P$ pounds is applied to its free left end. When the origin is taken at its free end, as shown in Figure 5.2.8, the deflection $y(x)$ of the beam can be shown to satisfy the differential equation

$$EIy'' = Py - w(x)\frac{x}{2}.$$

Find the deflection of the cantilever beam if $w(x) = w_0 x, 0 < x < L$, and $y(0) = 0, y'(L) = 0$.

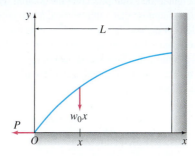

**FIGURE 5.2.8** Deflection of cantilever beam in Problem 7

**8.** When a compressive instead of a tensile force is applied at the free end of the beam in Problem 7, the differential equation of the deflection is

$$EIy'' = -Py - w(x)\frac{x}{2}.$$

Solve this equation if $w(x) = w_0 x, 0 < x < L$, and $y(0) = 0, y'(L) = 0$.

### Eigenvalues and Eigenfunctions

In Problems 9–20 find the eigenvalues and eigenfunctions for the given boundary-value problem.

**9.** $y'' + \lambda y = 0, \quad y(0) = 0, \quad y(\pi) = 0$

**10.** $y'' + \lambda y = 0, \quad y(0) = 0, \quad y(\pi/4) = 0$

**11.** $y'' + \lambda y = 0, \quad y'(0) = 0, \quad y(L) = 0$

**12.** $y'' + \lambda y = 0, \quad y(0) = 0, \quad y'(\pi/2) = 0$

**13.** $y'' + \lambda y = 0, \quad y'(0) = 0, \quad y'(\pi) = 0$

**14.** $y'' + \lambda y = 0, \quad y(-\pi) = 0, \quad y(\pi) = 0$

**15.** $y'' + 2y' + (\lambda + 1)y = 0, \quad y(0) = 0, \quad y(5) = 0$

**16.** $y'' + (\lambda + 1)y = 0, \quad y'(0) = 0, \quad y'(1) = 0$

**17.** $x^2 y'' + xy' + \lambda y = 0, \quad y(1) = 0, \quad y(e^\pi) = 0$

**18.** $x^2 y'' + xy' + \lambda y = 0, \quad y'(e^{-1}) = 0, \quad y(1) = 0$

**19.** $x^2 y'' + xy' + \lambda y = 0, \quad y'(1) = 0, \quad y'(e^2) = 0$

**20.** $x^2 y'' + xy' + \lambda y = 0, \quad y(1) = 0, \quad y'(e) = 0$

In Problems 21 and 22 find the eigenvalues and eigenfunctions for the given boundary-value problem. Consider only the case $\lambda = \alpha^4, \alpha > 0$. [*Hint*: Read (*ii*) in the *Remarks*.]

**21.** $y^{(4)} - \lambda y = 0, \quad y(0) = 0, \quad y''(0) = 0, \quad y(1) = 0, \quad y''(1) = 0$

**22.** $y^{(4)} - \lambda y = 0, \quad y'(0) = 0, \quad y'''(0) = 0, \quad y(\pi) = 0, \quad y''(\pi) = 0$

### Buckling of a Thin Column

**23.** Consider Figure 5.2.6. Where should physical restraints be placed on the column if we want the critical load to be $P_4$? Sketch the deflection curve corresponding to this load.

**24.** The critical loads of thin columns depend on the end conditions of the column. The value of the Euler load $P_1$ in Example 4 was derived under the assumption that the column was hinged at both ends. Suppose that a thin vertical homogeneous column is embedded at its base ($x = 0$) and free at its top ($x = L$) and that a constant axial load $P$ is applied to its free end. This load either causes a small deflection $\delta$ as shown in Figure 5.2.9 or does not cause such a deflection. In either case the differential equation for the deflection $y(x)$ is

$$EI\frac{d^2y}{dx^2} + Py = P\delta.$$

**FIGURE 5.2.9** Deflection of vertical column in Problem 24

**(a)** What is the predicted deflection when $\delta = 0$?

**(b)** When $\delta \neq 0$, show that the Euler load for this column is one-fourth of the Euler load for the hinged column in Example 4.

**25.** As was mentioned in Problem 24, the differential equation (5) that governs the deflection $y(x)$ of a thin elastic column subject to a constant compressive axial force $P$ is valid only when the ends of the column are hinged. In general, the differential equation governing the deflection of the column is given by

$$\frac{d^2}{dx^2}\left(EI\frac{d^2y}{dx^2}\right) + P\frac{d^2y}{dx^2} = 0.$$

Assume that the column is uniform ($EI$ is a constant) and that the ends of the column are hinged. Show that the solution of this fourth-order differential equation subject to the boundary conditions $y(0) = 0$, $y''(0) = 0$, $y(L) = 0$, $y''(L) = 0$ is equivalent to the analysis in Example 4.

**26.** Suppose that a uniform thin elastic column is hinged at the end $x = 0$ and embedded at the end $x = L$.

**(a)** Use the fourth-order differential equation given in Problem 25 to find the eigenvalues $\lambda_n$, the critical loads $P_n$, the Euler load $P_1$, and the deflections $y_n(x)$.

**(b)** Use a graphing utility to graph the first buckling mode.

## Rotating String

**27.** Consider the boundary-value problem introduced in the construction of the mathematical model for the shape of a rotating string:

$$T\frac{d^2y}{dx^2} + \rho\omega^2 y = 0, \quad y(0) = 0, \quad y(L) = 0.$$

For constant $T$ and $\rho$, define the **critical speeds** of angular rotation $\omega_n$ as the values of $\omega$ for which the boundary-value problem has nontrivial solutions. Find the critical speeds $\omega_n$ and the corresponding deflections $y_n(x)$.

**28.** When the magnitude of tension $T$ is not constant, then a model for the deflection curve or shape $y(x)$ assumed by a rotating string is given by

$$\frac{d}{dx}\left[T(x)\frac{dy}{dx}\right] + \rho\omega^2 y = 0.$$

Suppose that $1 < x < e$ and that $T(x) = x^2$.

**(a)** If $y(1) = 0$, $y(e) = 0$, and $\rho\omega^2 > 0.25$, show that the critical speeds of angular rotation are $\omega_n = \frac{1}{2}\sqrt{(4n^2\pi^2 + 1)/\rho}$ and the corresponding deflections are

$$y_n(x) = c_2 x^{-1/2} \sin(n\pi \ln x), \quad n = 1, 2, 3, \ldots.$$

**(b)** Use a graphing utility to graph the deflection curves on the interval $[1, e]$ for $n = 1, 2, 3$. Choose $c_2 = 1$.

## Additional Boundary-Value Problems

**29. Temperature in a Sphere**  Consider two concentric spheres of radius $r = a$ and $r = b$, $a < b$. See Figure 5.2.10. The temperature $u(r)$ in the region between the spheres is determined from the boundary-value problem

$$r\frac{d^2u}{dr^2} + 2\frac{du}{dr} = 0, \quad u(a) = u_0, \quad u(b) = u_1,$$

where $u_0$ and $u_1$ are constants. Solve for $u(r)$.

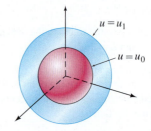

**FIGURE 5.2.10** Concentric spheres in Problem 29

**30. Temperature in a Ring**  The temperature $u(r)$ in the circular ring shown in Figure 5.2.11 is determined from the boundary-value problem

$$r\frac{d^2u}{dr^2} + \frac{du}{dr} = 0, \quad u(a) = u_0, \quad u(b) = u_1,$$

**FIGURE 5.2.11** Circular ring in Problem 30

where $u_0$ and $u_1$ are constants. Show that

$$u(r) = \frac{u_0 \ln(r/b) - u_1 \ln(r/a)}{\ln(a/b)}.$$

31. **Rotation of a Shaft** Suppose the $x$-axis on the interval $[0, L]$ is the geometric center of a long straight shaft, such as the propeller shaft of a ship. See Figure 5.2.12. When the shaft is rotating at a constant angular speed $\omega$ about this axis the deflection $y(x)$ of the shaft satisfies the differential equation

$$EI \frac{d^4 y}{dx^4} - \rho \omega^2 y = 0,$$

where $\rho$ is its density per unit length. If the shaft is simply supported, or hinged, at both ends the boundary conditions are then

$$y(0) = 0, \quad y''(0) = 0, \quad y(L) = 0, \quad y''(L) = 0.$$

(a) If $\lambda = \alpha^4 = \rho \omega^2 / EI$, then find the eigenvalues and eigenfunctions for this boundary-value problem.

(b) Use the eigenvalues $\lambda_n$ in part (a) to find corresponding angular speeds $\omega_n$. The values $\omega_n$ are called **critical speeds.** The value $\omega_1$ is called the **fundamental critical speed** and, analogous to Example 4, at this speed the shaft changes shape from $y = 0$ to a deflection given by $y_1(x)$.

**FIGURE 5.2.12** Propeller shaft of the battleship USS *Missouri*

National Archives and Records Administration

32. In Problem 31 suppose $L = 1$. If the shaft is fixed at both ends then the boundary conditions are

$$y(0) = 0, \quad y'(0) = 0, \quad y(1) = 0, \quad y'(1) = 0.$$

(a) Show that the eigenvalues $\lambda_n = \alpha_n^4$ are defined by the positive roots of $\cos \alpha \cosh \alpha = 1$. [*Hint*: See the instructions to Problems 21 and 22.]

(b) Show that the eigenfunctions are

$$y_n(x) = (-\sin \alpha_n + \sinh \alpha_n)(\cos \alpha_n x - \cosh \alpha_n x)$$
$$+ (\cos \alpha_n - \cosh \alpha_n)(\sin \alpha_n x - \sinh \alpha_n x).$$

### Discussion Problems

33. **Simple Harmonic Motion** The model $mx'' + kx = 0$ for simple harmonic motion, discussed in Section 5.1, can be related to Example 2 of this section.

Consider a free undamped spring/mass system for which the spring constant is, say, $k = 10$ lb/ft. Determine those masses $m_n$ that can be attached to the spring so that when each mass is released at the equilibrium position at $t = 0$ with a nonzero

velocity $v_0$, it will then pass through the equilibrium position at $t = 1$ second. How many times will each mass $m_n$ pass through the equilibrium position in the time interval $0 < t < 1$?

34. **Damped Motion** Assume that the model for the spring/mass system in Problem 33 is replaced by

$$mx'' + 2x' + kx = 0.$$

In other words, the system is free but is subjected to damping numerically equal to 2 times the instantaneous velocity. With the same initial conditions and spring constant as in Problem 33, investigate whether a mass $m$ can be found that will pass through the equilibrium position at $t = 1$ second.

In Problems 35 and 36 determine whether it is possible to find values $y_0$ and $y_1$ (Problem 35) and values of $L > 0$ (Problem 36) so that the given boundary-value problem has (**a**) precisely one nontrivial solution, (**b**) more than one solution, (**c**) no solution, (**d**) the trivial solution.

35. $y'' + 16y = 0, \quad y(0) = y_0, y(\pi/2) = y_1$

36. $y'' + 16y = 0, \quad y(0) = 1, y(L) = 1$

37. Consider the boundary-value problem

$$y'' + \lambda y = 0, \quad y(-\pi) = y(\pi), \quad y'(-\pi) = y'(\pi).$$

(a) The type of boundary conditions specified are called **periodic boundary conditions.** Give a geometric interpretation of these conditions.

(b) Find the eigenvalues and eigenfunctions of the problem.

(c) Use a graphing utility to graph some of the eigenfunctions. Verify your geometric interpretation of the boundary conditions given in part (a).

38. Show that the eigenvalues and eigenfunctions of the boundary-value problem

$$y'' + \lambda y = 0, \quad y(0) = 0, \quad y(1) + y'(1) = 0$$

are $\lambda_n = \alpha_n^2$ and $y_n(x) = \sin \alpha_n x$, respectively, where $\alpha_n, n = 1, 2, 3, \ldots$ are the consecutive positive roots of the equation $\tan \alpha = -\alpha$.

### Computer Lab Assignments

39. Use a CAS to plot graphs to convince yourself that the equation $\tan \alpha = -\alpha$ in Problem 38 has an infinite number of roots. Explain why the negative roots of the equation can be ignored. Explain why $\lambda = 0$ is not an eigenvalue even though $\alpha = 0$ is an obvious solution of the equation $\tan \alpha = -\alpha$.

40. Use a root-finding application of a CAS to approximate the first four eigenvalues $\lambda_1, \lambda_2, \lambda_3$, and $\lambda_4$ for the BVP in Problem 38.

41. Use a CAS to approximate the eigenvalues $\lambda_1, \lambda_2, \lambda_3$, and $\lambda_4$ of the boundary-value problem:

$$y'' + \lambda y = 0, \quad y(0) = 0, \quad y(1) - \tfrac{1}{2} y'(1) = 0.$$

Give the corresponding approximate eigenfunctions $y_1(x), y_2(x), y_3(x)$, and $y_4(x)$.

42. Use a CAS to approximate the eigenvalues $\lambda_1, \lambda_2, \lambda_3$, and $\lambda_4$ defined by the equation in part (a) of Problem 32.

## 5.3 Nonlinear Models

**INTRODUCTION** In this section we examine some nonlinear higher-order mathematical models. We are able to solve some of these models using the substitution method (leading to reduction of the order of the DE) introduced on page 189. In some cases in which the model cannot be solved, we show how a nonlinear DE can be replaced by a linear DE through a process called *linearization*.

**NONLINEAR SPRINGS** The mathematical model in (1) of Section 5.1 has the form

$$m\frac{d^2x}{dt^2} + F(x) = 0, \tag{1}$$

where $F(x) = kx$. Because $x$ denotes the displacement of the mass from its equilibrium position, $F(x) = kx$ is Hooke's law—that is, the force exerted by the spring that tends to restore the mass to the equilibrium position. A spring acting under a linear restoring force $F(x) = kx$ is naturally referred to as a **linear spring.** But springs are seldom perfectly linear. Depending on how it is constructed and the material that is used, a spring can range from "mushy," or soft, to "stiff," or hard, so its restorative force may vary from something below to something above that given by the linear law. In the case of free motion, if we assume that a nonaging spring has some nonlinear characteristics, then it might be reasonable to assume that the restorative force of a spring—that is, $F(x)$ in (1)—is proportional to, say, the cube of the displacement $x$ of the mass beyond its equilibrium position or that $F(x)$ is a linear combination of powers of the displacement such as that given by the nonlinear function $F(x) = kx + k_1x^3$. A spring whose mathematical model incorporates a nonlinear restorative force, such as

$$m\frac{d^2x}{dt^2} + kx^3 = 0 \qquad \text{or} \qquad m\frac{d^2x}{dt^2} + kx + k_1x^3 = 0, \tag{2}$$

is called a **nonlinear spring.** In addition, we examined mathematical models in which damping imparted to the motion was proportional to the instantaneous velocity $dx/dt$ and the restoring force of a spring was given by the linear function $F(x) = kx$. But these were simply assumptions; in more realistic situations damping could be proportional to some power of the instantaneous velocity $dx/dt$. The nonlinear differential equation

$$m\frac{d^2x}{dt^2} + \left|\frac{dx}{dt}\right|\frac{dx}{dt} + kx = 0 \tag{3}$$

is one model of a free spring/mass system in which the damping force is proportional to the square of the velocity. One can then envision other kinds of models: linear damping and nonlinear restoring force, nonlinear damping and nonlinear restoring force, and so on. The point is that nonlinear characteristics of a physical system lead to a mathematical model that is nonlinear.

Notice in (2) that both $F(x) = kx^3$ and $F(x) = kx + k_1x^3$ are odd functions of $x$. To see why a polynomial function containing only odd powers of $x$ provides a reasonable model for the restoring force, let us express $F$ as a power series centered at the equilibrium position $x = 0$:

$$F(x) = c_0 + c_1x + c_2x^2 + c_3x^3 + \cdots.$$

When the displacements $x$ are small, the values of $x^n$ are negligible for $n$ sufficiently large. If we truncate the power series with, say, the fourth term, then $F(x) = c_0 + c_1x + c_2x^2 + c_3x^3$. For the force at $x > 0$,

$$F(x) = c_0 + c_1x + c_2x^2 + c_3x^3,$$

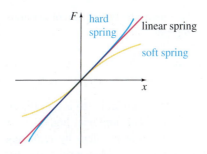

**FIGURE 5.3.1** Hard and soft springs

**(a)** hard spring

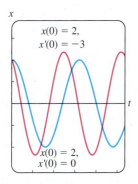

**(b)** soft spring

**FIGURE 5.3.2** Numerical solution curves

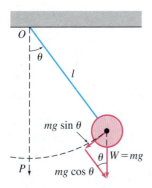

**FIGURE 5.3.3** Simple pendulum

and for the force at $-x < 0$,

$$F(-x) = c_0 - c_1x + c_2x^2 - c_3x^3$$

to have the same magnitude but act in the opposite direction, we must have $F(-x) = -F(x)$. Because this means that $F$ is an odd function, we must have $c_0 = 0$ and $c_2 = 0$, and so $F(x) = c_1x + c_3x^3$. Had we used only the first two terms in the series, the same argument yields the linear function $F(x) = c_1x$. A restoring force with mixed powers, such as $F(x) = c_1x + c_2x^2$, and the corresponding vibrations are said to be **unsymmetrical.** In the next discussion we shall write $c_1 = k$ and $c_3 = k_1$.

**HARD AND SOFT SPRINGS**    Let us take a closer look at the equation in (1) in the case in which the restoring force is given by $F(x) = kx + k_1x^3$, $k > 0$. The spring is said to be **hard** if $k_1 > 0$ and **soft** if $k_1 < 0$. Graphs of three types of restoring forces are illustrated in Figure 5.3.1. The next example illustrates these two special cases of the differential equation $md^2x/dt^2 + kx + k_1x^3 = 0$, $m > 0$, $k > 0$.

---

**EXAMPLE 1**    **Comparison of Hard and Soft Springs**

The differential equations

$$\frac{d^2x}{dt^2} + x + x^3 = 0 \tag{4}$$

and

$$\frac{d^2x}{dt^2} + x - x^3 = 0 \tag{5}$$

are special cases of the second equation in (2) and are models of a hard spring and a soft spring, respectively. Figure 5.3.2(a) shows two solutions of (4) and Figure 5.3.2(b) shows two solutions of (5) obtained from a numerical solver. The curves shown in red are solutions that satisfy the initial conditions $x(0) = 2$, $x'(0) = -3$; the two curves in blue are solutions that satisfy $x(0) = 2$, $x'(0) = 0$. These solution curves certainly suggest that the motion of a mass on the hard spring is oscillatory, whereas motion of a mass on the soft spring appears to be non-oscillatory. But we must be careful about drawing conclusions based on a couple of numerical solution curves. A more complete picture of the nature of the solutions of both of these equations can be obtained from the qualitative analysis discussed in Chapter 10. ■

**NONLINEAR PENDULUM**    Any object that swings back and forth is called a **physical pendulum.** The **simple pendulum** is a special case of the physical pendulum and consists of a rod of length $l$ to which a mass $m$ is attached at one end. In describing the motion of a simple pendulum in a vertical plane, we make the simplifying assumptions that the mass of the rod is negligible and that no external damping or driving forces act on the system. The displacement angle $\theta$ of the pendulum, measured from the vertical as shown in Figure 5.3.3, is considered positive when measured to the right of $OP$ and negative to the left of $OP$. Now recall the arc $s$ of a circle of radius $l$ is related to the central angle $\theta$ by the formula $s = l\theta$. Hence angular acceleration is

$$a = \frac{d^2s}{dt^2} = l\frac{d^2\theta}{dt^2}.$$

From Newton's second law we then have

$$F = ma = ml\frac{d^2\theta}{dt^2}.$$

From Figure 5.3.3 we see that the magnitude of the tangential component of the force due to the weight $W$ is $mg \sin \theta$. In direction this force is $-mg \sin \theta$ because it points to the left for $\theta > 0$ and to the right for $\theta < 0$. We equate the two different versions of the tangential force to obtain $ml\, d^2\theta/dt^2 = -mg \sin \theta$, or

$$\frac{d^2\theta}{dt^2} + \frac{g}{l} \sin \theta = 0. \tag{6}$$

**LINEARIZATION**   Because of the presence of $\sin \theta$, the model in (6) is non-linear. In an attempt to understand the behavior of the solutions of nonlinear higher-order differential equations, one sometimes tries to simplify the problem by replacing nonlinear terms by certain approximations. For example, the Maclaurin series for $\sin \theta$ is given by

$$\sin \theta = \theta - \frac{\theta^3}{3!} + \frac{\theta^5}{5!} - \cdots$$

so if we use the approximation $\sin \theta \approx \theta - \theta^3/6$, equation (6) becomes

$$\frac{d^2\theta}{dt^2} + \frac{g}{l}\theta - \frac{g}{6l}\theta^3 = 0.$$

Observe that this last equation is the same as the second nonlinear equation in (2) with $m = 1$, $k = g/l$, and $k_1 = -g/6l$. However, if we assume that the displacements $\theta$ are small enough to justify using the replacement $\sin \theta \approx \theta$, then (6) becomes

$$\frac{d^2\theta}{dt^2} + \frac{g}{l}\theta = 0. \tag{7}$$

See Problem 25 in Exercises 5.3. If we set $\omega^2 = g/l$, we recognize (7) as the differential equation (2) of Section 5.1 that is a model for the free undamped vibrations of a linear spring/mass system. In other words, (7) is again the basic linear equation $y'' + \lambda y = 0$ discussed on pages 215–216 of Section 5.2. As a consequence we say that equation (7) is a **linearization** of equation (6). Because the general solution of (7) is $\theta(t) = c_1 \cos \omega t + c_2 \sin \omega t$, this linearization suggests that for initial conditions amenable to small oscillations the motion of the pendulum described by (6) will be periodic.

**EXAMPLE 2**   Two Initial-Value Problems

The graphs in Figure 5.3.4(a) were obtained with the aid of a numerical solver and represent approximate or numerical solution curves of (6) when $\omega^2 = 1$. The blue curve depicts the solution of (6) that satisfies the initial conditions $\theta(0) = \frac{1}{2}$, $\theta'(0) = \frac{1}{2}$, whereas the red curve is the solution of (6) that satisfies $\theta(0) = \frac{1}{2}$, $\theta'(0) = 2$. The blue

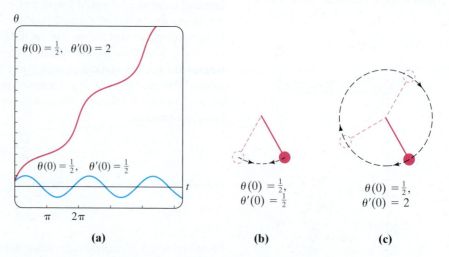

**FIGURE 5.3.4**   In Example 2, oscillating pendulum in (b); whirling pendulum in (c)

curve represents a periodic solution—the pendulum *oscillating* back and forth as shown in Figure 5.3.4(b) with an apparent amplitude $A \leq 1$. The red curve shows that $\theta$ increases without bound as time increases—the pendulum, starting from the same initial displacement, is given an initial velocity of magnitude great enough to send it over the top; in other words, the pendulum is *whirling* about its pivot as shown in Figure 5.3.4(c). In the absence of damping, the motion in each case is continued indefinitely.

**FIGURE 5.3.5** Shape of hanging telephone wires is a catenary

**TELEPHONE WIRES**  The first-order differential equation $dy/dx = W/T_1$ is equation (16) of Section 1.3. This differential equation, established with the aid of Figure 1.3.8 on page 28, serves as a mathematical model for the shape of a flexible cable suspended between two vertical supports when the cable is carrying a vertical load. In Section 2.2 we solved this simple DE under the assumption that the vertical load carried by the cables of a suspension bridge was the weight of a horizontal roadbed distributed evenly along the $x$-axis. With $W = \rho x$, $\rho$ the weight per unit length of the roadbed, the shape of each cable between the vertical supports turned out to be parabolic. We are now in a position to determine the shape of a uniform flexible cable hanging only under its own weight, such as a wire strung between two telephone posts. See Figure 5.3.5. The vertical load is now the wire itself, and so if $\rho$ is the linear density of the wire (measured, say, in pounds per feet) and $s$ is the length of the segment $P_1P_2$ in Figure 1.3.8 then $W = \rho s$. Hence

$$\frac{dy}{dx} = \frac{\rho s}{T_1}. \tag{8}$$

Since the arc length between points $P_1$ and $P_2$ is given by

$$s = \int_0^x \sqrt{1 + \left(\frac{dy}{dx}\right)^2}\, dx, \tag{9}$$

it follows from the Fundamental Theorem of Calculus that the derivative of (9) is

$$\frac{ds}{dx} = \sqrt{1 + \left(\frac{dy}{dx}\right)^2}. \tag{10}$$

Differentiating (8) with respect to $x$ and using (10) lead to the second-order equation

$$\frac{d^2y}{dx^2} = \frac{\rho}{T_1}\frac{ds}{dx} \quad \text{or} \quad \frac{d^2y}{dx^2} = \frac{\rho}{T_1}\sqrt{1 + \left(\frac{dy}{dx}\right)^2}. \tag{11}$$

In the example that follows we solve (11) and show that the curve assumed by the suspended cable is a **catenary.** Before proceeding, observe that the nonlinear second order differential equation (11) is one of those equations having the form $F(x, y', y'') = 0$ discussed in Section 4.10. Recall that we have a chance of solving an equation of this type by reducing the order of the equation by means of the substitution $u = y'$.

### EXAMPLE 3  A Solution of (11)

From the position of the $y$-axis in Figure 1.3.8 it is apparent that initial conditions associated with the second differential equation in (11) are $y(0) = a$ and $y'(0) = 0$.

If we substitute $u = y'$, then the equation in (11) becomes $\dfrac{du}{dx} = \dfrac{\rho}{T_1}\sqrt{1 + u^2}$. Separating variables, we find that

$$\int \frac{du}{\sqrt{1 + u^2}} = \frac{\rho}{T_1}\int dx \qquad \text{gives} \qquad \sinh^{-1}u = \frac{\rho}{T_1}x + c_1.$$

Now, $y'(0) = 0$ is equivalent to $u(0) = 0$. Since $\sinh^{-1} 0 = 0, c_1 = 0$, so $u = \sinh(\rho x/T_1)$. Finally, by integrating both sides of

$$\frac{dy}{dx} = \sinh\frac{\rho}{T_1}x, \qquad \text{we get} \qquad y = \frac{T_1}{\rho}\cosh\frac{\rho}{T_1}x + c_2.$$

Using $y(0) = a$, $\cosh 0 = 1$, the last equation implies that $c_2 = a - T_1/\rho$. Thus we see that the shape of the hanging wire is given by $y = (T_1/\rho)\cosh(\rho x/T_1) + a - T_1/\rho$. ∎

In Example 3, had we been clever enough at the start to choose $a = T_1/\rho$, then the solution of the problem would have been simply the hyperbolic cosine $y = (T_1/\rho)\cosh(\rho x/T_1)$.

**ROCKET MOTION**    In (12) of Section 1.3 we saw that the differential equation of a free-falling body of mass $m$ near the surface of the Earth is given by

$$m\frac{d^2s}{dt^2} = -mg \qquad \text{or simply} \qquad \frac{d^2s}{dt^2} = -g,$$

where $s$ represents the distance from the surface of the Earth to the object and the positive direction is considered to be upward. In other words, the underlying assumption here is that the distance $s$ to the object is small when compared with the radius $R$ of the Earth; put yet another way, the distance $y$ from the center of the Earth to the object is approximately the same as $R$. If, on the other hand, the distance $y$ to the object, such as a rocket or a space probe, is large when compared to $R$, then we combine Newton's second law of motion and his universal law of gravitation to derive a differential equation in the variable $y$.

Suppose a rocket is launched vertically upward from the ground as shown in Figure 5.3.6. If the positive direction is upward and air resistance is ignored, then the differential equation of motion after fuel burnout is

$$m\frac{d^2y}{dt^2} = -k\frac{Mm}{y^2} \qquad \text{or} \qquad \frac{d^2y}{dt^2} = -k\frac{M}{y^2}, \tag{12}$$

where $k$ is a constant of proportionality, $y$ is the distance from the center of the Earth to the rocket, $M$ is the mass of the Earth, and $m$ is the mass of the rocket. To determine the constant $k$, we use the fact that when $y = R$, $kMm/R^2 = mg$ or $k = gR^2/M$. Thus the last equation in (12) becomes

$$\frac{d^2y}{dt^2} = -g\frac{R^2}{y^2}. \tag{13}$$

See Problem 14 in Exercises 5.3.

**FIGURE 5.3.6** Distance to rocket is large compared to $R$.

**VARIABLE MASS**    Notice in the preceding discussion that we described the motion of the rocket after it has burned all its fuel, when presumably its mass $m$ is constant. Of course, during its powered ascent the total mass of the rocket varies as its fuel is being expended. We saw in (17) of Exercises 1.3 that the second law of motion, as originally advanced by Newton, states that when a body of mass $m$ moves through a force field with velocity $v$, the time rate of change of the momentum $mv$ of the body is equal to applied or net force $F$ acting on the body:

$$F = \frac{d}{dt}(mv). \tag{14}$$

If $m$ is constant, then (14) yields the more familiar form $F = m\, dv/dt = ma$, where $a$ is acceleration. We use the form of Newton's second law given in (14) in the next example, in which the mass $m$ of the body is variable.

5 lb
upward
force

$x(t)$

**FIGURE 5.3.7** Chain pulled upward by a constant force in Example 4

EXAMPLE 4 **Chain Pulled Upward by a Constant Force**

A uniform 10-foot-long chain is coiled loosely on the ground. One end of the chain is pulled vertically upward by means of constant force of 5 pounds. The chain weighs 1 pound per foot. Determine the height of the end above ground level at time $t$. See Figure 5.3.7.

**SOLUTION**  Let us suppose that $x = x(t)$ denotes the height of the end of the chain in the air at time $t$, $v = dx/dt$, and the positive direction is upward. For the portion of the chain that is in the air at time $t$ we have the following variable quantities:

$$weight: \quad W = (x \text{ ft}) \cdot (1 \text{ lb/ft}) = x,$$

$$mass: \quad m = W/g = x/32,$$

$$net\,force: \quad F = 5 - W = 5 - x.$$

Thus from (14) we have

Product Rule

$$\frac{d}{dt}\left(\frac{x}{32}v\right) = 5 - x \quad \text{or} \quad x\frac{dv}{dt} + v\frac{dx}{dt} = 160 - 32x. \qquad (15)$$

Because $v = dx/dt$, the last equation becomes

$$x\frac{d^2x}{dt^2} + \left(\frac{dx}{dt}\right)^2 + 32x = 160. \qquad (16)$$

The nonlinear second-order differential equation (16) has the form $F(x, x', x'') = 0$, which is the second of the two forms considered in Section 4.10 that can possibly be solved by reduction of order. To solve (16), we revert back to (15) and use $v = x'$ along with the Chain Rule. From $\dfrac{dv}{dt} = \dfrac{dv}{dx}\dfrac{dx}{dt} = v\dfrac{dv}{dx}$ the second equation in (15) can be rewritten as

$$xv\frac{dv}{dx} + v^2 = 160 - 32x. \qquad (17)$$

On inspection (17) might appear intractable, since it cannot be characterized as any of the first-order equations that were solved in Chapter 2. However, by rewriting (17) in differential form $M(x, v)\,dx + N(x, v)\,dv = 0$, we observe that although the equation

$$(v^2 + 32x - 160)\,dx + xv\,dv = 0 \qquad (18)$$

is not exact, it can be transformed into an exact equation by multiplying it by an integrating factor. From $(M_v - N_x)/N = 1/x$ we see from (13) of Section 2.4 that for $x > 0$ an integrating factor is $e^{\int dx/x} = e^{\ln x} = x$. When (18) is multiplied by $\mu(x) = x$, the resulting equation is exact (verify). By identifying $\partial f/\partial x = xv^2 + 32x^2 - 160x$, $\partial f/\partial v = x^2v$ and then proceeding as in Section 2.4, we obtain

$$\frac{1}{2}x^2v^2 + \frac{32}{3}x^3 - 80x^2 = c_1. \qquad (19)$$

Since we have assumed that all of the chain is on the floor initially, we have $x(0) = 0$. This last condition applied to (19) yields $c_1 = 0$. By solving the algebraic equation $\frac{1}{2}x^2v^2 + \frac{32}{3}x^3 - 80x^2 = 0$ for $v = dx/dt > 0$, we get another first-order differential equation,

$$\frac{dx}{dt} = \sqrt{160 - \frac{64}{3}x}.$$

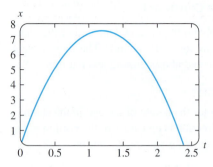

**FIGURE 5.3.8** Graph of (21) in Example 4

The last equation can be solved by separation of variables. You should verify that

$$-\frac{3}{32}\left(160 - \frac{64}{3}x\right)^{1/2} = t + c_2. \tag{20}$$

This time the initial condition $x(0) = 0$ implies that $c_2 = -3\sqrt{10}/8$. Finally, by squaring both sides of (20) and solving for $x$, we arrive at the desired result,

$$x(t) = \frac{15}{2} - \frac{15}{2}\left(1 - \frac{4\sqrt{10}}{15}t\right)^2. \tag{21}$$

The graph of (21) given in Figure 5.3.8 should not, on physical grounds, be taken at face value. See Problem 15 in Exercises 5.3.

---

## EXERCISES 5.3

*Answers to selected odd-numbered problems begin on page ANS-8.*

**To the Instructor**    In addition to Problems 24 and 25, all or portions of Problems 1–6, 8–13, 15, 20, and 21 could serve as Computer Lab Assignments.

### Nonlinear Springs

In Problems 1–4 the given differential equation is a model of an undamped spring/mass system in which the restoring force $F(x)$ in (1) is nonlinear. For each equation use a numerical solver to plot the solution curves that satisfy the given initial conditions. If the solutions appear to be periodic use the solution curve to estimate the period $T$ of oscillations.

**1.** $\dfrac{d^2x}{dt^2} + x^3 = 0,$

$x(0) = 1, x'(0) = 1;\quad x(0) = \frac{1}{2},\quad x'(0) = -1$

**2.** $\dfrac{d^2x}{dt^2} + 4x - 16x^3 = 0,$

$x(0) = 1, x'(0) = 1;\quad x(0) = -2,\quad x'(0) = 2$

**3.** $\dfrac{d^2x}{dt^2} + 2x - x^2 = 0,$

$x(0) = 1, x'(0) = 1;\quad x(0) = \frac{3}{2},\quad x'(0) = -1$

**4.** $\dfrac{d^2x}{dt^2} + xe^{0.01x} = 0,$

$x(0) = 1, x'(0) = 1;\quad x(0) = 3,\quad x'(0) = -1$

**5.** In Problem 3, suppose the mass is released from the initial position $x(0) = 1$ with an initial velocity $x'(0) = x_1$. Use a numerical solver to estimate the smallest value of $|x_1|$ at which the motion of the mass is nonperiodic.

**6.** In Problem 3, suppose the mass is released from an initial position $x(0) = x_0$ with the initial velocity $x'(0) = 1$. Use a numerical solver to estimate an interval $a \le x_0 \le b$ for which the motion is oscillatory.

**7.** Find a linearization of the differential equation in Problem 4.

**8.** Consider the model of an undamped nonlinear spring/mass system given by $x'' + 8x - 6x^3 + x^5 = 0$. Use a numerical

solver to discuss the nature of the oscillations of the system corresponding to the initial conditions:

$x(0) = 1,\quad x'(0) = 1;\qquad x(0) = -2,\quad x'(0) = \frac{1}{2};$

$x(0) = \sqrt{2},\quad x'(0) = 1;\quad x(0) = 2,\quad x'(0) = \frac{1}{2};$

$x(0) = 2,\quad x'(0) = 0;\qquad x(0) = -\sqrt{2},\quad x'(0) = -1.$

In Problems 9 and 10 the given differential equation is a model of a damped nonlinear spring/mass system. Predict the behavior of each system as $t \to \infty$. For each equation use a numerical solver to obtain the solution curves satisfying the given initial conditions.

**9.** $\dfrac{d^2x}{dt^2} + \dfrac{dx}{dt} + x + x^3 = 0,$

$x(0) = -3, x'(0) = 4;\quad x(0) = 0,\quad x'(0) = -8$

**10.** $\dfrac{d^2x}{dt^2} + \dfrac{dx}{dt} + x - x^3 = 0,$

$x(0) = 0, x'(0) = \frac{3}{2};\quad x(0) = -1,\quad x'(0) = 1$

**11.** The model $mx'' + kx + k_1x^3 = F_0\cos \omega t$ of an undamped periodically driven spring/mass system is called **Duffing's differential equation.** Consider the initial-value problem $x'' + x + k_1x^3 = 5 \cos t, x(0) = 1, x'(0) = 0$. Use a numerical solver to investigate the behavior of the system for values of $k_1 > 0$ ranging from $k_1 = 0.01$ to $k_1 = 100$. State your conclusions.

**12. (a)** Find values of $k_1 < 0$ for which the system in Problem 11 is oscillatory.

**(b)** Consider the initial-value problem

$$x'' + x + k_1x^3 = \cos \tfrac{3}{2}t,\quad x(0) = 0,\quad x'(0) = 0.$$

Find values for $k_1 < 0$ for which the system is oscillatory.

### Nonlinear Pendulum

**13.** Consider the model of the free damped nonlinear pendulum given by

$$\frac{d^2\theta}{dt^2} + 2\lambda\frac{d\theta}{dt} + \omega^2 \sin \theta = 0.$$

Use a numerical solver to investigate whether the motion in the two cases $\lambda^2 - \omega^2 > 0$ and $\lambda^2 - \omega^2 < 0$ corresponds, respectively, to the overdamped and underdamped cases discussed in Section 5.1 for spring/mass systems. For $\lambda^2 - \omega^2 > 0$, use $\lambda = 2$, $\omega = 1$, $\theta(0) = 1$, and $\theta'(0) = 2$. For $\lambda^2 - \omega^2 < 0$, use $\lambda = \frac{1}{3}$, $\omega = 1$, $\theta(0) = -2$, and $\theta'(0) = 4$.

### Rocket Motion

**14. (a)** Use the substitution $v = dy/dt$ to solve (13) for $v$ in terms of $y$. Assuming that the velocity of the rocket at burnout is $v = v_0$ and $y \approx R$ at that instant, show that the approximate value of the constant $c$ of integration is $c = -gR + \frac{1}{2}v_0^2$.

**(b)** Use the solution for $v$ in part (a) to show that the escape velocity of the rocket is given by $v_0 = \sqrt{2gR}$. [*Hint*: Take $y \to \infty$ and assume $v > 0$ for all time $t$.]

**(c)** The result in part (b) holds for any body in the Solar System. Use the values $g = 32$ ft/s$^2$ and $R = 4000$ mi to show that the escape velocity from the Earth is (approximately) $v_0 = 25{,}000$ mi/h.

**(d)** Find the escape velocity from the Moon if the acceleration of gravity is $0.165g$ and $R = 1080$ mi.

### Variable Mass

**15. (a)** In Example 4, how much of the chain would you intuitively expect the constant 5-pound force to be able to lift?

**(b)** What is the initial velocity of the chain?

**(c)** Why is the time interval corresponding to $x(t) \geq 0$ given in Figure 5.3.7 not the interval $I$ of definition of the solution (21)? Determine the interval $I$. How much chain is actually lifted? Explain any difference between this answer and your prediction in part (a).

**(d)** Why would you expect $x(t)$ to be a periodic solution?

**16.** A uniform chain of length $L$, measured in feet, is held vertically so that the lower end just touches the floor. The chain weighs 2 lb/ft. The upper end that is held is released from rest at $t = 0$ and the chain falls straight down. If $x(t)$ denotes the length of the chain on the floor at time $t$, air resistance is ignored, and the positive direction is taken to be downward, then

$$(L - x)\frac{d^2x}{dt^2} - \left(\frac{dx}{dt}\right)^2 = Lg.$$

**(a)** Solve for $v$ in terms of $x$. Solve for $x$ in terms of $t$. Express $v$ in terms of $t$.

**(b)** Determine how long it takes for the chain to fall completely to the ground.

**(c)** What velocity does the model in part (a) predict for the upper end of the chain as it hits the ground?

### Additional Mathematical Models

**17. Pursuit Curve** In a naval exercise a ship $S_1$ is pursued by a submarine $S_2$ as shown in Figure 5.3.9. Ship $S_1$ departs point $(0, 0)$ at $t = 0$ and proceeds along a straight-line course (the $y$-axis) at a constant speed $v_1$. The submarine $S_2$ keeps ship $S_1$ in visual contact, indicated by the straight dashed line $L$ in the figure,

while traveling at a constant speed $v_2$ along a curve $C$. Assume that ship $S_2$ starts at the point $(a, 0)$, $a > 0$, at $t = 0$ and that $L$ is tangent to $C$.

**(a)** Determine a mathematical model that describes the curve $C$.

**(b)** Find an explicit solution of the differential equation. For convenience define $r = v_1/v_2$.

**(c)** Determine whether the paths of $S_1$ and $S_2$ will ever intersect by considering the cases $r > 1$, $r < 1$, and $r = 1$. [*Hint*: $\dfrac{dt}{dx} = \dfrac{dt}{ds}\dfrac{ds}{dx}$, where $s$ is arc length measured along $C$.]

**FIGURE 5.3.9** Pursuit curve in Problem 17

**18. Pursuit Curve** In another naval exercise a destroyer $S_1$ pursues a submerged submarine $S_2$. Suppose that $S_1$ at $(9, 0)$ on the $x$-axis detects $S_2$ at $(0, 0)$ and that $S_2$ simultaneously detects $S_1$. The captain of the destroyer $S_1$ assumes that the submarine will take immediate evasive action and conjectures that its likely new course is the straight line indicated in Figure 5.3.10. When $S_1$ is at $(3, 0)$, it changes from its straight-line course toward the origin to a pursuit curve $C$. Assume that the speed of the destroyer is, at all times, a constant 30 mi/h and that the submarine's speed is a constant 15 mi/h.

**(a)** Explain why the captain waits until $S_1$ reaches $(3, 0)$ before ordering a course change to $C$.

**(b)** Using polar coordinates, find an equation $r = f(\theta)$ for the curve $C$.

**(c)** Let $T$ denote the time, measured from the initial detection, at which the destroyer intercepts the submarine. Find an upper bound for $T$.

**FIGURE 5.3.10** Pursuit curve in Problem 18

**19. The Ballistic Pendulum**   Historically, in order to maintain quality control over munitions (bullets) produced by an assembly line, the manufacturer would use a **ballistic pendulum** to determine the muzzle velocity of a gun, that is, the speed of a bullet as it leaves the barrel. Invented in 1742 by the English engineer Benjamin Robins, the ballistic pendulum is simply a plane pendulum consisting of a rod of negligible mass to which a block of wood of mass $m_w$ is attached. The system is set in motion by the impact of a bullet which is moving horizontally at the unknown velocity $v_b$; at the time of the impact, which we take as $t = 0$, the combined mass is $m_w + m_b$, where $m_b$ is the mass of the bullet imbedded in the wood. In (7) of this section, we saw that in the case of small oscillations, the angular displacement $\theta(t)$ of a plane pendulum shown in Figure 5.3.3 is given by the linear DE $\theta'' + (g/l)\theta = 0$, where $\theta > 0$ corresponds to motion to the right of vertical. The velocity $v_b$ can be found by measuring the height $h$ of the mass $m_w + m_b$ at the maximum displacement angle $\theta_{max}$ shown in Figure 5.3.11.

Intuitively, the horizontal velocity $V$ of the combined mass (wood plus bullet) after impact is only a fraction of the velocity $v_b$ of the bullet, that is,

$$V = \left(\frac{m_b}{m_w + m_b}\right) v_b.$$

Now recall, a distance $s$ traveled by a particle moving along a circular path is related to the radius $l$ and central angle $\theta$ by the formula $s = l\theta$. By differentiating the last formula with respect to time $t$, it follows that the angular velocity $\omega$ of the mass and its linear velocity $v$ are related by $v = l\omega$. Thus the initial angular velocity $\omega_0$ at the time $t$ at which the bullet impacts the wood block is related to $V$ by $V = l\omega_0$ or

$$\omega_0 = \left(\frac{m_b}{m_w + m_b}\right)\frac{v_b}{l}.$$

**(a)** Solve the initial-value problem

$$\frac{d^2\theta}{dt^2} + \frac{g}{l}\theta = 0, \quad \theta(0) = 0, \quad \theta'(0) = \omega_0.$$

**(b)** Use the result from part (a) to show that

$$v_b = \left(\frac{m_w + m_b}{m_b}\right)\sqrt{lg}\,\theta_{max}.$$

**(c)** Use Figure 5.3.11 to express $\cos\theta_{max}$ in terms of $l$ and $h$. Then use the first two terms of the Maclaurin series for $\cos\theta$ to express $\theta_{max}$ in terms of $l$ and $h$. Finally, show that $v_b$ is given (approximately) by

$$v_b = \left(\frac{m_w + m_b}{m_b}\right)\sqrt{2gh}.$$

**(d)** Use the result in part (c) to find $v_b$ and $m_b = 5$ g, $m_w = 1$ kg, and $h = 6$ cm.

**FIGURE 5.3.11** Ballistic pendulum in Problem 19

**20. Relief Supplies**   As shown in Figure 5.3.12, a plane flying horizontally at a constant speed $v_0$ drops a relief supply pack to a person on the ground. Assume the origin is the point where the supply pack is released and that the positive $x$-axis points forward and that positive $y$-axis points downward. Under the assumption that the horizontal and vertical components of the air resistance are proportional to $(dx/dt)^2$ and $(dy/dt)^2$, respectively, and if the position of the supply pack is given by $\mathbf{r}(t) = x(t)\mathbf{i} + y(t)\mathbf{j}$, then its velocity is $\mathbf{v}(t) = (dx/dt)\mathbf{i} + (dy/dt)\mathbf{j}$. Equating components in the vector form of Newton's second law of motion,

$$m\frac{d\mathbf{v}}{dt} = mg - k\left[\left(\frac{dx}{dt}\right)^2\mathbf{i} + \left(\frac{dy}{dt}\right)^2\mathbf{j}\right]$$

gives

$$m\frac{d^2x}{dt^2} = mg - k\left(\frac{dx}{dt}\right)^2, \quad x(0) = 0, \quad x'(0) = v_0$$

$$m\frac{d^2y}{dt^2} = mg - k\left(\frac{dy}{dt}\right)^2, \quad y(0) = 0, \quad y'(0) = 0.$$

**(a)** Solve both of the foregoing initial-value problems by means of the substitutions $u = dx/dt$, $w = dy/dt$, and separation of variables. [*Hint*: See the *Remarks* at the end of Section 3.2.]

**(b)** Suppose the plane files at an altitude of 1000 ft and that its constant speed is 300 mi/h. Assume that the constant of proportionality for air resistance is $k = 0.0053$ and that the supply pack weighs 256 lb. Use a root-finding application of a CAS or a graphic calculator to determine the horizontal distance the pack travels, measured from its point of release to the point where it hits the ground.

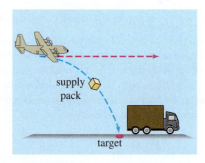

**FIGURE 5.3.12** Airplane drop in Problem 20

## Discussion Problems

**21.** Discuss why the damping term in equation (3) is written as

$$\beta\left|\frac{dx}{dt}\right|\frac{dx}{dt} \quad \text{instead of} \quad \beta\left(\frac{dx}{dt}\right)^2.$$

**22. (a)** Experiment with a calculator to find an interval $0 \le \theta < \theta_1$, where $\theta$ is measured in radians, for which you think $\sin\theta \approx \theta$ is a fairly good estimate. Then use a graphing utility to plot the graphs of $y = x$ and $y = \sin x$ on the same coordinate axes for $0 \le x \le \pi/2$. Do the graphs confirm your observations with the calculator?

**(b)** Use a numerical solver to plot the solution curves of the initial-value problems

$$\frac{d^2\theta}{dt^2} + \sin\theta = 0, \quad \theta(0) = \theta_0, \quad \theta'(0) = 0$$

and $\frac{d^2\theta}{dt^2} + \theta = 0, \quad \theta(0) = \theta_0, \quad \theta'(0) = 0$

for several values of $\theta_0$ in the interval $0 \le \theta < \theta_1$ found in part (a). Then plot solution curves of the initial-value problems for several values of $\theta_0$ for which $\theta_0 > \theta_1$.

**23. Pendulum Motion on the Moon** Does a pendulum of length $l$ oscillate faster on the Earth or on the Moon?

**(a)** Take $l = 3$ and $g = 32$ for the acceleration of gravity on Earth. Use a numerical solver to generate a numerical solution curve for the nonlinear model (6) subject to the initial conditions $\theta(0) = 1$, $\theta'(0) = 2$. Repeat using the same values but use $0.165g$ for the acceleration of gravity on the Moon.

**(b)** From the graphs in part (a), determine which pendulum oscillates faster. Which pendulum has the greater amplitude of motion?

**24. Pendulum Motion on the Moon-Continued** Repeat the two parts of Problem 23 this time using the linear model (7).

## Computer Lab Assignments

**25.** Consider the initial-value problem

$$\frac{d^2\theta}{dt^2} + \sin\theta = 0, \quad \theta(0) = \frac{\pi}{12}, \quad \theta'(0) = -\frac{1}{3}$$

for a nonlinear pendulum. Suppose we wish to determine the first time $t_1 > 0$ for which the pendulum in Figure 5.3.3, starting from its initial position to the right, reaches the position $OP$—that is, the first positive root of $\theta(t) = 0$. In this problem and the next we examine several ways to proceed.

**(a)** Approximate $t_1$ by solving the linear problem

$$\frac{d^2\theta}{dt^2} + \theta = 0, \quad \theta(0) = \frac{\pi}{12}, \quad \theta'(0) = -\frac{1}{3}.$$

**(b)** Use the method illustrated in Example 3 of Section 4.10 to find the first four nonzero terms of a Taylor series solution $\theta(t)$ centered at 0 for the nonlinear initial-value problem. Give the exact values of all coefficients.

**(c)** Use the first two terms of the Taylor series in part (b) to approximate $t_1$.

**(d)** Use the first three terms of the Taylor series in part (b) to approximate $t_1$.

**(e)** Use a root-finding application of a CAS or a graphic calculator and the first four terms of the Taylor series in part (b) to approximate $t_1$.

**(f)** In this part of the problem you are led through the commands in *Mathematica* that enable you to approximate the root $t_1$. The procedure is easily modified so that any root of $\theta(t) = 0$ can be approximated. (*If you do not have Mathematica, adapt the given procedure by finding the corresponding syntax for the CAS you have on hand.*) Reproduce and then, in turn, execute each line in the given sequence of commands.

```
sol = NDSolve [{y"[t] + Sin[y[t]] == 0,
               y[0] == Pi/12, y'[0] == −1/3},
               y, {t, 0, 5}]// Flatten
solution = y[t]/.sol
Clear[y]
y[t_]: = Evaluate[solution]
y[t]
gr1 = Plot[y[t], {t, 0, 5}]
root = FindRoot[y[t] == 0, {t, 1}]
```

**(g)** Appropriately modify the syntax in part (f) and find the next two positive roots of $\theta(t) = 0$.

**26.** Consider a pendulum that is released from rest from an initial displacement of $\theta_0$ radians. Solving the linear model (7) subject to the initial conditions $\theta(0) = \theta_0$, $\theta'(0) = 0$ gives $\theta(t) = \theta_0 \cos\sqrt{g/l}\,t$. The period of oscillations predicted by this model is given by the familiar formula $T = 2\pi/\sqrt{g/l} = 2\pi\sqrt{l/g}$. The interesting thing about this formula for $T$ is that it does not depend on the magnitude of the initial displacement $\theta_0$. In other words, the linear model predicts that the time it would take the pendulum to swing from an initial displacement of, say, $\theta_0 = \pi/2\,(= 90°)$ to $-\pi/2$ and back again would be exactly the same as the time it would take to cycle from, say, $\theta_0 = \pi/360\,(= 0.5°)$ to $-\pi/360$. This is intuitively unreasonable; the actual period must depend on $\theta_0$.

If we assume that $g = 32$ ft/s$^2$ and $l = 32$ ft, then the period of oscillation of the linear model is $T = 2\pi$ s. Let us compare this last number with the period predicted by the nonlinear model when $\theta_0 = \pi/4$. Using a numerical solver that is capable of generating hard data, approximate the solution of

$$\frac{d^2\theta}{dt^2} + \sin\theta = 0, \quad \theta(0) = \frac{\pi}{4}, \quad \theta'(0) = 0$$

on the interval $0 \le t \le 2$. As in Problem 25, if $t_1$ denotes the first time the pendulum reaches the position $OP$ in Figure 5.3.3, then the period of the nonlinear pendulum is $4t_1$. Here is another way of solving the equation $\theta(t) = 0$. Experiment with small step sizes and advance the time, starting at $t = 0$ and ending at $t = 2$. From your hard data observe the time $t_1$ when $\theta(t)$ changes, for the first time, from positive to negative. Use the value $t_1$ to determine the true value of the period of the nonlinear pendulum. Compute the percentage relative error in the period estimated by $T = 2\pi$.

# Chapter 5 **In Review**

*Answers to selected odd-numbered problems begin on page ANS-8.*

Answer Problems 1–8 without referring back to the text. Fill in the blank or answer true/false.

1. If a mass weighing 10 pounds stretches a spring 2.5 feet, a mass weighing 32 pounds will stretch it _____ feet.

2. The period of simple harmonic motion of mass weighing 8 pounds attached to a spring whose constant is 6.25 lb/ft is _____ seconds.

3. The differential equation of a spring/mass system is $x'' + 16x = 0$. If the mass is initially released from a point 1 meter above the equilibrium position with a downward velocity of 3 m/s, the amplitude of vibrations is _____ meters.

4. Pure resonance cannot take place in the presence of a damping force. _____

5. In the presence of a damping force, the displacements of a mass on a spring will always approach zero as $t \to \infty$. _____

6. A mass on a spring whose motion is critically damped can possibly pass through the equilibrium position twice. _____

7. At critical damping any increase in damping will result in an _____ system.

8. If simple harmonic motion is described by $x = (\sqrt{2}/2) \sin(2t + \phi)$, the phase angle $\phi$ is _____ when the initial conditions are $x(0) = -\frac{1}{2}$ and $x'(0) = 1$.

In Problems 9 and 10 the eigenvalues and eigenfunctions of the boundary-value problem $y'' + \lambda y = 0$, $y'(0) = 0$, $y'(\pi) = 0$ are $\lambda_n = n^2$, $n = 0, 1, 2,\ldots$, and $y = \cos nx$, respectively. Fill in the blanks.

9. A solution of the BVP when $\lambda = 8$ is $y =$ _____ because _____.

10. A solution of the BVP when $\lambda = 36$ is $y =$ _____ because _____.

11. A free undamped spring/mass system oscillates with a period of 3 seconds. When 8 pounds are removed from the spring, the system has a period of 2 seconds. What was the weight of the original mass on the spring?

12. A mass weighing 12 pounds stretches a spring 2 feet. The mass is initially released from a point 1 foot below the equilibrium position with an upward velocity of 4 ft/s.

    (a) Find the equation of motion.

    (b) What are the amplitude, period, and frequency of the simple harmonic motion?

    (c) At what times does the mass return to the point 1 foot below the equilibrium position?

    (d) At what times does the mass pass through the equilibrium position moving upward? Moving downward?

    (e) What is the velocity of the mass at $t = 3\pi/16$ s?

    (f) At what times is the velocity zero?

13. A force of 2 pounds stretches a spring 1 foot. With one end held fixed, a mass weighing 8 pounds is attached to the other end. The system lies on a table that imparts a frictional force numerically equal to $\frac{3}{2}$ times the instantaneous velocity. Initially, the mass is displaced 4 inches above the equilibrium position and released from rest. Find the equation of motion if the motion takes place along a horizontal straight line that is taken as the $x$-axis.

14. A mass weighing 32 pounds stretches a spring 6 inches. The mass moves through a medium offering a damping force that is numerically equal to $\beta$ times the instantaneous velocity. Determine the values of $\beta > 0$ for which the spring/mass system will exhibit oscillatory motion.

15. A spring with constant $k = 2$ is suspended in a liquid that offers a damping force numerically equal to 4 times the instantaneous velocity. If a mass $m$ is suspended from the spring, determine the values of $m$ for which the subsequent free motion is nonoscillatory.

16. The vertical motion of a mass attached to a spring is described by the initial-value problem

$$\tfrac{1}{4}x'' + x' + x = 0, \quad x(0) = 4, \quad x'(0) = 2.$$

Determine the maximum vertical displacement of the mass.

17. A mass weighing 4 pounds stretches a spring 18 inches. A periodic force equal to $f(t) = \cos \gamma t + \sin \gamma t$ is impressed on the system starting at $t = 0$. In the absence of a damping force, for what value of $\gamma$ will the system be in a state of pure resonance?

18. Find a particular solution for $x'' + 2\lambda x' + \omega^2 x = A$, where $A$ is a constant force.

19. A mass weighing 4 pounds is suspended from a spring whose constant is 3 lb/ft. The entire system is immersed in a fluid offering a damping force numerically equal to the instantaneous velocity. Beginning at $t = 0$, an external force equal to $f(t) = e^{-t}$ is impressed on the system. Determine the equation of motion if the mass is initially released from rest at a point 2 feet below the equilibrium position.

20. (a) A mass weighing $W$ pounds stretches a spring $\frac{1}{2}$ foot and stretches a different spring $\frac{1}{4}$ foot. The two springs are attached in series and the mass is then attached to the double spring as shown in Figure 5.1.6. Assume that the motion is free and that there is no damping force present. Determine the equation of motion if the mass is initially released at a point 1 foot below the equilibrium position with a downward velocity of $\frac{2}{3}$ ft/s.

    (b) Show that the maximum speed of the mass is $\frac{2}{3}\sqrt{3g + 1}$.

21. A series circuit contains an inductance of $L = 1$ h, a capacitance of $C = 10^{-4}$ f, and an electromotive force of $E(t) = 100 \sin 50t$ V. Initially, the charge $q$ and current $i$ are zero.

    (a) Determine the charge $q(t)$.

    (b) Determine the current $i(t)$.

    (c) Find the times for which the charge on the capacitor is zero.

**22. (a)** Show that the current $i(t)$ in an *LRC*-series circuit satisfies

$$L\frac{d^2i}{dt^2} + R\frac{di}{dt} + \frac{1}{C}i = E'(t),$$

where $E'(t)$ denotes the derivative of $E(t)$.

**(b)** Two initial conditions $i(0)$ and $i'(0)$ can be specified for the DE in part (a). If $i(0) = i_0$ and $q(0) = q_0$, what is $i'(0)$?

**23.** Consider the boundary-value problem

$$y'' + \lambda y = 0, \quad y(0) = y(2\pi), \quad y'(0) = y'(2\pi).$$

Show that except for the case $\lambda = 0$, there are two independent eigenfunctions corresponding to each eigenvalue.

**24.** A bead is constrained to slide along a frictionless rod of length $L$. The rod is rotating in a vertical plane with a constant angular velocity $\omega$ about a pivot $P$ fixed at the midpoint of the rod, but the design of the pivot allows the bead to move along the entire length of the rod. Let $r(t)$ denote the position of the bead relative to this rotating coordinate system as shown in Figure 5.R.1. To apply Newton's second law of motion to this rotating frame of reference, it is necessary to use the fact that the net force acting on the bead is the sum of the real forces (in this case, the force due to gravity) and the inertial forces (coriolis, transverse, and centripetal). The resulting differential equation for $r$ is:

$$m\frac{d^2r}{dt^2} = m\omega^2 r - mg\sin\omega t.$$

**(a)** Solve the foregoing DE subject to the initial conditions $r(0) = r_0, r'(0) = v_0$.

**(b)** Determine the initial conditions for which the bead exhibits simple harmonic motion. What is the minimum length $L$ of the rod for which it can accommodate simple harmonic motion of the bead?

**(c)** For initial conditions other than those obtained in part (b), the bead must eventually fly off the rod. Explain using the solution $r(t)$ in part (a).

**(d)** Suppose $\omega = 1$ rad/s. Use a graphing utility to graph the solution $r(t)$ for the initial conditions $r(0) = 0, r'(0) = v_0$, where $v_0$ is 0, 10, 15, 16, 16.1, and 17.

**(e)** Suppose the length of the rod is $L = 40$ ft. For each pair of initial conditions in part (d), use a root-finding application to find the total time that the bead stays on the rod.

**FIGURE 5.R.1** Rotating rod in Problem 24

**25.** Suppose a mass $m$ lying on a flat dry frictionless surface is attached to the free end of a spring whose constant is $k$. In Figure 5.R.2(a) the mass is shown at the equilibrium position $x = 0$, that is, the spring is neither stretched nor compressed. As shown in Figure 5.R.2(b), the displacement $x(t)$ of the mass to the right of the equilibrium position is positive and negative to the left. Determine a differential equation for the displacement $x(t)$ of the freely sliding mass. Discuss the difference between the derivation of this DE and the analysis leading to (1) of Section 5.1.

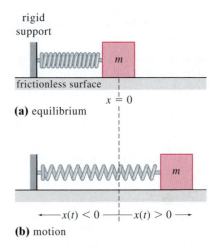

**(a)** equilibrium

**(b)** motion

**FIGURE 5.R.2** Sliding spring/mass system in Problem 25

**26.** Suppose the mass $m$ on the flat, dry, frictionless surface in Problem 25 is attached to two springs as shown Figure 5.R.3. If the spring constants are $k_1$ and $k_2$, determine a differential equation for the displacement $x(t)$ of the freely sliding mass.

**FIGURE 5.R.3** Double spring system in Problem 26

**27.** Suppose the mass $m$ in the spring/mass system in Problem 25 slides over a dry surface whose coefficient of sliding friction is $\mu > 0$. If the retarding force of kinetic friction has the constant magnitude $f_k = \mu mg$, where $mg$ is the weight of the mass, and acts opposite to the direction of motion, then it is known as **coulomb friction**. By using the **signum function**

$$\text{sgn}(x') = \begin{cases} -1, & x' < 0 \text{ (motion to left)} \\ 1, & x' > 0 \text{ (motion to right)} \end{cases}$$

determine a piecewise-defined differential equation for the displacement $x(t)$ of the damped sliding mass.

**28.** For simplicity, let us assume in Problem 27 that $m = 1, k = 1$, and $f_k = 1$.

**(a)** Find the displacement $x(t)$ of the mass if it is released from rest from a point 5.5 units to the right of the equilibrium position, that is, the initial conditions are $x(0) = 5.5$, $x'(0) = 0$. When released, intuitively the motion of the mass will be to the left. Give a time interval $[0, t_1]$ over which this solution is defined. Where is the mass at time $t_1$?

**(b)** For $t > t_1$ assume that the motion is now to the right. Using initial conditions at $t_1$, find $x(t)$ and give a time interval $[t_1, t_2]$ over which this solution is defined. Where is the mass at time $t_2$?

**(c)** For $t > t_2$ assume that the motion is now to the left. Using initial conditions at $t_2$, find $x(t)$ and give a time interval $[t_2, t_3]$ over which this solution is defined. Where is the mass at time $t_3$?

**(d)** Using initial conditions at $t_3$, show that the model predicts that there is no furthere motion for $t > t_3$.

**(e)** Graph the displacement $x(t)$ on the interval $[0, t_3]$.

29. Use a Maclaurin series to show that a power series solution of the initial-value problem

$$\frac{d^2\theta}{dt^2} + \frac{g}{l} \sin \theta = 0, \quad \theta(0) = \frac{\pi}{6}, \quad \theta'(0) = 0$$

is given by

$$\theta(t) = \frac{\pi}{6} - \frac{g}{4l} t^2 + \frac{\sqrt{3}g^2}{96l^2} t^4 + \cdots.$$

[*Hint*: See Example 3 in Section 4.10.]

30. **Spring Pendulum**   The rotational form of Newton's second law of motion is:

*The time rate of change of angular momentum about a point is equal to the moment of the resultant force (torque).*

In the absence of damping or other external forces, an analogue of (14) in Section 5.3 for the pendulum shown in Figure 5.3.3 is then

$$\frac{d}{dt}\left(ml^2 \frac{d\theta}{dt}\right) = -mgl \sin \theta. \tag{1}$$

**(a)** When $m$ and $l$ are constant show that (1) reduces to (6) of Section 5.3.

**(b)** Now suppose the rod in Figure 5.3.3 is replaced with a spring of negligible mass. When a mass $m$ is attached to its free end the spring hangs in the vertical equilibrium position shown in Figure 5.R.4 and has length $l_0$. When

**FIGURE 5.R.4** Spring pendulum in Problem 30

the spring pendulum is set in motion we assume that the motion takes place in a vertical plane and the spring is stiff enough not to bend. For $t > 0$ the length of the spring is then $l(t) = l_0 + x(t)$, where $x$ is the displacement from the equilibrium position. Find the differential equation for the displacement angle $\theta(t)$ defined by (1).

31. Suppose a pendulum is formed by attaching a mass $m$ to the end of a string of negligible mass and length $l$. At $t = 0$ the pendulum is released from rest at a small displacement angle $\theta_0 > 0$ to the right of the vertical equilibrium position $OP$. See Figure 5.R.5. At time $t_1 > 0$ the string hits a nail at a point $N$ on $OP$ a distance $\frac{3}{4}l$ from $O$, but the mass continues to the left as shown in the figure.

**(a)** Construct and solve a linear initial-value problem for the displacement angle $\theta_1(t)$ shown in the figure. Find the interval $[0, t_1]$ on which $\theta_1(t)$ is defined.

**(b)** Construct and solve a linear initial-value problem for the displacement angle $\theta_2(t)$ shown in the figure. Find the interval $[t_1, t_2]$ on which $\theta_2(t)$ is defined, where $t_2$ is the time that $m$ returns to the vertical line $NP$.

**FIGURE 5.R.5** Pendulum in Problem 31

32. **Galloping Gertie**   Bridges are good examples of vibrating mechanical systems that are constantly subjected to external forces, from cars driving on them, water pushing against their foundation, and wind blowing across their superstructure. On November 7, 1940, only four months after its grand opening, the Tacoma Narrows Suspension Bridge at Puget Sound in the state of Washington collapsed during a windstorm. See Figure 5.R.6. The crash came as no surprise since "Galloping Gertie," as the bridge was called by local residents, was famous for a vertical undulating motion of its roadway which gave many motorists a very exciting crossing. For many years it was conjectured that the poorly designed superstructure of the bridge caused the wind blowing across it to swirl in a periodic manner and that when the frequency of this force approached the natural frequency of the bridge, large upheavals of the lightweight roadway resulted. In other words, it was thought the bridge was a victim of mechanical resonance. But as we have seen on page 207, resonance is a linear phenomenon which can occur only in the complete absence of damping. In recent years the resonance theory has been replaced with mathematical models that can describe large oscillations even in the presence of damping. In his project, *The Collapse of the Tacoma Narrows Suspension Bridge*, that appeared in the last edition of

this text, Gilbert N. Lewis examines simple piecewise-defined models describing the driven oscillations of a mass (a portion of the roadway) attached to a spring (a vertical support cable) for which the amplitudes of oscillation increase over time. In this problem you are guided through the solution of one of the models discussed in that project.

The differential equation with a piecewise-defined restoring force,

$$\frac{d^2x}{dt^2} + F(x) = \sin 4t, \quad F(x) = \begin{cases} 4x, & x \geq 0 \\ x, & x < 0 \end{cases}$$

is a model for the displacement $x(t)$ of a unit mass in a driven spring/mass system. As in Section 5.1 we assume that the motion takes place along a vertical line, the equilibrium position is $x = 0$, and the positive direction is downward. The restoring force acts opposite to the direction of motion: a restoring force $4x$ when the mass is below ($x > 0$) the equilibrium position and a restoring force $x$ when the mass is above ($x < 0$) the equilibrium position.

**(a)** Solve the initial-value problem

$$\frac{d^2x}{dt^2} + 4x = \sin 4t, \quad x(0) = 0, \quad x'(0) = v_0 > 0. \qquad (2)$$

The initial conditions indicate that the mass is released from the equilibrium position with a downward velocity. Use the solution to determine the first time $t_1 > 0$ when $x(t) = 0$, that is, the first time that the mass returns to the equilibrium position after release. The solution of (2) is defined on the interval $[0, t_1]$. [*Hint*: The double-angle formula $\sin 4t = 2 \sin 2t \cos 2t$ will be helpful.]

**(b)** For a time interval on which $t > t_1$ the mass is above the equilibrium position and so we must now solve the new differential equation

$$\frac{d^2x}{dt^2} + x = \sin 4t. \qquad (3)$$

One initial condition is $x(t_1) = 0$. Find $x'(t_1)$ using the solution of (2) in part (a). Find a solution of equation (3) subject

to these new initial conditions. Use the solution to determine the second time $t_2 > t_1$ when $x(t) = 0$. The solution of (3) is defined on the interval $[t_1, t_2]$. [*Hint*: Use the double-angle formula for the sine function twice.]

**(c)** Construct and solve another initial-value problem to find $x(t)$ defined on the interval $[t_2, t_3]$, where $t_3 > t_2$ is the third time when $x(t) = 0$.

**(d)** Construct and solve another initial-value problem to find $x(t)$ defined on the interval $[t_3, t_4]$, where $t_4 > t_3$ is the fourth time when $x(t) = 0$.

**(e)** Because of the assumption that $v_0 > 0$ one *down-up* cycle of the mass is completed on the intervals $[0, t_2], [t_2, t_4], [t_4, t_6]$, and so on. Explain why the amplitudes of oscillation of the mass must increase over time. [*Hint*: Examine the velocity of the mass at the beginning of each cycle.]

**(f)** Assume in (2) that $v_0 = 0.01$. Use the four solutions on the intervals in parts (a), (b), (c), and (d) to construct a continuous piecewise-defined function $x(t)$ defined on the interval $[0, t_4]$. Use a graphing utility to obtain a graph of $x(t)$ on $[0, t_4]$.

**FIGURE 5.R.6** Collapse of the Tacoma Narrows Suspension Bridge

# 6

# Series Solutions of Linear Equations

Todd Dalton/Shutterstock.com

So far in our study of linear differential equations of order two (or higher) we have primarily solved equations with constant coefficients. The Cauchy-Euler equation in Section 4.7 was the only exception. In applications, linear high-order equations with variable coefficients are just as important, if not more than, DEs with constant coefficients. As pointed out in Section 4.7 even a simple linear second-order equation such as $y'' - xy = 0$ does not possess solutions that are elementary functions. But this is not to say we can't find two linearly independent solutions of $y'' - xy = 0$; we can. In Sections 6.2 and 6.4 we will see that solutions of this equation are functions that are defined by infinite series.

## 6.1 Review of Power Series

**INTRODUCTION** In Section 4.3 we saw that solving a homogeneous linear DE with constant coefficients was essentially a problem in algebra. By finding the roots of the auxiliary equation, we could write a general solution of the DE as a linear combination of the elementary functions $e^{\alpha x}$, $x^k e^{\alpha x}$, $x^k e^{\alpha x} \cos \beta x$, and $x^k e^{\alpha x} \sin \beta x$. But as was pointed out in the introduction to Section 4.7, most linear higher-order DEs with variable coefficients cannot be solved in terms of elementary functions. A usual course of action for equations of this sort is to assume a solution in the form of an infinite series and proceed in a manner similar to the method of undetermined coefficients (Section 4.4). In Section 6.2 we consider linear second-order DEs with variable coefficients that possess solutions in the form of a power series, and so it is appropriate that we begin this chapter with a review of that topic.

**POWER SERIES** Recall from calculus that **power series** in $x - a$ is an infinite series of the form

The index of summation need not start at $n = 0$.

$$\sum_{n=0}^{\infty} c_n(x - a)^n = c_0 + c_1(x - a) + c_2(x - a)^2 + \cdots.$$

Such a series is also said to be a **power series centered at $a$.** For example, the power series $\sum_{n=0}^{\infty}(x + 1)^n$ is centered at $a = -1$. In the next section we will be concerned principally with power series in $x$, in other words, power series that are centered at $a = 0$. For example,

$$\sum_{n=0}^{\infty} 2^n x^n = 1 + 2x + 4x^2 + \cdots$$

is a power series in $x$.

**IMPORTANT FACTS** The following bulleted list summarizes some important facts about power series $\sum_{n=0}^{\infty} c_n(x - a)^n$.

- **Convergence**  A power series is **convergent** at a specified value of $x$ if its sequence of partial sums $\{S_N(x)\}$ converges, that is, $\lim_{N \to \infty} S_N(x) = \lim_{N \to \infty} \sum_{n=0}^{N} c_n(x - a)^n$ exists. If the limit does not exist at $x$, then the series is said to be **divergent.**
- **Interval of Convergence**  Every power series has an **interval of convergence.** The interval of convergence is the set of *all* real numbers $x$ for which the series converges. The center of the interval of convergence is the center $a$ of the series.
- **Radius of Convergence**  The radius $R$ of the interval of convergence of a power series is called its **radius of convergence.** If $R > 0$, then a power series converges for $|x - a| < R$ and diverges for $|x - a| > R$. If the series converges only at its center $a$, then $R = 0$. If the series converges for all $x$, then we write $R = \infty$. Recall, the absolute-value inequality $|x - a| < R$ is equivalent to the simultaneous inequality $a - R < x < a + R$. A power series may or may not converge at the endpoints $a - R$ and $a + R$ of this interval.
- **Absolute Convergence**  In the interior of its interval of convergence a power series **converges absolutely.** In other words, if $x$ is in the interval of convergence and is not an endpoint of the interval, then the series of absolute values $\sum_{n=0}^{\infty} |c_n(x - a)^n|$ converges. See Figure 6.1.1.

**FIGURE 6.1.1** Absolute convergence within the interval of convergence and divergence outside of this interval

- **Ratio Test**　Convergence of power series can often be determined by the **ratio test.** Suppose $c_n \neq 0$ for all $n$ in $\sum_{n=0}^{\infty} c_n(x - a)^n$, and that

$$\lim_{n\to\infty} \left| \frac{c_{n+1}(x - a)^{n+1}}{c_n(x - a)^n} \right| = |x - a| \lim_{n\to\infty} \left| \frac{c_{n+1}}{c_n} \right| = L.$$

If $L < 1$, the series converges absolutely; if $L > 1$ the series diverges; and if $L = 1$ the test is inconclusive. The ratio test is always inconclusive at an endpoint $a \pm R$.

---

**EXAMPLE 1**　**Interval of Convergence**

Find the interval and radius of convergence for $\displaystyle\sum_{n=1}^{\infty} \frac{(x - 3)^n}{2^n n}$.

**SOLUTION**　The ratio test gives

$$\lim_{n\to\infty} \left| \frac{\dfrac{(x - 3)^{n+1}}{2^{n+1}(n + 1)}}{\dfrac{(x - 3)^n}{2^n n}} \right| = |x - 3| \lim_{n\to\infty} \frac{n + 1}{2n} = \frac{1}{2}|x - 3|.$$

The series converges absolutely for $\frac{1}{2}|x - 3| < 1$ or $|x - 3| < 2$ or $1 < x < 5$. This last inequality defines the *open* interval of convergence. The series diverges for $|x - 3| > 2$, that is, for $x > 5$ or $x < 1$. At the left endpoint $x = 1$ of the open interval of convergence, the series of constants $\sum_{n=1}^{\infty} ((-1)^n/n)$ is convergent by the alternating series test. At the right endpoint $x = 5$, the series $\sum_{n=1}^{\infty} (1/n)$ is the divergent harmonic series. The interval of convergence of the series is $[1, 5)$, and the radius of convergence is $R = 2$. ∎

- **A Power Series Defines a Function**　A power series defines a function, that is, $f(x) = \sum_{n=0}^{\infty} c_n(x - a)^n$ whose domain is the interval of convergence of the series. If the radius of convergence is $R > 0$ or $R = \infty$, then $f$ is continuous, differentiable, and integrable on the intervals $(a - R, a + R)$ or $(-\infty, \infty)$, respectively. Moreover, $f'(x)$ and $\int f(x)\,dx$ can be found by term-by-term differentiation and integration. Convergence at an endpoint may be either lost by differentiation or gained through integration. If

$$y = \sum_{n=1}^{\infty} c_n x^n = c_0 + c_1 x + c_2 x^2 + c_3 x^3 + \cdots$$

is a power series in $x$, then the first two derivatives are $y' = \sum_{n=0}^{\infty} n x^{n-1}$ and $y'' = \sum_{n=0}^{\infty} n(n - 1)x^{n-2}$. Notice that the first term in the first derivative and the first two terms in the second derivative are zero. We omit these zero terms and write

$$y' = \sum_{n=1}^{\infty} c_n n x^{n-1} = c_1 + 2c_2 x + 3c_3 x^2 + 4c_4 x^3 + \cdots$$

$$y'' = \sum_{n=2}^{\infty} c_n n(n - 1)x^{n-2} = 2c_2 + 6c_3 x + 12c_4 x^2 + \cdots. \qquad (1)$$

Be sure you understand the two results given in (1); especially note where the index of summation starts in each series. These results are important and will be used in all examples in the next section.

- **Identity Property**　If $\sum_{n=0}^{\infty} c_n(x - a)^n = 0$, $R > 0$, for all numbers $x$ in some open interval, then $c_n = 0$ for all $n$.
- **Analytic at a Point**　A function $f$ is said to be **analytic at a point** $a$ if it can be represented by a power series in $x - a$ with either a positive

or an infinite radius of convergence. In calculus it is seen that infinitely differentiable functions such as $e^x$, $\sin x$, $\cos x$, $e^x \ln(1+x)$, and so on, can be represented by Taylor series

$$\sum_{n=0}^{\infty} \frac{f^{(n)}(a)}{n!}(x-a)^n = f(a) + \frac{f'(a)}{1!}(x-a) + \frac{f''(a)}{1!}(x-a)^2 + \cdots$$

or by a Maclaurin series

$$\sum_{n=0}^{\infty} \frac{f^{(n)}(0)}{n!}x^n = f(0) + \frac{f'(0)}{1!}x + \frac{f''(0)}{1!}x^2 + \cdots.$$

You might remember some of the following Maclaurin series representations.

| Maclaurin Series | Interval of Convergence | |
|---|---|---|
| $e^x = 1 + \dfrac{x}{1!} + \dfrac{x^2}{2!} + \dfrac{x^3}{3!} + \cdots = \displaystyle\sum_{n=0}^{\infty} \frac{1}{n!}x^n$ | $(-\infty, \infty)$ | |
| $\cos x = 1 - \dfrac{x^2}{2!} + \dfrac{x^4}{4!} - \dfrac{x^6}{6!} + \cdots = \displaystyle\sum_{n=0}^{\infty} \frac{(-1)^n}{(2n)!}x^{2n}$ | $(-\infty, \infty)$ | |
| $\sin x = x - \dfrac{x^3}{3!} + \dfrac{x^5}{5!} - \dfrac{x^7}{7!} + \cdots = \displaystyle\sum_{n=0}^{\infty} \frac{(-1)^n}{(2n+1)!}x^{2n+1}$ | $(-\infty, \infty)$ | |
| $\tan^{-1} x = x - \dfrac{x^3}{3} + \dfrac{x^5}{5} - \dfrac{x^7}{7} + \cdots = \displaystyle\sum_{n=0}^{\infty} \frac{(-1)^n}{2n+1}x^{2n+1}$ | $[-1, 1]$ | (2) |
| $\cosh x = 1 + \dfrac{x^2}{2!} + \dfrac{x^4}{4!} + \dfrac{x^6}{6!} + \cdots = \displaystyle\sum_{n=0}^{\infty} \frac{1}{(2n)!}x^{2n}$ | $(-\infty, \infty)$ | |
| $\sinh x = x + \dfrac{x^3}{3!} + \dfrac{x^5}{5!} + \dfrac{x^7}{7!} + \cdots = \displaystyle\sum_{n=0}^{\infty} \frac{1}{(2n+1)!}x^{2n+1}$ | $(-\infty, \infty)$ | |
| $\ln(1+x) = x - \dfrac{x^2}{2} + \dfrac{x^3}{3} - \dfrac{x^4}{4} + \cdots = \displaystyle\sum_{n=1}^{\infty} \frac{(-1)^{n+1}}{n}x^n$ | $(-1, 1]$ | |
| $\dfrac{1}{1-x} = 1 + x + x^2 + x^3 + \cdots = \displaystyle\sum_{n=0}^{\infty} x^n$ | $(-1, 1)$ | |

These results can be used to obtain power series representations of other functions. For example, if we wish to find the Maclaurin series representation of, say, $e^{x^2}$ we need only replace $x$ in the Maclaurin series for $e^x$:

$$e^{x^2} = 1 + \frac{x^2}{1!} + \frac{x^4}{2!} + \frac{x^6}{3!} + \cdots = \sum_{n=0}^{\infty} \frac{1}{n!}x^{2n}.$$

Similarly, to obtain a Taylor series representation of $\ln x$ centered at $a = 1$ we replace $x$ by $x - 1$ in the Maclaurin series for $\ln(1 + x)$:

$$\ln x = \ln(1 + (x-1)) = (x-1) - \frac{(x-1)^2}{2} + \frac{(x-1)^3}{3} - \frac{(x-1)^4}{4} + \cdots = \sum_{n=1}^{\infty} \frac{(-1)^{n+1}}{n}(x-1)^n.$$

You can also verify that the interval of convergence is (0, 2] by using the ratio test.

The interval of convergence for the power series representation of $e^{x^2}$ is the same as that of $e^x$, that is, $(-\infty, \infty)$. But the interval of convergence of the Taylor series of $\ln x$ is now (0, 2]; this interval is $(-1, 1]$ shifted 1 unit to the right.

- **Arithmetic of Power Series** Power series can be combined through the operations of addition, multiplication, and division. The procedures for powers series are similar to the way in which two polynomials are added, multiplied, and divided—that is, we add coefficients of like powers of $x$, use the distributive law and collect like terms, and perform long division.

**EXAMPLE 2** **Multiplication of Power Series**

Find a power series representation of $e^x \sin x$.

**SOLUTION** We use the power series for $e^x$ and $\sin x$:

$$e^x \sin x = \left(1 + x + \frac{x^2}{2} + \frac{x^3}{6} + \frac{x^4}{24} + \cdots\right)\left(x - \frac{x^3}{6} + \frac{x^5}{120} - \frac{x^7}{5040} + \cdots\right)$$

$$= (1)x + (1)x^2 + \left(-\frac{1}{6} + \frac{1}{2}\right)x^3 + \left(-\frac{1}{6} + \frac{1}{6}\right)x^4 + \left(\frac{1}{120} - \frac{1}{12} + \frac{1}{24}\right)x^5 + \cdots$$

$$= x + x^2 + \frac{x^3}{3} - \frac{x^5}{30} + \cdots.$$

Since the power series of $e^x$ and $\sin x$ both converge on $(-\infty, \infty)$, the product series converges on the same interval. Problems involving multiplication or division of power series can be done with minimal fuss using a computer algebra system. ■

**SHIFTING THE SUMMATION INDEX**   For the three remaining sections of this chapter, it is crucial that you become adept at simplifying the sum of two or more power series, each series expressed in summation notation, to an expression with a single $\Sigma$. As the next example illustrates, combining two or more summations as a single summation often requires a reindexing, that is, a shift in the index of summation.

**EXAMPLE 3** **Addition of Power Series**

Write

$$\sum_{n=2}^{\infty} n(n-1)c_n x^{n-2} - \sum_{n=0}^{\infty} c_n x^{n+1}$$

as one power series.

**SOLUTION** In order to add the two series given in summation notation, it is necessary that both indices of summation start with the same number and that the powers of $x$ in each series be "in phase," in other words, if one series starts with a multiple of, say, $x$ to the first power, then we want the other series to start with the same power. Note that in the given problem, the first series starts with $x^0$ whereas the second series starts with $x^1$. By writing the first term of the first series outside of the summation notation,

series starts with $x$ for $n = 3$ ↓     series starts with $x$ for $n = 0$ ↓

$$\sum_{n=2}^{\infty} n(n-1)c_n x^{n-2} - \sum_{n=0}^{\infty} c_n x^{n+1} = 2 \cdot 1 c_2 x^0 + \sum_{n=3}^{\infty} n(n-1)c_n x^{n-2} - \sum_{n=0}^{\infty} c_n x^{n+1} \quad (3)$$

we see that both series on the right side start with the same power of $x$, namely, $x^1$. Now to get the same summation index we are inspired by the exponents of $x$; we let $k = n - 2$ in the first series and at the same time let $k = n + 1$ in the second series. For $n = 3$ in $k = n - 2$ we get $k = 1$, and for $n = 0$ in $k = n + 1$ we get $k = 1$, and so the right-hand side of (3) becomes

same

$$2c_2 + \sum_{k=1}^{\infty} (k+2)(k+1)c_{k+2} x^k - \sum_{k=1}^{\infty} c_{k-1} x^k. \quad (4)$$

same

Remember the summation index is a "dummy" variable; the fact that $k = n - 2$ in one case and $k = n + 1$ in the other should cause no confusion if you keep in mind that it is the *value* of the summation index that is important. In both cases $k$ takes on the same successive values $k = 1, 2, 3, \ldots$ when $n$ takes on the values $n = 2, 3, 4, \ldots$ for $k = n - 1$ and $n = 0, 1, 2, \ldots$ for $k = n + 1$. We are now in a position to add the series in (4) term-by-term:

$$\sum_{n=2}^{\infty} n(n-1)c_n x^{n-2} - \sum_{n=0}^{\infty} c_n x^{n+1} = 2c_2 + \sum_{k=1}^{\infty} [(k+2)(k+1)c_{k+2} - c_{k-1}]x^k. \quad (5) \quad \blacksquare$$

If you are not totally convinced of the result in (5), then write out a few terms on both sides of the equality.

**A PREVIEW**    The point of this section is to remind you of the salient facts about power series so that you are comfortable using power series in the next section to find solutions of linear second-order DEs. In the last example in this section we tie up many of the concepts just discussed; it also gives a preview of the method that will be used in Section 6.2. We purposely keep the example simple by solving a linear first-order equation. Also suspend, for the sake of illustration, the fact that you already know how to solve the given equation by the integrating-factor method in Section 2.3.

**EXAMPLE 4**    **A Power Series Solution**

Find a power series solution $y = \sum_{n=0}^{\infty} c_n x^n$ of the differential equation $y' + y = 0$.

**SOLUTION**    We break down the solution into a sequence of steps.

(*i*)    First calculate the derivative of the assumed solution:

$$y' = \sum_{n=1}^{\infty} c_n n x^{n-1} \quad \leftarrow \text{see the first line in (1)}$$

(*ii*)    Then substitute $y$ and $y'$ into the given DE:

$$y' + y = \sum_{n=1}^{\infty} c_n n x^{n-1} + \sum_{n=0}^{\infty} c_n x^n.$$

(*iii*)    Now shift the indices of summation. When the indices of summation have the same starting point and the powers of $x$ agree, combine the summations:

$$y' + y = \underbrace{\sum_{n=1}^{\infty} c_n n x^{n-1}}_{k = n-1} + \underbrace{\sum_{n=0}^{\infty} c_n x^n}_{k = n}$$

$$= \sum_{k=0}^{\infty} c_{k+1}(k+1)x^k + \sum_{k=0}^{\infty} c_k x^k$$

$$= \sum_{k=0}^{\infty} [c_{k+1}(k+1) + c_k]x^k.$$

(*iv*)    Because we want $y' + y = 0$ for all $x$ in some interval,

$$\sum_{k=0}^{\infty} [c_{k+1}(k+1) + c_k]x^k = 0$$

is an identity and so we must have $c_{k+1}(k+1) + c_k = 0$, or

$$c_{k+1} = -\frac{1}{k+1}c_k, \quad k = 0, 1, 2, \ldots.$$

(**v**) By letting $k$ take on successive integer values starting with $k = 0$, we find

$$c_1 = -\frac{1}{1}c_0 = -c_0$$

$$c_2 = -\frac{1}{2}c_1 = -\frac{1}{2}(-c_0) = \frac{1}{2}c_0$$

$$c_3 = -\frac{1}{3}c_2 = -\frac{1}{3}\left(\frac{1}{2}c_0\right) = -\frac{1}{3\cdot 2}c_0$$

$$c_4 = -\frac{1}{4}c_2 = -\frac{1}{4}\left(-\frac{1}{3\cdot 2}c_0\right) = \frac{1}{4\cdot 3\cdot 2}c_0$$

and so on, where $c_0$ is arbitrary.

(**vi**) Using the original assumed solution and the results in part (**v**) we obtain a formal power series solution

$$y = c_0 + c_1 x + c_2 x^2 + c_3 x^3 + c_4 x^4 + \cdots$$

$$= c_0 - c_0 x + \frac{1}{2}c_0 x^2 - c_0\frac{1}{3\cdot 2}x^3 + c_0\frac{1}{4\cdot 3\cdot 2}x^4 - \cdots$$

$$= c_0\left[1 - x + \frac{1}{2}x^2 - \frac{1}{3\cdot 2}x^3 + \frac{1}{4\cdot 3\cdot 2}x^4 - \cdots\right].$$

It should be fairly obvious that the pattern of the coefficients in part (**v**) is $c_k = c_0(-1)^k/k!$, $k = 0, 1, 2, \ldots$ so that in summation notation we can write

$$y = c_0\sum_{k=0}^{\infty}\frac{(-1)^k}{k!}x^k. \tag{8} \blacksquare$$

If desired we could switch back to  $n$ as the index of summation.

From the first power series representation in (2) the solution in (8) is recognized as $y = c_0e^{-x}$. Had you used the method of Section 2.3, you would have found that $y = ce^{-x}$ is a solution of $y' + y = 0$ on the interval $(-\infty, \infty)$. This interval is also the interval of convergence of the power series in (8).

## EXERCISES 6.1

Answers to selected odd-numbered problems begin on page ANS-9.

In Problems 1–10 find the interval and radius of convergence for the given power series.

**1.** $\displaystyle\sum_{n=1}^{\infty}\frac{(-1)^n}{n}x^n$

**2.** $\displaystyle\sum_{n=1}^{\infty}\frac{1}{n^2}x^n$

**3.** $\displaystyle\sum_{n=1}^{\infty}\frac{2^n}{n}x^n$

**4.** $\displaystyle\sum_{n=0}^{\infty}\frac{5^n}{n!}x^n$

**5.** $\displaystyle\sum_{k=1}^{\infty}\frac{(-1)^k}{10^k}(x-5)^k$

**6.** $\displaystyle\sum_{k=0}^{\infty}k!(x-1)^k$

**7.** $\displaystyle\sum_{k=1}^{\infty}\frac{1}{k^2+k}(3x-1)^k$

**8.** $\displaystyle\sum_{k=0}^{\infty}3^{-k}(4x-5)^k$

**9.** $\displaystyle\sum_{k=1}^{\infty}\frac{2^{5k}}{5^{2k}}\left(\frac{x}{3}\right)^k$

**10.** $\displaystyle\sum_{n=0}^{\infty}\frac{(-1)^n}{9^n}x^{2n+1}$

In Problems 11–16 use an appropriate series in (2) to find the Maclaurin series of the given function. Write your answer in summation notation.

**11.** $e^{-x/2}$

**12.** $xe^{3x}$

**13.** $\dfrac{1}{2+x}$

**14.** $\dfrac{x}{1+x^2}$

**15.** $\ln(1-x)$

**16.** $\sin x^2$

In Problems 17 and 18 use an appropriate series in (2) to find the Taylor series of the given function centered at the indicated value of $a$. Write your answer in summation notation.

**17.** $\sin x$, $a = 2\pi$    [*Hint*: Use periodicity.]

**18.** $\ln x$; $a = 2$    [*Hint*: $x = 2[1 + (x - 2)/2]$]

In Problems 19 and 20 the given function is analytic at $a = 0$. Use appropriate series in (2) and multiplication to find the first four nonzero terms of the Maclaurin series of the given function.

**19.** $\sin x\cos x$

**20.** $e^{-x}\cos x$

In Problems 21 and 22 the given function is analytic at $a = 0$. Use appropriate series in (2) and long division to find the first four nonzero terms of the Maclaurin series of the given function.

**21.** $\sec x$

**22.** $\tan x$

In Problems 23 and 24 use a substitution to shift the summation index so that the general term of given power series involves $x^k$.

**23.** $\displaystyle\sum_{n=1}^{\infty} nc_n x^{n+2}$

**24.** $\displaystyle\sum_{n=3}^{\infty} (2n - 1)c_n x^{n-3}$

In Problems 25–30 proceed as in Example 3 to rewrite the given expression using a single power series whose general term involves $x^k$.

**25.** $\displaystyle\sum_{n=1}^{\infty} nc_n x^{n-1} - \sum_{n=0}^{\infty} c_n x^n$

**26.** $\displaystyle\sum_{n=1}^{\infty} nc_n x^{n-1} + 3\sum_{n=0}^{\infty} c_n x^{n+2}$

**27.** $\displaystyle\sum_{n=1}^{\infty} 2nc_n x^{n-1} + \sum_{n=0}^{\infty} 6c_n x^{n+1}$

**28.** $\displaystyle\sum_{n=2}^{\infty} n(n - 1)c_n x^{n-2} + \sum_{n=0}^{\infty} c_n x^{n+2}$

**29.** $\displaystyle\sum_{n=2}^{\infty} n(n - 1)c_n x^{n-2} - 2\sum_{n=1}^{\infty} nc_n x^n + \sum_{n=0}^{\infty} c_n x^n$

**30.** $\displaystyle\sum_{n=2}^{\infty} n(n - 1)c_n x^n + 2\sum_{n=2}^{\infty} n(n - 1)c_n x^{n-2} + 3\sum_{n=1}^{\infty} nc_n x^n$

In Problems 31–34 verify by direct substitution that the given power series is a solution of the indicated differential equation. [*Hint*: For a power $x^{2n+1}$ let $k = n + 1$.]

**31.** $\displaystyle y = \sum_{n=0}^{\infty} \frac{(-1)^n}{n!} x^{2n}, \quad y' + 2xy = 0$

**32.** $\displaystyle y = \sum_{n=0}^{\infty} (-1)^n x^{2n}, \quad (1 + x^2)y' + 2xy = 0$

**33.** $\displaystyle y = \sum_{n=1}^{\infty} \frac{(-1)^{n+1}}{n} x^n, \quad (x + 1)y'' + y' = 0$

**34.** $\displaystyle y = \sum_{n=0}^{\infty} \frac{(-1)^n}{2^{2n}(n!)^2} x^{2n}, \quad xy'' + y' + xy = 0$

In Problems 35–38 proceed as in Example 4 and find a power series solution $y = \displaystyle\sum_{n=0}^{\infty} c_n x^n$ of the given linear first-order differential equation.

**35.** $y' - 5y = 0$       **36.** $4y' + y = 0$

**37.** $y' = xy$         **38.** $(1 + x)y' + y = 0$

### Discussion Problems

**39.** In Problem 19, find an easier way than multiplying two power series to obtain the Maclaurin series representation of $\sin x \cos x$.

**40.** In Problem 21, what do you think is the interval of convergence for the Maclaurin series of $\sec x$?

---

# 6.2 Solutions About Ordinary Points

**INTRODUCTION** In the last example of the preceding section we illustrated how to obtain a power series solution of a linear first-order differential equation. In this section we turn to the more important problem of finding power series solutions of linear second-order equations. More to the point, we are going to find solutions of linear second-order equations in the form of power series whose center is a number $x_0$ that is an **ordinary point** of the DE. We begin with the definition of an ordinary point.

**A DEFINITION** If we divide the homogeneous linear second-order differential equation

$$a_2(x)y'' + a_1(x)y' + a_0(x)y = 0 \tag{1}$$

by the lead coefficient $a_2(x)$ we obtain the standard form

$$y'' + P(x)y' + Q(x)y = 0. \tag{2}$$

We have the following definition.

### DEFINITION 6.2.1  Ordinary and Singular Points

A point $x = x_0$ is said to be an **ordinary point** of the differential of the differential equation (1) if both coefficients $P(x)$ and $Q(x)$ in the standard form (2) are analytic at $x_0$. A point that is *not* an ordinary point of (1) is said to be a **singular point** of the DE.

### EXAMPLE 1 Ordinary Points

**(a)** A homogeneous linear second-order differential equation with constant coefficients, such as

$$y'' + y = 0 \quad \text{and} \quad y'' + 3y' + 2y = 0,$$

can have no singular points. In other words, every finite value[*] of $x$ is an ordinary point of such equations.

**(b)** Every finite value of $x$ is an ordinary point of the differential equation

$$y'' + e^x y' + (\sin x)y = 0.$$

Specifically $x = 0$ is an ordinary point of the DE, because we have already seen in (2) of Section 6.1 that both $e^x$ and $\sin x$ are analytic at this point. ■

The negation of the second sentence in Definition 6.2.1 stipulates that if *at least one* of the coefficient functions $P(x)$ and $Q(x)$ in (2) fails to be analytic at $x_0$, then $x_0$ is a singular point.

### EXAMPLE 2 Singular Points

**(a)** The differential equation

$$y'' + xy' + (\ln x)y = 0$$

is already in standard form. The coefficient functions are

$$P(x) = x \quad \text{and} \quad Q(x) = \ln x.$$

Now $P(x) = x$ is analytic at every real number, and $Q(x) = \ln x$ is analytic at every *positive* real number. However, since $Q(x) = \ln x$ is discontinuous at $x = 0$ it cannot be represented by a power series in $x$, that is, a power series centered at 0. We conclude that $x = 0$ is a singular point of the DE.

**(b)** By putting $xy'' + y' + xy = 0$ in the standard form

$$y'' + \frac{1}{x}y' + y = 0,$$

we see that $P(x) = 1/x$ fails to be analytic at $x = 0$. Hence $x = 0$ is a singular point of the equation. ■

**POLYNOMIAL COEFFICIENTS**    We will primarily be interested in the case when the coefficients $a_2(x)$, $a_1(x)$, and $a_0(x)$ in (1) are polynomial functions with no common factors. A polynomial function is analytic at any value of $x$, and a rational function is analytic except at points where its denominator is zero. Thus, in (2) both coefficients

$$P(x) = \frac{a_1(x)}{a_2(x)} \quad \text{and} \quad Q(x) = \frac{a_0(x)}{a_2(x)}$$

---

[*]For our purposes, ordinary points and singular points will always be finite points. It is possible for an ODE to have, say, a singular point at infinity.

are analytic except at those numbers for which $a_2(x) = 0$. It follows, then, that

*A number $x = x_0$ is an ordinary point of (1) if $a_2(x_0) \neq 0$, whereas $x = x_0$ is a singular point of (1) if $a_2(x_0) = 0$.*

---

**EXAMPLE 3**  **Ordinary and Singular Points**

**(a)** The only singular points of the differential equation

$$(x^2 - 1)y'' + 2xy' + 6y = 0$$

are the solutions of $x^2 - 1 = 0$ or $x = \pm 1$. All other values of $x$ are ordinary points.

**(b)** Inspection of the Cauchy-Euler

$$\underset{\downarrow}{a_2(x) = x^2 = 0 \text{ at } x = 0}$$
$$x^2 y'' + y = 0$$

shows that it has a singular point at $x = 0$. All other values of $x$ are ordinary points.

**(c)** Singular points need not be real numbers. The equation

$$(x^2 + 1)y'' + xy' - y = 0$$

has singular points at the solutions of $x^2 + 1 = 0$—namely, $x = \pm i$. All other values of $x$, real or complex, are ordinary points. ∎

We state the following theorem about the existence of power series solutions without proof.

---

**THEOREM 6.2.1**  **Existence of Power Series Solutions**

If $x = x_0$ is an ordinary point of the differential equation (1), we can always find two linearly independent solutions in the form of a power series centered at $x_0$, that is,

$$y = \sum_{n=0}^{\infty} c_n(x - x_0)^n.$$

A power series solution converges at least on some interval defined by $|x - x_0| < R$, where $R$ is the distance from $x_0$ to the closest singular point.

---

A solution of the form $y = \sum_{n=0}^{\infty} c_n(x - x_0)^n$ is said to be a **solution about the ordinary point $x_0$**. The distance $R$ in Theorem 6.2.1 is the *minimum* value or *lower bound* for the radius of convergence.

---

**EXAMPLE 4**  **Minimum Radius of Convergence**

Find the minimum radius of convergence of a power series solution of the second-order differential equation

$$(x^2 - 2x + 5)y'' + xy' - y = 0$$

**(a)** about the ordinary point $x = 0$, **(b)** about the ordinary point $x = -1$.

**FIGURE 6.2.1** Distance from singular points to the ordinary point 0 in Example 4

**SOLUTION** By the quadratic formula we see from $x^2 - 2x + 5 = 0$ that the singular points of the given differential equation are the complex numbers $1 \pm 2i$.

**(a)** Because $x = 0$ is an ordinary point of the DE, Theorem 6.2.1 guarantees that we can find two power series solutions centered at 0. That is, solutions that look like $y = \sum_{n=0}^{\infty} c_n x^n$ and, moreover, we know without actually finding these solutions that each series must converge *at least* for $|x| < \sqrt{5}$, where $R = \sqrt{5}$ is the distance in the complex plane from either of the numbers $1 + 2i$ (the point $(1, 2)$) or $1 - 2i$ (the point $(1, -2)$) to the ordinary point 0 (the point $(0, 0)$). See Figure 6.2.1.

**(b)** Because $x = -1$ is an ordinary point of the DE, Theorem 6.2.1 guarantees that we can find two power series solutions that look like $y = \sum_{n=0}^{\infty} c_n (x + 1)^n$. Each of the power series converges at least for $|x + 1| < 2\sqrt{2}$ since the distance from each of the singular points to $-1$ (the point $(-1, 0)$) is $R = \sqrt{8} = 2\sqrt{2}$. ∎

In part (a) of Example 4, *one* of the two power series solutions centered at 0 of the differential equation is valid on an interval much larger than $(-\sqrt{5}, \sqrt{5})$; in actual fact this solution is valid on the interval $(-\infty, \infty)$ because it can be shown that one of the two solutions about 0 reduces to a polynomial.

**NOTE** In the examples that follow as well as in the problems of Exercises 6.2 we will, for the sake of simplicity, find only power series solutions about the ordinary point $x = 0$. If it is necessary to find power series solutions of an ODE about an ordinary point $x_0 \neq 0$, we can simply make the change of variable $t = x - x_0$ in the equation (this translates $x = x_0$ to $t = 0$), find solutions of the new equation of the form $y = \sum_{n=0}^{\infty} c_n t^n$, and then resubstitute $t = x - x_0$.

**FINDING A POWER SERIES SOLUTION** Finding a power series solution of a homogeneous linear second-order ODE has been accurately described as "the method of undetermined *series* coefficients" since the procedure is quite analogous to what we did in Section 4.4. In case you did not work through Example 4 of Section 6.1 here, in brief, is the idea. Substitute $y = \sum_{n=0}^{\infty} c_n x^n$ into the differential equation, combine series as we did in Example 3 of Section 6.1, and then equate the all coefficients to the right-hand side of the equation to determine the coefficients $c_n$. But because the right-hand side is zero, the last step requires, by the *identity property* in the bulleted list in Section 6.1, that all coefficients of $x$ must be equated to zero. No, this does *not* mean that all coefficients *are* zero; this would not make sense, after all Theorem 6.2.1 guarantees that we can find two solutions. We will see in Example 5 how the single assumption that $y = \sum_{n=0}^{\infty} c_n x^n = c_0 + c_1 x + c_2 x^2 + \cdots$ leads to two sets of coefficients so that we have two distinct power series $y_1(x)$ and $y_2(x)$, both expanded about the ordinary point $x = 0$. The general solution of the differential equation is $y = C_1 y_1(x) + C_2 y_2(x)$; indeed, it can be shown that $C_1 = c_0$ and $C_2 = c_1$.

---

**EXAMPLE 5**   **Power Series Solutions**

Solve $y'' - xy = 0$.

Before working through this example, we recommend that you reread Example 4 of Section 6.1.

**SOLUTION** Since there are no singular points, Theorem 6.2.1 guarantees two power series solutions centered at 0 that converge for $|x| < \infty$. Substituting $y = \sum_{n=0}^{\infty} c_n x^n$ and the second derivative $y'' = \sum_{n=2}^{\infty} n(n - 1) c_n x^{n-2}$ (see (1) in Section 6.1) into the differential equation give

$$y'' - xy = \sum_{n=2}^{\infty} c_n n(n - 1) x^{n-2} - x \sum_{n=0}^{\infty} c_n x^n = \sum_{n=2}^{\infty} c_n n(n - 1) x^{n-2} - \sum_{n=0}^{\infty} c_n x^{n+1}. \quad (3)$$

We have already added the last two series on the right-hand side of the equality in (3) by shifting the summation index. From the result given in (5) of Section 6.1

$$y'' - xy = 2c_2 + \sum_{k=1}^{\infty} [(k+1)(k+2)c_{k+2} - c_{k-1}]x^k = 0. \tag{4}$$

At this point we invoke the identity property. Since (4) is identically zero, it is necessary that the coefficient of each power of $x$ be set equal to zero—that is, $2c_2 = 0$ (it is the coefficient of $x^0$), and

$$(k+1)(k+2)c_{k+2} - c_{k-1} = 0, \qquad k = 1, 2, 3, \ldots. \tag{5}$$

Now $2c_2 = 0$ obviously dictates that $c_2 = 0$. But the expression in (5), called a **recurrence relation,** determines the $c_k$ in such a manner that we can choose a certain subset of the set of coefficients to be *nonzero.* Since $(k+1)(k+2) \neq 0$ for all values of $k$, we can solve (5) for $c_{k+2}$ in terms of $c_{k-1}$:

$$c_{k+2} = \frac{c_{k-1}}{(k+1)(k+2)}, \qquad k = 1, 2, 3, \ldots. \tag{6}$$

This relation generates consecutive coefficients of the assumed solution one at a time as we let $k$ take on the successive integers indicated in (6):

$$k = 1, \quad c_3 = \frac{c_0}{2 \cdot 3}$$

$$k = 2, \quad c_4 = \frac{c_1}{3 \cdot 4}$$

$$k = 3, \quad c_5 = \frac{c_2}{4 \cdot 5} = 0 \qquad\qquad\qquad\qquad \leftarrow c_2 \text{ is zero}$$

$$k = 4, \quad c_6 = \frac{c_3}{5 \cdot 6} = \frac{1}{2 \cdot 3 \cdot 5 \cdot 6} c_0$$

$$k = 5, \quad c_7 = \frac{c_4}{6 \cdot 7} = \frac{1}{3 \cdot 4 \cdot 6 \cdot 7} c_1$$

$$k = 6, \quad c_8 = \frac{c_5}{7 \cdot 8} = 0 \qquad\qquad\qquad\qquad \leftarrow c_5 \text{ is zero}$$

$$k = 7, \quad c_9 = \frac{c_6}{8 \cdot 9} = \frac{1}{2 \cdot 3 \cdot 5 \cdot 6 \cdot 8 \cdot 9} c_0$$

$$k = 8, \quad c_{10} = \frac{c_7}{9 \cdot 10} = \frac{1}{3 \cdot 4 \cdot 6 \cdot 7 \cdot 9 \cdot 10} c_1$$

$$k = 9, \quad c_{11} = \frac{c_8}{10 \cdot 11} = 0 \qquad\qquad\qquad \leftarrow c_8 \text{ is zero}$$

and so on. Now substituting the coefficients just obtained into the original assumption

$$y = c_0 + c_1x + c_2x^2 + c_3x^3 + c_4x^4 + c_5x^5 + c_6x^6 + c_7x^7 + c_8x^8 + c_9x^9 + c_{10}x^{10} + c_{11}x^{11} + \cdots,$$

we get

$$y = c_0 + c_1x + 0 + \frac{c_0}{2 \cdot 3}x^3 + \frac{c_1}{3 \cdot 4}x^4 + 0 + \frac{c_0}{2 \cdot 3 \cdot 5 \cdot 6}x^6$$

$$+ \frac{c_1}{3 \cdot 4 \cdot 6 \cdot 7}x^7 + 0 + \frac{c_0}{2 \cdot 3 \cdot 5 \cdot 6 \cdot 8 \cdot 9}x^9 + \frac{c_1}{3 \cdot 4 \cdot 6 \cdot 7 \cdot 9 \cdot 10}x^{10} + 0 + \cdots.$$

After grouping the terms containing $c_0$ and the terms containing $c_1$, we obtain $y = c_0 y_1(x) + c_1 y_2(x)$, where

$$y_1(x) = 1 + \frac{1}{2 \cdot 3}x^3 + \frac{1}{2 \cdot 3 \cdot 5 \cdot 6}x^6 + \frac{1}{2 \cdot 3 \cdot 5 \cdot 6 \cdot 8 \cdot 9}x^9 + \cdots = 1 + \sum_{k=1}^{\infty} \frac{1}{2 \cdot 3 \cdots (3k-1)(3k)}x^{3k}$$

$$y_2(x) = x + \frac{1}{3 \cdot 4}x^4 + \frac{1}{3 \cdot 4 \cdot 6 \cdot 7}x^7 + \frac{1}{3 \cdot 4 \cdot 6 \cdot 7 \cdot 9 \cdot 10}x^{10} + \cdots = x + \sum_{k=1}^{\infty} \frac{1}{3 \cdot 4 \cdots (3k)(3k+1)}x^{3k+1}.$$

Because the recursive use of (6) leaves $c_0$ and $c_1$ completely undetermined, they can be chosen arbitrarily. As was mentioned prior to this example, the linear combination $y = c_0 y_1(x) + c_1 y_2(x)$ actually represents the general solution of the differential equation. Although we know from Theorem 6.2.1 that each series solution converges for $|x| < \infty$, that is, on the interval $(-\infty, \infty)$. This fact can also be verified by the ratio test. ◼

The differential equation in Example 5 is called **Airy's equation** and is named after the English mathematician and astronomer **George Biddel Airy** (1801–1892). Airy's differential equation is encountered in the study of diffraction of light, diffraction of radio waves around the surface of the Earth, aerodynamics, and the deflection of a uniform thin vertical column that bends under its own weight. Other forms of Airy's equation are $y'' + xy = 0$ and $y'' \pm \alpha^2 xy = 0$. See Problem 43 in Exercises 6.4 for an application of the last equation.

**EXAMPLE 6** Power Series Solution

Solve $(x^2 + 1)y'' + xy' - y = 0$.

**SOLUTION** As we have already seen on page 245, the given differential equation has singular points at $x = \pm i$, and so a power series solution centered at 0 will converge at least for $|x| < 1$, where 1 is the distance in the complex plane from 0 to either $i$ or $-i$. The assumption $y = \sum_{n=0}^{\infty} c_n x^n$ and its first two derivatives lead to

$$(x^2 + 1) \sum_{n=2}^{\infty} n(n-1)c_n x^{n-2} + x \sum_{n=1}^{\infty} nc_n x^{n-1} - \sum_{n=0}^{\infty} c_n x^n$$

$$= \sum_{n=2}^{\infty} n(n-1)c_n x^n + \sum_{n=2}^{\infty} n(n-1)c_n x^{n-2} + \sum_{n=1}^{\infty} nc_n x^n - \sum_{n=0}^{\infty} c_n x^n$$

$$= 2c_2 x^0 - c_0 x^0 + 6c_3 x + c_1 x - c_1 x + \underbrace{\sum_{n=2}^{\infty} n(n-1)c_n x^n}_{k=n}$$

$$+ \underbrace{\sum_{n=4}^{\infty} n(n-1)c_n x^{n-2}}_{k=n-2} + \underbrace{\sum_{n=2}^{\infty} nc_n x^n}_{k=n} - \underbrace{\sum_{n=2}^{\infty} c_n x^n}_{k=n}$$

$$= 2c_2 - c_0 + 6c_3 x + \sum_{k=2}^{\infty} [k(k-1)c_k + (k+2)(k+1)c_{k+2} + kc_k - c_k]x^k$$

$$= 2c_2 - c_0 + 6c_3 x + \sum_{k=2}^{\infty} [(k+1)(k-1)c_k + (k+2)(k+1)c_{k+2}]x^k = 0.$$

From this identity we conclude that $2c_2 - c_0 = 0$, $6c_3 = 0$, and

$$(k+1)(k-1)c_k + (k+2)(k+1)c_{k+2} = 0.$$

Thus
$$c_2 = \frac{1}{2} c_0$$
$$c_3 = 0$$
$$c_{k+2} = \frac{1-k}{k+2} c_k, \qquad k = 2, 3, 4, \ldots.$$

Substituting $k = 2, 3, 4, \ldots$ into the last formula gives

$$c_4 = -\frac{1}{4} c_2 = -\frac{1}{2 \cdot 4} c_0 = -\frac{1}{2^2 2!} c_0$$

$$c_5 = -\frac{2}{5} c_3 = 0 \qquad \leftarrow c_3 \text{ is zero}$$

$$c_6 = -\frac{3}{6} c_4 = \frac{3}{2 \cdot 4 \cdot 6} c_0 = \frac{1 \cdot 3}{2^3 3!} c_0$$

$$c_7 = -\frac{4}{7} c_5 = 0 \qquad \leftarrow c_5 \text{ is zero}$$

$$c_8 = -\frac{5}{8} c_6 = -\frac{3 \cdot 5}{2 \cdot 4 \cdot 6 \cdot 8} c_0 = -\frac{1 \cdot 3 \cdot 5}{2^4 4!} c_0$$

$$c_9 = -\frac{6}{9} c_7 = 0, \qquad \leftarrow c_7 \text{ is zero}$$

$$c_{10} = -\frac{7}{10} c_8 = \frac{3 \cdot 5 \cdot 7}{2 \cdot 4 \cdot 6 \cdot 8 \cdot 10} c_0 = \frac{1 \cdot 3 \cdot 5 \cdot 7}{2^5 5!} c_0,$$

and so on. Therefore

$$y = c_0 + c_1 x + c_2 x^2 + c_3 x^3 + c_4 x^4 + c_5 x^5 + c_6 x^6 + c_7 x^7 + c_8 x^8 + c_9 x^9 + c_{10} x^{10} + \cdots$$

$$= c_0 \left[ 1 + \frac{1}{2} x^2 - \frac{1}{2^2 2!} x^4 + \frac{1 \cdot 3}{2^3 3!} x^6 - \frac{1 \cdot 3 \cdot 5}{2^4 4!} x^8 + \frac{1 \cdot 3 \cdot 5 \cdot 7}{2^5 5!} x^{10} - \cdots \right] + c_1 x$$

$$= c_0 y_1(x) + c_1 y_2(x).$$

The solutions are the polynomial $y_2(x) = x$ and the power series

$$y_1(x) = 1 + \frac{1}{2} x^2 + \sum_{n=2}^{\infty} (-1)^{n-1} \frac{1 \cdot 3 \cdot 5 \cdots (2n-3)}{2^n n!} x^{2n}, \qquad |x| < 1. \qquad \blacksquare$$

**EXAMPLE 7**  **Three-Term Recurrence Relation**

If we seek a power series solution $y = \sum_{n=0}^{\infty} c_n x^n$ for the differential equation

$$y'' - (1 + x) y = 0,$$

we obtain $c_2 = \frac{1}{2} c_0$ and the three-term recurrence relation

$$c_{k+2} = \frac{c_k + c_{k-1}}{(k+1)(k+2)}, \qquad k = 1, 2, 3, \ldots.$$

It follows from these two results that all coefficients $c_n$, for $n \geq 3$, are expressed in terms of *both* $c_0$ and $c_1$. To simplify life, we can first choose $c_0 \neq 0$, $c_1 = 0$; this yields coefficients for one solution expressed entirely in terms of $c_0$. Next, if we choose $c_0 = 0$, $c_1 \neq 0$, then coefficients for the other solution are expressed in terms of $c_1$. Using $c_2 = \frac{1}{2}c_0$ in both cases, the recurrence relation for $k = 1, 2, 3, \ldots$ gives

| $c_0 \neq 0, c_1 = 0$ | $c_0 = 0, c_1 \neq 0$ |
|---|---|
| $c_2 = \dfrac{1}{2}c_0$ | $c_2 = \dfrac{1}{2}c_0 = 0$ |
| $c_3 = \dfrac{c_1 + c_0}{2 \cdot 3} = \dfrac{c_0}{2 \cdot 3} = \dfrac{c_0}{6}$ | $c_3 = \dfrac{c_1 + c_0}{2 \cdot 3} = \dfrac{c_1}{2 \cdot 3} = \dfrac{c_1}{6}$ |
| $c_4 = \dfrac{c_2 + c_1}{3 \cdot 4} = \dfrac{c_0}{2 \cdot 3 \cdot 4} = \dfrac{c_0}{24}$ | $c_4 = \dfrac{c_2 + c_1}{3 \cdot 4} = \dfrac{c_1}{3 \cdot 4} = \dfrac{c_1}{12}$ |
| $c_5 = \dfrac{c_3 + c_2}{4 \cdot 5} = \dfrac{c_0}{4 \cdot 5}\left[\dfrac{1}{6} + \dfrac{1}{2}\right] = \dfrac{c_0}{30}$ | $c_5 = \dfrac{c_3 + c_2}{4 \cdot 5} = \dfrac{c_1}{4 \cdot 5 \cdot 6} = \dfrac{c_1}{120}$ |

and so on. Finally, we see that the general solution of the equation is $y = c_0 y_1(x) + c_1 y_2(x)$, where

$$y_1(x) = 1 + \frac{1}{2}x^2 + \frac{1}{6}x^3 + \frac{1}{24}x^4 + \frac{1}{30}x^5 + \cdots$$

and

$$y_2(x) = x + \frac{1}{6}x^3 + \frac{1}{12}x^4 + \frac{1}{120}x^5 + \cdots.$$

Each series converges for all finite values of $x$. ■

**NONPOLYNOMIAL COEFFICIENTS** The next example illustrates how to find a power series solution about the ordinary point $x_0 = 0$ of a differential equation when its coefficients are not polynomials. In this example we see an application of the multiplication of two power series.

**EXAMPLE 8** DE with Nonpolynomial Coefficients

Solve $y'' + (\cos x)y = 0$.

**SOLUTION** We see that $x = 0$ is an ordinary point of the equation because, as we have already seen, $\cos x$ is analytic at that point. Using the Maclaurin series for $\cos x$ given in (2) of Section 6.1, along with the usual assumption $y = \sum_{n=0}^{\infty} c_n x^n$ and the results in (1) of Section 6.1 we find

$$y'' + (\cos x)y = \sum_{n=2}^{\infty} n(n-1)c_n x^{n-2} + \left(1 - \frac{x^2}{2!} + \frac{x^4}{4!} - \frac{x^6}{6!} + \cdots\right)\sum_{n=0}^{\infty} c_n x^n$$

$$= 2c_2 + 6c_3 x + 12c_4 x^2 + 20c_5 x^3 + \cdots + \left(1 - \frac{x^2}{2!} + \frac{x^4}{4!} + \cdots\right)(c_0 + c_1 x + c_2 x^2 + c_3 x^3 + \cdots)$$

$$= 2c_2 + c_0 + (6c_3 + c_1)x + \left(12c_4 + c_2 - \frac{1}{2}c_0\right)x^2 + \left(20c_5 + c_3 - \frac{1}{2}c_1\right)x^3 + \cdots = 0.$$

It follows that

$$2c_2 + c_0 = 0, \qquad 6c_3 + c_1 = 0, \qquad 12c_4 + c_2 - \frac{1}{2}c_0 = 0, \qquad 20c_5 + c_3 - \frac{1}{2}c_1 = 0,$$

and so on. This gives $c_2 = -\frac{1}{2}c_0, c_3 = -\frac{1}{6}c_1, c_4 = \frac{1}{12}c_0, c_5 = \frac{1}{30}c_1, \ldots$ . By grouping terms, we arrive at the general solution $y = c_0 y_1(x) + c_1 y_2(x)$, where

$$y_1(x) = 1 - \frac{1}{2}x^2 + \frac{1}{12}x^4 - \cdots \qquad \text{and} \qquad y_2(x) = x - \frac{1}{6}x^3 + \frac{1}{30}x^5 - \cdots.$$

Because the differential equation has no finite singular points, both power series converge for $|x| < \infty$. ■

**SOLUTION CURVES** The approximate graph of a power series solution $y(x) = \sum_{n=0}^{\infty} c_n x^n$ can be obtained in several ways. We can always resort to graphing the terms in the sequence of partial sums of the series—in other words, the graphs of the polynomials $S_N(x) = \sum_{n=0}^{N} c_n x^n$. For large values of $N$, $S_N(x)$ should give us an indication of the behavior of $y(x)$ near the ordinary point $x = 0$. We can also obtain an approximate or numerical solution curve by using a solver as we did in Section 4.10. For example, if you carefully scrutinize the series solutions of Airy's equation in Example 5, you should see that $y_1(x)$ and $y_2(x)$ are, in turn, the solutions of the initial-value problems

**(a)** plot of $y_1(x)$

$$y'' - xy = 0, \quad y(0) = 1, \quad y'(0) = 0,$$
$$y'' - xy = 0, \quad y(0) = 0, \quad y'(0) = 1. \tag{7}$$

The specified initial conditions "pick out" the solutions $y_1(x)$ and $y_2(x)$ from $y = c_0 y_1(x) + c_1 y_2(x)$, since it should be apparent from our basic series assumption $y = \sum_{n=0}^{\infty} c_n x^n$ that $y(0) = c_0$ and $y'(0) = c_1$. Now if your numerical solver requires a system of equations, the substitution $y' = u$ in $y'' - xy = 0$ gives $y'' = u' = xy$, and so a system of two first-order equations equivalent to Airy's equation is

**(b)** plot of $y_2(x)$

$$y' = u$$
$$u' = xy. \tag{8}$$

**FIGURE 6.2.2** Numerical solution curves for Airy's DE

Initial conditions for the system in (8) are the two sets of initial conditions in (7) rewritten as $y(0) = 1$, $u(0) = 0$, and $y(0) = 0$, $u(0) = 1$. The graphs of $y_1(x)$ and $y_2(x)$ shown in Figure 6.2.2 were obtained with the aid of a numerical solver.

---

**REMARKS**

(*i*) In the problems that follow, do not expect to be able to write a solution in terms of summation notation in each case. Even though we can generate as many terms as desired in a series solution $y = \sum_{n=0}^{\infty} c_n x^n$ either through the use of a recurrence relation or, as in Example 8, by multiplication, it might not be possible to deduce any general term for the coefficients $c_n$. We might have to settle, as we did in Examples 7 and 8, for just writing out the first few terms of the series.

(*ii*) A point $x_0$ is an ordinary point of a *nonhomogeneous* linear second-order DE $y'' + P(x)y' + Q(x)y = f(x)$ if $P(x)$, $Q(x)$, and $f(x)$ are analytic at $x_0$. Moreover, Theorem 6.2.1 extends to such DEs; in other words, we can find power series solutions $y = \sum_{n=0}^{\infty} c_n(x - x_0)^n$ of nonhomogeneous linear DEs in the same manner as in Examples 5–8. See Problem 26 in Exercises 6.2.

---

## EXERCISES 6.2

*Answers to selected odd-numbered problems begin on page ANS-9.*

In Problems 1 and 2 without actually solving the given differential equation, find the minimum radius of convergence of power series solutions about the ordinary point $x = 0$. About the ordinary point $x = 1$.

**1.** $(x^2 - 25)y'' + 2xy' + y = 0$

**2.** $(x^2 - 2x + 10)y'' + xy' - 4y = 0$

In Problems 3–6 find two power series solutions of the given differential equation about the ordinary point $x = 0$. Compare the series solutions with the solutions of the differential equations obtained using the method of Section 4.3. Try to explain any differences between the two forms of the solutions.

**3.** $y'' + y = 0$      **4.** $y'' - y = 0$

**5.** $y'' - y' = 0$      **6.** $y'' + 2y' = 0$

In Problems 7–18 find two power series solutions of the given differential equation about the ordinary point $x = 0$.

**7.** $y'' + xy = 0$

**8.** $y'' + x^2y = 0$

**9.** $y'' - 2xy' + y = 0$

**10.** $y'' - xy' + 2y = 0$

**11.** $y'' + x^2y' + xy = 0$

**12.** $y'' + 2xy' + 2y = 0$

**13.** $(x - 1)y'' + y' = 0$

**14.** $(x + 2)y'' + xy' - y = 0$

**15.** $y'' - (x + 1)y' - y = 0$

**16.** $(x^2 + 1)y'' - 6y = 0$

**17.** $(x^2 + 2)y'' + 3xy' - y = 0$

**18.** $(x^2 - 1)y'' + xy' - y = 0$

In Problems 19–22 use the power series method to solve the given initial-value problem.

**19.** $(x - 1)y'' - xy' + y = 0$,  $y(0) = -2$,  $y'(0) = 6$

**20.** $(x + 1)y'' - (2 - x)y' + y = 0$,  $y(0) = 2$,  $y'(0) = -1$

**21.** $y'' - 2xy' + 8y = 0$,  $y(0) = 3$,  $y'(0) = 0$

**22.** $(x^2 + 1)y'' + 2xy' = 0$,  $y(0) = 0$,  $y'(0) = 1$

In Problems 23 and 24 use the procedure in Example 8 to find two power series solutions of the given differential equation about the ordinary point $x = 0$.

**23.** $y'' + (\sin x)y = 0$

**24.** $y'' + e^xy' - y = 0$

### Discussion Problems

**25.** Without actually solving the differential equation $(\cos x)y'' + y' + 5y = 0$, find the minimum radius of convergence of power series solutions about the ordinary point $x = 0$. About the ordinary point $x = 1$.

**26.** How can the power series method be used to solve the *nonhomogeneous* equation $y'' - xy = 1$ about the ordinary

point $x = 0$? Of $y'' - 4xy' - 4y = e^x$? Carry out your ideas by solving both DEs.

**27.** Is $x = 0$ an ordinary or a singular point of the differential equation $xy'' + (\sin x)y = 0$? Defend your answer with sound mathematics.   [*Hint*: Use the Maclaurin series of $\sin x$ and then examine $(\sin x)/x$.]

**28.** Is $x = 0$ an ordinary point of the differential equation $y'' + 5xy' + \sqrt{x}y = 0$?

### Computer Lab Assignments

**29. (a)** Find two power series solutions for $y'' + xy' + y = 0$ and express the solutions $y_1(x)$ and $y_2(x)$ in terms of summation notation.

**(b)** Use a CAS to graph the partial sums $S_N(x)$ for $y_1(x)$. Use $N = 2, 3, 5, 6, 8, 10$. Repeat using the partial sums $S_N(x)$ for $y_2(x)$.

**(c)** Compare the graphs obtained in part (b) with the curve obtained by using a numerical solver. Use the initial-conditions $y_1(0) = 1$, $y_1'(0) = 0$, and $y_2(0) = 0$, $y_2'(0) = 1$.

**(d)** Reexamine the solution $y_1(x)$ in part (a). Express this series as an elementary function. Then use (5) of Section 4.2 to find a second solution of the equation. Verify that this second solution is the same as the power series solution $y_2(x)$.

**30. (a)** Find one more nonzero term for each of the solutions $y_1(x)$ and $y_2(x)$ in Example 8.

**(b)** Find a series solution $y(x)$ of the initial-value problem $y'' + (\cos x)y = 0$,  $y(0) = 1$,  $y'(0) = 1$.

**(c)** Use a CAS to graph the partial sums $S_N(x)$ for the solution $y(x)$ in part (b). Use $N = 2, 3, 4, 5, 6, 7$.

**(d)** Compare the graphs obtained in part (c) with the curve obtained using a numerical solver for the initial-value problem in part (b).

## 6.3   Solutions About Singular Points

**INTRODUCTION**   The two differential equations

$$y'' - xy = 0 \qquad \text{and} \qquad xy'' + y = 0$$

are similar only in that they are both examples of simple linear second-order DEs with variable coefficients. That is all they have in common. Since $x = 0$ is an *ordinary point* of $y'' - xy = 0$, we saw in Section 6.2 that there was no problem in finding two distinct power series solutions centered at that point. In contrast, because $x = 0$ is a *singular point* of $xy'' + y = 0$, finding two infinite series—notice that we did not say *power series*—solutions of the equation about that point becomes a more difficult task.

The solution method that is discussed in this section does not always yield two infinite series solutions. When only one solution is found, we can use the formula given in (5) of Section 4.2 to find a second solution.

**A DEFINITION**    A singular point $x_0$ of a linear differential equation

$$a_2(x)y'' + a_1(x)y' + a_0(x)y = 0 \tag{1}$$

is further classified as either regular or irregular. The classification again depends on the functions $P$ and $Q$ in the standard form

$$y'' + P(x)y' + Q(x)y = 0. \tag{2}$$

---

**DEFINITION 6.3.1    Regular and Irregular Singular Points**

A singular point $x = x_0$ is said to be a **regular singular point** of the differential equation (1) if the functions $p(x) = (x - x_0)P(x)$ and $q(x) = (x - x_0)^2 Q(x)$ are both analytic at $x_0$. A singular point that is not regular is said to be an **irregular singular point** of the equation.

---

The second sentence in Definition 6.3.1 indicates that if one or both of the functions $p(x) = (x - x_0)P(x)$ and $q(x) = (x - x_0)^2 Q(x)$ fail to be analytic at $x_0$, then $x_0$ is an irregular singular point.

**POLYNOMIAL COEFFICIENTS**    As in Section 6.2, we are mainly interested in linear equations (1) where the coefficients $a_2(x)$, $a_1(x)$, and $a_0(x)$ are polynomials with no common factors. We have already seen that if $a_2(x_0) = 0$, then $x = x_0$ is a singular point of (1), since at least one of the rational functions $P(x) = a_1(x)/a_2(x)$ and $Q(x) = a_0(x)/a_2(x)$ in the standard form (2) fails to be analytic at that point. But since $a_2(x)$ is a polynomial and $x_0$ is one of its zeros, it follows from the Factor Theorem of algebra that $x - x_0$ is a factor of $a_2(x)$. This means that after $a_1(x)/a_2(x)$ and $a_0(x)/a_2(x)$ are reduced to lowest terms, the factor $x - x_0$ must remain, to some positive integer power, in one or both denominators. Now suppose that $x = x_0$ is a singular point of (1) but both the functions defined by the products $p(x) = (x - x_0)P(x)$ and $q(x) = (x - x_0)^2 Q(x)$ are analytic at $x_0$. We are led to the conclusion that multiplying $P(x)$ by $x - x_0$ and $Q(x)$ by $(x - x_0)^2$ has the effect (through cancellation) that $x - x_0$ no longer appears in either denominator. We can now determine whether $x_0$ is regular by a quick visual check of denominators:

*If $x - x_0$ appears at most to the first power in the denominator of $P(x)$ and at most to the second power in the denominator of $Q(x)$, then $x = x_0$ is a regular singular point.*

Moreover, observe that if $x = x_0$ is a regular singular point and we multiply (2) by $(x - x_0)^2$, then the original DE can be put into the form

$$(x - x_0)^2 y'' + (x - x_0)p(x)y' + q(x)y = 0, \tag{3}$$

where $p$ and $q$ are analytic at $x = x_0$.

---

**EXAMPLE 1    Classification of Singular Points**

It should be clear that $x = 2$ and $x = -2$ are singular points of

$$(x^2 - 4)^2 y'' + 3(x - 2)y' + 5y = 0.$$

After dividing the equation by $(x^2 - 4)^2 = (x - 2)^2(x + 2)^2$ and reducing the coefficients to lowest terms, we find that

$$P(x) = \frac{3}{(x - 2)(x + 2)^2} \quad \text{and} \quad Q(x) = \frac{5}{(x - 2)^2(x + 2)^2}.$$

We now test $P(x)$ and $Q(x)$ at each singular point.

For $x = 2$ to be a regular singular point, the factor $x - 2$ can appear at most to the first power in the denominator of $P(x)$ and at most to the second power in the denominator of $Q(x)$. A check of the denominators of $P(x)$ and $Q(x)$ shows that both these conditions are satisfied, so $x = 2$ is a regular singular point. Alternatively, we are led to the same conclusion by noting that both rational functions

$$p(x) = (x - 2)P(x) = \frac{3}{(x + 2)^2} \qquad \text{and} \qquad q(x) = (x - 2)^2 Q(x) = \frac{5}{(x + 2)^2}$$

are analytic at $x = 2$.

Now since the factor $x - (-2) = x + 2$ appears to the second power in the denominator of $P(x)$, we can conclude immediately that $x = -2$ is an irregular singular point of the equation. This also follows from the fact that

$$p(x) = (x + 2)P(x) = \frac{3}{(x - 2)(x + 2)}$$

is not analytic at $x = -2$.

In Example 1, notice that since $x = 2$ is a regular singular point, the original equation can be written as

$$\overset{\substack{p(x)\ \text{analytic} \\ \text{at } x = 2}}{\underset{\downarrow}{}} \qquad \overset{\substack{q(x)\ \text{analytic} \\ \text{at } x = 2}}{\underset{\downarrow}{}}$$

$$(x - 2)^2 y'' + (x - 2)\frac{3}{(x + 2)^2} y' + \frac{5}{(x + 2)^2} y = 0.$$

As another example, we can see that $x = 0$ is an irregular singular point of $x^3 y'' - 2xy' + 8y = 0$ by inspection of the denominators of $P(x) = -2/x^2$ and $Q(x) = 8/x^3$. On the other hand, $x = 0$ is a regular singular point of $xy'' - 2xy' + 8y = 0$, since $x - 0$ and $(x - 0)^2$ do not even appear in the respective denominators of $P(x) = -2$ and $Q(x) = 8/x$. For a singular point $x = x_0$ any nonnegative power of $x - x_0$ less than one (namely, zero) and any nonnegative power less than two (namely, zero and one) in the denominators of $P(x)$ and $Q(x)$, respectively, imply that $x_0$ is a regular singular point. A singular point can be a complex number. You should verify that $x = 3i$ and $x = -3i$ are two regular singular points of $(x^2 + 9)y'' - 3xy' + (1 - x)y = 0$.

**NOTE**    Any second-order Cauchy-Euler equation $ax^2 y'' + bxy' + cy = 0$, where $a$, $b$, and $c$ are real constants, has a regular singular point at $x = 0$. You should verify that two solutions of the Cauchy-Euler equation $x^2 y'' - 3xy' + 4y = 0$ on the interval $(0, \infty)$ are $y_1 = x^2$ and $y_2 = x^2 \ln x$. If we attempted to find a power series solution about the regular singular point $x = 0$ (namely, $y = \sum_{n=0}^{\infty} c_n x^n$), we would succeed in obtaining only the polynomial solution $y_1 = x^2$. The fact that we would not obtain the second solution is not surprising because $\ln x$ (and consequently $y_2 = x^2 \ln x$) is not analytic at $x = 0$—that is, $y_2$ does not possess a Taylor series expansion centered at $x = 0$.

**METHOD OF FROBENIUS**    To solve a differential equation (1) about a regular singular point, we employ the following theorem due to the eminent German mathematician **Ferdinand Georg Frobenius** (1849–1917).

---

**THEOREM 6.3.1**    **Frobenius' Theorem**

If $x = x_0$ is a regular singular point of the differential equation (1), then there exists at least one solution of the form

$$y = (x - x_0)^r \sum_{n=0}^{\infty} c_n (x - x_0)^n = \sum_{n=0}^{\infty} c_n (x - x_0)^{n+r}, \qquad (4)$$

where the number $r$ is a constant to be determined. The series will converge at least on some interval $0 < x - x_0 < R$.

Notice the words *at least* in the first sentence of Theorem 6.3.1. This means that in contrast to Theorem 6.2.1, Theorem 6.3.1 gives us no assurance that *two* series solutions of the type indicated in (4) can be found. The **method of Frobenius,** finding series solutions about a regular singular point $x_0$, is similar to the power-series method in the preceding section in that we substitute $y = \sum_{n=0}^{\infty} c_n(x - x_0)^{n+r}$ into the given differential equation and determine the unknown coefficients $c_n$ by a recurrence relation. However, we have an additional task in this procedure: Before determining the coefficients, we must find the unknown exponent $r$. If $r$ is found to be a number that is not a nonnegative integer, then the corresponding solution $y = \sum_{n=0}^{\infty} c_n(x - x_0)^{n+r}$ is not a power series.

As we did in the discussion of solutions about ordinary points, we shall always assume, for the sake of simplicity in solving differential equations, that the regular singular point is $x = 0$.

**EXAMPLE 2**  **Two Series Solutions**

Because $x = 0$ is a regular singular point of the differential equation

$$3xy'' + y' - y = 0, \tag{5}$$

we try to find a solution of the form $y = \sum_{n=0}^{\infty} c_n x^{n+r}$. Now

$$y' = \sum_{n=0}^{\infty} (n + r)c_n x^{n+r-1} \quad \text{and} \quad y'' = \sum_{n=0}^{\infty} (n + r)(n + r - 1)c_n x^{n+r-2},$$

so

$$3xy'' + y' - y = 3\sum_{n=0}^{\infty} (n + r)(n + r - 1)c_n x^{n+r-1} + \sum_{n=0}^{\infty} (n + r)c_n x^{n+r-1} - \sum_{n=0}^{\infty} c_n x^{n+r}$$

$$= \sum_{n=0}^{\infty} (n + r)(3n + 3r - 2)c_n x^{n+r-1} - \sum_{n=0}^{\infty} c_n x^{n+r}$$

$$= x^r \left[ r(3r - 2)c_0 x^{-1} + \underbrace{\sum_{n=1}^{\infty} (n + r)(3n + 3r - 2)c_n x^{n-1}}_{k = n - 1} - \underbrace{\sum_{n=0}^{\infty} c_n x^n}_{k = n} \right]$$

$$= x^r \left[ r(3r - 2)c_0 x^{-1} + \sum_{k=0}^{\infty} [(k + r + 1)(3k + 3r + 1)c_{k+1} - c_k]x^k \right] = 0,$$

which implies that $\qquad\qquad\qquad r(3r - 2)c_0 = 0$

and $\qquad (k + r + 1)(3k + 3r + 1)c_{k+1} - c_k = 0, \qquad k = 0, 1, 2, \ldots.$

Because nothing is gained by taking $c_0 = 0$, we must then have

$$r(3r - 2) = 0 \tag{6}$$

and $\qquad c_{k+1} = \dfrac{c_k}{(k + r + 1)(3k + 3r + 1)}, \qquad k = 0, 1, 2, \ldots. \tag{7}$

When substituted in (7), the two values of $r$ that satisfy the quadratic equation (6), $r_1 = \frac{2}{3}$ and $r_2 = 0$, give two different recurrence relations:

$$r_1 = \tfrac{2}{3}, \qquad c_{k+1} = \dfrac{c_k}{(3k + 5)(k + 1)}, \qquad k = 0, 1, 2, \ldots \tag{8}$$

$$r_2 = 0, \qquad c_{k+1} = \dfrac{c_k}{(k + 1)(3k + 1)}, \qquad k = 0, 1, 2, \ldots. \tag{9}$$

From (8) we find

$$c_1 = \frac{c_0}{5 \cdot 1}$$

$$c_2 = \frac{c_1}{8 \cdot 2} = \frac{c_0}{2!5 \cdot 8}$$

$$c_3 = \frac{c_2}{11 \cdot 3} = \frac{c_0}{3!5 \cdot 8 \cdot 11}$$

$$c_4 = \frac{c_3}{14 \cdot 4} = \frac{c_0}{4!5 \cdot 8 \cdot 11 \cdot 14}$$

$$\vdots$$

$$c_n = \frac{c_0}{n!5 \cdot 8 \cdot 11 \cdots (3n + 2)}.$$

From (9) we find

$$c_1 = \frac{c_0}{1 \cdot 1}$$

$$c_2 = \frac{c_1}{2 \cdot 4} = \frac{c_0}{2!1 \cdot 4}$$

$$c_3 = \frac{c_2}{3 \cdot 7} = \frac{c_0}{3!1 \cdot 4 \cdot 7}$$

$$c_4 = \frac{c_3}{4 \cdot 10} = \frac{c_0}{4!1 \cdot 4 \cdot 7 \cdot 10}$$

$$\vdots$$

$$c_n = \frac{c_0}{n!1 \cdot 4 \cdot 7 \cdots (3n - 2)}.$$

Here we encounter something that did not happen when we obtained solutions about an ordinary point; we have what looks to be two different sets of coefficients, but each set contains the *same* multiple $c_0$. If we omit this term, the series solutions are

$$y_1(x) = x^{2/3}\left[1 + \sum_{n=1}^{\infty} \frac{1}{n!5 \cdot 8 \cdot 11 \cdots (3n + 2)} x^n\right] \tag{10}$$

$$y_2(x) = x^0\left[1 + \sum_{n=1}^{\infty} \frac{1}{n!1 \cdot 4 \cdot 7 \cdots (3n - 2)} x^n\right]. \tag{11}$$

By the ratio test it can be demonstrated that both (10) and (11) converge for all values of $x$—that is, $|x| < \infty$. Also, it should be apparent from the form of these solutions that neither series is a constant multiple of the other, and therefore $y_1(x)$ and $y_2(x)$ are linearly independent on the entire $x$-axis. Hence by the superposition principle, $y = C_1 y_1(x) + C_2 y_2(x)$ is another solution of (5). On any interval that does not contain the origin, such as $(0, \infty)$, this linear combination represents the general solution of the differential equation. ∎

**INDICIAL EQUATION**    Equation (6) is called the **indicial equation** of the problem, and the values $r_1 = \frac{2}{3}$ and $r_2 = 0$ are called the **indicial roots,** or **exponents,** of the singularity $x = 0$. In general, after substituting $y = \sum_{n=0}^{\infty} c_n x^{n+r}$ into the given differential equation and simplifying, the indicial equation is a quadratic equation in $r$ that results from equating the *total coefficient of the lowest power of x to zero*. We solve for the two values of $r$ and substitute these values into a recurrence relation such as (7). Theorem 6.3.1 guarantees that at least one solution of the assumed series form can be found.

It is possible to obtain the indicial equation in advance of substituting $y = \sum_{n=0}^{\infty} c_n x^{n+r}$ into the differential equation. If $x = 0$ is a regular singular point of (1), then by Definition 6.3.1 both functions $p(x) = xP(x)$ and $q(x) = x^2 Q(x)$, where $P$ and $Q$ are defined by the standard form (2), are analytic at $x = 0$; that is, the power series expansions

$$p(x) = xP(x) = a_0 + a_1 x + a_2 x^2 + \cdots \quad \text{and} \quad q(x) = x^2 Q(x) = b_0 + b_1 x + b_2 x^2 + \cdots \tag{12}$$

are valid on intervals that have a positive radius of convergence. By multiplying (2) by $x^2$, we get the form given in (3):

$$x^2 y'' + x[xP(x)]y' + [x^2 Q(x)]y = 0. \tag{13}$$

After substituting $y = \sum_{n=0}^{\infty} c_n x^{n+r}$ and the two series in (12) into (13) and carrying out the multiplication of series, we find the general indicial equation to be

$$r(r-1) + a_0 r + b_0 = 0, \tag{14}$$

where $a_0$ and $b_0$ are as defined in (12). See Problems 13 and 14 in Exercises 6.3.

## EXAMPLE 3   Two Series Solutions

Solve $2xy'' + (1+x)y' + y = 0$.

**SOLUTION** Substituting $y = \sum_{n=0}^{\infty} c_n x^{n+r}$ gives

$$2xy'' + (1+x)y' + y = 2\sum_{n=0}^{\infty}(n+r)(n+r-1)c_n x^{n+r-1} + \sum_{n=0}^{\infty}(n+r)c_n x^{n+r-1}$$

$$+ \sum_{n=0}^{\infty}(n+r)c_n x^{n+r} + \sum_{n=0}^{\infty} c_n x^{n+r}$$

$$= \sum_{n=0}^{\infty}(n+r)(2n+2r-1)c_n x^{n+r-1} + \sum_{n=0}^{\infty}(n+r+1)c_n x^{n+r}$$

$$= x^r\left[ r(2r-1)c_0 x^{-1} + \underbrace{\sum_{n=1}^{\infty}(n+r)(2n+2r-1)c_n x^{n-1}}_{k=n-1} + \underbrace{\sum_{n=0}^{\infty}(n+r+1)c_n x^{n}}_{k=n} \right]$$

$$= x^r\left[ r(2r-1)c_0 x^{-1} + \sum_{k=0}^{\infty}[(k+r+1)(2k+2r+1)c_{k+1} + (k+r+1)c_k]x^k \right],$$

which implies that

$$r(2r-1) = 0 \tag{15}$$

and

$$(k+r+1)(2k+2r+1)c_{k+1} + (k+r+1)c_k = 0, \tag{16}$$

$k = 0, 1, 2, \ldots$. From (15) we see that the indicial roots are $r_1 = \frac{1}{2}$ and $r_2 = 0$. For $r_1 = \frac{1}{2}$ we can divide by $k + \frac{3}{2}$ in (16) to obtain

$$c_{k+1} = \frac{-c_k}{2(k+1)}, \qquad k = 0, 1, 2, \ldots, \tag{17}$$

whereas for $r_2 = 0$, (16) becomes

$$c_{k+1} = \frac{-c_k}{2k+1}, \qquad k = 0, 1, 2, \ldots. \tag{18}$$

| From (17) we find | From (18) we find |
|---|---|
| $c_1 = \dfrac{-c_0}{2 \cdot 1}$ | $c_1 = \dfrac{-c_0}{1}$ |
| $c_2 = \dfrac{-c_1}{2 \cdot 2} = \dfrac{c_0}{2^2 \cdot 2!}$ | $c_2 = \dfrac{-c_1}{3} = \dfrac{c_0}{1 \cdot 3}$ |
| $c_3 = \dfrac{-c_2}{2 \cdot 3} = \dfrac{-c_0}{2^3 \cdot 3!}$ | $c_3 = \dfrac{-c_2}{5} = \dfrac{-c_0}{1 \cdot 3 \cdot 5}$ |
| $c_4 = \dfrac{-c_3}{2 \cdot 4} = \dfrac{c_0}{2^4 \cdot 4!}$ | $c_4 = \dfrac{-c_3}{7} = \dfrac{c_0}{1 \cdot 3 \cdot 5 \cdot 7}$ |
| $\vdots$ | $\vdots$ |
| $c_n = \dfrac{(-1)^n c_0}{2^n n!}.$ | $c_n = \dfrac{(-1)^n c_0}{1 \cdot 3 \cdot 5 \cdot 7 \cdots (2n-1)}.$ |

Thus for the indicial root $r_1 = \frac{1}{2}$ we obtain the solution

$$y_1(x) = x^{1/2}\left[1 + \sum_{n=1}^{\infty} \frac{(-1)^n}{2^n n!} x^n\right] = \sum_{n=0}^{\infty} \frac{(-1)^n}{2^n n!} x^{n+1/2},$$

where we have again omitted $c_0$. The series converges for $x \geq 0$; as given, the series is not defined for negative values of $x$ because of the presence of $x^{1/2}$. For $r_2 = 0$ a second solution is

$$y_2(x) = 1 + \sum_{n=1}^{\infty} \frac{(-1)^n}{1 \cdot 3 \cdot 5 \cdot 7 \cdots (2n-1)} x^n, \qquad |x| < \infty.$$

On the interval $(0, \infty)$ the general solution is $y = C_1 y_1(x) + C_2 y_2(x)$.  ∎

---

### EXAMPLE 4    Only One Series Solution

Solve $xy'' + y = 0$.

**SOLUTION**  From $xP(x) = 0$, $x^2 Q(x) = x$ and the fact that $0$ and $x$ are their own power series centered at $0$ we conclude that $a_0 = 0$ and $b_0 = 0$, so from (14) the indicial equation is $r(r - 1) = 0$. You should verify that the two recurrence relations corresponding to the indicial roots $r_1 = 1$ and $r_2 = 0$ yield exactly the same set of coefficients. In other words, in this case the method of Frobenius produces only a single series solution

$$y_1(x) = \sum_{n=0}^{\infty} \frac{(-1)^n}{n!(n+1)!} x^{n+1} = x - \frac{1}{2} x^2 + \frac{1}{12} x^3 - \frac{1}{144} x^4 + \cdots.  \quad ∎$$

---

**THREE CASES**    For the sake of discussion let us again suppose that $x = 0$ is a regular singular point of equation (1) and that the indicial roots $r_1$ and $r_2$ of the singularity are real. When using the method of Frobenius, we distinguish three cases corresponding to the nature of the indicial roots $r_1$ and $r_2$. In the first two cases the symbol $r_1$ denotes the largest of two distinct roots, that is, $r_1 > r_2$. In the last case $r_1 = r_2$.

**CASE I:**   If $r_1$ and $r_2$ are distinct and the difference $r_1 - r_2$ is not a positive integer, then there exist two linearly independent solutions of equation (1) of the form

$$y_1(x) = \sum_{n=0}^{\infty} c_n x^{n+r_1}, \quad c_0 \neq 0, \qquad y_2(x) = \sum_{n=0}^{\infty} b_n x^{n+r_2}, \quad b_0 \neq 0.$$

This is the case illustrated in Examples 2 and 3.

Next we assume that the difference of the roots is $N$, where $N$ is a positive integer. In this case the second solution *may* contain a logarithm.

**CASE II:**   If $r_1$ and $r_2$ are distinct and the difference $r_1 - r_2$ is a positive integer, then there exist two linearly independent solutions of equation (1) of the form

$$y_1(x) = \sum_{n=0}^{\infty} c_n x^{n+r_1}, \qquad c_0 \neq 0, \tag{19}$$

$$y_2(x) = C y_1(x) \ln x + \sum_{n=0}^{\infty} b_n x^{n+r_2}, \qquad b_0 \neq 0, \tag{20}$$

where $C$ is a constant that could be zero.

Finally, in the last case, the case when $r_1 = r_2$, a second solution will *always* contain a logarithm. The situation is analogous to the solution of a Cauchy-Euler equation when the roots of the auxiliary equation are equal.

**CASE III:** If $r_1$ and $r_2$ are equal, then there always exist two linearly independent solutions of equation (1) of the form

$$y_1(x) = \sum_{n=0}^{\infty} c_n x^{n+r_1}, \qquad c_0 \neq 0, \tag{21}$$

$$y_2(x) = y_1(x) \ln x + \sum_{n=1}^{\infty} b_n x^{n+r_1}. \tag{22}$$

**FINDING A SECOND SOLUTION**    When the difference $r_1 - r_2$ is a positive integer (Case II), we *may* or *may not* be able to find two solutions having the form $y = \sum_{n=0}^{\infty} c_n x^{n+r}$. This is something that we do not know in advance but is determined after we have found the indicial roots and have carefully examined the recurrence relation that defines the coefficients $c_n$. We just may be lucky enough to find two solutions that involve only powers of $x$, that is, $y_1(x) = \sum_{n=0}^{\infty} c_n x^{n+r_1}$ (equation (19)) and $y_2(x) = \sum_{n=0}^{\infty} b_n x^{n+r_2}$ (equation (20) with $C = 0$). See Problem 31 in Exercises 6.3. On the other hand, in Example 4 we see that the difference of the indicial roots is a positive integer ($r_1 - r_2 = 1$) and the method of Frobenius failed to give a second series solution. In this situation equation (20), with $C \neq 0$, indicates what the second solution looks like. Finally, when the difference $r_1 - r_2$ is a zero (Case III), the method of Frobenius fails to give a second series solution; the second solution (22) always contains a logarithm and can be shown to be equivalent to (20) with $C = 1$. One way to obtain the second solution with the logarithmic term is to use the fact that

$$y_2(x) = y_1(x) \int \frac{e^{-\int P(x)\,dx}}{y_1^2(x)}\,dx \tag{23}$$

is also a solution of $y'' + P(x)y' + Q(x)y = 0$ whenever $y_1(x)$ is a known solution. We illustrate how to use (23) in the next example.

---

**EXAMPLE 5**    **Example 4 Revisited Using a CAS**

Find the general solution of $xy'' + y = 0$.

**SOLUTION**    From the known solution given in Example 4,

$$y_1(x) = x - \frac{1}{2}x^2 + \frac{1}{12}x^3 - \frac{1}{144}x^4 + \cdots,$$

we can construct a second solution $y_2(x)$ using formula (23). Those with the time, energy, and patience can carry out the drudgery of squaring a series, long division, and integration of the quotient by hand. But all these operations can be done with relative ease with the help of a CAS. We give the results:

$$y_2(x) = y_1(x) \int \frac{e^{-\int 0\,dx}}{[y_1(x)]^2}\,dx = y_1(x) \int \frac{dx}{\left[x - \dfrac{1}{2}x^2 + \dfrac{1}{12}x^3 - \dfrac{1}{144}x^4 + \cdots\right]^2}$$

$$= y_1(x) \int \frac{dx}{\left[x^2 - x^3 + \dfrac{5}{12}x^4 - \dfrac{7}{72}x^5 + \cdots\right]} \qquad \leftarrow \text{after squaring}$$

$$= y_1(x) \int \left[\frac{1}{x^2} + \frac{1}{x} + \frac{7}{12} + \frac{19}{72}x + \cdots\right]dx \qquad \leftarrow \text{after long division}$$

$$= y_1(x) \left[-\frac{1}{x} + \ln x + \frac{7}{12}x + \frac{19}{144}x^2 + \cdots\right] \qquad \leftarrow \text{after integrating}$$

$$= y_1(x) \ln x + y_1(x)\left[-\frac{1}{x} + \frac{7}{12}x + \frac{19}{144}x^2 + \cdots\right],$$

or    $y_2(x) = y_1(x) \ln x + \left[ -1 - \dfrac{1}{2}x + \dfrac{1}{2}x^2 + \cdots \right]$.    ← after multiplying out

On the interval $(0, \infty)$ the general solution is $y = C_1 y_1(x) + C_2 y_2(x)$.    ∎

Note that the final form of $y_2$ in Example 5 matches (20) with $C = 1$; the series in the brackets corresponds to the summation in (20) with $r_2 = 0$.

### REMARKS

(***i***) The three different forms of a linear second-order differential equation in (1), (2), and (3) were used to discuss various theoretical concepts. But on a practical level, when it comes to actually solving a differential equation using the method of Frobenius, it is advisable to work with the form of the DE given in (1).

(***ii***) When the difference of indicial roots $r_1 - r_2$ is a positive integer ($r_1 > r_2$), it sometimes pays to iterate the recurrence relation using the smaller root $r_2$ first. See Problems 31 and 32 in Exercises 6.3.

(***iii***) Because an indicial root $r$ is a solution of a quadratic equation, it could be complex. We shall not, however, investigate this case.

(***iv***) If $x = 0$ is an irregular singular point, then we might not be able to find *any* solution of the DE of form $y = \sum_{n=0}^{\infty} c_n x^{n+r}$.

## EXERCISES 6.3

*Answers to selected odd-numbered problems begin on page ANS-9.*

In Problems 1–10 determine the singular points of the given differential equation. Classify each singular point as regular or irregular.

**1.** $x^3 y'' + 4x^2 y' + 3y = 0$

**2.** $x(x + 3)^2 y'' - y = 0$

**3.** $(x^2 - 9)^2 y'' + (x + 3)y' + 2y = 0$

**4.** $y'' - \dfrac{1}{x}y' + \dfrac{1}{(x - 1)^3}y = 0$

**5.** $(x^3 + 4x)y'' - 2xy' + 6y = 0$

**6.** $x^2(x - 5)^2 y'' + 4xy' + (x^2 - 25)y = 0$

**7.** $(x^2 + x - 6)y'' + (x + 3)y' + (x - 2)y = 0$

**8.** $x(x^2 + 1)^2 y'' + y = 0$

**9.** $x^3(x^2 - 25)(x - 2)^2 y'' + 3x(x - 2)y' + 7(x + 5)y = 0$

**10.** $(x^3 - 2x^2 + 3x)^2 y'' + x(x - 3)^2 y' - (x + 1)y = 0$

In Problems 11 and 12 put the given differential equation into form (3) for each regular singular point of the equation. Identify the functions $p(x)$ and $q(x)$.

**11.** $(x^2 - 1)y'' + 5(x + 1)y' + (x^2 - x)y = 0$

**12.** $xy'' + (x + 3)y' + 7x^2 y = 0$

In Problems 13 and 14, $x = 0$ is a regular singular point of the given differential equation. Use the general form of the indicial equation in (14) to find the indicial roots of the singularity. Without solving, discuss the number of series solutions you would expect to find using the method of Frobenius.

**13.** $x^2 y'' + \left(\frac{5}{3}x + x^2\right)y' - \frac{1}{3}y = 0$

**14.** $xy'' + y' + 10y = 0$

In Problems 15–24, $x = 0$ is a regular singular point of the given differential equation. Show that the indicial roots of the singularity do not differ by an integer. Use the method of Frobenius to obtain two linearly independent series solutions about $x = 0$. Form the general solution on $(0, \infty)$.

**15.** $2xy'' - y' + 2y = 0$

**16.** $2xy'' + 5y' + xy = 0$

**17.** $4xy'' + \frac{1}{2}y' + y = 0$

**18.** $2x^2 y'' - xy' + (x^2 + 1)y = 0$

**19.** $3xy'' + (2 - x)y' - y = 0$

**20.** $x^2 y'' - \left(x - \frac{2}{9}\right)y = 0$

**21.** $2xy'' - (3 + 2x)y' + y = 0$

**22.** $x^2y'' + xy' + (x^2 - \frac{4}{9})y = 0$

**23.** $9x^2y'' + 9x^2y' + 2y = 0$

**24.** $2x^2y'' + 3xy' + (2x - 1)y = 0$

In Problems 25–30, $x = 0$ is a regular singular point of the given differential equation. Show that the indicial roots of the singularity differ by an integer. Use the method of Frobenius to obtain at least one series solution about $x = 0$. Use (23) where necessary and a CAS, if instructed, to find a second solution. Form the general solution on $(0, \infty)$.

**25.** $xy'' + 2y' - xy = 0$

**26.** $x^2y'' + xy' + (x^2 - \frac{1}{4})y = 0$

**27.** $xy'' - xy' + y = 0$    **28.** $y'' + \dfrac{3}{x}y' - 2y = 0$

**29.** $xy'' + (1 - x)y' - y = 0$    **30.** $xy'' + y' + y = 0$

In Problems 31 and 32, $x = 0$ is a regular singular point of the given differential equation. Show that the indicial roots of the singularity differ by an integer. Use the recurrence relation found by the method of Frobenius first with the larger root $r_1$. How many solutions did you find? Next use the recurrence relation with the smaller root $r_2$. How many solutions did you find?

**31.** $xy'' + (x - 6)y' - 3y = 0$

**32.** $x(x - 1)y'' + 3y' - 2y = 0$

**33. (a)** The differential equation $x^4y'' + \lambda y = 0$ has an irregular singular point at $x = 0$. Show that the substitution $t = 1/x$ yields the DE

$$\frac{d^2y}{dt^2} + \frac{2}{t}\frac{dy}{dt} + \lambda y = 0,$$

which now has a regular singular point at $t = 0$.

**(b)** Use the method of this section to find two series solutions of the second equation in part (a) about the regular singular point $t = 0$.

**(c)** Express each series solution of the original equation in terms of elementary functions.

### Mathematical Model

**34. Buckling of a Tapered Column**   In Example 4 of Section 5.2 we saw that when a constant vertical compressive force or load $P$ was applied to a thin column of uniform cross section, the deflection $y(x)$ was a solution of the boundary-value problem

$$EI\frac{d^2y}{dx^2} + Py = 0, \quad y(0) = 0, \quad y(L) = 0. \qquad (24)$$

The assumption here is that the column is hinged at both ends. The column will buckle or deflect only when the compressive force is a critical load $P_n$.

**(a)** In this problem let us assume that the column is of length $L$, is hinged at both ends, has circular cross sections, and is tapered as shown in Figure 6.3.1(a). If the column, a truncated cone, has a linear taper $y = cx$ as shown in cross section in Figure 6.3.1(b), the moment of inertia of a cross section with respect to an axis perpendicular to the $xy$-plane is $I = \frac{1}{4}\pi r^4$, where $r = y$ and $y = cx$. Hence we can write $I(x) = I_0(x/b)^4$, where $I_0 = I(b) = \frac{1}{4}\pi(cb)^4$. Substituting $I(x)$ into the differential equation in (24), we see that the deflection in this case is determined from the BVP

$$x^4\frac{d^2y}{dx^2} + \lambda y = 0, \quad y(a) = 0, \quad y(b) = 0,$$

where $\lambda = Pb^4/EI_0$. Use the results of Problem 33 to find the critical loads $P_n$ for the tapered column. Use an appropriate identity to express the buckling modes $y_n(x)$ as a single function.

**(b)** Use a CAS to plot the graph of the first buckling mode $y_1(x)$ corresponding to the Euler load $P_1$ when $b = 11$ and $a = 1$.

**FIGURE 6.3.1** Tapered column in Problem 34

### Discussion Problems

**35.** Discuss how you would define a regular singular point for the linear third-order differential equation

$$a_3(x)y''' + a_2(x)y'' + a_1(x)y' + a_0(x)y = 0.$$

**36.** Each of the differential equations

$$x^3y'' + y = 0 \quad \text{and} \quad x^2y'' + (3x - 1)y' + y = 0$$

has an irregular singular point at $x = 0$. Determine whether the method of Frobenius yields a series solution of each differential equation about $x = 0$. Discuss and explain your findings.

**37.** We have seen that $x = 0$ is a regular singular point of any Cauchy-Euler equation $ax^2y'' + bxy' + cy = 0$. Are the indicial equation (14) for a Cauchy-Euler equation and its auxiliary equation related? Discuss.

## 6.4 Special Functions

**INTRODUCTION**   In the *Remarks* at the end of Section 2.3 we mentioned the branch of mathematics called **special functions.** Perhaps a better title for this field of applied mathematics might be *named functions* because many of the functions studied bear proper names: Bessel functions, Legendre functions, Airy functions, Chebyshev polynomials, Hermite polynomials, Jacobi polynomials, Laguerre polynomials, Gauss' hypergeometric function, Mathieu functions, and so on. Historically, special functions were often the by-product of necessity: Someone needed a solution of a very specialized differential equation that arose from an attempt to solve a physical problem. In effect, a special function was determined or defined by the differential equation and many properties of the function could be discerned from the series form of the solution.

In this section we use the methods of Sections 6.2 and 6.3 to find solutions of two differential equations

$$x^2 y'' + xy' + (x^2 - \nu^2)y = 0 \tag{1}$$

$$(1 - x^2)y'' - 2xy' + n(n + 1)y = 0 \tag{2}$$

that arise in the advanced studies of applied mathematics, physics, and engineering. Equation (1) is called **Bessel's equation of order $\nu$** and is named after the noted German astronomer and mathematician **Friedrich Wilhelm Bessel** (1784–1846), who was the first person to determine the accurate distance from the Sun to another star. Bessel first encountered a special form of equation (1) in his study of elliptical planetary motion and eventually carried out a systematic study of the properties of the solutions of the general equation. Differential equation (2) is known as **Legendre's equation of order $n$** and is named after the French mathematician **Adrien-Marie Legendre** (1752–1833). When we solve (1) we will assume that $\nu \geq 0$; whereas in (2) we will consider only the case where $n$ is a nonnegative integer.

**SOLUTION OF BESSEL'S EQUATION**   Because $x = 0$ is a regular singular point of Bessel's equation, we know that there exists at least one solution of the form $y = \sum_{n=0}^{\infty} c_n x^{n+r}$. Substituting the last expression into (1) gives

$$x^2 y'' + xy' + (x^2 - \nu^2)y = \sum_{n=0}^{\infty} c_n(n + r)(n + r - 1)x^{n+r} + \sum_{n=0}^{\infty} c_n(n + r)x^{n+r} + \sum_{n=0}^{\infty} c_n x^{n+r+2} - \nu^2 \sum_{n=0}^{\infty} c_n x^{n+r}$$

$$= c_0(r^2 - r + r - \nu^2)x^r + x^r \sum_{n=1}^{\infty} c_n[(n + r)(n + r - 1) + (n + r) - \nu^2]x^n + x^r \sum_{n=0}^{\infty} c_n x^{n+2}$$

$$= c_0(r^2 - \nu^2)x^r + x^r \sum_{n=1}^{\infty} c_n[(n + r)^2 - \nu^2]x^n + x^r \sum_{n=0}^{\infty} c_n x^{n+2}. \tag{3}$$

From (3) we see that the indicial equation is $r^2 - \nu^2 = 0$, so the indicial roots are $r_1 = \nu$ and $r_2 = -\nu$. When $r_1 = \nu$, (3) becomes

$$x^\nu \sum_{n=1}^{\infty} c_n n(n + 2\nu)x^n + x^\nu \sum_{n=0}^{\infty} c_n x^{n+2}$$

$$= x^\nu \left[ (1 + 2\nu)c_1 x + \underbrace{\sum_{n=2}^{\infty} c_n n(n + 2\nu)x^n}_{k = n - 2} + \underbrace{\sum_{n=0}^{\infty} c_n x^{n+2}}_{k = n} \right]$$

$$= x^\nu \left[ (1 + 2\nu)c_1 x + \sum_{k=0}^{\infty} [(k + 2)(k + 2 + 2\nu)c_{k+2} + c_k]x^{k+2} \right] = 0.$$

Therefore by the usual argument we can write $(1 + 2\nu)c_1 = 0$ and

$$(k + 2)(k + 2 + 2\nu)c_{k+2} + c_k = 0$$

or $\quad c_{k+2} = \dfrac{-c_k}{(k + 2)(k + 2 + 2\nu)}, \quad k = 0, 1, 2, \ldots.$  $\qquad$ (4)

The choice $c_1 = 0$ in (4) implies that $c_3 = c_5 = c_7 = \cdots = 0$, so for $k = 0, 2, 4, \ldots$ we find, after letting $k + 2 = 2n$, $n = 1, 2, 3, \ldots$, that

$$c_{2n} = -\frac{c_{2n-2}}{2^2 n(n + \nu)}. \qquad (5)$$

Thus $\quad c_2 = -\dfrac{c_0}{2^2 \cdot 1 \cdot (1 + \nu)}$

$$c_4 = -\frac{c_2}{2^2 \cdot 2(2 + \nu)} = \frac{c_0}{2^4 \cdot 1 \cdot 2(1 + \nu)(2 + \nu)}$$

$$c_6 = -\frac{c_4}{2^2 \cdot 3(3 + \nu)} = -\frac{c_0}{2^6 \cdot 1 \cdot 2 \cdot 3(1 + \nu)(2 + \nu)(3 + \nu)}$$

$$\vdots$$

$$c_{2n} = \frac{(-1)^n c_0}{2^{2n} n!(1 + \nu)(2 + \nu) \cdots (n + \nu)}, \quad n = 1, 2, 3, \ldots. \qquad (6)$$

It is standard practice to choose $c_0$ to be a specific value, namely,

$$c_0 = \frac{1}{2^\nu \Gamma(1 + \nu)},$$

where $\Gamma(1 + \nu)$ is the gamma function. See Appendix A for a definition of the gamma function. Since this latter function possesses the convenient property $\Gamma(1 + \alpha) = \alpha\Gamma(\alpha)$, we can reduce the indicated product in the denominator of (6) to one term. For example,

$$\Gamma(1 + \nu + 1) = (1 + \nu)\Gamma(1 + \nu)$$

$$\Gamma(1 + \nu + 2) = (2 + \nu)\Gamma(2 + \nu) = (2 + \nu)(1 + \nu)\Gamma(1 + \nu).$$

Hence we can write (6) as

$$c_{2n} = \frac{(-1)^n}{2^{2n+\nu} n!(1 + \nu)(2 + \nu) \cdots (n + \nu)\Gamma(1 + \nu)} = \frac{(-1)^n}{2^{2n+\nu} n!\Gamma(1 + \nu + n)}$$

for $n = 0, 1, 2, \ldots.$

**BESSEL FUNCTIONS OF THE FIRST KIND** Using the coefficients $c_{2n}$ just obtained and $r = \nu$, a series solution of (1) is $y = \sum_{n=0}^{\infty} c_{2n} x^{2n+\nu}$. This solution is usually denoted by $J_\nu(x)$:

$$J_\nu(x) = \sum_{n=0}^{\infty} \frac{(-1)^n}{n!\Gamma(1 + \nu + n)} \left(\frac{x}{2}\right)^{2n+\nu}. \qquad (7)$$

If $\nu \geq 0$, the series converges at least on the interval $[0, \infty)$. Also, for the second exponent $r_2 = -\nu$ we obtain, in exactly the same manner,

$$J_{-\nu}(x) = \sum_{n=0}^{\infty} \frac{(-1)^n}{n!\Gamma(1 - \nu + n)} \left(\frac{x}{2}\right)^{2n-\nu}. \qquad (8)$$

The functions $J_\nu(x)$ and $J_{-\nu}(x)$ are called **Bessel functions of the first kind** of order $\nu$ and $-\nu$, respectively. Depending on the value of $\nu$, (8) may contain negative powers of $x$ and hence converges on $(0, \infty)$.*

Now some care must be taken in writing the general solution of (1). When $\nu = 0$, it is apparent that (7) and (8) are the same. If $\nu > 0$ and $r_1 - r_2 = \nu - (-\nu) = 2\nu$ is not a positive integer, it follows from Case I of Section 6.3 that $J_\nu(x)$ and $J_{-\nu}(x)$ are linearly independent solutions of (1) on $(0, \infty)$, and so the general solution on the interval is $y = c_1 J_\nu(x) + c_2 J_{-\nu}(x)$. But we also know from Case II of Section 6.3 that when $r_1 - r_2 = 2\nu$ is a positive integer, a second series solution of (1) *may* exist. In this second case we distinguish two possibilities. When $\nu = m =$ positive integer, $J_{-m}(x)$ defined by (8) and $J_m(x)$ are not linearly independent solutions. It can be shown that $J_{-m}$ is a constant multiple of $J_m$ (see Property (*i*) on page 267). In addition, $r_1 - r_2 = 2\nu$ can be a positive integer when $\nu$ is half an odd positive integer. It can be shown in this latter event that $J_\nu(x)$ and $J_{-\nu}(x)$ are linearly independent. In other words, the general solution of (1) on $(0, \infty)$ is

$$y = c_1 J_\nu(x) + c_2 J_{-\nu}(x), \qquad \nu \neq \text{integer}. \tag{9}$$

The graphs of $y = J_0(x)$ and $y = J_1(x)$ are given in Figure 6.4.1.

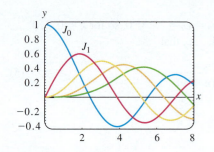

**FIGURE 6.4.1** Bessel functions of the first kind for $n = 0, 1, 2, 3, 4$

---

**EXAMPLE 1**   Bessel's Equation of Order $\frac{1}{2}$

By identifying $\nu^2 = \frac{1}{4}$ and $\nu = \frac{1}{2}$, we can see from (9) that the general solution of the equation $x^2 y'' + x y' + \left(x^2 - \frac{1}{4}\right) y = 0$ on $(0, \infty)$ is $y = c_1 J_{1/2}(x) + c_2 J_{-1/2}(x)$. ■

**BESSEL FUNCTIONS OF THE SECOND KIND**   If $\nu \neq$ integer, the function defined by the linear combination

$$Y_\nu(x) = \frac{\cos \nu\pi J_\nu(x) - J_{-\nu}(x)}{\sin \nu\pi} \tag{10}$$

and the function $J_\nu(x)$ are linearly independent solutions of (1). Thus another form of the general solution of (1) is $y = c_1 J_\nu(x) + c_2 Y_\nu(x)$, provided that $\nu \neq$ integer. As $\nu \to m$, $m$ an integer, (10) has the indeterminate form $0/0$. However, it can be shown by L'Hôpital's Rule that $\lim_{\nu \to m} Y_\nu(x)$ exists. Moreover, the function

$$Y_m(x) = \lim_{\nu \to m} Y_\nu(x)$$

and $J_m(x)$ are linearly independent solutions of $x^2 y'' + x y' + (x^2 - m^2) y = 0$. Hence for *any* value of $\nu$ the general solution of (1) on $(0, \infty)$ can be written as

$$y = c_1 J_\nu(x) + c_2 Y_\nu(x). \tag{11}$$

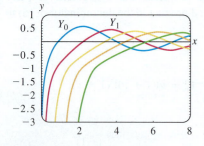

**FIGURE 6.4.2** Bessel functions of the second kind for $n = 0, 1, 2, 3, 4$

$Y_\nu(x)$ is called the **Bessel function of the second kind** of order $\nu$. Figure 6.4.2 shows the graphs of $Y_0(x)$ and $Y_1(x)$.

---

**EXAMPLE 2**   Bessel's Equation of Order 3

By identifying $\nu^2 = 9$ and $\nu = 3$, we see from (11) that the general solution of the equation $x^2 y'' + x y' + (x^2 - 9) y = 0$ on $(0, \infty)$ is $y = c_1 J_3(x) + c_2 Y_3(x)$. ■

**DEs SOLVABLE IN TERMS OF BESSEL FUNCTIONS**   Sometimes it is possible to transform a differential equation into equation (1) by means of a change of variable. We can then express the solution of the original equation in terms of Bessel functions. For example, if we let $t = \alpha x$, $\alpha > 0$, in

$$x^2 y'' + x y' + (\alpha^2 x^2 - \nu^2) y = 0, \tag{12}$$

---

*When we replace $x$ by $|x|$, the series given in (7) and (8) converge for $0 < |x| < \infty$.

then by the Chain Rule,

$$\frac{dy}{dx} = \frac{dy}{dt}\frac{dt}{dx} = \alpha\frac{dy}{dt} \qquad \text{and} \qquad \frac{d^2y}{dx^2} = \frac{d}{dt}\left(\frac{dy}{dx}\right)\frac{dt}{dx} = \alpha^2\frac{d^2y}{dt^2}.$$

Accordingly, (12) becomes

$$\left(\frac{t}{\alpha}\right)^2\alpha^2\frac{d^2y}{dt^2} + \left(\frac{t}{\alpha}\right)\alpha\frac{dy}{dt} + (t^2 - \nu^2)y = 0 \qquad \text{or} \qquad t^2\frac{d^2y}{dt^2} + t\frac{dy}{dt} + (t^2 - \nu^2)y = 0.$$

The last equation is Bessel's equation of order $\nu$ with solution $y = c_1J_\nu(t) + c_2Y_\nu(t)$. By resubstituting $t = \alpha x$ in the last expression, we find that the general solution of (12) is

$$y = c_1J_\nu(\alpha x) + c_2Y_\nu(\alpha x). \tag{13}$$

Equation (12), called the **parametric Bessel equation of order $\nu$,** and its general solution (13) are very important in the study of certain boundary-value problems involving partial differential equations that are expressed in cylindrical coordinates.

**MODIFIED BESSEL FUNCTIONS**    Another equation that bears a resemblance to (1) is the **modified Bessel equation of order $\nu$,**

$$x^2y'' + xy' - (x^2 + \nu^2)y = 0. \tag{14}$$

This DE can be solved in the manner just illustrated for (12). This time if we let $t = ix$, where $i^2 = -1$, then (14) becomes

$$t^2\frac{d^2y}{dt^2} + t\frac{dy}{dt} + (t^2 - \nu^2)y = 0.$$

Because solutions of the last DE are $J_\nu(t)$ and $Y_\nu(t)$, *complex-valued* solutions of (14) are $J_\nu(ix)$ and $Y_\nu(ix)$. A real-valued solution, called the **modified Bessel function of the first kind** of order $\nu$, is defined in terms of $J_\nu(ix)$:

$$I_\nu(x) = i^{-\nu}J_\nu(ix).$$

See Problem 21 in Exercises 6.4.

The general solution of (14) is

$$y = c_1I_\nu(x) + c_2I_{-\nu}(x), \quad \nu \neq \text{integer}. \tag{15}$$

When $\nu$ is an integer $n$ the functions $I_n(x)$ and $I_{-n}(x)$ are not linearly independent on the interval $(0, \infty)$. So analogous to (10) we define the **modified Bessel function of the second kind** of order $\nu \neq$ integer to be

$$K_\nu(x) = \frac{\pi}{2}\frac{I_{-\nu}(x) - I_\nu(x)}{\sin \nu\pi}, \tag{16}$$

and for integer $\nu = n$,

$$K_n(x) = \lim_{\nu \to n} K_\nu(x).$$

Because $I_\nu$ and $K_\nu$ are linearly independent on the interval $(0, \infty)$ for any value of $\nu$, the general solution of (14) on that interval is

$$y = c_1I_\nu(x) + c_2K_\nu(x). \tag{17}$$

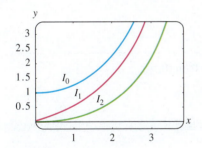

**FIGURE 6.4.3** Modified Bessel functions of the first kind for $n = 0, 1, 2$

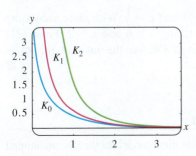

**FIGURE 6.4.4** Modified Bessel functions of the second kind for $n = 0, 1, 2$

The graphs of $y = I_0(x)$, $y = I_1(x)$, and $y = I_2(x)$ are given in Figure 6.4.3 and the graphs of $y = K_0(x)$, $y = K_1(x)$, and $y = K_2(x)$ are given in Figure 6.4.4. Unlike the Bessel functions of the first and second kinds, the modified Bessel functions of the first and second kind are not oscillatory. Figures 6.4.3 and 6.4.4 also illustrate the fact that the modified Bessel functions $I_n(x)$ and $K_n(x)$, $n = 0, 1, 2, \ldots$ have no real zeros in the interval $(0, \infty)$. Also notice that the modified Bessel functions of the second kind $K_n(x)$ like the Bessel functions of the second kind $Y_n(x)$ become unbounded as $x \to 0^+$.

A change of variable in (14) gives us the **parametric form of the modified Bessel equation of order $\nu$:**

$$x^2 y'' + xy' - (\alpha^2 x^2 + \nu^2)y = 0. \tag{18}$$

The general solution of the last equation on the interval $(0, \infty)$ is

$$y = c_1 I_\nu(\alpha x) + c_2 K_\nu(\alpha x). \tag{19}$$

### EXAMPLE 3  Parametric Modified Bessel Equation

By identifying $\alpha^2 = 25$, $\nu^2 = 4$, $\alpha = 5$, and $\nu = 2$ it follows from (19) that the general solution of the equation $x^2 y'' + xy' - (25x^2 + 4)y = 0$ on $(0, \infty)$ is $y = c_1 I_2(5x) + c_2 K_2(5x)$. ■

Yet another equation, important because many DEs fit into its form by appropriate choices of the parameters, is

$$y'' + \frac{1 - 2a}{x} y' + \left( b^2 c^2 x^{2c-2} + \frac{a^2 - p^2 c^2}{x^2} \right) y = 0, \quad p \geq 0. \tag{20}$$

Although we shall not supply the details, the general solution of (20),

$$y = x^a \left[ c_1 J_p(bx^c) + c_2 Y_p(bx^c) \right], \tag{21}$$

can be found by means of a change in both the independent and the dependent variables: $z = bx^c$, $y(x) = \left( \dfrac{z}{b} \right)^{a/c} w(z)$. If $p$ is not an integer, then $Y_p$ in (21) can be replaced by $J_{-p}$.

### EXAMPLE 4  Using (20)

Find the general solution of $xy'' + 3y' + 9y = 0$ on $(0, \infty)$.

**SOLUTION** By writing the given DE as

$$y'' + \frac{3}{x} y' + \frac{9}{x} y = 0,$$

we can make the following identifications with (20):

$$1 - 2a = 3, \qquad b^2 c^2 = 9, \qquad 2c - 2 = -1, \qquad \text{and} \qquad a^2 - p^2 c^2 = 0.$$

The first and third equations imply that $a = -1$ and $c = \frac{1}{2}$. With these values the second and fourth equations are satisfied by taking $b = 6$ and $p = 2$. From (21) we find that the general solution of the given DE on the interval $(0, \infty)$ is $y = x^{-1}[c_1 J_2(6x^{1/2}) + c_2 Y_2(6x^{1/2})]$. ■

### EXAMPLE 5  The Aging Spring Revisited

Recall that in Section 5.1 we saw that one mathematical model for the free undamped motion of a mass on an aging spring is given by $mx'' + ke^{-\alpha t}x = 0$, $\alpha > 0$. We are now in a position to find the general solution of the equation. It is left as a problem to show that the change of variables $s = \dfrac{2}{\alpha} \sqrt{\dfrac{k}{m}} e^{-\alpha t/2}$ transforms the differential equation of the aging spring into

$$s^2 \frac{d^2 x}{ds^2} + s \frac{dx}{ds} + s^2 x = 0.$$

The last equation is recognized as (1) with $\nu = 0$ and where the symbols $x$ and $s$ play the roles of $y$ and $x$, respectively. The general solution of the new equation is $x = c_1 J_0(s) + c_2 Y_0(s)$. If we resubstitute $s$, then the general solution of $mx'' + ke^{-\alpha t}x = 0$ is seen to be

$$x(t) = c_1 J_0\left(\frac{2}{\alpha} \sqrt{\frac{k}{m}} e^{-\alpha t/2}\right) + c_2 Y_0\left(\frac{2}{\alpha} \sqrt{\frac{k}{m}} e^{-\alpha t/2}\right).$$

See Problems 35 and 41 in Exercises 6.4.  ■

The other model that was discussed in Section 5.1 of a spring whose characteristics change with time was $mx'' + ktx = 0$. By dividing through by $m$, we see that the equation $x'' + \dfrac{k}{m}tx = 0$ is one form of Airy's differential equation. See page 248. The general solution of Airy's differential equation can also be written in terms of Bessel functions. See Problems 36, 37, and 42 in Exercises 6.4.

**PROPERTIES**  We list below a few of the more useful properties of Bessel functions of order $m$, $m = 0, 1, 2, \ldots$:

$(i)$  $J_{-m}(x) = (-1)^m J_m(x)$,  $\qquad$  $(ii)$  $J_m(-x) = (-1)^m J_m(x)$,

$(iii)$  $J_m(0) = \begin{cases} 0, & m > 0 \\ 1, & m = 0, \end{cases}$  $\qquad$  $(iv)$  $\displaystyle\lim_{x \to 0^+} Y_m(x) = -\infty.$

Note that Property $(ii)$ indicates that $J_m(x)$ is an even function if $m$ is an even integer and an odd function if $m$ is an odd integer. The graphs of $Y_0(x)$ and $Y_1(x)$ in Figure 6.4.2 illustrate Property $(iv)$, namely, $Y_m(x)$ is unbounded at the origin. This last fact is not obvious from (10). The solutions of the Bessel equation of order 0 can be obtained by using the solutions $y_1(x)$ in (21) and $y_2(x)$ in (22) of Section 6.3. It can be shown that (21) of Section 6.3 is $y_1(x) = J_0(x)$, whereas (22) of that section is

$$y_2(x) = J_0(x) \ln x - \sum_{k=1}^{\infty} \frac{(-1)^k}{(k!)^2}\left(1 + \frac{1}{2} + \cdots + \frac{1}{k}\right)\left(\frac{x}{2}\right)^{2k}.$$

The Bessel function of the second kind of order 0, $Y_0(x)$, is then defined to be the linear combination $Y_0(x) = \dfrac{2}{\pi}(\gamma - \ln 2)y_1(x) + \dfrac{2}{\pi}y_2(x)$ for $x > 0$. That is,

$$Y_0(x) = \frac{2}{\pi}J_0(x)\left[\gamma + \ln\frac{x}{2}\right] - \frac{2}{\pi}\sum_{k=1}^{\infty}\frac{(-1)^k}{(k!)^2}\left(1 + \frac{1}{2} + \cdots + \frac{1}{k}\right)\left(\frac{x}{2}\right)^{2k},$$

where $\gamma = 0.57721566\ldots$ is **Euler's constant.** Because of the presence of the logarithmic term, it is apparent that $Y_0(x)$ is discontinuous at $x = 0$.

**NUMERICAL VALUES**  The first five nonnegative zeros of $J_0(x)$, $J_1(x)$, $Y_0(x)$, and $Y_1(x)$ are given in Table 6.4.1. Some additional function values of these four functions are given in Table 6.4.2.

**TABLE 6.4.1**    Zeros of $J_0$, $J_1$, $Y_0$, and $Y_1$

| $J_0(x)$ | $J_1(x)$ | $Y_0(x)$ | $Y_1(x)$ |
|---|---|---|---|
| 2.4048 | 0.0000 | 0.8936 | 2.1971 |
| 5.5201 | 3.8317 | 3.9577 | 5.4297 |
| 8.6537 | 7.0156 | 7.0861 | 8.5960 |
| 11.7915 | 10.1735 | 10.2223 | 11.7492 |
| 14.9309 | 13.3237 | 13.3611 | 14.8974 |

**TABLE 6.4.2**    Numerical Values of $J_0$, $J_1$, $Y_0$, and $Y_1$

| $x$ | $J_0(x)$ | $J_1(x)$ | $Y_0(x)$ | $Y_1(x)$ |
|---|---|---|---|---|
| 0 | 1.0000 | 0.0000 | — | — |
| 1 | 0.7652 | 0.4401 | 0.0883 | −0.7812 |
| 2 | 0.2239 | 0.5767 | 0.5104 | −0.1070 |
| 3 | −0.2601 | 0.3391 | 0.3769 | 0.3247 |
| 4 | −0.3971 | −0.0660 | −0.0169 | 0.3979 |
| 5 | −0.1776 | −0.3276 | −0.3085 | 0.1479 |
| 6 | 0.1506 | −0.2767 | −0.2882 | −0.1750 |
| 7 | 0.3001 | −0.0047 | −0.0259 | −0.3027 |
| 8 | 0.1717 | 0.2346 | 0.2235 | −0.1581 |
| 9 | −0.0903 | 0.2453 | 0.2499 | 0.1043 |
| 10 | −0.2459 | 0.0435 | 0.0557 | 0.2490 |
| 11 | −0.1712 | −0.1768 | −0.1688 | 0.1637 |
| 12 | 0.0477 | −0.2234 | −0.2252 | −0.0571 |
| 13 | 0.2069 | −0.0703 | −0.0782 | −0.2101 |
| 14 | 0.1711 | 0.1334 | 0.1272 | −0.1666 |
| 15 | −0.0142 | 0.2051 | 0.2055 | 0.0211 |

**DIFFERENTIAL RECURRENCE RELATION**    Recurrence formulas that relate Bessel functions of different orders are important in theory and in applications. In the next example we derive a **differential recurrence relation.**

**EXAMPLE 6**    **Derivation Using the Series Definition**

Derive the formula $xJ_\nu'(x) = \nu J_\nu(x) - xJ_{\nu+1}(x)$.

**SOLUTION**    It follows from (7) that

$$xJ_\nu'(x) = \sum_{n=0}^{\infty} \frac{(-1)^n(2n+\nu)}{n!\,\Gamma(1+\nu+n)} \left(\frac{x}{2}\right)^{2n+\nu}$$

$$= \nu \sum_{n=0}^{\infty} \frac{(-1)^n}{n!\,\Gamma(1+\nu+n)} \left(\frac{x}{2}\right)^{2n+\nu} + 2 \sum_{n=0}^{\infty} \frac{(-1)^n n}{n!\,\Gamma(1+\nu+n)} \left(\frac{x}{2}\right)^{2n+\nu}$$

$$= \nu J_\nu(x) + x \underbrace{\sum_{n=1}^{\infty} \frac{(-1)^n}{(n-1)!\,\Gamma(1+\nu+n)} \left(\frac{x}{2}\right)^{2n+\nu-1}}_{k=n-1}$$

$$= \nu J_\nu(x) - x \sum_{k=0}^{\infty} \frac{(-1)^k}{k!\,\Gamma(2+\nu+k)} \left(\frac{x}{2}\right)^{2k+\nu+1} = \nu J_\nu(x) - xJ_{\nu+1}(x).$$

    The result in Example 6 can be written in an alternative form. Dividing $xJ_\nu'(x) - \nu J_\nu(x) = -xJ_{\nu+1}(x)$ by $x$ gives

$$J_\nu'(x) - \frac{\nu}{x} J_\nu(x) = -J_{\nu+1}(x).$$

This last expression is recognized as a linear first-order differential equation in $J_\nu(x)$. Multiplying both sides of the equality by the integrating factor $x^{-\nu}$ then yields

$$\frac{d}{dx}[x^{-\nu}J_\nu(x)] = -x^{-\nu}J_{\nu+1}(x). \tag{22}$$

It can be shown in a similar manner that

$$\frac{d}{dx}[x^\nu J_\nu(x)] = x^\nu J_{\nu-1}(x).  \quad (23)$$

See Problem 27 in Exercises 6.4. The differential recurrence relations (22) and (23) are also valid for the Bessel function of the second kind $Y_\nu(x)$. Observe that when $\nu = 0$, it follows from (22) that

$$\frac{d}{dx}J_0(x) = -J_1(x) \quad \text{and} \quad \frac{d}{dx}Y_0(x) = -Y_1(x).  \quad (24)$$

An application of these derivatives is given in Problem 41 in Exercises 6.4. Results similar to (24) also hold for the modified Bessel functions of the first and second kind of order $\nu = 0$:

$$\frac{d}{dx}I_0(x) = I_1(x) \quad \text{and} \quad \frac{d}{dx}K_0(x) = -K_1(x).  \quad (25)$$

See Problem 27 in Chapter 6 in Review for an application of the derivatives in (25).

**BESSEL FUNCTIONS OF HALF-INTEGRAL ORDER**  When the order is half an odd integer, that is, $\pm\frac{1}{2}, \pm\frac{3}{2}, \pm\frac{5}{2}, \ldots,$ Bessel functions of the first and second kinds can be expressed in terms of the elementary functions $\sin x$, $\cos x$, and powers of $x$. Let's consider the case when $\nu = \frac{1}{2}$. From (7)

$$J_{1/2}(x) = \sum_{n=0}^{\infty} \frac{(-1)^n}{n!\,\Gamma\!\left(1 + \frac{1}{2} + n\right)} \left(\frac{x}{2}\right)^{2n+1/2}.$$

In view of the property $\Gamma(1 + \alpha) = \alpha\Gamma(\alpha)$ and the fact that $\Gamma(\frac{1}{2}) = \sqrt{\pi}$ the values of $\Gamma(1 + \frac{1}{2} + n)$ for $n = 0, n = 1, n = 2,$ and $n = 3$ are, respectively,

$$\Gamma\!\left(\tfrac{3}{2}\right) = \Gamma\!\left(1 + \tfrac{1}{2}\right) = \tfrac{1}{2}\Gamma\!\left(\tfrac{1}{2}\right) = \tfrac{1}{2}\sqrt{\pi}$$

$$\Gamma\!\left(\tfrac{5}{2}\right) = \Gamma\!\left(1 + \tfrac{3}{2}\right) = \tfrac{3}{2}\Gamma\!\left(\tfrac{3}{2}\right) = \frac{3}{2^2}\sqrt{\pi}$$

$$\Gamma\!\left(\tfrac{7}{2}\right) = \Gamma\!\left(1 + \tfrac{5}{2}\right) = \tfrac{5}{2}\Gamma\!\left(\tfrac{5}{2}\right) = \frac{5\cdot3}{2^3}\sqrt{\pi} = \frac{5\cdot4\cdot3\cdot2\cdot1}{2^34\cdot2}\sqrt{\pi} = \frac{5!}{2^52!}\sqrt{\pi}$$

$$\Gamma\!\left(\tfrac{9}{2}\right) = \Gamma\!\left(1 + \tfrac{7}{2}\right) = \tfrac{7}{2}\Gamma\!\left(\tfrac{7}{2}\right) = \frac{7\cdot5}{2^6\cdot2!}\sqrt{\pi} = \frac{7\cdot6\cdot5!}{2^6\cdot6\cdot2!}\sqrt{\pi} = \frac{7!}{2^73!}\sqrt{\pi}.$$

In general,

$$\Gamma\!\left(1 + \tfrac{1}{2} + n\right) = \frac{(2n+1)!}{2^{2n+1}n!}\sqrt{\pi}.$$

Hence

$$J_{1/2}(x) = \sum_{n=0}^{\infty} \frac{(-1)^n}{n!\,\dfrac{(2n+1)!}{2^{2n+1}n!}\sqrt{\pi}} \left(\frac{x}{2}\right)^{2n+1/2} = \sqrt{\frac{2}{\pi x}} \sum_{n=0}^{\infty} \frac{(-1)^n}{(2n+1)!}\, x^{2n+1}.$$

From (2) of Section 6.1 the infinite series in the last line is recognized as the Maclaurin series for $\sin x$, and so we have shown that

$$J_{1/2}(x) = \sqrt{\frac{2}{\pi x}}\,\sin x.  \quad (26)$$

We leave it as an exercise to show that

$$J_{-1/2}(x) = \sqrt{\frac{2}{\pi x}}\,\cos x.  \quad (27)$$

See Figure 6.4.5 and Problems 31, 32, and 40 in Exercises 6.4.

**FIGURE 6.4.5** Bessel functions of order $\frac{1}{2}$ (blue) and order $-\frac{1}{2}$ (red)

If $n$ is an integer, then $\nu = n + \frac{1}{2}$ is half an odd integer. Because $\cos(n + \frac{1}{2})\pi = 0$ and $\sin(n + \frac{1}{2})\pi = \cos n\pi = (-1)^n$, we see from (10) that

$$Y_{n+1/2}(x) = (-1)^{n+1} J_{-(n+1/2)}(x).$$

For $n = 0$ and $n = -1$ we have, in turn, $Y_{1/2}(x) = -J_{-1/2}(x)$ and $Y_{-1/2}(x) = J_{1/2}(x)$. In view of (26) and (27) these results are the same as

$$Y_{1/2}(x) = -\sqrt{\frac{2}{\pi x}} \cos x \tag{28}$$

and

$$Y_{-1/2}(x) = \sqrt{\frac{2}{\pi x}} \sin x. \tag{29}$$

**SPHERICAL BESSEL FUNCTIONS**    Bessel functions of half-integral order are used to define two more important functions:

$$j_n(x) = \sqrt{\frac{\pi}{2x}} J_{n+1/2}(x) \quad \text{and} \quad y_n(x) = \sqrt{\frac{\pi}{2x}} Y_{n+1/2}(x). \tag{30}$$

The function $j_n(x)$ is called the **spherical Bessel function of the first kind** and $y_n(x)$ is the **spherical Bessel function of the second kind.** For example, for $n = 0$ the expressions in (30) become

$$j_0(x) = \sqrt{\frac{\pi}{2x}} J_{1/2}(x) = \sqrt{\frac{\pi}{2x}} \sqrt{\frac{2}{\pi x}} \sin x = \frac{\sin x}{x}$$

and

$$y_0(x) = \sqrt{\frac{\pi}{2x}} Y_{1/2}(x) = -\sqrt{\frac{\pi}{2x}} \sqrt{\frac{2}{\pi x}} \cos x = -\frac{\cos x}{x}.$$

It is apparent from (30) and Figure 6.4.2 for $n \geq 0$ the spherical Bessel of the second kind $y_n(x)$ becomes unbounded as $x \to 0^+$.

Spherical Bessel functions arise in the solution of a special partial differential equation expressed in spherical coordinates. See Problem 56 in Exercises 6.4 and Problem 13 in Exercises 13.3.

**SOLUTION OF LEGENDRE'S EQUATION**    Since $x = 0$ is an ordinary point of Legendre's equation (2), we substitute the series $y = \sum_{k=0}^{\infty} c_k x^k$, shift summation indices, and combine series to get

$$(1 - x^2)y'' - 2xy' + n(n + 1)y = [n(n + 1)c_0 + 2c_2] + [(n - 1)(n + 2)c_1 + 6c_3]x$$

$$+ \sum_{j=2}^{\infty} [(j + 2)(j + 1)c_{j+2} + (n - j)(n + j + 1)c_j]x^j = 0$$

which implies that

$$n(n + 1)c_0 + 2c_2 = 0$$

$$(n - 1)(n + 2)c_1 + 6c_3 = 0$$

$$(j + 2)(j + 1)c_{j+2} + (n - j)(n + j + 1)c_j = 0$$

or

$$c_2 = -\frac{n(n + 1)}{2!} c_0$$

$$c_3 = -\frac{(n - 1)(n + 2)}{3!} c_1$$

$$c_{j+2} = -\frac{(n - j)(n + j + 1)}{(j + 2)(j + 1)} c_j, \quad j = 2, 3, 4, \ldots. \tag{31}$$

If we let $j$ take on the values 2, 3, 4, . . . , the recurrence relation (31) yields

$$c_4 = -\frac{(n-2)(n+3)}{4 \cdot 3}c_2 = \frac{(n-2)n(n+1)(n+3)}{4!}c_0$$

$$c_5 = -\frac{(n-3)(n+4)}{5 \cdot 4}c_3 = \frac{(n-3)(n-1)(n+2)(n+4)}{5!}c_1$$

$$c_6 = -\frac{(n-4)(n+5)}{6 \cdot 5}c_4 = -\frac{(n-4)(n-2)n(n+1)(n+3)(n+5)}{6!}c_0$$

$$c_7 = -\frac{(n-5)(n+6)}{7 \cdot 6}c_5 = -\frac{(n-5)(n-3)(n-1)(n+2)(n+4)(n+6)}{7!}c_1$$

and so on. Thus for at least $|x| < 1$ we obtain two linearly independent power series solutions:

$$y_1(x) = c_0\left[1 - \frac{n(n+1)}{2!}x^2 + \frac{(n-2)n(n+1)(n+3)}{4!}x^4\right.$$

$$\left. - \frac{(n-4)(n-2)n(n+1)(n+3)(n+5)}{6!}x^6 + \cdots\right]$$

$$y_2(x) = c_1\left[x - \frac{(n-1)(n+2)}{3!}x^3 + \frac{(n-3)(n-1)(n+2)(n+4)}{5!}x^5\right.$$ (32)

$$\left. - \frac{(n-5)(n-3)(n-1)(n+2)(n+4)(n+6)}{7!}x^7 + \cdots\right].$$

Notice that if $n$ is an even integer, the first series terminates, whereas $y_2(x)$ is an infinite series. For example, if $n = 4$, then

$$y_1(x) = c_0\left[1 - \frac{4 \cdot 5}{2!}x^2 + \frac{2 \cdot 4 \cdot 5 \cdot 7}{4!}x^4\right] = c_0\left[1 - 10x^2 + \frac{35}{3}x^4\right].$$

Similarly, when $n$ is an odd integer, the series for $y_2(x)$ terminates with $x^n$; that is, *when $n$ is a nonnegative integer, we obtain an $n$th-degree polynomial solution* of Legendre's equation.

Because we know that a constant multiple of a solution of Legendre's equation is also a solution, it is traditional to choose specific values for $c_0$ or $c_1$, depending on whether $n$ is an even or odd positive integer, respectively. For $n = 0$ we choose $c_0 = 1$, and for $n = 2, 4, 6, \ldots$

$$c_0 = (-1)^{n/2}\frac{1 \cdot 3 \cdots (n-1)}{2 \cdot 4 \cdots n},$$

whereas for $n = 1$ we choose $c_1 = 1$, and for $n = 3, 5, 7, \ldots$

$$c_1 = (-1)^{(n-1)/2}\frac{1 \cdot 3 \cdots n}{2 \cdot 4 \cdots (n-1)}.$$

For example, when $n = 4$, we have

$$y_1(x) = (-1)^{4/2}\frac{1 \cdot 3}{2 \cdot 4}\left[1 - 10x^2 + \frac{35}{3}x^4\right] = \frac{1}{8}(35x^4 - 30x^2 + 3).$$

**LEGENDRE POLYNOMIALS**    These specific $n$th-degree polynomial solutions are called **Legendre polynomials** and are denoted by $P_n(x)$. From the series for $y_1(x)$ and $y_2(x)$ and from the above choices of $c_0$ and $c_1$ we find that the first six Legendre polynomials are

$$P_0(x) = 1, \qquad\qquad P_1(x) = x,$$

$$P_2(x) = \frac{1}{2}(3x^2 - 1), \qquad P_3(x) = \frac{1}{2}(5x^3 - 3x), \qquad (33)$$

$$P_4(x) = \frac{1}{8}(35x^4 - 30x^2 + 3), \qquad P_5(x) = \frac{1}{8}(63x^5 - 70x^3 + 15x).$$

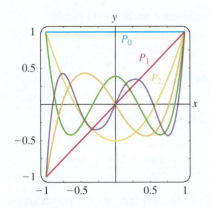

**FIGURE 6.4.6** Legendre polynomials for $n = 0, 1, 2, 3, 4, 5$

Remember, $P_0(x)$, $P_1(x)$, $P_2(x)$, $P_3(x)$, . . . are, in turn, particular solutions of the differential equations

$$n = 0: \quad (1 - x^2)y'' - 2xy' = 0,$$

$$n = 1: \quad (1 - x^2)y'' - 2xy' + 2y = 0,$$

$$n = 2: \quad (1 - x^2)y'' - 2xy' + 6y = 0, \tag{34}$$

$$n = 3: \quad (1 - x^2)y'' - 2xy' + 12y = 0,$$

$$\vdots \qquad \vdots$$

The graphs, on the interval $[-1, 1]$, of the six Legendre polynomials in (33) are given in Figure 6.4.6.

**PROPERTIES**   You are encouraged to verify the following properties using the Legendre polynomials in (33).

$$(\textbf{\textit{i}}) \;\; P_n(-x) = (-1)^n P_n(x)$$

$$(\textbf{\textit{ii}}) \;\; P_n(1) = 1 \qquad\qquad (\textbf{\textit{iii}}) \;\; P_n(-1) = (-1)^n$$

$$(\textbf{\textit{iv}}) \;\; P_n(0) = 0, \; n \text{ odd} \qquad (\textbf{\textit{v}}) \;\; P_n'(0) = 0, \; n \text{ even}$$

Property (*i*) indicates, as is apparent in Figure 6.4.6, that $P_n(x)$ is an even or odd function according to whether $n$ is even or odd.

**RECURRENCE RELATION**   Recurrence relations that relate Legendre polynomials of different degrees are also important in some aspects of their applications. We state, without proof, the three-term recurrence relation

$$(k + 1)P_{k+1}(x) - (2k + 1)xP_k(x) + kP_{k-1}(x) = 0, \tag{35}$$

which is valid for $k = 1, 2, 3, \ldots$ . In (33) we listed the first six Legendre polynomials. If, say, we wish to find $P_6(x)$, we can use (35) with $k = 5$. This relation expresses $P_6(x)$ in terms of the known $P_4(x)$ and $P_5(x)$. See Problem 47 in Exercises 6.4.

Another formula, although not a recurrence relation, can generate the Legendre polynomials by differentiation. **Rodrigues' formula** for these polynomials is

$$P_n(x) = \frac{1}{2^n n!} \frac{d^n}{dx^n} (x^2 - 1)^n, \qquad n = 0, 1, 2, \ldots . \tag{36}$$

See Problem 50 in Exercises 6.4.

---

**REMARKS**

Although we have assumed that the parameter $n$ in Legendre's differential equation $(1 - x^2)y'' - 2xy' + n(n + 1)y = 0$, represented a nonnegative integer, in a more general setting $n$ can represent any real number. Any solution of Legendre's equation is called a **Legendre function**. If $n$ is *not* a nonnegative integer, then both Legendre functions $y_1(x)$ and $y_2(x)$ given in (32) are infinite series convergent on the open interval $(-1, 1)$ and divergent (unbounded) at $x = \pm 1$. If $n$ is a nonnegative integer, then as we have just seen one of the Legendre functions in (32) is a polynomial and the other is an infinite series convergent for $-1 < x < 1$. You should be aware of the fact that Legendre's equation possesses solutions that are bounded on the *closed* interval $[-1, 1]$ only in the case when $n = 0, 1, 2, \ldots$. More to the point, the only Legendre functions that are bounded on the closed interval $[-1, 1]$ are the Legendre polynomials $P_n(x)$ or constant multiples of these polynomials. See Problem 49 in Exercises 6.4 and Problem 24 in Chapter 6 in Review.

## EXERCISES 6.4

Answers to selected odd-numbered problems begin on page ANS-10.

### Bessel's Equation

In Problems 1–6 use (1) to find the general solution of the given differential equation on $(0, \infty)$.

**1.** $x^2y'' + xy' + \left(x^2 - \frac{1}{9}\right)y = 0$

**2.** $x^2y'' + xy' + (x^2 - 1)y = 0$

**3.** $4x^2y'' + 4xy' + (4x^2 - 25)y = 0$

**4.** $16x^2y'' + 16xy' + (16x^2 - 1)y = 0$

**5.** $xy'' + y' + xy = 0$

**6.** $\dfrac{d}{dx}[xy'] + \left(x - \dfrac{4}{x}\right)y = 0$

In Problems 7 and 8 use (12) to find the general solution of the given differential equation on $(0, \infty)$.

**7.** $x^2y'' + xy' + (9x^2 - 4)y = 0$

**8.** $x^2y'' + xy' + \left(36x^2 - \frac{1}{4}\right)y = 0$

In Problems 9 and 10 use (18) to find the general solution of the given differential equation on $(0, \infty)$.

**9.** $x^2y'' + xy' - \left(16x^2 + \frac{4}{9}\right)y = 0$

**10.** $x^2y'' + xy' - (2x^2 + 64)y = 0$

In Problems 11 and 12 use the indicated change of variable to find the general solution of the given differential equation on $(0, \infty)$.

**11.** $x^2y'' + 2xy' + \alpha^2x^2y = 0; \quad y = x^{-1/2}v(x)$

**12.** $x^2y'' + \left(\alpha^2x^2 - v^2 + \frac{1}{4}\right)y = 0; \quad y = \sqrt{x}\,v(x)$

In Problems 13–20 use (20) to find the general solution of the given differential equation on $(0, \infty)$.

**13.** $xy'' + 2y' + 4y = 0$

**14.** $xy'' + 3y' + xy = 0$

**15.** $xy'' - y' + xy = 0$

**16.** $xy'' - 5y' + xy = 0$

**17.** $x^2y'' + (x^2 - 2)y = 0$

**18.** $4x^2y'' + (16x^2 + 1)y = 0$

**19.** $xy'' + 3y' + x^3y = 0$

**20.** $9x^2y'' + 9xy' + (x^6 - 36)y = 0$

**21.** Use the series in (7) to verify that $I_v(x) = i^{-v} J_v(ix)$ is a real function.

**22.** Assume that $b$ in equation (20) can be pure imaginary, that is, $b = \beta i, \beta > 0, i^2 = -1$. Use this assumption to express the general solution of the given differential equation in terms of the modified Bessel functions $I_n$ and $K_n$.

  **(a)** $y'' - x^2y = 0$

  **(b)** $xy'' + y' - 7x^3y = 0$

In Problems 23–26 first use (20) to express the general solution of the given differential equation in terms of Bessel functions. Then use (26) and (27) to express the general solution in terms of elementary functions.

**23.** $y'' + y = 0$

**24.** $x^2y'' + 4xy' + (x^2 + 2)y = 0$

**25.** $16x^2y'' + 32xy' + (x^4 - 12)y = 0$

**26.** $4x^2y'' - 4xy' + (16x^2 + 3)y = 0$

**27. (a)** Proceed as in Example 6 to show that
$$xJ_v'(x) = -vJ_v(x) + xJ_{v-1}(x).$$

  [*Hint*: Write $2n + v = 2(n + v) - v$.]

  **(b)** Use the result in part (a) to derive (23).

**28.** Use the formula obtained in Example 6 along with part (a) of Problem 27 to derive the recurrence relation
$$2vJ_v(x) = xJ_{v+1}(x) + xJ_{v-1}(x).$$

In Problems 29 and 30 use (22) or (23) to obtain the given result.

**29.** $\displaystyle\int_0^x rJ_0(r)\,dr = xJ_1(x)$    **30.** $J_0'(x) = J_{-1}(x) = -J_1(x)$

**31.** Proceed as on page 269 to derive the elementary form of $J_{-1/2}(x)$ given in (27).

**32.** Use the recurrence relation in Problem 28 along with (26) and (27) to express $J_{3/2}(x)$, $J_{-3/2}(x)$, $J_{5/2}(x)$, and $J_{-5/2}(x)$ in terms of $\sin x$, $\cos x$, and powers of $x$.

**33. (a)** Use the first formula in (30) and Problem 32 to find the spherical Bessel functions $j_1(x)$ and $j_2(x)$.

**(b)** Use a graphing utility to plot the graphs of $j_1(x)$ and $j_2(x)$ in the same coordinate plane.

**34. (a)** Use the second formula in (30) and Problem 32 to find the spherical Bessel functions $y_1(x)$ and $y_2(x)$.

**(b)** Use a graphing utility to plot the graphs of $y_1(x)$ and $y_2(x)$ in the same coordinate plane.

**35.** Use the change of variables $s = \dfrac{2}{\alpha}\sqrt{\dfrac{k}{m}}\,e^{-\alpha t/2}$ to show that the differential equation of the aging spring $mx'' + ke^{-\alpha t}x = 0$, $\alpha > 0$, becomes

$$s^2 \frac{d^2x}{ds^2} + s\frac{dx}{ds} + s^2x = 0.$$

**36.** Show that $y = x^{1/2}w\left(\tfrac{2}{3}\alpha x^{3/2}\right)$ is a solution of the given form of Airy's differential equation whenever $w$ is a solution of the indicated Bessel's equation. [*Hint*: After differentiating, substituting, and simplifying, then let $t = \tfrac{2}{3}\alpha x^{3/2}$.]

**(a)** $y'' + \alpha^2 xy = 0, x > 0; t^2w'' + tw' + (t^2 - \tfrac{1}{9})w = 0, t > 0$

**(b)** $y'' - \alpha^2 xy = 0, x > 0; t^2w'' + tw' - (t^2 + \tfrac{1}{9})w = 0, t > 0$

**37.** Use the result in parts (a) and (b) of Problem 36 to express the general solution on $(0, \infty)$ of each of the two forms of Airy's equation in terms of Bessel functions.

**38.** Use Table 6.4.1 to find the first three positive eigenvalues and corresponding eigenfunctions of the boundary-value problem

$$xy'' + y' + \lambda xy = 0,$$

$$y(x), y'(x) \text{ bounded as } x \to 0^+, \quad y(2) = 0.$$

[*Hint*: By identifying $\lambda = \alpha^2$, the DE is the parametric Bessel equation of order zero.]

**39. (a)** Use (20) to show that the general solution of the differential equation $xy'' + \lambda y = 0$ on the interval $(0, \infty)$ is

$$y = c_1\sqrt{x}J_1\left(2\sqrt{\lambda x}\right) + c_2\sqrt{x}Y_1\left(2\sqrt{\lambda x}\right).$$

**(b)** Verify by direct substitution that $y = \sqrt{x}J_1(2\sqrt{x})$ is a particular solution of the DE in the case $\lambda = 1$.

### Computer Lab Assignments

**40.** Use a CAS to graph $J_{3/2}(x)$, $J_{-3/2}(x)$, $J_{5/2}(x)$, and $J_{-5/2}(x)$.

**41. (a)** Use the general solution given in Example 5 to solve the IVP

$$4x'' + e^{-0.1t}x = 0, \quad x(0) = 1, \quad x'(0) = -\tfrac{1}{2}.$$

Also use $J_0'(x) = -J_1(x)$ and $Y_0'(x) = -Y_1(x)$ along with Table 6.4.1 or a CAS to evaluate coefficients.

**(b)** Use a CAS to graph the solution obtained in part (a) for $0 \le t < \infty$.

**42. (a)** Use the general solution obtained in Problem 37 to solve the IVP

$$4x'' + tx = 0, \quad x(0.1) = 1, \quad x'(0.1) = -\tfrac{1}{2}.$$

Use a CAS to evaluate coefficients.

**(b)** Use a CAS to graph the solution obtained in part (a) for $0 \le t \le 200$.

**43. Column Bending Under Its Own Weight** A uniform thin column of length $L$, positioned vertically with one end embedded in the ground, will deflect, or bend away, from the vertical under the influence of its own weight when its length or height exceeds a certain critical value. It can be shown that the angular deflection $\theta(x)$ of the column from the vertical at a point $P(x)$ is a solution of the boundary-value problem:

$$EI\frac{d^2\theta}{dx^2} + \delta g(L - x)\theta = 0, \quad \theta(0) = 0, \quad \theta'(L) = 0,$$

where $E$ is Young's modulus, $I$ is the cross-sectional moment of inertia, $\delta$ is the constant linear density, and $x$ is the distance along the column measured from its base. See Figure 6.4.7. The column will bend only for those values of $L$ for which the boundary-value problem has a nontrivial solution.

**(a)** Restate the boundary-value problem by making the change of variables $t = L - x$. Then use the results of a problem earlier in this exercise set to express the general solution of the differential equation in terms of Bessel functions.

**(b)** Use the general solution found in part (a) to find a solution of the BVP and an equation which defines the critical length $L$, that is, the smallest value of $L$ for which the column will start to bend.

**(c)** With the aid of a CAS, find the critical length $L$ of a solid steel rod of radius $r = 0.05$ in., $\delta g = 0.28\,A$ lb/in., $E = 2.6 \times 10^7$ lb/in.$^2$, $A = \pi r^2$, and $I = \tfrac{1}{4}\pi r^4$.

**FIGURE 6.4.7** Beam in Problem 43

**44. Buckling of a Thin Vertical Column** In Example 4 of Section 5.2 we saw that when a constant vertical compressive force, or load, $P$ was applied to a thin column of uniform cross section and hinged at both ends, the deflection $y(x)$ is a solution of the BVP:

$$EI\frac{d^2y}{dx^2} + Py = 0, \quad y(0) = 0, \quad y(L) = 0.$$

**(a)** If the bending stiffness factor $EI$ is proportional to $x$, then $EI(x) = kx$, where $k$ is a constant of proportionality. If $EI(L) = kL = M$ is the maximum stiffness factor, then $k = M/L$ and so $EI(x) = Mx/L$. Use the information in Problem 39 to find a solution of

$$M\frac{x}{L}\frac{d^2y}{dx^2} + Py = 0, \quad y(0) = 0, \quad y(L) = 0$$

if it is known that $\sqrt{x}Y_1(2\sqrt{\lambda x})$ is *not* zero at $x = 0$.

**(b)** Use Table 6.4.1 to find the Euler load $P_1$ for the column.

**(c)** Use a CAS to graph the first buckling mode $y_1(x)$ corresponding to the Euler load $P_1$. For simplicity assume that $c_1 = 1$ and $L = 1$.

**45. Pendulum of Varying Length**   For the simple pendulum described on page 223 of Section 5.3, suppose that the rod holding the mass $m$ at one end is replaced by a flexible wire and that the wire is fed by a pulley through the horizontal support at point $O$ in Figure 6.4.8. In this manner, while it is in motion in a vertical plane, the mass $m$ can be raised or lowered. In other words, the length $l(t)$ of the pendulum varies with time. Under the same assumptions leading to equation (6) in Section 5.3, it follows from (1) in Chapter 5 in Review that the differential equation for the displacement angle $\theta(t)$ is now

$$l\frac{d^2\theta}{dt^2} + 2\frac{dl}{dt}\frac{d\theta}{dt} + g\sin\theta = 0.$$

**(a)** If $l$ increases at a constant rate $v$ and $l(0) = l_0$, then show that a linearization of the foregoing differential equation is

$$(l_0 + vt)\frac{d^2\theta}{dt^2} + 2v\frac{d\theta}{dt} + g\theta = 0. \qquad (37)$$

**(b)** Make the change of variables $x = (l_0 + vt)/v$ and show that (37) becomes

$$\frac{d^2\theta}{dx^2} + \frac{2}{x}\frac{d\theta}{dx} + \frac{g}{vx}\theta = 0.$$

**(c)** Use part (b) and (20) to express the general solution of equation (37) in terms of Bessel functions.

**(d)** Use the general solution obtained in part (c) to solve the initial-value problem consisting of equation (37) and the initial conditions $\theta(0) = \theta_0$, $\theta'(0) = 0$. [*Hints*: To simplify calculations, use a further change of variable

$$u = \frac{2}{v}\sqrt{g(l_0 + vt)} = 2\sqrt{\frac{g}{v}}x^{1/2}.$$ Also, recall that (22)

holds for both $J_1(u)$ and $Y_1(u)$. Finally, the identity

$$J_1(u)Y_2(u) - J_2(u)Y_1(u) = -\frac{2}{\pi u}$$

will be helpful.]

**FIGURE 6.4.8** Pendulum of varying length in Problem 45

**(e)** Use a CAS to graph the solution $\theta(t)$ of the IVP in part (d) when $l_0 = 1$ ft, $\theta_0 = \frac{1}{10}$ radian, and $v = \frac{1}{60}$ ft/s. Experiment with the graph using different time intervals such as $[0, 10]$, $[0, 30]$, and so on.

**(f)** What do the graphs indicate about the displacement angle $\theta(t)$ as the length $l$ of the wire increases with time?

## Legendre's Equation

**46. (a)** Use the explicit solutions $y_1(x)$ and $y_2(x)$ of Legendre's equation given in (32) and the appropriate choice of $c_0$ and $c_1$ to find the Legendre polynomials $P_6(x)$ and $P_7(x)$.

**(b)** Write the differential equations for which $P_6(x)$ and $P_7(x)$ are particular solutions.

**47.** Use the recurrence relation (35) and $P_0(x) = 1$, $P_1(x) = x$, to generate the next six Legendre polynomials.

**48.** Show that the differential equation

$$\sin\theta\,\frac{d^2y}{d\theta^2} + \cos\theta\,\frac{dy}{d\theta} + n(n+1)(\sin\theta)y = 0$$

can be transformed into Legendre's equation by means of the substitution $x = \cos\theta$.

**49.** Find the first three positive values of $\lambda$ for which the problem

$$(1 - x^2)y'' - 2xy' + \lambda y = 0,$$

$$y(0) = 0, \quad y(x), \quad y'(x) \text{ bounded on } [-1, 1]$$

has nontrivial solutions.

## Computer Lab Assignments

**50.** For purposes of this problem ignore the list of Legendre polynomials given on page 271 and the graphs given in Figure 6.4.6. Use Rodrigues' formula (36) to generate the Legendre polynomials $P_1(x), P_2(x), \ldots, P_7(x)$. Use a CAS to carry out the differentiations and simplifications.

**51.** Use a CAS to graph $P_1(x), P_2(x), \ldots, P_7(x)$ on the interval $[-1, 1]$.

**52.** Use a root-finding application to find the zeros of $P_1(x), P_2(x), \ldots, P_7(x)$. If the Legendre polynomials are built-in functions of your CAS, find zeros of Legendre polynomials of higher degree. Form a conjecture about the location of the zeros of any Legendre polynomial $P_n(x)$, and then investigate to see whether it is true.

## Additional Differential Equations

**53.** The differential equation

$$y'' - 2xy' + 2\alpha y = 0$$

is known as **Hermite's equation of order $\alpha$** after the French mathematician **Charles Hermite** (1822–1901). Show that the general solution of the equation is $y(x) = c_0 y_1(x) + c_1 y_2(x)$, where

$$y_1(x) = 1 + \sum_{k=1}^{\infty}(-1)^k\frac{2^k\alpha(\alpha-2)\cdots(\alpha-2k+2)}{(2k)!}x^{2k}$$

$$y_2(x) = x + \sum_{k=1}^{\infty}(-1)^k\frac{2^k(\alpha-1)(\alpha-3)\cdots(\alpha-2k+1)}{(2k+1)!}x^{2k+1}$$

are power series solutions centered at the ordinary point 0.

**54. (a)** When $\alpha = n$ is a nonnegative integer, Hermite's differential equation always possesses a polynomial solution of degree $n$. Use $y_1(x)$, given in Problem 53, to find polynomial solutions for $n = 0$, $n = 2$, and $n = 4$. Then use $y_2(x)$ to find polynomial solutions for $n = 1$, $n = 3$, and $n = 5$.

**(b)** A **Hermite polynomial** $H_n(x)$ is defined to be the $n$th degree polynomial solution of Hermite's equation multiplied by an appropriate constant so that the coefficient of $x^n$ in $H_n(x)$ is $2^n$. Use the polynomial solutions in part (a) to show that the first six Hermite polynomials are

$$H_0(x) = 1$$
$$H_1(x) = 2x$$
$$H_2(x) = 4x^2 - 2$$
$$H_3(x) = 8x^3 - 12x$$
$$H_4(x) = 16x^4 - 48x^2 + 12$$
$$H_5(x) = 32x^5 - 160x^3 + 120x.$$

**55.** The differential equation

$$(1 - x^2)y'' - xy' + \alpha^2 y = 0,$$

where $\alpha$ is a parameter, is known as **Chebyshev's equation** after the Russian mathematician **Pafnuty Chebyshev** (1821–1894). When $\alpha = n$ is a nonnegative integer, Chebyshev's differential equation always possesses a polynomial solution of degree $n$. Find a fifth degree polynomial solution of this differential equation.

**56.** If $n$ is an integer, use the substitution $R(x) = (\alpha x)^{-1/2} Z(x)$ to show that the general solution of the differential equation

$$x^2 R'' + 2xR' + [\alpha^2 x^2 - n(n + 1)]R = 0$$

on the interval $(0, \infty)$ is $R(x) = c_1 j_n(\alpha x) + c_2 y_n(\alpha x)$, where $j_n(\alpha x)$ and $y_n(\alpha x)$ are the spherical Bessel functions of the first and second kind defined in (30).

# Chapter 6 In Review

*Answers to selected odd-numbered problems begin on page ANS-10.*

In Problems 1 and 2 answer true or false without referring back to the text.

**1.** The general solution of $x^2 y'' + xy' + (x^2 - 1)y = 0$ is $y = c_1 J_1(x) + c_2 J_{-1}(x)$. _____

**2.** Because $x = 0$ is an irregular singular point of $x^3 y'' - xy' + y = 0$, the DE possesses no solution that is analytic at $x = 0$. _____

**3.** Both power series solutions of $y'' + \ln(x + 1)y' + y = 0$ centered at the ordinary point $x = 0$ are guaranteed to converge for all $x$ in which *one* of the following intervals?

**(a)** $(-\infty, \infty)$  **(b)** $(-1, \infty)$

**(c)** $[-\frac{1}{2}, \frac{1}{2}]$  **(d)** $[-1, 1]$

**4.** $x = 0$ is an ordinary point of a certain linear differential equation. After the assumed solution $y = \sum_{n=0}^{\infty} c_n x^n$ is substituted into the DE, the following algebraic system is obtained by equating the coefficients of $x^0$, $x^1$, $x^2$, and $x^3$ to zero:

$$2c_2 + 2c_1 + c_0 = 0$$
$$6c_3 + 4c_2 + c_1 = 0$$
$$12c_4 + 6c_3 + c_2 - \tfrac{1}{3}c_1 = 0$$
$$20c_5 + 8c_4 + c_3 - \tfrac{2}{3}c_2 = 0.$$

Bearing in mind that $c_0$ and $c_1$ are arbitrary, write down the first five terms of two power series solutions of the differential equation.

**5.** Suppose the power series $\sum_{k=0}^{\infty} c_k(x - 4)^k$ is known to converge at $-2$ and diverge at $13$. Discuss whether the series converges at $-7, 0, 7, 10$, and $11$. Possible answers are *does, does not, might*.

**6.** Use the Maclaurin series for $\sin x$ and $\cos x$ along with long division to find the first three nonzero terms of a power series in $x$ for the function $f(x) = \dfrac{\sin x}{\cos x}$.

In Problems 7 and 8 construct a linear second-order differential equation that has the given properties.

**7.** A regular singular point at $x = 1$ and an irregular singular point at $x = 0$

**8.** Regular singular points at $x = 1$ and at $x = -3$

In Problems 9–14 use an appropriate infinite series method about $x = 0$ to find two solutions of the given differential equation.

**9.** $2xy'' + y' + y = 0$  **10.** $y'' - xy' - y = 0$

**11.** $(x - 1)y'' + 3y = 0$  **12.** $y'' - x^2 y' + xy = 0$

**13.** $xy'' - (x + 2)y' + 2y = 0$  **14.** $(\cos x)y'' + y = 0$

In Problems 15 and 16 solve the given initial-value problem.

**15.** $y'' + xy' + 2y = 0$,  $y(0) = 3$, $y'(0) = -2$

**16.** $(x + 2)y'' + 3y = 0$,  $y(0) = 0$, $y'(0) = 1$

**17.** Without actually solving the differential equation $(1 - 2\sin x)y'' + xy = 0$, find a lower bound for the radius of convergence of power series solutions about the ordinary point $x = 0$.

**18.** Even though $x = 0$ is an ordinary point of the differential equation, explain why it is not a good idea to try to find a solution of the IVP

$$y'' + xy' + y = 0, \quad y(1) = -6, \quad y'(1) = 3$$

of the form $y = \sum_{n=0}^{\infty} c_n x^n$. Using power series, find a better way to solve the problem.

In Problems 19 and 20 investigate whether $x = 0$ is an ordinary point, singular point, or irregular singular point of the given differential equation. [*Hint*: Recall the Maclaurin series for $\cos x$ and $e^x$.]

**19.** $xy'' + (1 - \cos x)y' + x^2 y = 0$

**20.** $(e^x - 1 - x)y'' + xy = 0$

**21.** Note that $x = 0$ is an ordinary point of the differential equation $y'' + x^2 y' + 2xy = 5 - 2x + 10x^3$. Use the assumption $y = \sum_{n=0}^{\infty} c_n x^n$ to find the general solution $y = y_c + y_p$ that consists of three power series centered at $x = 0$.

**22.** The first-order differential equation $dy/dx = x^2 + y^2$ cannot be solved in terms of elementary functions. However, a solution can be expressed in terms of Bessel functions.

(a) Show that the substitution $y = -\dfrac{1}{u}\dfrac{du}{dx}$ leads to the equation $u'' + x^2 u = 0$.

(b) Use (20) in Section 6.4 to find the general solution of $u'' + x^2 u = 0$.

(c) Use (22) and (23) in Section 6.4 in the forms
$$J_\nu'(x) = \frac{\nu}{x}J_\nu(x) - J_{\nu+1}(x)$$
and
$$J_\nu'(x) = -\frac{\nu}{x}J_\nu(x) + J_{\nu-1}(x)$$
as an aid to show that a one-parameter family of solutions of $dy/dx = x^2 + y^2$ is given by
$$y = x\frac{J_{3/4}\left(\frac{1}{2}x^2\right) - cJ_{-3/4}\left(\frac{1}{2}x^2\right)}{cJ_{1/4}\left(\frac{1}{2}x^2\right) + J_{-1/4}\left(\frac{1}{2}x^2\right)}.$$

**23.** (a) Use (10) of Section 6.4 and Problem 32 of Exercises 6.4 to show that
$$Y_{3/2}(x) = -\sqrt{\frac{2}{\pi x}}\left(\frac{\cos x}{x} + \sin x\right)$$

(b) Use the definition $I_\nu(x) = i^{-\nu}J_\nu(ix)$ to show that
$$I_{1/2}(x) = \sqrt{\frac{2}{\pi x}}\sinh x \quad \text{and} \quad I_{-1/2}(x) = \sqrt{\frac{2}{\pi x}}\cosh x.$$

(c) Use (16) of Section 6.4 and part (b) to show that
$$K_{1/2}(x) = \sqrt{\frac{\pi}{2x}}e^{-x}.$$

**24.** (a) From (33) and (34) of Section 6.4 we know that when $n = 0$, Legendre's differential equation $(1 - x^2)y'' - 2xy' = 0$ has the polynomial solution $y = P_0(x) = 1$. Use (5) of Section 4.2 to show that a second Legendre function satisfying the DE for $-1 < x < 1$ is
$$y = \frac{1}{2}\ln\left(\frac{1+x}{1-x}\right).$$

(b) We also know from (33) and (34) of Section 6.4 that when $n = 1$, Legendre's differential equation $(1 - x^2)y'' - 2xy' + 2y = 0$ possesses the polynomial solution $y = P_1(x) = x$. Use (5) of Section 4.2 to show that a second Legendre function satisfying the DE for $-1 < x < 1$ is
$$y = \frac{x}{2}\ln\left(\frac{1+x}{1-x}\right) - 1.$$

(c) Use a graphing utility to graph the logarithmic Legendre functions given in parts (a) and (b).

**25.** (a) Use binomial series to formally show that
$$(1 - 2xt + t^2)^{-1/2} = \sum_{n=0}^{\infty} P_n(x)t^n.$$

(b) Use the result obtained in part (a) to show that $P_n(1) = 1$ and $P_n(-1) = (-1)^n$. See Properties (*ii*) and (*iii*) on page 272.

**26.** Express the general solution of the given differential equation on the interval $(0, \infty)$ in terms of Bessel functions.

(a) $4x^2 y'' + 4xy' + (64x^2 - 25)y = 0$

(b) $x^2 y'' + xy' - (18x^2 + 9)y = 0$

**27. Cooling Fin** A cooling fin is an outward projection from a mechanical or electronic device from which heat can be radiated away from the device into the surrounding medium (such as air). See Figure 6.R.1. An annular, or ring-shaped, cooling fin is normally used on cylindrical surfaces such as a circular heating pipe. See Figure 6.R.2. In the latter case, let $r$ denote the radial distance measured from the center line of the pipe and $T(r)$ the temperature within the fin defined for $r_0 \le r \le r_1$. It can be shown that $T(r)$ satisfies the differential equation
$$\frac{d}{dr}\left(r\frac{dT}{dr}\right) = a^2 r(T - T_m),$$
where $a^2$ is a constant and $T_m$ is the constant air temperature. Suppose $r_0 = 1$, $r_1 = 3$, and $T_m = 70$. Use the substitution $w(r) = T(r) - 70$ to show that the solution of the given differential equation subject to the boundary conditions
$$T(1) = 160, \quad T'(3) = 0$$
is
$$T(r) = 70 + 90\frac{K_1(3a)I_0(ar) + I_1(3a)K_0(ar)}{K_1(3a)I_0(a) + I_1(3a)K_0(a)},$$
where and $I_0(x)$ and $K_0(x)$ are the modified Bessel functions of the first and second kind. You will also have to use the derivatives given in (25) of Section 6.4.

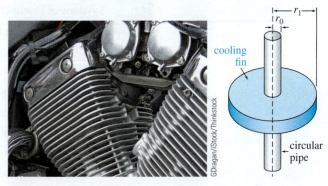

**FIGURE 6.R.1** Cooling fins on a motorcycle engine

**FIGURE 6.R.2** Annular cooling fin

**28.** Solve the differential equation in Problem 27 if the boundary conditions are
$$T(1) = 160, \quad T(3) = 90.$$

# 7

# The Laplace Transform

Raimundas/Shutterstock.com

In the linear mathematical models for a physical system such as a spring/mass system or a series electrical circuit, the right-hand member, or driving function, of the differential equations $mx'' + \beta x' + kx = f(t)$ or $Lq'' + Rq' + q/C = E(t)$ represents either an external force $f(t)$ or an impressed voltage $E(t)$. In Section 5.1 we solved problems in which the functions $f$ and $E$ were continuous. However, in practice discontinuous driving functions are not uncommon. Although we have solved piecewise-linear differential equations using the techniques of Chapters 2 and 4, the Laplace transform discussed in this chapter is an especially valuable tool that simplifies the solution of such equations.

## 7.1 Definition of the Laplace Transform

**INTRODUCTION** In elementary calculus you learned that differentiation and integration are *transforms*; this means, roughly speaking, that these operations transform a function into another function. For example, the function $f(x) = x^2$ is transformed, in turn, into a linear function and a family of cubic polynomial functions by the operations of differentiation and integration:

$$\frac{d}{dx}x^2 = 2x \qquad \text{and} \qquad \int x^2\, dx = \frac{1}{3}x^3 + c.$$

Moreover, these two transforms possess the **linearity property** that the transform of a linear combination of functions is a linear combination of the transforms. For $\alpha$ and $\beta$ constants

$$\frac{d}{dx}[\alpha f(x) + \beta\, g(x)] = \alpha f'(x) + \beta\, g'(x)$$

and

$$\int [\alpha f(x) + \beta\, g(x)]\, dx = \alpha \int f(x)\, dx + \beta \int g(x)\, dx$$

provided that each derivative and integral exists. In this section we will examine a special type of integral transform called the **Laplace transform.** In addition to possessing the linearity property the Laplace transform has many other interesting properties that make it very useful in solving linear initial-value problems.

**INTEGRAL TRANSFORM** If $f(x, y)$ is a function of two variables, then a definite integral of $f$ with respect to one of the variables leads to a function of the other variable. For example, by holding $y$ constant, we see that $\int_1^2 2xy^2\, dx = 3y^2$. Similarly, a definite integral such as $\int_a^b K(s, t) f(t)\, dt$ transforms a function $f$ of the variable $t$ into a function $F$ of the variable $s$. We are particularly interested in an **integral transform,** where the interval of integration is the unbounded interval $[0, \infty)$. If $f(t)$ is defined for $t \geq 0$, then the improper integral $\int_0^\infty K(s, t) f(t)\, dt$ is defined as a limit:

$$\int_0^\infty K(s, t) f(t)\, dt = \lim_{b \to \infty} \int_0^b K(s, t) f(t)\, dt. \tag{1}$$

If the limit in (1) exists, then we say that the integral exists or is **convergent;** if the limit does not exist, the integral does not exist and is **divergent.** The limit in (1) will, in general, exist for only certain values of the variable $s$.

 We will assume throughout that $s$ is a real variable.

**A DEFINITION** The function $K(s, t)$ in (1) is called the **kernel** of the transform. The choice $K(s, t) = e^{-st}$ as the kernel gives us an especially important integral transform.

---

**DEFINITION 7.1.1** Laplace Transform

Let $f$ be a function defined for $t \geq 0$. Then the integral

$$\mathcal{L}\{f(t)\} = \int_0^\infty e^{-st} f(t)\, dt \tag{2}$$

is said to be the **Laplace transform** of $f$, provided that the integral converges.

The Laplace transform is named in honor of the French mathematician and astronomer **Pierre-Simon Marquis de Laplace** (1749–1827).

When the defining integral (2) converges, the result is a function of $s$. In general discussion we shall use a lowercase letter to denote the function being transformed and the corresponding capital letter to denote its Laplace transform—for example,

$$\mathscr{L}\{f(t)\} = F(s), \qquad \mathscr{L}\{g(t)\} = G(s), \qquad \mathscr{L}\{y(t)\} = Y(s).$$

As the next four examples show, the domain of the function $F(s)$ depends on the function $f(t)$.

---

**EXAMPLE 1** Applying Definition 7.1.1
_____

Evaluate $\mathscr{L}\{1\}$.

**SOLUTION** From (2),

$$\mathscr{L}\{1\} = \int_0^\infty e^{-st}(1)\, dt = \lim_{b\to\infty} \int_0^b e^{-st}\, dt$$

$$= \lim_{b\to\infty} \frac{-e^{-st}}{s}\bigg|_0^b = \lim_{b\to\infty} \frac{-e^{-sb} + 1}{s} = \frac{1}{s}$$

provided that $s > 0$. In other words, when $s > 0$, the exponent $-sb$ is negative, and $e^{-sb} \to 0$ as $b \to \infty$. The integral diverges for $s < 0$. ∎

The use of the limit sign becomes somewhat tedious, so we shall adopt the notation $\big|_0^\infty$ as a shorthand for writing $\lim_{b\to\infty}(\ )\big|_0^b$. For example,

$$\mathscr{L}\{1\} = \int_0^\infty e^{-st}(1)\, dt = \frac{-e^{-st}}{s}\bigg|_0^\infty = \frac{1}{s}, \qquad s > 0.$$

At the upper limit, it is understood that we mean $e^{-st} \to 0$ as $t \to \infty$ for $s > 0$.

---

**EXAMPLE 2** Applying Definition 7.1.1
_____

Evaluate $\mathscr{L}\{t\}$.

**SOLUTION** From Definition 7.1.1 we have $\mathscr{L}\{t\} = \int_0^\infty e^{-st} t\, dt$. Integrating by parts and using $\lim_{t\to\infty} te^{-st} = 0$, $s > 0$, along with the result from Example 1, we obtain

$$\mathscr{L}\{t\} = \frac{-te^{-st}}{s}\bigg|_0^\infty + \frac{1}{s}\int_0^\infty e^{-st}\, dt = \frac{1}{s}\mathscr{L}\{1\} = \frac{1}{s}\left(\frac{1}{s}\right) = \frac{1}{s^2}. \qquad ∎$$

---

**EXAMPLE 3** Applying Definition 7.1.1
_____

Evaluate **(a)** $\mathscr{L}\{e^{-3t}\}$ **(b)** $\mathscr{L}\{e^{5t}\}$

**SOLUTION** In each case we use Definition 7.1.1.

**(a)**
$$\mathscr{L}\{e^{-3t}\} = \int_0^\infty e^{-3t} e^{-st}\, dt = \int_0^\infty e^{-(s+3)t}\, dt$$

$$= \frac{-e^{-(s+3)t}}{s+3}\bigg|_0^\infty$$

$$= \frac{1}{s+3}.$$

The last result is valid for $s > -3$ because in order to have $\lim_{t \to \infty} e^{-(s+3)t} = 0$ we must require that $s + 3 > 0$ or $s > -3$.

**(b)**
$$\mathcal{L}\{e^{5t}\} = \int_0^\infty e^{5t} e^{-st} \, dt = \int_0^\infty e^{-(s-5)t} \, dt$$
$$= \frac{-e^{-(s-5)t}}{s-5} \Big|_0^\infty$$
$$= \frac{1}{s-5}.$$

In contrast to part (a), this result is valid for $s > 5$ because $\lim_{t \to \infty} e^{-(s-5)t} = 0$ demands $s - 5 > 0$ or $s > 5$. ■

### EXAMPLE 4    Applying Definition 7.1.1

Evaluate $\mathcal{L}\{\sin 2t\}$.

**SOLUTION** From Definition 7.1.1 and two applications of integration by parts we obtain

$$\mathcal{L}\{\sin 2t\} = \int_0^\infty e^{-st} \sin 2t \, dt = \frac{-e^{-st} \sin 2t}{s} \Big|_0^\infty + \frac{2}{s} \int_0^\infty e^{-st} \cos 2t \, dt$$

$$= \frac{2}{s} \int_0^\infty e^{-st} \cos 2t \, dt, \qquad s > 0$$

$$\overset{\underset{\lim_{t \to \infty} e^{-st} \cos 2t = 0, \, s > 0}{\downarrow}}{\phantom{x}} \qquad \overset{\underset{\text{Laplace transform of } \sin 2t}{\downarrow}}{\phantom{x}}$$

$$= \frac{2}{s} \left[ \frac{-e^{-st} \cos 2t}{s} \Big|_0^\infty - \frac{2}{s} \int_0^\infty e^{-st} \sin 2t \, dt \right]$$

$$= \frac{2}{s^2} - \frac{4}{s^2} \mathcal{L}\{\sin 2t\}.$$

At this point we have an equation with $\mathcal{L}\{\sin 2t\}$ on both sides of the equality. Solving for that quantity yields the result

$$\mathcal{L}\{\sin 2t\} = \frac{2}{s^2 + 4}, \qquad s > 0. \qquad ■$$

**$\mathcal{L}$ IS A LINEAR TRANSFORM**    For a linear combination of functions we can write

$$\int_0^\infty e^{-st} [\alpha f(t) + \beta g(t)] \, dt = \alpha \int_0^\infty e^{-st} f(t) \, dt + \beta \int_0^\infty e^{-st} g(t) \, dt$$

whenever both integrals converge for $s > c$. Hence it follows that

$$\mathcal{L}\{\alpha f(t) + \beta g(t)\} = \alpha \mathcal{L}\{f(t)\} + \beta \mathcal{L}\{g(t)\} = \alpha F(s) + \beta G(s). \qquad (3)$$

Because of the property given in (3), $\mathcal{L}$ is said to be a **linear transform.**

### EXAMPLE 5    Linearity of the Laplace Transform

In this example we use the results of the preceding examples to illustrate the linearity of the Laplace transform.

**(a)** From Examples 1 and 2 we have for $s > 0$,

$$\mathcal{L}\{1 + 5t\} = \mathcal{L}\{1\} + 5\mathcal{L}\{t\} = \frac{1}{s} + \frac{5}{s^2}.$$

**(b)** From Examples 3 and 4 we have for $s > 5$,

$$\mathcal{L}\{4e^{5t} - 10 \sin 2t\} = 4\mathcal{L}\{e^{5t}\} - 10\mathcal{L}\{\sin 2t\} = \frac{4}{s-5} - \frac{20}{s^2 + 4}.$$

**(c)** From Examples 1, 2, and 3 we have for $s > 0$,

$$\mathscr{L}\{20e^{-3t} + 7t - 9\} = 20\mathscr{L}\{e^{-3t}\} + 7\mathscr{L}\{t\} - 9\mathscr{L}\{1\}$$

$$= \frac{20}{s+3} + \frac{7}{s^2} - \frac{9}{s}.$$

We state the generalization of some of the preceding examples by means of the next theorem. From this point on we shall also refrain from stating any restrictions on $s$; it is understood that $s$ is sufficiently restricted to guarantee the convergence of the appropriate Laplace transform.

---

**THEOREM 7.1.1** **Transforms of Some Basic Functions**

$$\text{(a) } \mathscr{L}\{1\} = \frac{1}{s}$$

$$\text{(b) } \mathscr{L}\{t^n\} = \frac{n!}{s^{n+1}}, \quad n = 1, 2, 3, \ldots \qquad \text{(c) } \mathscr{L}\{e^{at}\} = \frac{1}{s-a}$$

$$\text{(d) } \mathscr{L}\{\sin kt\} = \frac{k}{s^2 + k^2} \qquad \text{(e) } \mathscr{L}\{\cos kt\} = \frac{s}{s^2 + k^2}$$

$$\text{(f) } \mathscr{L}\{\sinh kt\} = \frac{k}{s^2 - k^2} \qquad \text{(g) } \mathscr{L}\{\cosh kt\} = \frac{s}{s^2 - k^2}$$

---

This result in (b) of Theorem 7.1.1 can be formally justified for $n$ a positive integer using integration by parts to first show that

$$\mathscr{L}\{t^n\} = \frac{n}{s}\,\mathscr{L}\{t^{n-1}\}.$$

Then for $n = 1, 2,$ and 3, we have, respectively,

$$\mathscr{L}\{t\} = \frac{1}{s} \cdot \mathscr{L}\{1\} = \frac{1}{s} \cdot \frac{1}{s} = \frac{1}{s^2}$$

$$\mathscr{L}\{t^2\} = \frac{2}{s} \cdot \mathscr{L}\{t\} = \frac{2}{s} \cdot \frac{1}{s^2} = \frac{2 \cdot 1}{s^3}$$

$$\mathscr{L}\{t^3\} = \frac{3}{s} \cdot \mathscr{L}\{t^2\} = \frac{3}{s} \cdot \frac{2 \cdot 1}{s^3} = \frac{3 \cdot 2 \cdot 1}{s^4}$$

If we carry on in this manner, you should be convinced that

$$\mathscr{L}\{t^n\} = \frac{n \cdots 3 \cdot 2 \cdot 1}{s^{n+1}} = \frac{n!}{s^{n+1}}.$$

**FIGURE 7.1.1** Piecewise continuous function

**SUFFICIENT CONDITIONS FOR EXISTENCE OF $\mathscr{L}\{f(t)\}$** The integral that defines the Laplace transform does not have to converge. For example, neither $\mathscr{L}\{1/t\}$ nor $\mathscr{L}\{e^{t^2}\}$ exists. Sufficient conditions guaranteeing the existence of $\mathscr{L}\{f(t)\}$ are that $f$ be piecewise continuous on $[0, \infty)$ and that $f$ be of exponential order for $t > T$. Recall that a function $f$ is **piecewise continuous** on $[0, \infty)$ if, in any interval $0 \le a \le t \le b$, there are at most a finite number of points $t_k$, $k = 1, 2, \ldots, n$ $(t_{k-1} < t_k)$ at which $f$ has finite discontinuities and is continuous on each open interval $(t_{k-1}, t_k)$. See Figure 7.1.1. The concept of **exponential order** is defined in the following manner.

---

**DEFINITION 7.1.2** **Exponential Order**

A function $f$ is said to be of **exponential order** if there exist constants $c$, $M > 0$, and $T > 0$ such that $|f(t)| \le Me^{ct}$ for all $t > T$.

**FIGURE 7.1.2** $f$ is of exponential order

If $f$ is an *increasing* function, then the condition $|f(t)| \le Me^{ct}$, $t > T$, simply states that the graph of $f$ on the interval $(T, \infty)$ does not grow faster than the graph of the exponential function $Me^{ct}$, where $c$ is a positive constant. See Figure 7.1.2. The functions $f(t) = t$, $f(t) = e^{-t}$, and $f(t) = 2 \cos t$ are all of exponential order because for $c = 1$, $M = 1$, $T = 0$ we have, respectively, for $t > 0$

$$|t| \le e^t, \qquad |e^{-t}| \le e^t, \qquad \text{and} \qquad |2 \cos t| \le 2e^t.$$

A comparison of the graphs on the interval $[0, \infty)$ is given in Figure 7.1.3.

**(a)**        **(b)**        **(c)**

**FIGURE 7.1.3** Three functions of exponential order

A positive integral power of $t$ is always of exponential order, since, for $c > 0$,

$$|t^n| \le Me^{ct} \qquad \text{or} \qquad \left|\frac{t^n}{e^{ct}}\right| \le M \quad \text{for } t > T$$

is equivalent to showing that $\lim_{t \to \infty} t^n / e^{ct}$ is finite for $n = 1, 2, 3, \ldots$. The result follows from $n$ applications of L'Hôpital's rule. A function such as $f(t) = e^{t^2}$ is not of exponential order since, as shown in Figure 7.1.4, $e^{t^2}$ grows faster than any positive linear power of $e$ for $t > c > 0$. This can also be seen from

$$\left|\frac{e^{t^2}}{e^{ct}}\right| = e^{t^2 - ct} = e^{t(t - c)} \to \infty \qquad \text{as} \quad t \to \infty$$

**FIGURE 7.1.4** $e^{t^2}$ is not of exponential order

for *any* value of $c$. By the same reasoning $e^{-st}e^{t^2} \to \infty$ as $t \to \infty$ for any $s$ and so the improper integral $\int_0^\infty e^{-st}e^{t^2}\, dt$ diverges. That is, $\mathscr{L}\{e^{t^2}\}$ does not exist.

---

**THEOREM 7.1.2** **Sufficient Conditions for Existence**

If $f$ is piecewise continuous on $[0, \infty)$ and of exponential order, then $\mathscr{L}\{f(t)\}$ exists for $s > c$.

---

**PROOF** By the additive interval property of definite integrals we can write

$$\mathscr{L}\{f(t)\} = \int_0^T e^{-st} f(t)\, dt + \int_T^\infty e^{-st} f(t)\, dt = I_1 + I_2.$$

The integral $I_1$ exists because it can be written as a sum of integrals over intervals on which $e^{-st} f(t)$ is continuous. Now since $f$ is of exponential order, there exist constants $c$, $M > 0$, $T > 0$ so that $|f(t)| \le Me^{ct}$ for $t > T$. We can then write

$$|I_2| \le \int_T^\infty |e^{-st}f(t)|\, dt \le M \int_T^\infty e^{-st}e^{ct}\, dt = M \int_T^\infty e^{-(s-c)t}\, dt = M\frac{e^{-(s-c)T}}{s - c}$$

for $s > c$. Since $\int_T^\infty Me^{-(s-c)t}\, dt$ converges, the integral $\int_T^\infty |e^{-st} f(t)|\, dt$ converges by the comparison test for improper integrals. This, in turn, implies that $I_2$ exists for $s > c$. The existence of $I_1$ and $I_2$ implies that $\mathscr{L}\{f(t)\} = \int_0^\infty e^{-st} f(t)\, dt$ exists for $s > c$. ∎

See (*i*) in the *Remarks*.

EXAMPLE 6 **Transform of a Piecewise Continuous Function**

Evaluate $\mathcal{L}\{f(t)\}$ where $f(t) = \begin{cases} 0, & 0 \leq t < 3 \\ 2, & t \geq 3. \end{cases}$

**SOLUTION** The function $f$, shown in Figure 7.1.5, is piecewise continuous and of exponential order for $t > 0$. Since $f$ is defined in two pieces, $\mathcal{L}\{f(t)\}$ is expressed as the sum of two integrals:

$$\mathcal{L}\{f(t)\} = \int_0^\infty e^{-st} f(t)\, dt = \int_0^3 e^{-st}(0)\, dt + \int_3^\infty e^{-st}(2)\, dt$$

$$= 0 + \frac{2e^{-st}}{-s}\bigg|_3^\infty$$

$$= \frac{2e^{-3s}}{s}, \qquad s > 0.$$

**FIGURE 7.1.5** Piecewise continuous function in Example 6

We conclude this section with an additional bit of theory related to the types of functions of $s$ that we will, generally, be working with. The next theorem indicates that not every arbitrary function of $s$ is a Laplace transform of a piecewise continuous function of exponential order.

**THEOREM 7.1.3** **Behavior of $F(s)$ as $s \to \infty$**

If $f$ is piecewise continuous on $[0, \infty)$ and of exponential order and $F(s) = \mathcal{L}\{f(t)\}$, then $\lim\limits_{s \to \infty} F(s) = 0$.

**PROOF** Since $f$ is of exponential order, there exist constants $\gamma$, $M_1 > 0$, and $T > 0$ so that $|f(t)| \leq M_1 e^{\gamma t}$ for $t > T$. Also, since $f$ is piecewise continuous for $0 \leq t \leq T$, it is necessarily bounded on the interval; that is, $|f(t)| \leq M_2 = M_2 e^{0t}$. If $M$ denotes the maximum of the set $\{M_1, M_2\}$ and $c$ denotes the maximum of $\{0, \gamma\}$, then

$$|F(s)| \leq \int_0^\infty e^{-st}|f(t)|\, dt \leq M \int_0^\infty e^{-st}e^{ct}\, dt = M \int_0^\infty e^{-(s-c)t}\, dt = \frac{M}{s-c}$$

for $s > c$. As $s \to \infty$, we have $|F(s)| \to 0$, and so $F(s) = \mathcal{L}\{f(t)\} \to 0$. ∎

**REMARKS**

(*i*) Throughout this chapter we shall be concerned primarily with functions that are both piecewise continuous and of exponential order. We note, however, that these two conditions are sufficient but not necessary for the existence of a Laplace transform. The function $f(t) = t^{-1/2}$ is not piecewise continuous on the interval $[0, \infty)$, but its Laplace transform exists. The function $f(t) = 2te^{t^2} \cos e^{t^2}$ is not of exponential order, but it can be shown that its Laplace transform exists. See Problems 43 and 53 in Exercises 7.1.

(*ii*) As a consequence of Theorem 7.1.3 we can say that functions of $s$ such as $F_1(s) = 1$ and $F_2(s) = s/(s + 1)$ are not the Laplace transforms of piecewise continuous functions of exponential order, since $F_1(s) \nrightarrow 0$ and $F_2(s) \nrightarrow 0$ as $s \to \infty$. But you should not conclude from this that $F_1(s)$ and $F_2(s)$ are *not* Laplace transforms. There are other kinds of functions.

In Problems 1–18 use Definition 7.1.1 to find $\mathscr{L}\{f(t)\}$.

**1.** $f(t) = \begin{cases} -1, & 0 \leq t < 1 \\ 1, & t \geq 1 \end{cases}$

**2.** $f(t) = \begin{cases} 4, & 0 \leq t < 2 \\ 0, & t \geq 2 \end{cases}$

**3.** $f(t) = \begin{cases} t, & 0 \leq t < 1 \\ 1, & t \geq 1 \end{cases}$

**4.** $f(t) = \begin{cases} 2t + 1, & 0 \leq t < 1 \\ 0, & t \geq 1 \end{cases}$

**5.** $f(t) = \begin{cases} \sin t, & 0 \leq t < \pi \\ 0, & t \geq \pi \end{cases}$

**6.** $f(t) = \begin{cases} 0, & 0 \leq t < \pi/2 \\ \cos t, & t \geq \pi/2 \end{cases}$

**7.** $f(t)$

**FIGURE 7.1.6** Graph for Problem 7

**8.** $f(t)$

**FIGURE 7.1.7** Graph for Problem 8

**9.** $f(t)$

**FIGURE 7.1.8** Graph for Problem 9

**10.** $f(t)$

**FIGURE 7.1.9** Graph for Problem 10

**11.** $f(t) = e^{t+7}$

**12.** $f(t) = e^{-2t-5}$

**13.** $f(t) = te^{4t}$

**14.** $f(t) = t^2 e^{-2t}$

**15.** $f(t) = e^{-t} \sin t$

**16.** $f(t) = e^t \cos t$

**17.** $f(t) = t \cos t$

**18.** $f(t) = t \sin t$

In Problems 19–36 use Theorem 7.1.1 to find $\mathscr{L}\{f(t)\}$.

**19.** $f(t) = 2t^4$

**20.** $f(t) = t^5$

**21.** $f(t) = 4t - 10$

**22.** $f(t) = 7t + 3$

**23.** $f(t) = t^2 + 6t - 3$

**24.** $f(t) = -4t^2 + 16t + 9$

**25.** $f(t) = (t + 1)^3$

**26.** $f(t) = (2t - 1)^3$

**27.** $f(t) = 1 + e^{4t}$

**28.** $f(t) = t^2 - e^{-9t} + 5$

**29.** $f(t) = (1 + e^{2t})^2$

**30.** $f(t) = (e^t - e^{-t})^2$

**31.** $f(t) = 4t^2 - 5 \sin 3t$

**32.** $f(t) = \cos 5t + \sin 2t$

**33.** $f(t) = \sinh kt$

**34.** $f(t) = \cosh kt$

**35.** $f(t) = e^t \sinh t$

**36.** $f(t) = e^{-t} \cosh t$

In Problems 37–40 find $\mathscr{L}\{f(t)\}$ by first using a trigonometric identity.

**37.** $f(t) = \sin 2t \cos 2t$

**38.** $f(t) = \cos^2 t$

**39.** $f(t) = \sin(4t + 5)$

**40.** $f(t) = 10 \cos\left(t - \dfrac{\pi}{6}\right)$

**41.** We have encountered the **gamma function** $\Gamma(\alpha)$ in our study of Bessel functions in Section 6.4 (page 263). One definition of this function is given by the improper integral

$$\Gamma(\alpha) = \int_0^\infty t^{\alpha-1} e^{-t}\, dt, \quad \alpha > 0.$$

Use this definition to show that $\Gamma(\alpha + 1) = \alpha\Gamma(\alpha)$. When $\alpha = n$ is a positive integer the last property can be used to show that $\Gamma(n + 1) = n!$. See Appendix A.

**42.** Use Problem 41 and the change of variable $u = st$ to obtain the generalization

$$\mathscr{L}\{t^\alpha\} = \frac{\Gamma(\alpha + 1)}{s^{\alpha+1}}, \quad \alpha > -1,$$

of the result in Theorem 7.1.1(b).

In Problems 43–46 use Problems 41 and 42 and the fact that $\Gamma\left(\frac{1}{2}\right) = \sqrt{\pi}$ to find the Laplace transform of the given function.

**43.** $f(t) = t^{-1/2}$

**44.** $f(t) = t^{1/2}$

**45.** $f(t) = t^{3/2}$

**46.** $f(t) = 2t^{1/2} + 8t^{5/2}$

### Discussion Problems

**47.** Suppose that $\mathscr{L}\{f_1(t)\} = F_1(s)$ for $s > c_1$ and that $\mathscr{L}\{f_2(t)\} = F_2(s)$ for $s > c_2$. When does

$$\mathscr{L}\{f_1(t) + f_2(t)\} = F_1(s) + F_2(s)?$$

**48.** Figure 7.1.4 suggests, but does not prove, that the function $f(t) = e^{t^2}$ is not of exponential order. How does the observation that $t^2 > \ln M + ct$, for $M > 0$ and $t$ sufficiently large, show that $e^{t^2} > Me^{ct}$ for any $c$?

**49.** Use part (c) of Theorem 7.1.1 to show that

$$\mathscr{L}\{e^{(a+ib)t}\} = \frac{s - a + ib}{(s - a)^2 + b^2},$$

where $a$ and $b$ are real and $i^2 = -1$. Show how Euler's formula (page 136) can then be used to deduce the results

$$\mathscr{L}\{e^{at} \cos bt\} = \frac{s - a}{(s - a)^2 + b^2}$$

$$\mathscr{L}\{e^{at} \sin bt\} = \frac{b}{(s - a)^2 + b^2}.$$

**50.** Under what conditions is a linear function $f(x) = mx + b$, $m \neq 0$, a linear transform?

**51.** Explain why the function

$$f(t) = \begin{cases} t, & 0 \leq t < 2 \\ 4, & 2 < t < 5 \\ 1/(t - 5), & t > 5 \end{cases}$$

is not piecewise continuous on $[0, \infty)$.

**52.** Show that the function $f(t) = 1/t^2$ does not possess a Laplace transform. [*Hint:* Write $\mathscr{L}\{1/t^2\}$ as two improper integrals:

$$\mathscr{L}\{1/t^2\} = \int_0^1 \frac{e^{-st}}{t^2}\, dt + \int_1^\infty \frac{e^{-st}}{t^2}\, dt = I_1 + I_2.$$

Show that $I_1$ diverges.]

**53.** The function $f(t) = 2te^{t^2} \cos t^2$ is not of exponential order. Nevertheless, show that the Laplace transform $\mathscr{L}\{2te^{t^2} \cos e^{t^2}\}$ exists. [*Hint*: Start with integration by parts.]

**54.** If $\mathscr{L}\{f(t)\} = F(s)$ and $a > 0$ is a constant, show that

$$\mathscr{L}\{f(at)\} = \frac{1}{a} F\left(\frac{s}{a}\right).$$

This result is known as the **change of scale theorem**.

In Problems 55–58 use the given Laplace transform and the result in Problem 54 to find the indicated Laplace transform. Assume that $a$ and $k$ are positive constants.

**55.** $\mathscr{L}\{e^t\} = \dfrac{1}{s-1};\quad \mathscr{L}\{e^{at}\}$   **56.** $\mathscr{L}\{\sin t\} = \dfrac{1}{s^2+1};\quad \mathscr{L}\{\sin kt\}$

**57.** $\mathscr{L}\{1 - \cos t\} = \dfrac{1}{s(s^2+1)};\quad \mathscr{L}\{1 - \cos kt\}$

**58.** $\mathscr{L}\{\sin t \sinh t\} = \dfrac{2s}{s^4+4};\quad \mathscr{L}\{\sin kt \sinh kt\}$

---

## 7.2 Inverse Transforms and Transforms of Derivatives

**INTRODUCTION**    In this section we take a few small steps into an investigation of how the Laplace transform can be used to solve certain types of equations for an unknown function. We begin the discussion with the concept of the inverse Laplace transform or, more precisely, the inverse of a Laplace transform $F(s)$. After some important preliminary background material on the Laplace transform of derivatives $f'(t), f''(t), \ldots$, we then illustrate how both the Laplace transform and the inverse Laplace transform come into play in solving some simple ordinary differential equations.

### 7.2.1    INVERSE TRANSFORMS

**THE INVERSE PROBLEM**    If $F(s)$ represents the Laplace transform of a function $f(t)$, that is, $\mathscr{L}\{f(t)\} = F(s)$, we then say $f(t)$ is the **inverse Laplace transform** of $F(s)$ and write $f(t) = \mathscr{L}^{-1}\{F(s)\}$. For example, from Examples 1, 2, and 3 of Section 7.1 we have, respectively,

| Transform | Inverse Transform |
|---|---|
| $\mathscr{L}\{1\} = \dfrac{1}{s}$ | $1 = \mathscr{L}^{-1}\left\{\dfrac{1}{s}\right\}$ |
| $\mathscr{L}\{t\} = \dfrac{1}{s^2}$ | $t = \mathscr{L}^{-1}\left\{\dfrac{1}{s^2}\right\}$ |
| $\mathscr{L}\{e^{-3t}\} = \dfrac{1}{s+3}$ | $e^{-3t} = \mathscr{L}^{-1}\left\{\dfrac{1}{s+3}\right\}$ |

We shall see shortly that in the application of the Laplace transform to equations we are not able to determine an unknown function $f(t)$ directly; rather, we are able to solve for the Laplace transform $F(s)$ of $f(t)$; but from that knowledge we ascertain $f$ by computing $f(t) = \mathscr{L}^{-1}\{F(s)\}$. The idea is simply this: Suppose $F(s) = \dfrac{-2s+6}{s^2+4}$ is a Laplace transform; find a function $f(t)$ such that $\mathscr{L}\{f(t)\} = F(s)$. We shall show how to solve this problem in Example 2.

For future reference the analogue of Theorem 7.1.1 for the inverse transform is presented as our next theorem.

> **THEOREM 7.2.1** Some Inverse Transforms
>
> $$\textbf{(a)} \quad 1 = \mathscr{L}^{-1}\left\{\frac{1}{s}\right\}$$
>
> $$\textbf{(b)} \quad t^n = \mathscr{L}^{-1}\left\{\frac{n!}{s^{n+1}}\right\}, \quad n = 1, 2, 3, \ldots \qquad \textbf{(c)} \quad e^{at} = \mathscr{L}^{-1}\left\{\frac{1}{s-a}\right\}$$
>
> $$\textbf{(d)} \quad \sin kt = \mathscr{L}^{-1}\left\{\frac{k}{s^2 + k^2}\right\} \qquad\qquad \textbf{(e)} \quad \cos kt = \mathscr{L}^{-1}\left\{\frac{s}{s^2 + k^2}\right\}$$
>
> $$\textbf{(f)} \quad \sinh kt = \mathscr{L}^{-1}\left\{\frac{k}{s^2 - k^2}\right\} \qquad\qquad \textbf{(g)} \quad \cosh kt = \mathscr{L}^{-1}\left\{\frac{s}{s^2 - k^2}\right\}$$

In evaluating inverse transforms, it often happens that a function of $s$ under consideration does not match *exactly* the form of a Laplace transform $F(s)$ given in a table. It may be necessary to "fix up" the function of $s$ by multiplying and dividing by an appropriate constant.

### EXAMPLE 1  Applying Theorem 7.2.1

Evaluate  **(a)** $\mathscr{L}^{-1}\left\{\dfrac{1}{s^5}\right\}$  **(b)** $\mathscr{L}^{-1}\left\{\dfrac{1}{s^2 + 7}\right\}$.

**SOLUTION (a)** To match the form given in part (b) of Theorem 7.2.1, we identify $n + 1 = 5$ or $n = 4$ and then multiply and divide by 4!:

$$\mathscr{L}^{-1}\left\{\frac{1}{s^5}\right\} = \frac{1}{4!}\mathscr{L}^{-1}\left\{\frac{4!}{s^5}\right\} = \frac{1}{24}t^4.$$

**(b)** To match the form given in part (d) of Theorem 7.2.1, we identify $k^2 = 7$, so $k = \sqrt{7}$. We fix up the expression by multiplying and dividing by $\sqrt{7}$:

$$\mathscr{L}^{-1}\left\{\frac{1}{s^2 + 7}\right\} = \frac{1}{\sqrt{7}}\,\mathscr{L}^{-1}\left\{\frac{\sqrt{7}}{s^2 + 7}\right\} = \frac{1}{\sqrt{7}}\sin\sqrt{7}t. \qquad \blacksquare$$

**$\mathscr{L}^{-1}$ IS A LINEAR TRANSFORM**  The inverse Laplace transform is also a linear transform; that is, for constants $\alpha$ and $\beta$

$$\mathscr{L}^{-1}\{\alpha F(s) + \beta G(s)\} = \alpha\,\mathscr{L}^{-1}\{F(s)\} + \beta\,\mathscr{L}^{-1}\{G(s)\}, \qquad (1)$$

where $F$ and $G$ are the transforms of some functions $f$ and $g$. Like (3) of Section 7.1, (1) extends to any finite linear combination of Laplace transforms.

### EXAMPLE 2  Termwise Division and Linearity

Evaluate  $\mathscr{L}^{-1}\left\{\dfrac{-2s + 6}{s^2 + 4}\right\}$.

**SOLUTION**  We first rewrite the given function of $s$ as two expressions by means of termwise division and then use (1):

termwise
division ↓

linearity and fixing
up constants ↓

$$\mathscr{L}^{-1}\left\{\frac{-2s + 6}{s^2 + 4}\right\} = \mathscr{L}^{-1}\left\{\frac{-2s}{s^2 + 4} + \frac{6}{s^2 + 4}\right\} = -2\ \mathscr{L}^{-1}\left\{\frac{s}{s^2 + 4}\right\} + \frac{6}{2}\ \mathscr{L}^{-1}\left\{\frac{2}{s^2 + 4}\right\} \quad (2)$$

$$= -2\cos 2t + 3\sin 2t. \quad \leftarrow \text{parts (e) and (d)}$$
of Theorem 7.2.1 with $k = 2$ ∎

**PARTIAL FRACTIONS**   Partial fractions play an important role in finding inverse Laplace transforms. The decomposition of a rational expression into component fractions can be done quickly by means of a single command on most computer algebra systems. Indeed, some CASs have packages that implement Laplace transform and inverse Laplace transform commands. But for those of you without access to such software, we will review in this and subsequent sections some of the basic algebra in the important cases in which the denominator of a Laplace transform $F(s)$ contains distinct linear factors, repeated linear factors, and quadratic polynomials with no real factors. Although we shall examine each of these cases as this chapter develops, it still might be a good idea for you to consult either a calculus text or a current precalculus text for a more comprehensive review of this theory.

The following example illustrates partial fraction decomposition in the case when the denominator of $F(s)$ is factorable into *distinct linear factors*.

**EXAMPLE 3**   **Partial Fractions: Distinct Linear Factors**

Evaluate $\mathscr{L}^{-1}\left\{\dfrac{s^2 + 6s + 9}{(s - 1)(s - 2)(s + 4)}\right\}$.

**SOLUTION**   There exist unique real constants $A$, $B$, and $C$ so that

$$\frac{s^2 + 6s + 9}{(s - 1)(s - 2)(s + 4)} = \frac{A}{s - 1} + \frac{B}{s - 2} + \frac{C}{s + 4}$$

$$= \frac{A(s - 2)(s + 4) + B(s - 1)(s + 4) + C(s - 1)(s - 2)}{(s - 1)(s - 2)(s + 4)}.$$

Since the denominators are identical, the numerators are identical:

$$s^2 + 6s + 9 = A(s - 2)(s + 4) + B(s - 1)(s + 4) + C(s - 1)(s - 2). \quad (3)$$

By comparing coefficients of powers of $s$ on both sides of the equality, we know that (3) is equivalent to a system of three equations in the three unknowns $A$, $B$, and $C$. However, there is a shortcut for determining these unknowns. If we set $s = 1$, $s = 2$, and $s = -4$ in (3), we obtain, respectively,

$$16 = A(-1)(5), \qquad 25 = B(1)(6), \qquad \text{and} \qquad 1 = C(-5)(-6),$$

and so $A = -\frac{16}{5}$, $B = \frac{25}{6}$, and $C = \frac{1}{30}$. Hence the partial fraction decomposition is

$$\frac{s^2 + 6s + 9}{(s - 1)(s - 2)(s + 4)} = -\frac{16/5}{s - 1} + \frac{25/6}{s - 2} + \frac{1/30}{s + 4}, \quad (4)$$

and thus, from the linearity of $\mathscr{L}^{-1}$ and part (c) of Theorem 7.2.1,

$$\mathscr{L}^{-1}\left\{\frac{s^2 + 6s + 9}{(s - 1)(s - 2)(s + 4)}\right\} = -\frac{16}{5}\mathscr{L}^{-1}\left\{\frac{1}{s - 1}\right\} + \frac{25}{6}\mathscr{L}^{-1}\left\{\frac{1}{s - 2}\right\} + \frac{1}{30}\mathscr{L}^{-1}\left\{\frac{1}{s + 4}\right\}$$

$$= -\frac{16}{5}e^t + \frac{25}{6}e^{2t} + \frac{1}{30}e^{-4t}. \quad (5) \ ∎$$

## 7.2.2 TRANSFORMS OF DERIVATIVES

**TRANSFORM A DERIVATIVE**   As was pointed out in the introduction to this chapter, our immediate goal is to use the Laplace transform to solve differential equations. To that end we need to evaluate quantities such as $\mathcal{L}\{dy/dt\}$ and $\mathcal{L}\{d^2y/dt^2\}$. For example, if $f'$ is continuous for $t \geq 0$, then integration by parts gives

$$\mathcal{L}\{f'(t)\} = \int_0^\infty e^{-st} f'(t)\, dt = e^{-st} f(t)\Big|_0^\infty + s\int_0^\infty e^{-st} f(t)\, dt$$

$$= -f(0) + s\mathcal{L}\{f(t)\}$$

or
$$\mathcal{L}\{f'(t)\} = sF(s) - f(0). \tag{6}$$

Here we have assumed that $e^{-st}f(t) \to 0$ as $t \to \infty$. Similarly, with the aid of (6),

$$\mathcal{L}\{f''(t)\} = \int_0^\infty e^{-st} f''(t)\, dt = e^{-st} f'(t)\Big|_0^\infty + s\int_0^\infty e^{-st} f'(t)\, dt$$

$$= -f'(0) + s\mathcal{L}\{f'(t)\}$$

$$= s[sF(s) - f(0)] - f'(0) \quad \leftarrow \text{from (6)}$$

or
$$\mathcal{L}\{f''(t)\} = s^2 F(s) - sf(0) - f'(0). \tag{7}$$

In like manner it can be shown that

$$\mathcal{L}\{f'''(t)\} = s^3 F(s) - s^2 f(0) - sf'(0) - f''(0). \tag{8}$$

The recursive nature of the Laplace transform of the derivatives of a function $f$ should be apparent from the results in (6), (7), and (8). The next theorem gives the Laplace transform of the $n$th derivative of $f$.

---

**THEOREM 7.2.2** Transform of a Derivative

If $f, f', \ldots, f^{(n-1)}$ are continuous on $[0, \infty)$ and are of exponential order and if $f^{(n)}(t)$ is piecewise continuous on $[0, \infty)$, then

$$\mathcal{L}\{f^{(n)}(t)\} = s^n F(s) - s^{n-1} f(0) - s^{n-2} f'(0) - \cdots - f^{(n-1)}(0),$$

where $F(s) = \mathcal{L}\{f(t)\}$.

---

**SOLVING LINEAR ODEs**   It is apparent from the general result given in Theorem 7.2.2 that $\mathcal{L}\{d^n y/dt^n\}$ depends on $Y(s) = \mathcal{L}\{y(t)\}$ and the $n - 1$ derivatives of $y(t)$ evaluated at $t = 0$. This property makes the Laplace transform ideally suited for solving linear initial-value problems in which the differential equation has *constant coefficients*. Such a differential equation is simply a linear combination of terms $y, y', y'', \ldots, y^{(n)}$:

$$a_n \frac{d^n y}{dt^n} + a_{n-1} \frac{d^{n-1} y}{dt^{n-1}} + \cdots + a_0 y = g(t),$$

$$y(0) = y_0, \, y'(0) = y_1, \, \ldots, \, y^{(n-1)}(0) = y_{n-1},$$

where the $a_i$, $i = 0, 1, \ldots, n$ and $y_0, y_1, \ldots, y_{n-1}$ are constants. By the linearity property the Laplace transform of this linear combination is a linear combination of Laplace transforms:

$$a_n \mathcal{L}\left\{\frac{d^n y}{dt^n}\right\} + a_{n-1}\mathcal{L}\left\{\frac{d^{n-1} y}{dt^{n-1}}\right\} + \cdots + a_0 \mathcal{L}\{y\} = \mathcal{L}\{g(t)\}. \tag{9}$$

From Theorem 7.2.2, (9) becomes

$$a_n[s^n Y(s) - s^{n-1}y(0) - \cdots - y^{(n-1)}(0)]$$
$$+ a_{n-1}[s^{n-1}Y(s) - s^{n-2}y(0) - \cdots - y^{(n-2)}(0)] + \cdots + a_0 Y(s) = G(s), \tag{10}$$

where $\mathscr{L}\{y(t)\} = Y(s)$ and $\mathscr{L}\{g(t)\} = G(s)$. In other words,

*The Laplace transform of a linear differential equation with constant coefficients becomes an algebraic equation in Y(s).*

If we solve the general transformed equation (10) for the symbol $Y(s)$, we first obtain $P(s)Y(s) = Q(s) + G(s)$ and then write

$$Y(s) = \frac{Q(s)}{P(s)} + \frac{G(s)}{P(s)}, \tag{11}$$

where $P(s) = a_n s^n + a_{n-1}s^{n-1} + \cdots + a_0$, $Q(s)$ is a polynomial in $s$ of degree less than or equal to $n - 1$ consisting of the various products of the coefficients $a_i$, $i = 1, \ldots, n$ and the prescribed initial conditions $y_0, y_1, \ldots, y_{n-1}$, and $G(s)$ is the Laplace transform of $g(t)$.* Typically, we put the two terms in (11) over the least common denominator and then decompose the expression into two or more partial fractions. Finally, the solution $y(t)$ of the original initial-value problem is $y(t) = \mathscr{L}^{-1}\{Y(s)\}$, where the inverse transform is done term by term. The procedure is summarized in the diagram in Figure 7.2.1.

**FIGURE 7.2.1**  Steps in solving an IVP by the Laplace transform

The next example illustrates the foregoing method of solving DEs, as well as partial fraction decomposition in the case when the denominator of $Y(s)$ contains a *quadratic polynomial with no real factors.*

**EXAMPLE 4**    Solving a First-Order IVP

Use the Laplace transform to solve the initial-value problem

$$\frac{dy}{dt} + 3y = 13 \sin 2t, \quad y(0) = 6.$$

**SOLUTION**  We first take the transform of each member of the differential equation:

$$\mathscr{L}\left\{\frac{dy}{dt}\right\} + 3\mathscr{L}\{y\} = 13\,\mathscr{L}\{\sin 2t\}. \tag{12}$$

From (6), $\mathscr{L}\{dy/dt\} = sY(s) - y(0) = sY(s) - 6$, and from part (d) of Theorem 7.1.1, $\mathscr{L}\{\sin 2t\} = 2/(s^2 + 4)$, so (12) is the same as

$$sY(s) - 6 + 3Y(s) = \frac{26}{s^2 + 4} \qquad \text{or} \qquad (s + 3)Y(s) = 6 + \frac{26}{s^2 + 4}.$$

---

*The polynomial $P(s)$ is the same as the $n$th-degree auxiliary polynomial in (12) in Section 4.3 with the usual symbol $m$ replaced by $s$.

Solving the last equation for $Y(s)$, we get

$$Y(s) = \frac{6}{s+3} + \frac{26}{(s+3)(s^2+4)} = \frac{6s^2+50}{(s+3)(s^2+4)}. \tag{13}$$

Since the quadratic polynomial $s^2 + 4$ does not factor using real numbers, its assumed numerator in the partial fraction decomposition is a linear polynomial in $s$:

$$\frac{6s^2+50}{(s+3)(s^2+4)} = \frac{A}{s+3} + \frac{Bs+C}{s^2+4}.$$

Putting the right-hand side of the equality over a common denominator and equating numerators gives $6s^2 + 50 = A(s^2 + 4) + (Bs + C)(s + 3)$. Setting $s = -3$ then immediately yields $A = 8$. Since the denominator has no more real zeros, we equate the coefficients of $s^2$ and $s$: $6 = A + B$ and $0 = 3B + C$. Using the value of $A$ in the first equation gives $B = -2$, and then using this last value in the second equation gives $C = 6$. Thus

$$Y(s) = \frac{6s^2+50}{(s+3)(s^2+4)} = \frac{8}{s+3} + \frac{-2s+6}{s^2+4}.$$

We are not quite finished because the last rational expression still has to be written as two fractions. This was done by termwise division in Example 2. From (2) of that example,

$$y(t) = 8\mathscr{L}^{-1}\left\{\frac{1}{s+3}\right\} - 2\mathscr{L}^{-1}\left\{\frac{s}{s^2+4}\right\} + 3\mathscr{L}^{-1}\left\{\frac{2}{s^2+4}\right\}.$$

It follows from parts (c), (d), and (e) of Theorem 7.2.1 that the solution of the initial-value problem is $y(t) = 8e^{-3t} - 2\cos 2t + 3\sin 2t$. ∎

---

### EXAMPLE 5  Solving a Second-Order IVP

Solve $y'' - 3y' + 2y = e^{-4t}$, $y(0) = 1$, $y'(0) = 5$.

**SOLUTION** Proceeding as in Example 4, we transform the DE. We take the sum of the transforms of each term, use (6) and (7), use the given initial conditions, use (c) of Theorem 7.1.1, and then solve for $Y(s)$:

$$\mathscr{L}\left\{\frac{d^2y}{dt^2}\right\} - 3\mathscr{L}\left\{\frac{dy}{dt}\right\} + 2\mathscr{L}\{y\} = \mathscr{L}\{e^{-4t}\}$$

$$s^2Y(s) - sy(0) - y'(0) - 3[sY(s) - y(0)] + 2Y(s) = \frac{1}{s+4}$$

$$(s^2 - 3s + 2)Y(s) = s + 2 + \frac{1}{s+4}$$

$$Y(s) = \frac{s+2}{s^2-3s+2} + \frac{1}{(s^2-3s+2)(s+4)} = \frac{s^2+6s+9}{(s-1)(s-2)(s+4)}. \tag{14}$$

The details of the partial fraction decomposition of $Y(s)$ in (14) have already been carried out in Example 3. In view of the results in (4) and (5) we have the solution of the initial-value problem

$$y(t) = \mathscr{L}^{-1}\{Y(s)\} = -\frac{16}{5}e^t + \frac{25}{6}e^{2t} + \frac{1}{30}e^{-4t}. \qquad ∎$$

Examples 4 and 5 illustrate the basic procedure for using the Laplace transform to solve a linear initial-value problem, but these examples may appear to demonstrate a method that is not much better than the approach to such problems outlined in Sections 2.3 and 4.3–4.6. Don't draw any negative conclusions from only two examples. Yes, there is a lot of algebra inherent in the use of the Laplace transform, *but* observe that we do not have to use variation of parameters or worry about the cases and algebra in the method of undetermined coefficients. Moreover, since the method incorporates the prescribed initial conditions directly into the solution, there is no need for the separate operation of applying the initial conditions to the general solution $y = c_1 y_1 + c_2 y_2 + \cdots + c_n y_n + y_p$ of the DE to find specific constants in a particular solution of the IVP.

The Laplace transform has many operational properties. In the sections that follow we will examine some of these properties and see how they enable us to solve problems of greater complexity.

## REMARKS

(*i*) The inverse Laplace transform of a function $F(s)$ may not be unique; in other words, it is possible that $\mathscr{L}\{f_1(t)\} = \mathscr{L}\{f_2(t)\}$ and yet $f_1 \neq f_2$. For our purposes this is not anything to be concerned about. If $f_1$ and $f_2$ are piecewise continuous on $[0, \infty)$ and of exponential order, then $f_1$ and $f_2$ are *essentially* the same. See Problem 50 in Exercises 7.2. However, if $f_1$ and $f_2$ are continuous on $[0, \infty)$ and $\mathscr{L}\{f_1(t)\} = \mathscr{L}\{f_2(t)\}$, then $f_1 = f_2$ on the interval.

(*ii*) This remark is for those of you who will be required to do partial fraction decompositions by hand. There is another way of determining the coefficients in a partial fraction decomposition in the special case when $\mathscr{L}\{f(t)\} = F(s)$ is a rational function of $s$ and the denominator of $F$ is a product of *distinct* linear factors. Let us illustrate by reexamining Example 3. Suppose we multiply both sides of the assumed decomposition

$$\frac{s^2 + 6s + 9}{(s-1)(s-2)(s+4)} = \frac{A}{s-1} + \frac{B}{s-2} + \frac{C}{s+4} \qquad (15)$$

by, say, $s - 1$, simplify, and then set $s = 1$. Since the coefficients of $B$ and $C$ on the right-hand side of the equality are zero, we get

$$\left. \frac{s^2 + 6s + 9}{(s-2)(s+4)} \right|_{s=1} = A \qquad \text{or} \qquad A = -\frac{16}{5}.$$

Written another way,

$$\left. \frac{s^2 + 6s + 9}{\boxed{(s-1)}(s-2)(s+4)} \right|_{s=1} = -\frac{16}{5} = A,$$

where we have shaded, or *covered up*, the factor that canceled when the left-hand side was multiplied by $s - 1$. Now to obtain $B$ and $C$, we simply evaluate the left-hand side of (15) while covering up, in turn, $s - 2$ and $s + 4$:

$$\left. \frac{s^2 + 6s + 9}{(s-1)\boxed{(s-2)}(s+4)} \right|_{s=2} = \frac{25}{6} = B$$

and

$$\left. \frac{s^2 + 6s + 9}{(s-1)(s-2)\boxed{(s+4)}} \right|_{s=-4} = \frac{1}{30} = C.$$

The desired decomposition (15) is given in (4). This special technique for determining coefficients is naturally known as the **cover-up method.**

(*iii*) In this remark we continue our introduction to the terminology of dynamical systems. Because of (9) and (10) the Laplace transform is well adapted to *linear* dynamical systems. The polynomial $P(s) = a_n s^n + a_{n-1} s^{n-1} + \cdots + a_0$ in (11) is the total coefficient of $Y(s)$ in (10) and is simply the left-hand side of the DE with the derivatives $d^k y/dt^k$ replaced by powers $s^k$, $k = 0, 1, \ldots, n$. It is usual practice to call the reciprocal of $P(s)$—namely, $W(s) = 1/P(s)$—the **transfer function** of the system and write (11) as

$$Y(s) = W(s)Q(s) + W(s)G(s). \tag{16}$$

In this manner we have separated, in an additive sense, the effects on the response that are due to the initial conditions (that is, $W(s)Q(s)$) from those due to the input function $g$ (that is, $W(s)G(s)$). See (13) and (14). Hence the response $y(t)$ of the system is a superposition of two responses:

$$y(t) = \mathscr{L}^{-1}\{W(s)Q(s)\} + \mathscr{L}^{-1}\{W(s)G(s)\} = y_0(t) + y_1(t).$$

If the input is $g(t) = 0$, then the solution of the problem is $y_0(t) = \mathscr{L}^{-1}\{W(s)Q(s)\}$. This solution is called the **zero-input response** of the system. On the other hand, the function $y_1(t) = \mathscr{L}^{-1}\{W(s)G(s)\}$ is the output due to the input $g(t)$. Now if the initial state of the system is the zero state (all the initial conditions are zero), then $Q(s) = 0$, and so the only solution of the initial-value problem is $y_1(t)$. The latter solution is called the **zero-state response** of the system. Both $y_0(t)$ and $y_1(t)$ are particular solutions: $y_0(t)$ is a solution of the IVP consisting of the associated homogeneous equation with the given initial conditions, and $y_1(t)$ is a solution of the IVP consisting of the nonhomogeneous equation with zero initial conditions. In Example 5 we see from (14) that the transfer function is $W(s) = 1/(s^2 - 3s + 2)$, the zero-input response is

$$y_0(t) = \mathscr{L}^{-1}\left\{ \frac{s+2}{(s-1)(s-2)} \right\} = -3e^t + 4e^{2t},$$

and the zero-state response is

$$y_1(t) = \mathscr{L}^{-1}\left\{ \frac{1}{(s-1)(s-2)(s+4)} \right\} = -\frac{1}{5}e^t + \frac{1}{6}e^{2t} + \frac{1}{30}e^{-4t}.$$

Verify that the sum of $y_0(t)$ and $y_1(t)$ is the solution $y(t)$ in Example 5 and that $y_0(0) = 1$, $y_0'(0) = 5$, whereas $y_1(0) = 0$, $y_1'(0) = 0$.

## EXERCISES 7.2

*Answers to selected odd-numbered problems begin on page ANS-11.*

### 7.2.1 Inverse Transforms

In Problems 1–30 use appropriate algebra and Theorem 7.2.1 to find the given inverse Laplace transform.

**1.** $\mathscr{L}^{-1}\left\{ \dfrac{1}{s^3} \right\}$

**2.** $\mathscr{L}^{-1}\left\{ \dfrac{1}{s^4} \right\}$

**3.** $\mathscr{L}^{-1}\left\{ \dfrac{1}{s^2} - \dfrac{48}{s^5} \right\}$

**4.** $\mathscr{L}^{-1}\left\{ \left( \dfrac{2}{s} - \dfrac{1}{s^3} \right)^2 \right\}$

**5.** $\mathscr{L}^{-1}\left\{ \dfrac{(s+1)^3}{s^4} \right\}$

**6.** $\mathscr{L}^{-1}\left\{ \dfrac{(s+2)^2}{s^3} \right\}$

**7.** $\mathscr{L}^{-1}\left\{ \dfrac{1}{s^2} - \dfrac{1}{s} + \dfrac{1}{s-2} \right\}$

**8.** $\mathscr{L}^{-1}\left\{ \dfrac{4}{s} + \dfrac{6}{s^5} - \dfrac{1}{s+8} \right\}$

**9.** $\mathscr{L}^{-1}\left\{ \dfrac{1}{4s+1} \right\}$

**10.** $\mathscr{L}^{-1}\left\{ \dfrac{1}{5s-2} \right\}$

**11.** $\mathscr{L}^{-1}\left\{ \dfrac{5}{s^2+49} \right\}$

**12.** $\mathscr{L}^{-1}\left\{ \dfrac{10s}{s^2+16} \right\}$

**13.** $\mathscr{L}^{-1}\left\{ \dfrac{4s}{4s^2+1} \right\}$

**14.** $\mathscr{L}^{-1}\left\{ \dfrac{1}{4s^2+1} \right\}$

**15.** $\mathscr{L}^{-1}\left\{ \dfrac{2s-6}{s^2+9} \right\}$

**16.** $\mathscr{L}^{-1}\left\{ \dfrac{s+1}{s^2+2} \right\}$

**17.** $\mathscr{L}^{-1}\left\{ \dfrac{1}{s^2+3s} \right\}$

**18.** $\mathscr{L}^{-1}\left\{ \dfrac{s+1}{s^2-4s} \right\}$

**19.** $\mathscr{L}^{-1}\left\{ \dfrac{s}{s^2+2s-3} \right\}$

**20.** $\mathscr{L}^{-1}\left\{ \dfrac{1}{s^2+s-20} \right\}$

**21.** $\mathscr{L}^{-1}\left\{\dfrac{0.9s}{(s-0.1)(s+0.2)}\right\}$

**22.** $\mathscr{L}^{-1}\left\{\dfrac{s-3}{(s-\sqrt{3})(s+\sqrt{3})}\right\}$

**23.** $\mathscr{L}^{-1}\left\{\dfrac{s}{(s-2)(s-3)(s-6)}\right\}$

**24.** $\mathscr{L}^{-1}\left\{\dfrac{s^2+1}{s(s-1)(s+1)(s-2)}\right\}$

**25.** $\mathscr{L}^{-1}\left\{\dfrac{1}{s^3+5s}\right\}$  **26.** $\mathscr{L}^{-1}\left\{\dfrac{s}{(s+2)(s^2+4)}\right\}$

**27.** $\mathscr{L}^{-1}\left\{\dfrac{2s-4}{(s^2+s)(s^2+1)}\right\}$  **28.** $\mathscr{L}^{-1}\left\{\dfrac{1}{s^4-9}\right\}$

**29.** $\mathscr{L}^{-1}\left\{\dfrac{1}{(s^2+1)(s^2+4)}\right\}$  **30.** $\mathscr{L}^{-1}\left\{\dfrac{6s+3}{s^4+5s^2+4}\right\}$

In Problems 31–34 find the given inverse Laplace transform by finding the Laplace transform of the indicated function $f$.

**31.** $\mathscr{L}^{-1}\left\{\dfrac{1}{(s-a)^2-b^2}\right\}$; $f(t)=e^{at}\sinh bt$

**32.** $\mathscr{L}^{-1}\left\{\dfrac{1}{s^2(s^2+a^2)}\right\}$; $f(t)=at-\sin at$

**33.** $\mathscr{L}^{-1}\left\{\dfrac{1}{(s^2+a^2)(s^2+b^2)}\right\}$; $f(t)=a\sin bt-b\sin at$

**34.** $\mathscr{L}^{-1}\left\{\dfrac{s}{(s^2+a^2)(s^2+b^2)}\right\}$; $f(t)=\cos bt-\cos at$

### 7.2.2 Transforms of Derivatives

In Problems 35–44 use the Laplace transform to solve the given initial-value problem.

**35.** $\dfrac{dy}{dt}-y=1$, $y(0)=0$

**36.** $2\dfrac{dy}{dt}+y=0$, $y(0)=-3$

**37.** $y'+6y=e^{4t}$, $y(0)=2$

**38.** $y'-y=2\cos 5t$, $y(0)=0$

**39.** $y''+5y'+4y=0$, $y(0)=1$, $y'(0)=0$

**40.** $y''-4y'=6e^{3t}-3e^{-t}$, $y(0)=1$, $y'(0)=-1$

**41.** $y''+y=\sqrt{2}\sin\sqrt{2}t$, $y(0)=10$, $y'(0)=0$

**42.** $y''+9y=e^t$, $y(0)=0$, $y'(0)=0$

**43.** $2y'''+3y''-3y'-2y=e^{-t}$, $y(0)=0$, $y'(0)=0$, $y''(0)=1$

**44.** $y'''+2y''-y'-2y=\sin 3t$, $y(0)=0$, $y'(0)=0$, $y''(0)=1$

The inverse forms of the results in Problem 49 in Exercises 7.1 are

$$\mathscr{L}^{-1}\left\{\dfrac{s-a}{(s-a)^2+b^2}\right\}=e^{at}\cos bt$$

$$\mathscr{L}^{-1}\left\{\dfrac{b}{(s-a)^2+b^2}\right\}=e^{at}\sin bt.$$

In Problems 45 and 46 use the Laplace transform and these inverses to solve the given initial-value problem.

**45.** $y'+y=e^{-3t}\cos 2t$, $y(0)=0$

**46.** $y''-2y'+5y=0$, $y(0)=1$, $y'(0)=3$

In Problems 47 and 48 use one of the inverse Laplace transforms found in Problems 31–34 to solve the given initial-value problem.

**47.** $y''+4y=10\cos 5t$, $y(0)=0$, $y'(0)=0$

**48.** $y''+2y=4t$, $y(0)=0$, $y'(0)=0$

### Discussion Problems

**49.** **(a)** With a slight change in notation the transform in (6) is the same as

$$\mathscr{L}\{f'(t)\}=s\mathscr{L}\{f(t)\}-f(0).$$

With $f(t)=te^{at}$, discuss how this result in conjunction with (c) of Theorem 7.1.1 can be used to evaluate $\mathscr{L}\{te^{at}\}$.

**(b)** Proceed as in part (a), but this time discuss how to use (7) with $f(t)=t\sin kt$ in conjunction with (d) and (e) of Theorem 7.1.1 to evaluate $\mathscr{L}\{t\sin kt\}$.

**50.** Make up two functions $f_1$ and $f_2$ that have the same Laplace transform. Do not think profound thoughts.

**51.** Reread (*iii*) in the *Remarks* on page 293. Find the zero-input and the zero-state response for the IVP in Problem 40.

**52.** Suppose $f(t)$ is a function for which $f'(t)$ is piecewise continuous and of exponential order $c$. Use results in this section and Section 7.1 to justify

$$f(0)=\lim_{s\to\infty}sF(s),$$

where $F(s)=\mathscr{L}\{f(t)\}$. Verify this result with $f(t)=\cos kt$.

## 7.3 Operational Properties I

**INTRODUCTION** It is not convenient to use Definition 7.1.1 each time we wish to find the Laplace transform of a function $f(t)$. For example, the integration by parts involved in evaluating, say, $\mathscr{L}\{e^t t^2 \sin 3t\}$ is formidable, to say the least. In this section and the next we present several labor-saving operational properties of the Laplace transform that enable us to build up a more extensive list of transforms (see the table in Appendix C) without having to resort to the basic definition and integration.

### 7.3.1 TRANSLATION ON THE s-AXIS

**A TRANSLATION** Evaluating transforms such as $\mathscr{L}\{e^{5t}t^3\}$ and $\mathscr{L}\{e^{-2t}\cos 4t\}$ is straightforward provided that we know (and we do) $\mathscr{L}\{t^3\}$ and $\mathscr{L}\{\cos 4t\}$. In general, if we know the Laplace transform of a function $f$, $\mathscr{L}\{f(t)\} = F(s)$, it is possible to compute the Laplace transform of an exponential multiple of $f$, that is, $\mathscr{L}\{e^{at}f(t)\}$, with no additional effort other than *translating*, or *shifting*, the transform $F(s)$ to $F(s - a)$. This result is known as the **first translation theorem** or **first shifting theorem.**

---

**THEOREM 7.3.1  First Translation Theorem**

If $\mathscr{L}\{f(t)\} = F(s)$ and $a$ is any real number, then

$$\mathscr{L}\{e^{at}f(t)\} = F(s - a).$$

---

**FIGURE 7.3.1** Shift on s-axis

**PROOF** The proof is immediate, since by Definition 7.1.1

$$\mathscr{L}\{e^{at}f(t)\} = \int_0^\infty e^{-st}e^{at}f(t)\,dt = \int_0^\infty e^{-(s-a)t}f(t)\,dt = F(s - a). \qquad \blacksquare$$

If we consider $s$ a real variable, then the graph of $F(s - a)$ is the graph of $F(s)$ shifted on the $s$-axis by the amount $|a|$. If $a > 0$, the graph of $F(s)$ is shifted $a$ units to the right, whereas if $a < 0$, the graph is shifted $|a|$ units to the left. See Figure 7.3.1.

For emphasis it is sometimes useful to use the symbolism

$$\mathscr{L}\{e^{at}f(t)\} = \mathscr{L}\{f(t)\}|_{s \to s-a},$$

where $s \to s - a$ means that in the Laplace transform $F(s)$ of $f(t)$ we replace the symbol $s$ wherever it appears by $s - a$.

---

**EXAMPLE 1    Using the First Translation Theorem**

Evaluate    **(a)** $\mathscr{L}\{e^{5t}t^3\}$    **(b)** $\mathscr{L}\{e^{-2t}\cos 4t\}$.

**SOLUTION** The results follow from Theorems 7.1.1 and 7.3.1.

**(a)** $\mathscr{L}\{e^{5t}t^3\} = \mathscr{L}\{t^3\}|_{s \to s-5} = \left.\dfrac{3!}{s^4}\right|_{s \to s-5} = \dfrac{6}{(s - 5)^4}$

**(b)** $\mathscr{L}\{e^{-2t}\cos 4t\} = \mathscr{L}\{\cos 4t\}|_{s \to s-(-2)} = \left.\dfrac{s}{s^2 + 16}\right|_{s \to s+2} = \dfrac{s + 2}{(s + 2)^2 + 16}$ ■

**INVERSE FORM OF THEOREM 7.3.1** To compute the inverse of $F(s - a)$, we must recognize $F(s)$, find $f(t)$ by taking the inverse Laplace transform of $F(s)$, and then multiply $f(t)$ by the exponential function $e^{at}$. This procedure can be summarized symbolically in the following manner:

$$\mathscr{L}^{-1}\{F(s - a)\} = \mathscr{L}^{-1}\{F(s)|_{s \to s-a}\} = e^{at}f(t), \tag{1}$$

where $f(t) = \mathscr{L}^{-1}\{F(s)\}$.

The first part of the next example illustrates partial fraction decomposition in the case when the denominator of $Y(s)$ contains *repeated linear factors*.

### EXAMPLE 2    Partial Fractions: Repeated Linear Factors

Evaluate   **(a)** $\mathscr{L}^{-1}\left\{\dfrac{2s + 5}{(s - 3)^2}\right\}$   **(b)** $\mathscr{L}^{-1}\left\{\dfrac{s/2 + 5/3}{s^2 + 4s + 6}\right\}$.

**SOLUTION   (a)** A repeated linear factor is a term $(s - a)^n$, where $a$ is a real number and $n$ is a positive integer $\geq 2$. Recall that if $(s - a)^n$ appears in the denominator of a rational expression, then the assumed decomposition contains $n$ partial fractions with constant numerators and denominators $s - a, (s - a)^2, \ldots, (s - a)^n$. Hence with $a = 3$ and $n = 2$ we write

$$\frac{2s + 5}{(s - 3)^2} = \frac{A}{s - 3} + \frac{B}{(s - 3)^2}.$$

By putting the two terms on the right-hand side over a common denominator, we obtain the numerator $2s + 5 = A(s - 3) + B$, and this identity yields $A = 2$ and $B = 11$. Therefore

$$\frac{2s + 5}{(s - 3)^2} = \frac{2}{s - 3} + \frac{11}{(s - 3)^2} \tag{2}$$

and

$$\mathscr{L}^{-1}\left\{\frac{2s + 5}{(s - 3)^2}\right\} = 2\mathscr{L}^{-1}\left\{\frac{1}{s - 3}\right\} + 11\mathscr{L}^{-1}\left\{\frac{1}{(s - 3)^2}\right\}. \tag{3}$$

Now $1/(s - 3)^2$ is $F(s) = 1/s^2$ shifted three units to the right. Since $\mathscr{L}^{-1}\{1/s^2\} = t$, it follows from (1) that

$$\mathscr{L}^{-1}\left\{\frac{1}{(s - 3)^2}\right\} = \mathscr{L}^{-1}\left\{\frac{1}{s^2}\bigg|_{s \to s-3}\right\} = e^{3t}t.$$

Finally, (3) is

$$\mathscr{L}^{-1}\left\{\frac{2s + 5}{(s - 3)^2}\right\} = 2e^{3t} + 11e^{3t}t. \tag{4}$$

**(b)** To start, observe that the quadratic polynomial $s^2 + 4s + 6$ has no real zeros and so has no real linear factors. In this situation we *complete the square*:

$$\frac{s/2 + 5/3}{s^2 + 4s + 6} = \frac{s/2 + 5/3}{(s + 2)^2 + 2}. \tag{5}$$

Our goal here is to recognize the expression on the right-hand side as some Laplace transform $F(s)$ in which $s$ has been replaced throughout by $s + 2$. What we are trying to do is analogous to working part (b) of Example 1 backwards. The denominator in (5) is already in the correct form—that is, $s^2 + 2$ with $s$ replaced by $s + 2$. However, we must fix up the numerator by manipulating the constants: $\frac{1}{2}s + \frac{5}{3} = \frac{1}{2}(s + 2) + \frac{5}{3} - \frac{2}{2} = \frac{1}{2}(s + 2) + \frac{2}{3}$.

Now by termwise division, the linearity of $\mathscr{L}^{-1}$, parts (d) and (e) of Theorem 7.2.1, and finally (1),

$$\frac{s/2 + 5/3}{(s + 2)^2 + 2} = \frac{\frac{1}{2}(s + 2) + \frac{2}{3}}{(s + 2)^2 + 2} = \frac{1}{2}\frac{s + 2}{(s + 2)^2 + 2} + \frac{2}{3}\frac{1}{(s + 2)^2 + 2}$$

$$\mathscr{L}^{-1}\left\{\frac{s/2 + 5/3}{s^2 + 4s + 6}\right\} = \frac{1}{2}\mathscr{L}^{-1}\left\{\frac{s + 2}{(s + 2)^2 + 2}\right\} + \frac{2}{3}\mathscr{L}^{-1}\left\{\frac{1}{(s + 2)^2 + 2}\right\}$$

$$= \frac{1}{2}\mathscr{L}^{-1}\left\{\frac{s}{s^2 + 2}\bigg|_{s \to s+2}\right\} + \frac{2}{3\sqrt{2}}\mathscr{L}^{-1}\left\{\frac{\sqrt{2}}{s^2 + 2}\bigg|_{s \to s+2}\right\} \tag{6}$$

$$= \frac{1}{2}e^{-2t}\cos\sqrt{2}t + \frac{\sqrt{2}}{3}e^{-2t}\sin\sqrt{2}t. \tag{7} \quad \blacksquare$$

EXAMPLE 3    An Initial-Value Problem

Solve $y'' - 6y' + 9y = t^2 e^{3t}$,  $y(0) = 2$,  $y'(0) = 17$.

**SOLUTION** Before transforming the DE, note that its right-hand side is similar to the function in part (a) of Example 1. After using linearity, Theorem 7.3.1, and the initial conditions, we simplify and then solve for $Y(s) = \mathcal{L}\{f(t)\}$:

$$\mathcal{L}\{y''\} - 6\mathcal{L}\{y'\} + 9\mathcal{L}\{y\} = \mathcal{L}\{t^2 e^{3t}\}$$

$$s^2 Y(s) - sy(0) - y'(0) - 6[sY(s) - y(0)] + 9Y(s) = \frac{2}{(s-3)^3}$$

$$(s^2 - 6s + 9)Y(s) = 2s + 5 + \frac{2}{(s-3)^3}$$

$$(s-3)^2 Y(s) = 2s + 5 + \frac{2}{(s-3)^3}$$

$$Y(s) = \frac{2s+5}{(s-3)^2} + \frac{2}{(s-3)^5}.$$

The first term on the right-hand side was already decomposed into individual partial fractions in (2) in part (a) of Example 2:

$$Y(s) = \frac{2}{s-3} + \frac{11}{(s-3)^2} + \frac{2}{(s-3)^5}.$$

Thus   $y(t) = 2\mathcal{L}^{-1}\left\{\frac{1}{s-3}\right\} + 11\mathcal{L}^{-1}\left\{\frac{1}{(s-3)^2}\right\} + \frac{2}{4!}\mathcal{L}^{-1}\left\{\frac{4!}{(s-3)^5}\right\}.$   (8)

From the inverse form (1) of Theorem 7.3.1, the last two terms in (8) are

$$\mathcal{L}^{-1}\left\{\frac{1}{s^2}\right\}\Big|_{s \to s-3} = te^{3t} \quad \text{and} \quad \mathcal{L}^{-1}\left\{\frac{4!}{s^5}\right\}\Big|_{s \to s-3} = t^4 e^{3t}.$$

Thus (8) is $y(t) = 2e^{3t} + 11te^{3t} + \frac{1}{12}t^4 e^{3t}.$ ∎

EXAMPLE 4    An Initial-Value Problem

Solve $y'' + 4y' + 6y = 1 + e^{-t}$,  $y(0) = 0$,  $y'(0) = 0$.

**SOLUTION**                $\mathcal{L}\{y''\} + 4\mathcal{L}\{y'\} + 6\mathcal{L}\{y\} = \mathcal{L}\{1\} + \mathcal{L}\{e^{-t}\}$

$$s^2 Y(s) - sy(0) - y'(0) + 4[sY(s) - y(0)] + 6Y(s) = \frac{1}{s} + \frac{1}{s+1}$$

$$(s^2 + 4s + 6)Y(s) = \frac{2s+1}{s(s+1)}$$

$$Y(s) = \frac{2s+1}{s(s+1)(s^2 + 4s + 6)}.$$

Since the quadratic term in the denominator does not factor into real linear factors, the partial fraction decomposition for $Y(s)$ is found to be

$$Y(s) = \frac{1/6}{s} + \frac{1/3}{s+1} - \frac{s/2 + 5/3}{s^2 + 4s + 6}.$$

Moreover, in preparation for taking the inverse transform we already manipulated the last term into the necessary form in part (b) of Example 2. So in view of the results in (6) and (7) we have the solution

$$y(t) = \frac{1}{6}\mathcal{L}^{-1}\left\{\frac{1}{s}\right\} + \frac{1}{3}\mathcal{L}^{-1}\left\{\frac{1}{s+1}\right\} - \frac{1}{2}\mathcal{L}^{-1}\left\{\frac{s+2}{(s+2)^2+2}\right\} - \frac{2}{3\sqrt{2}}\mathcal{L}^{-1}\left\{\frac{\sqrt{2}}{(s+2)^2+2}\right\}$$

$$= \frac{1}{6} + \frac{1}{3}e^{-t} - \frac{1}{2}e^{-2t}\cos\sqrt{2}t - \frac{\sqrt{2}}{3}e^{-2t}\sin\sqrt{2}t.$$ ∎

### 7.3.2 TRANSLATION ON THE *t*-AXIS

**UNIT STEP FUNCTION**    In engineering, one frequently encounters functions that are either "off" or "on." For example, an external force acting on a mechanical system or a voltage impressed on a circuit can be turned off after a period of time. It is convenient, then, to define a special function that is the number 0 (off) up to a certain time $t = a$ and then the number 1 (on) after that time. This function is called the **unit step function** or the **Heaviside function,** named after the English polymath **Oliver Heaviside** (1850–1925).

---

**DEFINITION 7.3.1**  Unit Step Function

The **unit step function** $\mathcal{U}(t - a)$ is defined to be

$$\mathcal{U}(t - a) = \begin{cases} 0, & 0 \le t < a \\ 1, & t \ge a. \end{cases}$$

---

**FIGURE 7.3.2**  Graph of unit step function

Notice that we define $\mathcal{U}(t - a)$ only on the nonnegative $t$-axis, since this is all that we are concerned with in the study of the Laplace transform. In a broader sense $\mathcal{U}(t - a) = 0$ for $t < a$. The graph of $\mathcal{U}(t - a)$ is given in Figure 7.3.2. In the case when $a = 0$, we take $\mathcal{U}(t) = 1$ for $t \ge 0$.

When a function $f$ defined for $t \ge 0$ is multiplied by $\mathcal{U}(t - a)$, the unit step function "turns off" a portion of the graph of that function. For example, consider the function $f(t) = 2t - 3$. To "turn off" the portion of the graph of $f$ for $0 \le t < 1$, we simply form the product $(2t - 3)\mathcal{U}(t - 1)$. See Figure 7.3.3. In general, the graph of $f(t)\mathcal{U}(t - a)$ is 0 (off) for $0 \le t < a$ and is the portion of the graph of $f$ (on) for $t \ge a$.

The unit step function can also be used to write piecewise-defined functions in a compact form. For example, if we consider $0 \le t < 2$, $2 \le t < 3$, and $t \ge 3$ and the corresponding values of $\mathcal{U}(t - 2)$ and $\mathcal{U}(t - 3)$, it should be apparent that the piecewise-defined function shown in Figure 7.3.4 is the same as $f(t) = 2 - 3\mathcal{U}(t - 2) + \mathcal{U}(t - 3)$. Also, a general piecewise-defined function of the type

$$f(t) = \begin{cases} g(t), & 0 \le t < a \\ h(t), & t \ge a \end{cases} \tag{9}$$

is the same as

$$f(t) = g(t) - g(t)\mathcal{U}(t - a) + h(t)\mathcal{U}(t - a). \tag{10}$$

Similarly, a function of the type

$$f(t) = \begin{cases} 0, & 0 \le t < a \\ g(t), & a \le t < b \\ 0, & t \ge b \end{cases} \tag{11}$$

**FIGURE 7.3.3**  Function is $f(t) = (2t - 3)\,\mathcal{U}(t - 1)$

can be written

$$f(t) = g(t)[\mathcal{U}(t - a) - \mathcal{U}(t - b)]. \tag{12}$$

**FIGURE 7.3.4**  Function is $f(t) = 2 - 3\mathcal{U}(t - 2) + \mathcal{U}(t - 3)$

---

**EXAMPLE 5**  A Piecewise-Defined Function

Express $f(t) = \begin{cases} 20t, & 0 \le t < 5 \\ 0, & t \ge 5 \end{cases}$ in terms of unit step functions. Graph.

**FIGURE 7.3.5** Function *f* in Example 5

**(a)** $f(t)$, $t \geq 0$

**(b)** $f(t - a)\,\mathcal{U}(t - a)$

**FIGURE 7.3.6** Shift on *t*-axis

**SOLUTION**   The graph of *f* is given in Figure 7.3.5. Now from (9) and (10) with $a = 5$, $g(t) = 20t$, and $h(t) = 0$ we get $f(t) = 20t - 20t\,\mathcal{U}(t - 5)$. ■

Consider a general function $y = f(t)$ defined for $t \geq 0$. The piecewise-defined function

$$f(t - a)\,\mathcal{U}(t - a) = \begin{cases} 0, & 0 \leq t < a \\ f(t - a), & t \geq a \end{cases} \tag{13}$$

plays a significant role in the discussion that follows. As shown in Figure 7.3.6, for $a > 0$ the graph of the function $y = f(t - a)\,\mathcal{U}(t - a)$ coincides with the graph of $y = f(t - a)$ for $t \geq a$ (which is the *entire* graph of $y = f(t)$, $t \geq 0$ shifted *a* units to the right on the *t*-axis), but is identically zero for $0 \leq t < a$.

We saw in Theorem 7.3.1 that an exponential multiple of $f(t)$ results in a translation of the transform $F(s)$ on the *s*-axis. As a consequence of the next theorem we see that whenever $F(s)$ is multiplied by an exponential function $e^{-as}$, $a > 0$, the inverse transform of the product $e^{-as}\,F(s)$ is the function *f* shifted along the *t*-axis in the manner illustrated in Figure 7.3.6(b). This result, presented next in its direct transform version, is called the **second translation theorem** or **second shifting theorem.**

---

**THEOREM 7.3.2**   **Second Translation Theorem**

If $F(s) = \mathcal{L}\{f(t)\}$ and $a > 0$, then

$$\mathcal{L}\{f(t - a)\,\mathcal{U}(t - a)\} = e^{-as}F(s).$$

---

**PROOF**   By the additive interval property of integrals,

$$\int_0^{\infty} e^{-st} f(t - a)\,\mathcal{U}(t - a)\, dt$$

can be written as two integrals:

$$\mathcal{L}\{f(t - a)\,\mathcal{U}(t - a)\} = \underbrace{\int_0^{a} e^{-st} f(t - a)\,\mathcal{U}(t - a)\, dt}_{\substack{\text{zero for} \\ 0 \leq t < a}} + \underbrace{\int_a^{\infty} e^{-st} f(t - a)\,\mathcal{U}(t - a)\, dt}_{\substack{\text{one for} \\ t \geq a}} = \int_a^{\infty} e^{-st} f(t - a)\, dt.$$

Now if we let $v = t - a$, $dv = dt$ in the last integral, then

$$\mathcal{L}\{f(t - a)\,\mathcal{U}(t - a)\} = \int_0^{\infty} e^{-s(v+a)} f(v)\, dv = e^{-as} \int_0^{\infty} e^{-sv} f(v)\, dv = e^{-as}\mathcal{L}\{f(t)\}. \quad ■$$

We often wish to find the Laplace transform of just a unit step function. This can be from either Definition 7.1.1 or Theorem 7.3.2. If we identify $f(t) = 1$ in Theorem 7.3.2, then $f(t - a) = 1$, $F(s) = \mathcal{L}\{1\} = 1/s$, and so

$$\mathcal{L}\{\mathcal{U}(t - a)\} = \frac{e^{-as}}{s}. \tag{14}$$

**EXAMPLE 6**   **Figure 7.3.4 Revisited**

Find the Laplace transform of the function *f* in Figure 7.3.4.

**SOLUTION**   We use *f* expressed in terms of the unit step function

$$f(t) = 2 - 3\mathcal{U}(t - 2) + \mathcal{U}(t - 3)$$

and the result given in (14):

$$\mathcal{L}\{f(t)\} = 2\mathcal{L}\{1\} - 3\mathcal{L}\{\mathcal{U}(t - 2)\} + \mathcal{L}\{\mathcal{U}(t - 3)\}$$

$$= \frac{2}{s} - 3\frac{e^{-2s}}{s} + \frac{e^{-3s}}{s}. \quad ■$$

**INVERSE FORM OF THEOREM 7.3.2** If $f(t) = \mathscr{L}^{-1}\{F(s)\}$, the inverse form of Theorem 7.3.2, $a > 0$, is

$$\mathscr{L}^{-1}\{e^{-as}F(s)\} = f(t - a)\,\mathcal{U}(t - a). \tag{15}$$

---

### EXAMPLE 7 Using Formula (15)

Evaluate **(a)** $\mathscr{L}^{-1}\left\{\dfrac{1}{s-4}e^{-2s}\right\}$ **(b)** $\mathscr{L}^{-1}\left\{\dfrac{s}{s^2+9}e^{-\pi s/2}\right\}$.

**SOLUTION (a)** With the three identifications $a = 2$, $F(s) = 1/(s - 4)$, and $\mathscr{L}^{-1}\{F(s)\} = e^{4t}$, we have from (15)

$$\mathscr{L}^{-1}\left\{\frac{1}{s-4}e^{-2s}\right\} = e^{4(t-2)}\,\mathcal{U}(t - 2).$$

**(b)** With $a = \pi/2$, $F(s) = s/(s^2 + 9)$, and $\mathscr{L}^{-1}\{F(s)\} = \cos 3t$, (15) yields

$$\mathscr{L}^{-1}\left\{\frac{s}{s^2+9}e^{-\pi s/2}\right\} = \cos 3\left(t - \frac{\pi}{2}\right)\mathcal{U}\left(t - \frac{\pi}{2}\right).$$

The last expression can be simplified somewhat by using the addition formula for the cosine. Verify that the result is the same as $-\sin 3t\,\mathcal{U}\left(t - \dfrac{\pi}{2}\right)$. ■

**ALTERNATIVE FORM OF THEOREM 7.3.2** We are frequently confronted with the problem of finding the Laplace transform of a product of a function $g$ and a unit step function $\mathcal{U}(t - a)$ where the function $g$ lacks the precise shifted form $f(t - a)$ in Theorem 7.3.2. To find the Laplace transform of $g(t)\mathcal{U}(t - a)$, it is possible to fix up $g(t)$ into the required form $f(t - a)$ by algebraic manipulations. For example, if we wanted to use Theorem 7.3.2 to find the Laplace transform of $t^2\mathcal{U}(t - 2)$, we would have to force $g(t) = t^2$ into the form $f(t - 2)$. You should work through the details and verify that $t^2 = (t - 2)^2 + 4(t - 2) + 4$ is an identity. Therefore

$$\mathscr{L}\{t^2\mathcal{U}(t - 2)\} = \mathscr{L}\{(t - 2)^2\,\mathcal{U}(t - 2) + 4(t - 2)\,\mathcal{U}(t - 2) + 4\mathcal{U}(t - 2)\},$$

where each term on the right-hand side can now be evaluated by Theorem 7.3.2. But since these manipulations are time consuming and often not obvious, it is simpler to devise an alternative version of Theorem 7.3.2. Using Definition 7.1.1, the definition of $\mathcal{U}(t - a)$, and the substitution $u = t - a$, we obtain

$$\mathscr{L}\{g(t)\,\mathcal{U}(t - a)\} = \int_a^\infty e^{-st}g(t)\,dt = \int_0^\infty e^{-s(u+a)}\,g(u + a)\,du.$$

That is, $$\mathscr{L}\{g(t)\mathcal{U}(t - a)\} = e^{-as}\mathscr{L}\{g(t + a)\}. \tag{16}$$

---

### EXAMPLE 8 Second Translation Theorem—Alternative Form

Evaluate $\mathscr{L}\{\cos t\,\mathcal{U}(t - \pi)\}$.

**SOLUTION** With $g(t) = \cos t$ and $a = \pi$, then $g(t + \pi) = \cos(t + \pi) = -\cos t$ by the addition formula for the cosine function. Hence by (16),

$$\mathscr{L}\{\cos t\,\mathcal{U}(t - \pi)\} = -e^{-\pi s}\mathscr{L}\{\cos t\} = -\frac{s}{s^2+1}e^{-\pi s}. \quad ■$$

In the next two examples we solve, in turn, an initial-value problem and a boundary-value problem involving a piecewise-linear differential equation.

EXAMPLE 9    **An Initial-Value Problem**

Solve $y' + y = f(t)$, $y(0) = 5$, where $f(t) = \begin{cases} 0, & 0 \le t < \pi \\ 3\cos t, & t \ge \pi. \end{cases}$

**SOLUTION** The function $f$ can be written as $f(t) = 3\cos t\, \mathcal{U}(t - \pi)$, so by linearity, the results of Example 7, and the usual partial fractions, we have

$$\mathcal{L}\{y'\} + \mathcal{L}\{y\} = 3\mathcal{L}\{\cos t\, \mathcal{U}(t - \pi)\}$$

$$sY(s) - y(0) + Y(s) = -3\frac{s}{s^2 + 1}e^{-\pi s}$$

$$(s + 1)Y(s) = 5 - \frac{3s}{s^2 + 1}e^{-\pi s}$$

$$Y(s) = \frac{5}{s + 1} - \frac{3}{2}\left[-\frac{1}{s + 1}e^{-\pi s} + \frac{1}{s^2 + 1}e^{-\pi s} + \frac{s}{s^2 + 1}e^{-\pi s}\right]. \qquad (17)$$

Now proceeding as we did in Example 7, it follows from (15) with $a = \pi$ that the inverses of the terms inside the brackets are

$$\mathcal{L}^{-1}\left\{\frac{1}{s + 1}e^{-\pi s}\right\} = e^{-(t - \pi)}\mathcal{U}(t - \pi), \qquad \mathcal{L}^{-1}\left\{\frac{1}{s^2 + 1}e^{-\pi s}\right\} = \sin(t - \pi)\mathcal{U}(t - \pi),$$

and $$\mathcal{L}^{-1}\left\{\frac{s}{s^2 + 1}e^{-\pi s}\right\} = \cos(t - \pi)\mathcal{U}(t - \pi).$$

Thus the inverse of (17) is

$$y(t) = 5e^{-t} + \frac{3}{2}e^{-(t - \pi)}\mathcal{U}(t - \pi) - \frac{3}{2}\sin(t - \pi)\mathcal{U}(t - \pi) - \frac{3}{2}\cos(t - \pi)\mathcal{U}(t - \pi)$$

$$= 5e^{-t} + \frac{3}{2}[e^{-(t - \pi)} + \sin t + \cos t]\,\mathcal{U}(t - \pi) \qquad \leftarrow \text{trigonometric identities}$$

$$= \begin{cases} 5e^{-t}, & 0 \le t < \pi \\ 5e^{-t} + \frac{3}{2}e^{-(t - \pi)} + \frac{3}{2}\sin t + \frac{3}{2}\cos t, & t \ge \pi. \end{cases} \qquad (18)$$

We obtained the graph of (18) shown in Figure 7.3.7 by using a graphing utility.  ■

**FIGURE 7.3.7** Graph of function (18) in Example 9

**BEAMS** In Section 5.2 we saw that the static deflection $y(x)$ of a uniform beam of length $L$ carrying load $w(x)$ per unit length is found from the linear fourth-order differential equation

$$EI\frac{d^4y}{dx^4} = w(x), \qquad (19)$$

where $E$ is Young's modulus of elasticity and $I$ is a moment of inertia of a cross section of the beam. The Laplace transform is particularly useful in solving (19) when $w(x)$ is piecewise-defined. However, to use the Laplace transform, we must tacitly assume that $y(x)$ and $w(x)$ are defined on $(0, \infty)$ rather than on $(0, L)$. Note, too, that the next example is a boundary-value problem rather than an initial-value problem.

EXAMPLE 10    **A Boundary-Value Problem**

A beam of length $L$ is embedded at both ends, as shown in Figure 7.3.8. Find the deflection of the beam when the load is given by

$$w(x) = \begin{cases} w_0\left(1 - \dfrac{2}{L}x\right), & 0 < x < L/2 \\ 0, & L/2 < x < L. \end{cases}$$

$w(x)$

wall

$L$

**FIGURE 7.3.8** Embedded beam with variable load in Example 10

**SOLUTION** Recall that because the beam is embedded at both ends, the boundary conditions are $y(0) = 0$, $y'(0) = 0$, $y(L) = 0$, $y'(L) = 0$. Now by (10) we can express $w(x)$ in terms of the unit step function:

$$w(x) = w_0\left(1 - \frac{2}{L}x\right) - w_0\left(1 - \frac{2}{L}x\right)\mathcal{U}\left(x - \frac{L}{2}\right)$$

$$= \frac{2w_0}{L}\left[\frac{L}{2} - x + \left(x - \frac{L}{2}\right)\mathcal{U}\left(x - \frac{L}{2}\right)\right].$$

Transforming (19) with respect to the variable $x$ gives

$$EI(s^4 Y(s) - s^3 y(0) - s^2 y'(0) - s y''(0) - y'''(0)) = \frac{2w_0}{L}\left[\frac{L/2}{s} - \frac{1}{s^2} + \frac{1}{s^2}e^{-Ls/2}\right]$$

or

$$s^4 Y(s) - s y''(0) - y'''(0) = \frac{2w_0}{EIL}\left[\frac{L/2}{s} - \frac{1}{s^2} + \frac{1}{s^2}e^{-Ls/2}\right].$$

If we let $c_1 = y''(0)$ and $c_2 = y'''(0)$, then

$$Y(s) = \frac{c_1}{s^3} + \frac{c_2}{s^4} + \frac{2w_0}{EIL}\left[\frac{L/2}{s^5} - \frac{1}{s^6} + \frac{1}{s^6}e^{-Ls/2}\right],$$

and consequently

$$y(x) = \frac{c_1}{2!}\mathcal{L}^{-1}\left\{\frac{2!}{s^3}\right\} + \frac{c_2}{3!}\mathcal{L}^{-1}\left\{\frac{3!}{s^4}\right\} + \frac{2w_0}{EIL}\left[\frac{L/2}{4!}\mathcal{L}^{-1}\left\{\frac{4!}{s^5}\right\} - \frac{1}{5!}\mathcal{L}^{-1}\left\{\frac{5!}{s^6}\right\} + \frac{1}{5!}\mathcal{L}^{-1}\left\{\frac{5!}{s^6}e^{-Ls/2}\right\}\right]$$

$$= \frac{c_1}{2}x^2 + \frac{c_2}{6}x^3 + \frac{w_0}{60\,EIL}\left[\frac{5L}{2}x^4 - x^5 + \left(x - \frac{L}{2}\right)^5\mathcal{U}\left(x - \frac{L}{2}\right)\right].$$

Applying the conditions $y(L) = 0$ and $y'(L) = 0$ to the last result yields a system of equations for $c_1$ and $c_2$:

$$c_1\frac{L^2}{2} + c_2\frac{L^3}{6} + \frac{49w_0 L^4}{1920EI} = 0$$

$$c_1 L + c_2\frac{L^2}{2} + \frac{85w_0 L^3}{960EI} = 0.$$

Solving, we find $c_1 = 23w_0 L^2/(960EI)$ and $c_2 = -9w_0 L/(40EI)$. Thus the deflection is given by

$$y(x) = \frac{23w_0 L^2}{1920EI}x^2 - \frac{3w_0 L}{80EI}x^3 + \frac{w_0}{60EIL}\left[\frac{5L}{2}x^4 - x^5 + \left(x - \frac{L}{2}\right)^5\mathcal{U}\left(x - \frac{L}{2}\right)\right]. \quad\blacksquare$$

### REMARKS

Outside the discussion of the Laplace transform, the unit step function is defined on the interval $(-\infty, \infty)$, that is,

$$\mathcal{U}(t - a) = \begin{cases} 0, & t < a \\ 1, & t \geq a. \end{cases}$$

Using this slight modification of Definition 7.3.1, a special case of (12) when $g(t) = 1$ is sometimes called the **boxcar function** and denoted by

$$\Pi(t) = \mathcal{U}(t - a) - \mathcal{U}(t - b).$$

See Figure 7.3.9.

**FIGURE 7.3.9** Boxcar function

# EXERCISES 7.3

*Answers to selected odd-numbered problems begin on page ANS-11.*

## 7.3.1 Translation on the $s$-Axis

In Problems 1–20 find either $F(s)$ or $f(t)$, as indicated.

**1.** $\mathscr{L}\{te^{10t}\}$

**2.** $\mathscr{L}\{te^{-6t}\}$

**3.** $\mathscr{L}\{t^3e^{-2t}\}$

**4.** $\mathscr{L}\{t^{10}e^{-7t}\}$

**5.** $\mathscr{L}\{t(e^t + e^{2t})^2\}$

**6.** $\mathscr{L}\{e^{2t}(t - 1)^2\}$

**7.** $\mathscr{L}\{e^t \sin 3t\}$

**8.** $\mathscr{L}\{e^{-2t} \cos 4t\}$

**9.** $\mathscr{L}\{(1 - e^t + 3e^{-4t}) \cos 5t\}$

**10.** $\mathscr{L}\left\{e^{3t}\left(9 - 4t + 10 \sin \dfrac{t}{2}\right)\right\}$

**11.** $\mathscr{L}^{-1}\left\{\dfrac{1}{(s + 2)^3}\right\}$

**12.** $\mathscr{L}^{-1}\left\{\dfrac{1}{(s - 1)^4}\right\}$

**13.** $\mathscr{L}^{-1}\left\{\dfrac{1}{s^2 - 6s + 10}\right\}$

**14.** $\mathscr{L}^{-1}\left\{\dfrac{1}{s^2 + 2s + 5}\right\}$

**15.** $\mathscr{L}^{-1}\left\{\dfrac{s}{s^2 + 4s + 5}\right\}$

**16.** $\mathscr{L}^{-1}\left\{\dfrac{2s + 5}{s^2 + 6s + 34}\right\}$

**17.** $\mathscr{L}^{-1}\left\{\dfrac{s}{(s + 1)^2}\right\}$

**18.** $\mathscr{L}^{-1}\left\{\dfrac{5s}{(s - 2)^2}\right\}$

**19.** $\mathscr{L}^{-1}\left\{\dfrac{2s - 1}{s^2(s + 1)^3}\right\}$

**20.** $\mathscr{L}^{-1}\left\{\dfrac{(s + 1)^2}{(s + 2)^4}\right\}$

In Problems 21–30 use the Laplace transform to solve the given initial-value problem.

**21.** $y' + 4y = e^{-4t}, \quad y(0) = 2$

**22.** $y' - y = 1 + te^t, \quad y(0) = 0$

**23.** $y'' + 2y' + y = 0, \quad y(0) = 1, \quad y'(0) = 1$

**24.** $y'' - 4y' + 4y = t^3e^{2t}, \quad y(0) = 0, \quad y'(0) = 0$

**25.** $y'' - 6y' + 9y = t, \quad y(0) = 0, \quad y'(0) = 1$

**26.** $y'' - 4y' + 4y = t^3, \quad y(0) = 1, \quad y'(0) = 0$

**27.** $y'' - 6y' + 13y = 0, \quad y(0) = 0, \quad y'(0) = -3$

**28.** $2y'' + 20y' + 51y = 0, \quad y(0) = 2, \quad y'(0) = 0$

**29.** $y'' - y' = e^t \cos t, \quad y(0) = 0, \quad y'(0) = 0$

**30.** $y'' - 2y' + 5y = 1 + t, \quad y(0) = 0, \quad y'(0) = 4$

In Problems 31 and 32 use the Laplace transform and the procedure outlined in Example 10 to solve the given boundary-value problem.

**31.** $y'' + 2y' + y = 0, \quad y'(0) = 2, \quad y(1) = 2$

**32.** $y'' + 8y' + 20y = 0, \quad y(0) = 0, \quad y'(\pi) = 0$

**33.** A 4-pound weight stretches a spring 2 feet. The weight is released from rest 18 inches above the equilibrium position, and the resulting motion takes place in a medium offering a

damping force numerically equal to $\frac{7}{8}$ times the instantaneous velocity. Use the Laplace transform to find the equation of motion $x(t)$.

**34.** Recall that the differential equation for the instantaneous charge $q(t)$ on the capacitor in an $LRC$-series circuit is given by

$$L\frac{d^2q}{dt^2} + R\frac{dq}{dt} + \frac{1}{C}q = E(t). \qquad (20)$$

See Section 5.1. Use the Laplace transform to find $q(t)$ when $L = 1$ h, $R = 20 \, \Omega$, $C = 0.005$ f, $E(t) = 150$ V, $t > 0$, $q(0) = 0$, and $i(0) = 0$. What is the current $i(t)$?

**35.** Consider a battery of constant voltage $E_0$ that charges the capacitor shown in Figure 7.3.10. Divide equation (20) by $L$ and define $2\lambda = R/L$ and $\omega^2 = 1/LC$. Use the Laplace transform to show that the solution $q(t)$ of $q'' + 2\lambda q' + \omega^2 q = E_0/L$ subject to $q(0) = 0, i(0) = 0$ is

$$q(t) = \begin{cases} E_0C\left[1 - e^{-\lambda t}\left(\cosh \sqrt{\lambda^2 - \omega^2}\,t \right.\right. \\ \qquad \left.\left. + \dfrac{\lambda}{\sqrt{\lambda^2 - \omega^2}} \sinh \sqrt{\lambda^2 - \omega^2}\,t\right)\right], \quad \lambda > \omega, \\ E_0C[1 - e^{-\lambda t}(1 + \lambda t)], \qquad\qquad\quad \lambda = \omega, \\ E_0C\left[1 - e^{-\lambda t}\left(\cos \sqrt{\omega^2 - \lambda^2}\,t \right.\right. \\ \qquad \left.\left. + \dfrac{\lambda}{\sqrt{\omega^2 - \lambda^2}} \sin \sqrt{\omega^2 - \lambda^2}\,t\right)\right], \quad \lambda < \omega. \end{cases}$$

**FIGURE 7.3.10** Series circuit in Problem 35

**36.** Use the Laplace transform to find the charge $q(t)$ in an $RC$ series circuit when $q(0) = 0$ and $E(t) = E_0e^{-kt}, k > 0$. Consider two cases: $k \neq 1/RC$ and $k = 1/RC$.

## 7.3.2 Translation on the $t$-Axis

In Problems 37–48 find either $F(s)$ or $f(t)$, as indicated.

**37.** $\mathscr{L}\{(t - 1)\mathscr{U}(t - 1)\}$

**38.** $\mathscr{L}\{e^{2-t}\mathscr{U}(t - 2)\}$

**39.** $\mathscr{L}\{t\,\mathscr{U}(t - 2)\}$

**40.** $\mathscr{L}\{(3t + 1)\,\mathscr{U}(t - 1)\}$

**41.** $\mathscr{L}\{\cos 2t\,\mathscr{U}(t - \pi)\}$

**42.** $\mathscr{L}\left\{\sin t\,\mathscr{U}\left(t - \dfrac{\pi}{2}\right)\right\}$

**43.** $\mathscr{L}^{-1}\left\{\dfrac{e^{-2s}}{s^3}\right\}$

**44.** $\mathscr{L}^{-1}\left\{\dfrac{(1 + e^{-2s})^2}{s + 2}\right\}$

**45.** $\mathscr{L}^{-1}\left\{\dfrac{e^{-\pi s}}{s^2 + 1}\right\}$

**46.** $\mathscr{L}^{-1}\left\{\dfrac{se^{-\pi s/2}}{s^2 + 4}\right\}$

**47.** $\mathscr{L}^{-1}\left\{\dfrac{e^{-s}}{s(s + 1)}\right\}$

**48.** $\mathscr{L}^{-1}\left\{\dfrac{e^{-2s}}{s^2(s - 1)}\right\}$

In Problems 49–54 match the given graph with one of the functions in (a)–(f). The graph of $f(t)$ is given in Figure 7.3.11.

**(a)** $f(t) - f(t)\,\mathcal{U}(t - a)$

**(b)** $f(t - b)\,\mathcal{U}(t - b)$

**(c)** $f(t)\,\mathcal{U}(t - a)$

**(d)** $f(t) - f(t)\,\mathcal{U}(t - b)$

**(e)** $f(t)\,\mathcal{U}(t - a) - f(t)\,\mathcal{U}(t - b)$

**(f)** $f(t - a)\,\mathcal{U}(t - a) - f(t - a)\,\mathcal{U}(t - b)$

**FIGURE 7.3.11** Graph for Problems 49–54

**49.**

**FIGURE 7.3.12** Graph for Problem 49

**50.**

**FIGURE 7.3.13** Graph for Problem 50

**51.**

**FIGURE 7.3.14** Graph for Problem 51

**52.**

**FIGURE 7.3.15** Graph for Problem 52

**53.**

**FIGURE 7.3.16** Graph for Problem 53

**54.**

**FIGURE 7.3.17** Graph for Problem 54

In Problems 55–62 write each function in terms of unit step functions. Find the Laplace transform of the given function.

**55.** $f(t) = \begin{cases} 2, & 0 \le t < 3 \\ -2, & t \ge 3 \end{cases}$

**56.** $f(t) = \begin{cases} 1, & 0 \le t < 4 \\ 0, & 4 \le t < 5 \\ 1, & t \ge 5 \end{cases}$

**57.** $f(t) = \begin{cases} 0, & 0 \le t < 1 \\ t^2, & t \ge 1 \end{cases}$

**58.** $f(t) = \begin{cases} 0, & 0 \le t < 3\pi/2 \\ \sin t, & t \ge 3\pi/2 \end{cases}$

**59.** $f(t) = \begin{cases} t, & 0 \le t < 2 \\ 0, & t \ge 2 \end{cases}$

**60.** $f(t) = \begin{cases} \sin t, & 0 \le t < 2\pi \\ 0, & t \ge 2\pi \end{cases}$

**61.**

rectangular pulse

**FIGURE 7.3.18** Graph for Problem 61

**62.**

staircase function

**FIGURE 7.3.19** Graph for Problem 62

In Problems 63–70 use the Laplace transform to solve the given initial-value problem.

**63.** $y' + y = f(t)$, $y(0) = 0$, where

$$f(t) = \begin{cases} 0, & 0 \le t < 1 \\ 5, & t \ge 1 \end{cases}$$

**64.** $y' + y = f(t)$, $y(0) = 0$, where

$$f(t) = \begin{cases} 1, & 0 \le t < 1 \\ -1, & t \ge 1 \end{cases}$$

**65.** $y' + 2y = f(t), \quad y(0) = 0$, where

$$f(t) = \begin{cases} t, & 0 \le t < 1 \\ 0, & t \ge 1 \end{cases}$$

**66.** $y'' + 4y = f(t), \quad y(0) = 0, \quad y'(0) = -1$, where

$$f(t) = \begin{cases} 1, & 0 \le t < 1 \\ 0, & t \ge 1 \end{cases}$$

**67.** $y'' + 4y = \sin t \, \mathcal{U}(t - 2\pi), \quad y(0) = 1, \quad y'(0) = 0$

**68.** $y'' - 5y' + 6y = \mathcal{U}(t - 1), \quad y(0) = 0, \quad y'(0) = 1$

**69.** $y'' + y = f(t), \quad y(0) = 0, \quad y'(0) = 1$, where

$$f(t) = \begin{cases} 0, & 0 \le t < \pi \\ 1, & \pi \le t < 2\pi \\ 0, & t \ge 2\pi \end{cases}$$

**70.** $y'' + 4y' + 3y = 1 - \mathcal{U}(t - 2) - \mathcal{U}(t - 4) + \mathcal{U}(t - 6),$
$y(0) = 0, \quad y'(0) = 0$

**71.** Suppose a 32-pound weight stretches a spring 2 feet. If the weight is released from rest at the equilibrium position, find the equation of motion $x(t)$ if an impressed force $f(t) = 20t$ acts on the system for $0 \le t < 5$ and is then removed (see Example 5). Ignore any damping forces. Use a graphing utility to graph $x(t)$ on the interval $[0, 10]$.

**72.** Solve Problem 71 if the impressed force $f(t) = \sin t$ acts on the system for $0 \le t < 2\pi$ and is then removed.

In Problems 73 and 74 use the Laplace transform to find the charge $q(t)$ on the capacitor in an *RC*-series circuit subject to the given conditions.

**73.** $q(0) = 0, R = 2.5 \ \Omega, C = 0.08$ f, $E(t)$ given in Figure 7.3.20

**FIGURE 7.3.20** $E(t)$ in Problem 73

**74.** $q(0) = q_0, R = 10 \ \Omega, C = 0.1$ f, $E(t)$ given in Figure 7.3.21

**FIGURE 7.3.21** $E(t)$ in Problem 74

**75. (a)** Use the Laplace transform to find the current $i(t)$ in a single-loop *LR*-series circuit when $i(0) = 0$, $L = 1$ h, $R = 10 \ \Omega$, and $E(t)$ is as given in Figure 7.3.22.

**(b)** Use a graphing utility to graph $i(t)$ for $0 \le t \le 6$. Use the graph to estimate $i_{max}$ and $i_{min}$, the maximum and minimum values of the current.

**FIGURE 7.3.22** $E(t)$ in Problem 75

**76. (a)** Use the Laplace transform to find the charge $q(t)$ on the capacitor in an *RC*-series circuit when $q(0) = 0, R = 50 \ \Omega$, $C = 0.01$ f, and $E(t)$ is as given in Figure 7.3.23.

**(b)** Assume that $E_0 = 100$ V. Use a graphing utility to graph $q(t)$ for $0 \le t \le 6$. Use the graph to estimate $q_{max}$, the maximum value of the charge.

**FIGURE 7.3.23** $E(t)$ in Problem 76

**77.** A cantilever beam is embedded at its left end and free at its right end. Use the Laplace transform to find the deflection $y(x)$ when the load is given by

$$w(x) = \begin{cases} w_0, & 0 < x < L/2 \\ 0, & L/2 \le x < L. \end{cases}$$

**78.** Solve Problem 77 when the load is given by

$$w(x) = \begin{cases} 0, & 0 < x < L/3 \\ w_0, & L/3 < x < 2L/3 \\ 0, & 2L/3 < x < L. \end{cases}$$

**79.** Find the deflection $y(x)$ of a cantilever beam embedded at its left end and free at its right end when the load is as given in Example 10.

**80.** A beam is embedded at its left end and simply supported at its right end. Find the deflection $y(x)$ when the load is as given in Problem 77.

### Mathematical Model

**81. Cake Inside an Oven**   Reread Example 4 in Section 3.1 on the cooling of a cake that is taken out of an oven.

**(a)** Devise a mathematical model for the temperature of a cake while it is *inside* the oven based on the following assumptions: At $t = 0$ the cake mixture is at the room temperature of 70°; the oven is not preheated, so at $t = 0$, when the cake mixture is placed into the oven, the temperature inside the oven is also 70°; the temperature of the oven increases linearly until $t = 4$ minutes, when the desired temperature

of 300° is attained; the oven temperature is a constant 300°
for $t \geq 4$.

(b) Use the Laplace transform to solve the initial-value
problem in part (a).

## Discussion Problems

**82.** Discuss how you would fix up each of the following functions so
that Theorem 7.3.2 could be used directly to find the given Laplace
transform. Check your answers using (16) of this section.

(a) $\mathscr{L}\{(2t + 1)\mathcal{U}(t - 1)\}$

(b) $\mathscr{L}\{e^t \mathcal{U}(t - 5)\}$

(c) $\mathscr{L}\{\cos t\, \mathcal{U}(t - \pi)\}$

(d) $\mathscr{L}\{(t^2 - 3t)\mathcal{U}(t - 2)\}$

**83. (a)** Assume that Theorem 7.3.1 holds when the symbol $a$ is
replaced by $ki$, where $k$ is a real number and $i^2 = -1$. Show
that $\mathscr{L}\{te^{kti}\}$ can be used to deduce

$$\mathscr{L}\{t \cos kt\} = \frac{s^2 - k^2}{(s^2 + k^2)^2}$$

$$\mathscr{L}\{t \sin kt\} = \frac{2ks}{(s^2 + k^2)^2}.$$

**(b)** Now use the Laplace transform to solve the initial-
value problem $x'' + \omega^2 x = \cos \omega t$, $x(0) = 0$, $x'(0) = 0$.

---

## 7.4 Operational Properties II

**INTRODUCTION**  In this section we develop several more operational properties
of the Laplace transform. Specifically, we shall see how to find the transform of a
function $f(t)$ that is multiplied by a monomial $t^n$, the transform of a special type of
integral, and the transform of a periodic function. The last two transform properties
allow us to solve some equations that we have not encountered up to this point:
Volterra integral equations, integrodifferential equations, and ordinary differential
equations in which the input function is a periodic piecewise-defined function.

### 7.4.1 DERIVATIVES OF A TRANSFORM

**MULTIPLYING A FUNCTION BY $t^n$**  The Laplace transform of the product of a
function $f(t)$ with $t$ can be found by differentiating the Laplace transform of $f(t)$. To
motivate this result, let us assume that $F(s) = \mathscr{L}\{f(t)\}$ exists and that it is possible
to interchange the order of differentiation and integration. Then

$$\frac{d}{ds}F(s) = \frac{d}{ds}\int_0^\infty e^{-st}f(t)\,dt = \int_0^\infty \frac{\partial}{\partial s}[e^{-st}f(t)]\,dt = -\int_0^\infty e^{-st}\,tf(t)\,dt = -\mathscr{L}\{tf(t)\};$$

that is,
$$\mathscr{L}\{tf(t)\} = -\frac{d}{ds}\mathscr{L}\{f(t)\}.$$

We can use the last result to find the Laplace transform of $t^2 f(t)$:

$$\mathscr{L}\{t^2 f(t)\} = \mathscr{L}\{t \cdot tf(t)\} = -\frac{d}{ds}\mathscr{L}\{tf(t)\} = -\frac{d}{ds}\left(-\frac{d}{ds}\mathscr{L}\{f(t)\}\right) = \frac{d^2}{ds^2}\mathscr{L}\{f(t)\}.$$

The preceding two cases suggest the general result for $\mathscr{L}\{t^n f(t)\}$.

---

**THEOREM 7.4.1**  **Derivatives of Transforms**

If $F(s) = \mathscr{L}\{f(t)\}$ and $n = 1, 2, 3, \ldots$, then

$$\mathscr{L}\{t^n f(t)\} = (-1)^n \frac{d^n}{ds^n}F(s).$$

## EXAMPLE 1 Using Theorem 7.4.1

Evaluate $\mathscr{L}\{t \sin kt\}$.

**SOLUTION** With $f(t) = \sin kt$, $F(s) = k/(s^2 + k^2)$, and $n = 1$, Theorem 7.4.1 gives

$$\mathscr{L}\{t \sin kt\} = -\frac{d}{ds}\mathscr{L}\{\sin kt\} = -\frac{d}{ds}\left(\frac{k}{s^2 + k^2}\right) = \frac{2ks}{(s^2 + k^2)^2}.$$    ∎

If we want to evaluate $\mathscr{L}\{t^2 \sin kt\}$ and $\mathscr{L}\{t^3 \sin kt\}$, all we need do, in turn, is take the negative of the derivative with respect to $s$ of the result in Example 1 and then take the negative of the derivative with respect to $s$ of $\mathscr{L}\{t^2 \sin kt\}$.

**NOTE** To find transforms of functions $t^n e^{at}$ we can use either Theorem 7.3.1 or Theorem 7.4.1. For example,

Theorem 7.3.1: $\mathscr{L}\{te^{3t}\} = \mathscr{L}\{t\}_{s \to s-3} = \left.\frac{1}{s^2}\right|_{s \to s-3} = \frac{1}{(s-3)^2}.$

Theorem 7.4.1: $\mathscr{L}\{te^{3t}\} = -\frac{d}{ds}\mathscr{L}\{e^{3t}\} = -\frac{d}{ds}\frac{1}{s-3} = (s-3)^{-2} = \frac{1}{(s-3)^2}.$

## EXAMPLE 2 An Initial-Value Problem

Solve $x'' + 16x = \cos 4t$, $x(0) = 0$, $x'(0) = 1$.

**SOLUTION** The initial-value problem could describe the forced, undamped, and resonant motion of a mass on a spring. The mass starts with an initial velocity of 1 ft/s in the downward direction from the equilibrium position.

Transforming the differential equation gives

$$(s^2 + 16)X(s) = 1 + \frac{s}{s^2 + 16} \qquad \text{or} \qquad X(s) = \frac{1}{s^2 + 16} + \frac{s}{(s^2 + 16)^2}.$$

Now we just saw in Example 1 that

$$\mathscr{L}^{-1}\left\{\frac{2ks}{(s^2 + k^2)^2}\right\} = t \sin kt, \tag{1}$$

and so with the identification $k = 4$ in (1) and in part (d) of Theorem 7.2.1, we obtain

$$x(t) = \frac{1}{4}\mathscr{L}^{-1}\left\{\frac{4}{s^2 + 16}\right\} + \frac{1}{8}\mathscr{L}^{-1}\left\{\frac{8s}{(s^2 + 16)^2}\right\}$$

$$= \frac{1}{4}\sin 4t + \frac{1}{8}t \sin 4t.$$    ∎

### 7.4.2 TRANSFORMS OF INTEGRALS

**CONVOLUTION** If functions $f$ and $g$ are piecewise continuous on the interval $[0, \infty)$, then the **convolution** of $f$ and $g$, denoted by the symbol $f * g$, is a function defined by the integral

$$f * g = \int_0^t f(\tau)g(t - \tau)\,d\tau. \tag{2}$$

Because we are integrating in (2) with respect to the variable $\tau$ (the lower case Greek letter *tau*), the convolution $f * g$ is a function of $t$. To emphasize this fact, (2) is also written $(f * g)(t)$. As the notation $f * g$ suggests, the convolution (2) is often interpreted as a *generalized product* of two functions $f$ and $g$.

### EXAMPLE 3 Convolution of Two Functions

Evaluate **(a)** $e^t * \sin t$ **(b)** $\mathcal{L}\{e^t * \sin t\}$.

**SOLUTION** **(a)** With the identifications

$$f(t) = e^t,\ g(t) = \sin t \qquad \text{and} \qquad f(\tau) = e^\tau,\ g(t - \tau) = \sin(t - \tau),$$

it follows from (2) and integration by parts that

$$e^t * \sin t = \int_0^t e^\tau \sin(t - \tau)\,d\tau$$

$$= \frac{1}{2}[e^\tau \sin(t - \tau) + e^\tau \cos(t - \tau)]_0^t$$

$$= \frac{1}{2}(-\sin t - \cos t + e^t). \tag{3}$$

**(b)** Then from (3) and parts (c), (d), and (e) of Theorem 7.1.1 we find

$$\mathcal{L}\{e^t * \sin t\} = -\frac{1}{2}\mathcal{L}\{\sin t\} - \frac{1}{2}\mathcal{L}\{\cos t\} + \frac{1}{2}\mathcal{L}\{e^t\}$$

$$= -\frac{1}{2}\frac{1}{s^2 + 1} - \frac{1}{2}\frac{s}{s^2 + 1} + \frac{1}{2}\frac{1}{s - 1}$$

$$= \frac{1}{(s - 1)(s^2 + 1)}.$$

It is left as an exercise to show that

$$\int_0^t f(\tau)g(t - \tau)\,d\tau = \int_0^t f(t - \tau)g(\tau)\,d\tau,$$

that is, $f * g = g * f$. In other words, the convolution of two functions is commutative.

**CONVOLUTION THEOREM** We have seen that if $f$ and $g$ are both piecewise continuous for $t \geq 0$, then the Laplace transform of a sum $f + g$ is the sum of the individual Laplace transforms. While it is *not* true that the Laplace transform of the product $fg$ is the product of the Laplace transforms, we see in the next theorem—called the **convolution theorem**—that the Laplace transform of the generalized product $f * g$ is the product of the Laplace transforms of $f$ and $g$.

### THEOREM 7.4.2 Convolution Theorem

If $f(t)$ and $g(t)$ are piecewise continuous on $[0, \infty)$ and of exponential order, then

$$\mathcal{L}\{f * g\} = \mathcal{L}\{f(t)\}\,\mathcal{L}\{g(t)\} = F(s)G(s).$$

**PROOF** Let

$$F(s) = \mathcal{L}\{f(t)\} = \int_0^\infty e^{-s\tau} f(\tau)\,d\tau$$

and

$$G(s) = \mathcal{L}\{g(t)\} = \int_0^\infty e^{-s\beta} g(\beta)\,d\beta.$$

**FIGURE 7.4.1** Changing order of integration from $t$ first to $\tau$ first

Proceeding formally, we have

$$F(s)G(s) = \left(\int_0^\infty e^{-s\tau}f(\tau)\,d\tau\right)\left(\int_0^\infty e^{-s\beta}g(\beta)\,d\beta\right)$$

$$= \int_0^\infty \int_0^\infty e^{-s(\tau+\beta)}f(\tau)g(\beta)\,d\tau\,d\beta$$

$$= \int_0^\infty f(\tau)\,d\tau \int_0^\infty e^{-s(\tau+\beta)}g(\beta)\,d\beta.$$

Holding $\tau$ fixed, we let $t = \tau + \beta$, $dt = d\beta$, so that

$$F(s)G(s) = \int_0^\infty f(\tau)\,d\tau \int_\tau^\infty e^{-st}g(t-\tau)\,dt.$$

In the $t\tau$-plane we are integrating over the shaded region in Figure 7.4.1. Since $f$ and $g$ are piecewise continuous on $[0, \infty)$ and of exponential order, it is possible to interchange the order of integration:

$$F(s)\,G(s) = \int_0^\infty e^{-st}dt \int_0^t f(\tau)g(t-\tau)\,d\tau = \int_0^\infty e^{-st}\left\{\int_0^t f(\tau)g(t-\tau)\,d\tau\right\}dt = \mathscr{L}\{f * g\}. \blacksquare$$

Theorem 7.4.2 shows that we can find the Laplace transform of the convolution $f * g$ of two functions without actually evaluating the definite integral $\int_0^t f(\tau)g(t-\tau)\,d\tau$ as we did in (3). The next example illustrates the idea.

**EXAMPLE 4**    **Using Theorem 7.4.2**

Evaluate $\mathscr{L}\left\{\int_0^t e^\tau \sin(t - \tau)\,d\tau\right\}$.

**SOLUTION** This is the same as the transform $\mathscr{L}\{e^t * \sin t\}$ that we found in part (b) of Example 3. This time we use Theorem 7.4.2 that the Laplace transform of the convolution of $f$ and $g$ is the product of their Laplace transforms:

$$\mathscr{L}\left\{\int_0^t e^\tau \sin(t - \tau)\,d\tau\right\} = \mathscr{L}\{e^t * \sin t\}$$

$$= \mathscr{L}\{e^t\} \cdot \mathscr{L}\{\sin t\}$$

$$= \frac{1}{s - 1} \cdot \frac{1}{s^2 + 1}$$

$$= \frac{1}{(s - 1)(s^2 + 1)}. \blacksquare$$

**INVERSE FORM OF THEOREM 7.4.2**    The convolution theorem is sometimes useful in finding the inverse Laplace transform of the product of two Laplace transforms. From Theorem 7.4.2 we have

$$\mathscr{L}^{-1}\{F(s)G(s)\} = f * g. \tag{4}$$

Many of the results in the table of Laplace transforms in Appendix C can be derived using (4). For example, in the next example we obtain entry 25 of the table:

$$\mathscr{L}\{\sin kt - kt \cos kt\} = \frac{2k^3}{(s^2 + k^2)^2}. \tag{5}$$

**EXAMPLE 5**    **Inverse Transform as a Convolution**

Evaluate $\mathscr{L}^{-1}\left\{\dfrac{1}{(s^2 + k^2)^2}\right\}$.

**SOLUTION** Let $F(s) = G(s) = \dfrac{1}{s^2 + k^2}$ so that

$$f(t) = g(t) = \frac{1}{k}\mathscr{L}^{-1}\left\{\frac{k}{s^2 + k^2}\right\} = \frac{1}{k}\sin kt.$$

In this case (4) gives

$$\mathscr{L}^{-1}\left\{\frac{1}{(s^2 + k^2)^2}\right\} = \frac{1}{k^2}\int_0^t \sin k\tau \sin k(t - \tau)\, d\tau. \qquad (6)$$

With the aid of the product-to-sum trigonometric identity

$$\sin A \sin B = \frac{1}{2}[\cos(A - B) - \cos(A + B)]$$

and the substitutions $A = k\tau$ and $B = k(t - \tau)$ we can carry out the integration in (6):

$$\mathscr{L}^{-1}\left\{\frac{1}{(s^2 + k^2)^2}\right\} = \frac{1}{2k^2}\int_0^t [\cos k(2\tau - t) - \cos kt]\, d\tau$$

$$= \frac{1}{2k^2}\left[\frac{1}{2k}\sin k(2\tau - t) - \tau \cos kt\right]_0^t$$

$$= \frac{\sin kt - kt \cos kt}{2k^3}.$$

Multiplying both sides by $2k^3$ gives the inverse form of (5).  ◼

**TRANSFORM OF AN INTEGRAL**   When $g(t) = 1$ and $\mathscr{L}\{g(t)\} = G(s) = 1/s$, the convolution theorem implies that the Laplace transform of the integral of $f$ is

$$\mathscr{L}\left\{\int_0^t f(\tau)\, d\tau\right\} = \frac{F(s)}{s}. \qquad (7)$$

The inverse form of (7),

$$\int_0^t f(\tau)\, d\tau = \mathscr{L}^{-1}\left\{\frac{F(s)}{s}\right\}, \qquad (8)$$

can be used in lieu of partial fractions when $s^n$ is a factor of the denominator and $f(t) = \mathscr{L}^{-1}\{F(s)\}$ is easy to integrate. For example, we know for $f(t) = \sin t$ that $F(s) = 1/(s^2 + 1)$, and so by (8)

$$\mathscr{L}^{-1}\left\{\frac{1}{s(s^2 + 1)}\right\} = \mathscr{L}^{-1}\left\{\frac{1/(s^2 + 1)}{s}\right\} = \int_0^t \sin \tau\, d\tau = 1 - \cos t$$

$$\mathscr{L}^{-1}\left\{\frac{1}{s^2(s^2 + 1)}\right\} = \mathscr{L}^{-1}\left\{\frac{1/s(s^2 + 1)}{s}\right\} = \int_0^t (1 - \cos \tau)\, d\tau = t - \sin t$$

$$\mathscr{L}^{-1}\left\{\frac{1}{s^3(s^2 + 1)}\right\} = \mathscr{L}^{-1}\left\{\frac{1/s^2(s^2 + 1)}{s}\right\} = \int_0^t (\tau - \sin \tau)\, d\tau = \tfrac{1}{2}t^2 - 1 + \cos t$$

and so on.

**VOLTERRA INTEGRAL EQUATION**   The convolution theorem and the result in (7) are useful in solving other types of equations in which an unknown function appears under an integral sign. In the next example we solve a **Volterra integral equation** for $f(t)$,

$$f(t) = g(t) + \int_0^t f(\tau)\, h(t - \tau)\, d\tau. \qquad (9)$$

The functions $g(t)$ and $h(t)$ are known. Notice that the integral in (9) has the convolution form (2) with the symbol $h$ playing the part of $g$.

**EXAMPLE 6**   **An Integral Equation**

Solve $f(t) = 3t^2 - e^{-t} - \displaystyle\int_0^t f(\tau)\, e^{t-\tau}\, d\tau$ for $f(t)$.

**SOLUTION**  In the integral we identify $h(t - \tau) = e^{t-\tau}$ so that $h(t) = e^t$. We take the Laplace transform of each term; in particular, by Theorem 7.4.2 the transform of the integral is the product of $\mathcal{L}\{f(t)\} = F(s)$ and $\mathcal{L}\{e^t\} = 1/(s - 1)$:

$$F(s) = 3 \cdot \frac{2}{s^3} - \frac{1}{s + 1} - F(s) \cdot \frac{1}{s - 1}.$$

After solving the last equation for $F(s)$ and carrying out the partial fraction decomposition, we find

$$F(s) = \frac{6}{s^3} - \frac{6}{s^4} + \frac{1}{s} - \frac{2}{s + 1}.$$

The inverse transform then gives

$$f(t) = 3\mathcal{L}^{-1}\left\{\frac{2!}{s^3}\right\} - \mathcal{L}^{-1}\left\{\frac{3!}{s^4}\right\} + \mathcal{L}^{-1}\left\{\frac{1}{s}\right\} - 2\mathcal{L}^{-1}\left\{\frac{1}{s + 1}\right\}$$

$$= 3t^2 - t^3 + 1 - 2e^{-t}. \qquad\blacksquare$$

**SERIES CIRCUITS**   In a single-loop or series circuit, Kirchhoff's second law states that the sum of the voltage drops across an inductor, resistor, and capacitor is equal to the impressed voltage $E(t)$. Now it is known that the voltage drops across an inductor, resistor, and capacitor are, respectively,

$$L\frac{di}{dt}, \qquad R\,i(t), \quad \text{and} \quad \frac{1}{C}\int_0^t i(\tau)\,d\tau,$$

where $i(t)$ is the current and $L$, $R$, and $C$ are constants. It follows that the current in a circuit, such as that shown in Figure 7.4.2, is governed by the **integrodifferential equation**

$$L\frac{di}{dt} + R\,i(t) + \frac{1}{C}\int_0^t i(\tau)\,d\tau = E(t). \tag{10}$$

**FIGURE 7.4.2** *LRC*-series circuit

**EXAMPLE 7**   **An Integrodifferential Equation**

Determine the current $i(t)$ in a single-loop *LRC*-series circuit when $L = 0.1$ h, $R = 2$ Ω, $C = 0.1$ f, $i(0) = 0$, and the impressed voltage is

$$E(t) = 120t - 120t\,\mathcal{U}(t - 1).$$

**SOLUTION**  With the given data equation (10) becomes

$$0.1\frac{di}{dt} + 2i + 10\int_0^t i(\tau)\,d\tau = 120t - 120t\,\mathcal{U}(t - 1).$$

Now by (7), $\mathcal{L}\{\int_0^t i(\tau)\,d\tau\} = I(s)/s$, where $I(s) = \mathcal{L}\{i(t)\}$. Thus the Laplace transform of the integrodifferential equation is

$$0.1sI(s) + 2I(s) + 10\frac{I(s)}{s} = 120\left[\frac{1}{s^2} - \frac{1}{s^2}e^{-s} - \frac{1}{s}e^{-s}\right]. \qquad \leftarrow \text{by (16) of Section 7.3}$$

Multiplying this equation by $10s$, using $s^2 + 20s + 100 = (s + 10)^2$, and then solving for $I(s)$ gives

$$I(s) = 1200\left[\frac{1}{s(s + 10)^2} - \frac{1}{s(s + 10)^2}e^{-s} - \frac{1}{(s + 10)^2}e^{-s}\right].$$

By partial fractions,

$$I(s) = 1200\left[\frac{1/100}{s} - \frac{1/100}{s + 10} - \frac{1/10}{(s + 10)^2} - \frac{1/100}{s}e^{-s}\right.$$

$$\left. + \frac{1/100}{s + 10}e^{-s} + \frac{1/10}{(s + 10)^2}e^{-s} - \frac{1}{(s + 10)^2}e^{-s}\right].$$

From the inverse form of the second translation theorem, (15) of Section 7.3, we finally obtain

$$i(t) = 12[1 - \mathcal{U}(t - 1)] - 12[e^{-10t} - e^{-10(t-1)}\mathcal{U}(t - 1)]$$
$$- 120te^{-10t} - 1080(t - 1)e^{-10(t-1)}\mathcal{U}(t - 1).$$

Written as a piecewise-defined function, the current is

$$i(t) = \begin{cases} 12 - 12e^{-10t} - 120te^{-10t}, & 0 \le t < 1 \\ -12e^{-10t} + 12e^{-10(t-1)} - 120te^{-10t} - 1080(t - 1)e^{-10(t-1)}, & t \ge 1. \end{cases}$$

FIGURE 7.4.3 Graph of current $i(t)$ in Example 7

Using this last expression and a CAS, we graph $i(t)$ on each of the two intervals and then combine the graphs. Note in Figure 7.4.3 that even though the input $E(t)$ is discontinuous, the output or response $i(t)$ is a continuous function. ■

Optional material if Section 4.8 was covered.

**POST SCRIPT—GREEN'S FUNCTIONS REDUX**  By applying the Laplace transform to the initial-value problem

$$y'' + ay' + by = f(t), \quad y(0) = 0, \quad y'(0) = 0,$$

where $a$ and $b$ are constants, we find that the transform of $y(t)$ is

$$Y(s) = \frac{F(s)}{s^2 + as + b},$$

where $F(s) = \mathcal{L}\{f(t)\}$. By rewriting the foregoing transform as the product

$$Y(s) = \frac{1}{s^2 + as + b}F(s)$$

we can use the inverse form of the convolution theorem (4) to write the solution of the IVP as

$$y(t) = \int_0^t g(t - \tau) f(\tau) \, d\tau, \tag{11}$$

where $\mathcal{L}^{-1}\left\{\dfrac{1}{s^2 + as + b}\right\} = g(t)$ and $\mathcal{L}^{-1}\{F(s)\} = f(t)$. On the other hand, we know from (10) of Section 4.8 that the solution of the IVP is also given by

$$y(t) = \int_0^t G(t, \tau) f(\tau) \, d\tau, \tag{12}$$

where $G(t, \tau)$ is the Green's function for the differential equation.

By comparing (11) and (12) we see that the Green's function for the differential equation is related to $\mathcal{L}^{-1}\left\{\dfrac{1}{s^2 + as + b}\right\} = g(t)$ by

$$G(t, \tau) = g(t - \tau). \tag{13}$$

For example, for the initial-value problem $y'' + 4y = f(t)$, $y(0) = 0$, $y'(0) = 0$ we find

$$\mathcal{L}^{-1}\left\{\frac{1}{s^2 + 4}\right\} = \tfrac{1}{2} \sin 2t = g(t).$$

In Example 4 of Section 4.8, the roles of the symbols $x$ and $t$ are played by $t$ and $\tau$ in this discussion.

Thus from (13) we see that the Green's function for the DE $y'' + 4y = f(t)$ is $G(t, \tau) = g(t - \tau) = \tfrac{1}{2} \sin 2(t - \tau)$. See Example 4 in Section 4.8.

## REMARKS

Although the Laplace transform was known for a long time prior to the twentieth century, it was not used to solve differential equations. The fact that we use the Laplace transform today to solve a variety of equations is due to Oliver Heaviside (see page 298). In 1893 Heaviside invented an operational calculus for solving differential equations encountered in electrical engineering. Heaviside was no mathematician, and his procedures for solving differential equations were formal manipulations or procedures lacking mathematical justification. Nonetheless these procedures worked. In an attempt to put his operational calculus on a sound foundation, mathematicians discovered that the rules of his calculus matched many properties of the Laplace transform. Over time, Heaviside's operation calculus disappeared to be replaced by the theory and applications of the Laplace transform.

You should verify either by substitution in the equation or by the methods of Section 2.3 that $y(t) = e^{-t}\int_0^t e^{u^2+u}\,du$ is a perfectly good solution of the linear initial-value problem $y' + y = e^{t^2}$, $y(0) = 0$. We now solve the same equation with a formal application of the Laplace transform. If we denote $\mathscr{L}\{y(t)\} = Y(s)$ and $\mathscr{L}\{e^{t^2}\} = F(s)$, then the transform of the equation is

$$sY(s) - y(0) + Y(s) = F(s) \quad \text{or} \quad Y(s) = \frac{F(s)}{s + 1}.$$

Using $\mathscr{L}^{-1}\{F(s)\} = e^{t^2}$ and $\mathscr{L}^{-1}\left\{\dfrac{1}{s+1}\right\} = e^{-t}$ it follows from the inverse form (4) of the convolution theorem that the solution of the initial-value problem is

$$y(t) = \mathscr{L}^{-1}\left\{F(s) \cdot \frac{1}{s+1}\right\} = \int_0^t e^{\tau^2} \cdot e^{-(t-\tau)}\,d\tau = e^{-t}\int_0^t e^{\tau^2+\tau}\,d\tau.$$

With $\tau$ playing the part of $u$, this is the solution as first given. What's wrong here?

### 7.4.3 TRANSFORM OF A PERIODIC FUNCTION

**PERIODIC FUNCTION**    If a periodic function has period $T$, $T > 0$, then $f(t + T) = f(t)$. The next theorem shows that the Laplace transform of a periodic function can be obtained by integration over one period.

---

**THEOREM  7.4.3  Transform of a Periodic Function**

If $f(t)$ is piecewise continuous on $[0, \infty)$, of exponential order, and periodic with period $T$, then

$$\mathscr{L}\{f(t)\} = \frac{1}{1 - e^{-sT}}\int_0^T e^{-st}f(t)\,dt.$$

---

**PROOF**    Write the Laplace transform of $f$ as two integrals:

$$\mathscr{L}\{f(t)\} = \int_0^T e^{-st}f(t)\,dt + \int_T^\infty e^{-st}f(t)\,dt.$$

When we let $t = u + T$, the last integral becomes

$$\int_T^\infty e^{-st}f(t)\,dt = \int_0^\infty e^{-s(u+T)}f(u + T)\,du = e^{-sT}\int_0^\infty e^{-su}f(u)\,du = e^{-sT}\mathscr{L}\{f(t)\}.$$

Therefore
$$\mathscr{L}\{f(t)\} = \int_0^T e^{-st}f(t)\,dt + e^{-sT}\mathscr{L}\{f(t)\}.$$

Solving the equation in the last line for $\mathscr{L}\{f(t)\}$ proves the theorem. ■

FIGURE 7.4.4　Square wave in Example 8

### EXAMPLE 8　Transform of a Periodic Function

Find the Laplace transform of the periodic function shown in Figure 7.4.4.

**SOLUTION** The function $E(t)$ is called a square wave and has period $T = 2$. For $0 \le t < 2$, $E(t)$ can be defined by

$$E(t) = \begin{cases} 1, & 0 \le t < 1 \\ 0, & 1 \le t < 2 \end{cases}$$

and outside the interval by $E(t + 2) = E(t)$. Now from Theorem 7.4.3

$$\mathscr{L}\{E(t)\} = \frac{1}{1 - e^{-2s}} \int_0^2 e^{-st}E(t)\,dt = \frac{1}{1 - e^{-2s}}\left[\int_0^1 e^{-st} \cdot 1\,dt + \int_1^2 e^{-st} \cdot 0\,dt\right]$$

$$= \frac{1}{1 - e^{-2s}} \frac{1 - e^{-s}}{s} \qquad \leftarrow 1 - e^{-2s} = (1 + e^{-s})(1 - e^{-s})$$

$$= \frac{1}{s(1 + e^{-s})}. \tag{14}$$ ■

### EXAMPLE 9　A Periodic Impressed Voltage

The differential equation for the current $i(t)$ in a single-loop $LR$-series circuit is

$$L\frac{di}{dt} + Ri = E(t). \tag{15}$$

Determine the current $i(t)$ when $i(0) = 0$ and $E(t)$ is the square wave function given in Figure 7.4.4.

**SOLUTION** If we use the result in (14) of the preceding example, the Laplace transform of the DE is

$$LsI(s) + RI(s) = \frac{1}{s(1 + e^{-s})} \qquad \text{or} \qquad I(s) = \frac{1/L}{s(s + R/L)} \cdot \frac{1}{1 + e^{-s}}. \tag{16}$$

To find the inverse Laplace transform of the last function, we first make use of geometric series. With the identification $x = e^{-s}$, $s > 0$, the geometric series

$$\frac{1}{1 + x} = 1 - x + x^2 - x^3 + \cdots \qquad \text{becomes} \qquad \frac{1}{1 + e^{-s}} = 1 - e^{-s} + e^{-2s} - e^{-3s} + \cdots.$$

From
$$\frac{1}{s(s + R/L)} = \frac{L/R}{s} - \frac{L/R}{s + R/L}$$

we can then rewrite (16) as

$$I(s) = \frac{1}{R}\left(\frac{1}{s} - \frac{1}{s + R/L}\right)(1 - e^{-s} + e^{-2s} - e^{-3s} + \cdots)$$

$$= \frac{1}{R}\left(\frac{1}{s} - \frac{e^{-s}}{s} + \frac{e^{-2s}}{s} - \frac{e^{-3s}}{s} + \cdots\right) - \frac{1}{R}\left(\frac{1}{s + R/L} - \frac{1}{s + R/L}e^{-s} + \frac{e^{-2s}}{s + R/L} - \frac{e^{-3s}}{s + R/L} + \cdots\right).$$

By applying the form of the second translation theorem to each term of both series, we obtain

$$i(t) = \frac{1}{R}(1 - \mathcal{U}(t - 1) + \mathcal{U}(t - 2) - \mathcal{U}(t - 3) + \cdots)$$

$$-\frac{1}{R}(e^{-Rt/L} - e^{-R(t-1)/L}\,\mathcal{U}(t - 1) + e^{-R(t-2)/L}\,\mathcal{U}(t - 2) - e^{-R(t-3)/L}\,\mathcal{U}(t - 3) + \cdots)$$

or, equivalently,

$$i(t) = \frac{1}{R}(1 - e^{-Rt/L}) + \frac{1}{R}\sum_{n=1}^{\infty}(-1)^n(1 - e^{-R(t-n)/L})\,\mathcal{U}(t - n).$$

To interpret the solution, let us assume for the sake of illustration that $R = 1$, $L = 1$, and $0 \le t < 4$. In this case

$$i(t) = 1 - e^{-t} - (1 - e^{t-1})\,\mathcal{U}(t - 1) + (1 - e^{-(t-2)})\,\mathcal{U}(t - 2) - (1 - e^{-(t-3)})\,\mathcal{U}(t - 3);$$

in other words,

$$i(t) = \begin{cases} 1 - e^{-t}, & 0 \le t < 1 \\ -e^{-t} + e^{-(t-1)}, & 1 \le t < 2 \\ 1 - e^{-t} + e^{-(t-1)} - e^{-(t-2)}, & 2 \le t < 3 \\ -e^{-t} + e^{-(t-1)} - e^{-(t-2)} + e^{-(t-3)}, & 3 \le t < 4. \end{cases}$$

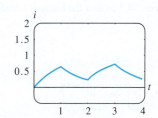

**FIGURE 7.4.5** Graph of current $i(t)$ in Example 9

The graph of $i(t)$ for $0 \le t < 4$, given in Figure 7.4.5, was obtained with the help of a CAS.

---

## EXERCISES 7.4

Answers to selected odd-numbered problems begin on page ANS-12.

### 7.4.1 Derivatives of a Transform

In Problems 1–8 use Theorem 7.4.1 to evaluate the given Laplace transform.

**1.** $\mathcal{L}\{te^{-10t}\}$

**2.** $\mathcal{L}\{t^3e^t\}$

**3.** $\mathcal{L}\{t\cos 2t\}$

**4.** $\mathcal{L}\{t\sinh 3t\}$

**5.** $\mathcal{L}\{t^2\sinh t\}$

**6.** $\mathcal{L}\{t^2\cos t\}$

**7.** $\mathcal{L}\{te^{2t}\sin 6t\}$

**8.** $\mathcal{L}\{te^{-3t}\cos 3t\}$

In Problems 9–14 use the Laplace transform to solve the given initial-value problem. Use the table of Laplace transforms in Appendix C as needed.

**9.** $y' + y = t\sin t, \quad y(0) = 0$

**10.** $y' - y = te^t\sin t, \quad y(0) = 0$

**11.** $y'' + 9y = \cos 3t, \quad y(0) = 2, \quad y'(0) = 5$

**12.** $y'' + y = \sin t, \quad y(0) = 1, \quad y'(0) = -1$

**13.** $y'' + 16y = f(t), \quad y(0) = 0, \quad y'(0) = 1$, where

$$f(t) = \begin{cases} \cos 4t, & 0 \le t < \pi \\ 0, & t \ge \pi \end{cases}$$

**14.** $y'' + y = f(t), \quad y(0) = 1, \quad y'(0) = 0$, where

$$f(t) = \begin{cases} 1, & 0 \le t < \pi/2 \\ \sin t, & t \ge \pi/2 \end{cases}$$

In Problems 15 and 16 use a graphing utility to graph the indicated solution.

**15.** $y(t)$ of Problem 13 for $0 \le t < 2\pi$

**16.** $y(t)$ of Problem 14 for $0 \le t < 3\pi$

In some instances the Laplace transform can be used to solve linear differential equations with variable monomial coefficients. In Problems 17 and 18 use Theorem 7.4.1 to reduce the given differential equation to a linear first-order DE in the transformed function $Y(s) = \mathcal{L}\{y(t)\}$. Solve the first-order DE for $Y(s)$ and then find $y(t) = \mathcal{L}^{-1}\{Y(s)\}$.

**17.** $ty'' - y' = 2t^2, \quad y(0) = 0$

**18.** $2y'' + ty' - 2y = 10, \quad y(0) = y'(0) = 0$

### 7.4.2 Transforms of Integrals

In Problems 19–22 proceed as in Example 3 and find the convolution $f * g$ of the given functions. After integrating find the Laplace transform of $f * g$.

**19.** $f(t) = 4t, \quad g(t) = 3t^2$

**20.** $f(t) = t, \quad g(t) = e^{-t}$

**21.** $f(t) = e^{-t}, \quad g(t) = e^t$

**22.** $f(t) = \cos 2t, \quad g(t) = e^t$

In Problems 23–34 proceed as in Example 4 and find the Laplace transform of $f * g$ using Theorem 7.4.2. Do not evaluate the convolution integral before transforming.

**23.** $\mathcal{L}\{1 * t^3\}$

**24.** $\mathcal{L}\{t^2 * te^t\}$

**25.** $\mathcal{L}\{e^{-t} * e^t \cos t\}$

**26.** $\mathcal{L}\{e^{2t} * \sin t\}$

**27.** $\mathscr{L}\left\{\displaystyle\int_0^t e^\tau \, d\tau\right\}$

**28.** $\mathscr{L}\left\{\displaystyle\int_0^t \cos \tau \, d\tau\right\}$

**29.** $\mathscr{L}\left\{\displaystyle\int_0^t e^{-\tau} \cos \tau \, d\tau\right\}$

**30.** $\mathscr{L}\left\{\displaystyle\int_0^t \tau \sin \tau \, d\tau\right\}$

**31.** $\mathscr{L}\left\{\displaystyle\int_0^t \tau \, e^{t-\tau} \, d\tau\right\}$

**32.** $\mathscr{L}\left\{\displaystyle\int_0^t \sin \tau \cos (t - \tau) \, d\tau\right\}$

**33.** $\mathscr{L}\left\{t \displaystyle\int_0^t \sin \tau \, d\tau\right\}$

**34.** $\mathscr{L}\left\{t \displaystyle\int_0^t \tau \, e^{-\tau} \, d\tau\right\}$

In Problems 35–38 use (8) to evaluate the given inverse transform.

**35.** $\mathscr{L}^{-1}\left\{\dfrac{1}{s(s-1)}\right\}$

**36.** $\mathscr{L}^{-1}\left\{\dfrac{1}{s^2(s-1)}\right\}$

**37.** $\mathscr{L}^{-1}\left\{\dfrac{1}{s^3(s-1)}\right\}$

**38.** $\mathscr{L}^{-1}\left\{\dfrac{1}{s(s-a)^2}\right\}$

**39.** The table in Appendix C does not contain an entry for

$$\mathscr{L}^{-1}\left\{\dfrac{8k^3 s}{(s^2 + k^2)^3}\right\}.$$

  **(a)** Use (4) along with the results in (5) to evaluate this inverse transform. Use a CAS as an aid in evaluating the convolution integral.

  **(b)** Reexamine your answer to part (a). Could you have obtained the result in a different manner?

**40.** Use the Laplace transform and the results of Problem 39 to solve the initial-value problem

$$y'' + y = \sin t + t \sin t, \quad y(0) = 0, \quad y'(0) = 0.$$

Use a graphing utility to graph the solution.

In Problems 41–50 use the Laplace transform to solve the given integral equation or integrodifferential equation.

**41.** $f(t) + \displaystyle\int_0^t (t - \tau) f(\tau) \, d\tau = t$

**42.** $f(t) = 2t - 4 \displaystyle\int_0^t \sin \tau \, f(t - \tau) \, d\tau$

**43.** $f(t) = t e^t + \displaystyle\int_0^t \tau \, f(t - \tau) \, d\tau$

**44.** $f(t) + 2 \displaystyle\int_0^t f(\tau) \cos (t - \tau) \, d\tau = 4e^{-t} + \sin t$

**45.** $f(t) + \displaystyle\int_0^t f(\tau) \, d\tau = 1$

**46.** $f(t) = \cos t + \displaystyle\int_0^t e^{-\tau} f(t - \tau) \, d\tau$

**47.** $f(t) = 1 + t - \dfrac{8}{3} \displaystyle\int_0^t (\tau - t)^3 f(\tau) \, d\tau$

**48.** $t - 2f(t) = \displaystyle\int_0^t (e^\tau - e^{-\tau}) f(t - \tau) \, d\tau$

**49.** $y'(t) = 1 - \sin t - \displaystyle\int_0^t y(\tau) \, d\tau, \quad y(0) = 0$

**50.** $\dfrac{dy}{dt} + 6y(t) + 9 \displaystyle\int_0^t y(\tau) \, d\tau = 1, \quad y(0) = 0$

In Problems 51 and 52 solve equation (10) subject to $i(0) = 0$ with $L$, $R$, $C$, and $E(t)$ as given. Use a graphing utility to graph the solution for $0 \le t \le 3$.

**51.** $L = 0.1$ h, $R = 3$ $\Omega$, $C = 0.05$ f,
$\quad E(t) = 100[\mathscr{U}(t - 1) - \mathscr{U}(t - 2)]$

**52.** $L = 0.005$ h, $R = 1$ $\Omega$, $C = 0.02$ f,
$\quad E(t) = 100[t - (t - 1)\mathscr{U}(t - 1)]$

## 7.4.3 Transform of a Periodic Function

In Problems 53–58 use Theorem 7.4.3 to find the Laplace transform of the given periodic function.

**53.**

meander function

**FIGURE 7.4.6** Graph for Problem 53

**54.**

square wave

**FIGURE 7.4.7** Graph for Problem 54

**55.**

sawtooth function

**FIGURE 7.4.8** Graph for Problem 55

**56.**

triangular wave

**FIGURE 7.4.9** Graph for Problem 56

**57.**

full-wave rectification of sin $t$

**FIGURE 7.4.10** Graph for Problem 57

**58.**

half-wave rectification of sin $t$

**FIGURE 7.4.11** Graph for Problem 58

In Problems 59 and 60 solve equation (15) subject to $i(0) = 0$ with $E(t)$ as given. Use a graphing utility to graph the solution for $0 \le t \le 4$ in the case when $L = 1$ and $R = 1$.

**59.** $E(t)$ is the meander function in Problem 53 with amplitude 1 and $a = 1$.

**60.** $E(t)$ is the sawtooth function in Problem 55 with amplitude 1 and $b = 1$.

In Problems 61 and 62 solve the model for a driven spring/mass system with damping

$$m\frac{d^2x}{dt^2} + \beta\frac{dx}{dt} + kx = f(t), \quad x(0) = 0, \quad x'(0) = 0,$$

where the driving function $f$ is as specified. Use a graphing utility to graph $x(t)$ for the indicated values of $t$.

**61.** $m = \frac{1}{2}, \beta = 1, k = 5, f$ is the meander function in Problem 53 with amplitude 10, and $a = \pi, 0 \le t \le 2\pi$.

**62.** $m = 1, \beta = 2, k = 1, f$ is the square wave in Problem 54 with amplitude 5, and $a = \pi, 0 \le t \le 4\pi$.

## Discussion Problems

**63.** Discuss how Theorem 7.4.1 can be used to find

$$\mathcal{L}^{-1}\left\{\ln\frac{s-3}{s+1}\right\}.$$

**64.** In Section 6.4 we saw that $ty'' + y' + ty = 0$ is Bessel's equation of order $\nu = 0$. In view of (24) of that section and Table 6.4.1 a solution of the initial-value problem $ty'' + y' + ty = 0, y(0) = 1, y'(0) = 0$, is $y = J_0(t)$. Use this result and the procedure outlined in the instructions to Problems 17 and 18 to show that

$$\mathcal{L}\{J_0(t)\} = \frac{1}{\sqrt{s^2+1}}.$$

[*Hint*: You might need to use Problem 52 in Exercises 7.2.]

**65. (a) Laguerre's differential equation**

$$ty'' + (1-t)y' + ny = 0$$

is known to possess polynomial solutions when $n$ is a nonnegative integer. These solutions are naturally called **Laguerre polynomials** and are denoted by $L_n(t)$. Find $y = L_n(t)$, for $n = 0, 1, 2, 3, 4$ if it is known that $L_n(0) = 1$.

**(b)** Show that

$$\mathcal{L}\left\{\frac{e^t}{n!}\frac{d^n}{dt^n}t^n e^{-t}\right\} = Y(s),$$

where $Y(s) = \mathcal{L}\{y\}$ and $y = L_n(t)$ is a polynomial solution of the DE in part (a). Conclude that

$$L_n(t) = \frac{e^t}{n!}\frac{d^n}{dt^n}t^n e^{-t}, \quad n = 0, 1, 2, \ldots.$$

This last relation for generating the Laguerre polynomials is the analogue of Rodrigues' formula for the Legendre polynomials. See (36) in Section 6.4.

**66.** The Laplace transform $\mathcal{L}\{e^{-t^2}\}$ exists, but without finding it solve the initial-value problem $y'' + y = e^{-t^2}, y(0) = 0, y'(0) = 0$.

**67.** Solve the integral equation

$$f(t) = e^t + e^t\int_0^t e^{-\tau}f(\tau)\,d\tau.$$

**68. (a)** Show that the square wave function $E(t)$ given in Figure 7.4.4 can be written

$$E(t) = \sum_{k=0}^{\infty} (-1)^k \mathcal{U}(t-k).$$

**(b)** Obtain (14) of this section by taking the Laplace transform of each term in the series in part (a).

**69.** Use the Laplace transform as an aide in evaluating the improper integral $\int_0^\infty te^{-2t}\sin 4t\,dt$.

**70.** If we assume that $\mathcal{L}\{f(t)/t\}$ exists and $\mathcal{L}\{f(t)\} = F(s)$, then

$$\mathcal{L}\left\{\frac{f(t)}{t}\right\} = \int_s^\infty F(u)\,du.$$

Use this result to find the Laplace transform of the given function. The symbols $a$ and $k$ are positive constants.

**(a)** $f(t) = \dfrac{\sin at}{t}$    **(b)** $f(t) = \dfrac{2(1 - \cos kt)}{t}$

**71. Transform of the Logarithm**   Because $f(t) = \ln t$ has an infinite discontinuity at $t = 0$ it might be assumed that $\mathcal{L}\{\ln t\}$ does not exist; however, this is incorrect. The point of this problem is to guide you through the formal steps leading to the Laplace transform of $f(t) = \ln t, t > 0$.

**(a)** Use integration by parts to show that

$$\mathcal{L}\{\ln t\} = s\,\mathcal{L}\{t\ln t\} - \frac{1}{s}.$$

**(b)** If $\mathcal{L}\{\ln t\} = Y(s)$, use Theorem 7.4.1 with $n = 1$ to show that part (a) becomes

$$s\frac{dY}{ds} + Y = -\frac{1}{s}.$$

Find an explicit solution $Y(s)$ of the foregoing differential equation.

**(c)** Finally, the integral definition of **Euler's constant** (sometimes called the **Euler-Mascheroni constant**) is $\gamma = -\int_0^\infty e^{-t}\ln t\,dt$, where $\gamma = 0.5772156649\ldots$. Use $Y(1) = -\gamma$ in the solution in part (b) to show that

$$\mathcal{L}\{\ln t\} = -\frac{\gamma}{s} - \frac{\ln s}{s}, \quad s > 0.$$

## Computer Lab Assignments

**72.** In this problem you are led through the commands in *Mathematica* that enable you to obtain the symbolic Laplace transform of a differential equation and the solution of the initial-value problem by finding the inverse transform. In *Mathematica* the Laplace transform of a function $y(t)$ is obtained using **LaplaceTransform [y[t], t, s]**. In line two of the syntax we replace **LaplaceTransform [y[t], t, s]** by the symbol **Y**. (*If you do not have* Mathematica, *then adapt the given procedure by finding the corresponding syntax for the CAS you have on hand.*)

Consider the initial-value problem

$$y'' + 6y' + 9y = t \sin t, \quad y(0) = 2, \quad y'(0) = -1.$$

Load the Laplace transform package. Precisely reproduce and then, in turn, execute each line in the following sequence of commands. Either copy the output by hand or print out the results.

```
diffequat = y''[t] + 6y'[t] + 9y[t] == t Sin[t]
transformdeq = LaplaceTransform [diffequat, t, s] /.
    {y[0] − > 2, y'[0] − > −1,
    LaplaceTransform [y[t], t, s] − > Y}
soln = Solve[transformdeq, Y]//Flatten
Y = Y/.soln
InverseLaplaceTransform[ Y, s, t]
```

**73.** Appropriately modify the procedure of Problem 72 to find a solution of

$$y''' + 3y' - 4y = 0,$$
$$y(0) = 0, \quad y'(0) = 0, \quad y''(0) = 1.$$

**74.** The charge $q(t)$ on a capacitor in an *LC*-series circuit is given by

$$\frac{d^2q}{dt^2} + q = 1 - 4\mathcal{U}(t - \pi) + 6\mathcal{U}(t - 3\pi),$$
$$q(0) = 0, \quad q'(0) = 0.$$

Appropriately modify the procedure of Problem 72 to find $q(t)$. Graph your solution.

---

## 7.5 The Dirac Delta Function

**INTRODUCTION**    In the last paragraph on page 284, we indicated that as an immediate consequence of Theorem 7.1.3, $F(s) = 1$ cannot be the Laplace transform of a function $f$ that is piecewise continuous on $[0, \infty)$ and of exponential order. In the discussion that follows we are going to introduce a function that is very different from the kinds that you have studied in previous courses. We shall see that there does indeed exist a function—or, more precisely, a *generalized function*—whose Laplace transform is $F(s) = 1$.

**UNIT IMPULSE**    Mechanical systems are often acted on by an external force (or electromotive force in an electrical circuit) of large magnitude that acts only for a very short period of time. For example, a vibrating airplane wing could be struck by lightning, a mass on a spring could be given a sharp blow by a ball peen hammer, and a ball (baseball, golf ball, tennis ball) could be sent soaring when struck violently by some kind of club (baseball bat, golf club, tennis racket). See Figure 7.5.1. The graph of the piecewise-defined function

**FIGURE 7.5.1** A golf club applies a force of large magnitude on the ball for a very short period of time

$$\delta_a(t - t_0) = \begin{cases} 0, & 0 \leq t < t_0 - a \\ \dfrac{1}{2a}, & t_0 - a \leq t < t_0 + a \\ 0, & t \geq t_0 + a, \end{cases} \tag{1}$$

$a > 0$, $t_0 > 0$, shown in Figure 7.5.2(a), could serve as a model for such a force. For a small value of $a$, $\delta_a(t - t_0)$ is essentially a constant function of large magnitude that is "on" for just a very short period of time, around $t_0$. The behavior of $\delta_a(t - t_0)$ as $a \to 0$ is illustrated in Figure 7.5.2(b). The function $\delta_a(t - t_0)$ is called a **unit impulse,** because it possesses the integration property $\int_0^\infty \delta_a(t - t_0)\, dt = 1$.

**(a)** graph of $\delta_a(t - t_0)$

**(b)** behavior of $\delta_a$ as $a \to 0$

**FIGURE 7.5.2** Unit impulse

**DIRAC DELTA FUNCTION**    In practice it is convenient to work with another type of unit impulse, a "function" that approximates $\delta_a(t - t_0)$ and is defined by the limit

$$\delta(t - t_0) = \lim_{a \to 0} \delta_a(t - t_0). \tag{2}$$

The latter expression, which is not a function at all, can be characterized by the two properties

$$(i) \; \delta(t - t_0) = \begin{cases} \infty, & t = t_0 \\ 0, & t \neq t_0 \end{cases} \quad \text{and} \quad (ii) \int_0^\infty \delta(t - t_0) \, dt = 1.$$

The unit impulse $\delta(t - t_0)$ is called the **Dirac delta function.**

It is possible to obtain the Laplace transform of the Dirac delta function by the formal assumption that $\mathscr{L}\{\delta(t - t_0)\} = \lim_{a \to 0} \mathscr{L}\{\delta_a(t - t_0)\}$.

**THEOREM 7.5.1**    **Transform of the Dirac Delta Function**

For $t_0 > 0$, $\qquad\qquad \mathscr{L}\{\delta(t - t_0)\} = e^{-st_0}. \tag{3}$

**PROOF**    To begin, we can write $\delta_a(t - t_0)$ in terms of the unit step function by virtue of (11) and (12) of Section 7.3:

$$\delta_a(t - t_0) = \frac{1}{2a} [\mathscr{U}(t - (t_0 - a)) - \mathscr{U}(t - (t_0 + a))].$$

By linearity and (14) of Section 7.3 the Laplace transform of this last expression is

$$\mathscr{L}\{\delta_a(t - t_0)\} = \frac{1}{2a} \left[ \frac{e^{-s(t_0 - a)}}{s} - \frac{e^{-s(t_0 + a)}}{s} \right] = e^{-st_0} \left( \frac{e^{sa} - e^{-sa}}{2sa} \right). \tag{4}$$

Since (4) has the indeterminate form $0/0$ as $a \to 0$, we apply L'Hôpital's Rule:

$$\mathscr{L}\{\delta(t - t_0)\} = \lim_{a \to 0} \mathscr{L}\{\delta_a(t - t_0)\} = e^{-st_0} \lim_{a \to 0} \left( \frac{e^{sa} - e^{-sa}}{2sa} \right) = e^{-st_0}. \quad\blacksquare$$

Now when $t_0 = 0$, it seems plausible to conclude from (3) that

$$\mathscr{L}\{\delta(t)\} = 1.$$

The last result emphasizes the fact that $\delta(t)$ is not the usual type of function that we have been considering, since we expect from Theorem 7.1.3 that $\mathscr{L}\{f(t)\} \to 0$ as $s \to \infty$.

**EXAMPLE 1**    **Two Initial-Value Problems**

Solve $y'' + y = 4 \, \delta(t - 2\pi)$ subject to

**(a)** $y(0) = 1, \quad y'(0) = 0$    **(b)** $y(0) = 0, \quad y'(0) = 0$.

The two initial-value problems could serve as models for describing the motion of a mass on a spring moving in a medium in which damping is negligible. At $t = 2\pi$ the mass is given a sharp blow. In (a) the mass is released from rest 1 unit below the equilibrium position. In (b) the mass is at rest in the equilibrium position.

**SOLUTION    (a)** From (3) the Laplace transform of the differential equation is

$$s^2 Y(s) - s + Y(s) = 4e^{-2\pi s} \quad \text{or} \quad Y(s) = \frac{s}{s^2 + 1} + \frac{4e^{-2\pi s}}{s^2 + 1}.$$

**FIGURE 7.5.3** Mass is struck at $t = 2\pi$ in part (a) of Example 1

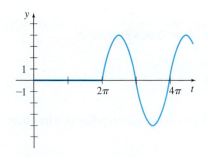

**FIGURE 7.5.4** No motion until mass is struck at $t = 2\pi$ in part (b) of Example 1

Using the inverse form of the second translation theorem, (15) of Section 7.3, we find

$$y(t) = \cos t + 4 \sin (t - 2\pi) \, \mathcal{U}(t - 2\pi).$$

Since $\sin(t - 2\pi) = \sin t$, the foregoing solution can be written as

$$y(t) = \begin{cases} \cos t, & 0 \le t < 2\pi \\ \cos t + 4 \sin t, & t \ge 2\pi. \end{cases} \tag{5}$$

In Figure 7.5.3 we see from the graph of (5) that the mass is exhibiting simple harmonic motion until it is struck at $t = 2\pi$. The influence of the unit impulse is to increase the amplitude of vibration to $\sqrt{17}$ for $t > 2\pi$.

**(b)** In this case the transform of the equation is simply

$$Y(s) = \frac{4e^{-2\pi s}}{s^2 + 1},$$

and so

$$y(t) = 4 \sin (t - 2\pi) \, \mathcal{U}(t - 2\pi)$$

$$= \begin{cases} 0, & 0 \le t < 2\pi \\ 4 \sin t, & t \ge 2\pi. \end{cases} \tag{6}$$

The graph of (6) in Figure 7.5.4 shows, as we would expect from the initial conditions that the mass exhibits no motion until it is struck at $t = 2\pi$.    ∎

### REMARKS

(*i*) If $\delta(t - t_0)$ were a function in the usual sense, then property (*i*) on page 319 would imply $\int_0^{\infty} \delta(t - t_0) \, dt = 0$ rather than $\int_0^{\infty} \delta(t - t_0) \, dt = 1$. Because the Dirac delta function did not "behave" like an ordinary function, even though its users produced correct results, it was met initially with great scorn by mathematicians. However, in the 1940s Dirac's controversial function was put on a rigorous footing by the French mathematician Laurent Schwartz in his book *Théorie des distributions*, and this, in turn, led to an entirely new branch of mathematics known as the **theory of distributions** or **generalized functions.** In this theory (2) is not an accepted definition of $\delta(t - t_0)$, nor does one speak of a function whose values are either $\infty$ or 0. Although we shall not pursue this topic any further, suffice it to say that the Dirac delta function is best characterized by its effect on other functions. If *f* is a continuous function, then

$$\int_0^{\infty} f(t) \, \delta(t - t_0) \, dt = f(t_0) \tag{7}$$

can be taken as the *definition* of $\delta(t - t_0)$. This result is known as the **sifting property,** since $\delta(t - t_0)$ has the effect of sifting the value $f(t_0)$ out of the set of values of *f* on $[0, \infty)$. Note that property (*ii*) (with $f(t) = 1$) and (3) (with $f(t) = e^{-st}$) are consistent with (7).

(*ii*) In (*iii*) in the *Remarks* at the end of Section 7.2 we indicated that the transfer function of a general linear *n*th-order differential equation with constant coefficients is $W(s) = 1/P(s)$, where $P(s) = a_n s^n + a_{n-1} s^{n-1} + \cdots + a_0$. The transfer function is the Laplace transform of function $w(t)$, called the **weight function** of a linear system. But $w(t)$ can also be characterized in terms of the discussion at hand. For simplicity let us consider a second-order linear system in which the input is a unit impulse at $t = 0$:

$$a_2 y'' + a_1 y' + a_0 y = \delta(t), \quad y(0) = 0, \quad y'(0) = 0.$$

Applying the Laplace transform and using $\mathcal{L}\{\delta(t)\} = 1$ shows that the transform of the response $y$ in this case is the transfer function

$$Y(s) = \frac{1}{a_2 s^2 + a_1 s + a_0} = \frac{1}{P(s)} = W(s) \quad \text{and} \quad \text{so} \quad y = \mathcal{L}^{-1}\left\{\frac{1}{P(s)}\right\} = w(t).$$

From this we can see, in general, that the weight function $y = w(t)$ of an $n$th-order linear system is the zero-state response of the system to a unit impulse. For this reason $w(t)$ is also called the **impulse response** of the system.

---

## EXERCISES 7.5

*Answers to selected odd-numbered problems begin on page ANS-12.*

In Problems 1–12 use the Laplace transform to solve the given initial-value problem.

**1.** $y' - 3y = \delta(t - 2), \quad y(0) = 0$

**2.** $y' + y = \delta(t - 1), \quad y(0) = 2$

**3.** $y'' + y = \delta(t - 2\pi), \quad y(0) = 0, \quad y'(0) = 1$

**4.** $y'' + 16y = \delta(t - 2\pi), \quad y(0) = 0, \quad y'(0) = 0$

**5.** $y'' + y = \delta\left(t - \frac{1}{2}\pi\right) + \delta\left(t - \frac{3}{2}\pi\right), \quad y(0) = 0, \quad y'(0) = 0$

**6.** $y'' + y = \delta(t - 2\pi) + \delta(t - 4\pi), \quad y(0) = 1, \quad y'(0) = 0$

**7.** $y'' + 2y' = \delta(t - 1), \quad y(0) = 0, \quad y'(0) = 1$

**8.** $y'' - 2y' = 1 + \delta(t - 2), \quad y(0) = 0, \quad y'(0) = 1$

**9.** $y'' + 4y' + 5y = \delta(t - 2\pi), \quad y(0) = 0, \quad y'(0) = 0$

**10.** $y'' + 2y' + y = \delta(t - 1), \quad y(0) = 0, \quad y'(0) = 0$

**11.** $y'' + 4y' + 13y = \delta(t - \pi) + \delta(t - 3\pi), \quad y(0) = 1, \quad y'(0) = 0$

**12.** $y'' - 7y' + 6y = e^t + \delta(t - 2) + \delta(t - 4), \quad y(0) = 0, \quad y'(0) = 0$

In Problems 13 and 14 use the Laplace transform to solve the given initial-value problem. Graph your solution on the interval $[0, 8\pi]$.

**13.** $y'' + y = \displaystyle\sum_{k=1}^{\infty} \delta(t - k\pi), \quad y(0) = 0, \quad y'(0) = 1$

**14.** $y'' + y = \displaystyle\sum_{k=1}^{\infty} \delta(t - 2k\pi), \quad y(0) = 0, \quad y'(0) = 1$

In Problems 15 and 16 a uniform beam of length $L$ carries a concentrated load $w_0$ at $x = \frac{1}{2}L$. See Figure 7.5.5 (Problem 15) and Figure 7.5.6 (Problem 16). Use the Laplace transform to solve the differential equation

$$EI \frac{d^4 y}{dx^4} = w_0 \delta\left(x - \tfrac{1}{2}L\right), \quad 0 < x < L,$$

subject to the given boundary conditions.

**15.** $y(0) = 0, \quad y'(0) = 0, \quad y''(L) = 0, \quad y'''(L) = 0$

**FIGURE 7.5.5** Beam embedded at its left end and free at its right end

**16.** $y(0) = 0, \quad y'(0) = 0, \quad y(L) = 0, \quad y'(L) = 0$

**FIGURE 7.5.6** Beam embedded at both ends

## Discussion Problems

**17.** Someone tells you that the solutions of the two IVPs

$$y'' + 2y' + 10y = 0, \qquad y(0) = 0, \quad y'(0) = 1$$
$$y'' + 2y' + 10y = \delta(t), \qquad y(0) = 0, \quad y'(0) = 0$$

are exactly the same. Do you agree or disagree? Defend your answer.

**18.** Reread (*i*) in the *Remarks* at the end of this section. Then use the Laplace transform to solve the initial-value problem:

$$y'' + 4y' + 3y = e^t \delta(t - 1), \quad y(0) = 0, \quad y'(0) = 2.$$

Use a graphing utility to graph $y(t)$ for $0 \le t \le 5$.

# 7.6 Systems of Linear Differential Equations

**INTRODUCTION**   When initial conditions are specified, the Laplace transform of each equation in a system of linear differential equations with constant coefficients reduces the system of DEs to a set of simultaneous algebraic equations in the transformed functions. We solve the system of algebraic equations for each of the transformed functions and then find the inverse Laplace transforms in the usual manner.

**COUPLED SPRINGS**   Two masses $m_1$ and $m_2$ are connected to two springs $A$ and $B$ of negligible mass having spring constants $k_1$ and $k_2$, respectively. In turn the two springs are attached as shown in Figure 7.6.1. Let $x_1(t)$ and $x_2(t)$ denote the vertical displacements of the masses from their equilibrium positions. When the system is in motion, spring $B$ is subject to both an elongation and a compression; hence its net elongation is $x_2 - x_1$. Therefore it follows from Hooke's law that springs $A$ and $B$ exert forces $-k_1x_1$ and $k_2(x_2 - x_1)$, respectively, on $m_1$. If no external force is impressed on the system and if no damping force is present, then the net force on $m_1$ is $-k_1x_1 + k_2(x_2 - x_1)$. By Newton's second law we can write

$$m_1 \frac{d^2x_1}{dt^2} = -k_1 x_1 + k_2(x_2 - x_1).$$

Similarly, the net force exerted on mass $m_2$ is due solely to the net elongation of $B$; that is, $-k_2(x_2 - x_1)$. Hence we have

$$m_2 \frac{d^2x_2}{dt^2} = -k_2(x_2 - x_1).$$

In other words, the motion of the coupled system is represented by the system of simultaneous second-order differential equations

$$m_1 x_1'' = -k_1 x_1 + k_2(x_2 - x_1) \tag{1}$$
$$m_2 x_2'' = -k_2(x_2 - x_1).$$

In the next example we solve (1) under the assumptions that $k_1 = 6$, $k_2 = 4$, $m_1 = 1$, $m_2 = 1$, and that the masses start from their equilibrium positions with opposite unit velocities.

**(a)** equilibrium   **(b)** motion   **(c)** forces

**FIGURE 7.6.1** Coupled spring/mass system

---

**EXAMPLE 1   Coupled Springs**

Solve

$$x_1'' + 10x_1 \quad\quad - 4x_2 = 0$$
$$-4x_1 + x_2'' + 4x_2 = 0 \tag{2}$$

subject to $x_1(0) = 0$, $x_1'(0) = 1$, $x_2(0) = 0$, $x_2'(0) = -1$.

**SOLUTION**   The Laplace transform of each equation is

$$s^2 X_1(s) - s x_1(0) - x_1'(0) + 10X_1(s) - 4X_2(s) = 0$$
$$-4X_1(s) + s^2 X_2(s) - s x_2(0) - x_2'(0) + 4X_2(s) = 0,$$

where $X_1(s) = \mathscr{L}\{x_1(t)\}$ and $X_2(s) = \mathscr{L}\{x_2(t)\}$. The preceding system is the same as

$$(s^2 + 10)X_1(s) - \quad\quad 4X_2(s) = 1$$
$$-4X_1(s) + (s^2 + 4)X_2(s) = -1. \tag{3}$$

Solving (3) for $X_1(s)$ and using partial fractions on the result yields

$$X_1(s) = \frac{s^2}{(s^2 + 2)(s^2 + 12)} = -\frac{1/5}{s^2 + 2} + \frac{6/5}{s^2 + 12},$$

and therefore

$$x_1(t) = -\frac{1}{5\sqrt{2}} \mathscr{L}^{-1}\left\{\frac{\sqrt{2}}{s^2 + 2}\right\} + \frac{6}{5\sqrt{12}} \mathscr{L}^{-1}\left\{\frac{\sqrt{12}}{s^2 + 12}\right\}$$

$$= -\frac{\sqrt{2}}{10} \sin \sqrt{2}t + \frac{\sqrt{3}}{5} \sin 2\sqrt{3}t.$$

Substituting the expression for $X_1(s)$ into the first equation of (3) gives

$$X_2(s) = -\frac{s^2 + 6}{(s^2 + 2)(s^2 + 12)} = -\frac{2/5}{s^2 + 2} - \frac{3/5}{s^2 + 12}$$

and

$$x_2(t) = -\frac{2}{5\sqrt{2}} \mathscr{L}^{-1}\left\{\frac{\sqrt{2}}{s^2 + 2}\right\} - \frac{3}{5\sqrt{12}} \mathscr{L}^{-1}\left\{\frac{\sqrt{12}}{s^2 + 12}\right\}$$

$$= -\frac{\sqrt{2}}{5} \sin \sqrt{2}t - \frac{\sqrt{3}}{10} \sin 2\sqrt{3}t.$$

Finally, the solution to the given system (2) is

$$x_1(t) = -\frac{\sqrt{2}}{10} \sin \sqrt{2}t + \frac{\sqrt{3}}{5} \sin 2\sqrt{3}t$$

$$x_2(t) = -\frac{\sqrt{2}}{5} \sin \sqrt{2}t - \frac{\sqrt{3}}{10} \sin 2\sqrt{3}t.$$

(4)

**(a)** plot of $x_1(t)$

**(b)** plot of $x_2(t)$

**FIGURE 7.6.2** Displacements of the two masses in Example 1

The graphs of $x_1$ and $x_2$ in Figure 7.6.2 reveal the complicated oscillatory motion of each mass. ∎

**NETWORKS**    In (18) of Section 3.3 we saw the currents $i_1(t)$ and $i_2(t)$ in the network shown in Figure 7.6.3, containing an inductor, a resistor, and a capacitor, were governed by the system of first-order differential equations

$$L\frac{di_1}{dt} + Ri_2 = E(t)$$

$$RC\frac{di_2}{dt} + i_2 - i_1 = 0.$$

(5)

**FIGURE 7.6.3** Electrical network

We solve this system by the Laplace transform in the next example.

---

**EXAMPLE 2**   **An Electrical Network**

Solve the system in (5) under the conditions $E(t) = 60$ V, $L = 1$ h, $R = 50\ \Omega$, $C = 10^{-4}$ f, and the currents $i_1$ and $i_2$ are initially zero.

**SOLUTION** We must solve

$$\frac{di_1}{dt} + 50i_2 = 60$$

$$50(10^{-4})\frac{di_2}{dt} + i_2 - i_1 = 0$$

subject to $i_1(0) = 0$, $i_2(0) = 0$.

Applying the Laplace transform to each equation of the system and simplifying gives

$$sI_1(s) + \qquad 50I_2(s) = \frac{60}{s}$$

$$-200I_1(s) + (s + 200)I_2(s) = 0,$$

where $I_1(s) = \mathscr{L}\{i_1(t)\}$ and $I_2(s) = \mathscr{L}\{i_2(t)\}$. Solving the system for $I_1$ and $I_2$ and decomposing the results into partial fractions gives

$$I_1(s) = \frac{60s + 12{,}000}{s(s + 100)^2} = \frac{6/5}{s} - \frac{6/5}{s + 100} - \frac{60}{(s + 100)^2}$$

$$I_2(s) = \frac{12{,}000}{s(s + 100)^2} = \frac{6/5}{s} - \frac{6/5}{s + 100} - \frac{120}{(s + 100)^2}.$$

Taking the inverse Laplace transform, we find the currents to be

$$i_1(t) = \frac{6}{5} - \frac{6}{5}e^{-100t} - 60te^{-100t}$$

$$i_2(t) = \frac{6}{5} - \frac{6}{5}e^{-100t} - 120te^{-100t}. \qquad \blacksquare$$

Note that both $i_1(t)$ and $i_2(t)$ in Example 2 tend toward the value $E/R = \frac{6}{5}$ as $t \to \infty$. Furthermore, since the current through the capacitor is $i_3(t) = i_1(t) - i_2(t) = 60te^{-100t}$, we observe that $i_3(t) \to 0$ as $t \to \infty$.

**FIGURE 7.6.4** Double pendulum

**DOUBLE PENDULUM**   Consider the double-pendulum system consisting of a pendulum attached to a pendulum shown in Figure 7.6.4. We assume that the system oscillates in a vertical plane under the influence of gravity, that the mass of each rod is negligible, and that no damping forces act on the system. Figure 7.6.4 also shows that the displacement angle $\theta_1$ is measured (in radians) from a vertical line extending downward from the pivot of the system and that $\theta_2$ is measured from a vertical line extending downward from the center of mass $m_1$. The positive direction is to the right; the negative direction is to the left. As we might expect from the analysis leading to equation (6) of Section 5.3, the system of differential equations describing the motion is nonlinear:

$$(m_1 + m_2)l_1^2\theta_1'' + m_2l_1l_2\theta_2'' \cos(\theta_1 - \theta_2) + m_2l_1l_2(\theta_2')^2 \sin(\theta_1 - \theta_2) + (m_1 + m_2)l_1g \sin\theta_1 = 0$$
$$m_2l_2^2\theta_2'' + m_2l_1l_2\theta_1''\cos(\theta_1 - \theta_2) - m_2l_1l_2(\theta_1')^2 \sin(\theta_1 - \theta_2) + m_2l_2g \sin\theta_2 = 0. \tag{6}$$

But if the displacements $\theta_1(t)$ and $\theta_2(t)$ are assumed to be small, then the approximations $\cos(\theta_1 - \theta_2) \approx 1$, $\sin(\theta_1 - \theta_2) \approx 0$, $\sin\theta_1 \approx \theta_1$, $\sin\theta_2 \approx \theta_2$ enable us to replace system (6) by the linearization

$$(m_1 + m_2)l_1^2\theta_1'' + m_2l_1l_2\theta_2'' + (m_1 + m_2)l_1g\theta_1 = 0$$
$$m_2l_2^2\theta_2'' + m_2l_1l_2\theta_1'' + m_2l_2g\theta_2 = 0. \tag{7}$$

**EXAMPLE 3**   Double Pendulum

It is left as an exercise to fill in the details of using the Laplace transform to solve system (7) when $m_1 = 3$, $m_2 = 1$, $l_1 = l_2 = 16$, $\theta_1(0) = 1$, $\theta_2(0) = -1$, $\theta_1'(0) = 0$, and $\theta_2'(0) = 0$. You should find that

$$\theta_1(t) = \frac{1}{4}\cos\frac{2}{\sqrt{3}}t + \frac{3}{4}\cos 2t$$
$$\theta_2(t) = \frac{1}{2}\cos\frac{2}{\sqrt{3}}t - \frac{3}{2}\cos 2t. \tag{8}$$

With the aid of a CAS the positions of the two masses at $t = 0$ and at subsequent times are shown in Figure 7.6.5. See Problem 21 in Exercises 7.6.

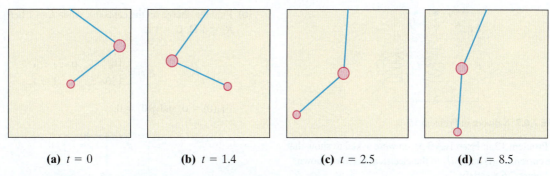

**(a)** $t = 0$        **(b)** $t = 1.4$        **(c)** $t = 2.5$        **(d)** $t = 8.5$

**FIGURE 7.6.5** Positions of masses on double pendulum at various times in Example 3

## EXERCISES 7.6

*Answers to selected odd-numbered problems begin on page ANS-12.*

In Problems 1–12 use the Laplace transform to solve the given system of differential equations.

**1.** $\dfrac{dx}{dt} = -x + y$

$\dfrac{dy}{dt} = 2x$

$x(0) = 0, \quad y(0) = 1$

**2.** $\dfrac{dx}{dt} = 2y + e^t$

$\dfrac{dy}{dt} = 8x - t$

$x(0) = 1, \quad y(0) = 1$

**3.** $\dfrac{dx}{dt} = x - 2y$

$\dfrac{dy}{dt} = 5x - y$

$x(0) = -1, \quad y(0) = 2$

**4.** $\dfrac{dx}{dt} + 3x + \dfrac{dy}{dt} = 1$

$\dfrac{dx}{dt} - x + \dfrac{dy}{dt} - y = e^t$

$x(0) = 0, \quad y(0) = 0$

**5.** $2\dfrac{dx}{dt} + \dfrac{dy}{dt} - 2x = 1$

$\dfrac{dx}{dt} + \dfrac{dy}{dt} - 3x - 3y = 2$

$x(0) = 0, \quad y(0) = 0$

**6.** $\dfrac{dx}{dt} + x - \dfrac{dy}{dt} + y = 0$

$\dfrac{dx}{dt} + \dfrac{dy}{dt} + 2y = 0$

$x(0) = 0, \quad y(0) = 1$

**7.** $\dfrac{d^2x}{dt^2} + x - y = 0$

$\dfrac{d^2y}{dt^2} + y - x = 0$

$x(0) = 0, \quad x'(0) = -2,$

$y(0) = 0, \quad y'(0) = 1$

**8.** $\dfrac{d^2x}{dt^2} + \dfrac{dx}{dt} + \dfrac{dy}{dt} = 0$

$\dfrac{d^2y}{dt^2} + \dfrac{dy}{dt} - 4\dfrac{dx}{dt} = 0$

$x(0) = 1, \quad x'(0) = 0,$

$y(0) = -1, \quad y'(0) = 5$

**9.** $\dfrac{d^2x}{dt^2} + \dfrac{d^2y}{dt^2} = t^2$

$\dfrac{d^2x}{dt^2} - \dfrac{d^2y}{dt^2} = 4t$

$x(0) = 8, \quad x'(0) = 0,$

$y(0) = 0, \quad y'(0) = 0$

**10.** $\dfrac{dx}{dt} - 4x + \dfrac{d^3y}{dt^3} = 6 \sin t$

$\dfrac{dx}{dt} + 2x - 2\dfrac{d^3y}{dt^3} = 0$

$x(0) = 0, \quad y(0) = 0,$

$y'(0) = 0, \quad y''(0) = 0$

**11.** $\dfrac{d^2x}{dt^2} + 3\dfrac{dy}{dt} + 3y = 0$

$\dfrac{d^2x}{dt^2} + 3y = te^{-t}$

$x(0) = 0, \quad x'(0) = 2, \quad y(0) = 0$

**12.** $\dfrac{dx}{dt} = 4x - 2y + 2\mathcal{U}(t - 1)$

$\dfrac{dy}{dt} = 3x - y + \mathcal{U}(t - 1)$

$x(0) = 0, \quad y(0) = \tfrac{1}{2}$

**13.** Solve system (1) when $k_1 = 3$, $k_2 = 2$, $m_1 = 1$, $m_2 = 1$ and $x_1(0) = 0$, $x_1'(0) = 1$, $x_2(0) = 1$, $x_2'(0) = 0$.

**14.** Derive the system of differential equations describing the straight-line vertical motion of the coupled springs shown in Figure 7.6.6. Use the Laplace transform to solve the system when $k_1 = 1$, $k_2 = 1$, $k_3 = 1$, $m_1 = 1$, $m_2 = 1$ and $x_1(0) = 0$, $x_1'(0) = -1$, $x_2(0) = 0$, $x_2'(0) = 1$.

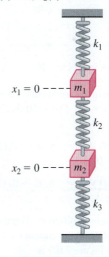

**FIGURE 7.6.6** Coupled springs in Problem 14

**15. (a)** Show that the system of differential equations for the currents $i_2(t)$ and $i_3(t)$ in the electrical network shown in Figure 7.6.7 is

$$L_1 \frac{di_2}{dt} + Ri_2 + Ri_3 = E(t)$$

$$L_2 \frac{di_3}{dt} + Ri_2 + Ri_3 = E(t).$$

**(b)** Solve the system in part (a) if $R = 5\ \Omega$, $L_1 = 0.01$ h, $L_2 = 0.0125$ h, $E = 100$ V, $i_2(0) = 0$, and $i_3(0) = 0$.

**(c)** Determine the current $i_1(t)$.

**FIGURE 7.6.7** Network in Problem 15

**16. (a)** In Problem 12 in Exercises 3.3 you were asked to show that the currents $i_2(t)$ and $i_3(t)$ in the electrical network shown in Figure 7.6.8 satisfy

$$L\frac{di_2}{dt} + L\frac{di_3}{dt} + R_1 i_2 = E(t)$$

$$-R_1\frac{di_2}{dt} + R_2\frac{di_3}{dt} + \frac{1}{C}i_3 = 0.$$

Solve the system if $R_1 = 10\ \Omega$, $R_2 = 5\ \Omega$, $L = 1$ h, $C = 0.2$ f,

$$E(t) = \begin{cases} 120, & 0 \le t < 2 \\ 0, & t \ge 2, \end{cases}$$

$i_2(0) = 0$, and $i_3(0) = 0$.

**(b)** Determine the current $i_1(t)$.

**FIGURE 7.6.8** Network in Problem 16

**17.** Solve the system given in (17) of Section 3.3 when $R_1 = 6\ \Omega$, $R_2 = 5\ \Omega$, $L_1 = 1$ h, $L_2 = 1$ h, $E(t) = 50 \sin t$ V, $i_2(0) = 0$, and $i_3(0) = 0$.

**18.** Solve (5) when $E = 60$ V, $L = \frac{1}{2}$ h, $R = 50\ \Omega$, $C = 10^{-4}$ f, $i_1(0) = 0$, and $i_2(0) = 0$.

**19.** Solve (5) when $E = 60$ V, $L = 2$ h, $R = 50\ \Omega$, $C = 10^{-4}$ f, $i_1(0) = 0$, and $i_2(0) = 0$.

**20. (a)** Show that the system of differential equations for the charge on the capacitor $q(t)$ and the current $i_3(t)$ in the electrical network shown in Figure 7.6.9 is

$$R_1\frac{dq}{dt} + \frac{1}{C}q + R_1 i_3 = E(t)$$

$$L\frac{di_3}{dt} + R_2 i_3 - \frac{1}{C}q = 0.$$

**(b)** Find the charge on the capacitor when $L = 1$ h, $R_1 = 1\ \Omega$, $R_2 = 1\ \Omega$, $C = 1$ f,

$$E(t) = \begin{cases} 0, & 0 < t < 1 \\ 50e^{-t}, & t \ge 1, \end{cases}$$

$i_3(0) = 0$, and $q(0) = 0$.

**FIGURE 7.6.9** Network in Problem 20

## Computer Lab Assignments

**21. (a)** Use the Laplace transform and the information given in Example 3 to obtain the solution (8) of the system given in (7).

**(b)** Use a graphing utility to graph $\theta_1(t)$ and $\theta_2(t)$ in the $t\theta$-plane. Which mass has extreme displacements of greater magnitude? Use the graphs to estimate the first time that each mass passes through its equilibrium position. Discuss whether the motion of the pendulums is periodic.

**(c)** Graph $\theta_1(t)$ and $\theta_2(t)$ in the $\theta_1\theta_2$-plane as parametric equations. The curve defined by these parametric equations is called a **Lissajous curve.**

**(d)** The positions of the masses at $t = 0$ are given in Figure 7.6.5(a). Note that we have used 1 radian $\approx 57.3°$. Use a calculator or a table application in a CAS to construct a table of values of the angles $\theta_1$ and $\theta_2$ for $t = 1, 2, \ldots, 10$ s. Then plot the positions of the two masses at these times.

**(e)** Use a CAS to find the first time that $\theta_1(t) = \theta_2(t)$ and compute the corresponding angular value. Plot the positions of the two masses at these times.

**(f)** Utilize the CAS to draw appropriate lines to simulate the pendulum rods, as in Figure 7.6.5. Use the animation capability of your CAS to make a "movie" of the motion of the double pendulum from $t = 0$ to $t = 10$ using a time increment of 0.1. [*Hint*: Express the coordinates $(x_1(t), y_1(t))$ and $(x_2(t), y_2(t))$ of the masses $m_1$ and $m_2$, respectively, in terms of $\theta_1(t)$ and $\theta_2(t)$.]

## Chapter 7 **In Review**

*Answers to selected odd-numbered problems begin on page ANS-13.*

In Problems 1 and 2 use the definition of the Laplace transform to find $\mathscr{L}\{f(t)\}$.

**1.** $f(t) = \begin{cases} t, & 0 \le t < 1 \\ 2 - t, & t \ge 1 \end{cases}$

**2.** $f(t) = \begin{cases} 0, & 0 \le t < 2 \\ 1, & 2 \le t < 4 \\ 0, & t \ge 4 \end{cases}$

In Problems 3–24 fill in the blanks or answer true or false.

**3.** If $f$ is not piecewise continuous on $[0, \infty)$, then $\mathscr{L}\{f(t)\}$ will not exist. _____

**4.** The function $f(t) = (e^t)^{10}$ is not of exponential order. _____

**5.** $F(s) = s^2/(s^2 + 4)$ is not the Laplace transform of a function that is piecewise continuous and of exponential order. _____

**6.** If $\mathscr{L}\{f(t)\} = F(s)$ and $\mathscr{L}\{g(t)\} = G(s)$, then
$\mathscr{L}^{-1}\{F(s)G(s)\} = f(t)g(t)$. _____

**7.** $\mathscr{L}\{e^{-7t}\} = $ _____

**8.** $\mathscr{L}\{te^{-7t}\} = $ _____

**9.** $\mathscr{L}\{\sin 2t\} = $ _____

**10.** $\mathscr{L}\{e^{-3t}\sin 2t\} = $ _____

**11.** $\mathscr{L}\{t \sin 2t\} = $ _____

**12.** $\mathscr{L}\{\sin 2t\, \mathscr{U}(t - \pi)\} = $ _____

**13.** $\mathscr{L}^{-1}\left\{\dfrac{20}{s^6}\right\} = $ _____

**14.** $\mathscr{L}^{-1}\left\{\dfrac{1}{3s - 1}\right\} = $ _____

**15.** $\mathscr{L}^{-1}\left\{\dfrac{1}{(s - 5)^3}\right\} = $ _____

**16.** $\mathscr{L}^{-1}\left\{\dfrac{1}{s^2 - 5}\right\} = $ _____

**17.** $\mathscr{L}^{-1}\left\{\dfrac{s}{s^2 - 10s + 29}\right\} = $ _____

**18.** $\mathscr{L}^{-1}\left\{\dfrac{e^{-5s}}{s^2}\right\} = $ _____

**19.** $\mathscr{L}^{-1}\left\{\dfrac{s + \pi}{s^2 + \pi^2}e^{-s}\right\} = $ _____

**20.** $\mathscr{L}^{-1}\left\{\dfrac{1}{L^2 s^2 + n^2 \pi^2}\right\} = $ _____

**21.** $\mathscr{L}\{e^{-5t}\}$ exists for $s > $ _____.

**22.** If $\mathscr{L}\{f(t)\} = F(s)$, then $\mathscr{L}\{te^{8t}f(t)\} = $ _____.

**23.** If $\mathscr{L}\{f(t)\} = F(s)$ and $k > 0$, then
$\mathscr{L}\{e^{at}f(t - k)\mathscr{U}(t - k)\} = $ _____.

**24.** $\mathscr{L}\{\int_0^t e^{a\tau} f(\tau)\, d\tau\} = $ _____ whereas
$\mathscr{L}\{e^{at}\int_0^t f(\tau)\, d\tau\} = $ _____.

In Problems 25–28 use the unit step function to find an equation for each graph in terms of the function $y = f(t)$, whose graph is given in Figure 7.R.1.

**FIGURE 7.R.1** Graph for Problems 25–28

**25.**

**FIGURE 7.R.2** Graph for Problem 25

**26.**

**FIGURE 7.R.3** Graph for Problem 26

**27.**

**FIGURE 7.R.4** Graph for Problem 27

**28.**

**FIGURE 7.R.5** Graph for Problem 28

In Problems 29–32 express $f$ in terms of unit step functions. Find $\mathscr{L}\{f(t)\}$ and $\mathscr{L}\{e^t f(t)\}$.

**29.**

**FIGURE 7.R.6** Graph for Problem 29

**30.**

**FIGURE 7.R.7**  Graph for Problem 30

**31.**

**FIGURE 7.R.8**  Graph for Problem 31

**32.**

**FIGURE 7.R.9**  Graph for Problem 32

In Problems 33 and 34 sketch the graph of the given function. Find $\mathscr{L}\{f(t)\}$.

**33.** $f(t) = -1 + 2 \sum_{k=1}^{\infty} (-1)^{k+1} \mathcal{U}(t - k)$

**34.** $f(t) = \sum_{k=0}^{\infty} (2k + 1 - t)[\mathcal{U}(t - 2k) - \mathcal{U}(t - 2k - 1)]$

In Problems 35–42 use the Laplace transform to solve the given equation.

**35.** $y'' - 2y' + y = e^t, \quad y(0) = 0, \quad y'(0) = 5$

**36.** $y'' - 8y' + 20y = te^t, \quad y(0) = 0, \quad y'(0) = 0$

**37.** $y'' + 6y' + 5y = t - t\mathcal{U}(t - 2), \quad y(0) = 1, \quad y'(0) = 0$

**38.** $y' - 5y = f(t)$, where

$$f(t) = \begin{cases} t^2, & 0 \le t < 1 \\ 0, & t \ge 1 \end{cases}, \quad y(0) = 1$$

**39.** $y' + 2y = f(t), \quad y(0) = 1$, where $f(t)$ is given in Figure 7.R.10.

**FIGURE 7.R.10**  Graph for Problem 39

**40.** $y'' + 5y' + 4y = f(t), \quad y(0) = 0, \quad y'(0) = 3$, where

$$f(t) = 12 \sum_{k=0}^{\infty} (-1)^k \mathcal{U}(t - k).$$

**41.** $y'(t) = \cos t + \int_0^t y(\tau) \cos(t - \tau)\, d\tau, \quad y(0) = 1$

**42.** $\int_0^t f(\tau) f(t - \tau)\, d\tau = 6t^3$

In Problems 43 and 44 use the Laplace transform to solve each system.

**43.**  $x' + y = t$
   $4x + y' = 0$
   $x(0) = 1, \quad y(0) = 2$

**44.**  $x'' + y'' = e^{2t}$
   $2x' + y'' = -e^{2t}$
   $x(0) = 0, \quad y(0) = 0,$
   $x'(0) = 0, \quad y'(0) = 0$

**45.** The current $i(t)$ in an $RC$-series circuit can be determined from the integral equation

$$Ri + \frac{1}{C} \int_0^t i(\tau)\, d\tau = E(t),$$

where $E(t)$ is the impressed voltage. Determine $i(t)$ when $R = 10\ \Omega$, $C = 0.5$ f, and $E(t) = 2(t^2 + t)$.

**46.** A series circuit contains an inductor, a resistor, and a capacitor for which $L = \frac{1}{2}$ h, $R = 10\ \Omega$, and $C = 0.01$ f, respectively. The voltage

$$E(t) = \begin{cases} 10, & 0 \le t < 5 \\ 0, & t \ge 5 \end{cases}$$

is applied to the circuit. Determine the instantaneous charge $q(t)$ on the capacitor for $t > 0$ if $q(0) = 0$ and $q'(0) = 0$.

**47.** A uniform cantilever beam of length $L$ is embedded at its left end ($x = 0$) and free at its right end. Find the deflection $y(x)$ if the load per unit length is given by

$$w(x) = \frac{2w_0}{L} \left[ \frac{L}{2} - x + \left( x - \frac{L}{2} \right) \mathcal{U}\left( x - \frac{L}{2} \right) \right].$$

**48.** When a uniform beam is supported by an elastic foundation, the differential equation for its deflection $y(x)$ is

$$EI \frac{d^4 y}{dx^4} + ky = w(x),$$

where $k$ is the modulus of the foundation and $-ky$ is the restoring force of the foundation that acts in the direction opposite to that of the load $w(x)$. See Figure 7.R.11. For algebraic convenience suppose that the differential equation is written as

$$\frac{d^4 y}{dx^4} + 4a^4 y = \frac{w(x)}{EI},$$

where $a = (k/4EI)^{1/4}$. Assume $L = \pi$ and $a = 1$. Find the deflection $y(x)$ of a beam that is supported on an elastic foundation when

**(a)** the beam is simply supported at both ends and a constant load $w_0$ is uniformly distributed along its length,

**(b)** the beam is embedded at both ends and $w(x)$ is a concentrated load $w_0$ applied at $x = \pi/2$. [*Hint*: In both parts of this problem use the table of Laplace transforms in Appendix C and the fact that $s^4 + 4 = (s^2 - 2s + 2)(s^2 + 2s + 2)$.]

**FIGURE 7.R.11**  Beam on elastic foundation in Problem 48

**49. (a)** Suppose two identical pendulums are coupled by means of a spring with constant $k$. See Figure 7.R.12. Under the same assumptions made in the discussion preceding Example 3 in Section 7.6, it can be shown that when the displacement angles $\theta_1(t)$ and $\theta_2(t)$ are small, the system of linear differential equations describing the motion is

$$\theta_1'' + \frac{g}{l}\theta_1 = -\frac{k}{m}(\theta_1 - \theta_2)$$

$$\theta_2'' + \frac{g}{l}\theta_2 = \frac{k}{m}(\theta_1 - \theta_2).$$

Use the Laplace transform to solve the system when $\theta_1(0) = \theta_0$, $\theta_1'(0) = 0$, $\theta_2(0) = \psi_0$, $\theta_2'(0) = 0$, where $\theta_0$ and $\psi_0$ are constants. For convenience let $\omega^2 = g/l$, $K = k/m$.

**(b)** Use the solution in part (a) to discuss the motion of the coupled pendulums in the special case when the initial conditions are $\theta_1(0) = \theta_0$, $\theta_1'(0) = 0$, $\theta_2(0) = \theta_0$, $\theta_2'(0) = 0$. When the initial conditions are $\theta_1(0) = \theta_0$, $\theta_1'(0) = 0$, $\theta_2(0) = -\theta_0$, $\theta_2'(0) = 0$.

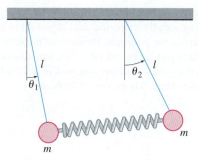

**FIGURE 7.R.12** Coupled pendulums in Problem 49

**50. Coulomb Friction Revisited** In Problem 27 in Chapter 5 in Review we examined a spring/mass system in which a mass $m$ slides over a dry horizontal surface whose coefficient of kinetic friction is a constant $\mu$. The constant retarding force $f_k = \mu mg$ of the dry surface that acts opposite to the direction of motion is called Coulomb friction after the French physicist **Charles-Augustin de Coulomb** (1736–1806). You were asked to show that the piecewise-linear differential equation for the displacement $x(t)$ of the mass is given by

$$m\frac{d^2x}{dt^2} + kx = \begin{cases} f_k, & x' < 0 \text{ (motion to left)} \\ -f_k, & x' > 0 \text{ (motion to right).} \end{cases}$$

**(a)** Suppose that the mass is released from rest from a point $x(0) = x_0 > 0$ and that there are no other external forces. Then the differential equations describing the motion of the mass $m$ are

$$x'' + \omega^2 x = F, \quad 0 < t < T/2$$

$$x'' + \omega^2 x = -F, \quad T/2 < t < T$$

$$x'' + \omega^2 x = F, \quad T < t < 3T/2,$$

and so on, where $\omega^2 = k/m$, $F = f_k/m = \mu g$, $g = 32$, and $T = 2\pi/\omega$. Show that the times $0, T/2, T, 3T/2, \dots$ correspond to $x'(t) = 0$.

**(b)** Explain why, in general, the initial displacement must satisfy $\omega^2|x_0| > F$.

**(c)** Explain why the interval $-F/\omega^2 \le x \le F/\omega^2$ is appropriately called the "dead zone" of the system.

**(d)** Use the Laplace transform and the concept of the meander function to solve for the displacement $x(t)$ for $t \ge 0$.

**(e)** Show that in the case $m = 1$, $k = 1$, $f_k = 1$, and $x_0 = 5.5$ that on the interval $[0, 2\pi)$ your solution agrees with parts (a) and (b) of Problem 28 in Chapter 5 in Review.

**(f)** Show that each successive oscillation is $2F/\omega^2$ shorter than the preceding one.

**(g)** Predict the long-term behavior of the system.

**51. Range of a Projectile—No Air Resistance (a)** A projectile, such as the canon ball shown in Figure 7.R.13, has weight $w = mg$ and initial velocity $\mathbf{v}_0$ that is tangent to its path of motion. If air resistance and all other forces except its weight are ignored, we saw in Problem 23 of Exercises 4.9 that motion of the projectile is described by the system of linear differential equations

$$m\frac{d^2x}{dt^2} = 0$$

$$m\frac{d^2y}{dt^2} = -mg.$$

Use the Laplace transform to solve this system subject to the initial conditions

$$x(0) = 0, \, x'(0) = v_0\cos\theta, \, y(0) = 0, \, y'(0) = v_0\sin\theta,$$

where $v_0 = |\mathbf{v}_0|$ is constant and $\theta$ is the constant angle of elevation shown in Figure 7.R.13 on page 330. The solutions $x(t)$ and $y(t)$ are parametric equations of the trajectory of the projectile.

**(b)** Use $x(t)$ in part (a) to eliminate the parameter $t$ in $y(t)$. Use the resulting equation for $y$ to show that the horizontal range $R$ of the projectile is given by

$$R = \frac{v_0^2}{g}\sin 2\theta.$$

**(c)** From the formula in part (b), we see that $R$ is a maximum when $\sin 2\theta = 1$ or when $\theta = \pi/4$. Show that the same range—less than the maximum—can be attained by firing the gun at either of two complementary angles $\theta$ and $\pi/2 - \theta$. The only difference is that the smaller angle results in a low trajectory whereas the larger angle gives a high trajectory.

**(d)** Suppose $g = 32$ ft/s$^2$, $\theta = 38°$, and $v_0 = 300$ ft/s. Use part (b) to find the horizontal range of the projectile. Find the time when the projectile hits the ground.

**(e)** Use the parametric equations $x(t)$ and $y(t)$ in part (a) along with the numerical data in part (d) to plot the ballistic curve of the projectile. Repeat with $\theta = 52°$ and $v_0 = 300$ ft/s. Superimpose both curves on the same coordinate system.

**FIGURE 7.R.13** Projectile in Problem 51

52. **Range of a Projectile—With Air Resistance** (a) Now suppose that air resistance is a retarding force tangent to the path but acts opposite to the motion. If we take air resistance to be proportional to the velocity of the projectile, then we saw in Problem 24 of Exercises 4.9 that motion of the projectile is describe by the system of linear differential equations

$$m\frac{d^2x}{dt^2} = -\beta\frac{dx}{dt}$$

$$m\frac{d^2y}{dt^2} = -mg - \beta\frac{dy}{dt},$$

where $\beta > 0$. Use the Laplace transform to solve this system subject to the initial conditions $x(0) = 0$, $x'(0) = v_0\cos\theta$, $y(0) = 0$, $y'(0) = v_0\sin\theta$, where $v_0 = |\mathbf{v}_0|$ and $\theta$ are constant.

(b) Suppose $m = \frac{1}{4}$ slug, $g = 32\,\text{ft/s}^2$, $\beta = 0.02$, $\theta = 38°$, and $v_0 = 300\,\text{ft/s}$. Use a CAS to find the time when the projectile hits the ground and then compute its corresponding horizontal range.

(c) Repeat part (b) using the complementary angle $\theta = 52°$ and compare the range with that found in part (b). Does the property in part (c) of Problem 51 hold?

(d) Use the parametric equations $x(t)$ and $y(t)$ in part (a) along with the numerical data in part (b) to plot the ballistic curve of the projectile. Repeat with the same numerical data in part (b) but take $\theta = 52°$. Superimpose both curves on the same coordinate system. Compare these curves with those obtained in part (e) of Problem 51.

# Systems of Linear First-Order Differential Equations

Pavel L Photo and Video/Shutterstock.com

We encountered systems of ordinary differential equations in Sections 3.3, 4.9, and 7.6 and were able to solve some systems using either systematic elimination or the Laplace transform. In this chapter we focus only on systems of linear first-order differential equations. We will see that the general theory of systems of linear DEs and the solution procedure is similar to that of linear higher-order equations considered in Chapter 4. This material is fundamental to the analysis of systems of nonlinear first-order equations in Chapter 10.

Matrix notation and properties are used extensively throughout this chapter. If you are unfamiliar with these concepts, review Appendix B or a linear algebra text.

# 8.1 Preliminary Theory—Linear Systems

**INTRODUCTION** Recall that in Section 4.9 we illustrated how to solve systems of $n$ linear differential equations in $n$ unknowns of the form

$$
\begin{aligned}
P_{11}(D)x_1 + P_{12}(D)x_2 + \cdots + P_{1n}(D)x_n &= b_1(t) \\
P_{21}(D)x_1 + P_{22}(D)x_2 + \cdots + P_{2n}(D)x_n &= b_2(t) \\
&\vdots \\
P_{n1}(D)x_1 + P_{n2}(D)x_2 + \cdots + P_{nn}(D)x_n &= b_n(t),
\end{aligned}
\tag{1}
$$

where the $P_{ij}$ were polynomials of various degrees in the differential operator $D$. In this chapter we confine our study to systems of first-order DEs that are special cases of systems that have the normal form

$$
\begin{aligned}
\frac{dx_1}{dt} &= g_1(t, x_1, x_2, \ldots, x_n) \\[4pt]
\frac{dx_2}{dt} &= g_2(t, x_1, x_2, \ldots, x_n) \\
\vdots \qquad & \qquad \vdots \\
\frac{dx_n}{dt} &= g_n(t, x_1, x_2, \ldots, x_n).
\end{aligned}
\tag{2}
$$

A system such as (2) of $n$ first-order equations is called a **first-order system.**

**LINEAR SYSTEMS** When each of the functions $g_1, g_2, \ldots, g_n$ in (2) is linear in the dependent variables $x_1, x_2, \ldots, x_n$, we get the **normal form** of a first-order system of linear equations:

$$
\begin{aligned}
\frac{dx_1}{dt} &= a_{11}(t)x_1 + a_{12}(t)x_2 + \cdots + a_{1n}(t)x_n + f_1(t) \\[4pt]
\frac{dx_2}{dt} &= a_{21}(t)x_1 + a_{22}(t)x_2 + \cdots + a_{2n}(t)x_n + f_2(t) \\
\vdots \qquad & \qquad \vdots \\
\frac{dx_n}{dt} &= a_{n1}(t)x_1 + a_{n2}(t)x_2 + \cdots + a_{nn}(t)x_n + f_n(t).
\end{aligned}
\tag{3}
$$

We refer to a system of the form given in (3) simply as a **linear system.** We assume that the coefficients $a_{ij}$ as well as the functions $f_i$ are continuous on a common interval $I$. When $f_i(t) = 0$, $i = 1, 2, \ldots, n$, the linear system (3) is said to be **homogeneous;** otherwise, it is **nonhomogeneous.**

**MATRIX FORM OF A LINEAR SYSTEM**  If $\mathbf{X}$, $\mathbf{A}(t)$, and $\mathbf{F}(t)$ denote the respective matrices

$$
\mathbf{X} = \begin{pmatrix} x_1(t) \\ x_2(t) \\ \vdots \\ x_n(t) \end{pmatrix}, \qquad
\mathbf{A}(t) = \begin{pmatrix} a_{11}(t) & a_{12}(t) & \cdots & a_{1n}(t) \\ a_{21}(t) & a_{22}(t) & \cdots & a_{2n}(t) \\ \vdots & & & \vdots \\ a_{n1}(t) & a_{n2}(t) & \cdots & a_{nn}(t) \end{pmatrix}, \qquad
\mathbf{F}(t) = \begin{pmatrix} f_1(t) \\ f_2(t) \\ \vdots \\ f_n(t) \end{pmatrix},
$$

then the system of linear first-order differential equations (3) can be written as

$$\frac{d}{dt}\begin{pmatrix} x_1 \\ x_2 \\ \vdots \\ x_n \end{pmatrix} = \begin{pmatrix} a_{11}(t) & a_{12}(t) & \cdots & a_{1n}(t) \\ a_{21}(t) & a_{22}(t) & \cdots & a_{2n}(t) \\ \vdots & & & \vdots \\ a_{n1}(t) & a_{n2}(t) & \cdots & a_{nn}(t) \end{pmatrix}\begin{pmatrix} x_1 \\ x_2 \\ \vdots \\ x_n \end{pmatrix} + \begin{pmatrix} f_1(t) \\ f_2(t) \\ \vdots \\ f_n(t) \end{pmatrix}$$

or simply
$$\mathbf{X}' = \mathbf{AX} + \mathbf{F}. \tag{4}$$

If the system is homogeneous, its matrix form is then

$$\mathbf{X}' = \mathbf{AX}. \tag{5}$$

### EXAMPLE 1   Systems Written in Matrix Notation

**(a)** If $\mathbf{X} = \begin{pmatrix} x \\ y \end{pmatrix}$, then the matrix form of the homogeneous linear system

$$\begin{aligned} \frac{dx}{dt} &= 3x + 4y \\ \frac{dy}{dt} &= 5x - 7y \end{aligned} \qquad \text{is} \quad \mathbf{X}' = \begin{pmatrix} 3 & 4 \\ 5 & -7 \end{pmatrix}\mathbf{X}.$$

**(b)** If $\mathbf{X} = \begin{pmatrix} x \\ y \\ z \end{pmatrix}$, then the matrix form of the nonhomogeneous linear system

$$\begin{aligned} \frac{dx}{dt} &= 6x + \phantom{7}y + z + \phantom{1}t \\ \frac{dy}{dt} &= 8x + 7y - z + 10t \\ \frac{dz}{dt} &= 2x + 9y - z + \phantom{1}6t \end{aligned} \qquad \text{is} \quad \mathbf{X}' = \begin{pmatrix} 6 & 1 & 1 \\ 8 & 7 & -1 \\ 2 & 9 & -1 \end{pmatrix}\mathbf{X} + \begin{pmatrix} t \\ 10t \\ 6t \end{pmatrix}.$$

### DEFINITION 8.1.1   Solution Vector

A **solution vector** on an interval $I$ is any column matrix

$$\mathbf{X} = \begin{pmatrix} x_1(t) \\ x_2(t) \\ \vdots \\ x_n(t) \end{pmatrix}$$

whose entries are differentiable functions satisfying the system (4) on the interval.

A solution vector of (4) is, of course, equivalent to $n$ scalar equations $x_1 = \phi_1(t)$, $x_2 = \phi_2(t)$, ..., $x_n = \phi_n(t)$ and can be interpreted geometrically as a set of parametric equations of a space curve. In the important case $n = 2$ the equations $x_1 = \phi_1(t)$, $x_2 = \phi_2(t)$ represent a curve in the $x_1x_2$-plane. It is common practice to call a curve in the plane a **trajectory** and to call the $x_1x_2$-plane the **phase plane.** We will come back to these concepts and illustrate them in the next section.

EXAMPLE 2   **Verification of Solutions**

Verify that on the interval $(-\infty, \infty)$

$$\mathbf{X}_1 = \begin{pmatrix} 1 \\ -1 \end{pmatrix} e^{-2t} = \begin{pmatrix} e^{-2t} \\ -e^{-2t} \end{pmatrix} \quad \text{and} \quad \mathbf{X}_2 = \begin{pmatrix} 3 \\ 5 \end{pmatrix} e^{6t} = \begin{pmatrix} 3e^{6t} \\ 5e^{6t} \end{pmatrix}$$

are solutions of
$$\mathbf{X}' = \begin{pmatrix} 1 & 3 \\ 5 & 3 \end{pmatrix} \mathbf{X}. \tag{6}$$

**SOLUTION** From $\mathbf{X}'_1 = \begin{pmatrix} -2e^{-2t} \\ 2e^{-2t} \end{pmatrix}$ and $\mathbf{X}'_2 = \begin{pmatrix} 18e^{6t} \\ 30e^{6t} \end{pmatrix}$ we see that

$$\mathbf{AX}_1 = \begin{pmatrix} 1 & 3 \\ 5 & 3 \end{pmatrix} \begin{pmatrix} e^{-2t} \\ -e^{-2t} \end{pmatrix} = \begin{pmatrix} e^{-2t} - 3e^{-2t} \\ 5e^{-2t} - 3e^{-2t} \end{pmatrix} = \begin{pmatrix} -2e^{-2t} \\ 2e^{-2t} \end{pmatrix} = \mathbf{X}'_1,$$

and
$$\mathbf{AX}_2 = \begin{pmatrix} 1 & 3 \\ 5 & 3 \end{pmatrix} \begin{pmatrix} 3e^{6t} \\ 5e^{6t} \end{pmatrix} = \begin{pmatrix} 3e^{6t} + 15e^{6t} \\ 15e^{6t} + 15e^{6t} \end{pmatrix} = \begin{pmatrix} 18e^{6t} \\ 30e^{6t} \end{pmatrix} = \mathbf{X}'_2. \quad \blacksquare$$

Much of the theory of systems of $n$ linear first-order differential equations is similar to that of linear $n$th-order differential equations.

**INITIAL-VALUE PROBLEM**   Let $t_0$ denote a number in an interval $I$ and

$$\mathbf{X}(t_0) = \begin{pmatrix} x_1(t_0) \\ x_2(t_0) \\ \vdots \\ x_n(t_0) \end{pmatrix} \quad \text{and} \quad \mathbf{X}_0 = \begin{pmatrix} \gamma_1 \\ \gamma_2 \\ \vdots \\ \gamma_n \end{pmatrix},$$

where the $\gamma_i$, $i = 1, 2, \ldots, n$ are given constants. Then the problem

$$\begin{array}{ll} \textit{Solve:} & \mathbf{X}' = \mathbf{A}(t)\mathbf{X} + \mathbf{F}(t) \\ \textit{Subject to:} & \mathbf{X}(t_0) = \mathbf{X}_0 \end{array} \tag{7}$$

is an **initial-value problem** on the interval.

---

### THEOREM 8.1.1   Existence of a Unique Solution

Let the entries of the matrices $\mathbf{A}(t)$ and $\mathbf{F}(t)$ be functions continuous on a common interval $I$ that contains the point $t_0$. Then there exists a unique solution of the initial-value problem (7) on the interval.

---

**HOMOGENEOUS SYSTEMS**   In the next several definitions and theorems we are concerned only with homogeneous systems. Without stating it, we shall always assume that the $a_{ij}$ and the $f_i$ are continuous functions of $t$ on some common interval $I$.

**SUPERPOSITION PRINCIPLE**   The following result is a **superposition principle** for solutions of linear systems.

---

### THEOREM 8.1.2   Superposition Principle

Let $\mathbf{X}_1, \mathbf{X}_2, \ldots, \mathbf{X}_k$ be a set of solution vectors of the homogeneous system (5) on an interval $I$. Then the linear combination

$$\mathbf{X} = c_1\mathbf{X}_1 + c_2\mathbf{X}_2 + \cdots + c_k\mathbf{X}_k,$$

where the $c_i$, $i = 1, 2, \ldots, k$ are arbitrary constants, is also a solution of the system on the interval.

It follows from Theorem 8.1.2 that a constant multiple of any solution vector of a homogeneous system of linear first-order differential equations is also a solution.

---

### EXAMPLE 3    Using the Superposition Principle

You should practice by verifying that the two vectors

$$\mathbf{X}_1 = \begin{pmatrix} \cos t \\ -\frac{1}{2}\cos t + \frac{1}{2}\sin t \\ -\cos t - \sin t \end{pmatrix} \quad \text{and} \quad \mathbf{X}_2 = \begin{pmatrix} 0 \\ e^t \\ 0 \end{pmatrix}$$

are solutions of the system

$$\mathbf{X}' = \begin{pmatrix} 1 & 0 & 1 \\ 1 & 1 & 0 \\ -2 & 0 & -1 \end{pmatrix} \mathbf{X}. \tag{8}$$

By the superposition principle the linear combination

$$\mathbf{X} = c_1 \mathbf{X}_1 + c_2 \mathbf{X}_2 = c_1 \begin{pmatrix} \cos t \\ -\frac{1}{2}\cos t + \frac{1}{2}\sin t \\ -\cos t - \sin t \end{pmatrix} + c_2 \begin{pmatrix} 0 \\ e^t \\ 0 \end{pmatrix}$$

is yet another solution of the system.  ∎

**LINEAR DEPENDENCE AND LINEAR INDEPENDENCE**    We are primarily interested in linearly independent solutions of the homogeneous system (5).

---

### DEFINITION 8.1.2   Linear Dependence/Independence

Let $\mathbf{X}_1, \mathbf{X}_2, \ldots, \mathbf{X}_k$ be a set of solution vectors of the homogeneous system (5) on an interval $I$. We say that the set is **linearly dependent** on the interval if there exist constants $c_1, c_2, \ldots, c_k$, not all zero, such that

$$c_1 \mathbf{X}_1 + c_2 \mathbf{X}_2 + \cdots + c_k \mathbf{X}_k = \mathbf{0}$$

for every $t$ in the interval. If the set of vectors is not linearly dependent on the interval, it is said to be **linearly independent.**

---

The case when $k = 2$ should be clear; two solution vectors $\mathbf{X}_1$ and $\mathbf{X}_2$ are linearly dependent if one is a constant multiple of the other, and conversely. For $k > 2$ a set of solution vectors is linearly dependent if we can express at least one solution vector as a linear combination of the remaining vectors.

**WRONSKIAN**    As in our earlier consideration of the theory of a single ordinary differential equation, we can introduce the concept of the **Wronskian** determinant as a test for linear independence. We state the following theorem without proof.

---

### THEOREM 8.1.3   Criterion for Linearly Independent Solutions

Let $\qquad \mathbf{X}_1 = \begin{pmatrix} x_{11} \\ x_{21} \\ \vdots \\ x_{n1} \end{pmatrix}, \quad \mathbf{X}_2 = \begin{pmatrix} x_{12} \\ x_{22} \\ \vdots \\ x_{n2} \end{pmatrix}, \quad \ldots, \quad \mathbf{X}_n = \begin{pmatrix} x_{1n} \\ x_{2n} \\ \vdots \\ x_{nn} \end{pmatrix}$

be $n$ solution vectors of the homogeneous system (5) on an interval $I$. Then the set of solution vectors is linearly independent on $I$ if and only if the **Wronskian**

$$W(\mathbf{X}_1, \mathbf{X}_2, \ldots, \mathbf{X}_n) = \begin{vmatrix} x_{11} & x_{12} & \ldots & x_{1n} \\ x_{21} & x_{22} & \ldots & x_{2n} \\ \vdots & & & \vdots \\ x_{n1} & x_{n2} & \ldots & x_{nn} \end{vmatrix} \neq 0 \tag{9}$$

for every $t$ in the interval.

It can be shown that if $\mathbf{X}_1, \mathbf{X}_2, \ldots, \mathbf{X}_n$ are solution vectors of (5), then for every $t$ in $I$ either $W(\mathbf{X}_1, \mathbf{X}_2, \ldots, \mathbf{X}_n) \neq 0$ or $W(\mathbf{X}_1, \mathbf{X}_2, \ldots, \mathbf{X}_n) = 0$. Thus if we can show that $W \neq 0$ for some $t_0$ in $I$, then $W \neq 0$ for every $t$, and hence the solutions are linearly independent on the interval.

Notice that, unlike our definition of the Wronskian in Section 4.1, here the definition of the determinant (9) does not involve differentiation.

### EXAMPLE 4 Linearly Independent Solutions

In Example 2 we saw that $\mathbf{X}_1 = \begin{pmatrix} 1 \\ -1 \end{pmatrix} e^{-2t}$ and $\mathbf{X}_2 = \begin{pmatrix} 3 \\ 5 \end{pmatrix} e^{6t}$ are solutions of system (6). Clearly, $\mathbf{X}_1$ and $\mathbf{X}_2$ are linearly independent on the interval $(-\infty, \infty)$, since neither vector is a constant multiple of the other. In addition, we have

$$W(\mathbf{X}_1, \mathbf{X}_2) = \begin{vmatrix} e^{-2t} & 3e^{6t} \\ -e^{-2t} & 5e^{6t} \end{vmatrix} = 8e^{4t} \neq 0$$

for all real values of $t$. ∎

### DEFINITION 8.1.3 Fundamental Set of Solutions

Any set $\mathbf{X}_1, \mathbf{X}_2, \ldots, \mathbf{X}_n$ of $n$ linearly independent solution vectors of the homogeneous system (5) on an interval $I$ is said to be a **fundamental set of solutions** on the interval.

### THEOREM 8.1.4 Existence of a Fundamental Set

There exists a fundamental set of solutions for the homogeneous system (5) on an interval $I$.

The next two theorems are the linear system equivalents of Theorems 4.1.5 and 4.1.6.

### THEOREM 8.1.5 General Solution—Homogeneous Systems

Let $\mathbf{X}_1, \mathbf{X}_2, \ldots, \mathbf{X}_n$ be a fundamental set of solutions of the homogeneous system (5) on an interval $I$. Then the **general solution** of the system on the interval is

$$\mathbf{X} = c_1 \mathbf{X}_1 + c_2 \mathbf{X}_2 + \cdots + c_n \mathbf{X}_n,$$

where the $c_i, i = 1, 2, \ldots, n$ are arbitrary constants.

**EXAMPLE 5** General Solution of System (6)

From Example 2 we know that $\mathbf{X}_1 = \begin{pmatrix} 1 \\ -1 \end{pmatrix} e^{-2t}$ and $\mathbf{X}_2 = \begin{pmatrix} 3 \\ 5 \end{pmatrix} e^{6t}$ are linearly independent solutions of (6) on $(-\infty, \infty)$. Hence $\mathbf{X}_1$ and $\mathbf{X}_2$ form a fundamental set of solutions on the interval. The general solution of the homogeneous system on the interval is then

$$\mathbf{X} = c_1 \mathbf{X}_1 + c_2 \mathbf{X}_2 = c_1 \begin{pmatrix} 1 \\ -1 \end{pmatrix} e^{-2t} + c_2 \begin{pmatrix} 3 \\ 5 \end{pmatrix} e^{6t}. \qquad (10) \quad \blacksquare$$

**EXAMPLE 6** General Solution of System (8)

The vectors

$$\mathbf{X}_1 = \begin{pmatrix} \cos t \\ -\frac{1}{2}\cos t + \frac{1}{2}\sin t \\ -\cos t - \sin t \end{pmatrix}, \qquad \mathbf{X}_2 = \begin{pmatrix} 0 \\ 1 \\ 0 \end{pmatrix} e^t, \qquad \mathbf{X}_3 = \begin{pmatrix} \sin t \\ -\frac{1}{2}\sin t - \frac{1}{2}\cos t \\ -\sin t + \cos t \end{pmatrix}$$

are solutions of the system (8) in Example 3 (see Problem 16 in Exercises 8.1). Now

$$W(\mathbf{X}_1, \mathbf{X}_2, \mathbf{X}_3) = \begin{vmatrix} \cos t & 0 & \sin t \\ -\frac{1}{2}\cos t + \frac{1}{2}\sin t & e^t & -\frac{1}{2}\sin t - \frac{1}{2}\cos t \\ -\cos t - \sin t & 0 & -\sin t + \cos t \end{vmatrix} = e^t \neq 0$$

for all real values of $t$. We conclude that $\mathbf{X}_1$, $\mathbf{X}_2$, and $\mathbf{X}_3$ form a fundamental set of solutions on $(-\infty, \infty)$. Thus the general solution of the homogeneous system on the interval is the linear combination $\mathbf{X} = c_1 \mathbf{X}_1 + c_2 \mathbf{X}_2 + c_3 \mathbf{X}_3$; that is,

$$\mathbf{X} = c_1 \begin{pmatrix} \cos t \\ -\frac{1}{2}\cos t + \frac{1}{2}\sin t \\ -\cos t - \sin t \end{pmatrix} + c_2 \begin{pmatrix} 0 \\ 1 \\ 0 \end{pmatrix} e^t + c_3 \begin{pmatrix} \sin t \\ -\frac{1}{2}\sin t - \frac{1}{2}\cos t \\ -\sin t + \cos t \end{pmatrix}. \quad \blacksquare$$

**NONHOMOGENEOUS SYSTEMS** For nonhomogeneous systems a **particular solution** $\mathbf{X}_p$ on an interval $I$ is any vector, free of arbitrary parameters, whose entries are functions that satisfy the system (4).

**THEOREM 8.1.6** General Solution — Nonhomogeneous Systems

Let $\mathbf{X}_p$ be a given solution of the nonhomogeneous system (4) on an interval $I$ and let

$$\mathbf{X}_c = c_1 \mathbf{X}_1 + c_2 \mathbf{X}_2 + \cdots + c_n \mathbf{X}_n$$

denote the general solution on the same interval of the associated homogeneous system (5). Then the **general solution** of the nonhomogeneous system on the interval is

$$\mathbf{X} = \mathbf{X}_c + \mathbf{X}_p.$$

The general solution $\mathbf{X}_c$ of the associated homogeneous system (5) is called the **complementary function** of the nonhomogeneous system (4).

**EXAMPLE 7**  General Solution—Nonhomogeneous System

The vector $\mathbf{X}_p = \begin{pmatrix} 3t - 4 \\ -5t + 6 \end{pmatrix}$ is a particular solution of the nonhomogeneous system

$$\mathbf{X}' = \begin{pmatrix} 1 & 3 \\ 5 & 3 \end{pmatrix} \mathbf{X} + \begin{pmatrix} 12t - 11 \\ -3 \end{pmatrix} \qquad (11)$$

on the interval $(-\infty, \infty)$. (Verify this.) The complementary function of (11) on the same interval, or the general solution of $\mathbf{X}' = \begin{pmatrix} 1 & 3 \\ 5 & 3 \end{pmatrix} \mathbf{X}$, was seen in (10) of Example 5 to be $\mathbf{X}_c = c_1 \begin{pmatrix} 1 \\ -1 \end{pmatrix} e^{-2t} + c_2 \begin{pmatrix} 3 \\ 5 \end{pmatrix} e^{6t}$. Hence by Theorem 8.1.6

$$\mathbf{X} = \mathbf{X}_c + \mathbf{X}_p = c_1 \begin{pmatrix} 1 \\ -1 \end{pmatrix} e^{-2t} + c_2 \begin{pmatrix} 3 \\ 5 \end{pmatrix} e^{6t} + \begin{pmatrix} 3t - 4 \\ -5t + 6 \end{pmatrix}$$

is the general solution of the nonhomogeneous system (11) on $(-\infty, \infty)$.  ∎

---

## EXERCISES 8.1

*Answers to selected odd-numbered problems begin on page ANS-13.*

In Problems 1–6 write the given linear system in matrix form.

**1.** $\dfrac{dx}{dt} = 3x - 5y$

$\dfrac{dy}{dt} = 4x + 8y$

**2.** $\dfrac{dx}{dt} = 4x - 7y$

$\dfrac{dy}{dt} = 5x$

**3.** $\dfrac{dx}{dt} = -3x + 4y - 9z$

$\dfrac{dy}{dt} = 6x - y$

$\dfrac{dz}{dt} = 10x + 4y + 3z$

**4.** $\dfrac{dx}{dt} = x - y$

$\dfrac{dy}{dt} = x + 2z$

$\dfrac{dz}{dt} = -x + z$

**5.** $\dfrac{dx}{dt} = x - y + z + t - 1$

$\dfrac{dy}{dt} = 2x + y - z - 3t^2$

$\dfrac{dz}{dt} = x + y + z + t^2 - t + 2$

**6.** $\dfrac{dx}{dt} = -3x + 4y + e^{-t} \sin 2t$

$\dfrac{dy}{dt} = 5x + 9z + 4e^{-t} \cos 2t$

$\dfrac{dz}{dt} = y + 6z - e^{-t}$

In Problems 7–10 write the given linear system without the use of matrices.

**7.** $\mathbf{X}' = \begin{pmatrix} 4 & 2 \\ -1 & 3 \end{pmatrix} \mathbf{X} + \begin{pmatrix} 1 \\ -1 \end{pmatrix} e^t$

**8.** $\mathbf{X}' = \begin{pmatrix} 7 & 5 & -9 \\ 4 & 1 & 1 \\ 0 & -2 & 3 \end{pmatrix} \mathbf{X} + \begin{pmatrix} 0 \\ 2 \\ 1 \end{pmatrix} e^{5t} - \begin{pmatrix} 8 \\ 0 \\ 3 \end{pmatrix} e^{-2t}$

**9.** $\dfrac{d}{dt} \begin{pmatrix} x \\ y \\ z \end{pmatrix} = \begin{pmatrix} 1 & -1 & 2 \\ 3 & -4 & 1 \\ -2 & 5 & 6 \end{pmatrix} \begin{pmatrix} x \\ y \\ z \end{pmatrix} + \begin{pmatrix} 1 \\ 2 \\ 2 \end{pmatrix} e^{-t} - \begin{pmatrix} 3 \\ -1 \\ 1 \end{pmatrix} t$

**10.** $\dfrac{d}{dt} \begin{pmatrix} x \\ y \end{pmatrix} = \begin{pmatrix} 3 & -7 \\ 1 & 1 \end{pmatrix} \begin{pmatrix} x \\ y \end{pmatrix} + \begin{pmatrix} 4 \\ 8 \end{pmatrix} \sin t + \begin{pmatrix} t - 4 \\ 2t + 1 \end{pmatrix} e^{4t}$

In Problems 11–16 verify that the vector $\mathbf{X}$ is a solution of the given homogeneous linear system.

**11.** $\dfrac{dx}{dt} = 3x - 4y$

$\dfrac{dy}{dt} = 4x - 7y; \quad \mathbf{X} = \begin{pmatrix} 1 \\ 2 \end{pmatrix} e^{-5t}$

**12.** $\dfrac{dx}{dt} = -2x + 5y$

$\dfrac{dy}{dt} = -2x + 4y; \quad \mathbf{X} = \begin{pmatrix} 5 \cos t \\ 3 \cos t - \sin t \end{pmatrix} e^t$

**13.** $\mathbf{X}' = \begin{pmatrix} -1 & \frac{1}{4} \\ 1 & -1 \end{pmatrix} \mathbf{X}; \quad \mathbf{X} = \begin{pmatrix} -1 \\ 2 \end{pmatrix} e^{-3t/2}$

**14.** $\mathbf{X}' = \begin{pmatrix} 2 & 1 \\ -1 & 0 \end{pmatrix} \mathbf{X}; \quad \mathbf{X} = \begin{pmatrix} 1 \\ 3 \end{pmatrix} e^t + \begin{pmatrix} 4 \\ -4 \end{pmatrix} t e^t$

**15.** $\mathbf{X}' = \begin{pmatrix} 1 & 2 & 1 \\ 6 & -1 & 0 \\ -1 & -2 & -1 \end{pmatrix} \mathbf{X}; \quad \mathbf{X} = \begin{pmatrix} 1 \\ 6 \\ -13 \end{pmatrix}$

**16.** $\mathbf{X}' = \begin{pmatrix} 1 & 0 & 1 \\ 1 & 1 & 0 \\ -2 & 0 & -1 \end{pmatrix} \mathbf{X}; \quad \mathbf{X} = \begin{pmatrix} \sin t \\ -\frac{1}{2} \sin t - \frac{1}{2} \cos t \\ -\sin t + \cos t \end{pmatrix}$

In Problems 17–20 the given vectors are solutions of a system $X' = AX$. Determine whether the vectors form a fundamental set on the interval $(-\infty, \infty)$.

**17.** $X_1 = \begin{pmatrix} 1 \\ 1 \end{pmatrix} e^{-2t}, \quad X_2 = \begin{pmatrix} 1 \\ -1 \end{pmatrix} e^{-6t}$

**18.** $X_1 = \begin{pmatrix} 1 \\ -1 \end{pmatrix} e^t, \quad X_2 = \begin{pmatrix} 2 \\ 6 \end{pmatrix} e^t + \begin{pmatrix} 8 \\ -8 \end{pmatrix} te^t$

**19.** $X_1 = \begin{pmatrix} 1 \\ -2 \\ 4 \end{pmatrix} + t \begin{pmatrix} 1 \\ 2 \\ 2 \end{pmatrix}, \quad X_2 = \begin{pmatrix} 1 \\ -2 \\ 4 \end{pmatrix},$

$X_3 = \begin{pmatrix} 3 \\ -6 \\ 12 \end{pmatrix} + t \begin{pmatrix} 2 \\ 4 \\ 4 \end{pmatrix}$

**20.** $X_1 = \begin{pmatrix} 1 \\ 6 \\ -13 \end{pmatrix}, \quad X_2 = \begin{pmatrix} 1 \\ -2 \\ -1 \end{pmatrix} e^{-4t}, \quad X_3 = \begin{pmatrix} 2 \\ 3 \\ -2 \end{pmatrix} e^{3t}$

In Problems 21–24 verify that the vector $X_p$ is a particular solution of the given nonhomogeneous linear system.

**21.** $\dfrac{dx}{dt} = x + 4y + 2t - 7$

$\dfrac{dy}{dt} = 3x + 2y - 4t - 18; \quad X_p = \begin{pmatrix} 2 \\ -1 \end{pmatrix} t + \begin{pmatrix} 5 \\ 1 \end{pmatrix}$

**22.** $X' = \begin{pmatrix} 2 & 1 \\ 1 & -1 \end{pmatrix} X + \begin{pmatrix} -5 \\ 2 \end{pmatrix}; \quad X_p = \begin{pmatrix} 1 \\ 3 \end{pmatrix}$

**23.** $X' = \begin{pmatrix} 2 & 1 \\ 3 & 4 \end{pmatrix} X - \begin{pmatrix} 1 \\ 7 \end{pmatrix} e^t; \quad X_p = \begin{pmatrix} 1 \\ 1 \end{pmatrix} e^t + \begin{pmatrix} 1 \\ -1 \end{pmatrix} te^t$

**24.** $X' = \begin{pmatrix} 1 & 2 & 3 \\ -4 & 2 & 0 \\ -6 & 1 & 0 \end{pmatrix} X + \begin{pmatrix} -1 \\ 4 \\ 3 \end{pmatrix} \sin 3t; \quad X_p = \begin{pmatrix} \sin 3t \\ 0 \\ \cos 3t \end{pmatrix}$

**25.** Prove that the general solution of the homogeneous linear system

$$X' = \begin{pmatrix} 0 & 6 & 0 \\ 1 & 0 & 1 \\ 1 & 1 & 0 \end{pmatrix} X$$

on the interval $(-\infty, \infty)$ is

$$X = c_1 \begin{pmatrix} 6 \\ -1 \\ -5 \end{pmatrix} e^{-t} + c_2 \begin{pmatrix} -3 \\ 1 \\ 1 \end{pmatrix} e^{-2t} + c_3 \begin{pmatrix} 2 \\ 1 \\ 1 \end{pmatrix} e^{3t}.$$

**26.** Prove that the general solution of the nonhomogeneous linear system

$$X' = \begin{pmatrix} -1 & -1 \\ -1 & 1 \end{pmatrix} X + \begin{pmatrix} 1 \\ 1 \end{pmatrix} t^2 + \begin{pmatrix} 4 \\ -6 \end{pmatrix} t + \begin{pmatrix} -1 \\ 5 \end{pmatrix}$$

on the interval $(-\infty, \infty)$ is

$$X = c_1 \begin{pmatrix} 1 \\ -1 - \sqrt{2} \end{pmatrix} e^{\sqrt{2}t} + c_2 \begin{pmatrix} 1 \\ -1 + \sqrt{2} \end{pmatrix} e^{-\sqrt{2}t}$$

$$+ \begin{pmatrix} 1 \\ 0 \end{pmatrix} t^2 + \begin{pmatrix} -2 \\ 4 \end{pmatrix} t + \begin{pmatrix} 1 \\ 0 \end{pmatrix}.$$

---

### 8.2 Homogeneous Linear Systems

**INTRODUCTION** We saw in Example 5 of Section 8.1 that the general solution of the homogeneous system $X' = \begin{pmatrix} 1 & 3 \\ 5 & 3 \end{pmatrix} X$ is

$$X = c_1 X_1 + c_2 X_2 = c_1 \begin{pmatrix} 1 \\ -1 \end{pmatrix} e^{-2t} + c_2 \begin{pmatrix} 3 \\ 5 \end{pmatrix} e^{6t}.$$

Because the solution vectors $X_1$ and $X_2$ have the form

$$X_i = \begin{pmatrix} k_1 \\ k_2 \end{pmatrix} e^{\lambda_i t}, \qquad i = 1, 2,$$

where $k_1$, $k_2$, $\lambda_1$, and $\lambda_2$ are constants, we are prompted to ask whether we can always find a solution of the form

$$X = \begin{pmatrix} k_1 \\ k_2 \\ \vdots \\ k_n \end{pmatrix} e^{\lambda t} = K e^{\lambda t} \tag{1}$$

for the general homogeneous linear first-order system

$$X' = AX, \tag{2}$$

where $A$ is an $n \times n$ matrix of constants.

**EIGENVALUES AND EIGENVECTORS**    If (1) is to be a solution vector of the homogeneous linear system (2), then $\mathbf{X}' = \mathbf{K}\lambda e^{\lambda t}$, so the system becomes $\mathbf{K}\lambda e^{\lambda t} = \mathbf{AK}e^{\lambda t}$. After dividing out $e^{\lambda t}$ and rearranging, we obtain $\mathbf{AK} = \lambda\mathbf{K}$ or $\mathbf{AK} - \lambda\mathbf{K} = \mathbf{0}$. Since $\mathbf{K} = \mathbf{IK}$, the last equation is the same as

$$(\mathbf{A} - \lambda\mathbf{I})\mathbf{K} = \mathbf{0}. \tag{3}$$

The matrix equation (3) is equivalent to the simultaneous algebraic equations

$$
\begin{aligned}
(a_{11} - \lambda)k_1 + \quad & a_{12}k_2 + \cdots + \quad && a_{1n}k_n = 0 \\
a_{21}k_1 + (a_{22} &- \lambda)k_2 + \cdots + \quad && a_{2n}k_n = 0 \\
& \quad\vdots && \quad\vdots \\
a_{n1}k_1 + \quad & a_{n2}k_2 + \cdots + (a_{nn} &&- \lambda)k_n = 0.
\end{aligned}
$$

Thus to find a nontrivial solution $\mathbf{X}$ of (2), we must first find a nontrivial solution of the foregoing system; in other words, we must find a nontrivial vector $\mathbf{K}$ that satisfies (3). But for (3) to have solutions other than the obvious solution $k_1 = k_2 = \cdots = k_n = 0$, we must have

$$\det(\mathbf{A} - \lambda\mathbf{I}) = 0.$$

This polynomial equation in $\lambda$ is called the **characteristic equation** of the matrix $\mathbf{A}$; its solutions are the **eigenvalues** of $\mathbf{A}$. A solution $\mathbf{K} \neq \mathbf{0}$ of (3) corresponding to an eigenvalue $\lambda$ is called an **eigenvector** of $\mathbf{A}$. A solution of the homogeneous system (2) is then $\mathbf{X} = \mathbf{K}e^{\lambda t}$.

In the discussion that follows we examine three cases: real and distinct eigenvalues (that is, no eigenvalues are equal), repeated eigenvalues, and, finally, complex eigenvalues.

## 8.2.1 DISTINCT REAL EIGENVALUES

When the $n \times n$ matrix $\mathbf{A}$ possesses $n$ distinct real eigenvalues $\lambda_1, \lambda_2, \ldots, \lambda_n$, then a set of $n$ linearly independent eigenvectors $\mathbf{K}_1, \mathbf{K}_2, \ldots, \mathbf{K}_n$ can always be found, and

$$\mathbf{X}_1 = \mathbf{K}_1 e^{\lambda_1 t}, \qquad \mathbf{X}_2 = \mathbf{K}_2 e^{\lambda_2 t}, \qquad \ldots, \qquad \mathbf{X}_n = \mathbf{K}_n e^{\lambda_n t}$$

is a fundamental set of solutions of (2) on the interval $(-\infty, \infty)$.

---

**THEOREM 8.2.1**  General Solution—Homogeneous Systems

Let $\lambda_1, \lambda_2, \ldots, \lambda_n$ be $n$ distinct real eigenvalues of the coefficient matrix $\mathbf{A}$ of the homogeneous system (2) and let $\mathbf{K}_1, \mathbf{K}_2, \ldots, \mathbf{K}_n$ be the corresponding eigenvectors. Then the **general solution** of (2) on the interval $(-\infty, \infty)$ is given by

$$\mathbf{X} = c_1\mathbf{K}_1 e^{\lambda_1 t} + c_2\mathbf{K}_2 e^{\lambda_2 t} + \cdots + c_n\mathbf{K}_n e^{\lambda_n t}.$$

---

**EXAMPLE 1**  Distinct Eigenvalues

Solve

$$
\begin{aligned}
\frac{dx}{dt} &= 2x + 3y \\[2mm]
\frac{dy}{dt} &= 2x + \ y.
\end{aligned}
\tag{4}
$$

**SOLUTION**  We first find the eigenvalues and eigenvectors of the matrix of coefficients.

From the characteristic equation

$$\det(\mathbf{A} - \lambda\mathbf{I}) = \begin{vmatrix} 2 - \lambda & 3 \\ 2 & 1 - \lambda \end{vmatrix} = \lambda^2 - 3\lambda - 4 = (\lambda + 1)(\lambda - 4) = 0$$

we see that the eigenvalues are $\lambda_1 = -1$ and $\lambda_2 = 4$.

Now for $\lambda_1 = -1$, (3) is equivalent to

$$3k_1 + 3k_2 = 0$$
$$2k_1 + 2k_2 = 0.$$

Thus $k_1 = -k_2$. When $k_2 = -1$, the related eigenvector is

$$\mathbf{K}_1 = \begin{pmatrix} 1 \\ -1 \end{pmatrix}.$$

For $\lambda_2 = 4$ we have

$$-2k_1 + 3k_2 = 0$$
$$2k_1 - 3k_2 = 0$$

so $k_1 = \frac{3}{2}k_2$; therefore with $k_2 = 2$ the corresponding eigenvector is

$$\mathbf{K}_2 = \begin{pmatrix} 3 \\ 2 \end{pmatrix}.$$

Since the matrix of coefficients $\mathbf{A}$ is a $2 \times 2$ matrix and since we have found two linearly independent solutions of (4),

$$\mathbf{X}_1 = \begin{pmatrix} 1 \\ -1 \end{pmatrix} e^{-t} \quad \text{and} \quad \mathbf{X}_2 = \begin{pmatrix} 3 \\ 2 \end{pmatrix} e^{4t},$$

we conclude that the general solution of the system is

$$\mathbf{X} = c_1\mathbf{X}_1 + c_2\mathbf{X}_2 = c_1\begin{pmatrix} 1 \\ -1 \end{pmatrix} e^{-t} + c_2\begin{pmatrix} 3 \\ 2 \end{pmatrix} e^{4t}. \tag{5} \blacksquare$$

**PHASE PORTRAIT** You should keep firmly in mind that writing a solution of a system of linear first-order differential equations in terms of matrices is simply an alternative to the method that we employed in Section 4.9, that is, listing the individual functions and the relationship between the constants. If we add the vectors on the right-hand side of (5) and then equate the entries with the corresponding entries in the vector on the left-hand side, we obtain the more familiar statement

$$x = c_1 e^{-t} + 3c_2 e^{4t}, \qquad y = -c_1 e^{-t} + 2c_2 e^{4t}.$$

As was pointed out in Section 8.1, we can interpret these equations as parametric equations of curves in the $xy$-plane or **phase plane**. Each curve, corresponding to specific choices for $c_1$ and $c_2$, is called a **trajectory**. For the choice of constants $c_1 = c_2 = 1$ in the solution (5) we see in Figure 8.2.1 the graph of $x(t)$ in the

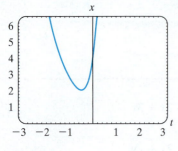

**(a)** graph of $x = e^{-t} + 3e^{4t}$

**(b)** graph of $y = -e^{-t} + 2e^{4t}$

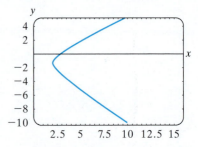

**(c)** trajectory defined by $x = e^{-t} + 3e^{4t}$, $y = -e^{-t} + 2e^{4t}$ in the phase plane

**FIGURE 8.2.1** A solution from (5) yields three different curves in three different planes

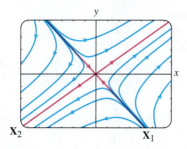

**FIGURE 8.2.2** A phase portrait of system (4)

*tx*-plane, the graph of $y(t)$ in the *ty*-plane, and the trajectory consisting of the points $(x(t), y(t))$ in the phase plane. A collection of representative trajectories in the phase plane, as shown in Figure 8.2.2, is said to be a **phase portrait** of the given linear system. What appears to be *two* red lines in Figure 8.2.2 are actually *four* red half-lines defined parametrically in the first, second, third, and fourth quadrants by the solutions $\mathbf{X}_2$, $-\mathbf{X}_1$, $-\mathbf{X}_2$, and $\mathbf{X}_1$, respectively. For example, the Cartesian equations $y = \frac{2}{3}x$, $x > 0$, and $y = -x$, $x > 0$, of the half-lines in the first and fourth quadrants were obtained by eliminating the parameter $t$ in the solutions $x = 3e^{4t}$, $y = 2e^{4t}$, and $x = e^{-t}$, $y = -e^{-t}$, respectively. Moreover, each eigenvector can be visualized as a two-dimensional vector lying along one of these half-lines. The eigenvector $\mathbf{K}_2 = \begin{pmatrix} 3 \\ 2 \end{pmatrix}$ lies along $y = \frac{2}{3}x$ in the first quadrant, and $\mathbf{K}_1 = \begin{pmatrix} 1 \\ -1 \end{pmatrix}$ lies along $y = -x$ in the fourth quadrant. Each vector starts at the origin; $\mathbf{K}_2$ terminates at the point $(2, 3)$, and $\mathbf{K}_1$ terminates at $(1, -1)$.

The origin is not only a constant solution $x = 0$, $y = 0$ of every $2 \times 2$ homogeneous linear system $\mathbf{X}' = \mathbf{AX}$, but also an important point in the qualitative study of such systems. If we think in physical terms, the arrowheads on each trajectory in Figure 8.2.2 indicate the direction that a particle with coordinates $(x(t), y(t))$ on that trajectory at time $t$ moves as time increases. Observe that the arrowheads, with the exception of only those on the half-lines in the second and fourth quadrants, indicate that a particle moves away from the origin as time $t$ increases. If we imagine time ranging from $-\infty$ to $\infty$, then inspection of the solution $x = c_1 e^{-t} + 3c_2 e^{4t}$, $y = -c_1 e^{-t} + 2c_2 e^{4t}$, $c_1 \neq 0$, $c_2 \neq 0$ shows that a trajectory, or moving particle, "starts" asymptotic to one of the half-lines defined by $\mathbf{X}_1$ or $-\mathbf{X}_1$ (since $e^{4t}$ is negligible for $t \to -\infty$) and "finishes" asymptotic to one of the half-lines defined by $\mathbf{X}_2$ and $-\mathbf{X}_2$ (since $e^{-t}$ is negligible for $t \to \infty$).

We note in passing that Figure 8.2.2 represents a phase portrait that is typical of *all* $2 \times 2$ homogeneous linear systems $\mathbf{X}' = \mathbf{AX}$ with real eigenvalues of opposite signs. See Problem 19 in Exercises 8.2. Moreover, phase portraits in the two cases when distinct real eigenvalues have the same algebraic sign are typical of all such $2 \times 2$ linear systems; the only difference is that the arrowheads indicate that a particle moves away from the origin on any trajectory as $t \to \infty$ when both $\lambda_1$ and $\lambda_2$ are positive and moves toward the origin on any trajectory when both $\lambda_1$ and $\lambda_2$ are negative. Consequently, we call the origin a **repeller** in the case $\lambda_1 > 0$, $\lambda_2 > 0$ and an **attractor** in the case $\lambda_1 < 0$, $\lambda_2 < 0$. See Problem 20 in Exercises 8.2. The origin in Figure 8.2.2 is neither a repeller nor an attractor. Investigation of the remaining case when $\lambda = 0$ is an eigenvalue of a $2 \times 2$ homogeneous linear system is left as an exercise. See Problem 53 in Exercises 8.2.

---

**EXAMPLE 2** **Distinct Eigenvalues**

Solve

$$\frac{dx}{dt} = -4x + y + z$$

$$\frac{dx}{dt} = x + 5y - z \tag{6}$$

$$\frac{dz}{dt} = y - 3z.$$

**SOLUTION** Using the cofactors of the third row, we find

$$\det(\mathbf{A} - \lambda\mathbf{I}) = \begin{vmatrix} -4 - \lambda & 1 & 1 \\ 1 & 5 - \lambda & -1 \\ 0 & 1 & -3 - \lambda \end{vmatrix} = -(\lambda + 3)(\lambda + 4)(\lambda - 5) = 0,$$

and so the eigenvalues are $\lambda_1 = -3$, $\lambda_2 = -4$, and $\lambda_3 = 5$.

For $\lambda_1 = -3$ Gauss-Jordan elimination gives

$$(\mathbf{A} + 3\mathbf{I}|\mathbf{0}) = \begin{pmatrix} -1 & 1 & 1 & | & 0 \\ 1 & 8 & -1 & | & 0 \\ 0 & 1 & 0 & | & 0 \end{pmatrix} \xrightarrow[\text{operations}]{\text{row}} \begin{pmatrix} 1 & 0 & -1 & | & 0 \\ 0 & 1 & 0 & | & 0 \\ 0 & 0 & 0 & | & 0 \end{pmatrix}.$$

Therefore $k_1 = k_3$ and $k_2 = 0$. The choice $k_3 = 1$ gives an eigenvector and corresponding solution vector

$$\mathbf{K}_1 = \begin{pmatrix} 1 \\ 0 \\ 1 \end{pmatrix}, \qquad \mathbf{X}_1 = \begin{pmatrix} 1 \\ 0 \\ 1 \end{pmatrix} e^{-3t}. \tag{7}$$

Similarly, for $\lambda_2 = -4$

$$(\mathbf{A} + 4\mathbf{I}|\mathbf{0}) = \begin{pmatrix} 0 & 1 & 1 & | & 0 \\ 1 & 9 & -1 & | & 0 \\ 0 & 1 & 1 & | & 0 \end{pmatrix} \xrightarrow[\text{operations}]{\text{row}} \begin{pmatrix} 1 & 0 & -10 & | & 0 \\ 0 & 1 & 1 & | & 0 \\ 0 & 0 & 0 & | & 0 \end{pmatrix}$$

implies that $k_1 = 10k_3$ and $k_2 = -k_3$. Choosing $k_3 = 1$, we get a second eigenvector and solution vector

$$\mathbf{K}_2 = \begin{pmatrix} 10 \\ -1 \\ 1 \end{pmatrix}, \qquad \mathbf{X}_2 = \begin{pmatrix} 10 \\ -1 \\ 1 \end{pmatrix} e^{-4t}. \tag{8}$$

Finally, when $\lambda_3 = 5$, the augmented matrices

$$(\mathbf{A} - 5\mathbf{I}|\mathbf{0}) = \begin{pmatrix} -9 & 1 & 1 & | & 0 \\ 1 & 0 & -1 & | & 0 \\ 0 & 1 & -8 & | & 0 \end{pmatrix} \xrightarrow[\text{operations}]{\text{row}} \begin{pmatrix} 1 & 0 & -1 & | & 0 \\ 0 & 1 & -8 & | & 0 \\ 0 & 0 & 0 & | & 0 \end{pmatrix}$$

yield

$$\mathbf{K}_3 = \begin{pmatrix} 1 \\ 8 \\ 1 \end{pmatrix}, \qquad \mathbf{X}_3 = \begin{pmatrix} 1 \\ 8 \\ 1 \end{pmatrix} e^{5t}. \tag{9}$$

The general solution of (6) is a linear combination of the solution vectors in (7), (8), and (9):

$$\mathbf{X} = c_1 \begin{pmatrix} 1 \\ 0 \\ 1 \end{pmatrix} e^{-3t} + c_2 \begin{pmatrix} 10 \\ -1 \\ 1 \end{pmatrix} e^{-4t} + c_3 \begin{pmatrix} 1 \\ 8 \\ 1 \end{pmatrix} e^{5t}. \qquad ■$$

**USE OF COMPUTERS** Software packages such as MATLAB, *Mathematica*, and *Maple* can be real time savers in finding eigenvalues and eigenvectors of a matrix **A**.

## 8.2.2 REPEATED EIGENVALUES

Of course, not all of the $n$ eigenvalues $\lambda_1, \lambda_2, \ldots, \lambda_n$ of an $n \times n$ matrix **A** need be distinct; that is, some of the eigenvalues may be repeated. For example, the characteristic equation of the coefficient matrix in the system

$$\mathbf{X}' = \begin{pmatrix} 3 & -18 \\ 2 & -9 \end{pmatrix} \mathbf{X} \tag{10}$$

is readily shown to be $(\lambda + 3)^2 = 0$, and therefore $\lambda_1 = \lambda_2 = -3$ is a root of *multiplicity two*. For this value we find the single eigenvector

$$\mathbf{K}_1 = \begin{pmatrix} 3 \\ 1 \end{pmatrix}, \quad \text{so} \quad \mathbf{X}_1 = \begin{pmatrix} 3 \\ 1 \end{pmatrix} e^{-3t} \tag{11}$$

is one solution of (10). But since we are obviously interested in forming the general solution of the system, we need to pursue the question of finding a second solution.

In general, if $m$ is a positive integer and $(\lambda - \lambda_1)^m$ is a factor of the characteristic equation while $(\lambda - \lambda_1)^{m+1}$ is not a factor, then $\lambda_1$ is said to be an **eigenvalue of multiplicity $m$.** The next three examples illustrate the following cases:

**(*i*)** For some $n \times n$ matrices $\mathbf{A}$ it may be possible to find $m$ linearly independent eigenvectors $\mathbf{K}_1, \mathbf{K}_2, \ldots, \mathbf{K}_m$ corresponding to an eigenvalue $\lambda_1$ of multiplicity $m \leq n$. In this case the general solution of the system contains the linear combination

$$c_1 \mathbf{K}_1 e^{\lambda_1 t} + c_2 \mathbf{K}_2 e^{\lambda_1 t} + \cdots + c_m \mathbf{K}_m e^{\lambda_1 t}.$$

**(*ii*)** If there is only one eigenvector corresponding to the eigenvalue $\lambda_1$ of multiplicity $m$, then $m$ linearly independent solutions of the form

$$\mathbf{X}_1 = \mathbf{K}_{11} e^{\lambda_1 t}$$
$$\mathbf{X}_2 = \mathbf{K}_{21} t e^{\lambda_1 t} + \mathbf{K}_{22} e^{\lambda_1 t}$$
$$\vdots$$
$$\mathbf{X}_m = \mathbf{K}_{m1} \frac{t^{m-1}}{(m-1)!} e^{\lambda_1 t} + \mathbf{K}_{m2} \frac{t^{m-2}}{(m-2)!} e^{\lambda_1 t} + \cdots + \mathbf{K}_{mm} e^{\lambda_1 t},$$

where $\mathbf{K}_{ij}$ are column vectors, can always be found.

**EIGENVALUE OF MULTIPLICITY TWO**  We begin by considering eigenvalues of multiplicity two. In the first example we illustrate a matrix for which we can find two distinct eigenvectors corresponding to a double eigenvalue.

**EXAMPLE 3**  Repeated Eigenvalues

Solve $\mathbf{X}' = \begin{pmatrix} 1 & -2 & 2 \\ -2 & 1 & -2 \\ 2 & -2 & 1 \end{pmatrix} \mathbf{X}.$

**SOLUTION**  Expanding the determinant in the characteristic equation

$$\det(\mathbf{A} - \lambda \mathbf{I}) = \begin{vmatrix} 1 - \lambda & -2 & 2 \\ -2 & 1 - \lambda & -2 \\ 2 & -2 & 1 - \lambda \end{vmatrix} = 0$$

yields $-(\lambda + 1)^2 (\lambda - 5) = 0$. We see that $\lambda_1 = \lambda_2 = -1$ and $\lambda_3 = 5$.

For $\lambda_1 = -1$ Gauss-Jordan elimination immediately gives

$$(\mathbf{A} + \mathbf{I} \,|\, \mathbf{0}) = \begin{pmatrix} 2 & -2 & 2 & | & 0 \\ -2 & 2 & -2 & | & 0 \\ 2 & -2 & 2 & | & 0 \end{pmatrix} \xrightarrow[\text{operations}]{\text{row}} \begin{pmatrix} 1 & -1 & 1 & | & 0 \\ 0 & 0 & 0 & | & 0 \\ 0 & 0 & 0 & | & 0 \end{pmatrix}.$$

The first row of the last matrix means $k_1 - k_2 + k_3 = 0$ or $k_1 = k_2 - k_3$. The choices $k_2 = 1$, $k_3 = 0$ and $k_2 = 1$, $k_3 = 1$ yield, in turn, $k_1 = 1$ and $k_1 = 0$. Thus two eigenvectors corresponding to $\lambda_1 = -1$ are

$$\mathbf{K}_1 = \begin{pmatrix} 1 \\ 1 \\ 0 \end{pmatrix} \quad \text{and} \quad \mathbf{K}_2 = \begin{pmatrix} 0 \\ 1 \\ 1 \end{pmatrix}.$$

Since neither eigenvector is a constant multiple of the other, we have found two linearly independent solutions,

$$\mathbf{X}_1 = \begin{pmatrix} 1 \\ 1 \\ 0 \end{pmatrix} e^{-t} \quad \text{and} \quad \mathbf{X}_2 = \begin{pmatrix} 0 \\ 1 \\ 1 \end{pmatrix} e^{-t},$$

corresponding to the same eigenvalue. Lastly, for $\lambda_3 = 5$ the reduction

$$(\mathbf{A} - 5\mathbf{I} | \mathbf{0}) = \begin{pmatrix} -4 & -2 & 2 & | & 0 \\ -2 & -4 & -2 & | & 0 \\ 2 & -2 & -4 & | & 0 \end{pmatrix} \xrightarrow[\text{operations}]{\text{row}} \begin{pmatrix} 1 & 0 & -1 & | & 0 \\ 0 & 1 & 1 & | & 0 \\ 0 & 0 & 0 & | & 0 \end{pmatrix}$$

implies that $k_1 = k_3$ and $k_2 = -k_3$. Picking $k_3 = 1$ gives $k_1 = 1, k_2 = -1$; thus a third eigenvector is

$$\mathbf{K}_3 = \begin{pmatrix} 1 \\ -1 \\ 1 \end{pmatrix}.$$

We conclude that the general solution of the system is

$$\mathbf{X} = c_1 \begin{pmatrix} 1 \\ 1 \\ 0 \end{pmatrix} e^{-t} + c_2 \begin{pmatrix} 0 \\ 1 \\ 1 \end{pmatrix} e^{-t} + c_3 \begin{pmatrix} 1 \\ -1 \\ 1 \end{pmatrix} e^{5t}. \qquad \blacksquare$$

The matrix of coefficients $\mathbf{A}$ in Example 3 is a special kind of matrix known as a symmetric matrix. An $n \times n$ matrix $\mathbf{A}$ is said to be **symmetric** if its transpose $\mathbf{A}^T$ (where the rows and columns are interchanged) is the same as $\mathbf{A}$—that is, if $\mathbf{A}^T = \mathbf{A}$. It can be proved that if the matrix $\mathbf{A}$ in the system $\mathbf{X}' = \mathbf{AX}$ is symmetric and has real entries, then we can always find $n$ linearly independent eigenvectors $\mathbf{K}_1, \mathbf{K}_2, \ldots, \mathbf{K}_n$, and the general solution of such a system is as given in Theorem 8.2.1. As illustrated in Example 3, this result holds even when some of the eigenvalues are repeated.

**SECOND SOLUTION**    Now suppose that $\lambda_1$ is an eigenvalue of multiplicity two and that there is only one eigenvector associated with this value. A second solution can be found of the form

$$\mathbf{X}_2 = \mathbf{K}te^{\lambda_1 t} + \mathbf{P}e^{\lambda_1 t}, \qquad (12)$$

where

$$\mathbf{K} = \begin{pmatrix} k_1 \\ k_2 \\ \vdots \\ k_n \end{pmatrix} \quad \text{and} \quad \mathbf{P} = \begin{pmatrix} p_1 \\ p_2 \\ \vdots \\ p_n \end{pmatrix}.$$

To see this, we substitute (12) into the system $\mathbf{X}' = \mathbf{AX}$ and simplify:

$$(\mathbf{AK} - \lambda_1 \mathbf{K})te^{\lambda_1 t} + (\mathbf{AP} - \lambda_1 \mathbf{P} - \mathbf{K})e^{\lambda_1 t} = \mathbf{0}.$$

Since this last equation is to hold for all values of $t$, we must have

$$(\mathbf{A} - \lambda_1 \mathbf{I})\mathbf{K} = \mathbf{0} \qquad (13)$$

and

$$(\mathbf{A} - \lambda_1 \mathbf{I})\mathbf{P} = \mathbf{K}. \qquad (14)$$

Equation (13) simply states that $\mathbf{K}$ must be an eigenvector of $\mathbf{A}$ associated with $\lambda_1$. By solving (13), we find one solution $\mathbf{X}_1 = \mathbf{K}e^{\lambda_1 t}$. To find the second solution $\mathbf{X}_2$, we need only solve the additional system (14) for the vector $\mathbf{P}$.

**EXAMPLE 4**  Repeated Eigenvalues

Find the general solution of the system given in (10).

**SOLUTION** From (11) we know that $\lambda_1 = -3$ and that one solution is $X_1 = \begin{pmatrix} 3 \\ 1 \end{pmatrix} e^{-3t}$. Identifying $K = \begin{pmatrix} 3 \\ 1 \end{pmatrix}$ and $P = \begin{pmatrix} p_1 \\ p_2 \end{pmatrix}$, we find from (14) that we must now solve

$$(A + 3I)P = K \qquad \text{or} \qquad \begin{array}{l} 6p_1 - 18p_2 = 3 \\ 2p_1 - 6p_2 = 1. \end{array}$$

Since this system is obviously equivalent to one equation, we have an infinite number of choices for $p_1$ and $p_2$. For example, by choosing $p_1 = 1$, we find $p_2 = \frac{1}{6}$. However, for simplicity we shall choose $p_1 = \frac{1}{2}$ so that $p_2 = 0$. Hence $P = \begin{pmatrix} \frac{1}{2} \\ 0 \end{pmatrix}$. Thus from (12) we find $X_2 = \begin{pmatrix} 3 \\ 1 \end{pmatrix} te^{-3t} + \begin{pmatrix} \frac{1}{2} \\ 0 \end{pmatrix} e^{-3t}$. The general solution of (10) is then $X = c_1 X_1 + c_2 X_2$ or

$$X = c_1 \begin{pmatrix} 3 \\ 1 \end{pmatrix} e^{-3t} + c_2 \left[ \begin{pmatrix} 3 \\ 1 \end{pmatrix} te^{-3t} + \begin{pmatrix} \frac{1}{2} \\ 0 \end{pmatrix} e^{-3t} \right].$$ ∎

By assigning various values to $c_1$ and $c_2$ in the solution in Example 4, we can plot trajectories of the system in (10). A phase portrait of (10) is given in Figure 8.2.3. The solutions $X_1$ and $-X_1$ determine two half-lines $y = \frac{1}{3}x$, $x > 0$ and $y = \frac{1}{3}x$, $x < 0$, respectively, shown in red in the figure. Because the single eigenvalue is negative and $e^{-3t} \to 0$ as $t \to \infty$ on *every* trajectory, we have $(x(t), y(t)) \to (0, 0)$ as $t \to \infty$. This is why the arrowheads in Figure 8.2.3 indicate that a particle on any trajectory moves toward the origin as time increases and why the origin is an attractor in this case. Moreover, a moving particle or trajectory $x = 3c_1 e^{-3t} + c_2(3te^{-3t} + \frac{1}{2}e^{-3t})$, $y = c_1 e^{-3t} + c_2 te^{-3t}$, $c_2 \neq 0$, approaches $(0, 0)$ tangentially to one of the half-lines as $t \to \infty$. In contrast, when the repeated eigenvalue is positive, the situation is reversed and the origin is a repeller. See Problem 23 in Exercises 8.2. Analogous to Figure 8.2.2, Figure 8.2.3 is typical of *all* $2 \times 2$ homogeneous linear systems $X' = AX$ that have two repeated negative eigenvalues. See Problem 34 in Exercises 8.2.

**FIGURE 8.2.3** A phase portrait of system (10)

**EIGENVALUE OF MULTIPLICITY THREE** When the coefficient matrix $A$ has only one eigenvector associated with an eigenvalue $\lambda_1$ of multiplicity three, we can find a second solution of the form (12) and a third solution of the form

$$X_3 = K \frac{t^2}{2} e^{\lambda_1 t} + Pte^{\lambda_1 t} + Qe^{\lambda_1 t}, \qquad (15)$$

where

$$K = \begin{pmatrix} k_1 \\ k_2 \\ \vdots \\ k_n \end{pmatrix}, \qquad P = \begin{pmatrix} p_1 \\ p_2 \\ \vdots \\ p_n \end{pmatrix}, \qquad \text{and} \qquad Q = \begin{pmatrix} q_1 \\ q_2 \\ \vdots \\ q_n \end{pmatrix}.$$

By substituting (15) into the system $X' = AX$, we find that the column vectors $K$, $P$, and $Q$ must satisfy

$$(A - \lambda_1 I)K = 0 \qquad (16)$$

$$(A - \lambda_1 I)P = K \qquad (17)$$

and

$$(A - \lambda_1 I)Q = P \qquad (18)$$

Of course, the solutions of (16) and (17) can be used in forming the solutions $X_1$ and $X_2$.

EXAMPLE 5    **Repeated Eigenvalues**

Solve $\mathbf{X}' = \begin{pmatrix} 2 & 1 & 6 \\ 0 & 2 & 5 \\ 0 & 0 & 2 \end{pmatrix} \mathbf{X}$.

**SOLUTION**  The characteristic equation $(\lambda - 2)^3 = 0$ shows that $\lambda_1 = 2$ is an eigenvalue of multiplicity three. By solving $(\mathbf{A} - 2\mathbf{I})\mathbf{K} = \mathbf{0}$, we find the single eigenvector

$$\mathbf{K} = \begin{pmatrix} 1 \\ 0 \\ 0 \end{pmatrix}.$$

We next solve the systems $(\mathbf{A} - 2\mathbf{I})\mathbf{P} = \mathbf{K}$ and $(\mathbf{A} - 2\mathbf{I})\mathbf{Q} = \mathbf{P}$ in succession and find that

$$\mathbf{P} = \begin{pmatrix} 0 \\ 1 \\ 0 \end{pmatrix} \quad \text{and} \quad \mathbf{Q} = \begin{pmatrix} 0 \\ -\frac{6}{5} \\ \frac{1}{5} \end{pmatrix}.$$

Using (12) and (15), we see that the general solution of the system is

$$\mathbf{X} = c_1 \begin{pmatrix} 1 \\ 0 \\ 0 \end{pmatrix} e^{2t} + c_2 \left[ \begin{pmatrix} 1 \\ 0 \\ 0 \end{pmatrix} te^{2t} + \begin{pmatrix} 0 \\ 1 \\ 0 \end{pmatrix} e^{2t} \right] + c_3 \left[ \begin{pmatrix} 1 \\ 0 \\ 0 \end{pmatrix} \frac{t^2}{2} e^{2t} + \begin{pmatrix} 0 \\ 1 \\ 0 \end{pmatrix} te^{2t} + \begin{pmatrix} 0 \\ -\frac{6}{5} \\ \frac{1}{5} \end{pmatrix} e^{2t} \right]. \quad \blacksquare$$

### REMARKS

When an eigenvalue $\lambda_1$ has multiplicity $m$, either we can find $m$ linearly independent eigenvectors or the number of corresponding eigenvectors is less than $m$. Hence the two cases listed on page 345 are not all the possibilities under which a repeated eigenvalue can occur. It can happen, say, that a $5 \times 5$ matrix has an eigenvalue of multiplicity five and there exist three corresponding linearly independent eigenvectors. See Problems 33 and 54 in Exercises 8.2.

## 8.2.3  COMPLEX EIGENVALUES

If $\lambda_1 = \alpha + \beta i$ and $\lambda_2 = \alpha - \beta i$, $\beta > 0$, $i^2 = -1$ are complex eigenvalues of the coefficient matrix $\mathbf{A}$, we can then certainly expect their corresponding eigenvectors to also have complex entries.[*]

For example, the characteristic equation of the system

$$\frac{dx}{dt} = 6x - y$$

$$\frac{dy}{dt} = 5x + 4y \tag{19}$$

is          $\det(\mathbf{A} - \lambda\mathbf{I}) = \begin{vmatrix} 6 - \lambda & -1 \\ 5 & 4 - \lambda \end{vmatrix} = \lambda^2 - 10\lambda + 29 = 0.$

From the quadratic formula we find $\lambda_1 = 5 + 2i$, $\lambda_2 = 5 - 2i$.

---

[*]When the characteristic equation has real coefficients, complex eigenvalues always appear in conjugate pairs.

Now for $\lambda_1 = 5 + 2i$ we must solve

$$(1 - 2i)k_1 - \qquad k_2 = 0$$
$$5k_1 - (1 + 2i)\,k_2 = 0.$$

Since $k_2 = (1 - 2i)k_1,$[*] the choice $k_1 = 1$ gives the following eigenvector and corresponding solution vector:

$$\mathbf{K}_1 = \begin{pmatrix} 1 \\ 1 - 2i \end{pmatrix}, \qquad \mathbf{X}_1 = \begin{pmatrix} 1 \\ 1 - 2i \end{pmatrix} e^{(5 + 2i)t}.$$

In like manner, for $\lambda_2 = 5 - 2i$ we find

$$\mathbf{K}_2 = \begin{pmatrix} 1 \\ 1 + 2i \end{pmatrix}, \qquad \mathbf{X}_2 = \begin{pmatrix} 1 \\ 1 + 2i \end{pmatrix} e^{(5 - 2i)t}.$$

We can verify by means of the Wronskian that these solution vectors are linearly independent, and so the general solution of (19) is

$$\mathbf{X} = c_1 \begin{pmatrix} 1 \\ 1 - 2i \end{pmatrix} e^{(5+2i)t} + c_2 \begin{pmatrix} 1 \\ 1 + 2i \end{pmatrix} e^{(5-2i)t}. \tag{20}$$

Note that the entries in $\mathbf{K}_2$ corresponding to $\lambda_2$ are the conjugates of the entries in $\mathbf{K}_1$ corresponding to $\lambda_1$. The conjugate of $\lambda_1$ is, of course, $\lambda_2$. We write this as $\lambda_2 = \bar{\lambda}_1$ and $\mathbf{K}_2 = \bar{\mathbf{K}}_1$. We have illustrated the following general result.

> **THEOREM 8.2.2  Solutions Corresponding to a Complex Eigenvalue**
>
> Let $\mathbf{A}$ be the coefficient matrix having real entries of the homogeneous system (2), and let $\mathbf{K}_1$ be an eigenvector corresponding to the complex eigenvalue $\lambda_1 = \alpha + i\beta$, $\alpha$ and $\beta$ real. Then
>
> $$\mathbf{K}_1 e^{\lambda_1 t} \qquad \text{and} \qquad \bar{\mathbf{K}}_1 e^{\bar{\lambda}_1 t}$$
>
> are solutions of (2).

It is desirable and relatively easy to rewrite a solution such as (20) in terms of real functions. To this end we first use Euler's formula to write

$$e^{(5 + 2i)t} = e^{5t} e^{2ti} = e^{5t}(\cos 2t + i \sin 2t)$$
$$e^{(5 - 2i)t} = e^{5t} e^{-2ti} = e^{5t}(\cos 2t - i \sin 2t).$$

Then, after we multiply complex numbers, collect terms, and replace $c_1 + c_2$ by $C_1$ and $(c_1 - c_2)i$ by $C_2$, (20) becomes

$$\mathbf{X} = C_1 \mathbf{X}_1 + C_2 \mathbf{X}_2, \tag{21}$$

where

$$\mathbf{X}_1 = \left[ \begin{pmatrix} 1 \\ 1 \end{pmatrix} \cos 2t - \begin{pmatrix} 0 \\ -2 \end{pmatrix} \sin 2t \right] e^{5t}$$

and

$$\mathbf{X}_2 = \left[ \begin{pmatrix} 0 \\ -2 \end{pmatrix} \cos 2t + \begin{pmatrix} 1 \\ 1 \end{pmatrix} \sin 2t \right] e^{5t}.$$

It is now important to realize that the vectors $\mathbf{X}_1$ and $\mathbf{X}_2$ in (21) constitute a linearly independent set of *real* solutions of the original system. Consequently, we are justified in ignoring the relationship between $C_1$, $C_2$ and $c_1$, $c_2$, and we can regard $C_1$ and

---

[*]Note that the second equation is simply $(1 + 2i)$ times the first.

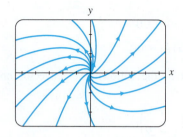

**FIGURE 8.2.4** A phase portrait of system (19)

$C_2$ as completely arbitrary and real. In other words, the linear combination (21) is an alternative general solution of (19). Moreover, with the real form given in (21) we are able to obtain a phase portrait of the system in (19). From (21) we find $x(t)$ and $y(t)$ to be

$$x = C_1 e^{5t} \cos 2t + C_2 e^{5t} \sin 2t$$

$$y = (C_1 - 2C_2)e^{5t} \cos 2t + (2C_1 + C_2)e^{5t} \sin 2t.$$

By plotting the trajectories $(x(t), y(t))$ for various values of $C_1$ and $C_2$, we obtain the phase portrait of (19) shown in Figure 8.2.4. Because the real part of $\lambda_1$ is $5 > 0$, $e^{5t} \to \infty$ as $t \to \infty$. This is why the arrowheads in Figure 8.2.4 point away from the origin; a particle on any trajectory spirals away from the origin as $t \to \infty$. The origin is a repeller.

The process by which we obtained the real solutions in (21) can be generalized. Let $\mathbf{K}_1$ be an eigenvector of the coefficient matrix $\mathbf{A}$ (with real entries) corresponding to the complex eigenvalue $\lambda_1 = \alpha + i\beta$. Then the solution vectors in Theorem 8.2.2 can be written as

$$\mathbf{K}_1 e^{\lambda_1 t} = \mathbf{K}_1 e^{\alpha t} e^{i\beta t} = \mathbf{K}_1 e^{\alpha t}(\cos \beta t + i \sin \beta t)$$

$$\overline{\mathbf{K}}_1 e^{\overline{\lambda}_1 t} = \overline{\mathbf{K}}_1 e^{\alpha t} e^{-i\beta t} = \overline{\mathbf{K}}_1 e^{\alpha t}(\cos \beta t - i \sin \beta t).$$

By the superposition principle, Theorem 8.1.2, the following vectors are also solutions:

$$\mathbf{X}_1 = \frac{1}{2}(\mathbf{K}_1 e^{\lambda_1 t} + \overline{\mathbf{K}}_1 e^{\overline{\lambda}_1 t}) = \frac{1}{2}(\mathbf{K}_1 + \overline{\mathbf{K}}_1)e^{\alpha t} \cos \beta t - \frac{i}{2}(-\mathbf{K}_1 + \overline{\mathbf{K}}_1)e^{\alpha t} \sin \beta t$$

$$\mathbf{X}_2 = \frac{i}{2}(-\mathbf{K}_1 e^{\lambda_1 t} + \overline{\mathbf{K}}_1 e^{\overline{\lambda}_1 t}) = \frac{i}{2}(-\mathbf{K}_1 + \overline{\mathbf{K}}_1)e^{\alpha t} \cos \beta t + \frac{1}{2}(\mathbf{K}_1 + \overline{\mathbf{K}}_1)e^{\alpha t} \sin \beta t.$$

Both $\frac{1}{2}(z + \overline{z}) = a$ and $\frac{1}{2}i(-z + \overline{z}) = b$ are *real* numbers for *any* complex number $z = a + ib$. Therefore, the entries in the column vectors $\frac{1}{2}(\mathbf{K}_1 + \overline{\mathbf{K}}_1)$ and $\frac{1}{2}i(-\mathbf{K}_1 + \overline{\mathbf{K}}_1)$ are real numbers. By defining

$$\mathbf{B}_1 = \frac{1}{2}(\mathbf{K}_1 + \overline{\mathbf{K}}_1) \qquad \text{and} \qquad \mathbf{B}_2 = \frac{i}{2}(-\mathbf{K}_1 + \overline{\mathbf{K}}_1), \qquad (22)$$

we are led to the following theorem.

---

**THEOREM 8.2.3** **Real Solutions Corresponding to a Complex Eigenvalue**

Let $\lambda_1 = \alpha + i\beta$ be a complex eigenvalue of the coefficient matrix $\mathbf{A}$ in the homogeneous system (2) and let $\mathbf{B}_1$ and $\mathbf{B}_2$ denote the column vectors defined in (22). Then

$$\mathbf{X}_1 = [\mathbf{B}_1 \cos \beta t - \mathbf{B}_2 \sin \beta t]e^{\alpha t}$$
$$\mathbf{X}_2 = [\mathbf{B}_2 \cos \beta t + \mathbf{B}_1 \sin \beta t]e^{\alpha t} \qquad (23)$$

are linearly independent solutions of (2) on $(-\infty, \infty)$.

---

The matrices $\mathbf{B}_1$ and $\mathbf{B}_2$ in (22) are often denoted by

$$\mathbf{B}_1 = \text{Re}(\mathbf{K}_1) \qquad \text{and} \qquad \mathbf{B}_2 = \text{Im}(\mathbf{K}_1) \qquad (24)$$

since these vectors are, respectively, the *real* and *imaginary* parts of the eigenvector $\mathbf{K}_1$. For example, (21) follows from (23) with

$$\mathbf{K}_1 = \begin{pmatrix} 1 \\ 1 - 2i \end{pmatrix} = \begin{pmatrix} 1 \\ 1 \end{pmatrix} + i\begin{pmatrix} 0 \\ -2 \end{pmatrix},$$

$$\mathbf{B}_1 = \text{Re}(\mathbf{K}_1) = \begin{pmatrix} 1 \\ 1 \end{pmatrix} \qquad \text{and} \qquad \mathbf{B}_2 = \text{Im}(\mathbf{K}_1) = \begin{pmatrix} 0 \\ -2 \end{pmatrix}.$$

**EXAMPLE 6** Complex Eigenvalues

Solve the initial-value problem

$$\mathbf{X}' = \begin{pmatrix} 2 & 8 \\ -1 & -2 \end{pmatrix} \mathbf{X}, \quad \mathbf{X}(0) = \begin{pmatrix} 2 \\ -1 \end{pmatrix}. \tag{25}$$

**SOLUTION** First we obtain the eigenvalues from

$$\det(\mathbf{A} - \lambda\mathbf{I}) = \begin{vmatrix} 2 - \lambda & 8 \\ -1 & -2 - \lambda \end{vmatrix} = \lambda^2 + 4 = 0.$$

The eigenvalues are $\lambda_1 = 2i$ and $\lambda_2 = \overline{\lambda_1} = -2i$. For $\lambda_1$ the system

$$(2 - 2i)k_1 + 8k_2 = 0$$
$$-k_1 + (-2 - 2i)k_2 = 0$$

gives $k_1 = -(2 + 2i)k_2$. By choosing $k_2 = -1$, we get

$$\mathbf{K}_1 = \begin{pmatrix} 2 + 2i \\ -1 \end{pmatrix} = \begin{pmatrix} 2 \\ -1 \end{pmatrix} + i\begin{pmatrix} 2 \\ 0 \end{pmatrix}.$$

Now from (24) we form

$$\mathbf{B}_1 = \text{Re}(\mathbf{K}_1) = \begin{pmatrix} 2 \\ -1 \end{pmatrix} \quad \text{and} \quad \mathbf{B}_2 = \text{Im}(\mathbf{K}_1) = \begin{pmatrix} 2 \\ 0 \end{pmatrix}.$$

Since $\alpha = 0$, it follows from (23) that the general solution of the system is

$$\mathbf{X} = c_1 \left[ \begin{pmatrix} 2 \\ -1 \end{pmatrix} \cos 2t - \begin{pmatrix} 2 \\ 0 \end{pmatrix} \sin 2t \right] + c_2 \left[ \begin{pmatrix} 2 \\ 0 \end{pmatrix} \cos 2t + \begin{pmatrix} 2 \\ -1 \end{pmatrix} \sin 2t \right]$$
$$= c_1 \begin{pmatrix} 2\cos 2t - 2\sin 2t \\ -\cos 2t \end{pmatrix} + c_2 \begin{pmatrix} 2\cos 2t + 2\sin 2t \\ -\sin 2t \end{pmatrix}. \tag{26}$$

Some graphs of the curves or trajectories defined by solution (26) of the system are illustrated in the phase portrait in Figure 8.2.5. Now the initial condition $\mathbf{X}(0) = \begin{pmatrix} 2 \\ -1 \end{pmatrix}$ or, equivalently, $x(0) = 2$ and $y(0) = -1$ yields the algebraic system $2c_1 + 2c_2 = 2$, $-c_1 = -1$, whose solution is $c_1 = 1$, $c_2 = 0$. Thus the solution to the problem is $\mathbf{X} = \begin{pmatrix} 2\cos 2t - 2\sin 2t \\ -\cos 2t \end{pmatrix}$. The specific trajectory defined parametrically by the particular solution $x = 2\cos 2t - 2\sin 2t$, $y = -\cos 2t$ is the red curve in Figure 8.2.5. Note that this curve passes through $(2, -1)$. ∎

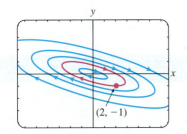

**FIGURE 8.2.5** A phase portrait of (25) in Example 6

---

**REMARKS**

In this section we have examined exclusively homogeneous first-order systems of linear equations in normal form $\mathbf{X}' = \mathbf{AX}$. But often the mathematical model of a dynamical physical system is a homogeneous second-order system whose normal form is $\mathbf{X}'' = \mathbf{AX}$. For example, the model for the coupled springs in (1) of Section 7.6,

$$m_1 x_1'' = -k_1 x_1 + k_2(x_2 - x_1)$$
$$m_2 x_2'' = -k_2(x_2 - x_1), \tag{27}$$

can be written as $\qquad \mathbf{MX}'' = \mathbf{KX},$

*(continued)*

where

$$\mathbf{M} = \begin{pmatrix} m_1 & 0 \\ 0 & m_2 \end{pmatrix}, \quad \mathbf{K} = \begin{pmatrix} -k_1 - k_2 & k_2 \\ k_2 & -k_2 \end{pmatrix}, \quad \text{and} \quad \mathbf{X} = \begin{pmatrix} x_1(t) \\ x_2(t) \end{pmatrix}.$$

Since $\mathbf{M}$ is nonsingular, we can solve for $\mathbf{X}''$ as $\mathbf{X}'' = \mathbf{A}\mathbf{X}$, where $\mathbf{A} = \mathbf{M}^{-1}\mathbf{K}$. Thus (27) is equivalent to

$$\mathbf{X}'' = \begin{pmatrix} -\dfrac{k_1}{m_1} - \dfrac{k_2}{m_1} & \dfrac{k_2}{m_1} \\ \dfrac{k_2}{m_2} & -\dfrac{k_2}{m_2} \end{pmatrix} \mathbf{X}. \tag{28}$$

The methods of this section can used to solve (28) by transforming it into a first-order system by means of substitutions. If we let $x_1' = x_3$ and $x_2' = x_4$, then $x_3' = x_1''$ and $x_4' = x_2''$ and so (28) is equivalent to a system of *four* linear first-order DEs:

$$\begin{aligned} x_1' &= x_3 \\ x_2' &= x_4 \\ x_3' &= -\left(\dfrac{k_1}{m_1} + \dfrac{k_2}{m_1}\right)x_1 + \dfrac{k_2}{m_1}x_2 \quad \text{or} \quad \mathbf{X}' = \begin{pmatrix} 0 & 0 & 1 & 0 \\ 0 & 0 & 0 & 1 \\ -\dfrac{k_1}{m_1} - \dfrac{k_2}{m_1} & \dfrac{k_2}{m_1} & 0 & 0 \\ \dfrac{k_2}{m_2} & -\dfrac{k_2}{m_2} & 0 & 0 \end{pmatrix} \mathbf{X}. \tag{29} \\ x_4' &= \dfrac{k_2}{m_2}x_1 - \dfrac{k_2}{m_2}x_2 \end{aligned}$$

By finding the eigenvalues and eigenvectors of the coefficient matrix $\mathbf{A}$ in (29), we see that the solution of this first-order system gives the complete state of the physical system—the positions of the masses relative to the equilibrium positions ($x_1$ and $x_2$) as well as the velocities of the masses ($x_3$ and $x_4$) at time $t$. See Problem 52 in Exercises 8.2.

## EXERCISES 8.2

*Answers to selected odd-numbered problems begin on page ANS-13.*

### 8.2.1 Distinct Real Eigenvalues

In Problems 1–12 find the general solution of the given system.

**1.** $\dfrac{dx}{dt} = x + 2y$

$\dfrac{dy}{dt} = 4x + 3y$

**2.** $\dfrac{dx}{dt} = 2x + 2y$

$\dfrac{dy}{dt} = x + 3y$

**3.** $\dfrac{dx}{dt} = -4x + 2y$

$\dfrac{dy}{dt} = -\dfrac{5}{2}x + 2y$

**4.** $\dfrac{dx}{dt} = -\dfrac{5}{2}x + 2y$

$\dfrac{dy}{dt} = \dfrac{3}{4}x - 2y$

**5.** $\mathbf{X}' = \begin{pmatrix} 10 & -5 \\ 8 & -12 \end{pmatrix} \mathbf{X}$

**6.** $\mathbf{X}' = \begin{pmatrix} -6 & 2 \\ -3 & 1 \end{pmatrix} \mathbf{X}$

**7.** $\dfrac{dx}{dt} = x + y - z$

$\dfrac{dy}{dt} = 2y$

$\dfrac{dz}{dt} = y - z$

**8.** $\dfrac{dx}{dt} = 2x - 7y$

$\dfrac{dy}{dt} = 5x + 10y + 4z$

$\dfrac{dz}{dt} = 5y + 2z$

**9.** $\mathbf{X}' = \begin{pmatrix} -1 & 1 & 0 \\ 1 & 2 & 1 \\ 0 & 3 & -1 \end{pmatrix} \mathbf{X}$

**10.** $\mathbf{X}' = \begin{pmatrix} 1 & 0 & 1 \\ 0 & 1 & 0 \\ 1 & 0 & 1 \end{pmatrix} \mathbf{X}$

**11.** $\mathbf{X}' = \begin{pmatrix} -1 & -1 & 0 \\ \frac{3}{4} & -\frac{3}{2} & 3 \\ \frac{1}{8} & \frac{1}{4} & -\frac{1}{2} \end{pmatrix} \mathbf{X}$

**12.** $\mathbf{X}' = \begin{pmatrix} -1 & 4 & 2 \\ 4 & -1 & -2 \\ 0 & 0 & 6 \end{pmatrix} \mathbf{X}$

In Problems 13 and 14 solve the given initial-value problem.

**13.** $\mathbf{X}' = \begin{pmatrix} \frac{1}{2} & 0 \\ 1 & -\frac{1}{2} \end{pmatrix} \mathbf{X}, \quad \mathbf{X}(0) = \begin{pmatrix} 3 \\ 5 \end{pmatrix}$

**14.** $\mathbf{X}' = \begin{pmatrix} 1 & 1 & 4 \\ 0 & 2 & 0 \\ 1 & 1 & 1 \end{pmatrix} \mathbf{X}, \quad \mathbf{X}(0) = \begin{pmatrix} 1 \\ 3 \\ 0 \end{pmatrix}$

**15.** As shown in Figure 8.2.6 two large connected mixing tanks $A$ and $B$ initially contain 100 gallons of brine. Liquid is pumped in and out of the tanks as indicated in the figure; the mixture pumped between and out of a tank is assumed to be well-stirred.

(a) Construct a mathematical model in the form of a linear system of first-order differential equations for the number of pounds $x_1(t)$ and $x_2(t)$ of salt in tanks $A$ and $B$, respectively, at time $t$. Write the system in matrix form. [*Hint*: Review Section 3.3.]

(b) Use the eigenvalue method of this section to solve the linear system in part (a) subject to $x_1(0) = 20$, $x_2(0) = 5$.

(c) Use a graphing utility or CAS to plot the graphs of $x_1(t)$ and $x_2(t)$ in the same coordinate plane.

(d) Suppose the system of mixing tanks is to be turned off when the number of pounds of salt in tank $B$ equals that in tank $A$. Use a root-finding application of a CAS or calculator to approximate that time.

pure water
2 gal/min

mixture
1 gal/min

mixture
1 gal/min

mixture
2 gal/min

mixture
1 gal/min

**FIGURE 8.2.6** Mixing tanks in Problem 15

**16.** In Problem 27 of Exercises 4.9 you were asked to solve the following linear system

$$\frac{dx_1}{dt} = -\frac{1}{50}x_1$$

$$\frac{dx_2}{dt} = \frac{1}{50}x_1 - \frac{2}{75}x_2$$

$$\frac{dx_3}{dt} = \frac{2}{75}x_2 - \frac{1}{25}x_3$$

using elimination techniques. This system is a mathematical model for the number of pounds of salt $x_1(t)$, $x_2(t)$, and $x_3(t)$ in the connected mixing tanks $A$, $B$, and $C$ shown in Figure 3.3.8 on page 112.

(a) Use the eigenvalue method of this section to solve the system subject to $x_1(0) = 15$, $x_2(0) = 10$, $x_3(0) = 5$.

(b) What are $\lim_{t\to\infty} x_1(t)$, $\lim_{t\to\infty} x_2(t)$, and $\lim_{t\to\infty} x_3(t)$? Interpret this result.

**Computer Lab Assignments**

In Problems 17 and 18 use a CAS or linear algebra software as an aid in finding the general solution of the given system.

**17.** $\mathbf{X}' = \begin{pmatrix} 0.9 & 2.1 & 3.2 \\ 0.7 & 6.5 & 4.2 \\ 1.1 & 1.7 & 3.4 \end{pmatrix} \mathbf{X}$

**18.** $\mathbf{X}' = \begin{pmatrix} 1 & 0 & 2 & -1.8 & 0 \\ 0 & 5.1 & 0 & -1 & 3 \\ 1 & 2 & -3 & 0 & 0 \\ 0 & 1 & -3.1 & 4 & 0 \\ -2.8 & 0 & 0 & 1.5 & 1 \end{pmatrix} \mathbf{X}$

**19.** (a) Use computer software to obtain the phase portrait of the system in Problem 5. If possible, include arrowheads as in Figure 8.2.2. Also include four half-lines in your phase portrait.

(b) Obtain the Cartesian equations of each of the four half-lines in part (a).

(c) Draw the eigenvectors on your phase portrait of the system.

**20.** Find phase portraits for the systems in Problems 2 and 4. For each system find any half-line trajectories and include these lines in your phase portrait.

### 8.2.2 Repeated Eigenvalues

In Problems 21–30 find the general solution of the given system.

**21.** $\dfrac{dx}{dt} = 3x - y$

$\dfrac{dy}{dt} = 9x - 3y$

**22.** $\dfrac{dx}{dt} = -6x + 5y$

$\dfrac{dy}{dt} = -5x + 4y$

**23.** $\mathbf{X}' = \begin{pmatrix} -1 & 3 \\ -3 & 5 \end{pmatrix} \mathbf{X}$

**24.** $\mathbf{X}' = \begin{pmatrix} 12 & -9 \\ 4 & 0 \end{pmatrix} \mathbf{X}$

**25.** $\dfrac{dx}{dt} = 3x - y - z$

$\dfrac{dy}{dt} = x + y - z$

$\dfrac{dz}{dt} = x - y + z$

**26.** $\dfrac{dx}{dt} = 3x + 2y + 4z$

$\dfrac{dy}{dt} = 2x + 2z$

$\dfrac{dz}{dt} = 4x + 2y + 3z$

**27.** $\mathbf{X}' = \begin{pmatrix} 5 & -4 & 0 \\ 1 & 0 & 2 \\ 0 & 2 & 5 \end{pmatrix} \mathbf{X}$

**28.** $\mathbf{X}' = \begin{pmatrix} 1 & 0 & 0 \\ 0 & 3 & 1 \\ 0 & -1 & 1 \end{pmatrix} \mathbf{X}$

**29.** $\mathbf{X}' = \begin{pmatrix} 1 & 0 & 0 \\ 2 & 2 & -1 \\ 0 & 1 & 0 \end{pmatrix} \mathbf{X}$

**30.** $\mathbf{X}' = \begin{pmatrix} 4 & 1 & 0 \\ 0 & 4 & 1 \\ 0 & 0 & 4 \end{pmatrix} \mathbf{X}$

In Problems 31 and 32 solve the given initial-value problem.

**31.** $\mathbf{X}' = \begin{pmatrix} 2 & 4 \\ -1 & 6 \end{pmatrix} \mathbf{X}, \quad \mathbf{X}(0) = \begin{pmatrix} -1 \\ 6 \end{pmatrix}$

**32.** $\mathbf{X}' = \begin{pmatrix} 0 & 0 & 1 \\ 0 & 1 & 0 \\ 1 & 0 & 0 \end{pmatrix} \mathbf{X}, \quad \mathbf{X}(0) = \begin{pmatrix} 1 \\ 2 \\ 5 \end{pmatrix}$

**33.** Show that the $5 \times 5$ matrix

$$\mathbf{A} = \begin{pmatrix} 2 & 1 & 0 & 0 & 0 \\ 0 & 2 & 0 & 0 & 0 \\ 0 & 0 & 2 & 0 & 0 \\ 0 & 0 & 0 & 2 & 1 \\ 0 & 0 & 0 & 0 & 2 \end{pmatrix}$$

has an eigenvalue $\lambda_1$ of multiplicity 5. Show that three linearly independent eigenvectors corresponding to $\lambda_1$ can be found.

## Computer Lab Assignments

**34.** Find phase portraits for the systems in Problems 22 and 23. For each system find any half-line trajectories and include these lines in your phase portrait.

### 8.2.3  Complex Eigenvalues

In Problems 35–46 find the general solution of the given system.

**35.** $\dfrac{dx}{dt} = 6x - y$

$\dfrac{dy}{dt} = 5x + 2y$

**36.** $\dfrac{dx}{dt} = x + y$

$\dfrac{dy}{dt} = -2x - y$

**37.** $\dfrac{dx}{dt} = 5x + y$

$\dfrac{dy}{dt} = -2x + 3y$

**38.** $\dfrac{dx}{dt} = 4x + 5y$

$\dfrac{dy}{dt} = -2x + 6y$

**39.** $\mathbf{X}' = \begin{pmatrix} 4 & -5 \\ 5 & -4 \end{pmatrix}\mathbf{X}$

**40.** $\mathbf{X}' = \begin{pmatrix} 1 & -8 \\ 1 & -3 \end{pmatrix}\mathbf{X}$

**41.** $\dfrac{dx}{dt} = z$

$\dfrac{dy}{dt} = -z$

$\dfrac{dz}{dt} = y$

**42.** $\dfrac{dx}{dt} = 2x + y + 2z$

$\dfrac{dy}{dt} = 3x + 6z$

$\dfrac{dz}{dt} = -4x - 3z$

**43.** $\mathbf{X}' = \begin{pmatrix} 1 & -1 & 2 \\ -1 & 1 & 0 \\ -1 & 0 & 1 \end{pmatrix}\mathbf{X}$

**44.** $\mathbf{X}' = \begin{pmatrix} 4 & 0 & 1 \\ 0 & 6 & 0 \\ -4 & 0 & 4 \end{pmatrix}\mathbf{X}$

**45.** $\mathbf{X}' = \begin{pmatrix} 2 & 5 & 1 \\ -5 & -6 & 4 \\ 0 & 0 & 2 \end{pmatrix}\mathbf{X}$

**46.** $\mathbf{X}' = \begin{pmatrix} 2 & 4 & 4 \\ -1 & -2 & 0 \\ -1 & 0 & -2 \end{pmatrix}\mathbf{X}$

In Problems 47 and 48 solve the given initial-value problem.

**47.** $\mathbf{X}' = \begin{pmatrix} 1 & -12 & -14 \\ 1 & 2 & -3 \\ 1 & 1 & -2 \end{pmatrix}\mathbf{X}, \quad \mathbf{X}(0) = \begin{pmatrix} 4 \\ 6 \\ -7 \end{pmatrix}$

**48.** $\mathbf{X}' = \begin{pmatrix} 6 & -1 \\ 5 & 4 \end{pmatrix}\mathbf{X}, \quad \mathbf{X}(0) = \begin{pmatrix} -2 \\ 8 \end{pmatrix}$

**49.** The system of mixing tanks shown in Figure 8.2.7 is a closed system. The tanks $A$, $B$, and $C$ initially contain the number of gallons of brine indicated in the figure.

    **(a)** Construct a mathematical model in the form of a linear system of first-order differential equations for the number of pounds of salt $x_1(t)$, $x_2(t)$, and $x_3(t)$ in the tanks $A$, $B$, and $C$ at time $t$, respectively. Write the system in matrix form.

    **(b)** Use the eigenvalue method of this section to solve the linear system in part (a) subject to $x_1(0) = 30$, $x_2(0) = 20$, $x_3(0) = 5$.

FIGURE 8.2.7  Mixing tanks in Problem 49

**50.** For the linear system in Problem 49:

    **(a)** Show that $x_1(t) + x_2(t) + x_3(t) = 55$. Interpret this result.

    **(b)** What are $\lim_{t\to\infty} x_1(t)$, $\lim_{t\to\infty} x_2(t)$, and $\lim_{t\to\infty} x_3(t)$. Interpret this result.

## Computer Lab Assignments

**51.** Find phase portraits for the systems in Problems 38, 39, and 40.

**52.** Solve (2) of Section 7.6 using the method outlined in the *Remarks* (page 351)—that is, express (2) of Section 7.6 as a first-order system of four linear equations. Use a CAS or linear algebra software as an aid in finding eigenvalues and eigenvectors of a $4 \times 4$ matrix. Then apply the initial conditions to your general solution to obtain (4) of Section 7.6.

## Discussion Problems

**53.** Solve each of the following linear systems.

    **(a)** $\mathbf{X}' = \begin{pmatrix} 1 & 1 \\ 1 & 1 \end{pmatrix}\mathbf{X}$     **(b)** $\mathbf{X}' = \begin{pmatrix} 1 & 1 \\ -1 & -1 \end{pmatrix}\mathbf{X}$

Find a phase portrait of each system. What is the geometric significance of the line $y = -x$ in each portrait?

**54.** Consider the $5 \times 5$ matrix given in Problem 33. Solve the system $\mathbf{X}' = \mathbf{AX}$ without the aid of matrix methods, but write the general solution using matrix notation. Use the general solution as a basis for a discussion of how the system can be solved using the matrix methods of this section. Carry out your ideas.

**55.** Obtain a Cartesian equation of the curve defined parametrically by the solution of the linear system in Example 6. Identify the curve passing through $(2, -1)$ in Figure 8.2.5. [*Hint:* Compute $x^2$, $y^2$, and $xy$.]

**56.** Examine your phase portraits in Problem 51. Under what conditions will the phase portrait of a $2 \times 2$ homogeneous linear system with complex eigenvalues consist of a family of closed curves? Consist of a family of spirals? Under what conditions is the origin $(0, 0)$ a repeller? An attractor?

# Nonhomogeneous Linear Systems

**INTRODUCTION** In Section 8.1 we saw that the general solution of a nonhomogeneous linear system $\mathbf{X}' = \mathbf{AX} + \mathbf{F}(t)$ on an interval $I$ is $\mathbf{X} = \mathbf{X}_c + \mathbf{X}_p$, where $\mathbf{X}_c = c_1\mathbf{X}_1 + c_2\mathbf{X}_2 + \cdots + c_n\mathbf{X}_n$ is the **complementary function** or general solution of the associated homogeneous linear system $\mathbf{X}' = \mathbf{AX}$ and $\mathbf{X}_p$ is any **particular solution** of the nonhomogeneous system. In Section 8.2 we saw how to obtain $\mathbf{X}_c$ when the coefficient matrix $\mathbf{A}$ was an $n \times n$ matrix of constants. In the present section we consider two methods for obtaining $\mathbf{X}_p$.

The methods of **undetermined coefficients** and **variation of parameters** used in Chapter 4 to find particular solutions of nonhomogeneous linear ODEs can both be adapted to the solution of nonhomogeneous linear systems $\mathbf{X}' = \mathbf{AX} + \mathbf{F}(t)$. Of the two methods, variation of parameters is the more powerful technique. However, there are instances when the method of undetermined coefficients provides a quick means of finding a particular solution.

## 8.3.1 UNDETERMINED COEFFICIENTS

**THE ASSUMPTIONS** As in Section 4.4, the method of undetermined coefficients consists of making an educated guess about the form of a particular solution vector $\mathbf{X}_p$; the guess is motivated by the types of functions that make up the entries of the column matrix $\mathbf{F}(t)$. Not surprisingly, the matrix version of undetermined coefficients is applicable to $\mathbf{X}' = \mathbf{AX} + \mathbf{F}(t)$ only when the entries of $\mathbf{A}$ are constants and the entries of $\mathbf{F}(t)$ are constants, polynomials, exponential functions, sines and cosines, or finite sums and products of these functions.

**EXAMPLE 1** Undetermined Coefficients

Solve the system $\mathbf{X}' = \begin{pmatrix} -1 & 2 \\ -1 & 1 \end{pmatrix}\mathbf{X} + \begin{pmatrix} -8 \\ 3 \end{pmatrix}$ on $(-\infty, \infty)$.

**SOLUTION** We first solve the associated homogeneous system

$$\mathbf{X}' = \begin{pmatrix} -1 & 2 \\ -1 & 1 \end{pmatrix}\mathbf{X}.$$

The characteristic equation of the coefficient matrix $\mathbf{A}$,

$$\det(\mathbf{A} - \lambda\mathbf{I}) = \begin{vmatrix} -1 - \lambda & 2 \\ -1 & 1 - \lambda \end{vmatrix} = \lambda^2 + 1 = 0,$$

yields the complex eigenvalues $\lambda_1 = i$ and $\lambda_2 = \overline{\lambda_1} = -i$. By the procedures of Section 8.2 we find

$$\mathbf{X}_c = c_1\begin{pmatrix} \cos t + \sin t \\ \cos t \end{pmatrix} + c_2\begin{pmatrix} \cos t - \sin t \\ -\sin t \end{pmatrix}.$$

Now since $\mathbf{F}(t)$ is a constant vector, we assume a constant particular solution vector $\mathbf{X}_p = \begin{pmatrix} a_1 \\ b_1 \end{pmatrix}$. Substituting this latter assumption into the original system and equating entries leads to

$$0 = -a_1 + 2b_1 - 8$$
$$0 = -a_1 + b_1 + 3.$$

Solving this algebraic system gives $a_1 = 14$ and $b_1 = 11$, and so a particular solution is $\mathbf{X}_p = \begin{pmatrix} 14 \\ 11 \end{pmatrix}$. The general solution of the original system of DEs on the interval $(-\infty, \infty)$ is then $\mathbf{X} = \mathbf{X}_c + \mathbf{X}_p$ or

$$\mathbf{X} = c_1 \begin{pmatrix} \cos t + \sin t \\ \cos t \end{pmatrix} + c_2 \begin{pmatrix} \cos t - \sin t \\ -\sin t \end{pmatrix} + \begin{pmatrix} 14 \\ 11 \end{pmatrix}. \qquad \blacksquare$$

### EXAMPLE 2  Undetermined Coefficients

Solve the system $\mathbf{X}' = \begin{pmatrix} 6 & 1 \\ 4 & 3 \end{pmatrix} \mathbf{X} + \begin{pmatrix} 6t \\ -10t + 4 \end{pmatrix}$ on $(-\infty, \infty)$.

**SOLUTION** The eigenvalues and corresponding eigenvectors of the associated homogeneous system $\mathbf{X}' = \begin{pmatrix} 6 & 1 \\ 4 & 3 \end{pmatrix} \mathbf{X}$ are found to be $\lambda_1 = 2$, $\lambda_2 = 7$, $\mathbf{K}_1 = \begin{pmatrix} 1 \\ -4 \end{pmatrix}$, and $\mathbf{K}_2 = \begin{pmatrix} 1 \\ 1 \end{pmatrix}$. Hence the complementary function is

$$\mathbf{X}_c = c_1 \begin{pmatrix} 1 \\ -4 \end{pmatrix} e^{2t} + c_2 \begin{pmatrix} 1 \\ 1 \end{pmatrix} e^{7t}.$$

Now because $\mathbf{F}(t)$ can be written $\mathbf{F}(t) = \begin{pmatrix} 6 \\ -10 \end{pmatrix} t + \begin{pmatrix} 0 \\ 4 \end{pmatrix}$, we shall try to find a particular solution of the system that possesses the *same* form:

$$\mathbf{X}_p = \begin{pmatrix} a_2 \\ b_2 \end{pmatrix} t + \begin{pmatrix} a_1 \\ b_1 \end{pmatrix}.$$

Substituting this last assumption into the given system yields

$$\begin{pmatrix} a_2 \\ b_2 \end{pmatrix} = \begin{pmatrix} 6 & 1 \\ 4 & 3 \end{pmatrix} \left[ \begin{pmatrix} a_2 \\ b_2 \end{pmatrix} t + \begin{pmatrix} a_1 \\ b_1 \end{pmatrix} \right] + \begin{pmatrix} 6 \\ -10 \end{pmatrix} t + \begin{pmatrix} 0 \\ 4 \end{pmatrix}$$

or
$$\begin{pmatrix} 0 \\ 0 \end{pmatrix} = \begin{pmatrix} (6a_2 + b_2 + 6)t + 6a_1 + b_1 - a_2 \\ (4a_2 + 3b_2 - 10)t + 4a_1 + 3b_1 - b_2 + 4 \end{pmatrix}.$$

From the last identity we obtain four algebraic equations in four unknowns

$$\begin{array}{ccc} 6a_2 + b_2 + 6 = 0 & & 6a_1 + b_1 - a_2 = 0 \\ 4a_2 + 3b_2 - 10 = 0 & \text{and} & 4a_1 + 3b_1 - b_2 + 4 = 0. \end{array}$$

Solving the first two equations simultaneously yields $a_2 = -2$ and $b_2 = 6$. We then substitute these values into the last two equations and solve for $a_1$ and $b_1$. The results are $a_1 = -\frac{4}{7}$, $b_1 = \frac{10}{7}$. It follows, therefore, that a particular solution vector is

$$\mathbf{X}_p = \begin{pmatrix} -2 \\ 6 \end{pmatrix} t + \begin{pmatrix} -\frac{4}{7} \\ \frac{10}{7} \end{pmatrix}.$$

The general solution of the system on $(-\infty, \infty)$ is $\mathbf{X} = \mathbf{X}_c + \mathbf{X}_p$ or

$$\mathbf{X} = c_1 \begin{pmatrix} 1 \\ -4 \end{pmatrix} e^{2t} + c_2 \begin{pmatrix} 1 \\ 1 \end{pmatrix} e^{7t} + \begin{pmatrix} -2 \\ 6 \end{pmatrix} t + \begin{pmatrix} -\frac{4}{7} \\ \frac{10}{7} \end{pmatrix}. \qquad \blacksquare$$

**EXAMPLE 3** Form of $\mathbf{X}_p$

Determine the form of a particular solution vector $\mathbf{X}_p$ for the system

$$\frac{dx}{dt} = 5x + 3y - 2e^{-t} + 1$$

$$\frac{dy}{dt} = -x + y + e^{-t} - 5t + 7.$$

**SOLUTION** Because $\mathbf{F}(t)$ can be written in matrix terms as

$$\mathbf{F}(t) = \begin{pmatrix} -2 \\ 1 \end{pmatrix} e^{-t} + \begin{pmatrix} 0 \\ -5 \end{pmatrix} t + \begin{pmatrix} 1 \\ 7 \end{pmatrix}$$

a natural assumption for a particular solution would be

$$\mathbf{X}_p = \begin{pmatrix} a_3 \\ b_3 \end{pmatrix} e^{-t} + \begin{pmatrix} a_2 \\ b_2 \end{pmatrix} t + \begin{pmatrix} a_1 \\ b_1 \end{pmatrix}.$$

■

### REMARKS

The method of undetermined coefficients for linear systems is not as straightforward as the last three examples would seem to indicate. In Section 4.4 the form of a particular solution $y_p$ was predicated on prior knowledge of the complementary function $y_c$. The same is true for the formation of $\mathbf{X}_p$. But there are further difficulties: The special rules governing the form of $y_p$ in Section 4.4 do not *quite* carry to the formation of $\mathbf{X}_p$. For example, if $\mathbf{F}(t)$ is a constant vector, as in Example 1, and $\lambda = 0$ is an eigenvalue of multiplicity one, then $\mathbf{X}_c$ contains a constant vector. Under the Multiplication Rule on page 148 we would ordinarily try a particular solution of the form $\mathbf{X}_p = \begin{pmatrix} a_1 \\ b_1 \end{pmatrix} t$. This is not the proper assumption for linear systems; it should be $\mathbf{X}_p = \begin{pmatrix} a_2 \\ b_2 \end{pmatrix} t + \begin{pmatrix} a_1 \\ b_1 \end{pmatrix}$.

Similarly, in Example 3, if we replace $e^{-t}$ in $\mathbf{F}(t)$ by $e^{2t}$ ($\lambda = 2$ is an eigenvalue), then the correct form of the particular solution vector is

$$\mathbf{X}_p = \begin{pmatrix} a_4 \\ b_4 \end{pmatrix} t e^{2t} + \begin{pmatrix} a_3 \\ b_3 \end{pmatrix} e^{2t} + \begin{pmatrix} a_2 \\ b_2 \end{pmatrix} t + \begin{pmatrix} a_1 \\ b_1 \end{pmatrix}.$$

Rather than delving into these difficulties, we turn instead to the method of variation of parameters.

## 8.3.2 VARIATION OF PARAMETERS

**A FUNDAMENTAL MATRIX** If $\mathbf{X}_1, \mathbf{X}_2, \dots, \mathbf{X}_n$ is a fundamental set of solutions of the homogeneous system $\mathbf{X}' = \mathbf{A}\mathbf{X}$ on an interval $I$, then its general solution on the interval is the linear combination $\mathbf{X} = c_1\mathbf{X}_1 + c_2\mathbf{X}_2 + \cdots + c_n\mathbf{X}_n$ or

$$\mathbf{X} = c_1 \begin{pmatrix} x_{11} \\ x_{21} \\ \vdots \\ x_{n1} \end{pmatrix} + c_2 \begin{pmatrix} x_{12} \\ x_{22} \\ \vdots \\ x_{n2} \end{pmatrix} + \cdots + c_n \begin{pmatrix} x_{1n} \\ x_{2n} \\ \vdots \\ x_{nn} \end{pmatrix} = \begin{pmatrix} c_1 x_{11} + c_2 x_{12} + \cdots + c_n x_{1n} \\ c_1 x_{21} + c_2 x_{22} + \cdots + c_n x_{2n} \\ \vdots \\ c_1 x_{n1} + c_2 x_{n2} + \cdots + c_n x_{nn} \end{pmatrix}. \quad (1)$$

The last matrix in (1) is recognized as the product of an $n \times n$ matrix with an $n \times 1$ matrix. In other words, the general solution (1) can be written as the product

$$\mathbf{X} = \mathbf{\Phi}(t)\mathbf{C}, \tag{2}$$

where $\mathbf{C}$ is an $n \times 1$ column vector of arbitrary constants $c_1, c_2, \ldots, c_n$ and the $n \times n$ matrix, whose columns consist of the entries of the solution vectors of the system $\mathbf{X}' = \mathbf{AX}$,

$$\mathbf{\Phi}(t) = \begin{pmatrix} x_{11} & x_{12} & \cdots & x_{1n} \\ x_{21} & x_{22} & \cdots & x_{2n} \\ \vdots & & & \vdots \\ x_{n1} & x_{n2} & \cdots & x_{nn} \end{pmatrix},$$

is called a **fundamental matrix** of the system on the interval.

In the discussion that follows we need to use two properties of a fundamental matrix:

- A fundamental matrix $\mathbf{\Phi}(t)$ is nonsingular.
- If $\mathbf{\Phi}(t)$ is a fundamental matrix of the system $\mathbf{X}' = \mathbf{AX}$, then

$$\mathbf{\Phi}'(t) = \mathbf{A}\mathbf{\Phi}(t). \tag{3}$$

A reexamination of (9) of Theorem 8.1.3 shows that $\det \mathbf{\Phi}(t)$ is the same as the Wronskian $W(\mathbf{X}_1, \mathbf{X}_2, \ldots, \mathbf{X}_n)$. Hence the linear independence of the columns of $\mathbf{\Phi}(t)$ on the interval $I$ guarantees that $\det \mathbf{\Phi}(t) \neq 0$ for every $t$ in the interval. Since $\mathbf{\Phi}(t)$ is nonsingular, the multiplicative inverse $\mathbf{\Phi}^{-1}(t)$ exists for every $t$ in the interval. The result given in (3) follows immediately from the fact that every column of $\mathbf{\Phi}(t)$ is a solution vector of $\mathbf{X}' = \mathbf{AX}$.

**VARIATION OF PARAMETERS** Analogous to the procedure in Section 4.6 we ask whether it is possible to replace the matrix of constants $\mathbf{C}$ in (2) by a column matrix of functions

$$\mathbf{U}(t) = \begin{pmatrix} u_1(t) \\ u_2(t) \\ \vdots \\ u_n(t) \end{pmatrix} \qquad \text{so} \qquad \mathbf{X}_p = \mathbf{\Phi}(t)\mathbf{U}(t) \tag{4}$$

is a particular solution of the nonhomogeneous system

$$\mathbf{X}' = \mathbf{AX} + \mathbf{F}(t). \tag{5}$$

By the Product Rule the derivative of the last expression in (4) is

$$\mathbf{X}_p' = \mathbf{\Phi}(t)\mathbf{U}'(t) + \mathbf{\Phi}'(t)\mathbf{U}(t). \tag{6}$$

Note that the order of the products in (6) is very important. Since $\mathbf{U}(t)$ is a column matrix, the products $\mathbf{U}'(t)\mathbf{\Phi}(t)$ and $\mathbf{U}(t)\mathbf{\Phi}'(t)$ are not defined. Substituting (4) and (6) into (5) gives

$$\mathbf{\Phi}(t)\mathbf{U}'(t) + \mathbf{\Phi}'(t)\mathbf{U}(t) = \mathbf{A}\mathbf{\Phi}(t)\mathbf{U}(t) + \mathbf{F}(t). \tag{7}$$

Now if we use (3) to replace $\mathbf{\Phi}'(t)$, (7) becomes

$$\mathbf{\Phi}(t)\mathbf{U}'(t) + \mathbf{A}\mathbf{\Phi}(t)\mathbf{U}(t) = \mathbf{A}\mathbf{\Phi}(t)\mathbf{U}(t) + \mathbf{F}(t)$$

or

$$\mathbf{\Phi}(t)\mathbf{U}'(t) = \mathbf{F}(t). \tag{8}$$

Multiplying both sides of equation (8) by $\mathbf{\Phi}^{-1}(t)$ gives

$$\mathbf{U}'(t) = \mathbf{\Phi}^{-1}(t)\mathbf{F}(t) \qquad \text{and so} \qquad \mathbf{U}(t) = \int \mathbf{\Phi}^{-1}(t)\mathbf{F}(t)\, dt.$$

Since $\mathbf{X}_p = \mathbf{\Phi}(t)\mathbf{U}(t)$, we conclude that a particular solution of (5) is

$$\mathbf{X}_p = \mathbf{\Phi}(t) \int \mathbf{\Phi}^{-1}(t)\mathbf{F}(t)\, dt. \tag{9}$$

To calculate the indefinite integral of the column matrix $\mathbf{\Phi}^{-1}(t)\mathbf{F}(t)$ in (9), we integrate each entry. Thus the general solution of the system (5) is $\mathbf{X} = \mathbf{X}_c + \mathbf{X}_p$ or

$$\mathbf{X} = \mathbf{\Phi}(t)\mathbf{C} + \mathbf{\Phi}(t) \int \mathbf{\Phi}^{-1}(t)\mathbf{F}(t)\, dt. \tag{10}$$

Note that it is not necessary to use a constant of integration in the evaluation of $\int \mathbf{\Phi}^{-1}(t)\mathbf{F}(t)\, dt$ for the same reasons stated in the discussion of variation of parameters in Section 4.6.

---

### EXAMPLE 4 Variation of Parameters

Solve the system

$$\mathbf{X}' = \begin{pmatrix} -3 & 1 \\ 2 & -4 \end{pmatrix}\mathbf{X} + \begin{pmatrix} 3t \\ e^{-t} \end{pmatrix} \tag{11}$$

on $(-\infty, \infty)$.

**SOLUTION** We first solve the associated homogeneous system

$$\mathbf{X}' = \begin{pmatrix} -3 & 1 \\ 2 & -4 \end{pmatrix}\mathbf{X}. \tag{12}$$

The characteristic equation of the coefficient matrix is

$$\det(\mathbf{A} - \lambda\mathbf{I}) = \begin{vmatrix} -3 - \lambda & 1 \\ 2 & -4 - \lambda \end{vmatrix} = (\lambda + 2)(\lambda + 5) = 0,$$

so the eigenvalues are $\lambda_1 = -2$ and $\lambda_2 = -5$. By the usual method we find that the eigenvectors corresponding to $\lambda_1$ and $\lambda_2$ are, respectively, $\mathbf{K}_1 = \begin{pmatrix} 1 \\ 1 \end{pmatrix}$ and $\mathbf{K}_2 = \begin{pmatrix} 1 \\ -2 \end{pmatrix}$. The solution vectors of the homogeneous system (12) are then

$$\mathbf{X}_1 = \begin{pmatrix} 1 \\ 1 \end{pmatrix}e^{-2t} = \begin{pmatrix} e^{-2t} \\ e^{-2t} \end{pmatrix} \quad \text{and} \quad \mathbf{X}_2 = \begin{pmatrix} 1 \\ -2 \end{pmatrix}e^{-5t} = \begin{pmatrix} e^{-5t} \\ -2e^{-5t} \end{pmatrix}.$$

The entries in $\mathbf{X}_1$ form the first column of $\mathbf{\Phi}(t)$, and the entries in $\mathbf{X}_2$ form the second column of $\mathbf{\Phi}(t)$. Hence

$$\mathbf{\Phi}(t) = \begin{pmatrix} e^{-2t} & e^{-5t} \\ e^{-2t} & -2e^{-5t} \end{pmatrix} \quad \text{and} \quad \mathbf{\Phi}^{-1}(t) = \begin{pmatrix} \frac{2}{3}e^{2t} & \frac{1}{3}e^{2t} \\ \frac{1}{3}e^{5t} & -\frac{1}{3}e^{5t} \end{pmatrix}.$$

From (9) we obtain the particular solution

$$\mathbf{X}_p = \mathbf{\Phi}(t) \int \mathbf{\Phi}^{-1}(t)\mathbf{F}(t)\, dt = \begin{pmatrix} e^{-2t} & e^{-5t} \\ e^{-2t} & -2e^{-5t} \end{pmatrix} \int \begin{pmatrix} \frac{2}{3}e^{2t} & \frac{1}{3}e^{2t} \\ \frac{1}{3}e^{5t} & -\frac{1}{3}e^{5t} \end{pmatrix}\begin{pmatrix} 3t \\ e^{-t} \end{pmatrix} dt$$

$$= \begin{pmatrix} e^{-2t} & e^{-5t} \\ e^{-2t} & -2e^{-5t} \end{pmatrix} \int \begin{pmatrix} 2te^{2t} + \frac{1}{3}e^{t} \\ te^{5t} - \frac{1}{3}e^{4t} \end{pmatrix} dt$$

$$= \begin{pmatrix} e^{-2t} & e^{-5t} \\ e^{-2t} & -2e^{-5t} \end{pmatrix}\begin{pmatrix} te^{2t} - \frac{1}{2}e^{2t} + \frac{1}{3}e^{t} \\ \frac{1}{5}te^{5t} - \frac{1}{25}e^{5t} - \frac{1}{12}e^{4t} \end{pmatrix}$$

$$= \begin{pmatrix} \frac{6}{5}t - \frac{27}{50} + \frac{1}{4}e^{-t} \\ \frac{3}{5}t - \frac{21}{50} + \frac{1}{2}e^{-t} \end{pmatrix}.$$

Hence from (10) the general solution of (11) on the interval $(-\infty, \infty)$ is

$$
\mathbf{X} = \begin{pmatrix} e^{-2t} & e^{-5t} \\ e^{-2t} & -2e^{-5t} \end{pmatrix} \begin{pmatrix} c_1 \\ c_2 \end{pmatrix} + \begin{pmatrix} \frac{6}{5}t - \frac{27}{50} + \frac{1}{4}e^{-t} \\ \frac{3}{5}t - \frac{21}{50} + \frac{1}{2}e^{-t} \end{pmatrix}
$$

$$
= c_1 \begin{pmatrix} 1 \\ 1 \end{pmatrix} e^{-2t} + c_2 \begin{pmatrix} 1 \\ -2 \end{pmatrix} e^{-5t} + \begin{pmatrix} \frac{6}{5} \\ \frac{3}{5} \end{pmatrix} t - \begin{pmatrix} \frac{27}{50} \\ \frac{21}{50} \end{pmatrix} + \begin{pmatrix} \frac{1}{4} \\ \frac{1}{2} \end{pmatrix} e^{-t}.
$$ ■

**INITIAL-VALUE PROBLEM**   The general solution of (5) on an interval can be written in the alternative manner

$$
\mathbf{X} = \mathbf{\Phi}(t)\mathbf{C} + \mathbf{\Phi}(t) \int_{t_0}^{t} \mathbf{\Phi}^{-1}(s)\mathbf{F}(s)\, ds, \tag{13}
$$

where $t$ and $t_0$ are points in the interval. This last form is useful in solving (5) subject to an initial condition $\mathbf{X}(t_0) = \mathbf{X}_0$, because the limits of integration are chosen so that the particular solution vanishes at $t = t_0$. Substituting $t = t_0$ into (13) yields $\mathbf{X}_0 = \mathbf{\Phi}(t_0)\mathbf{C}$ from which we get $\mathbf{C} = \mathbf{\Phi}^{-1}(t_0)\mathbf{X}_0$. Substituting this last result into (13) gives the following solution of the initial-value problem:

$$
\mathbf{X} = \mathbf{\Phi}(t)\mathbf{\Phi}^{-1}(t_0)\mathbf{X}_0 + \mathbf{\Phi}(t) \int_{t_0}^{t} \mathbf{\Phi}^{-1}(s)\mathbf{F}(s)\, ds. \tag{14}
$$

## EXERCISES 8.3

Answers to selected odd-numbered problems begin on page ANS-14.

### 8.3.1  Undetermined Coefficients

In Problems 1–8 use the method of undetermined coefficients to solve the given nonhomogeneous system.

1. $\dfrac{dx}{dt} = 2x + 3y - 7$

   $\dfrac{dy}{dt} = -x - 2y + 5$

2. $\dfrac{dx}{dt} = 5x + 9y + 2$

   $\dfrac{dy}{dt} = -x + 11y + 6$

3. $\mathbf{X}' = \begin{pmatrix} 1 & 3 \\ 3 & 1 \end{pmatrix} \mathbf{X} + \begin{pmatrix} -2t^2 \\ t + 5 \end{pmatrix}$

4. $\mathbf{X}' = \begin{pmatrix} 1 & -4 \\ 4 & 1 \end{pmatrix} \mathbf{X} + \begin{pmatrix} 4t + 9e^{6t} \\ -t + e^{6t} \end{pmatrix}$

5. $\mathbf{X}' = \begin{pmatrix} 4 & \frac{1}{3} \\ 9 & 6 \end{pmatrix} \mathbf{X} + \begin{pmatrix} -3 \\ 10 \end{pmatrix} e^{t}$

6. $\mathbf{X}' = \begin{pmatrix} -1 & 5 \\ -1 & 1 \end{pmatrix} \mathbf{X} + \begin{pmatrix} \sin t \\ -2 \cos t \end{pmatrix}$

7. $\mathbf{X}' = \begin{pmatrix} 1 & 1 & 1 \\ 0 & 2 & 3 \\ 0 & 0 & 5 \end{pmatrix} \mathbf{X} + \begin{pmatrix} 1 \\ -1 \\ 2 \end{pmatrix} e^{4t}$

8. $\mathbf{X}' = \begin{pmatrix} 0 & 0 & 5 \\ 0 & 5 & 0 \\ 5 & 0 & 0 \end{pmatrix} \mathbf{X} + \begin{pmatrix} 5 \\ -10 \\ 40 \end{pmatrix}$

In Problems 9 and 10, solve the given initial-value problem.

9. $\mathbf{X}' = \begin{pmatrix} -1 & -2 \\ 3 & 4 \end{pmatrix} \mathbf{X} + \begin{pmatrix} 3 \\ 3 \end{pmatrix}, \quad \mathbf{X}(0) = \begin{pmatrix} -4 \\ 5 \end{pmatrix}$

10. $\mathbf{X}' = \begin{pmatrix} 1 & -1 \\ 1 & 3 \end{pmatrix} \mathbf{X} + \begin{pmatrix} t \\ t+1 \end{pmatrix}, \quad \mathbf{X}(0) = \begin{pmatrix} 3 \\ 2 \end{pmatrix}$

11. Consider the large mixing tanks shown in Figure 8.3.1. Suppose that both tanks $A$ and $B$ initially contain 100 gallons of brine. Liquid is pumped in and out of the tanks as indicated in the figure; the mixture pumped between and out of the tanks is assumed to be well-stirred.

    (a) Construct a mathematical model in the form of a linear system of first-order differential equations for the number of pounds $x_1(t)$ and $x_2(t)$ of salt in tanks $A$ and $B$, respectively, at time $t$. Write the system in matrix form. [*Hint*: Review Section 3.3.]

    (b) Use the method of undetermined coefficients to solve the linear system in part (a) subject to $x_1(0) = 60$, $x_2(0) = 10$.

    (c) What are $\lim\limits_{t\to\infty} x_1(t)$ and $\lim\limits_{t\to\infty} x_2(t)$? Interpret this result.

    (d) Use a graphing utility to plot the graphs of $x_1(t)$ and $x_2(t)$ on the same coordinate axes.

**FIGURE 8.3.1**  Mixing tanks in Problem 11

**12. (a)** The system of differential equations for the currents $i_2(t)$ and $i_3(t)$ in the electrical network shown in Figure 8.3.2 is

$$\frac{d}{dt}\begin{pmatrix} i_2 \\ i_3 \end{pmatrix} = \begin{pmatrix} -R_1/L_1 & -R_1/L_1 \\ -R_1/L_2 & -(R_1+R_2)/L_2 \end{pmatrix}\begin{pmatrix} i_2 \\ i_3 \end{pmatrix} + \begin{pmatrix} E/L_1 \\ E/L_2 \end{pmatrix}.$$

Use the method of undetermined coefficients to solve the system if $R_1 = 2\ \Omega$, $R_2 = 3\ \Omega$, $L_1 = 1$ h, $L_2 = 1$ h, $E = 60$ V, $i_2(0) = 0$, and $i_3(0) = 0$.

**(b)** Determine the current $i_1(t)$.

**FIGURE 8.3.2** Network in Problem 12

### 8.3.2 Variation of Parameters

In Problems 13–32 use variation of parameters to solve the given nonhomogeneous system.

**13.** $\dfrac{dx}{dt} = 3x - 3y + 4$

$\dfrac{dy}{dt} = 2x - 2y - 1$

**14.** $\dfrac{dx}{dt} = 2x - y$

$\dfrac{dy}{dt} = 3x - 2y + 4t$

**15.** $\mathbf{X}' = \begin{pmatrix} 3 & -5 \\ \frac{3}{4} & -1 \end{pmatrix}\mathbf{X} + \begin{pmatrix} 1 \\ -1 \end{pmatrix}e^{t/2}$

**16.** $\mathbf{X}' = \begin{pmatrix} 2 & -1 \\ 4 & 2 \end{pmatrix}\mathbf{X} + \begin{pmatrix} \sin 2t \\ 2\cos 2t \end{pmatrix}e^{2t}$

**17.** $\mathbf{X}' = \begin{pmatrix} 0 & 2 \\ -1 & 3 \end{pmatrix}\mathbf{X} + \begin{pmatrix} 1 \\ -1 \end{pmatrix}e^{t}$

**18.** $\mathbf{X}' = \begin{pmatrix} 0 & 2 \\ -1 & 3 \end{pmatrix}\mathbf{X} + \begin{pmatrix} 2 \\ e^{-3t} \end{pmatrix}$

**19.** $\mathbf{X}' = \begin{pmatrix} 1 & 8 \\ 1 & 1 \end{pmatrix}\mathbf{X} + \begin{pmatrix} 12 \\ 12 \end{pmatrix}t$

**20.** $\mathbf{X}' = \begin{pmatrix} 1 & 8 \\ 1 & -1 \end{pmatrix}\mathbf{X} + \begin{pmatrix} e^{-t} \\ te^{t} \end{pmatrix}$

**21.** $\mathbf{X}' = \begin{pmatrix} 3 & 2 \\ -2 & -1 \end{pmatrix}\mathbf{X} + \begin{pmatrix} 2e^{-t} \\ e^{-t} \end{pmatrix}$

**22.** $\mathbf{X}' = \begin{pmatrix} 3 & 2 \\ -2 & -1 \end{pmatrix}\mathbf{X} + \begin{pmatrix} 1 \\ 1 \end{pmatrix}$

**23.** $\mathbf{X}' = \begin{pmatrix} 0 & -1 \\ 1 & 0 \end{pmatrix}\mathbf{X} + \begin{pmatrix} \sec t \\ 0 \end{pmatrix}$

**24.** $\mathbf{X}' = \begin{pmatrix} 1 & -1 \\ 1 & 1 \end{pmatrix}\mathbf{X} + \begin{pmatrix} 3 \\ 3 \end{pmatrix}e^{t}$

**25.** $\mathbf{X}' = \begin{pmatrix} 1 & -1 \\ 1 & 1 \end{pmatrix}\mathbf{X} + \begin{pmatrix} \cos t \\ \sin t \end{pmatrix}e^{t}$

**26.** $\mathbf{X}' = \begin{pmatrix} 2 & -2 \\ 8 & -6 \end{pmatrix}\mathbf{X} + \begin{pmatrix} 1 \\ 3 \end{pmatrix}\dfrac{e^{-2t}}{t}$

**27.** $\mathbf{X}' = \begin{pmatrix} 0 & 1 \\ -1 & 0 \end{pmatrix}\mathbf{X} + \begin{pmatrix} 0 \\ \sec t \tan t \end{pmatrix}$

**28.** $\mathbf{X}' = \begin{pmatrix} 0 & 1 \\ -1 & 0 \end{pmatrix}\mathbf{X} + \begin{pmatrix} 1 \\ \cot t \end{pmatrix}$

**29.** $\mathbf{X}' = \begin{pmatrix} 1 & 2 \\ -\frac{1}{2} & 1 \end{pmatrix}\mathbf{X} + \begin{pmatrix} \csc t \\ \sec t \end{pmatrix}e^{t}$

**30.** $\mathbf{X}' = \begin{pmatrix} 1 & -2 \\ 1 & -1 \end{pmatrix}\mathbf{X} + \begin{pmatrix} \tan t \\ 1 \end{pmatrix}$

**31.** $\mathbf{X}' = \begin{pmatrix} 1 & 1 & 0 \\ 1 & 1 & 0 \\ 0 & 0 & 3 \end{pmatrix}\mathbf{X} + \begin{pmatrix} e^{t} \\ e^{2t} \\ te^{3t} \end{pmatrix}$

**32.** $\mathbf{X}' = \begin{pmatrix} 3 & -1 & -1 \\ 1 & 1 & -1 \\ 1 & -1 & 1 \end{pmatrix}\mathbf{X} + \begin{pmatrix} 0 \\ t \\ 2e^{t} \end{pmatrix}$

In Problems 33 and 34 use (14) to solve the given initial-value problem.

**33.** $\mathbf{X}' = \begin{pmatrix} 3 & -1 \\ -1 & 3 \end{pmatrix}\mathbf{X} + \begin{pmatrix} 4e^{2t} \\ 4e^{4t} \end{pmatrix}, \quad \mathbf{X}(0) = \begin{pmatrix} 1 \\ 1 \end{pmatrix}$

**34.** $\mathbf{X}' = \begin{pmatrix} 1 & -1 \\ 1 & -1 \end{pmatrix}\mathbf{X} + \begin{pmatrix} 1/t \\ 1/t \end{pmatrix}, \quad \mathbf{X}(1) = \begin{pmatrix} 2 \\ -1 \end{pmatrix}$

**35.** The system of differential equations for the currents $i_1(t)$ and $i_2(t)$ in the electrical network shown in Figure 8.3.3 is

$$\frac{d}{dt}\begin{pmatrix} i_1 \\ i_2 \end{pmatrix} = \begin{pmatrix} -(R_1+R_2)/L_2 & R_2/L_2 \\ R_2/L_1 & -R_2/L_1 \end{pmatrix}\begin{pmatrix} i_1 \\ i_2 \end{pmatrix} + \begin{pmatrix} E/L_2 \\ 0 \end{pmatrix}.$$

Use variation of parameters to solve the system if $R_1 = 8\ \Omega$, $R_2 = 3\ \Omega$, $L_1 = 1$ h, $L_2 = 1$ h, $E(t) = 100\sin t$ V, $i_1(0) = 0$, and $i_2(0) = 0$.

**FIGURE 8.3.3** Network in Problem 35

### Discussion Problems

**36.** If $y_1$ and $y_2$ are linearly independent solutions of the associated homogeneous DE for $y'' + P(x)y' + Q(x)y = f(x)$, show in the case of a nonhomogeneous linear second-order DE that (9) reduces to the form of variation of parameters discussed in Section 4.6.

## Computer Lab Assignments

**37.** Solving a nonhomogeneous linear system $X' = AX + F(t)$ by variation of parameters when $A$ is a $3 \times 3$ (or larger) matrix is almost an impossible task to do by hand. Consider the system

$$X' = \begin{pmatrix} 2 & -2 & 2 & 1 \\ -1 & 3 & 0 & 3 \\ 0 & 0 & 4 & -2 \\ 0 & 0 & 2 & -1 \end{pmatrix} X + \begin{pmatrix} te^t \\ e^{-t} \\ e^{2t} \\ 1 \end{pmatrix}.$$

(a) Use a CAS or linear algebra software to find the eigenvalues and eigenvectors of the coefficient matrix.

(b) Form a fundamental matrix $\Phi(t)$ and use the computer to find $\Phi^{-1}(t)$.

(c) Use the computer to carry out the computations of:
$\Phi^{-1}(t)F(t)$, $\int \Phi^{-1}(t)F(t)\,dt$, $\Phi(t)\int \Phi^{-1}(t)F(t)\,dt$, $\Phi(t)C$, and $\Phi(t)C + \int \Phi^{-1}(t)F(t)\,dt$, where $C$ is a column matrix of constants $c_1$, $c_2$, $c_3$, and $c_4$.

(d) Rewrite the computer output for the general solution of the system in the form $X = X_c + X_p$, where $X_c = c_1X_1 + c_2X_2 + c_3X_3 + c_4X_4$.

# 8.4 Matrix Exponential

**INTRODUCTION** Matrices can be used in an entirely different manner to solve a system of linear first-order differential equations. Recall that the simple linear first-order differential equation $x' = ax$, where $a$ is constant, has the general solution $x = ce^{at}$, where $c$ is a constant. It seems natural then to ask whether we can define a matrix exponential function $e^{At}$, where $A$ is a matrix of constants, so that a solution of the linear system $X' = AX$ is $e^{At}$.

**HOMOGENEOUS SYSTEMS** We shall now see that it is possible to define a matrix exponential $e^{At}$ so that

$$X = e^{At}C \tag{1}$$

is a solution of the homogeneous system $X' = AX$. Here $A$ is an $n \times n$ matrix of constants, and $C$ is an $n \times 1$ column matrix of arbitrary constants. Note in (1) that the matrix $C$ post multiplies $e^{At}$ because we want $e^{At}$ to be an $n \times n$ matrix. While the complete development of the meaning and theory of the matrix exponential would require a thorough knowledge of matrix algebra, one way of defining $e^{At}$ is inspired by the power series representation of the scalar exponential function $e^{at}$:

$$e^{at} = 1 + at + \frac{(at)^2}{2!} + \cdots + \frac{(at)^k}{k!} + \cdots$$

$$= 1 + at + a^2\frac{t^2}{2!} + \cdots + \alpha^k\frac{t^k}{k!} + \cdots = \sum_{k=0}^{\infty} \alpha^k \frac{t^k}{k!}. \tag{2}$$

The series in (2) converges for all $t$. Using this series, with 1 replaced by the identity matrix $I$ and the constant $a$ replaced by an $n \times n$ matrix $A$ of constants, we arrive at a definition for the $n \times n$ matrix $e^{At}$.

### DEFINITION 8.4.1 Matrix Exponential

For any $n \times n$ matrix $A$,

$$e^{At} = I + At + A^2\frac{t^2}{2!} + \cdots + A^k\frac{t^k}{k!} + \cdots = \sum_{k=0}^{\infty} A^k \frac{t^k}{k!}. \tag{3}$$

It can be shown that the series given in (3) converges to an $n \times n$ matrix for every value of $t$. Also, $A^2 = AA$, $A^3 = A(A^2)$, and so on.

## EXAMPLE 1  Matrix Exponential Using (3)

Compute $e^{\mathbf{A}t}$ for the matrix

$$\mathbf{A} = \begin{pmatrix} 2 & 0 \\ 0 & 3 \end{pmatrix}.$$

**SOLUTION**  From the various powers

$$\mathbf{A}^2 = \begin{pmatrix} 2^2 & 0 \\ 0 & 3^2 \end{pmatrix}, \mathbf{A}^3 = \begin{pmatrix} 2^3 & 0 \\ 0 & 3^3 \end{pmatrix}, \mathbf{A}^4 = \begin{pmatrix} 2^4 & 0 \\ 0 & 3^4 \end{pmatrix}, \ldots, \mathbf{A}^n = \begin{pmatrix} 2^n & 0 \\ 0 & 3^n \end{pmatrix}, \ldots,$$

we see from (3) that

$$e^{\mathbf{A}t} = \mathbf{I} + \mathbf{A}t + \mathbf{A}^2 \frac{t^2}{2!} + \cdots$$

$$= \begin{pmatrix} 1 & 0 \\ 0 & 1 \end{pmatrix} + \begin{pmatrix} 2 & 0 \\ 0 & 3 \end{pmatrix} t + \begin{pmatrix} 2^2 & 0 \\ 0 & 3^2 \end{pmatrix} \frac{t^2}{2!} + \cdots + \begin{pmatrix} 2^n & 0 \\ 0 & 3^n \end{pmatrix} \frac{t^n}{n!} + \cdots$$

$$= \begin{pmatrix} 1 + 2t + 2^2 \dfrac{t^2}{2!} + \cdots & 0 \\ 0 & 1 + 3t + 3^2 \dfrac{t^2}{2!} + \cdots \end{pmatrix}.$$

In view of (2) and the identifications $a = 2$ and $a = 3$, the power series in the first and second rows of the last matrix represent, respectively, $e^{2t}$ and $e^{3t}$ and so we have

$$e^{\mathbf{A}t} = \begin{pmatrix} e^{2t} & 0 \\ 0 & e^{3t} \end{pmatrix}.$$

The matrix in Example 1 is an example of a $2 \times 2$ diagonal matrix. In general, an $n \times n$ matrix $\mathbf{A}$ is a **diagonal matrix** if all its entries off the main diagonal are zero, that is,

$$\mathbf{A} = \begin{pmatrix} a_{11} & 0 & \cdots & 0 \\ 0 & a_{22} & \cdots & 0 \\ \vdots & \vdots & & \vdots \\ 0 & 0 & \cdots & a_{nn} \end{pmatrix}.$$

Hence if $\mathbf{A}$ is any $n \times n$ diagonal matrix it follows from Example 1 that

$$e^{\mathbf{A}t} = \begin{pmatrix} e^{a_{11}t} & 0 & \cdots & 0 \\ 0 & e^{a_{22}t} & \cdots & 0 \\ \vdots & \vdots & & \vdots \\ 0 & 0 & \cdots & e^{a_{nn}t} \end{pmatrix}.$$

**DERIVATIVE OF $e^{\mathbf{A}t}$**  The derivative of the matrix exponential is analogous to the differentiation property of the scalar exponential $\dfrac{d}{dt} e^{at} = ae^{at}$. To justify

$$\frac{d}{dt} e^{\mathbf{A}t} = \mathbf{A}e^{\mathbf{A}t}, \tag{4}$$

we differentiate (3) term by term:

$$\frac{d}{dt} e^{\mathbf{A}t} = \frac{d}{dt} \left[ \mathbf{I} + \mathbf{A}t + \mathbf{A}^2 \frac{t^2}{2!} + \cdots + \mathbf{A}^k \frac{t^k}{k!} + \cdots \right] = \mathbf{A} + \mathbf{A}^2 t + \frac{1}{2!} \mathbf{A}^3 t^2 + \cdots$$

$$= \mathbf{A} \left[ \mathbf{I} + \mathbf{A}t + \mathbf{A}^2 \frac{t^2}{2!} + \cdots \right] = \mathbf{A}e^{\mathbf{A}t}.$$

Because of (4), we can now prove that (1) is a solution of $\mathbf{X}' = \mathbf{A}\mathbf{X}$ for every $n \times 1$ vector $\mathbf{C}$ of constants:

$$\mathbf{X}' = \frac{d}{dt} e^{\mathbf{A}t} \mathbf{C} = \mathbf{A}e^{\mathbf{A}t}\mathbf{C} = \mathbf{A}(e^{\mathbf{A}t}\mathbf{C}) = \mathbf{A}\mathbf{X}.$$

**$e^{At}$ IS A FUNDAMENTAL MATRIX** If we denote the matrix exponential $e^{At}$ by the symbol $\Psi(t)$, then (4) is equivalent to the matrix differential equation $\Psi'(t) = A\Psi(t)$ (see (3) of Section 8.3). In addition, it follows immediately from Definition 8.4.1 that $\Psi(0) = e^{A0} = I$, and so $\det \Psi(0) \neq 0$. It turns out that these two properties are sufficient for us to conclude that $\Psi(t)$ is a fundamental matrix of the system $X' = AX$.

**NONHOMOGENEOUS SYSTEMS** We saw in (4) of Section 2.3 that the general solution of the single linear first-order differential equation $x' = ax + f(t)$, where $a$ is a constant, can be expressed as

$$x = ce^{at} + e^{at}\int_{t_0}^{t} e^{-as}f(s)\, ds.$$

For a nonhomogeneous system of linear first-order differential equations it can be shown that the general solution of $X' = AX + F(t)$, where $A$ is an $n \times n$ matrix of constants, is

$$X = e^{At}C + e^{At}\int_{t_0}^{t} e^{-As}F(s)\, ds. \tag{5}$$

Since the matrix exponential $e^{At}$ is a fundamental matrix, it is always nonsingular and $e^{-As} = (e^{As})^{-1}$. In practice, $e^{-As}$ can be obtained from $e^{At}$ by simply replacing $t$ by $-s$.

**COMPUTATION OF $e^{At}$** The definition of $e^{At}$ given in (3) can, of course, always be used to compute $e^{At}$. However, the practical utility of (3) is limited by the fact that the entries in $e^{At}$ are power series in $t$. With a natural desire to work with simple and familiar things, we then try to recognize whether these series define a closed-form function. Fortunately, there are many alternative ways of computing $e^{At}$; the following discussion shows how the Laplace transform can be used.

**USE OF THE LAPLACE TRANSFORM** We saw in (5) that $X = e^{At}$ is a solution of $X' = AX$. Indeed, since $e^{A0} = I$, $X = e^{At}$ is a solution of the initial-value problem

$$X' = AX, \quad X(0) = I. \tag{6}$$

If $x(s) = \mathscr{L}\{X(t)\} = \mathscr{L}\{e^{At}\}$, then the Laplace transform of (6) is

$$sx(s) - X(0) = Ax(s) \quad \text{or} \quad (sI - A)x(s) = I.$$

Multiplying the last equation by $(sI - A)^{-1}$ implies that $x(s) = (sI - A)^{-1}I = (sI - A)^{-1}$. In other words, $\mathscr{L}\{e^{At}\} = (sI - A)^{-1}$ or

$$e^{At} = \mathscr{L}^{-1}\{(sI - A)^{-1}\}. \tag{7}$$

---

**EXAMPLE 2** Matrix Exponential Using (7)

Use the Laplace transform to compute $e^{At}$ for $A = \begin{pmatrix} 1 & -1 \\ 2 & -2 \end{pmatrix}$.

**SOLUTION** First we compute the matrix $sI - A$ and find its inverse:

$$sI - A = \begin{pmatrix} s - 1 & 1 \\ -2 & s + 2 \end{pmatrix},$$

$$(sI - A)^{-1} = \begin{pmatrix} s - 1 & 1 \\ -2 & s + 2 \end{pmatrix}^{-1} = \begin{pmatrix} \dfrac{s + 2}{s(s + 1)} & \dfrac{-1}{s(s + 1)} \\[2mm] \dfrac{2}{s(s + 1)} & \dfrac{s - 1}{s(s + 1)} \end{pmatrix}.$$

Then we decompose the entries of the last matrix into partial fractions:

$$(s\mathbf{I} - \mathbf{A})^{-1} = \begin{pmatrix} \dfrac{2}{s} - \dfrac{1}{s+1} & -\dfrac{1}{s} + \dfrac{1}{s+1} \\ \dfrac{2}{s} - \dfrac{2}{s+1} & -\dfrac{1}{s} + \dfrac{2}{s+1} \end{pmatrix}. \tag{8}$$

It follows from (7) that the inverse Laplace transform of (8) gives the desired result,

$$e^{\mathbf{A}t} = \begin{pmatrix} 2 - e^{-t} & -1 + e^{-t} \\ 2 - 2e^{-t} & -1 + 2e^{-t} \end{pmatrix}. \qquad \blacksquare$$

**USE OF COMPUTERS**    For those who are willing to momentarily trade understanding for speed of solution, $e^{\mathbf{A}t}$ can be computed with the aid of computer software. See Problems 27 and 28 in Exercises 8.4.

---

## EXERCISES 8.4

In Problems 1 and 2 use (3) to compute $e^{\mathbf{A}t}$ and $e^{-\mathbf{A}t}$.

**1.** $\mathbf{A} = \begin{pmatrix} 1 & 0 \\ 0 & 2 \end{pmatrix}$    **2.** $\mathbf{A} = \begin{pmatrix} 0 & 1 \\ 1 & 0 \end{pmatrix}$

In Problems 3 and 4 use (3) to compute $e^{\mathbf{A}t}$.

**3.** $\mathbf{A} = \begin{pmatrix} 1 & 1 & 1 \\ 1 & 1 & 1 \\ -2 & -2 & -2 \end{pmatrix}$

**4.** $\mathbf{A} = \begin{pmatrix} 0 & 0 & 0 \\ 3 & 0 & 0 \\ 5 & 1 & 0 \end{pmatrix}$

In Problems 5–8 use (1) to find the general solution of the given system.

**5.** $\mathbf{X}' = \begin{pmatrix} 1 & 0 \\ 0 & 2 \end{pmatrix}\mathbf{X}$    **6.** $\mathbf{X}' = \begin{pmatrix} 0 & 1 \\ 1 & 0 \end{pmatrix}\mathbf{X}$

**7.** $\mathbf{X}' = \begin{pmatrix} 1 & 1 & 1 \\ 1 & 1 & 1 \\ -2 & -2 & -2 \end{pmatrix}\mathbf{X}$    **8.** $\mathbf{X}' = \begin{pmatrix} 0 & 0 & 0 \\ 3 & 0 & 0 \\ 5 & 1 & 0 \end{pmatrix}\mathbf{X}$

In Problems 9–12 use (5) to find the general solution of the given system.

**9.** $\mathbf{X}' = \begin{pmatrix} 1 & 0 \\ 0 & 2 \end{pmatrix}\mathbf{X} + \begin{pmatrix} 3 \\ -1 \end{pmatrix}$

**10.** $\mathbf{X}' = \begin{pmatrix} 1 & 0 \\ 0 & 2 \end{pmatrix}\mathbf{X} + \begin{pmatrix} t \\ e^{4t} \end{pmatrix}$

**11.** $\mathbf{X}' = \begin{pmatrix} 0 & 1 \\ 1 & 0 \end{pmatrix}\mathbf{X} + \begin{pmatrix} 1 \\ 1 \end{pmatrix}$

**12.** $\mathbf{X}' = \begin{pmatrix} 0 & 1 \\ 1 & 0 \end{pmatrix}\mathbf{X} + \begin{pmatrix} \cosh t \\ \sinh t \end{pmatrix}$

**13.** Solve the system in Problem 7 subject to the initial condition

$$\mathbf{X}(0) = \begin{pmatrix} 1 \\ -4 \\ 6 \end{pmatrix}.$$

**14.** Solve the system in Problem 9 subject to the initial condition

$$\mathbf{X}(0) = \begin{pmatrix} 4 \\ 3 \end{pmatrix}.$$

In Problems 15–18 use the method of Example 2 to compute $e^{\mathbf{A}t}$ for the coefficient matrix. Use (1) to find the general solution of the given system.

**15.** $\mathbf{X}' = \begin{pmatrix} 4 & 3 \\ -4 & -4 \end{pmatrix}\mathbf{X}$    **16.** $\mathbf{X}' = \begin{pmatrix} 4 & -2 \\ 1 & 1 \end{pmatrix}\mathbf{X}$

**17.** $\mathbf{X}' = \begin{pmatrix} 5 & -9 \\ 1 & -1 \end{pmatrix}\mathbf{X}$    **18.** $\mathbf{X}' = \begin{pmatrix} 0 & 1 \\ -2 & -2 \end{pmatrix}\mathbf{X}$

Let $\mathbf{P}$ denote a matrix whose columns are eigenvectors $\mathbf{K}_1, \mathbf{K}_2, \ldots, \mathbf{K}_n$ corresponding to distinct eigenvalues $\lambda_1, \lambda_2, \ldots, \lambda_n$ of an $n \times n$ matrix $\mathbf{A}$. Then it can be shown that $\mathbf{A} = \mathbf{PDP}^{-1}$, where $\mathbf{D}$ is a diagonal matrix defined by

$$\mathbf{D} = \begin{pmatrix} \lambda_1 & 0 & \cdots & 0 \\ 0 & \lambda_2 & \cdots & 0 \\ \vdots & \vdots & & \vdots \\ 0 & 0 & \cdots & \lambda_n \end{pmatrix}. \tag{9}$$

In Problems 19 and 20 verify the foregoing result for the given matrix.

**19.** $\mathbf{A} = \begin{pmatrix} 2 & 1 \\ -3 & 6 \end{pmatrix}$    **20.** $\mathbf{A} = \begin{pmatrix} 2 & 1 \\ 1 & 2 \end{pmatrix}$

**21.** Suppose $\mathbf{A} = \mathbf{P}\mathbf{D}\mathbf{P}^{-1}$, where $\mathbf{D}$ is defined as in (9). Use (3) to show that $e^{\mathbf{A}t} = \mathbf{P}e^{\mathbf{D}t}\mathbf{P}^{-1}$.

**22.** If $\mathbf{D}$ is defined as in (9), then find $e^{\mathbf{D}t}$.

In Problems 23 and 24 use the results of Problems 19–22 to solve the given system.

**23.** $\mathbf{X}' = \begin{pmatrix} 2 & 1 \\ -3 & 6 \end{pmatrix}\mathbf{X}$  **24.** $\mathbf{X}' = \begin{pmatrix} 2 & 1 \\ 1 & 2 \end{pmatrix}\mathbf{X}$

### Discussion Problems

**25.** Reread the discussion leading to the result given in (7). Does the matrix $s\mathbf{I} - \mathbf{A}$ always have an inverse? Discuss.

**26.** A matrix $\mathbf{A}$ is said to be **nilpotent** if there exists some positive integer $m$ such that $\mathbf{A}^m = \mathbf{0}$. Verify that

$$\mathbf{A} = \begin{pmatrix} -1 & 1 & 1 \\ -1 & 0 & 1 \\ -1 & 1 & 1 \end{pmatrix}$$

is nilpotent. Discuss why it is relatively easy to compute $e^{\mathbf{A}t}$ when $\mathbf{A}$ is nilpotent. Compute $e^{\mathbf{A}t}$ and then use (1) to solve the system $\mathbf{X}' = \mathbf{A}\mathbf{X}$.

### Computer Lab Assignments

**27. (a)** Use (1) to find the general solution of $\mathbf{X}' = \begin{pmatrix} 4 & 2 \\ 3 & 3 \end{pmatrix}\mathbf{X}$.

Use a CAS to find $e^{\mathbf{A}t}$. Then use the computer to find eigenvalues and eigenvectors of the coefficient matrix

$\mathbf{A} = \begin{pmatrix} 4 & 2 \\ 3 & 3 \end{pmatrix}$ and form the general solution in the manner of Section 8.2. Finally, reconcile the two forms of the general solution of the system.

**(b)** Use (1) to find the general solution of $\mathbf{X}' = \begin{pmatrix} -3 & -1 \\ 2 & -1 \end{pmatrix}\mathbf{X}$.

Use a CAS to find $e^{\mathbf{A}t}$. In the case of complex output, utilize the software to do the simplification; for example, in *Mathematica*, if $\mathbf{m} = \mathbf{MatrixExp[A\ t]}$ has complex entries, then try the command **Simplify[ComplexExpand[m]]**.

**28.** Use (1) to find the general solution of

$$\mathbf{X}' = \begin{pmatrix} -4 & 0 & 6 & 0 \\ 0 & -5 & 0 & -4 \\ -1 & 0 & 1 & 0 \\ 0 & 3 & 0 & 2 \end{pmatrix}\mathbf{X}.$$

Use MATLAB or a CAS to find $e^{\mathbf{A}t}$.

## Chapter 8 In Review

*Answers to selected odd-numbered problems begin on page ANS-15.*

In Problems 1 and 2 fill in the blanks.

**1.** The vector $\mathbf{X} = k\begin{pmatrix} 4 \\ 5 \end{pmatrix}$ is a solution of

$$\mathbf{X}' = \begin{pmatrix} 1 & 4 \\ 2 & -1 \end{pmatrix}\mathbf{X} - \begin{pmatrix} 8 \\ 1 \end{pmatrix}$$

for $k = $ _____.

**2.** The vector $\mathbf{X} = c_1\begin{pmatrix} -1 \\ 1 \end{pmatrix}e^{-9t} + c_2\begin{pmatrix} 5 \\ 3 \end{pmatrix}e^{7t}$ is a solution of

the initial-value problem $\mathbf{X}' = \begin{pmatrix} 1 & 10 \\ 6 & -3 \end{pmatrix}\mathbf{X}, \mathbf{X}(0) = \begin{pmatrix} 2 \\ 0 \end{pmatrix}$

for $c_1 = $ _____ and $c_2 = $ _____.

**3.** Consider the linear system $\mathbf{X}' = \begin{pmatrix} 4 & 6 & 6 \\ 1 & 3 & 2 \\ -1 & -4 & -3 \end{pmatrix}\mathbf{X}$.

Without attempting to solve the system, determine which one of the vectors

$$\mathbf{K}_1 = \begin{pmatrix} 0 \\ 1 \\ 1 \end{pmatrix}, \quad \mathbf{K}_2 = \begin{pmatrix} 1 \\ 1 \\ -1 \end{pmatrix}, \quad \mathbf{K}_3 = \begin{pmatrix} 3 \\ 1 \\ -1 \end{pmatrix}, \quad \mathbf{K}_4 = \begin{pmatrix} 6 \\ 2 \\ -5 \end{pmatrix}$$

is an eigenvector of the coefficient matrix. What is the solution of the system corresponding to this eigenvector?

**4.** Consider the linear system $\mathbf{X}' = \mathbf{A}\mathbf{X}$ of two differential equations, where $\mathbf{A}$ is a real coefficient matrix. What is the general solution of the system if it is known that

$\lambda_1 = 1 + 2i$ is an eigenvalue and $\mathbf{K}_1 = \begin{pmatrix} 1 \\ i \end{pmatrix}$ is a corresponding eigenvector?

In Problems 5–14 solve the given linear system.

**5.** $\dfrac{dx}{dt} = 2x + y$
$\dfrac{dy}{dt} = -x$

**6.** $\dfrac{dx}{dt} = -4x + 2y$
$\dfrac{dy}{dt} = 2x - 4y$

**7.** $\mathbf{X}' = \begin{pmatrix} 1 & 2 \\ -2 & 1 \end{pmatrix}\mathbf{X}$

**8.** $\mathbf{X}' = \begin{pmatrix} -2 & 5 \\ -2 & 4 \end{pmatrix}\mathbf{X}$

**9.** $\mathbf{X}' = \begin{pmatrix} 1 & -1 & 1 \\ 0 & 1 & 3 \\ 4 & 3 & 1 \end{pmatrix}\mathbf{X}$

**10.** $\mathbf{X}' = \begin{pmatrix} 0 & 2 & 1 \\ 1 & 1 & -2 \\ 2 & 2 & -1 \end{pmatrix}\mathbf{X}$

**11.** $\mathbf{X}' = \begin{pmatrix} 2 & 8 \\ 0 & 4 \end{pmatrix}\mathbf{X} + \begin{pmatrix} 2 \\ 16t \end{pmatrix}$

**12.** $\mathbf{X}' = \begin{pmatrix} 1 & 2 \\ -\frac{1}{2} & 1 \end{pmatrix}\mathbf{X} + \begin{pmatrix} 0 \\ e^t \tan t \end{pmatrix}$

**13.** $\mathbf{X}' = \begin{pmatrix} -1 & 1 \\ -2 & 1 \end{pmatrix}\mathbf{X} + \begin{pmatrix} 1 \\ \cot t \end{pmatrix}$

**14.** $\mathbf{X}' = \begin{pmatrix} 3 & 1 \\ -1 & 1 \end{pmatrix}\mathbf{X} + \begin{pmatrix} -2 \\ 1 \end{pmatrix}e^{2t}$

**15. (a)** Consider the linear system $\mathbf{X}' = \mathbf{AX}$ of three first-order differential equations, where the coefficient matrix is

$$\mathbf{A} = \begin{pmatrix} 5 & 3 & 3 \\ 3 & 5 & 3 \\ -5 & -5 & -3 \end{pmatrix}$$

and $\lambda = 2$ is known to be an eigenvalue of multiplicity two. Find two different solutions of the system corresponding to this eigenvalue without using a special formula (such as (12) of Section 8.2).

**(b)** Use the procedure of part (a) to solve

$$\mathbf{X}' = \begin{pmatrix} 1 & 1 & 1 \\ 1 & 1 & 1 \\ 1 & 1 & 1 \end{pmatrix}\mathbf{X}.$$

**16.** Verify that $\mathbf{X} = \begin{pmatrix} c_1 \\ c_2 \end{pmatrix} e^t$ is a solution of the linear system

$$\mathbf{X}' = \begin{pmatrix} 1 & 0 \\ 0 & 1 \end{pmatrix}\mathbf{X}$$

for arbitrary constants $c_1$ and $c_2$. By hand, draw a phase portrait of the system.

# Numerical Solutions of Ordinary Differential Equations

Paul B. Moore/Shutterstock.com

Even if a solution of a differential equation exists, we may not be able to exhibit it in explicit or implicit form. In many instances we have to be content with an approximation of the solution. If a solution exists, it represents a set of points in the Cartesian plane. In this chapter we continue to explore the basic idea introduced in Section 2.6, that is, using the differential equation to construct an algorithm to approximate the *y*-coordinates of points on the actual solution curve. Our concentration in this chapter is primarily on first-order initial-value problems, but it concludes with a method for approximating solutions of linear second-order boundary-value problems.

## 9.1 Euler Methods and Error Analysis

**INTRODUCTION**  In Chapter 2 we examined one of the simplest numerical methods for approximating solutions of first-order initial-value problems $y' = f(x, y)$, $y(x_0) = y_0$. Recall that the backbone of **Euler's method** is the formula

$$y_{n+1} = y_n + hf(x_n, y_n), \tag{1}$$

where $f$ is the function obtained from the differential equation $y' = f(x, y)$. The recursive use of (1) for $n = 0, 1, 2, \ldots$ yields the $y$-coordinates $y_1, y_2, y_3, \ldots$ of points on successive "tangent lines" to the solution curve at $x_1, x_2, x_3, \ldots$ or $x_n = x_0 + nh$, where $h$ is a constant and is the size of the step between $x_n$ and $x_{n+1}$. The values $y_1, y_2, y_3, \ldots$ approximate the values of a solution $y(x)$ of the IVP at $x_1, x_2, x_3, \ldots$. But whatever advantage (1) has in its simplicity is lost in the crudeness of its approximations.

**A COMPARISON**  In Problem 4 in Exercises 2.6 you were asked to use Euler's method to obtain the approximate value of $y(1.5)$ for the solution of the initial-value problem $y' = 2xy$, $y(1) = 1$. You should have obtained the analytic solution $y = e^{x^2-1}$ and results similar to those given in Tables 9.1.1 and 9.1.2.

**TABLE 9.1.1**  Euler's Method with $h = 0.1$

| $x_n$ | $y_n$ | Actual value | Abs. error | % Rel. error |
|-------|-------|--------------|------------|--------------|
| 1.00 | 1.0000 | 1.0000 | 0.0000 | 0.00 |
| 1.10 | 1.2000 | 1.2337 | 0.0337 | 2.73 |
| 1.20 | 1.4640 | 1.5527 | 0.0887 | 5.71 |
| 1.30 | 1.8154 | 1.9937 | 0.1784 | 8.95 |
| 1.40 | 2.2874 | 2.6117 | 0.3244 | 12.42 |
| 1.50 | 2.9278 | 3.4903 | 0.5625 | 16.12 |

**TABLE 9.1.2**  Euler's Method with $h = 0.05$

| $x_n$ | $y_n$ | Actual value | Abs. error | % Rel. error |
|-------|-------|--------------|------------|--------------|
| 1.00 | 1.0000 | 1.0000 | 0.0000 | 0.00 |
| 1.05 | 1.1000 | 1.1079 | 0.0079 | 0.72 |
| 1.10 | 1.2155 | 1.2337 | 0.0182 | 1.47 |
| 1.15 | 1.3492 | 1.3806 | 0.0314 | 2.27 |
| 1.20 | 1.5044 | 1.5527 | 0.0483 | 3.11 |
| 1.25 | 1.6849 | 1.7551 | 0.0702 | 4.00 |
| 1.30 | 1.8955 | 1.9937 | 0.0982 | 4.93 |
| 1.35 | 2.1419 | 2.2762 | 0.1343 | 5.90 |
| 1.40 | 2.4311 | 2.6117 | 0.1806 | 6.92 |
| 1.45 | 2.7714 | 3.0117 | 0.2403 | 7.98 |
| 1.50 | 3.1733 | 3.4903 | 0.3171 | 9.08 |

In this case, with a step size $h = 0.1$ a 16% relative error in the calculation of the approximation to $y(1.5)$ is totally unacceptable. At the expense of doubling the number of calculations, some improvement in accuracy is obtained by halving the step size to $h = 0.05$.

**ERRORS IN NUMERICAL METHODS**  In choosing and using a numerical method for the solution of an initial-value problem, we must be aware of the various sources of errors. For some kinds of computation the accumulation of errors might reduce the accuracy of an approximation to the point of making the computation useless. On the other hand, depending on the use to which a numerical solution may be put, extreme accuracy might not be worth the added expense and complication.

One source of error that is always present in calculations is **round-off error.** This error results from the fact that any calculator or computer can represent numbers using only a finite number of digits. Suppose, for the sake of illustration, that we have a calculator that uses base 10 arithmetic and carries four digits, so that $\frac{1}{3}$ is represented in the calculator as 0.3333 and $\frac{1}{9}$ is represented as 0.1111. If we use this calculator to compute $\left(x^2 - \frac{1}{9}\right) / \left(x - \frac{1}{3}\right)$ for $x = 0.3334$, we obtain

$$\frac{(0.3334)^2 - 0.1111}{0.3334 - 0.3333} = \frac{0.1112 - 0.1111}{0.3334 - 0.3333} = 1.$$

With the help of a little algebra, however, we see that

$$\frac{x^2 - \frac{1}{9}}{x - \frac{1}{3}} = \frac{\left(x - \frac{1}{3}\right)\left(x + \frac{1}{3}\right)}{x - \frac{1}{3}} = x + \frac{1}{3},$$

so when $x = 0.3334$, $\left(x^2 - \frac{1}{9}\right) / \left(x - \frac{1}{3}\right) \approx 0.3334 + 0.3333 = 0.6667$. This example shows that the effects of round-off error can be quite serious unless some care is taken. One way to reduce the effect of round-off error is to minimize the number of calculations. Another technique on a computer is to use double-precision arithmetic to check the results. In general, round-off error is unpredictable and difficult to analyze, and we will neglect it in the error analysis that follows. We will concentrate on investigating the error introduced by using a formula or algorithm to approximate the values of the solution.

**TRUNCATION ERRORS FOR EULER'S METHOD** In the sequence of values $y_1, y_2, y_3, \ldots$ generated from (1), usually the value of $y_1$ will not agree with the actual solution at $x_1$—namely, $y(x_1)$—because the algorithm gives only a straight-line approximation to the solution. See Figure 2.6.2. The error is called the **local truncation error, formula error,** or **discretization error.** It occurs at each step; that is, if we assume that $y_n$ is accurate, then $y_{n+1}$ will contain local truncation error.

To derive a formula for the local truncation error for Euler's method, we use Taylor's formula with remainder. If a function $y(x)$ possesses $k + 1$ derivatives that are continuous on an open interval containing $a$ and $x$, then

$$y(x) = y(a) + y'(a)\frac{x - a}{1!} + \cdots + y^{(k)}(a)\frac{(x - a)^k}{k!} + y^{(k+1)}(c)\frac{(x - a)^{k+1}}{(k + 1)!},$$

where $c$ is some point between $a$ and $x$. Setting $k = 1$, $a = x_n$, and $x = x_{n+1} = x_n + h$, we get

$$y(x_{n+1}) = y(x_n) + y'(x_n)\frac{h}{1!} + y''(c)\frac{h^2}{2!}$$

or

$$y(x_{n+1}) = \underbrace{y_n + hf(x_n, y_n)}_{y_{n+1}} + y''(c)\frac{h^2}{2!}.$$

Euler's method (1) is the last formula without the last term; hence the local truncation error in $y_{n+1}$ is

$$y''(c)\frac{h^2}{2!}, \quad \text{where} \quad x_n < c < x_{n+1}.$$

The value of $c$ is usually unknown (it exists theoretically), so the *exact* error cannot be calculated, but an upper bound on the absolute value of the error is $Mh^2/2!$, where $M = \max_{x_n < x < x_{n+1}} |y''(x)|$.

In discussing errors that arise from the use of numerical methods, it is helpful to use the notation $O(h^n)$. To define this concept, we let $e(h)$ denote the error in a numerical calculation depending on $h$. Then $e(h)$ is said to be of order $h^n$, denoted by $O(h^n)$, if there exist a constant $C$ and a positive integer $n$ such that $|e(h)| \leq Ch^n$ for $h$ sufficiently small. Thus the local truncation error for Euler's method is $O(h^2)$. We note that, in general, if $e(h)$ in a numerical method is of order $h^n$ and $h$ is halved, the new error is approximately $C(h/2)^n = Ch^n/2^n$; that is, the error is reduced by a factor of $1/2^n$.

**EXAMPLE 1    Bound for Local Truncation Errors**

Find a bound for the local truncation errors for Euler's method applied to $y' = 2xy, y(1) = 1$.

**SOLUTION** From the solution $y = e^{x^2-1}$ we get $y'' = (2 + 4x^2)e^{x^2-1}$, so the local truncation error is

$$y''(c)\frac{h^2}{2} = (2 + 4c^2)e^{(c^2-1)}\frac{h^2}{2},$$

where $c$ is between $x_n$ and $x_n + h$. In particular, for $h = 0.1$ we can get an upper bound on the local truncation error for $y_1$ by replacing $c$ by 1.1:

$$[2 + (4)(1.1)^2]e^{((1.1)^2-1)}\frac{(0.1)^2}{2} = 0.0422.$$

From Table 9.1.1 we see that the error after the first step is 0.0337, less than the value given by the bound.

Similarly, we can get a bound for the local truncation error for any of the five steps given in Table 9.1.1 by replacing $c$ by 1.5 (this value of $c$ gives the largest value of $y''(c)$ for any of the steps and may be too generous for the first few steps). Doing this gives

$$[2 + (4)(1.5)^2]e^{((1.5)^2-1)}\frac{(0.1)^2}{2} = 0.1920 \tag{2}$$

as an upper bound for the local truncation error in each step. ◼

Note that if $h$ is halved to 0.05 in Example 1, then the error bound is 0.0480, about one-fourth as much as shown in (2). This is expected because the local truncation error for Euler's method is $O(h^2)$.

In the above analysis we assumed that the value of $y_n$ was exact in the calculation of $y_{n+1}$, but it is not because it contains local truncation errors from previous steps. The total error in $y_{n+1}$ is an accumulation of the errors in each of the previous steps. This total error is called the **global truncation error.** A complete analysis of the global truncation error is beyond the scope of this text, but it can be shown that the global truncation error for Euler's method is $O(h)$.

We expect that, for Euler's method, if the step size is halved the error will be approximately halved as well. This is borne out in Tables 9.1.1 and 9.1.2 where the absolute error at $x = 1.50$ with $h = 0.1$ is 0.5625 and with $h = 0.05$ is 0.3171, approximately half as large.

In general it can be shown that if a method for the numerical solution of a differential equation has local truncation error $O(h^{\alpha+1})$, then the global truncation error is $O(h^\alpha)$.

For the remainder of this section and in the subsequent sections we study methods that give significantly greater accuracy than does Euler's method.

**IMPROVED EULER'S METHOD**    The numerical method defined by the formula

$$y_{n+1} = y_n + h\frac{f(x_n, y_n) + f(x_{n+1}, y_{n-1}^*)}{2}, \tag{3}$$

where

$$y_{n+1}^* = y_n + hf(x_n, y_n) \tag{4}$$

**FIGURE 9.1.1** Slope of red dashed line is the average of $m_0$ and $m_1$

is commonly known as the **improved Euler's method.** To compute $y_{n+1}$ for $n = 0, 1, 2, \ldots$ from (3), we must, at each step, first use Euler's method (4) to obtain an initial estimate $y_{n+1}^*$. For example, with $n = 0$, (4) gives $y_1^* = y_0 + hf(x_0, y_0)$, and then, knowing this value, we use (3) to get $y_1 = y_0 + h\dfrac{f(x_0, y_0) + f(x_1, y_1^*)}{2}$, where $x_1 = x_0 + h$. These equations can be readily visualized. In Figure 9.1.1, observe that $m_0 = f(x_0, y_0)$ and $m_1 = f(x_1, y_1^*)$ are slopes of the solid straight lines shown passing through the points $(x_0, y_0)$ and $(x_1, y_1^*)$, respectively. By taking an average of these slopes, that is, $m_{ave} = \dfrac{f(x_0, y_0) + f(x_1, y_1^*)}{2}$, we obtain the slope of the parallel dashed skew lines. With the first step, rather than advancing along the line through $(x_0, y_0)$ with slope $f(x_0, y_0)$ to the point with $y$-coordinate $y_1^*$ obtained by Euler's method, we advance instead along the red dashed line through $(x_0, y_0)$ with slope $m_{ave}$ until we reach $x_1$. It seems plausible from inspection of the figure that $y_1$ is an improvement over $y_1^*$.

In general, the improved Euler's method is an example of a **predictor-corrector method.** The value of $y_{n+1}^*$ given by (4) predicts a value of $y(x_n)$, whereas the value of $y_{n+1}$ defined by formula (3) corrects this estimate.

### EXAMPLE 2    Improved Euler's Method

Use the improved Euler's method to obtain the approximate value of $y(1.5)$ for the solution of the initial-value problem $y' = 2xy$, $y(1) = 1$. Compare the results for $h = 0.1$ and $h = 0.05$.

**SOLUTION**  With $x_0 = 1$, $y_0 = 1$, $f(x_n, y_n) = 2x_n y_n$, $n = 0$, and $h = 0.1$, we first compute (4):

$$y_1^* = y_0 + (0.1)(2x_0 y_0) = 1 + (0.1)2(1)(1) = 1.2.$$

We use this last value in (3) along with $x_1 = 1 + h = 1 + 0.1 = 1.1$:

$$y_1 = y_0 + (0.1)\frac{2x_0 y_0 + 2x_1 y_1^*}{2} = 1 + (0.1)\frac{2(1)(1) + 2(1.1)(1.2)}{2} = 1.232.$$

The comparative values of the calculations for $h = 0.1$ and $h = 0.05$ are given in Tables 9.1.3 and 9.1.4, respectively.

**TABLE 9.1.3**    Improved Euler's Method with $h = 0.1$

| $x_n$ | $y_n$ | Actual value | Abs. error | % Rel. error |
|-------|-------|--------------|------------|--------------|
| 1.00 | 1.0000 | 1.0000 | 0.0000 | 0.00 |
| 1.10 | 1.2320 | 1.2337 | 0.0017 | 0.14 |
| 1.20 | 1.5479 | 1.5527 | 0.0048 | 0.31 |
| 1.30 | 1.9832 | 1.9937 | 0.0106 | 0.53 |
| 1.40 | 2.5908 | 2.6117 | 0.0209 | 0.80 |
| 1.50 | 3.4509 | 3.4904 | 0.0394 | 1.13 |

**TABLE 9.1.4**    Improved Euler's Method with $h = 0.05$

| $x_n$ | $y_n$ | Actual value | Abs. error | % Rel. error |
|-------|-------|--------------|------------|--------------|
| 1.00 | 1.0000 | 1.0000 | 0.0000 | 0.00 |
| 1.05 | 1.1077 | 1.1079 | 0.0002 | 0.02 |
| 1.10 | 1.2332 | 1.2337 | 0.0004 | 0.04 |
| 1.15 | 1.3798 | 1.3806 | 0.0008 | 0.06 |
| 1.20 | 1.5514 | 1.5527 | 0.0013 | 0.08 |
| 1.25 | 1.7531 | 1.7551 | 0.0020 | 0.11 |
| 1.30 | 1.9909 | 1.9937 | 0.0029 | 0.14 |
| 1.35 | 2.2721 | 2.2762 | 0.0041 | 0.18 |
| 1.40 | 2.6060 | 2.6117 | 0.0057 | 0.22 |
| 1.45 | 3.0038 | 3.0117 | 0.0079 | 0.26 |
| 1.50 | 3.4795 | 3.4904 | 0.0108 | 0.31 |

A brief word of caution is in order here. We cannot compute all the values of $y_n^*$ first and then substitute these values into formula (3). In other words, we cannot use the data in Table 9.1.1 to help construct the values in Table 9.1.3. Why not?

**TRUNCATION ERRORS FOR THE IMPROVED EULER'S METHOD**   The local truncation error for the improved Euler's method is $O(h^3)$. The derivation of this result is similar to the derivation of the local truncation error for Euler's method. Since the local truncation error for the improved Euler's method is $O(h^3)$, the global truncation error is $O(h^2)$. This can be seen in Example 2; when the step size is halved from $h = 0.1$ to $h = 0.05$, the absolute error at $x = 1.50$ is reduced from 0.0394 to 0.0108, a reduction of approximately $\left(\frac{1}{2}\right)^2 = \frac{1}{4}$.

---

## EXERCISES 9.1

*Answers to selected odd-numbered problems begin on page ANS-15.*

In Problems 1–10 use the improved Euler's method to obtain a four-decimal approximation of the indicated value. First use $h = 0.1$ and then use $h = 0.05$.

**1.** $y' = 2x - 3y + 1$,   $y(1) = 5$;   $y(1.5)$

**2.** $y' = 4x - 2y$,   $y(0) = 2$;   $y(0.5)$

**3.** $y' = 1 + y^2$,   $y(0) = 0$;   $y(0.5)$

**4.** $y' = x^2 + y^2$,   $y(0) = 1$;   $y(0.5)$

**5.** $y' = e^{-y}$,   $y(0) = 0$;   $y(0.5)$

**6.** $y' = x + y^2$,   $y(0) = 0$;   $y(0.5)$

**7.** $y' = (x - y)^2$,   $y(0) = 0.5$;   $y(0.5)$

**8.** $y' = xy + \sqrt{y}$,   $y(0) = 1$;   $y(0.5)$

**9.** $y' = xy^2 - \dfrac{y}{x}$,   $y(1) = 1$;   $y(1.5)$

**10.** $y' = y - y^2$,   $y(0) = 0.5$;   $y(0.5)$

**11.** Consider the initial-value problem $y' = (x + y - 1)^2$, $y(0) = 2$. Use the improved Euler's method with $h = 0.1$ and $h = 0.05$ to obtain approximate values of the solution at $x = 0.5$. At each step compare the approximate value with the actual value of the analytic solution.

**12.** Although it might not be obvious from the differential equation, its solution could "behave badly" near a point $x$ at which we wish to approximate $y(x)$. Numerical procedures may give widely differing results near this point. Let $y(x)$ be the solution of the initial-value problem $y' = x^2 + y^3$, $y(1) = 1$.

  **(a)** Use a numerical solver to graph the solution on the interval [1, 1.4].

  **(b)** Using the step size $h = 0.1$, compare the results obtained from Euler's method with the results from the improved Euler's method in the approximation of $y(1.4)$.

**13.** Consider the initial-value problem $y' = 2y$, $y(0) = 1$. The analytic solution is $y = e^{2x}$.

  **(a)** Approximate $y(0.1)$ using one step and Euler's method.

  **(b)** Find a bound for the local truncation error in $y_1$.

  **(c)** Compare the error in $y_1$ with your error bound.

  **(d)** Approximate $y(0.1)$ using two steps and Euler's method.

  **(e)** Verify that the global truncation error for Euler's method is $O(h)$ by comparing the errors in parts (a) and (d).

**14.** Repeat Problem 13 using the improved Euler's method. Its global truncation error is $O(h^2)$.

**15.** Repeat Problem 13 using the initial-value problem $y' = x - 2y$, $y(0) = 1$. The analytic solution is

$$y(x) = \tfrac{1}{2}x - \tfrac{1}{4} + \tfrac{5}{4}e^{-2x}.$$

**16.** Repeat Problem 15 using the improved Euler's method. Its global truncation error is $O(h^2)$.

**17.** Consider the initial-value problem $y' = 2x - 3y + 1$, $y(1) = 5$. The analytic solution is

$$y(x) = \tfrac{1}{9} + \tfrac{2}{3}x + \tfrac{38}{9}e^{-3(x-1)}.$$

  **(a)** Find a formula involving $c$ and $h$ for the local truncation error in the $n$th step if Euler's method is used.

  **(b)** Find a bound for the local truncation error in each step if $h = 0.1$ is used to approximate $y(1.5)$.

  **(c)** Approximate $y(1.5)$ using $h = 0.1$ and $h = 0.05$ with Euler's method. See Problem 1 in Exercises 2.6.

  **(d)** Calculate the errors in part (c) and verify that the global truncation error of Euler's method is $O(h)$.

**18.** Repeat Problem 17 using the improved Euler's method, which has a global truncation error $O(h^2)$. See Problem 1. You might need to keep more than four decimal places to see the effect of reducing the order of the error.

**19.** Repeat Problem 17 for the initial-value problem $y' = e^{-y}$, $y(0) = 0$. The analytic solution is $y(x) = \ln(x + 1)$. Approximate $y(0.5)$. See Problem 5 in Exercises 2.6.

**20.** Repeat Problem 19 using the improved Euler's method, which has global truncation error $O(h^2)$. See Problem 5. You might need to keep more than four decimal places to see the effect of reducing the order of error.

### Discussion Problems

**21.** Answer the question "Why not?" that follows the three sentences after Example 2 on page 372.

## 9.2 Runge-Kutta Methods

**INTRODUCTION** Probably one of the more popular as well as most accurate numerical procedures used in obtaining approximate solutions to a first-order initial-value problem $y' = f(x, y)$, $y(x_0) = y_0$ is the **fourth-order Runge-Kutta method.** As the name suggests, there are Runge-Kutta methods of different orders.

**RUNGE-KUTTA METHODS** Fundamentally, all Runge-Kutta methods are generalizations of the basic Euler formula (1) of Section 9.1 in that the slope function $f$ is replaced by a weighted average of slopes over the interval $x_n \leq x \leq x_{n+1}$. That is,

$$\overbrace{\phantom{w_1k_1 + w_2k_2 + \cdots + w_mk_m}}^{\text{weighted average}}$$
$$y_{n+1} = y_n + h\,(w_1k_1 + w_2k_2 + \cdots + w_mk_m). \tag{1}$$

Here the weights $w_i$, $i = 1, 2, \ldots, m$, are constants that generally satisfy $w_1 + w_2 + \cdots + w_m = 1$, and each $k_i$, $i = 1, 2, \ldots, m$, is the function $f$ evaluated at a selected point $(x, y)$ for which $x_n \leq x \leq x_{n+1}$. We shall see that the $k_i$ are defined recursively. The number $m$ is called the **order** of the method. Observe that by taking $m = 1$, $w_1 = 1$, and $k_1 = f(x_n, y_n)$, we get the familiar Euler formula $y_{n+1} = y_n + hf(x_n, y_n)$. Hence Euler's method is said to be a **first-order Runge-Kutta method.**

The average in (1) is not formed willy-nilly, but parameters are chosen so that (1) agrees with a Taylor polynomial of degree $m$. As we saw in the preceding section, if a function $y(x)$ possesses $k + 1$ derivatives that are continuous on an open interval containing $a$ and $x$, then we can write

$$y(x) = y(a) + y'(a)\frac{x - a}{1!} + y''(a)\frac{(x - a)^2}{2!} + \cdots + y^{(k+1)}(c)\frac{(x - a)^{k+1}}{(k + 1)!},$$

where $c$ is some number between $a$ and $x$. If we replace $a$ by $x_n$ and $x$ by $x_{n+1} = x_n + h$, then the foregoing formula becomes

$$y(x_{n+1}) = y(x_n + h) = y(x_n) + hy'(x_n) + \frac{h^2}{2!}y''(x_n) + \cdots + \frac{h^{k+1}}{(k + 1)!}y^{(k+1)}(c),$$

where $c$ is now some number between $x_n$ and $x_{n+1}$. When $y(x)$ is a solution of $y' = f(x, y)$ in the case $k = 1$ and the remainder $\frac{1}{2}h^2y''(c)$ is close to 0, we see that a Taylor polynomial $y(x_{n+1}) = y(x_n) + hy'(x_n)$ of degree one agrees with the approximation formula of Euler's method

$$y_{n+1} = y_n + hy'_n = y_n + hf(x_n, y_n).$$

**A SECOND-ORDER RUNGE-KUTTA METHOD** To further illustrate (1), we consider now a **second-order Runge-Kutta procedure.** This consists of finding constants or parameters $w_1$, $w_2$, $\alpha$, and $\beta$ so that the formula

$$y_{n+1} = y_n + h(w_1k_1 + w_2k_2), \tag{2}$$

where
$$k_1 = f(x_n, y_n)$$
$$k_2 = f(x_n + \alpha h, y_n + \beta hk_1),$$

agrees with a Taylor polynomial of degree two. For our purposes it suffices to say that this can be done whenever the constants satisfy

$$w_1 + w_2 = 1, \qquad w_2\alpha = \frac{1}{2}, \qquad \text{and} \qquad w_2\beta = \frac{1}{2}. \tag{3}$$

This is an algebraic system of three equations in four unknowns and has infinitely many solutions:

$$w_1 = 1 - w_2, \qquad \alpha = \frac{1}{2w_2}, \qquad \text{and} \qquad \beta = \frac{1}{2w_2}, \tag{4}$$

where $w_2 \neq 0$. For example, the choice $w_2 = \frac{1}{2}$ yields $w_1 = \frac{1}{2}$, $\alpha = 1$, and $\beta = 1$, and so (2) becomes

$$y_{n+1} = y_n + \frac{h}{2}(k_1 + k_2),$$

where $\qquad k_1 = f(x_n, y_n) \qquad$ and $\qquad k_2 = f(x_n + h, y_n + hk_1).$

Since $x_n + h = x_{n+1}$ and $y_n + hk_1 = y_n + hf(x_n, y_n)$, the foregoing result is recognized to be the improved Euler's method that is summarized in (3) and (4) of Section 9.1.

In view of the fact that $w_2 \neq 0$ can be chosen arbitrarily in (4), there are many possible second-order Runge-Kutta methods. See Problem 2 in Exercises 9.2.

We shall skip any discussion of third-order methods in order to come to the principal point of discussion in this section.

### A FOURTH-ORDER RUNGE-KUTTA METHOD   A **fourth-order Runge-Kutta procedure** consists of finding parameters so that the formula

$$y_{n+1} = y_n + h(w_1 k_1 + w_2 k_2 + w_3 k_3 + w_4 k_4), \tag{5}$$

where
$$k_1 = f(x_n, y_n)$$

$$k_2 = f(x_n + \alpha_1 h, y_n + \beta_1 hk_1)$$

$$k_3 = f(x_n + \alpha_2 h, y_n + \beta_2 hk_1 + \beta_3 hk_2)$$

$$k_4 = f(x_n + \alpha_3 h, y_n + \beta_4 hk_1 + \beta_5 hk_2 + \beta_6 hk_3),$$

agrees with a Taylor polynomial of degree four. This results in a system of 11 equations in 13 unknowns. The most commonly used set of values for the parameters yields the following result:

$$y_{n+1} = y_n + \frac{h}{6}(k_1 + 2k_2 + 2k_3 + k_4),$$

$$k_1 = f(x_n, y_n)$$

$$k_2 = f\left(x_n + \tfrac{1}{2}h, y_n + \tfrac{1}{2}hk_1\right) \tag{6}$$

$$k_3 = f\left(x_n + \tfrac{1}{2}h, y_n + \tfrac{1}{2}hk_2\right)$$

$$k_4 = f(x_n + h, y_n + hk_3).$$

While other fourth-order formulas are easily derived, the algorithm summarized in (6) is so widely used and recognized as a valuable computational tool it is often referred to as *the* fourth-order Runge-Kutta method or *the classical* Runge-Kutta method. It is (6) that we have in mind, hereafter, when we use the abbreviation the **RK4 method.**

You are advised to look carefully at the formulas in (6); note that $k_2$ depends on $k_1$, $k_3$ depends on $k_2$, and $k_4$ depends on $k_3$. Also, $k_2$ and $k_3$ involve approximations to the slope at the midpoint $x_n + \frac{1}{2}h$ of the interval defined by $x_n \leq x \leq x_{n+1}$.

### EXAMPLE 1   RK4 Method

Use the RK4 method with $h = 0.1$ to obtain an approximation to $y(1.5)$ for the solution of $y' = 2xy$, $y(1) = 1$.

**SOLUTION** For the sake of illustration let us compute the case when $n = 0$. From (6) we find

$$k_1 = f(x_0, y_0) = 2x_0 y_0 = 2$$

$$k_2 = f\left(x_0 + \tfrac{1}{2}(0.1), y_0 + \tfrac{1}{2}(0.1)2\right)$$

$$= 2\left(x_0 + \tfrac{1}{2}(0.1)\right)\left(y_0 + \tfrac{1}{2}(0.2)\right) = 2.31$$

$$k_3 = f\left(x_0 + \tfrac{1}{2}(0.1), y_0 + \tfrac{1}{2}(0.1)2.31\right)$$

$$= 2\left(x_0 + \tfrac{1}{2}(0.1)\right)\left(y_0 + \tfrac{1}{2}(0.231)\right) = 2.34255$$

$$k_4 = f(x_0 + (0.1), y_0 + (0.1)2.34255)$$

$$= 2(x_0 + 0.1)(y_0 + 0.234255) = 2.715361$$

and therefore

$$y_1 = y_0 + \frac{0.1}{6}(k_1 + 2k_2 + 2k_3 + k_4)$$

$$= 1 + \frac{0.1}{6}(2 + 2(2.31) + 2(2.34255) + 2.715361) = 1.23367435.$$

The remaining calculations are summarized in Table 9.2.1, whose entries are rounded to four decimal places.  ■

Inspection of Table 9.2.1 shows why the fourth-order Runge-Kutta method is so popular. If four-decimal-place accuracy is all that we desire, there is no need to use a smaller step size. Table 9.2.2 compares the results of applying Euler's, the improved Euler's, and the fourth-order Runge-Kutta methods to the initial-value problem $y' = 2xy$, $y(1) = 1$. (See Tables 9.1.1–9.1.4.)

**TABLE 9.2.1**  RK4 Method with $h = 0.1$

| $x_n$ | $y_n$ | Actual value | Abs. error | % Rel. error |
|-------|-------|--------------|-----------|--------------|
| 1.00 | 1.0000 | 1.0000 | 0.0000 | 0.00 |
| 1.10 | 1.2337 | 1.2337 | 0.0000 | 0.00 |
| 1.20 | 1.5527 | 1.5527 | 0.0000 | 0.00 |
| 1.30 | 1.9937 | 1.9937 | 0.0000 | 0.00 |
| 1.40 | 2.6116 | 2.6117 | 0.0001 | 0.00 |
| 1.50 | 3.4902 | 3.4904 | 0.0001 | 0.00 |

**TABLE 9.2.2**  $y' = 2xy$, $y(1) = 1$

| | Comparison of numerical methods with $h = 0.1$ | | | | | Comparison of numerical methods with $h = 0.05$ | | | |
|-------|-------|------------------|-------|--------------|-------|-------|------------------|-------|--------------|
| $x_n$ | Euler | Improved Euler | RK4 | Actual value | $x_n$ | Euler | Improved Euler | RK4 | Actual value |
| 1.00 | 1.0000 | 1.0000 | 1.0000 | 1.0000 | 1.00 | 1.0000 | 1.0000 | 1.0000 | 1.0000 |
| 1.10 | 1.2000 | 1.2320 | 1.2337 | 1.2337 | 1.05 | 1.1000 | 1.1077 | 1.1079 | 1.1079 |
| 1.20 | 1.4640 | 1.5479 | 1.5527 | 1.5527 | 1.10 | 1.2155 | 1.2332 | 1.2337 | 1.2337 |
| 1.30 | 1.8154 | 1.9832 | 1.9937 | 1.9937 | 1.15 | 1.3492 | 1.3798 | 1.3806 | 1.3806 |
| 1.40 | 2.2874 | 2.5908 | 2.6116 | 2.6117 | 1.20 | 1.5044 | 1.5514 | 1.5527 | 1.5527 |
| 1.50 | 2.9278 | 3.4509 | 3.4902 | 3.4904 | 1.25 | 1.6849 | 1.7531 | 1.7551 | 1.7551 |
| | | | | | 1.30 | 1.8955 | 1.9909 | 1.9937 | 1.9937 |
| | | | | | 1.35 | 2.1419 | 2.2721 | 2.2762 | 2.2762 |
| | | | | | 1.40 | 2.4311 | 2.6060 | 2.6117 | 2.6117 |
| | | | | | 1.45 | 2.7714 | 3.0038 | 3.0117 | 3.0117 |
| | | | | | 1.50 | 3.1733 | 3.4795 | 3.4903 | 3.4904 |

**TRUNCATION ERRORS FOR THE RK4 METHOD**  In Section 9.1 we saw that global truncation errors for Euler's method and for the improved Euler's method are, respectively, $O(h)$ and $O(h^2)$. Because the first equation in (6) agrees with a Taylor polynomial of degree four, the local truncation error for this method is $y^{(5)}(c)\, h^5/5!$ or $O(h^5)$, and the global truncation error is thus $O(h^4)$. It is now obvious why Euler's method, the improved Euler's method, and (6) are *first-*, *second-*, and *fourth-order* Runge-Kutta methods, respectively.

**EXAMPLE 2**  Bound for Local Truncation Errors

Find a bound for the local truncation errors for the RK4 method applied to $y' = 2xy$, $y(1) = 1$.

**SOLUTION** By computing the fifth derivative of the known solution $y(x) = e^{x^2-1}$, we get

$$y^{(5)}(c)\frac{h^5}{5!} = (120c + 160c^3 + 32c^5)e^{c^2-1}\frac{h^5}{5!}. \tag{7}$$

Thus with $c = 1.5$, (7) yields a bound of 0.00028 on the local truncation error for each of the five steps when $h = 0.1$. Note that in Table 9.2.1 the error in $y_1$ is much less than this bound.

Table 9.2.3 gives the approximations to the solution of the initial-value problem at $x = 1.5$ that are obtained from the RK4 method. By computing the value of the analytic solution at $x = 1.5$, we can find the error in these approximations. Because the method is so accurate, many decimal places must be used in the numerical solution to see the effect of halving the step size. Note that when $h$ is halved, from $h = 0.1$ to $h = 0.05$, the error is divided by a factor of about $2^4 = 16$, as expected. ■

**TABLE 9.2.3** RK4 Method

| $h$ | Approx. | Error |
|------|-----------|------------------------------|
| 0.1 | 3.49021064 | $1.32321089 \times 10^{-4}$ |
| 0.05 | 3.49033382 | $9.13776090 \times 10^{-6}$ |

**ADAPTIVE METHODS**   We have seen that the accuracy of a numerical method for approximating solutions of differential equations can be improved by decreasing the step size $h$. Of course, this enhanced accuracy is usually obtained at a cost—namely, increased computation time and greater possibility of round-off error. In general, over the interval of approximation there may be subintervals where a relatively large step size suffices and other subintervals where a smaller step is necessary to keep the truncation error within a desired limit. Numerical methods that use a variable step size are called **adaptive methods.** One of the more popular of the adaptive routines is the **Runge-Kutta-Fehlberg method.** Because Fehlberg employed two Runge-Kutta methods of differing orders, a fourth- and a fifth-order method, this algorithm is frequently denoted as the **RKF45 method.**[*]

---

[*]The Runge-Kutta method of order four used in RKF45 is *not* the same as that given in (6).

---

## EXERCISES 9.2

Answers to selected odd-numbered problems begin on page ANS-16.

1. Use the RK4 method with $h = 0.1$ to approximate $y(0.5)$, where $y(x)$ is the solution of the initial-value problem $y' = (x + y - 1)^2$, $y(0) = 2$. Compare this approximate value with the actual value obtained in Problem 11 in Exercises 9.1.

2. Assume that $w_2 = \frac{3}{4}$ in (4). Use the resulting second-order Runge-Kutta method to approximate $y(0.5)$, where $y(x)$ is the solution of the initial-value problem in Problem 1. Compare this approximate value with the approximate value obtained in Problem 11 in Exercises 9.1.

In Problems 3–12 use the RK4 method with $h = 0.1$ to obtain a four-decimal approximation of the indicated value.

3. $y' = 2x - 3y + 1$, $y(1) = 5$; $y(1.5)$

4. $y' = 4x - 2y$, $y(0) = 2$; $y(0.5)$

5. $y' = 1 + y^2$, $y(0) = 0$; $y(0.5)$

6. $y' = x^2 + y^2$, $y(0) = 1$; $y(0.5)$

7. $y' = e^{-y}$, $y(0) = 0$; $y(0.5)$

8. $y' = x + y^2$, $y(0) = 0$; $y(0.5)$

9. $y' = (x - y)^2$, $y(0) = 0.5$; $y(0.5)$

10. $y' = xy + \sqrt{y}$, $y(0) = 1$; $y(0.5)$

11. $y' = xy^2 - \dfrac{y}{x}$, $y(1) = 1$; $y(1.5)$

12. $y' = y - y^2$, $y(0) = 0.5$; $y(0.5)$

13. If air resistance is proportional to the square of the instantaneous velocity, then the velocity $v$ of a mass $m$ dropped from a given height is determined from

$$m\frac{dv}{dt} = mg - kv^2, \qquad k > 0.$$

Let $v(0) = 0$, $k = 0.125$, $m = 5$ slugs, and $g = 32$ ft/s$^2$.

(a) Use the RK4 method with $h = 1$ to approximate the velocity $v(5)$.

(b) Use a numerical solver to graph the solution of the IVP on the interval $[0, 6]$.

(c) Use separation of variables to solve the IVP and then find the actual value $v(5)$.

14. A mathematical model for the area $A$ (in cm$^2$) that a colony of bacteria (*B. dendroides*) occupies is given by

$$\frac{dA}{dt} = A(2.128 - 0.0432A).[*]$$

---

[*]See V. A. Kostitzin, *Mathematical Biology* (London: Harrap, 1939).

Suppose that the initial area is 0.24 cm$^2$.

(a) Use the RK4 method with $h = 0.5$ to complete the following table:

| t (days) | 1 | 2 | 3 | 4 | 5 |
|---|---|---|---|---|---|
| A (observed) | 2.78 | 13.53 | 36.30 | 47.50 | 49.40 |
| A (approximated) | | | | | |

(b) Use a numerical solver to graph the solution of the initial-value problem. Estimate the values $A(1), A(2), A(3), A(4)$, and $A(5)$ from the graph.

(c) Use separation of variables to solve the initial-value problem and compute the actual values $A(1), A(2), A(3), A(4)$, and $A(5)$.

15. Consider the initial-value problem $y' = x^2 + y^3, y(1) = 1$. See Problem 12 in Exercises 9.1.

(a) Compare the results obtained from using the RK4 method over the interval $[1, 1.4]$ with step sizes $h = 0.1$ and $h = 0.05$.

(b) Use a numerical solver to graph the solution of the initial-value problem on the interval $[1, 1.4]$.

16. Consider the initial-value problem $y' = 2y, y(0) = 1$. The analytic solution is $y(x) = e^{2x}$.

(a) Approximate $y(0.1)$ using one step and the RK4 method.

(b) Find a bound for the local truncation error in $y_1$.

(c) Compare the error in $y_1$ with your error bound.

(d) Approximate $y(0.1)$ using two steps and the RK4 method.

(e) Verify that the global truncation error for the RK4 method is $O(h^4)$ by comparing the errors in parts (a) and (d).

17. Repeat Problem 16 using the initial-value problem $y' = -2y + x, \quad y(0) = 1$. The analytic solution is

$$y(x) = \tfrac{1}{2}x - \tfrac{1}{4} + \tfrac{5}{4}e^{-2x}.$$

18. Consider the initial-value problem $y' = 2x - 3y + 1, y(1) = 5$. The analytic solution is

$$y(x) = \tfrac{1}{9} + \tfrac{2}{3}x + \tfrac{38}{9}e^{-3(x-1)}.$$

(a) Find a formula involving $c$ and $h$ for the local truncation error in the $n$th step if the RK4 method is used.

(b) Find a bound for the local truncation error in each step if $h = 0.1$ is used to approximate $y(1.5)$.

(c) Approximate $y(1.5)$ using the RK4 method with $h = 0.1$ and $h = 0.05$. See Problem 3. You will need to carry more than six decimal places to see the effect of reducing the step size.

19. Repeat Problem 18 for the initial-value problem $y' = e^{-y}, y(0) = 0$. The analytic solution is $y(x) = \ln(x + 1)$. Approximate $y(0.5)$. See Problem 7.

### Discussion Problems

20. A count of the number of evaluations of the function $f$ used in solving the initial-value problem $y' = f(x, y), y(x_0) = y_0$ is used as a measure of the computational complexity of a numerical method. Determine the number of evaluations of $f$ required for each step of Euler's, the improved Euler's, and the RK4 methods. By considering some specific examples, compare the accuracy of these methods when used with comparable computational complexities.

### Computer Lab Assignments

21. The RK4 method for solving an initial-value problem over an interval $[a, b]$ results in a finite set of points that are supposed to approximate points on the graph of the exact solution. To expand this set of discrete points to an approximate solution defined at all points on the interval $[a, b]$, we can use an **interpolating function.** This is a function, supported by most computer algebra systems, that agrees with the given data exactly and assumes a smooth transition between data points. These interpolating functions may be polynomials or sets of polynomials joined together smoothly. In *Mathematica* the command **y = Interpolation[data]** can be used to obtain an interpolating function through the points **data = $\{\{x_0, y_0\}, \{x_1, y_1\}, \ldots, \{x_n, y_n\}\}$**. The interpolating function **y[x]** can now be treated like any other function built into the computer algebra system.

(a) Find the analytic solution of the initial-value problem $y' = -y + 10 \sin 3x; y(0) = 0$ on the interval $[0, 2]$. Graph this solution and find its positive roots.

(b) Use the RK4 method with $h = 0.1$ to approximate a solution of the initial-value problem in part (a). Obtain an interpolating function and graph it. Find the positive roots of the interpolating function of the interval $[0, 2]$.

---

## 9.3 Multistep Methods

**INTRODUCTION** Euler's method, the improved Euler's method, and the Runge-Kutta methods are examples of **single-step** or **starting methods.** In these methods each successive value $y_{n+1}$ is computed based only on information about the immediately preceding value $y_n$. On the other hand, **multistep** or **continuing methods** use the values from several computed steps to obtain the value of $y_{n+1}$. There are a large number of multistep method formulas for approximating solutions of DEs, but since it is not our intention to survey the vast field of numerical procedures, we will consider only one such method here.

**ADAMS-BASHFORTH-MOULTON METHOD** The multistep method that is discussed in this section is called the fourth-order **Adams-Bashforth-Moulton method.** Like the improved Euler's method it is a predictor-corrector method—that is, one formula is used to predict a value $y_{n+1}^*$, which in turn is used to obtain a corrected value $y_{n+1}$. The predictor in this method is the Adams-Bashforth formula

$$y_{n+1}^* = y_n + \frac{h}{24}(55y_n' - 59y_{n-1}' + 37y_{n-2}' - 9y_{n-3}'), \tag{1}$$

$$y_n' = f(x_n, y_n)$$
$$y_{n-1}' = f(x_{n-1}, y_{n-1})$$
$$y_{n-2}' = f(x_{n-2}, y_{n-2})$$
$$y_{n-3}' = f(x_{n-3}, y_{n-3})$$

for $n \geq 3$. The value of $y_{n+1}^*$ is then substituted into the Adams-Moulton corrector

$$y_{n+1} = y_n + \frac{h}{24}(9y_{n+1}' + 19y_n' - 5y_{n-1}' + y_{n-2}') \tag{2}$$

$$y_{n+1}' = f(x_{n+1}, y_{n+1}^*).$$

Notice that formula (1) requires that we know the values of $y_0$, $y_1$, $y_2$, and $y_3$ to obtain $y_4$. The value of $y_0$ is, of course, the given initial condition. The local truncation error of the Adams-Bashforth-Moulton method is $O(h^5)$, the values of $y_1$, $y_2$, and $y_3$ are generally computed by a method with the same error property, such as the fourth-order Runge-Kutta method.

### EXAMPLE 1 Adams-Bashforth-Moulton Method

Use the Adams-Bashforth-Moulton method with $h = 0.2$ to obtain an approximation to $y(0.8)$ for the solution of

$$y' = x + y - 1, \quad y(0) = 1.$$

**SOLUTION** With a step size of $h = 0.2$, $y(0.8)$ will be approximated by $y_4$. To get started, we use the RK4 method with $x_0 = 0$, $y_0 = 1$, and $h = 0.2$ to obtain

$$y_1 = 1.02140000, \qquad y_2 = 1.09181796, \qquad y_3 = 1.22210646.$$

Now with the identifications $x_0 = 0$, $x_1 = 0.2$, $x_2 = 0.4$, $x_3 = 0.6$, and $f(x, y) = x + y - 1$, we find

$$y_0' = f(x_0, y_0) = (0) + (1) - 1 = 0$$
$$y_1' = f(x_1, y_1) = (0.2) + (1.02140000) - 1 = 0.22140000$$
$$y_2' = f(x_2, y_2) = (0.4) + (1.09181796) - 1 = 0.49181796$$
$$y_3' = f(x_3, y_3) = (0.6) + (1.22210646) - 1 = 0.82210646.$$

With the foregoing values the predictor (1) then gives

$$y_4^* = y_3 + \frac{0.2}{24}(55y_3' - 59y_2' + 37y_1' - 9y_0') = 1.42535975.$$

To use the corrector (2), we first need

$$y_4' = f(x_4, y_4^*) = 0.8 + 1.42535975 - 1 = 1.22535975.$$

Finally, (2) yields

$$y_4 = y_3 + \frac{0.2}{24}(9y_4' + 19y_3' - 5y_2' + y_1') = 1.42552788.$$

You should verify that the actual value of $y(0.8)$ in Example 1 is $y(0.8) = 1.42554093$. See Problem 1 in Exercises 9.3.

**STABILITY OF NUMERICAL METHODS**    An important consideration in using numerical methods to approximate the solution of an initial-value problem is the stability of the method. Simply stated, a numerical method is **stable** if small changes in the initial condition result in only small changes in the computed solution. A numerical method is said to be **unstable** if it is not stable. The reason that stability considerations are important is that in each step after the first step of a numerical technique we are essentially starting over again with a new initial-value problem, where the initial condition is the approximate solution value computed in the preceding step. Because of the presence of round-off error, this value will almost certainly vary at least slightly from the true value of the solution. Besides round-off error, another common source of error occurs in the initial condition itself; in physical applications the data are often obtained by imprecise measurements.

One possible method for detecting instability in the numerical solution of a specific initial-value problem is to compare the approximate solutions obtained when decreasing step sizes are used. If the numerical method is unstable, the error may actually increase with smaller step sizes. Another way of checking stability is to observe what happens to solutions when the initial condition is slightly perturbed (for example, change $y(0) = 1$ to $y(0) = 0.999$).

For a more detailed and precise discussion of stability, consult a numerical analysis text. In general, all of the methods that we have discussed in this chapter have good stability characteristics.

**ADVANTAGES AND DISADVANTAGES OF MULTISTEP METHODS**    Many considerations enter into the choice of a method to solve a differential equation numerically. Single-step methods, particularly the RK4 method, are often chosen because of their accuracy and the fact that they are easy to program. However, a major drawback is that the right-hand side of the differential equation must be evaluated many times at each step. For instance, the RK4 method requires four function evaluations for each step. On the other hand, if the function evaluations in the previous step have been calculated and stored, a multistep method requires only one new function evaluation for each step. This can lead to great savings in time and expense.

As an example, solving $y' = f(x, y)$, $y(x_0) = y_0$ numerically using $n$ steps by the fourth-order Runge-Kutta method requires $4n$ function evaluations. The Adams-Bashforth multistep method requires 16 function evaluations for the Runge-Kutta fourth-order starter and $n - 4$ for the $n$ Adams-Bashforth steps, giving a total of $n + 12$ function evaluations for this method. In general the Adams-Bashforth multistep method requires slightly more than a quarter of the number of function evaluations required for the RK4 method. If the evaluation of $f(x, y)$ is complicated, the multistep method will be more efficient.

Another issue that is involved with multistep methods is how many times the Adams-Moulton corrector formula should be repeated in each step. Each time the corrector is used, another function evaluation is done, and so the accuracy is increased at the expense of losing an advantage of the multistep method. In practice, the corrector is calculated once, and if the value of $y_{n+1}$ is changed by a large amount, the entire problem is restarted using a smaller step size. This is often the basis of the variable step size methods, whose discussion is beyond the scope of this text.

**1.** Find the analytic solution of the initial-value problem in Example 1. Compare the actual values of $y(0.2)$, $y(0.4)$, $y(0.6)$, and $y(0.8)$ with the approximations $y_1$, $y_2$, $y_3$, and $y_4$.

**2.** Write a computer program to implement the Adams-Bashforth-Moulton method.

In Problems 3 and 4 use the Adams-Bashforth-Moulton method to approximate $y(0.8)$, where $y(x)$ is the solution of the given initial-value problem. Use $h = 0.2$ and the RK4 method to compute $y_1$, $y_2$, and $y_3$.

**3.** $y' = 2x - 3y + 1$,   $y(0) = 1$

**4.** $y' = 4x - 2y$,   $y(0) = 2$

In Problems 5–8 use the Adams-Bashforth-Moulton method to approximate $y(1.0)$, where $y(x)$ is the solution of the given initial-value problem. First use $h = 0.2$ and then use $h = 0.1$. Use the RK4 method to compute $y_1$, $y_2$, and $y_3$.

**5.** $y' = 1 + y^2$,   $y(0) = 0$

**6.** $y' = y + \cos x$,   $y(0) = 1$

**7.** $y' = (x - y)^2$,   $y(0) = 0$

**8.** $y' = xy + \sqrt{y}$,   $y(0) = 1$

---

## 9.4 Higher-Order Equations and Systems

**INTRODUCTION** So far, we have focused on numerical techniques that can be used to approximate the solution of a first-order initial-value problem $y' = f(x, y)$, $y(x_0) = y_0$. In order to approximate the solution of a second-order initial-value problem, we must express a second-order DE as a system of two first-order DEs. To do this, we begin by writing the second-order DE in normal form by solving for $y''$ in terms of $x$, $y$, and $y'$.

**SECOND-ORDER IVPs**   A second-order initial-value problem

$$y'' = f(x, y, y'), \quad y(x_0) = y_0, \quad y'(x_0) = u_0 \tag{1}$$

can be expressed as an initial-value problem for a system of first-order differential equations. If we let $y' = u$, the differential equation in (1) becomes the system

$$\begin{aligned} y' &= u \\ u' &= f(x, y, u). \end{aligned} \tag{2}$$

Since $y'(x_0) = u(x_0)$, the corresponding initial conditions for (2) are then $y(x_0) = y_0$, $u(x_0) = u_0$. The system (2) can now be solved numerically by simply applying a particular numerical method to each first-order differential equation in the system. For example, **Euler's method** applied to the system (2) would be

$$\begin{aligned} y_{n+1} &= y_n + hu_n \\ u_{n+1} &= u_n + hf(x_n, y_n, u_n), \end{aligned} \tag{3}$$

whereas the **fourth-order Runge-Kutta method,** or **RK4 method,** would be

$$\begin{aligned} y_{n+1} &= y_n + \frac{h}{6}(m_1 + 2m_2 + 2m_3 + m_4) \\ u_{n+1} &= u_n + \frac{h}{6}(k_1 + 2k_2 + 2k_3 + k_4) \end{aligned} \tag{4}$$

where

$$\begin{aligned} m_1 &= u_n & k_1 &= f(x_n, y_n, u_n) \\ m_2 &= u_n + \tfrac{1}{2}hk_1 & k_2 &= f\left(x_n + \tfrac{1}{2}h, y_n + \tfrac{1}{2}hm_1, u_n + \tfrac{1}{2}hk_1\right) \\ m_3 &= u_n + \tfrac{1}{2}hk_2 & k_3 &= f\left(x_n + \tfrac{1}{2}h, y_n + \tfrac{1}{2}hm_2, u_n + \tfrac{1}{2}hk_2\right) \\ m_4 &= u_n + hk_3 & k_4 &= f(x_n + h, y_n + hm_3, u_n + hk_3). \end{aligned}$$

In general, we can express every $n$th-order differential equation $y^{(n)} = f(x, y, y', \ldots, y^{(n-1)})$ as a system of $n$ first-order equations using the substitutions $y = u_1, y' = u_2, y'' = u_3, \ldots, y^{(n-1)} = u_n$.

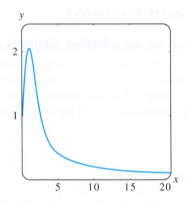

**(a)** Euler's method (red) and the RK4 method (blue)

**(b)** RK4 method

**FIGURE 9.4.1** Numerical solution curves generated by different methods

### EXAMPLE 1 Euler's Method

Use Euler's method to obtain the approximate value of $y(0.2)$, where $y(x)$ is the solution of the initial-value problem

$$y'' + xy' + y = 0, \quad y(0) = 1, \quad y'(0) = 2. \tag{5}$$

**SOLUTION** In terms of the substitution $y' = u$, the equation is equivalent to the system

$$y' = u$$
$$u' = -xu - y.$$

Thus from (3) we obtain

$$y_{n+1} = y_n + hu_n$$
$$u_{n+1} = u_n + h[-x_n u_n - y_n].$$

Using the step size $h = 0.1$ and $y_0 = 1, u_0 = 2$, we find

$$y_1 = y_0 + (0.1)u_0 = 1 + (0.1)2 = 1.2$$
$$u_1 = u_0 + (0.1)[-x_0 u_0 - y_0] = 2 + (0.1)[-(0)(2) - 1] = 1.9$$
$$y_2 = y_1 + (0.1)u_1 = 1.2 + (0.1)(1.9) = 1.39$$
$$u_2 = u_1 + (0.1)[-x_1 u_1 - y_1] = 1.9 + (0.1)[-(0.1)(1.9) - 1.2] = 1.761.$$

In other words, $y(0.2) \approx 1.39$ and $y'(0.2) \approx 1.761$. ■

With the aid of the graphing feature of a numerical solver, in Figure 9.4.1(a) we compare the solution curve of (5) generated by Euler's method ($h = 0.1$) on the interval $[0, 3]$ with the solution curve generated by the RK4 method ($h = 0.1$). From Figure 9.4.1(b) it appears that the solution $y(x)$ of (4) has the property that $y(x) \to 0$ and $x \to \infty$.

If desired, we can use the method of Section 6.2 to obtain two power series solutions of the differential equation in (5). But unless this method reveals that the DE possesses an elementary solution, we will still only be able to approximate $y(0.2)$ using a partial sum. Reinspection of the infinite series solutions of Airy's differential equation $y'' - xy = 0$, given on page 246, does not reveal the oscillatory behavior of the solutions $y_1(x)$ and $y_2(x)$ exhibited in the graphs in Figure 6.2.2. Those graphs were obtained from a numerical solver using the RK4 method with a step size of $h = 0.1$.

**SYSTEMS REDUCED TO FIRST-ORDER SYSTEMS** Using a procedure similar to that just discussed for second-order equations, we can often reduce a system of higher-order differential equations to a system of first-order equations by first solving for the highest-order derivative of each dependent variable and then making appropriate substitutions for the lower-order derivatives.

### EXAMPLE 2 A System Rewritten as a First-Order System

Write
$$x'' - x' + 5x + 2y'' = e^t$$
$$-2x + y'' + 2y = 3t^2$$

as a system of first-order differential equations.

**SOLUTION** Write the system as

$$x'' + 2y'' = e^t - 5x + x'$$
$$y'' = 3t^2 + 2x - 2y$$

and then eliminate $y''$ by multiplying the second equation by 2 and subtracting. This gives

$$x'' = -9x + 4y + x' + e^t - 6t^2.$$

Since the second equation of the system already expresses the highest-order derivative of $y$ in terms of the remaining functions, we are now in a position to introduce new variables. If we let $x' = u$ and $y' = v$, the expressions for $x''$ and $y''$ become, respectively,

$$u' = x'' = -9x + 4y + u + e^t - 6t^2$$
$$v' = y'' = 2x - 2y + 3t^2.$$

The original system can then be written in the form

$$x' = u$$
$$y' = v$$
$$u' = -9x + 4y + u + e^t - 6t^2$$
$$v' = 2x - 2y + 3t^2. \qquad \blacksquare$$

It might not always be possible to carry out the reductions illustrated in Example 2.

### NUMERICAL SOLUTION OF A SYSTEM   The solution of a system of the form

$$\frac{dx_1}{dt} = g_1(t, x_1, x_2, \ldots, x_n)$$

$$\frac{dx_2}{dt} = g_2(t, x_1, x_2, \ldots, x_n)$$

$$\vdots \qquad\qquad \vdots$$

$$\frac{dx_n}{dt} = g_n(t, x_1, x_2, \ldots, x_n)$$

can be approximated by a version of Euler's, the Runge-Kutta, or the Adams-Bashforth-Moulton method adapted to the system. For instance, the RK4 method applied to the system

$$x' = f(t, x, y)$$
$$y' = g(t, x, y) \qquad\qquad (6)$$
$$x(t_0) = x_0, \quad y(t_0) = y_0,$$

looks like this:

$$x_{n+1} = x_n + \frac{h}{6}(m_1 + 2m_2 + 2m_3 + m_4)$$

$$\qquad\qquad\qquad\qquad\qquad\qquad (7)$$

$$y_{n+1} = y_n + \frac{h}{6}(k_1 + 2k_2 + 2k_3 + k_4),$$

where

$$m_1 = f(t_n, x_n, y_n) \qquad\qquad k_1 = g(t_n, x_n, y_n)$$

$$m_2 = f\left(t_n + \tfrac{1}{2}h, x_n + \tfrac{1}{2}hm_1, y_n + \tfrac{1}{2}hk_1\right) \qquad k_2 = g\left(t_n + \tfrac{1}{2}h, x_n + \tfrac{1}{2}hm_1, y_n + \tfrac{1}{2}hk_1\right)$$

$$m_3 = f\left(t_n + \tfrac{1}{2}h, x_n + \tfrac{1}{2}hm_2, y_n + \tfrac{1}{2}hk_2\right) \qquad k_3 = g\left(t_n + \tfrac{1}{2}h, x_n + \tfrac{1}{2}hm_2, y_n + \tfrac{1}{2}hk_2\right)$$

$$m_4 = f(t_n + h, x_n + hm_3, y_n + hk_3) \qquad k_4 = g(t_n + h, x_n + hm_3, y_n + hk_3).$$

$$(8)$$

EXAMPLE 3 RK4 Method

Consider the initial-value problem

$$x' = 2x + 4y$$

$$y' = -x + 6y$$

$$x(0) = -1, \quad y(0) = 6.$$

Use the RK4 method to approximate $x(0.6)$ and $y(0.6)$. Compare the results for $h = 0.2$ and $h = 0.1$.

**SOLUTION** We illustrate the computations of $x_1$ and $y_1$ with step size $h = 0.2$. With the identifications $f(t, x, y) = 2x + 4y$, $g(t, x, y) = -x + 6y$, $t_0 = 0$, $x_0 = -1$, and $y_0 = 6$ we see from (8) that

$$m_1 = f(t_0, x_0, y_0) = f(0, -1, 6) = 2(-1) + 4(6) = 22$$

$$k_1 = g(t_0, x_0, y_0) = g(0, -1, 6) = -1(-1) + 6(6) = 37$$

$$m_2 = f\left(t_0 + \tfrac{1}{2}h, x_0 + \tfrac{1}{2}hm_1, y_0 + \tfrac{1}{2}hk_1\right) = f(0.1, 1.2, 9.7) = 41.2$$

$$k_2 = g\left(t_0 + \tfrac{1}{2}h, x_0 + \tfrac{1}{2}hm_1, y_0 + \tfrac{1}{2}hk_1\right) = g(0.1, 1.2, 9.7) = 57$$

$$m_3 = f\left(t_0 + \tfrac{1}{2}h, x_0 + \tfrac{1}{2}hm_2, y_0 + \tfrac{1}{2}hk_2\right) = f(0.1, 3.12, 11.7) = 53.04$$

$$k_3 = g\left(t_0 + \tfrac{1}{2}h, x_0 + \tfrac{1}{2}hm_2, y_0 + \tfrac{1}{2}hk_2\right) = g(0.1, 3.12, 11.7) = 67.08$$

$$m_4 = f(t_0 + h, x_0 + hm_3, y_0 + hk_3) = f(0.2, 9.608, 19.416) = 96.88$$

$$k_4 = g(t_0 + h, x_0 + hm_3, y_0 + hk_3) = g(0.2, 9.608, 19.416) = 106.888.$$

Therefore from (7) we get

$$x_1 = x_0 + \frac{0.2}{6}(m_1 + 2m_2 + 2m_3 + m_4)$$

$$= -1 + \frac{0.2}{6}(22 + 2(41.2) + 2(53.04) + 96.88) = 9.2453$$

$$y_1 = y_0 + \frac{0.2}{6}(k_1 + 2k_2 + 2k_3 + k_4)$$

$$= 6 + \frac{0.2}{6}(37 + 2(57) + 2(67.08) + 106.888) = 19.0683,$$

where, as usual, the computed values of $x_1$ and $y_1$ are rounded to four decimal places. These numbers give us the approximation $x_1 \approx x(0.2)$ and $y_1 \approx y(0.2)$. The subsequent values, obtained with the aid of a computer, are summarized in Tables 9.4.1 and 9.4.2. ■

You should verify that the solution of the initial-value problem in Example 3 is given by $x(t) = (26t - 1)e^{4t}$, $y(t) = (13t + 6)e^{4t}$. From these equations we see that the actual values $x(0.6) = 160.9384$ and $y(0.6) = 152.1198$ compare favorably with the entries in the last line of Table 9.4.2. The graph of the solution in a neighborhood of $t = 0$ is shown in Figure 9.4.2; the graph was obtained from a numerical solver using the RK4 method with $h = 0.1$.

**TABLE 9.4.1** $h = 0.2$

| $t_n$ | $x_n$ | $y_n$ |
|-------|-------|-------|
| 0.00 | −1.0000 | 6.0000 |
| 0.20 | 9.2453 | 19.0683 |
| 0.40 | 46.0327 | 55.1203 |
| 0.60 | 158.9430 | 150.8192 |

**TABLE 9.4.2** $h = 0.1$

| $t_n$ | $x_n$ | $y_n$ |
|-------|-------|-------|
| 0.00 | −1.0000 | 6.0000 |
| 0.10 | 2.3840 | 10.8883 |
| 0.20 | 9.3379 | 19.1332 |
| 0.30 | 22.5541 | 32.8539 |
| 0.40 | 46.5103 | 55.4420 |
| 0.50 | 88.5729 | 93.3006 |
| 0.60 | 160.7563 | 152.0025 |

**FIGURE 9.4.2** Numerical solution curves for IVP in Example 3

In conclusion, we state Euler's method for the general system (6):

$$x_{n+1} = x_n + hf(t_n, x_n, y_n)$$

$$y_{n+1} = y_n + hg(t_n, x_n, y_n).$$

## EXERCISES 9.4

Answers to selected odd-numbered problems begin on page ANS-16.

**1.** Use Euler's method to approximate $y(0.2)$, where $y(x)$ is the solution of the initial-value problem

$$y'' - 4y' + 4y = 0, \quad y(0) = -2, \quad y'(0) = 1.$$

Use $h = 0.1$. Find the analytic solution of the problem, and compare the actual value of $y(0.2)$ with $y_2$.

**2.** Use Euler's method to approximate $y(1.2)$, where $y(x)$ is the solution of the initial-value problem

$$x^2 y'' - 2xy' + 2y = 0, \quad y(1) = 4, \quad y'(1) = 9,$$

where $x > 0$. Use $h = 0.1$. Find the analytic solution of the problem, and compare the actual value of $y(1.2)$ with $y_2$.

In Problems 3 and 4 repeat the indicated problem using the RK4 method. First use $h = 0.2$ and then use $h = 0.1$.

**3.** Problem 1           **4.** Problem 2

**5.** Use the RK4 method to approximate $y(0.2)$, where $y(x)$ is the solution of the initial-value problem

$$y'' - 2y' + 2y = e^t \cos t, \quad y(0) = 1, \quad y'(0) = 2.$$

First use $h = 0.2$ and then use $h = 0.1$.

**6.** When $E = 100$ V, $R = 10$ $\Omega$, and $L = 1$ h, the system of differential equations for the currents $i_1(t)$ and $i_3(t)$ in the electrical network given in Figure 9.4.3 is

$$\frac{di_1}{dt} = -20i_1 + 10i_3 + 100$$

$$\frac{di_3}{dt} = 10i_1 - 20i_3,$$

where $i_1(0) = 0$ and $i_3(0) = 0$. Use the RK4 method to approximate $i_1(t)$ and $i_3(t)$ at $t = 0.1, 0.2, 0.3, 0.4$, and $0.5$. Use $h = 0.1$. Use a numerical solver to graph the solution for $0 \le t \le 5$. Use the graphs to predict the behavior of $i_1(t)$ and $i_3(t)$ as $t \to \infty$.

**FIGURE 9.4.3** Network in Problem 6

In Problems 7–12 use the Runge-Kutta method to approximate $x(0.2)$ and $y(0.2)$. First use $h = 0.2$ and then use $h = 0.1$. Use a numerical solver and $h = 0.1$ to graph the solution in a neighborhood of $t = 0$.

**7.** $x' = 2x - y$
$y' = x$
$x(0) = 6, \quad y(0) = 2$

**8.** $x' = x + 2y$
$y' = 4x + 3y$
$x(0) = 1, \quad y(0) = 1$

**9.** $x' = -y + t$
$y' = x - t$
$x(0) = -3, \quad y(0) = 5$

**10.** $x' = 6x + y + 6t$
$y' = 4x + 3y - 10t + 4$
$x(0) = 0.5, \quad y(0) = 0.2$

**11.** $x' + 4x - y' = 7t$
$x' + y' - 2y = 3t$
$x(0) = 1, \quad y(0) = -2$

**12.** $\quad\quad x' + y' = 4t$
$-x' + y' + y = 6t^2 + 10$
$x(0) = 3, \quad y(0) = -1$

---

## 9.5 Second-Order Boundary-Value Problems

**INTRODUCTION**   We just saw in Section 9.4 how to approximate the solution of a *second-order initial-value problem*

$$y'' = f(x, y, y'), \quad y(x_0) = y_0, \quad y'(x_0) = u_0.$$

In this section we are going to examine two methods for approximating a solution of a *second-order boundary-value problem*

$$y'' = f(x, y, y'), \quad y(a) = \alpha, \quad y(b) = \beta.$$

Unlike the procedures that are used with second-order initial-value problems, the methods of second-order boundary-value problems do not require writing the second-order DE as a system of first-order DEs.

**FINITE DIFFERENCE APPROXIMATIONS** The Taylor series expansion, centered at a point $a$, of a function $y(x)$ is

$$y(x) = y(a) + y'(a)\frac{x - a}{1!} + y''(a)\frac{(x - a)^2}{2!} + y'''(a)\frac{(x - a)^3}{3!} + \cdots.$$

If we set $h = x - a$, then the preceding line is the same as

$$y(x) = y(a) + y'(a)\frac{h}{1!} + y''(a)\frac{h^2}{2!} + y'''(a)\frac{h^3}{3!} + \cdots.$$

For the subsequent discussion it is convenient then to rewrite this last expression in two alternative forms:

$$y(x + h) = y(x) + y'(x)h + y''(x)\frac{h^2}{2} + y'''(x)\frac{h^3}{6} + \cdots \tag{1}$$

and

$$y(x - h) = y(x) - y'(x)h + y''(x)\frac{h^2}{2} - y'''(x)\frac{h^3}{6} + \cdots. \tag{2}$$

If $h$ is small, we can ignore terms involving $h^4$, $h^5$, ... since these values are negligible. Indeed, if we ignore all terms involving $h^2$ and higher, then solving (1) and (2), in turn, for $y'(x)$ yields the following approximations for the first derivative:

$$y'(x) \approx \frac{1}{h}[y(x + h) - y(x)] \tag{3}$$

$$y'(x) \approx \frac{1}{h}[y(x) - y(x - h)]. \tag{4}$$

Subtracting (1) and (2) also gives

$$y'(x) \approx \frac{1}{2h}[y(x + h) - y(x - h)]. \tag{5}$$

On the other hand, if we ignore terms involving $h^3$ and higher, then by adding (1) and (2), we obtain an approximation for the second derivative $y''(x)$:

$$y''(x) \approx \frac{1}{h^2}[y(x + h) - 2y(x) + y(x - h)]. \tag{6}$$

The right-hand sides of (3), (4), (5), and (6) are called **difference quotients.** The expressions

$$y(x + h) - y(x), \quad y(x) - y(x - h), \quad y(x + h) - y(x - h),$$

and

$$y(x + h) - 2y(x) + y(x - h)$$

are called **finite differences.** Specifically, $y(x + h) - y(x)$ is called a **forward difference,** $y(x) - y(x - h)$ is a **backward difference,** and both $y(x + h) - y(x - h)$ and $y(x + h) - 2y(x) + y(x - h)$ are called **central differences.** The results given in (5) and (6) are referred to as **central difference approximations** for the derivatives $y'$ and $y''$.

**FINITE DIFFERENCE METHOD** Consider now a linear second-order boundary-value problem

$$y'' + P(x)y' + Q(x)y = f(x), \qquad y(a) = \alpha, \qquad y(b) = \beta. \tag{7}$$

Suppose $a = x_0 < x_1 < x_2 < \cdots < x_{n-1} < x_n = b$ represents a regular partition of the interval $[a, b]$, that is, $x_i = a + ih$, where $i = 0, 1, 2, \ldots, n$ and $h = (b - a)/n$. The points

$$x_1 = a + h, \qquad x_2 = a + 2h, \ldots, \qquad x_{n-1} = a + (n - 1)h$$

are called **interior mesh points** of the interval $[a, b]$. If we let

$$y_i = y(x_i), \qquad P_i = P(x_i), \qquad Q_i = Q(x_i), \qquad \text{and} \qquad f_i = f(x_i)$$

and if $y''$ and $y'$ in (7) are replaced by the central difference approximations (5) and (6), we get

$$\frac{y_{i+1} - 2y_i + y_{i-1}}{h^2} + P_i \frac{y_{i+1} - y_{i-1}}{2h} + Q_i y_i = f_i$$

or, after simplifying,

$$\left(1 + \frac{h}{2} P_i\right) y_{i+1} + (-2 + h^2 Q_i) y_i + \left(1 - \frac{h}{2} P_i\right) y_{i-1} = h^2 f_i. \qquad (8)$$

The last equation, known as a **finite difference equation,** is an approximation to the differential equation. It enables us to approximate the solution $y(x)$ of (7) at the interior mesh points $x_1, x_2, \ldots, x_{n-1}$ of the interval $[a, b]$. By letting $i$ take on the values $1, 2, \ldots, n - 1$ in (8), we obtain $n - 1$ equations in the $n - 1$ unknowns $y_1, y_2, \ldots, y_{n-1}$. Bear in mind that we know $y_0$ and $y_n$, since these are the prescribed boundary conditions $y_0 = y(x_0) = y(a) = \alpha$ and $y_n = y(x_n) = y(b) = \beta$.

In Example 1 we consider a boundary-value problem for which we can compare the approximate values that we find with the actual values of an explicit solution.

---

**EXAMPLE 1**   **Using the Finite Difference Method**

Use the difference equation (8) with $n = 4$ to approximate the solution of the boundary-value problem $y'' - 4y = 0$, $y(0) = 0$, $y(1) = 5$.

**SOLUTION** To use (8), we identify $P(x) = 0$, $Q(x) = -4$, $f(x) = 0$, and $h = (1 - 0)/4 = \frac{1}{4}$. Hence the difference equation is

$$y_{i+1} - 2.25 y_i + y_{i-1} = 0. \qquad (9)$$

Now the interior points are $x_1 = 0 + \frac{1}{4}$, $x_2 = 0 + \frac{2}{4}$, $x_3 = 0 + \frac{3}{4}$, so for $i = 1$, 2, and 3, (9) yields the following system for the corresponding $y_1$, $y_2$, and $y_3$:

$$y_2 - 2.25 y_1 + y_0 = 0$$
$$y_3 - 2.25 y_2 + y_1 = 0$$
$$y_4 - 2.25 y_3 + y_2 = 0.$$

With the boundary conditions $y_0 = 0$ and $y_4 = 5$ the foregoing system becomes

$$-2.25 y_1 + \qquad y_2 \qquad\qquad = 0$$
$$y_1 - 2.25 y_2 + \qquad y_3 = 0$$
$$y_2 - 2.25 y_3 = -5.$$

Solving the system gives $y_1 = 0.7256$, $y_2 = 1.6327$, and $y_3 = 2.9479$.

Now the general solution of the given differential equation is $y = c_1 \cosh 2x + c_2 \sinh 2x$. The condition $y(0) = 0$ implies that $c_1 = 0$. The other boundary condition gives $c_2$. In this way we see that a solution of the boundary-value problem is $y(x) = (5 \sinh 2x)/\sinh 2$. Thus the actual values (rounded to four decimal places) of this solution at the interior points are as follows: $y(0.25) = 0.7184$, $y(0.5) = 1.6201$, and $y(0.75) = 2.9354$. ∎

The accuracy of the approximations in Example 1 can be improved by using a smaller value of $h$. Of course, the trade-off here is that a smaller value of $h$ necessitates solving a larger system of equations. It is left as an exercise to show that with $h = \frac{1}{8}$, approximations to $y(0.25)$, $y(0.5)$, and $y(0.75)$ are 0.7202, 1.6233, and 2.9386, respectively. See Problem 11 in Exercises 9.5.

---

### EXAMPLE 2  Using the Finite Difference Method

Use the difference equation (8) with $n = 10$ to approximate the solution of

$$y'' + 3y' + 2y = 4x^2, \quad y(1) = 1, \quad y(2) = 6.$$

**SOLUTION** In this case we identify $P(x) = 3$, $Q(x) = 2$, $f(x) = 4x^2$, and $h = (2 - 1)/10 = 0.1$, and so (8) becomes

$$1.15y_{i+1} - 1.98y_i + 0.85y_{i-1} = 0.04x_i^2. \tag{10}$$

Now the interior points are $x_1 = 1.1$, $x_2 = 1.2$, $x_3 = 1.3$, $x_4 = 1.4$, $x_5 = 1.5$, $x_6 = 1.6$, $x_7 = 1.7$, $x_8 = 1.8$, and $x_9 = 1.9$. For $i = 1, 2, \ldots, 9$ and $y_0 = 1$, $y_{10} = 6$, (10) gives a system of nine equations and nine unknowns:

$$1.15y_2 - 1.98y_1 \qquad\qquad\qquad = -0.8016$$
$$1.15y_3 - 1.98y_2 + 0.85y_1 = 0.0576$$
$$1.15y_4 - 1.98y_3 + 0.85y_2 = 0.0676$$
$$1.15y_5 - 1.98y_4 + 0.85y_3 = 0.0784$$
$$1.15y_6 - 1.98y_5 + 0.85y_4 = 0.0900$$
$$1.15y_7 - 1.98y_6 + 0.85y_5 = 0.1024$$
$$1.15y_8 - 1.98y_7 + 0.85y_6 = 0.1156$$
$$1.15y_9 - 1.98y_8 + 0.85y_7 = 0.1296$$
$$\qquad\qquad\quad - 1.98y_9 + 0.85y_8 = -6.7556.$$

We can solve this large system using Gaussian elimination or, with relative ease, by means of a computer algebra system. The result is found to be $y_1 = 2.4047$, $y_2 = 3.4432$, $y_3 = 4.2010$, $y_4 = 4.7469$, $y_5 = 5.1359$, $y_6 = 5.4124$, $y_7 = 5.6117$, $y_8 = 5.7620$, and $y_9 = 5.8855$. ■

**SHOOTING METHOD** Another way of approximating a solution of a boundary-value problem $y'' = f(x, y, y')$, $y(a) = \alpha$, $y(b) = \beta$ is called the **shooting method.** The starting point in this method is the replacement of the second-order boundary-value problem by a second-order initial-value problem

$$y'' = f(x, y, y'), \quad y(a) = \alpha, \quad y'(a) = m_1. \tag{11}$$

The number $m_1$ in (11) is simply a guess for the unknown slope of the solution curve at the known point $(a, y(a))$. We then apply one of the step-by-step numerical techniques to the second-order equation in (11) to find an approximation $\beta_1$ for the value of $y(b)$. If $\beta_1$ agrees with the given value $y(b) = \beta$ to some preassigned tolerance, we stop; otherwise, the calculations are repeated, starting with a different guess $y'(a) = m_2$ to obtain a second approximation $\beta_2$ for $y(b)$. This method can be continued in a trial-and-error manner, or the subsequent slopes $m_3$, $m_4, \ldots$ can be adjusted in some systematic way; linear interpolation is particularly successful when the differential equation in (11) is linear. The procedure is analogous to shooting (the "aim" is the choice of the initial slope) at a target until the bullseye $y(b)$ is hit. See Problem 14 in Exercises 9.5.

Of course, underlying the use of these numerical methods is the assumption, which we know is not always warranted, that a solution of the boundary-value problem exists.

The approximation method using finite differences can be extended to boundary-value problems in which the first derivative is specified at a boundary—for example, a problem such as $y'' = f(x, y, y')$, $y'(a) = \alpha$, $y(b) = \beta$. See Problem 13 in Exercises 9.5.

## EXERCISES 9.5

Answers to selected odd-numbered problems begin on page ANS-16.

In Problems 1–10 use the finite difference method and the indicated value of $n$ to approximate the solution of the given boundary-value problem.

**1.** $y'' + 9y = 0$,   $y(0) = 4$,   $y(2) = 1$;   $n = 4$

**2.** $y'' - y = x^2$,   $y(0) = 0$,   $y(1) = 0$;   $n = 4$

**3.** $y'' + 2y' + y = 5x$,   $y(0) = 0$,   $y(1) = 0$;   $n = 5$

**4.** $y'' - 10y' + 25y = 1$,   $y(0) = 1$,   $y(1) = 0$;   $n = 5$

**5.** $y'' - 4y' + 4y = (x + 1)e^{2x}$,
$y(0) = 3$,   $y(1) = 0$;   $n = 6$

**6.** $y'' + 5y' = 4\sqrt{x}$,   $y(1) = 1$,   $y(2) = -1$;   $n = 6$

**7.** $x^2y'' + 3xy' + 3y = 0$,   $y(1) = 5$,   $y(2) = 0$;   $n = 8$

**8.** $x^2y'' - xy' + y = \ln x$,   $y(1) = 0$,   $y(2) = -2$;   $n = 8$

**9.** $y'' + (1 - x)y' + xy = x$,   $y(0) = 0$,   $y(1) = 2$;   $n = 10$

**10.** $y'' + xy' + y = x$,   $y(0) = 1$,   $y(1) = 0$;   $n = 10$

**11.** Rework Example 1 using $n = 8$.

**12.** The electrostatic potential $u$ between two concentric spheres of radius $r = 1$ and $r = 4$ is determined from

$$\frac{d^2u}{dr^2} + \frac{2}{r}\frac{du}{dr} = 0, \quad u(1) = 50, \quad u(4) = 100.$$

Use the method of this section with $n = 6$ to approximate the solution of this boundary-value problem.

**13.** Consider the boundary-value problem $y'' + xy = 0$, $y'(0) = 1$, $y(1) = -1$.

(a) Find the difference equation corresponding to the differential equation. Show that for $i = 0, 1, 2, \ldots, n - 1$ the difference equation yields $n$ equations in $n + 1$ unknowns $y_{-1}, y_0, y_1, y_2, \ldots, y_{n-1}$. Here $y_{-1}$ and $y_0$ are unknowns, since $y_{-1}$ represents an approximation to $y$ at the exterior point $x = -h$ and $y_0$ is not specified at $x = 0$.

(b) Use the central difference approximation (5) to show that $y_1 - y_{-1} = 2h$. Use this equation to eliminate $y_{-1}$ from the system in part (a).

(c) Use $n = 5$ and the system of equations found in parts (a) and (b) to approximate the solution of the original boundary-value problem.

## Computer Lab Assignments

**14.** Consider the boundary-value problem $y'' = y' - \sin(xy)$, $y(0) = 1$, $y(1) = 1.5$. Use the shooting method to approximate the solution of this problem. (The approximation can be obtained using a numerical technique—say, the RK4 method with $h = 0.1$; or, even better, if you have access to a CAS such as *Mathematica* or *Maple*, the **NDSolve** function can be used.)

## Chapter 9 **In Review**

Answers to selected odd-numbered problems begin on page ANS-17.

In Problems 1–4 construct a table comparing the indicated values of $y(x)$ using Euler's method, the improved Euler's method, and the RK4 method. Compute to four rounded decimal places. First use $h = 0.1$ and then use $h = 0.05$.

**1.** $y' = 2 \ln xy$,   $y(1) = 2$;
$y(1.1)$,   $y(1.2)$,   $y(1.3)$,   $y(1.4)$,   $y(1.5)$

**2.** $y' = \sin x^2 + \cos y^2$,   $y(0) = 0$;
$y(0.1)$,   $y(0.2)$,   $y(0.3)$,   $y(0.4)$,   $y(0.5)$

**3.** $y' = \sqrt{x + y}$,   $y(0.5) = 0.5$;
$y(0.6)$,   $y(0.7)$,   $y(0.8)$,   $y(0.9)$,   $y(1.0)$

**4.** $y' = xy + y^2$,   $y(1) = 1$;
$y(1.1)$,   $y(1.2)$,   $y(1.3)$,   $y(1.4)$,   $y(1.5)$

**5.** Use Euler's method to approximate $y(0.2)$, where $y(x)$ is the solution of the initial-value problem

$y'' - (2x + 1)y = 1$, $y(0) = 3$, $y'(0) = 1$. First use one step with $h = 0.2$ and then repeat the calculations using two steps with $h = 0.1$.

**6.** Use the Adams-Bashforth-Moulton method to approximate $y(0.4)$, where $y(x)$ is the solution of the initial-value problem $y' = 4x - 2y$, $y(0) = 2$. Use $h = 0.1$ and the RK4 method to compute $y_1, y_2$, and $y_3$.

**7.** Use Euler's method with $h = 0.1$ to approximate $x(0.2)$ and $y(0.2)$, where $x(t), y(t)$ is the solution of the initial-value problem

$$x' = x + y$$
$$y' = x - y$$
$$x(0) = 1, \quad y(0) = 2.$$

**8.** Use the finite difference method with $n = 10$ to approximate the solution of the boundary-value problem

$$y'' + 6.55(1 + x)y = 1, \quad y(0) = 0, \quad y(1) = 0.$$

# Appendices

A | Integral-Defined Functions
B | Matrices
C | Laplace Transforms

# Appendix A  Integral-Defined Functions

**INTRODUCTION**   Many important functions encountered in applied mathematics are defined in terms of an integral. There are several ways of doing this. In Sections 1.1, 2.3, and 4.2 we have already seen that a solution of a differential equation may lead to a function of the form

$$F(x) = \int_a^x g(t)\,dt. \tag{1}$$

An integral-defined function can also be of the form

$$F(x) = \int_a^b g(x, t)\,dt, \tag{2}$$

where it is understood that $x$ is a parameter, that is, $x$ is treated as a constant in the $t$-integration. In many cases, especially for functions with a name, the defining integral is nonelementary (review page 11). The integrand $g$ in (2) could depend on several parameters, say, $g(x, y, t)$. In this case the integral (2) defines a function of *two* variables $x$ and $y$. Moreover, (1) and (2) can also be an improper integral. For example, in (1) the lower limit of integration $a$ could be $-\infty$ or the interval of integration could be chosen to be $[x, \infty)$. In (2), one or both of the limits of integration could be infinite, or the integrand $g(x, t)$ could have an infinite discontinuity in the interval of integration $[a, b]$. The convergence of an improper integral will, in general, depend on values of the parameter $x$. See Examples 1, 2, and 3 in Section 7.1.

In Sections 2.3 and 14.1 we saw that the **error function** and **complementary error function**,

$$\text{erf}(x) = \frac{2}{\sqrt{\pi}} \int_0^x e^{-t^2}\,dt \qquad \text{and} \qquad \text{erfc}(x) = \frac{2}{\sqrt{\pi}} \int_x^\infty e^{-t^2}\,dt \tag{3}$$

are both of the type given in (1) with the identification $g(t) = (2/\sqrt{\pi})e^{-t^2}$. As we have seen in Chapter 7, the Laplace transform of a function $f(t)$ is the function $F(s) = \int_0^\infty e^{-st}f(t)\,dt$. The last integral is of the type given in (2) with the symbol $s$ playing the part of $x$ so that $g(s, t) = e^{-st}f(t)$. In Chapter 14, the Fourier transform $F(\alpha) = \int_{-\infty}^\infty f(x)e^{i\alpha x}\,dx$ is also of the type given in (2) with the symbols $x$ and $\alpha$ replacing $t$ and $x$, respectively, and $g(\alpha, x) = f(x)e^{i\alpha x}$.

We begin our discussion of specific integral-defined functions with one of the form given in (2).

**THE GAMMA FUNCTION**   On a short list of important functions in the study of special functions (such as the Bessel function) the **gamma function** would appear near the top. The integral definition of this function,

$$\Gamma(x) = \int_0^\infty t^{x-1}e^{-t}\,dt, \tag{4}$$

was first given by the Swiss mathematician **Leonhard Euler** (1707–1783) in his text *Institutiones calculi integralis* published in 1768.

**FIGURE A.1** Graph of the gamma function

**GRAPHS AND PROPERTIES**   Convergence of the improper integral (4) requires that $x - 1 > -1$ or $x > 0$. Although the integral (4) does not converge for $x \le 0$, it can be shown by alternative definitions that the domain of the gamma function can be expanded to the set of real numbers *except* the nonpositive integers: $x = -n$, $n = 0, 1, 2, \ldots$. See Problem 28 in Exercises for Appendix A. Considered as a function of a real variable $x$, the graph of $\Gamma(x)$ is given in Figure A.1. As seen in the figure the dashed lines $x = 0$, $x = -1$, $x = -2$, $\ldots$ are vertical asymptotes of the graph.

There is a remarkably simple and extremely useful connection between the value of the gamma function at a number $x$ and its value at $x + 1$. To obtain this relationship we replace $x$ by $x + 1$ in (4) and integrate by parts:

$$\Gamma(x + 1) = \int_0^\infty t^{(x+1)-1} e^{-t}\, dt = \int_0^\infty t^x e^{-t}\, dt$$

$$= \lim_{b \to \infty} \left[ -t^x e^{-t} \Big|_0^b + x \int_0^b t^{x-1} e^{-t}\, dt \right] \tag{5}$$

$$= x \lim_{b \to \infty} \int_0^b t^{x-1} e^{-t}\, dt. \tag{6}$$

In (5), $\lim_{b \to \infty} t^x e^{-t} \big|_0^b = 0$ follows from the assumption that $x > 0$ and repeated applications of L'Hôpital's rule. The limit in (6), $\lim_{b \to \infty} \int_0^b t^{x-1} e^{-t}\, dt$, is the definition of the improper integral (4). Hence we have shown that

$$\Gamma(x + 1) = x\Gamma(x). \tag{7}$$

Equation (7) is called a **recursion formula.**

It is easy to find the value of $\Gamma(x)$ when $x$ is a positive integer. For example, when $x = 1$ we see that (4) is an elementary integral:

$$\Gamma(1) = \int_0^\infty e^{-t}\, dt = \lim_{b \to \infty} \int_0^b e^{-t}\, dt = \lim_{b \to \infty} (-e^{-b} + 1) = 1.$$

From this single numerical value the repeated application of formula (7) gives:

$$\begin{aligned}
\Gamma(2) &= \Gamma(1 + 1) = 1 \cdot \Gamma(1) = 1 \\
\Gamma(3) &= \Gamma(2 + 1) = 2 \cdot \Gamma(2) = 2 \cdot 1 \\
\Gamma(4) &= \Gamma(3 + 1) = 3 \cdot \Gamma(3) = 3 \cdot 2 \cdot 1 \\
\Gamma(5) &= \Gamma(4 + 1) = 4 \cdot \Gamma(4) = 4 \cdot 3 \cdot 2 \cdot 1
\end{aligned} \tag{8}$$

and so on. In this manner it is seen that when $n$ is a positive integer,

$$\Gamma(n + 1) = n!. \tag{9}$$

Recall, $n!$ (read "$n$ factorial") is defined as the product of consecutive integers from 1 to $n$, that is, $n! = n(n - 1) \cdots 3 \cdot 2 \cdot 1$ and is called the **factorial function.** For convenience it is customary to define $0! = 1$. Because $n!$ is a special case of (4), the gamma function is sometimes referred to as the **generalized factorial function.**

In the discussion of the Bessel functions of half-integral order in Section 6.4 we used the value of $\Gamma\left(\frac{1}{2}\right) = \sqrt{\pi}$. With $x = \frac{1}{2}$ we see that (4) becomes

$$\Gamma\left(\tfrac{1}{2}\right) = \int_0^\infty t^{-1/2} e^{-t}\, dt. \tag{10}$$

Note that the foregoing integral is improper for two reasons: an infinite limit of integration and the integrand has an infinite discontinuity at 0. Nonetheless (10) can be explicitly evaluated. To that end, we begin with the substitutions $t = u^2$ and $dt = 2u\, du$:

$$\Gamma\left(\tfrac{1}{2}\right) = \int_0^\infty u^{-1} e^{-u^2} (2u\, du) = 2 \int_0^\infty e^{-u^2}\, du. \tag{11}$$

But $\int_0^\infty e^{-u^2}\, du = \int_0^\infty e^{-v^2}\, dv$, so

$$\left[ \Gamma\left(\tfrac{1}{2}\right) \right]^2 = \left( 2 \int_0^\infty e^{-u^2} du \right)\left( 2 \int_0^\infty e^{-v^2} dv \right) = 4 \int_0^\infty \int_0^\infty e^{-(u^2 + v^2)}\, du\, dv.$$

Switching to polar coordinates $u = r\cos\theta$, $v = r\sin\theta$ enables us to evaluate the double integral:

$$4\int_0^\infty \int_0^\infty e^{-(u^2+v^2)}\,du\,dv = 4\int_0^{\pi/2}\int_0^\infty e^{-r^2}r\,dr\,d\theta = \pi.$$

Hence

$$\left[\Gamma\!\left(\tfrac{1}{2}\right)\right]^2 = \pi \quad\text{or}\quad \Gamma\!\left(\tfrac{1}{2}\right) = \sqrt{\pi}. \tag{12}$$

We used (11), $2\int_0^\infty e^{-u^2}\,du = \sqrt{\pi}$, in the form $(2/\sqrt{\pi})\int_0^\infty e^{-u^2}\,du = 1$ to derive the fundamental identity that relates the error function and complementary error function: $\operatorname{erf}(x) + \operatorname{erfc}(x) = 1$. See Section 2.3.

Because of (7) and (12) the value of the gamma function for $x$ equal to one half an odd integer, that is, $x = n + \tfrac{1}{2}$ where $n = 0, \pm1, \pm2, \ldots$, can be expressed in terms of $\Gamma\!\left(\tfrac{1}{2}\right) = \sqrt{\pi}$. The recursion formula (7) written as

$$\Gamma(x) = \frac{\Gamma(x+1)}{x} \tag{13}$$

is one way to extend the definition of the gamma function to negative real numbers with the negative integers being the only exception. Part (b) of the next example illustrates this idea.

### EXAMPLE 1 Using (7), (12), and (13)

Evaluate **(a)** $\Gamma\!\left(\tfrac{3}{2}\right)$  **(b)** $\Gamma\!\left(-\tfrac{1}{2}\right)$.

**SOLUTION**

**(a)** With $x = \tfrac{1}{2}$ and the result in (12) the recursion formula (7) yields

$$\Gamma\!\left(\tfrac{3}{2}\right) = \Gamma\!\left(\tfrac{1}{2}+1\right) = \tfrac{1}{2}\Gamma\!\left(\tfrac{1}{2}\right) = \tfrac{1}{2}\sqrt{\pi}.$$

**(b)** If we choose $x = -\tfrac{1}{2}$, then formula (13) gives

$$\Gamma\!\left(-\tfrac{1}{2}\right) = \frac{\Gamma\!\left(-\tfrac{1}{2}+1\right)}{-\tfrac{1}{2}} = -2\Gamma\!\left(\tfrac{1}{2}\right) = -2\sqrt{\pi}. \qquad\blacksquare$$

**THE BETA FUNCTION** Analogous to (2) an integral of the type $\int_a^b g(x, y, t)\,dt$ defines a function of two variables $x$ and $y$. An example of this kind of integral is

$$B(x, y) = \int_0^1 t^{x-1}(1-t)^{y-1}\,dt. \tag{14}$$

The foregoing integral is known as the **beta function** and is closely related to the gamma function. The integrand of (14) can be discontinuous at 0 and at 1 for particular choices of $x$ and $y$, but it can be shown that of the integral exists for $x > 0$, $y > 0$. Although we are not going to prove it, the beta function can be expressed in terms of the gamma function:

$$B(x, y) = \frac{\Gamma(x)\Gamma(y)}{\Gamma(x+y)}, \quad x > 0, \quad y > 0. \tag{15}$$

### EXAMPLE 2 Using (12) and (15)

Note that the integrand has an infinite discontinuity at $t = 1$.

Evaluate $\displaystyle\int_0^1 \sqrt{\frac{t}{1-t}}\,dt$.

**SOLUTION** Even though the given integral is improper we can evaluate it straight away using (15). By rewriting the integral in the equivalent form

$$\int_0^1 t^{1/2}(1-t)^{-1/2}\,dt$$

we recognize it as (14) in the case $x = \frac{3}{2}$ and $y = \frac{1}{2}$. From (15) it follows that,

$$\int_0^1 t^{1/2}(1-t)^{-1/2}\,dt = B\left(\tfrac{3}{2}, \tfrac{1}{2}\right) = \frac{\Gamma\left(\tfrac{3}{2}\right)\Gamma\left(\tfrac{1}{2}\right)}{\Gamma(2)} = \frac{\pi}{2},$$

where we have used the previously obtained values $\Gamma\left(\tfrac{3}{2}\right) = \tfrac{1}{2}\sqrt{\pi}$, $\Gamma\left(\tfrac{1}{2}\right) = \sqrt{\pi}$, and $\Gamma(2) = 1$. ■

**OTHER INTEGRAL-DEFINED FUNCTIONS** We have barely touched the subject of integral-defined functions. Here are a few more functions that occur in mathematics, physics, and engineering:

**Sine and cosine integral functions:**

$$\text{Si}(x) = \int_0^x \frac{\sin t}{t}\,dt, \quad \text{Ci}(x) = -\int_x^\infty \frac{\cos t}{t}\,dt \tag{16}$$

**Fresnel sine and cosine integral functions:**

$$S(x) = \int_0^x \sin\left(\tfrac{1}{2}\pi t^2\right)dt, \quad C(x) = \int_0^x \cos\left(\tfrac{1}{2}\pi t^2\right)dt \tag{17}$$

**Exponential integral function:**

$$\text{Ei}(x) = \int_{-\infty}^x \frac{e^t}{t}\,dt \tag{18}$$

**Logarithmic integral function:**

$$\text{Li}(x) = \int_2^x \frac{1}{\ln t}\,dt \tag{19}$$

**Airy functions of first and second kind:**

$$\text{Ai}(x) = \frac{1}{\pi}\int_0^\infty \cos\left(xt + \tfrac{1}{3}t^3\right)dt, \quad \text{Bi}(x) = \frac{1}{\pi}\int_0^\infty \left[e^{xt - \frac{1}{3}t^3} + \sin\left(xt + \tfrac{1}{3}t^3\right)\right]dt \tag{20}$$

The definitions of the functions in (16)–(20) vary slightly throughout the literature and websites. For example, the Fresnel integrals, which first appeared in the study of optics, are often defined without the factor $\frac{1}{2}\pi$ in the integrands. In the computer algebra system *Mathematica*, the exponential integral is written $\text{Ei}(x) = -\int_{-\infty}^\infty (e^{-t}/t)\,dt$. Using the substitution $t = -u$ it is easily shown this latter form is equivalent to (18). Finally, it should be noted that the sine integral function $\text{Si}(x) = \int_0^x \frac{\sin t}{t}\,dt$ is not regarded as an improper integral because the integrand $f(t) = (\sin t)/t$ has a removable discontinuity at $t = 0$, that is, the integrand should be interpreted as

$$f(t) = \begin{cases} \dfrac{\sin t}{t}, & t > 0 \\ 1, & t = 0. \end{cases}$$

The graphs of the functions (16)–(20) are given in Figure A.2.

**(a)** sine and cosine integral functions

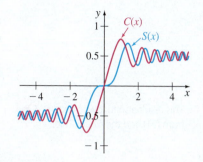

**(b)** Fresnel sine and cosine integral functions

**(c)** exponential integral function

**(d)** logarithmic integral function

**(e)** Airy functions of first and second kind

**FIGURE A.2**   Graphs of Si(x), Ci(x), S(x), C(x), Ei(x), Li(x), Ai(x), and Bi(x)

**DERIVATIVE OF AN INTEGRAL-DEFINED FUNCTION**   The integral (1) and its derivative are encountered in the first semester of a course in calculus. Recall, one form of the form of the Fundamental Theorem of Calculus states that if $g$ is continuous on an interval and $a$ is a number in the interval, then $F(x) = \int_a^x g(t)\,dt$ is a differentiable function and its derivative is

$$F'(x) = \frac{d}{dx}\int_a^x g(t)\,dt = g(x). \tag{21}$$

In the most general case of (2) where the limits of integration are functions of $x$,

$$F(x) = \int_{u(x)}^{v(x)} g(x, t)\,dt,$$

the derivative is given by **Leibniz's rule:**

$$F'(x) = \frac{d}{dx}\int_{u(x)}^{v(x)} g(x, t)\,dt = g(x, v(x))\frac{dv}{dx} - g(x, u(x))\frac{du}{dx} + \int_{u(x)}^{v(x)} \frac{\partial}{\partial x} g(x, t)\,dt. \tag{22}$$

Without going into any details, it is usually assumed that $u(x)$, $v(x)$, $u'(x)$, and $v'(x)$ are continuous on an interval and $g(x, t)$, and $g_x(x, t)$ are continuous on some region of the $xt$-plane.

Two special cases of (22) are of particular interest. First, if the integrand $g(x, t)$ is replaced by $g(t)$, then $(\partial/\partial x)g(t) = 0$ and (22) reduces to

This result also follows directly from (21) and the Chain Rule.

$$F'(x) = \frac{d}{dx}\int_{u(x)}^{v(x)} g(t)\,dt = g(v(x))\frac{dv}{dx} - g(u(x))\frac{du}{dx}. \tag{23}$$

Second, if $u = a$ and $v = b$ denote constants, then $u' = 0$ and $v' = 0$ and (22) gives a rule for differentiating under an integral sign:

$$F'(x) = \frac{d}{dx}\int_a^b g(x, t)\,dt = \int_a^b \frac{\partial}{\partial x} g(x, t)\,dt. \tag{24}$$

Under slightly more stringent conditions, the result in (24) is also valid when the interval of integration is $[a, \infty)$. The last result leads us back to differential equations. See Problems 49–52 in Exercises for Appendix A.

---

**EXAMPLE 3**    Using (22), (23), and (24)

Compute the derivative of the functions

**(a)** $F(x) = \int_{x^2}^{x^3} \sin(4xt^2)\,dt$    **(b)** $F(x) = \int_{x^2}^{x^3} \sin(4t^2)\,dt$    **(c)** $F(x) = \int_0^1 \sin(4xt^2)\,dt$

**SOLUTION**

**(a)** From (22) with $u = x^2$ and $v = x^3$ we have

$$F'(x) = \frac{d}{dx}\int_{x^2}^{x^3} \sin(4xt^2)\,dt$$

$$= \sin(4x(x^3)^2) \cdot 3x^2 - \sin(4x(x^2)^2) \cdot 2x + 4\int_{x^2}^{x^3} t^2 \cos(4xt^2)\,dt$$

$$= 3x^2\sin(4x^7) - 2x\sin(4x^5) + 4\int_{x^2}^{x^3} t^2\cos(4xt^2)\,dt.$$

**(b)** From (23),

$$F'(x) = \frac{d}{dx}\int_{x^2}^{x^3} \sin(4t^2)\,dt$$

$$= \sin(4(x^3)^2) \cdot 3x^2 - \sin(4(x^2)^2) \cdot 2x$$

$$= 3x^2\sin(4x^6) - 2x\sin(4x^4).$$

**(c)** From (24),

$$F'(x) = \frac{d}{dx}\int_0^1 \sin(4xt^2)\,dt = \int_0^1 \frac{\partial}{\partial x}\sin(4xt^2)\,dt = 4\int_0^1 t^2\cos(4xt^2)\,dt. \qquad \blacksquare$$

The result in (24) can sometimes be an aid in evaluating an integral that at first glance seems intractable. But the procedure usually involves formal manipulations and cleverness that cannot be taught. Problems 53–56 in Exercises for Appendix A give you a small sample of the idea.

## The Gamma Function

In Problems 1 and 2 evaluate the given quantity.

**1.** $\Gamma(6)$

**2.** $\Gamma(8)$

In Problems 3–6 use (7) and the known value $\Gamma\left(\frac{1}{2}\right) = \sqrt{\pi}$ to evaluate the given quantity.

**3.** $\Gamma\left(\frac{5}{2}\right)$

**4.** $\Gamma\left(\frac{7}{2}\right)$

**5.** $\Gamma\left(\frac{9}{2}\right)$

**6.** $\Gamma\left(\frac{11}{2}\right)$

In Problems 7–10 use (13) and the known value $\Gamma\left(-\frac{1}{2}\right) = -2\sqrt{\pi}$ to evaluate the given quantity.

**7.** $\Gamma\left(-\frac{3}{2}\right)$

**8.** $\Gamma\left(-\frac{5}{2}\right)$

**9.** $\Gamma\left(-\frac{7}{2}\right)$

**10.** $\Gamma\left(-\frac{9}{2}\right)$

In Problems 11–14 use (7) and the given numerical value to evaluate the indicated quantity.

**11.** $\Gamma(0.2) = 4.5908;\ \Gamma(1.2)$   **12.** $\Gamma(-0.1) = -10.6863;\ \Gamma(0.9)$

**13.** $\Gamma(-0.3) = -4.3269;\ \Gamma(1.7)$   **14.** $\Gamma(1.5) = 0.8862;\ \Gamma(3.5)$

**15.** Express $\Gamma(x + 4)$ in terms of $\Gamma(x)$.

**16.** For $c > 0$, $x > 0$, use a substitution to show that

$$\int_0^\infty t^{x-1}e^{-ct}\,dt = c^{-x}\,\Gamma(x).$$

In Problems 17–20 use the result in Problem 16 to express the given improper integral as a gamma function. Then evaluate the integral.

**17.** $\displaystyle\int_0^\infty \frac{e^{-2t}}{\sqrt{t}}\,dt$

**18.** $\displaystyle\int_0^\infty t^{3/2}e^{-4t}\,dt$

**19.** $\displaystyle\int_0^\infty t^4 e^{-6t}\,dt$

**20.** $\displaystyle\int_0^\infty t^3 e^{-t/3}\,dt$

In Problems 21–24 use the substitution $t = -\ln x$ and the result in Problem 16 to express the given improper integral as a gamma function. Then evaluate the integral.

**21.** $\displaystyle\int_0^1 x^3\left(\ln\frac{1}{x}\right)^3 dx$

**22.** $\displaystyle\int_0^1 x^4\left(\ln\frac{1}{x}\right)^2 dx$

**23.** $\displaystyle\int_0^1 \left(\frac{1}{x}\ln\frac{1}{x}\right)^{1/2} dx$

**24.** $\displaystyle\int_0^1 x^{1/2}\left(\ln\frac{1}{x}\right)^{-1/2} dx$

In Problems 25 and 26 use the indicated substitution to express the given improper integral as a gamma function. Then evaluate the integral using the indicated numerical value.

**25.** $\displaystyle\int_0^\infty x^5 e^{-x^5}\,dx;\ \ t = x^5,\ \Gamma\left(\frac{6}{5}\right) = 0.9182$

**26.** $\displaystyle\int_0^\infty x^4 e^{-x^3}\,dx;\ \ t = x^3,\ \Gamma\left(\frac{5}{3}\right) = 0.9027$

**27. (a)** For $x > 0$ the gamma function possesses derivatives of all orders. Use Leibniz's rule in the form given in (24) to show that

$$\Gamma'(x) = \int_0^\infty t^{x-1}e^{-t}\ln t\,dt$$

$$\Gamma''(x) = \int_0^\infty t^{x-1}e^{-t}(\ln t)^2\,dt.$$

**(b)** For $x > 0$, $\Gamma(x) > 0$ and $\Gamma''(x) > 0$. Without looking back at Figure A.1, what does this say about the graph of $\Gamma(x)$?

**28.** A definition of the gamma function given by Carl Friedrich Gauss in 1811 that is valid for all real numbers, except $x = 0, x = -1, x = -2, \ldots$, is

$$\Gamma(x) = \lim_{n\to\infty} \frac{n!\,n^x}{x(x + 1)(x + 2)\cdots(x + n)}.$$

Use this definition to give an alternative derivation of the recursion formula (7).

## The Beta Function

In Problems 29 and 30 express the given improper integral as a beta function. Then evaluate the integral using (15) and known values of the gamma function.

**29.** $\displaystyle\int_0^1 \frac{1}{\sqrt{t(1 - t)}}\,dt$

**30.** $\displaystyle\int_0^1 \sqrt{t}\sqrt{(1 - t)^3}\,dt$

In Problems 31 and 32 use the indicated substitution to express the given improper integral as a beta function. Then evaluate the integral using (15) and known values of the gamma function.

**31.** $\displaystyle\int_0^1 \frac{x^2}{\sqrt{1 - x^2}}\,dx;\ \ t = x^2$   **32.** $\displaystyle\int_0^1 \frac{x^5}{\sqrt{1 - x^4}}\,dx;\ \ t = x^4$

In Problems 33 and 34 use the substitution $t = x/2$ to express the given improper integral as a beta function. Then evaluate the integral using (15) and known values of the gamma function.

**33.** $\displaystyle\int_0^2 x^{-1/2}(2 - x)^{5/2}\,dx$   **34.** $\displaystyle\int_0^2 \frac{x^2}{\sqrt{2 - x}}\,dx$

**35.** Show that the beta function is symmetric in $x$ and $y$, that is, $B(x, y) = B(y, x)$.

**36.** Show that if $m + 1$ and $n + 1$ are positive integers, then (15) becomes

$$B(m + 1, n + 1) = \frac{m!\,n!}{(m + n + 1)!}.$$

**37.** Use the substitution $t = \sin^2\theta$ to show that the beta function (14) can be expressed as

$$B(x, y) = 2\int_0^{\pi/2} \sin^{2x-1}\theta\,\cos^{2y-1}\theta\,d\theta.$$

**38.** If $m = 2x - 1$ and $n = 2y - 1$, then show that the integral in Problem 37 can be written

$$\int_0^{\pi/2} \sin^m \theta \cos^n \theta \, d\theta = \frac{\Gamma\left(\dfrac{m+1}{2}\right)\Gamma\left(\dfrac{n+1}{2}\right)}{2\Gamma\left(\dfrac{m+n+2}{2}\right)}, \quad m > -1, n > -1.$$

In Problems 39–42 use the result in Problem 38 to evaluate the given integral.

**39.** $\displaystyle\int_0^{\pi/2} \sin^3 \theta \cos^3 \theta \, d\theta$    **40.** $\displaystyle\int_0^{\pi/2} \sin^6 \theta \cos^5 \theta \, d\theta$

**41.** $\displaystyle\int_0^{\pi/2} \cos^5 \theta \, d\theta$    **42.** $\displaystyle\int_0^{\pi/2} \sin^7 \theta \, d\theta$

## Leibniz's Rule

In Problems 43–48 use Leibniz's rule in the forms given in (22) and (23) to find the derivative of the given function.

**43.** $\displaystyle F(x) = \int_x^{2x^3} e^{-x^3 t^2} dt$    **44.** $\displaystyle F(x) = \int_0^{x^2} \frac{\sin(x\sqrt{t})}{t} dt$

**45.** $\displaystyle F(x) = \int_{x^2}^{\cos x} (x^2 + t^2)^{10} dt$    **46.** $\displaystyle F(x) = \int_{\sqrt{3x}}^{\sin 2x} \ln(x^2 + t^2) dt$

**47.** $\displaystyle F(x) = \int_{\pi/2}^{\pi - x} e^{\sin t} \sin(\cos t) dt$    **48.** $\displaystyle F(x) = \int_{\sqrt{x}}^{4\sqrt{x}} e^{t^2} dt$

In Problems 49 and 50 use Leibniz's rule in the form given in (24) to show that the indicated integral-defined function is a solution of the given second-order differential equation. [*Hint*: After computing the first and second derivatives find a term in $y'$ that can be evaluated using integration by parts.]

**49.** $xy'' + y' + xy = 0; \quad y = \displaystyle\int_0^{\pi} \cos(x \cos t) dt$

**50.** $xy'' + y' - xy = 0; \quad y = \displaystyle\int_0^{\pi} e^{x \cos t} dt$

**51.** Assume that Leibniz's rule (24) is valid for the improper integral

$$y = e^{x^2} \int_0^{\infty} t^2 e^{-t^2} \cos 2xt \, dt.$$

Show that this integral-defined function is a solution of the second-order differential equation $y'' + 2xy' + 4y = 0$. [*Hint*: After computing the first and second derivatives find a term in $y''$ that can be evaluated using integration by parts.]

**52. (a)** Find the derivative of the error function $\text{erf}(x)$. Then use the identity $\text{erf}(x) + \text{erf} \, x(x) = 1$ to find the derivative of the complementary error function $\text{erfc}(x)$.

**(b)** Use the last result in part (a) to show that

$$y = e^{-x^2} - 2x \int_x^{\infty} e^{-t^2} dt$$

is a solution of the differential equation $y'' + 2xy' - 2y = 0$.

**53. (a)** Show that for $x > 0$,

$$\int_0^{\infty} \frac{1}{x^2 + t^2} dt = \frac{\pi}{2x}.$$

[*Hint*: See formula 43 in the table of integrals given on the right inside page of the front cover.]

**(b)** Use the result in part (a) and (24) to show that

$$\int_0^{\infty} \frac{1}{(x^2 + t^2)^2} dt = \frac{\pi}{4x^3}.$$

**54.** Consider the function

$$F(x) = \int_0^1 \frac{t^x - 1}{\ln t} dt, \quad x \geq 0.$$

**(a)** Use (24) to find $F'(x)$ as an integral. Evaluate this integral.

**(b)** Use the result in part (a) to find $F(x) = \int F'(x) dx + C$, where $C$ is a constant.

**(c)** Use the result in part (b) and $F(0) = 0$ to find an explicit function $F(x)$.

**55.** Consider the function

$$F(x) = \int_0^{\infty} \frac{1 - e^{-xt^2}}{t^2} dt, \quad x \geq 0.$$

**(a)** Use (24) to find $F'(x)$ as an integral. Evaluate this integral using a substitution and the known result $\int_0^{\infty} e^{-u^2} du = \sqrt{\pi}/2$. (See page APP-5.)

**(b)** Use the result in part (a) to find $F(x) = \int F'(x) dx + C$, where $C$ is a constant.

**(c)** Use the result in part (b) and $F(0) = 0$ to find an explicit function $F(x)$.

**56.** Consider the function

$$F(x) = \int_0^{\infty} e^{-tx} \frac{\sin t}{t} dt, \quad x \geq 0.$$

**(a)** Use (24) to find $F'(x)$ as an integral. Evaluate this integral using integration by parts.

**(b)** Use the result in part (a) to find $F(x) = \int F'(x) dx + C$, where $C$ is a constant.

**(c)** Use the result in part (b) and $F(0) = \pi/2$ to find an explicit function $F(x)$.

# Appendix B Matrices

## B.1 BASIC DEFINITIONS AND THEORY

### DEFINITION B.1 Matrix

A **matrix A** is any rectangular array of numbers or functions:

$$\mathbf{A} = \begin{pmatrix} a_{11} & a_{12} & \cdots & a_{1n} \\ a_{21} & a_{22} & \cdots & a_{2n} \\ \vdots & & & \vdots \\ a_{m1} & a_{m2} & \cdots & a_{mn} \end{pmatrix}. \tag{1}$$

If a matrix has $m$ rows and $n$ columns, we say that its **size** is $m$ by $n$ (written $m \times n$). An $n \times n$ matrix is called a **square matrix** of order $n$.

The **entry** in the $i$th row and $j$th column of an $m \times n$ matrix $\mathbf{A}$ is written $a_{ij}$. An $m \times n$ matrix $\mathbf{A}$ is then abbreviated as $\mathbf{A} = (a_{ij})_{m \times n}$ or simply $\mathbf{A} = (a_{ij})$. A $1 \times 1$ matrix is simply one constant or function.

### DEFINITION B.2 Equality of Matrices

Two $m \times n$ matrices $\mathbf{A}$ and $\mathbf{B}$ are **equal** if $a_{ij} = b_{ij}$ for each $i$ and $j$.

### DEFINITION B.3 Column Matrix

A **column matrix X** is any matrix having $n$ rows and one column:

$$\mathbf{X} = \begin{pmatrix} b_{11} \\ b_{21} \\ \vdots \\ b_{n1} \end{pmatrix} = (b_{i1})_{n \times 1}.$$

A column matrix is also called a **column vector** or simply a **vector.**

### DEFINITION B.4 Multiples of Matrices

A **multiple** of a matrix $\mathbf{A}$ is defined to be

$$k\mathbf{A} = \begin{pmatrix} ka_{11} & ka_{12} & \cdots & ka_{1n} \\ ka_{21} & ka_{22} & \cdots & ka_{2n} \\ \vdots & & & \vdots \\ ka_{m1} & ka_{m2} & \cdots & ka_{mn} \end{pmatrix} = (ka_{ij})_{m \times n},$$

where $k$ is a constant or a function.

### EXAMPLE 1 Multiples of Matrices

(a) $5 \begin{pmatrix} 2 & -3 \\ 4 & -1 \\ \frac{1}{5} & 6 \end{pmatrix} = \begin{pmatrix} 10 & -15 \\ 20 & -5 \\ 1 & 30 \end{pmatrix}$     (b) $e^t \begin{pmatrix} 1 \\ -2 \\ 4 \end{pmatrix} = \begin{pmatrix} e^t \\ -2e^t \\ 4e^t \end{pmatrix}$ ■

We note in passing that, for any matrix **A**, the product $k\mathbf{A}$ is the same as $\mathbf{A}k$. For example,

$$e^{-3t} \begin{pmatrix} 2 \\ 5 \end{pmatrix} = \begin{pmatrix} 2e^{-3t} \\ 5e^{-3t} \end{pmatrix} = \begin{pmatrix} 2 \\ 5 \end{pmatrix} e^{-3t}.$$

### DEFINITION B.5 Addition of Matrices

The **sum** of two $m \times n$ matrices **A** and **B** is defined to be the matrix

$$\mathbf{A} + \mathbf{B} = (a_{ij} + b_{ij})_{m \times n}.$$

In other words, when adding two matrices of the same size, we add the corresponding entries.

### EXAMPLE 2 Matrix Addition

The sum of $\mathbf{A} = \begin{pmatrix} 2 & -1 & 3 \\ 0 & 4 & 6 \\ -6 & 10 & -5 \end{pmatrix}$ and $\mathbf{B} = \begin{pmatrix} 4 & 7 & -8 \\ 9 & 3 & 5 \\ 1 & -1 & 2 \end{pmatrix}$ is

$$\mathbf{A} + \mathbf{B} = \begin{pmatrix} 2+4 & -1+7 & 3+(-8) \\ 0+9 & 4+3 & 6+5 \\ -6+1 & 10+(-1) & -5+2 \end{pmatrix} = \begin{pmatrix} 6 & 6 & -5 \\ 9 & 7 & 11 \\ -5 & 9 & -3 \end{pmatrix}.$$ ■

### EXAMPLE 3 A Matrix Written as a Sum of Column Matrices

The single matrix $\begin{pmatrix} 3t^2 - 2e^t \\ t^2 + 7t \\ 5t \end{pmatrix}$ can be written as the sum of three column vectors:

$$\begin{pmatrix} 3t^2 - 2e^t \\ t^2 + 7t \\ 5t \end{pmatrix} = \begin{pmatrix} 3t^2 \\ t^2 \\ 0 \end{pmatrix} + \begin{pmatrix} 0 \\ 7t \\ 5t \end{pmatrix} + \begin{pmatrix} -2e^t \\ 0 \\ 0 \end{pmatrix} = \begin{pmatrix} 3 \\ 1 \\ 0 \end{pmatrix} t^2 + \begin{pmatrix} 0 \\ 7 \\ 5 \end{pmatrix} t + \begin{pmatrix} -2 \\ 0 \\ 0 \end{pmatrix} e^t.$$ ■

The difference of two $m \times n$ matrices is defined in the usual manner: $\mathbf{A} - \mathbf{B} = \mathbf{A} + (-\mathbf{B})$, where $-\mathbf{B} = (-1)\mathbf{B}$.

### DEFINITION B.6  Multiplication of Matrices

Let $\mathbf{A}$ be a matrix having $m$ rows and $n$ columns and $\mathbf{B}$ be a matrix having $n$ rows and $p$ columns. We define the **product AB** to be the $m \times p$ matrix

$$\mathbf{AB} = \begin{pmatrix} a_{11} & a_{12} & \cdots & a_{1n} \\ a_{21} & a_{22} & \cdots & a_{2n} \\ \vdots & & & \vdots \\ a_{m1} & a_{m2} & \cdots & a_{mn} \end{pmatrix} \begin{pmatrix} b_{11} & b_{12} & \cdots & b_{1p} \\ b_{21} & b_{22} & \cdots & b_{2p} \\ \vdots & & & \vdots \\ b_{n1} & b_{n2} & \cdots & b_{np} \end{pmatrix}$$

$$= \begin{pmatrix} a_{11}b_{11} + a_{12}b_{21} + \cdots + a_{1n}b_{n1} & \cdots & a_{11}b_{1p} + a_{12}b_{2p} + \cdots + a_{1n}b_{np} \\ a_{21}b_{11} + a_{22}b_{21} + \cdots + a_{2n}b_{n1} & \cdots & a_{21}b_{1p} + a_{22}b_{2p} + \cdots + a_{2n}b_{np} \\ \vdots & & \vdots \\ a_{m1}b_{11} + a_{m2}b_{21} + \cdots + a_{mn}b_{n1} & \cdots & a_{m1}b_{1p} + a_{m2}b_{2p} + \cdots + a_{mn}b_{np} \end{pmatrix}$$

$$= \left( \sum_{k=1}^{n} a_{ik}b_{kj} \right)_{m \times p}.$$

Note carefully in Definition B.6 that the product $\mathbf{AB} = \mathbf{C}$ is defined only when the number of columns in the matrix $\mathbf{A}$ is the same as the number of rows in $\mathbf{B}$. The size of the product can be determined from

$$\mathbf{A}_{m \times n} \mathbf{B}_{n \times p} = \mathbf{C}_{m \times p}.$$

Also, you might recognize that the entries in, say, the $i$th row of the final matrix $\mathbf{AB}$ are formed by using the component definition of the inner, or dot, product of the $i$th row of $\mathbf{A}$ with each of the columns of $\mathbf{B}$.

### EXAMPLE 4  Multiplication of Matrices

**(a)** For $\mathbf{A} = \begin{pmatrix} 4 & 7 \\ 3 & 5 \end{pmatrix}$ and $\mathbf{B} = \begin{pmatrix} 9 & -2 \\ 6 & 8 \end{pmatrix}$,

$$\mathbf{AB} = \begin{pmatrix} 4 \cdot 9 + 7 \cdot 6 & 4 \cdot (-2) + 7 \cdot 8 \\ 3 \cdot 9 + 5 \cdot 6 & 3 \cdot (-2) + 5 \cdot 8 \end{pmatrix} = \begin{pmatrix} 78 & 48 \\ 57 & 34 \end{pmatrix}.$$

**(b)** For $\mathbf{A} = \begin{pmatrix} 5 & 8 \\ 1 & 0 \\ 2 & 7 \end{pmatrix}$ and $\mathbf{B} = \begin{pmatrix} -4 & -3 \\ 2 & 0 \end{pmatrix}$,

$$\mathbf{AB} = \begin{pmatrix} 5 \cdot (-4) + 8 \cdot 2 & 5 \cdot (-3) + 8 \cdot 0 \\ 1 \cdot (-4) + 0 \cdot 2 & 1 \cdot (-3) + 0 \cdot 0 \\ 2 \cdot (-4) + 7 \cdot 2 & 2 \cdot (-3) + 7 \cdot 0 \end{pmatrix} = \begin{pmatrix} -4 & -15 \\ -4 & -3 \\ 6 & -6 \end{pmatrix}. \qquad \blacksquare$$

In general, *matrix multiplication is not commutative*; that is, $\mathbf{AB} \neq \mathbf{BA}$.

Observe in part (a) of Example 4 that $\mathbf{BA} = \begin{pmatrix} 30 & 53 \\ 48 & 82 \end{pmatrix}$, whereas in part (b) the product $\mathbf{BA}$ is not defined, since Definition B.6 requires that the first matrix (in this case $\mathbf{B}$) have the same number of columns as the second matrix has rows.

We are particularly interested in the product of a square matrix and a column vector.

**Multiplication of Matrices**

**(a)** $\begin{pmatrix} 2 & -1 & 3 \\ 0 & 4 & 5 \\ 1 & -7 & 9 \end{pmatrix} \begin{pmatrix} -3 \\ 6 \\ 4 \end{pmatrix} = \begin{pmatrix} 2 \cdot (-3) + (-1) \cdot 6 + 3 \cdot 4 \\ 0 \cdot (-3) + 4 \cdot 6 + 5 \cdot 4 \\ 1 \cdot (-3) + (-7) \cdot 6 + 9 \cdot 4 \end{pmatrix} = \begin{pmatrix} 0 \\ 44 \\ -9 \end{pmatrix}$

**(b)** $\begin{pmatrix} -4 & 2 \\ 3 & 8 \end{pmatrix} \begin{pmatrix} x \\ y \end{pmatrix} = \begin{pmatrix} -4x + 2y \\ 3x + 8y \end{pmatrix}$

**MULTIPLICATIVE IDENTITY**    For a given positive integer $n$ the $n \times n$ matrix

$$\mathbf{I} = \begin{pmatrix} 1 & 0 & 0 & \cdots & 0 \\ 0 & 1 & 0 & \cdots & 0 \\ \vdots & & & & \vdots \\ 0 & 0 & 0 & \cdots & 1 \end{pmatrix}$$

is called the **multiplicative identity matrix.** It follows from Definition B.6 that for any $n \times n$ matrix $\mathbf{A}$.

$$\mathbf{AI} = \mathbf{IA} = \mathbf{A}.$$

Also, it is readily verified that, if $\mathbf{X}$ is an $n \times 1$ column matrix, then $\mathbf{IX} = \mathbf{X}$.

**ZERO MATRIX**    A matrix consisting of all zero entries is called a **zero matrix** and is denoted by $\mathbf{0}$. For example,

$$\mathbf{0} = \begin{pmatrix} 0 \\ 0 \end{pmatrix}, \qquad \mathbf{0} = \begin{pmatrix} 0 & 0 \\ 0 & 0 \end{pmatrix}, \qquad \mathbf{0} = \begin{pmatrix} 0 & 0 \\ 0 & 0 \\ 0 & 0 \end{pmatrix},$$

and so on. If $\mathbf{A}$ and $\mathbf{0}$ are $m \times n$ matrices, then

$$\mathbf{A} + \mathbf{0} = \mathbf{0} + \mathbf{A} = \mathbf{A}.$$

**ASSOCIATIVE LAW**    Although we shall not prove it, matrix multiplication is **associative.** If $\mathbf{A}$ is an $m \times p$ matrix, $\mathbf{B}$ a $p \times r$ matrix, and $\mathbf{C}$ an $r \times n$ matrix, then

$$\mathbf{A}(\mathbf{BC}) = (\mathbf{AB})\mathbf{C}$$

is an $m \times n$ matrix.

**DISTRIBUTIVE LAW**    If all products are defined, multiplication is **distributive** over addition:

$$\mathbf{A}(\mathbf{B} + \mathbf{C}) = \mathbf{AB} + \mathbf{AC} \qquad \text{and} \qquad (\mathbf{B} + \mathbf{C})\mathbf{A} = \mathbf{BA} + \mathbf{CA}.$$

**DETERMINANT OF A MATRIX**    Associated with every *square* matrix $\mathbf{A}$ of constants is a number called the **determinant of the matrix,** which is denoted by det $\mathbf{A}$.

**Determinant of a Square Matrix**

For $\mathbf{A} = \begin{pmatrix} 3 & 6 & 2 \\ 2 & 5 & 1 \\ -1 & 2 & 4 \end{pmatrix}$ we expand det $\mathbf{A}$ by cofactors of the first row:

$$\det \mathbf{A} = \begin{vmatrix} 3 & 6 & 2 \\ 2 & 5 & 1 \\ -1 & 2 & 4 \end{vmatrix} = 3 \begin{vmatrix} 5 & 1 \\ 2 & 4 \end{vmatrix} - 6 \begin{vmatrix} 2 & 1 \\ -1 & 4 \end{vmatrix} + 2 \begin{vmatrix} 2 & 5 \\ -1 & 2 \end{vmatrix}$$

$$= 3(20 - 2) - 6(8 + 1) + 2(4 + 5) = 18.$$

It can be proved that a determinant det **A** can be expanded by cofactors using any row or column. If det **A** has a row (or a column) containing many zero entries, then wisdom dictates that we expand the determinant by that row (or column).

---

**DEFINITION B.7    Transpose of a Matrix**

The **transpose** of the $m \times n$ matrix (1) is the $n \times m$ matrix $\mathbf{A}^T$ given by

$$\mathbf{A}^T = \begin{pmatrix} a_{11} & a_{21} & \cdots & a_{m1} \\ a_{12} & a_{22} & \cdots & a_{m2} \\ \vdots & & & \vdots \\ a_{1n} & a_{2n} & \cdots & a_{mn} \end{pmatrix}.$$

In other words, the rows of a matrix **A** become the columns of its transpose $\mathbf{A}^T$.

---

**EXAMPLE 7    Transpose of a Matrix**

(a) The transpose of $\mathbf{A} = \begin{pmatrix} 3 & 6 & 2 \\ 2 & 5 & 1 \\ -1 & 2 & 4 \end{pmatrix}$ is $\mathbf{A}^T = \begin{pmatrix} 3 & 2 & -1 \\ 6 & 5 & 2 \\ 2 & 1 & 4 \end{pmatrix}.$

(b) If $\mathbf{X} = \begin{pmatrix} 5 \\ 0 \\ 3 \end{pmatrix}$, then $\mathbf{X}^T = (5 \quad 0 \quad 3)$. ∎

---

**DEFINITION B.8    Multiplicative Inverse of a Matrix**

Let **A** be an $n \times n$ matrix. If there exists an $n \times n$ matrix **B** such that

$$\mathbf{AB} = \mathbf{BA} = \mathbf{I},$$

where **I** is the multiplicative identity, then **B** is said to be the **multiplicative inverse of A** and is denoted by $\mathbf{B} = \mathbf{A}^{-1}$.

---

**DEFINITION B.9    Nonsingular/Singular Matrices**

Let **A** be an $n \times n$ matrix. If det $\mathbf{A} \neq 0$, then **A** is said to be **nonsingular.** If det $\mathbf{A} = 0$, then **A** is said to be **singular.**

---

The following theorem gives a necessary and sufficient condition for a square matrix to have a multiplicative inverse.

---

**THEOREM B.1    Nonsingularity Implies A Has an Inverse**

An $n \times n$ matrix **A** has a multiplicative inverse $\mathbf{A}^{-1}$ if and only if **A** is nonsingular.

---

The following theorem gives one way of finding the multiplicative inverse for a nonsingular matrix.

### THEOREM  B.2   A Formula for the Inverse of a Matrix

Let $\mathbf{A}$ be an $n \times n$ nonsingular matrix and let $C_{ij} = (-1)^{i+j} M_{ij}$, where $M_{ij}$ is the determinant of the $(n - 1) \times (n - 1)$ matrix obtained by deleting the $i$th row and $j$th column from $\mathbf{A}$. Then

$$\mathbf{A}^{-1} = \frac{1}{\det \mathbf{A}} (C_{ij})^T. \tag{2}$$

Each $C_{ij}$ in Theorem B.2 is simply the **cofactor** (signed minor) of the corresponding entry $a_{ij}$ in $\mathbf{A}$. Note that the transpose is utilized in formula (2).

For future reference we observe in the case of a $2 \times 2$ nonsingular matrix

$$\mathbf{A} = \begin{pmatrix} a_{11} & a_{12} \\ a_{21} & a_{22} \end{pmatrix}$$

that $C_{11} = a_{22}$, $C_{12} = -a_{21}$, $C_{21} = -a_{12}$, and $C_{22} = a_{11}$. Thus

$$\mathbf{A}^{-1} = \frac{1}{\det \mathbf{A}} \begin{pmatrix} a_{22} & -a_{21} \\ -a_{12} & a_{11} \end{pmatrix}^T = \frac{1}{\det \mathbf{A}} \begin{pmatrix} a_{22} & -a_{12} \\ -a_{21} & a_{11} \end{pmatrix}. \tag{3}$$

For a $3 \times 3$ nonsingular matrix

$$\mathbf{A} = \begin{pmatrix} a_{11} & a_{12} & a_{13} \\ a_{21} & a_{22} & a_{23} \\ a_{31} & a_{32} & a_{33} \end{pmatrix},$$

$$C_{11} = \begin{vmatrix} a_{22} & a_{23} \\ a_{32} & a_{33} \end{vmatrix}, \qquad C_{12} = -\begin{vmatrix} a_{21} & a_{23} \\ a_{31} & a_{33} \end{vmatrix}, \qquad C_{13} = \begin{vmatrix} a_{21} & a_{22} \\ a_{31} & a_{32} \end{vmatrix},$$

and so on. Carrying out the transposition gives

$$\mathbf{A}^{-1} = \frac{1}{\det \mathbf{A}} \begin{pmatrix} C_{11} & C_{21} & C_{31} \\ C_{12} & C_{22} & C_{32} \\ C_{13} & C_{23} & C_{33} \end{pmatrix}. \tag{4}$$

### EXAMPLE 8   Inverse of a 2 × 2 Matrix

Find the multiplicative inverse for $\mathbf{A} = \begin{pmatrix} 1 & 4 \\ 2 & 10 \end{pmatrix}$.

**SOLUTION** Since $\det \mathbf{A} = 10 - 8 = 2 \neq 0$, $\mathbf{A}$ is nonsingular. It follows from Theorem B.1 that $\mathbf{A}^{-1}$ exists. From (3) we find

$$\mathbf{A}^{-1} = \frac{1}{2} \begin{pmatrix} 10 & -4 \\ -2 & 1 \end{pmatrix} = \begin{pmatrix} 5 & -2 \\ -1 & \frac{1}{2} \end{pmatrix}. \qquad \blacksquare$$

Not every square matrix has a multiplicative inverse. The matrix $\mathbf{A} = \begin{pmatrix} 2 & 2 \\ 3 & 3 \end{pmatrix}$ is singular, since $\det \mathbf{A} = 0$. Hence $\mathbf{A}^{-1}$ does not exist.

### EXAMPLE 9   Inverse of a 3 × 3 Matrix

Find the multiplicative inverse for $\mathbf{A} = \begin{pmatrix} 2 & 2 & 0 \\ -2 & 1 & 1 \\ 3 & 0 & 1 \end{pmatrix}$.

**SOLUTION** Since det $\mathbf{A} = 12 \neq 0$, the given matrix is nonsingular. The cofactors corresponding to the entries in each row of det $\mathbf{A}$ are

$$C_{11} = \begin{vmatrix} 1 & 1 \\ 0 & 1 \end{vmatrix} = 1 \qquad C_{12} = -\begin{vmatrix} -2 & 1 \\ 3 & 1 \end{vmatrix} = 5 \qquad C_{13} = \begin{vmatrix} -2 & 1 \\ 3 & 0 \end{vmatrix} = -3$$

$$C_{21} = -\begin{vmatrix} 2 & 0 \\ 0 & 1 \end{vmatrix} = -2 \qquad C_{22} = \begin{vmatrix} 2 & 0 \\ 3 & 1 \end{vmatrix} = 2 \qquad C_{23} = -\begin{vmatrix} 2 & 2 \\ 3 & 0 \end{vmatrix} = 6$$

$$C_{31} = \begin{vmatrix} 2 & 0 \\ 1 & 1 \end{vmatrix} = 2 \qquad C_{32} = -\begin{vmatrix} 2 & 0 \\ -2 & 1 \end{vmatrix} = -2 \qquad C_{33} = \begin{vmatrix} 2 & 2 \\ -2 & 1 \end{vmatrix} = 6.$$

If follows from (4) that

$$\mathbf{A}^{-1} = \frac{1}{12}\begin{pmatrix} 1 & -2 & 2 \\ 5 & 2 & -2 \\ -3 & 6 & 6 \end{pmatrix} = \begin{pmatrix} \frac{1}{12} & -\frac{1}{6} & \frac{1}{6} \\ \frac{5}{12} & \frac{1}{6} & -\frac{1}{6} \\ -\frac{1}{4} & \frac{1}{2} & \frac{1}{2} \end{pmatrix}.$$

You are urged to verify that $\mathbf{A}^{-1}\mathbf{A} = \mathbf{A}\mathbf{A}^{-1} = \mathbf{I}$. ∎

Formula (2) presents obvious difficulties for nonsingular matrices larger than $3 \times 3$. For example, to apply (2) to a $4 \times 4$ matrix, we would have to calculate *sixteen* $3 \times 3$ determinants.* In the case of a large matrix there are more efficient ways of finding $\mathbf{A}^{-1}$. The curious reader is referred to any text in linear algebra.

Since our goal is to apply the concept of a matrix to systems of linear first-order differential equations, we need the following definitions.

---

**DEFINITION B.10** Derivative of a Matrix of Functions

If $\mathbf{A}(t) = (a_{ij}(t))_{m \times n}$ is a matrix whose entries are functions differentiable on a common interval, then

$$\frac{d\mathbf{A}}{dt} = \left(\frac{d}{dt}a_{ij}\right)_{m \times n}.$$

---

**DEFINITION B.11** Integral of a Matrix of Functions

If $\mathbf{A}(t) = (a_{ij}(t))_{m \times n}$ is a matrix whose entries are functions continuous on a common interval containing $t$ and $t_0$, then

$$\int_{t_0}^{t} \mathbf{A}(s)\, ds = \left(\int_{t_0}^{t} a_{ij}(s)\, ds\right)_{m \times n}.$$

---

To differentiate (integrate) a matrix of functions, we simply differentiate (integrate) each entry. The derivative of a matrix is also denoted by $\mathbf{A}'(t)$.

**EXAMPLE 10** Derivative/Integral of a Matrix

If $\quad \mathbf{X}(t) = \begin{pmatrix} \sin 2t \\ e^{3t} \\ 8t - 1 \end{pmatrix}, \quad$ then $\quad \mathbf{X}'(t) = \begin{pmatrix} \dfrac{d}{dt}\sin 2t \\[4pt] \dfrac{d}{dt}e^{3t} \\[4pt] \dfrac{d}{dt}(8t - 1) \end{pmatrix} = \begin{pmatrix} 2\cos 2t \\ 3e^{3t} \\ 8 \end{pmatrix}$

---

*Strictly speaking, a determinant is a number, but it is sometimes convenient to refer to a determinant as if it were an array.

and
$$\int_0^t \mathbf{X}(s)\,ds = \begin{pmatrix} \int_0^t \sin 2s\,ds \\ \int_0^t e^{3s}\,ds \\ \int_0^t (8s-1)\,ds \end{pmatrix} = \begin{pmatrix} -\frac{1}{2}\cos 2t + \frac{1}{2} \\ \frac{1}{3}e^{3t} - \frac{1}{3} \\ 4t^2 - t \end{pmatrix}.$$

## B.2 GAUSSIAN AND GAUSS-JORDAN ELIMINATION

Matrices are an invaluable aid in solving algebraic systems of $n$ linear equations in $n$ variables or unknowns,

$$\begin{aligned}
a_{11}x_1 + a_{12}x_2 + \cdots + a_{1n}x_n &= b_1 \\
a_{21}x_1 + a_{22}x_2 + \cdots + a_{2n}x_n &= b_2 \\
&\;\;\vdots \qquad\qquad\qquad \vdots \\
a_{n1}x_1 + a_{n2}x_2 + \cdots + a_{nn}x_n &= b_n.
\end{aligned} \tag{5}$$

If $\mathbf{A}$ denotes the matrix of coefficients in (5), we know that Cramer's rule could be used to solve the system whenever $\det \mathbf{A} \neq 0$. However, that rule requires a herculean effort if $\mathbf{A}$ is larger than $3 \times 3$. The procedure that we shall now consider has the distinct advantage of being not only an efficient way of handling large systems, but also a means of solving consistent systems (5) in which $\det \mathbf{A} = 0$ and a means of solving $m$ linear equations in $n$ unknowns.

---

**DEFINITION B.12  Augmented Matrix**

The **augmented matrix** of the system (5) is the $n \times (n+1)$ matrix

$$\begin{pmatrix}
a_{11} & a_{12} & \cdots & a_{1n} & \big| & b_1 \\
a_{21} & a_{22} & \cdots & a_{2n} & \big| & b_2 \\
\vdots & & & & \big| & \vdots \\
a_{n1} & a_{n2} & \cdots & a_{nn} & \big| & b_n
\end{pmatrix}.$$

---

If $\mathbf{B}$ is the column matrix of the $b_i$, $i = 1, 2, \ldots, n$, the augmented matrix of (5) is denoted by $(\mathbf{A}|\mathbf{B})$.

**ELEMENTARY ROW OPERATIONS**   Recall from algebra that we can transform an algebraic system of equations into an equivalent system (that is, one having the same solution) by multiplying an equation by a nonzero constant, interchanging the positions of any two equations in a system, and adding a nonzero constant multiple of an equation to another equation. These operations on equations in a system are, in turn, equivalent to **elementary row operations** on an augmented matrix:

*(i)*    Multiply a row by a nonzero constant.
*(ii)*   Interchange any two rows.
*(iii)*  Add a nonzero constant multiple of one row to any other row.

**ELIMINATION METHODS**   To solve a system such as (5) using an augmented matrix, we use either **Gaussian elimination** or the **Gauss-Jordan elimination method.** In the former method, we carry out a succession of elementary row operations until we arrive at an augmented matrix in **row-echelon form:**

*(i)*    The first nonzero entry in a nonzero row is 1.
*(ii)*   In consecutive nonzero rows the first entry 1 in the lower row appears to the right of the first 1 in the higher row.
*(iii)*  Rows consisting of all 0's are at the bottom of the matrix.

In the Gauss-Jordan method the row operations are continued until we obtain an augmented matrix that is in **reduced row-echelon form.** A reduced row-echelon matrix has the same three properties listed above in addition to the following one:

(*iv*)  A column containing a first entry 1 has 0's everywhere else.

### EXAMPLE 11  Row-Echelon/Reduced Row-Echelon Form

**(a)**  The augmented matrices

$$\begin{pmatrix} 1 & 5 & 0 & | & 2 \\ 0 & 1 & 0 & | & -1 \\ 0 & 0 & 0 & | & 0 \end{pmatrix} \quad \text{and} \quad \begin{pmatrix} 0 & 0 & 1 & -6 & 2 & | & 2 \\ 0 & 0 & 0 & 0 & 1 & | & 4 \end{pmatrix}$$

are in row-echelon form. You should verify that the three criteria are satisfied.

**(b)**  The augmented matrices

$$\begin{pmatrix} 1 & 0 & 0 & | & 7 \\ 0 & 1 & 0 & | & -1 \\ 0 & 0 & 0 & | & 0 \end{pmatrix} \quad \text{and} \quad \begin{pmatrix} 0 & 0 & 1 & -6 & 0 & | & -6 \\ 0 & 0 & 0 & 0 & 1 & | & 4 \end{pmatrix}$$

are in reduced row-echelon form. Note that the remaining entries in the columns containing a leading entry 1 are all 0's. ■

Note that in Gaussian elimination we stop once we have obtained *an* augmented matrix in row-echelon form. In other words, by using different sequences of row operations we may arrive at different row-echelon forms. This method then requires the use of back-substitution. In Gauss-Jordan elimination we stop when we have obtained *the* augmented matrix in reduced row-echelon form. Any sequence of row operations will lead to the same augmented matrix in reduced row-echelon form. This method does not require back-substitution; the solution of the system will be apparent by inspection of the final matrix. In terms of the equations of the original system, our goal in both methods is simply to make the coefficient of $x_1$ in the first equation* equal to 1 and then use multiples of that equation to eliminate $x_1$ from other equations. The process is repeated on the other variables.

To keep track of the row operations on an augmented matrix, we utilize the following notation:

| Symbol | Meaning |
|---|---|
| $R_{ij}$ | Interchange rows $i$ and $j$ |
| $cR_i$ | Multiply the $i$th row by the nonzero constant $c$ |
| $cR_i + R_j$ | Multiply the $i$th row by $c$ and add to the $j$th row |

### EXAMPLE 12  Solution by Elimination

Solve

$$2x_1 + 6x_2 + x_3 = 7$$

$$x_1 + 2x_2 - x_3 = -1$$

$$5x_1 + 7x_2 - 4x_3 = 9$$

using **(a)** Gaussian elimination and **(b)** Gauss-Jordan elimination.

---

*We can always interchange equations so that the first equation contains the variable $x_1$.

**SOLUTION** (a) Using row operations on the augmented matrix of the system, we obtain

$$\begin{pmatrix} 2 & 6 & 1 & 7 \\ 1 & 2 & -1 & -1 \\ 5 & 7 & -4 & 9 \end{pmatrix} \xrightarrow{R_{12}} \begin{pmatrix} 1 & 2 & -1 & -1 \\ 2 & 6 & 1 & 7 \\ 5 & 7 & -4 & 9 \end{pmatrix} \xrightarrow[-5R_1 + R_3]{-2R_1 + R_2} \begin{pmatrix} 1 & 2 & -1 & -1 \\ 0 & 2 & 3 & 9 \\ 0 & -3 & 1 & 14 \end{pmatrix}$$

$$\xrightarrow{\frac{1}{2}R_2} \begin{pmatrix} 1 & 2 & -1 & -1 \\ 0 & 1 & \frac{3}{2} & \frac{9}{2} \\ 0 & -3 & 1 & 14 \end{pmatrix} \xrightarrow{3R_2 + R_3} \begin{pmatrix} 1 & 2 & -1 & -1 \\ 0 & 1 & \frac{3}{2} & \frac{9}{2} \\ 0 & 0 & \frac{11}{2} & \frac{55}{2} \end{pmatrix} \xrightarrow{\frac{2}{11}R_3} \begin{pmatrix} 1 & 2 & -1 & -1 \\ 0 & 1 & \frac{3}{2} & \frac{9}{2} \\ 0 & 0 & 1 & 5 \end{pmatrix}.$$

The last matrix is in row-echelon form and represents the system

$$x_1 + 2x_2 - x_3 = -1$$
$$x_2 + \frac{3}{2}x_3 = \frac{9}{2}$$
$$x_3 = 5.$$

Substituting $x_3 = 5$ into the second equation then gives $x_2 = -3$. Substituting both these values back into the first equation finally yields $x_1 = 10$.

(b) We start with the last matrix above. Since the first entries in the second and third rows are 1's, we must, in turn, make the remaining entries in the second and third columns 0's:

$$\begin{pmatrix} 1 & 2 & -1 & -1 \\ 0 & 1 & \frac{3}{2} & \frac{9}{2} \\ 0 & 0 & 1 & 5 \end{pmatrix} \xrightarrow{-2R_2 + R_1} \begin{pmatrix} 1 & 0 & -4 & -10 \\ 0 & 1 & \frac{3}{2} & \frac{9}{2} \\ 0 & 0 & 1 & 5 \end{pmatrix} \xrightarrow[-\frac{3}{2}R_3 + R_2]{4R_3 + R_1} \begin{pmatrix} 1 & 0 & 0 & 10 \\ 0 & 1 & 0 & -3 \\ 0 & 0 & 1 & 5 \end{pmatrix}.$$

The last matrix is now in reduced row-echelon form. Because of what the matrix means in terms of equations, it is evident that the solution of the system is $x_1 = 10, x_2 = -3, x_3 = 5$. ■

## EXAMPLE 13 Gauss-Jordan Elimination

Solve

$$x + 3y - 2z = -7$$
$$4x + y + 3z = 5$$
$$2x - 5y + 7z = 19.$$

**SOLUTION** We solve the system using Gauss-Jordan elimination:

$$\begin{pmatrix} 1 & 3 & -2 & -7 \\ 4 & 1 & 3 & 5 \\ 2 & -5 & 7 & 19 \end{pmatrix} \xrightarrow[-2R_1 + R_3]{-4R_1 + R_2} \begin{pmatrix} 1 & 3 & -2 & -7 \\ 0 & -11 & 11 & 33 \\ 0 & -11 & 11 & 33 \end{pmatrix}$$

$$\xrightarrow[-\frac{1}{11}R_3]{-\frac{1}{11}R_2} \begin{pmatrix} 1 & 3 & -2 & -7 \\ 0 & 1 & -1 & -3 \\ 0 & 1 & -1 & -3 \end{pmatrix} \xrightarrow[-R_2 + R_3]{-3R_2 + R_1} \begin{pmatrix} 1 & 0 & 1 & 1 \\ 0 & 1 & -1 & -3 \\ 0 & 0 & 0 & 0 \end{pmatrix}.$$

In this case, the last matrix in reduced row-echelon form implies that the original system of three equations in three unknowns is really equivalent to two equations in three unknowns. Since only $z$ is common to both equations (the nonzero rows), we can assign its values arbitrarily. If we let $z = t$, where $t$ represents any real number, then we see that the system has infinitely many solutions: $x = 2 - t$, $y = -3 + t$, $z = t$. Geometrically, these equations are the parametric equations for the line of intersection of the planes $x + 0y + z = 2$ and $0x + y - z = 3$. ◼

**USING ROW OPERATIONS TO FIND AN INVERSE**  Because of the number of determinants that must be evaluated, formula (2) in Theorem B.2 is seldom used to find the inverse when the matrix **A** is large. In the case of $3 \times 3$ or larger matrices the method described in the next theorem is a particularly efficient means for finding $\mathbf{A}^{-1}$.

---

**THEOREM B.3**  **Finding $\mathbf{A}^{-1}$ Using Elementary Row Operations**

If an $n \times n$ matrix **A** can be transformed into the $n \times n$ identity **I** by a sequence of elementary row operations, then **A** is nonsingular. The same sequence of operations that transforms **A** into the identity **I** will also transform **I** into $\mathbf{A}^{-1}$.

---

It is convenient to carry out these row operations on **A** and **I** simultaneously by means of an $n \times 2n$ matrix obtained by augmenting **A** with the identity **I** as shown here:

$$(\mathbf{A} \mid \mathbf{I}) = \begin{pmatrix} a_{11} & a_{12} & \cdots & a_{1n} & 1 & 0 & \cdots & 0 \\ a_{21} & a_{22} & \cdots & a_{2n} & 1 & 0 & \cdots & 0 \\ \vdots & & & \vdots & \vdots & & & \vdots \\ a_{n1} & a_{n2} & \cdots & a_{nn} & 0 & 0 & \cdots & 1 \end{pmatrix}.$$

The procedure for finding $\mathbf{A}^{-1}$ is outlined in the following diagram:

Perform row operations on **A** until **I** is obtained. This means that **A** is nonsingular.

$(\mathbf{A} \mid \mathbf{I}) \longrightarrow (\mathbf{I} \mid \mathbf{A}^{-1}).$

By simultaneously applying the same row operations to **I**, we get $\mathbf{A}^{-1}$.

---

**EXAMPLE 14**  **Inverse by Elementary Row Operations**

Find the multiplicative inverse for $\mathbf{A} = \begin{pmatrix} 2 & 0 & 1 \\ -2 & 3 & 4 \\ -5 & 5 & 6 \end{pmatrix}$.

**SOLUTION**  We shall use the same notation as we did when we reduced an augmented matrix to reduced row-echelon form:

$$\begin{pmatrix} 2 & 0 & 1 & 1 & 0 & 0 \\ -2 & 3 & 4 & 0 & 1 & 0 \\ -5 & 5 & 6 & 0 & 0 & 1 \end{pmatrix} \xrightarrow{\frac{1}{2}R_1} \begin{pmatrix} 1 & 0 & \frac{1}{2} & \frac{1}{2} & 0 & 0 \\ -2 & 3 & 4 & 0 & 1 & 0 \\ -5 & 5 & 6 & 0 & 0 & 1 \end{pmatrix} \xrightarrow[5R_1 + R_3]{2R_1 + R_2} \begin{pmatrix} 1 & 0 & \frac{1}{2} & \frac{1}{2} & 0 & 0 \\ 0 & 3 & 5 & 1 & 1 & 0 \\ 0 & 5 & \frac{17}{2} & \frac{5}{2} & 0 & 1 \end{pmatrix}$$

$$\xrightarrow[\frac{1}{5}R_3]{\frac{1}{3}R_2} \left( \begin{array}{ccc|ccc} 1 & 0 & \frac{1}{2} & \frac{1}{2} & 0 & 0 \\ 0 & 1 & \frac{5}{3} & \frac{1}{3} & \frac{1}{3} & 0 \\ 0 & 1 & \frac{17}{10} & \frac{1}{2} & 0 & \frac{1}{5} \end{array} \right) \xrightarrow{-R_2 + R_3} \left( \begin{array}{ccc|ccc} 1 & 0 & \frac{1}{2} & \frac{1}{2} & 0 & 0 \\ 0 & 1 & \frac{5}{3} & \frac{1}{3} & \frac{1}{3} & 0 \\ 0 & 0 & \frac{1}{30} & \frac{1}{6} & -\frac{1}{3} & \frac{1}{5} \end{array} \right)$$

$$\xrightarrow{30R_3} \left( \begin{array}{ccc|ccc} 1 & 0 & \frac{1}{2} & \frac{1}{2} & 0 & 0 \\ 0 & 1 & \frac{5}{3} & \frac{1}{3} & \frac{1}{3} & 0 \\ 0 & 0 & 1 & 5 & -10 & 6 \end{array} \right) \xrightarrow[-\frac{5}{3}R_3 + R_2]{-\frac{1}{2}R_3 + R_1} \left( \begin{array}{ccc|ccc} 1 & 0 & 0 & -2 & 5 & -3 \\ 0 & 1 & 0 & -8 & 17 & -10 \\ 0 & 0 & 1 & 5 & -10 & 6 \end{array} \right).$$

Because $\mathbf{I}$ appears to the left of the vertical line, we conclude that the matrix to the right of the line is

$$\mathbf{A}^{-1} = \left( \begin{array}{ccc} -2 & 5 & -3 \\ -8 & 17 & -10 \\ 5 & -10 & 6 \end{array} \right).$$ ▪

If row reduction of $(\mathbf{A}|\mathbf{I})$ leads to the situation

$$(\mathbf{A} \mid \mathbf{I}) \xrightarrow[\text{operations}]{\text{row}} (\mathbf{B} \mid \mathbf{C}),$$

where the matrix $\mathbf{B}$ contains a row of zeros, then necessarily $\mathbf{A}$ is singular. Since further reduction of $\mathbf{B}$ always yields another matrix with a row of zeros, we can never transform $\mathbf{A}$ into $\mathbf{I}$.

## B.3   THE EIGENVALUE PROBLEM

Gauss-Jordan elimination can be used to find the eigenvectors of a square matrix.

---

### DEFINITION B.13   Eigenvalues and Eigenvectors

Let $\mathbf{A}$ be an $n \times n$ matrix. A number $\lambda$ is said to be an **eigenvalue** of $\mathbf{A}$ if there exists a *nonzero* solution vector $\mathbf{K}$ of the linear system

$$\mathbf{AK} = \lambda\mathbf{K}. \tag{6}$$

The solution vector $\mathbf{K}$ is said to be an **eigenvector** corresponding to the eigenvalue $\lambda$.

---

The word *eigenvalue* is a combination of German and English terms adapted from the German word *eigenwert*, which, translated literally, is "proper value." Eigenvalues and eigenvectors are also called **characteristic values** and **characteristic vectors,** respectively.

### EXAMPLE 15   Eigenvector of a Matrix

Verify that $\mathbf{K} = \left( \begin{array}{c} 1 \\ -1 \\ 1 \end{array} \right)$ is an eigenvector of the matrix

$$\mathbf{A} = \left( \begin{array}{ccc} 0 & -1 & -3 \\ 2 & 3 & 3 \\ -2 & 1 & 1 \end{array} \right).$$

**SOLUTION** By carrying out the multiplication $\mathbf{AK}$, we see that

$$\mathbf{AK} = \begin{pmatrix} 0 & -1 & -3 \\ 2 & 3 & 3 \\ -2 & 1 & 1 \end{pmatrix} \begin{pmatrix} 1 \\ -1 \\ 1 \end{pmatrix} = \begin{pmatrix} -2 \\ 2 \\ -2 \end{pmatrix} = (-2) \begin{pmatrix} 1 \\ -1 \\ 1 \end{pmatrix} = \overset{\text{eigenvalue}}{(-2)}\mathbf{K}.$$

We see from the preceding line and Definition B.13 that $\lambda = -2$ is an eigenvalue of $\mathbf{A}$. ◼

Using properties of matrix algebra, we can write (6) in the alternative form

$$(\mathbf{A} - \lambda\mathbf{I})\mathbf{K} = \mathbf{0}, \tag{7}$$

where $\mathbf{I}$ is the multiplicative identity. If we let

$$\mathbf{K} = \begin{pmatrix} k_1 \\ k_2 \\ \vdots \\ k_n \end{pmatrix},$$

then (7) is the same as

$$\begin{aligned} (a_{11} - \lambda)k_1 + \quad & a_{12}k_2 + \cdots + \quad & a_{1n}k_n = 0 \\ a_{21}k_1 + (a_{22} - \lambda)k_2 + \cdots + \quad & a_{2n}k_n = 0 \\ & \vdots \qquad\qquad\qquad \vdots \\ a_{n1}k_1 + \quad & a_{n2}k_2 + \cdots + (a_{nn} - \lambda)k_n = 0. \end{aligned} \tag{8}$$

Although an obvious solution of (8) is $k_1 = 0$, $k_2 = 0$, ..., $k_n = 0$, we are seeking only nontrivial solutions. It is known that a **homogeneous** system of $n$ linear equations in $n$ unknowns (that is, $b_i = 0$, $i = 1, 2, \ldots, n$ in (5)) has a nontrivial solution if and only if the determinant of the coefficient matrix is equal to zero. Thus to find a nonzero solution $\mathbf{K}$ for (7), we must have

$$\det(\mathbf{A} - \lambda\mathbf{I}) = 0. \tag{9}$$

Inspection of (8) shows that the expansion of $\det(\mathbf{A} - \lambda\mathbf{I})$ by cofactors results in an $n$th-degree polynomial in $\lambda$. The equation (9) is called the **characteristic equation** of $\mathbf{A}$. Thus *the eigenvalues of $\mathbf{A}$ are the roots of the characteristic equation.* To find an eigenvector corresponding to an eigenvalue $\lambda$, we simply solve the system of equations $(\mathbf{A} - \lambda\mathbf{I})\mathbf{K} = \mathbf{0}$ by applying Gauss-Jordan elimination to the augmented matrix $(\mathbf{A} - \lambda\mathbf{I}|\mathbf{0})$.

### EXAMPLE 16 Eigenvalues/Eigenvectors

Find the eigenvalues and eigenvectors of $\mathbf{A} = \begin{pmatrix} 1 & 2 & 1 \\ 6 & -1 & 0 \\ -1 & -2 & -1 \end{pmatrix}$.

**SOLUTION** To expand the determinant in the characteristic equation, we use the cofactors of the second row:

$$\det(\mathbf{A} - \lambda\mathbf{I}) = \begin{vmatrix} 1 - \lambda & 2 & 1 \\ 6 & -1 - \lambda & 0 \\ -1 & -2 & -1 - \lambda \end{vmatrix} = -\lambda^3 - \lambda^2 + 12\lambda = 0.$$

From $-\lambda^3 - \lambda^2 + 12\lambda = -\lambda(\lambda + 4)(\lambda - 3) = 0$ we see that the eigenvalues are $\lambda_1 = 0$, $\lambda_2 = -4$, and $\lambda_3 = 3$. To find the eigenvectors, we must now reduce $(\mathbf{A} - \lambda\mathbf{I}|\mathbf{0})$ three times corresponding to the three distinct eigenvalues.

For $\lambda_1 = 0$ we have

$$(A - 0I \mid 0) = \begin{pmatrix} 1 & 2 & 1 & \mid & 0 \\ 6 & -1 & 0 & \mid & 0 \\ -1 & -2 & -1 & \mid & 0 \end{pmatrix} \xrightarrow[R_1 + R_3]{-6R_1 + R_2} \begin{pmatrix} 1 & 2 & 1 & \mid & 0 \\ 0 & -13 & -6 & \mid & 0 \\ 0 & 0 & 0 & \mid & 0 \end{pmatrix}$$

$$\xrightarrow{-\frac{1}{13}R_2} \begin{pmatrix} 1 & 2 & 1 & \mid & 0 \\ 0 & 1 & \frac{6}{13} & \mid & 0 \\ 0 & 0 & 0 & \mid & 0 \end{pmatrix} \xrightarrow{-2R_2 + R_1} \begin{pmatrix} 1 & 0 & \frac{1}{13} & \mid & 0 \\ 0 & 1 & \frac{6}{13} & \mid & 0 \\ 0 & 0 & 0 & \mid & 0 \end{pmatrix}.$$

Thus we see that $k_1 = -\frac{1}{13}k_3$ and $k_2 = -\frac{6}{13}k_3$. Choosing $k_3 = -13$, we get the eigenvector[*]

$$K_1 = \begin{pmatrix} 1 \\ 6 \\ -13 \end{pmatrix}.$$

For $\lambda_2 = -4$,

$$(A + 4I \mid 0) = \begin{pmatrix} 5 & 2 & 1 & \mid & 0 \\ 6 & 3 & 0 & \mid & 0 \\ -1 & -2 & 3 & \mid & 0 \end{pmatrix} \xrightarrow[R_{31}]{-R_3} \begin{pmatrix} 1 & 2 & -3 & \mid & 0 \\ 6 & 3 & 0 & \mid & 0 \\ 5 & 2 & 1 & \mid & 0 \end{pmatrix}$$

$$\xrightarrow[-5R_1 + R_3]{-6R_1 + R_2} \begin{pmatrix} 1 & 2 & -3 & \mid & 0 \\ 0 & -9 & 18 & \mid & 0 \\ 0 & -8 & 16 & \mid & 0 \end{pmatrix} \xrightarrow[-\frac{1}{8}R_3]{-\frac{1}{9}R_2} \begin{pmatrix} 1 & 2 & -3 & \mid & 0 \\ 0 & 1 & -2 & \mid & 0 \\ 0 & 1 & -2 & \mid & 0 \end{pmatrix} \xrightarrow[-R_2 + R_3]{-2R_2 + R_1} \begin{pmatrix} 1 & 0 & 1 & \mid & 0 \\ 0 & 1 & -2 & \mid & 0 \\ 0 & 0 & 0 & \mid & 0 \end{pmatrix}$$

implies that $k_1 = -k_3$ and $k_2 = 2k_3$. Choosing $k_3 = 1$ then yields the second eigenvector

$$K_2 = \begin{pmatrix} -1 \\ 2 \\ 1 \end{pmatrix}.$$

Finally, for $\lambda_3 = 3$ Gauss-Jordan elimination gives

$$(A - 3I \mid 0) = \begin{pmatrix} -2 & 2 & 1 & \mid & 0 \\ 6 & -4 & 0 & \mid & 0 \\ -1 & -2 & -4 & \mid & 0 \end{pmatrix} \xrightarrow{\text{row operations}} \begin{pmatrix} 1 & 0 & 1 & \mid & 0 \\ 0 & 1 & \frac{3}{2} & \mid & 0 \\ 0 & 0 & 0 & \mid & 0 \end{pmatrix},$$

so $k_1 = -k_3$ and $k_2 = -\frac{3}{2}k_3$. The choice of $k_3 = -2$ leads to the third eigenvector:

$$K_3 = \begin{pmatrix} 2 \\ 3 \\ -2 \end{pmatrix}. \qquad \blacksquare$$

When an $n \times n$ matrix $A$ possesses $n$ distinct eigenvalues $\lambda_1, \lambda_2, \ldots, \lambda_n$, it can be proved that a set of $n$ linearly independent[†] eigenvectors $K_1, K_2, \ldots, K_n$ can be found. However, when the characteristic equation has repeated roots, it may not be possible to find $n$ linearly independent eigenvectors for $A$. If an $n \times n$ matrix $A$ possesses less than $n$ linearly independent eigenvectors, it is said to be **defective.**

---

[*]Of course, $k_3$ could be chosen as any nonzero number. In other words, a nonzero constant multiple of an eigenvector is also an eigenvector.

[†]Linear independence of column vectors is defined in exactly the same manner as for functions.

EXAMPLE 17    **Eigenvalues/Eigenvectors**

Find the eigenvalues and eigenvectors of $\mathbf{A} = \begin{pmatrix} 3 & 4 \\ -1 & 7 \end{pmatrix}$.

**SOLUTION**  From the characteristic equation

$$\det(\mathbf{A} - \lambda\mathbf{I}) = \begin{vmatrix} 3 - \lambda & 4 \\ -1 & 7 - \lambda \end{vmatrix} = (\lambda - 5)^2 = 0$$

we see that $\lambda_1 = \lambda_2 = 5$ is an eigenvalue of multiplicity two. In the case of a $2 \times 2$ matrix there is no need to use Gauss-Jordan elimination. To find the eigenvector(s) corresponding to $\lambda_1 = 5$, we resort to the system $(\mathbf{A} - 5\mathbf{I}\,|\,\mathbf{0})$ in its equivalent form

$$-2k_1 + 4k_2 = 0$$
$$-k_1 + 2k_2 = 0.$$

It is apparent from this system that $k_1 = 2k_2$. Thus if we choose $k_2 = 1$, we find the single eigenvector

$$\mathbf{K}_1 = \begin{pmatrix} 2 \\ 1 \end{pmatrix}.$$

The matrix $\mathbf{A}$ is defective.  ∎

EXAMPLE 18    **Eigenvalues/Eigenvectors**

Find the eigenvalues and eigenvectors of $\mathbf{A} = \begin{pmatrix} 9 & 1 & 1 \\ 1 & 9 & 1 \\ 1 & 1 & 9 \end{pmatrix}$.

**SOLUTION**  The characteristic equation

$$\det(\mathbf{A} - \lambda\mathbf{I}) = \begin{vmatrix} 9 - \lambda & 1 & 1 \\ 1 & 9 - \lambda & 1 \\ 1 & 1 & 9 - \lambda \end{vmatrix} = -(\lambda - 11)(\lambda - 8)^2 = 0$$

shows that $\lambda_1 = 11$ and that $\lambda_2 = \lambda_3 = 8$ is an eigenvalue of multiplicity two.

For $\lambda_1 = 11$ Gauss-Jordan elimination gives

$$(\mathbf{A} - 11\mathbf{I}\,|\,\mathbf{0}) = \begin{pmatrix} -2 & 1 & 1 & | & 0 \\ 1 & -2 & 1 & | & 0 \\ 1 & 1 & -2 & | & 0 \end{pmatrix} \xrightarrow[\text{operations}]{\text{row}} \begin{pmatrix} 1 & 0 & -1 & | & 0 \\ 0 & 1 & -1 & | & 0 \\ 0 & 0 & 0 & | & 0 \end{pmatrix}.$$

Hence $k_1 = k_3$ and $k_2 = k_3$. If $k_3 = 1$, then

$$\mathbf{K}_1 = \begin{pmatrix} 1 \\ 1 \\ 1 \end{pmatrix}.$$

Now for $\lambda_2 = 8$ we have

$$(\mathbf{A} - 8\mathbf{I}\,|\,\mathbf{0}) = \begin{pmatrix} 1 & 1 & 1 & | & 0 \\ 1 & 1 & 1 & | & 0 \\ 1 & 1 & 1 & | & 0 \end{pmatrix} \xrightarrow[\text{operations}]{\text{row}} \begin{pmatrix} 1 & 1 & 1 & | & 0 \\ 0 & 0 & 0 & | & 0 \\ 0 & 0 & 0 & | & 0 \end{pmatrix}.$$

In the equation $k_1 + k_2 + k_3 = 0$, we are free to select two of the variables arbitrarily. Choosing, on the one hand, $k_2 = 1$, $k_3 = 0$ and, on the other, $k_2 = 0$, $k_3 = 1$, we obtain two linearly independent eigenvectors

$$\mathbf{K}_2 = \begin{pmatrix} -1 \\ 1 \\ 0 \end{pmatrix} \quad \text{and} \quad \mathbf{K}_3 = \begin{pmatrix} -1 \\ 0 \\ 1 \end{pmatrix}.$$

The matrix $\mathbf{A}$ is *not* defective.

## EXERCISES FOR APPENDIX B

Answers to selected odd-numbered problems begin on page ANS-30.

### B.1  Basic Definitions and Theory

**1.** If $\mathbf{A} = \begin{pmatrix} 4 & 5 \\ -6 & 9 \end{pmatrix}$ and $\mathbf{B} = \begin{pmatrix} -2 & 6 \\ 8 & -10 \end{pmatrix}$, find

(a) $\mathbf{A} + \mathbf{B}$   (b) $\mathbf{B} - \mathbf{A}$   (c) $2\mathbf{A} + 3\mathbf{B}$

**2.** If $\mathbf{A} = \begin{pmatrix} -2 & 0 \\ 4 & 1 \\ 7 & 3 \end{pmatrix}$ and $\mathbf{B} = \begin{pmatrix} 3 & -1 \\ 0 & 2 \\ -4 & -2 \end{pmatrix}$, find

(a) $\mathbf{A} - \mathbf{B}$   (b) $\mathbf{B} - \mathbf{A}$   (c) $2(\mathbf{A} + \mathbf{B})$

**3.** If $\mathbf{A} = \begin{pmatrix} 2 & -3 \\ -5 & 4 \end{pmatrix}$ and $\mathbf{B} = \begin{pmatrix} -1 & 6 \\ 3 & 2 \end{pmatrix}$, find

(a) $\mathbf{AB}$   (b) $\mathbf{BA}$   (c) $\mathbf{A}^2 = \mathbf{AA}$   (d) $\mathbf{B}^2 = \mathbf{BB}$

**4.** If $\mathbf{A} = \begin{pmatrix} 1 & 4 \\ 5 & 10 \\ 8 & 12 \end{pmatrix}$ and $\mathbf{B} = \begin{pmatrix} -4 & 6 & -3 \\ 1 & -3 & 2 \end{pmatrix}$, find

(a) $\mathbf{AB}$   (b) $\mathbf{BA}$

**5.** If $\mathbf{A} = \begin{pmatrix} 1 & -2 \\ -2 & 4 \end{pmatrix}$, $\mathbf{B} = \begin{pmatrix} 6 & 3 \\ 2 & 1 \end{pmatrix}$, and $\mathbf{C} = \begin{pmatrix} 0 & 2 \\ 3 & 4 \end{pmatrix}$, find

(a) $\mathbf{BC}$   (b) $\mathbf{A}(\mathbf{BC})$   (c) $\mathbf{C}(\mathbf{BA})$   (d) $\mathbf{A}(\mathbf{B} + \mathbf{C})$

**6.** If $\mathbf{A} = (5 \quad -6 \quad 7)$, $\mathbf{B} = \begin{pmatrix} 3 \\ 4 \\ -1 \end{pmatrix}$, and

$\mathbf{C} = \begin{pmatrix} 1 & 2 & 4 \\ 0 & 1 & -1 \\ 3 & 2 & 1 \end{pmatrix}$, find

(a) $\mathbf{AB}$   (b) $\mathbf{BA}$   (c) $(\mathbf{BA})\mathbf{C}$   (d) $(\mathbf{AB})\mathbf{C}$

**7.** If $\mathbf{A} = \begin{pmatrix} 4 \\ 8 \\ -10 \end{pmatrix}$ and $\mathbf{B} = (2 \quad 4 \quad 5)$, find

(a) $\mathbf{A}^T\mathbf{A}$   (b) $\mathbf{B}^T\mathbf{B}$   (c) $\mathbf{A} + \mathbf{B}^T$

**8.** If $\mathbf{A} = \begin{pmatrix} 1 & 2 \\ 2 & 4 \end{pmatrix}$ and $\mathbf{B} = \begin{pmatrix} -2 & 3 \\ 5 & 7 \end{pmatrix}$, find

(a) $\mathbf{A} + \mathbf{B}^T$   (b) $2\mathbf{A}^T - \mathbf{B}^T$   (c) $\mathbf{A}^T(\mathbf{A} - \mathbf{B})$

**9.** If $\mathbf{A} = \begin{pmatrix} 3 & 4 \\ 8 & 1 \end{pmatrix}$ and $\mathbf{B} = \begin{pmatrix} 5 & 10 \\ -2 & -5 \end{pmatrix}$, find

(a) $(\mathbf{AB})^T$   (b) $\mathbf{B}^T\mathbf{A}^T$

**10.** If $\mathbf{A} = \begin{pmatrix} 5 & 9 \\ -4 & 6 \end{pmatrix}$ and $\mathbf{B} = \begin{pmatrix} -3 & 11 \\ -7 & 2 \end{pmatrix}$, find

(a) $\mathbf{A}^T + \mathbf{B}^T$   (b) $(\mathbf{A} + \mathbf{B})^T$

In Problems 11–14 write the given sum as a single column matrix.

**11.** $4\begin{pmatrix} -1 \\ 2 \end{pmatrix} - 2\begin{pmatrix} 2 \\ 8 \end{pmatrix} + 3\begin{pmatrix} -2 \\ 3 \end{pmatrix}$

**12.** $3t\begin{pmatrix} 2 \\ t \\ -1 \end{pmatrix} + (t - 1)\begin{pmatrix} -1 \\ -t \\ 3 \end{pmatrix} - 2\begin{pmatrix} 3t \\ 4 \\ -5t \end{pmatrix}$

**13.** $\begin{pmatrix} 2 & -3 \\ 1 & 4 \end{pmatrix}\begin{pmatrix} -2 \\ 5 \end{pmatrix} - \begin{pmatrix} -1 & 6 \\ -2 & 3 \end{pmatrix}\begin{pmatrix} -7 \\ 2 \end{pmatrix}$

**14.** $\begin{pmatrix} 1 & -3 & 4 \\ 2 & 5 & -1 \\ 0 & -4 & -2 \end{pmatrix}\begin{pmatrix} t \\ 2t-1 \\ -t \end{pmatrix} + \begin{pmatrix} -t \\ 1 \\ 4 \end{pmatrix} - \begin{pmatrix} 2 \\ 8 \\ -6 \end{pmatrix}$

In Problems 15–22 determine whether the given matrix is singular or nonsingular. If it is nonsingular, find $\mathbf{A}^{-1}$ using Theorem B.2.

**15.** $\mathbf{A} = \begin{pmatrix} -3 & 6 \\ -2 & 4 \end{pmatrix}$   **16.** $\mathbf{A} = \begin{pmatrix} 2 & 5 \\ 1 & 4 \end{pmatrix}$

**17.** $\mathbf{A} = \begin{pmatrix} 4 & 8 \\ -3 & -5 \end{pmatrix}$   **18.** $\mathbf{A} = \begin{pmatrix} 7 & 10 \\ 2 & 2 \end{pmatrix}$

**19.** $\mathbf{A} = \begin{pmatrix} 2 & 1 & 0 \\ -1 & 2 & 1 \\ 1 & 2 & 1 \end{pmatrix}$   **20.** $\mathbf{A} = \begin{pmatrix} 3 & 2 & 1 \\ 4 & 1 & 0 \\ -2 & 5 & -1 \end{pmatrix}$

**21.** $A = \begin{pmatrix} 2 & 1 & 1 \\ 1 & -2 & -3 \\ 3 & 2 & 4 \end{pmatrix}$    **22.** $A = \begin{pmatrix} 4 & 1 & -1 \\ 6 & 2 & -3 \\ -2 & -1 & 2 \end{pmatrix}$

In Problems 23 and 24 show that the given matrix is nonsingular for every real value of $t$. Find $A^{-1}(t)$ using Theorem B.2.

**23.** $A(t) = \begin{pmatrix} 2e^{-t} & e^{4t} \\ 4e^{-t} & 3e^{4t} \end{pmatrix}$

**24.** $A(t) = \begin{pmatrix} 2e^t \sin t & -2e^t \cos t \\ e^t \cos t & e^t \sin t \end{pmatrix}$

In Problems 25–28 find $dX/dt$.

**25.** $X = \begin{pmatrix} 5e^{-t} \\ 2e^{-t} \\ -7e^{-t} \end{pmatrix}$    **26.** $X = \begin{pmatrix} \frac{1}{2}\sin 2t - 4\cos 2t \\ -3\sin 2t + 5\cos 2t \end{pmatrix}$

**27.** $X = 2\begin{pmatrix} 1 \\ -1 \end{pmatrix}e^{2t} + 4\begin{pmatrix} 2 \\ 1 \end{pmatrix}e^{-3t}$    **28.** $X = \begin{pmatrix} 5te^{2t} \\ t\sin 3t \end{pmatrix}$

**29.** Let $A(t) = \begin{pmatrix} e^{4t} & \cos \pi t \\ 2t & 3t^2 - 1 \end{pmatrix}$. Find

    **(a)** $\dfrac{dA}{dt}$    **(b)** $\displaystyle\int_0^2 A(t)\,dt$    **(c)** $\displaystyle\int_0^t A(s)\,ds$

**30.** Let $A(t) = \begin{pmatrix} \dfrac{1}{t^2+1} & 3t \\ t^2 & t \end{pmatrix}$ and $B(t) = \begin{pmatrix} 6t & 2 \\ 1/t & 4t \end{pmatrix}$. Find

    **(a)** $\dfrac{dA}{dt}$              **(b)** $\dfrac{dB}{dt}$

    **(c)** $\displaystyle\int_0^1 A(t)\,dt$    **(d)** $\displaystyle\int_1^2 B(t)\,dt$

    **(e)** $A(t)B(t)$      **(f)** $\dfrac{d}{dt}A(t)B(t)$

    **(g)** $\displaystyle\int_1^t A(s)B(s)\,ds$

## B.2 Gaussian and Gauss-Jordan Elimination

In Problems 31–38 solve the given system of equations by either Gaussian elimination or Gauss-Jordan elimination.

**31.** $\begin{aligned} x + y - 2z &= 14 \\ 2x - y + z &= 0 \\ 6x + 3y + 4z &= 1 \end{aligned}$    **32.** $\begin{aligned} 5x - 2y + 4z &= 10 \\ x + y + z &= 9 \\ 4x - 3y + 3z &= 1 \end{aligned}$

**33.** $\begin{aligned} y + z &= -5 \\ 5x + 4y - 16z &= -10 \\ x - y - 5z &= 7 \end{aligned}$    **34.** $\begin{aligned} 3x + y + z &= 4 \\ 4x + 2y - z &= 7 \\ x + y - 3z &= 6 \end{aligned}$

**35.** $\begin{aligned} 2x + y + z &= 4 \\ 10x - 2y + 2z &= -1 \\ 6x - 2y + 4z &= 8 \end{aligned}$    **36.** $\begin{aligned} x + 2z &= 8 \\ x + 2y - 2z &= 4 \\ 2x + 5y - 6z &= 6 \end{aligned}$

**37.** $\begin{aligned} x_1 + x_2 - x_3 - x_4 &= -1 \\ x_1 + x_2 + x_3 + x_4 &= 3 \\ x_1 - x_2 + x_3 - x_4 &= 3 \\ 4x_1 + x_2 - 2x_3 + x_4 &= 0 \end{aligned}$    **38.** $\begin{aligned} 2x_1 + x_2 + x_3 &= 0 \\ x_1 + 3x_2 + x_3 &= 0 \\ 7x_1 + x_2 + 3x_3 &= 0 \end{aligned}$

In Problems 39 and 40 use Gauss-Jordan elimination to demonstrate that the given system of equations has no solution.

**39.** $\begin{aligned} x + 2y + 4z &= 2 \\ 2x + 4y + 3z &= 1 \\ x + 2y - z &= 7 \end{aligned}$    **40.** $\begin{aligned} x_1 + x_2 - x_3 + 3x_4 &= 1 \\ x_2 - x_3 - 4x_4 &= 0 \\ x_1 + 2x_2 - 2x_3 - x_4 &= 6 \\ 4x_1 + 7x_2 - 7x_3 &= 9 \end{aligned}$

In Problems 41–46 use Theorem B.3 to find $A^{-1}$ for the given matrix or show that no inverse exists.

**41.** $A = \begin{pmatrix} 4 & 2 & 3 \\ 2 & 1 & 0 \\ -1 & -2 & 0 \end{pmatrix}$    **42.** $A = \begin{pmatrix} 2 & 4 & -2 \\ 4 & 2 & -2 \\ 8 & 10 & -6 \end{pmatrix}$

**43.** $A = \begin{pmatrix} -1 & 3 & 0 \\ 1 & -2 & 1 \\ 0 & 1 & 2 \end{pmatrix}$    **44.** $A = \begin{pmatrix} 1 & 2 & 3 \\ 0 & 1 & 4 \\ 0 & 0 & 8 \end{pmatrix}$

**45.** $A = \begin{pmatrix} 1 & 2 & 3 & 1 \\ -1 & 0 & 2 & 1 \\ 2 & 1 & -3 & 0 \\ 1 & 1 & 2 & 1 \end{pmatrix}$    **46.** $A = \begin{pmatrix} 1 & 0 & 0 & 0 \\ 0 & 0 & 1 & 0 \\ 0 & 0 & 0 & 1 \\ 0 & 1 & 0 & 0 \end{pmatrix}$

## B.3 The Eigenvalue Problem

In Problems 47–54 find the eigenvalues and eigenvectors of the given matrix.

**47.** $\begin{pmatrix} -1 & 2 \\ -7 & 8 \end{pmatrix}$    **48.** $\begin{pmatrix} 2 & 1 \\ 2 & 1 \end{pmatrix}$

**49.** $\begin{pmatrix} -8 & -1 \\ 16 & 0 \end{pmatrix}$    **50.** $\begin{pmatrix} 1 & 1 \\ \frac{1}{4} & 1 \end{pmatrix}$

**51.** $\begin{pmatrix} 5 & -1 & 0 \\ 0 & -5 & 9 \\ 5 & -1 & 0 \end{pmatrix}$    **52.** $\begin{pmatrix} 3 & 0 & 0 \\ 0 & 2 & 0 \\ 4 & 0 & 1 \end{pmatrix}$

**53.** $\begin{pmatrix} 0 & 4 & 0 \\ -1 & -4 & 0 \\ 0 & 0 & -2 \end{pmatrix}$    **54.** $\begin{pmatrix} 1 & 6 & 0 \\ 0 & 2 & 1 \\ 0 & 1 & 2 \end{pmatrix}$

In Problems 55 and 56 show that the given matrix has complex eigenvalues. Find the eigenvectors of the matrix.

**55.** $\begin{pmatrix} -1 & 2 \\ -5 & 1 \end{pmatrix}$    **56.** $\begin{pmatrix} 2 & -1 & 0 \\ 5 & 2 & 4 \\ 0 & 1 & 2 \end{pmatrix}$

## Miscellaneous Problems

**57.** If $\mathbf{A}(t)$ is a $2 \times 2$ matrix of differentiable functions and $\mathbf{X}(t)$ is a $2 \times 1$ column matrix of differentiable functions, prove the product rule

$$\frac{d}{dt}[\mathbf{A}(t)\mathbf{X}(t)] = \mathbf{A}(t)\mathbf{X}'(t) + \mathbf{A}'(t)\mathbf{X}(t).$$

**58.** Derive formula (3). [*Hint*: Find a matrix

$$\mathbf{B} = \begin{pmatrix} b_{11} & b_{12} \\ b_{21} & b_{22} \end{pmatrix}$$

for which $\mathbf{AB} = \mathbf{I}$. Solve for $b_{11}, b_{12}, b_{21}$, and $b_{22}$. Then show that $\mathbf{BA} = \mathbf{I}$.]

**59.** If $\mathbf{A}$ is nonsingular and $\mathbf{AB} = \mathbf{AC}$, show that $\mathbf{B} = \mathbf{C}$.

**60.** If $\mathbf{A}$ and $\mathbf{B}$ are nonsingular, show that $(\mathbf{AB})^{-1} = \mathbf{B}^{-1}\mathbf{A}^{-1}$.

**61.** Let $\mathbf{A}$ and $\mathbf{B}$ be $n \times n$ matrices. In general, is

$$(\mathbf{A} + \mathbf{B})^2 = \mathbf{A}^2 + 2\mathbf{AB} + \mathbf{B}^2?$$

**62.** A square matrix $\mathbf{A}$ is said to be a **diagonal matrix** if all its entries off the main diagonal are zero—that is, $a_{ij} = 0, i \neq j$. The entries $a_{ii}$ on the main diagonal may or may not be zero. The multiplicative identity matrix $\mathbf{I}$ is an example of a diagonal matrix.

(a) Find the inverse of the $2 \times 2$ diagonal matrix

$$\mathbf{A} = \begin{pmatrix} a_{11} & 0 \\ 0 & a_{22} \end{pmatrix}$$

when $a_{11} \neq 0, a_{22} \neq 0$.

(b) Find the inverse of a $3 \times 3$ diagonal matrix $\mathbf{A}$ whose main diagonal entries $a_{ii}$ are all nonzero.

(c) In general, what is the inverse of an $n \times n$ diagonal matrix $\mathbf{A}$ whose main diagonal entries $a_{ii}$ are all nonzero?

# Appendix C Laplace Transforms

| $f(t)$ | $\mathscr{L}\{f(t)\} = F(s)$ |
|---|---|
| **1.** $1$ | $\dfrac{1}{s}$ |
| **2.** $t$ | $\dfrac{1}{s^2}$ |
| **3.** $t^n$ | $\dfrac{n!}{s^{n+1}}$,  $n$ a positive integer |
| **4.** $t^{-1/2}$ | $\sqrt{\dfrac{\pi}{s}}$ |
| **5.** $t^{1/2}$ | $\dfrac{\sqrt{\pi}}{2s^{3/2}}$ |
| **6.** $t^{\alpha}$ | $\dfrac{\Gamma(\alpha + 1)}{s^{\alpha+1}}$,  $\alpha > -1$ |
| **7.** $\sin kt$ | $\dfrac{k}{s^2 + k^2}$ |
| **8.** $\cos kt$ | $\dfrac{s}{s^2 + k^2}$ |
| **9.** $\sin^2 kt$ | $\dfrac{2k^2}{s(s^2 + 4k^2)}$ |
| **10.** $\cos^2 kt$ | $\dfrac{s^2 + 2k^2}{s(s^2 + 4k^2)}$ |
| **11.** $e^{at}$ | $\dfrac{1}{s - a}$ |
| **12.** $\sinh kt$ | $\dfrac{k}{s^2 - k^2}$ |
| **13.** $\cosh kt$ | $\dfrac{s}{s^2 - k^2}$ |
| **14.** $\sinh^2 kt$ | $\dfrac{2k^2}{s(s^2 - 4k^2)}$ |
| **15.** $\cosh^2 kt$ | $\dfrac{s^2 - 2k^2}{s(s^2 - 4k^2)}$ |
| **16.** $te^{at}$ | $\dfrac{1}{(s - a)^2}$ |
| **17.** $t^n e^{at}$ | $\dfrac{n!}{(s - a)^{n+1}}$,  $n$ a positive integer |
| **18.** $e^{at} \sin kt$ | $\dfrac{k}{(s - a)^2 + k^2}$ |
| **19.** $e^{at} \cos kt$ | $\dfrac{s - a}{(s - a)^2 + k^2}$ |
| **20.** $e^{at} \sinh kt$ | $\dfrac{k}{(s - a)^2 - k^2}$ |
| **21.** $e^{at} \cosh kt$ | $\dfrac{s - a}{(s - a)^2 - k^2}$ |

| $f(t)$ | $\mathscr{L}\{f(t)\} = F(s)$ |
|---|---|
| **22.** $t \sin kt$ | $\dfrac{2ks}{(s^2 + k^2)^2}$ |
| **23.** $t \cos kt$ | $\dfrac{s^2 - k^2}{(s^2 + k^2)^2}$ |
| **24.** $\sin kt + kt \cos kt$ | $\dfrac{2ks^2}{(s^2 + k^2)^2}$ |
| **25.** $\sin kt - kt \cos kt$ | $\dfrac{2k^3}{(s^2 + k^2)^2}$ |
| **26.** $t \sinh kt$ | $\dfrac{2ks}{(s^2 - k^2)^2}$ |
| **27.** $t \cosh kt$ | $\dfrac{s^2 + k^2}{(s^2 - k^2)^2}$ |
| **28.** $\dfrac{e^{at} - e^{bt}}{a - b}$ | $\dfrac{1}{(s - a)(s - b)}$ |
| **29.** $\dfrac{ae^{at} - be^{bt}}{a - b}$ | $\dfrac{s}{(s - a)(s - b)}$ |
| **30.** $1 - \cos kt$ | $\dfrac{k^2}{s(s^2 + k^2)}$ |
| **31.** $kt - \sin kt$ | $\dfrac{k^3}{s^2(s^2 + k^2)}$ |
| **32.** $\dfrac{a \sin bt - b \sin at}{ab(a^2 - b^2)}$ | $\dfrac{1}{(s^2 + a^2)(s^2 + b^2)}$ |
| **33.** $\dfrac{\cos bt - \cos at}{a^2 - b^2}$ | $\dfrac{s}{(s^2 + a^2)(s^2 + b^2)}$ |
| **34.** $\sin kt \sinh kt$ | $\dfrac{2k^2 s}{s^4 + 4k^4}$ |
| **35.** $\sin kt \cosh kt$ | $\dfrac{k(s^2 + 2k^2)}{s^4 + 4k^4}$ |
| **36.** $\cos kt \sinh kt$ | $\dfrac{k(s^2 - 2k^2)}{s^4 + 4k^4}$ |
| **37.** $\cos kt \cosh kt$ | $\dfrac{s^3}{s^4 + 4k^4}$ |
| **38.** $\sin kt \cosh kt + \cos kt \sinh kt$ | $\dfrac{2ks^2}{s^4 + 4k^4}$ |
| **39.** $\sin kt \cosh kt - \cos kt \sinh kt$ | $\dfrac{4k^3}{s^4 + 4k^4}$ |
| **40.** $\sinh kt - \sin kt$ | $\dfrac{2k^3}{s^4 - k^4}$ |
| **41.** $\cosh kt - \cos kt$ | $\dfrac{2k^2 s}{s^4 - k^4}$ |
| **42.** $J_0(kt)$ | $\dfrac{1}{\sqrt{s^2 + k^2}}$ |

| $f(t)$ | $\mathscr{L}\{f(t)\} = F(s)$ |
|---|---|
| **43.** $\dfrac{e^{bt} - e^{at}}{t}$ | $\ln\dfrac{s-a}{s-b}$ |
| **44.** $\dfrac{2(1 - \cos kt)}{t}$ | $\ln\dfrac{s^2 + k^2}{s^2}$ |
| **45.** $\dfrac{2(1 - \cosh kt)}{t}$ | $\ln\dfrac{s^2 - k^2}{s^2}$ |
| **46.** $\dfrac{\sin at}{t}$ | $\arctan\left(\dfrac{a}{s}\right)$ |
| **47.** $\dfrac{\sin at \cos bt}{t}$ | $\dfrac{1}{2}\arctan\dfrac{a+b}{s} + \dfrac{1}{2}\arctan\dfrac{a-b}{s}$ |
| **48.** $\dfrac{1}{\sqrt{\pi t}}\,e^{-a^2/4t}$ | $\dfrac{e^{-a\sqrt{s}}}{\sqrt{s}}$ |
| **49.** $\dfrac{a}{2\sqrt{\pi t^3}}\,e^{-a^2/4t}$ | $e^{-a\sqrt{s}}$ |
| **50.** $\operatorname{erfc}\left(\dfrac{a}{2\sqrt{t}}\right)$ | $\dfrac{e^{-a\sqrt{s}}}{s}$ |
| **51.** $2\sqrt{\dfrac{t}{\pi}}\,e^{-a^2/4t} - a\operatorname{erfc}\left(\dfrac{a}{2\sqrt{t}}\right)$ | $\dfrac{e^{-a\sqrt{s}}}{s\sqrt{s}}$ |
| **52.** $e^{ab}e^{b^2t}\operatorname{erfc}\left(b\sqrt{t} + \dfrac{a}{2\sqrt{t}}\right)$ | $\dfrac{e^{-a\sqrt{s}}}{\sqrt{s}(\sqrt{s}+b)}$ |
| **53.** $-e^{ab}e^{b^2t}\operatorname{erfc}\left(b\sqrt{t} + \dfrac{a}{2\sqrt{t}}\right)$ $+ \operatorname{erfc}\left(\dfrac{a}{2\sqrt{t}}\right)$ | $\dfrac{be^{-a\sqrt{s}}}{s(\sqrt{s}+b)}$ |
| **54.** $e^{at}f(t)$ | $F(s-a)$ |
| **55.** $\mathscr{U}(t-a)$ | $\dfrac{e^{-as}}{s}$ |
| **56.** $f(t-a)\,\mathscr{U}(t-a)$ | $e^{-as}F(s)$ |
| **57.** $g(t)\,\mathscr{U}(t-a)$ | $e^{-as}\mathscr{L}\{g(t+a)\}$ |
| **58.** $f^{(n)}(t)$ | $s^n F(s) - s^{(n-1)}f(0) - \cdots - f^{(n-1)}(0)$ |
| **59.** $t^n f(t)$ | $(-1)^n\dfrac{d^n}{ds^n}F(s)$ |
| **60.** $\displaystyle\int_0^t f(\tau)g(t-\tau)\,d\tau$ | $F(s)G(s)$ |
| **61.** $\delta(t)$ | $1$ |
| **62.** $\delta(t - t_0)$ | $e^{-st_0}$ |

# Answers for Selected Odd-Numbered Problems

## EXERCISES 1.1 (PAGE 12)

**1.** linear, second order    **3.** linear, fourth order

**5.** nonlinear, second order    **7.** linear, third order

**9.** linear in $x$ but nonlinear in $y$

**15.** domain of function is $[-2, \infty)$; largest interval of definition for solution is $(-2, \infty)$

**17.** domain of function is the set of real numbers except $x = 2$ and $x = -2$; largest intervals of definition for solution are $(-\infty, -2)$, $(-2, 2)$, or $(2, \infty)$

**19.** $X = \dfrac{e^t - 1}{e^t - 2}$ defined on $(-\infty, \ln 2)$ or on $(\ln 2, \infty)$

**31.** $m = -2$    **33.** $m = 2, m = 3$    **35.** $m = 0, m = -1$

**37.** $y = 2$    **39.** no constant solutions

## EXERCISES 1.2 (PAGE 19)

**1.** $y = 1/(1 - 4e^{-x})$

**3.** $y = 1/(x^2 - 1)$; $(1, \infty)$

**5.** $y = 1/(x^2 + 1)$; $(-\infty, \infty)$

**7.** $x = -\cos t + 8 \sin t$

**9.** $x = \dfrac{\sqrt{3}}{4} \cos t + \dfrac{1}{4} \sin t$    **11.** $y = \dfrac{3}{2} e^x - \dfrac{1}{2} e^{-x}$

**13.** $y = 5e^{-x-1}$    **15.** $y = 0, y = x^3$

**17.** half-planes defined by either $y > 0$ or $y < 0$

**19.** half-planes defined by either $x > 0$ or $x < 0$

**21.** the regions defined by $y > 2, y < -2$, or $-2 < y < 2$

**23.** any region not containing $(0, 0)$

**25.** yes

**27.** no

**29.** (a) $y = cx$

    (b) any rectangular region not touching the $y$-axis

    (c) No, the function is not differentiable at $x = 0$.

**31.** (b) $y = 1/(1 - x)$ on $(-\infty, 1)$;
    $y = -1/(x + 1)$ on $(-1, \infty)$;

    (c) $y = 0$ on $(-\infty, \infty)$

**39.** $y = 3 \sin 2x$

**41.** $y = 0$

**43.** no solution

## EXERCISES 1.3 (PAGE 30)

**1.** $\dfrac{dP}{dt} = kP + r$; $\dfrac{dP}{dt} = kP - r$

**3.** $\dfrac{dP}{dt} = k_1 P - k_2 P^2$

**7.** $\dfrac{dx}{dt} = kx(1000 - x)$

**9.** $\dfrac{dA}{dt} + \dfrac{1}{100} A = 0$; $A(0) = 50$

**11.** $\dfrac{dA}{dt} + \dfrac{7}{600 - t} A = 6$    **13.** $\dfrac{dh}{dt} = -\dfrac{c\pi}{450} \sqrt{h}$

**15.** $L\dfrac{di}{dt} + Ri = E(t)$    **17.** $m\dfrac{dv}{dt} = mg - kv^2$

**19.** $m\dfrac{d^2x}{dt^2} = -kx$

**21.** $m\dfrac{dv}{dt} + v\dfrac{dm}{dt} + kv = -mg + R$

**23.** $\dfrac{d^2r}{dt^2} + \dfrac{gR^2}{r^2} = 0$    **25.** $\dfrac{dA}{dt} = k(M - A), k > 0$

**27.** $\dfrac{dx}{dt} + kx = r, k > 0$    **29.** $\dfrac{dy}{dx} = \dfrac{-x + \sqrt{x^2 + y^2}}{y}$

## CHAPTER 1 IN REVIEW (PAGE 34)

**1.** $\dfrac{dy}{dx} = 10y$    **3.** $y'' + k^2 y = 0$

**5.** $y'' - 2y' + y = 0$    **7.** (a), (d)

**9.** (b)    **11.** (b)

**13.** $y = c_1$ and $y = c_2 e^x$, $c_1$ and $c_2$ constants

**15.** $y' = x^2 + y^2$

**17.** (a) The domain is the set of all real numbers.

    (b) either $(-\infty, 0)$ or $(0, \infty)$

**19.** For $x_0 = -1$ the interval is $(-\infty, 0)$, and for $x_0 = 2$ the interval is $(0, \infty)$.

**21.** (c) $y = \begin{cases} -x^2, & x < 0 \\ x^2, & x \geq 0 \end{cases}$    **23.** $(-\infty, \infty)$

**25.** $(0, \infty)$    **35.** $y = \dfrac{1}{2} e^{3x} - \dfrac{1}{2} e^{-x} - 2x$

**37.** $y = \dfrac{3}{2} e^{3x-3} + \dfrac{9}{2} e^{-x+1} - 2x$.

**39.** $y_0 = -3, y_1 = 0$

## EXERCISES 2.1 (PAGE 44)

**21.** 0 is asymptotically stable (attractor); 3 is unstable (repeller).

**23.** 2 is semi-stable.

**25.** $-2$ is unstable (repeller); 0 is semi-stable; 2 is asymptotically stable (attractor).

**27.** $-1$ is asymptotically stable (attractor); 0 is unstable (repeller).

**39.** $0 < P_0 < h/k$

**41.** $\sqrt{mg/k}$

## EXERCISES 2.2 (PAGE 52)

**1.** $y = -\dfrac{1}{5} \cos 5x + c$    **3.** $y = \dfrac{1}{3} e^{-3x} + c$

**5.** $y = cx^4$    **7.** $-3e^{-2y} = 2e^{3x} + c$

**9.** $\dfrac{1}{3} x^3 \ln x - \dfrac{1}{9} x^3 = \dfrac{1}{2} y^2 + 2y + \ln|y| + c$

**11.** $4 \cos y = 2x + \sin 2x + c$

**13.** $(e^x + 1)^{-2} + 2(e^y + 1)^{-1} = c$

**15.** $S = ce^{kr}$    **17.** $P = \dfrac{ce^t}{1 + ce^t}$

**19.** $(y + 3)^5 e^x = c(x + 4)^5 e^y$    **21.** $y = \sin\!\left(\dfrac{1}{2} x^2 + c\right)$

**23.** $x = \tan\left(4t - \frac{3}{4}\pi\right)$

**25.** $y = \dfrac{e^{-(1+1/x)}}{x}$

**27.** $y = \frac{1}{2}x + \frac{\sqrt{3}}{2}\sqrt{1 - x^2}$

**29.** $y = e^{\int_4^x e^{-t^2}\,dt}$

**31.** $y = -\sqrt{x^2 + x - 1};\ \left(-\infty, -\frac{1+\sqrt{5}}{2}\right)$

**33.** $y = -\ln(2 - e^x);\quad (-\infty, \ln 2)$

**35. (a)** $y = 2,\ y = -2,\ y = 2\,\dfrac{3 - e^{4x-1}}{3 + e^{4x-1}}$

**37.** $y = -1$ and $y = 1$ are singular solutions of Problem 21; $y = 0$ of Problem 22

**39.** $y = 1$

**41.** $y = 1 + \frac{1}{10}\tan\left(\frac{1}{10}x\right)$

**45.** $y = \tan x - \sec x + c$

**47.** $y = [-1 + c(1 + \sqrt{x})]^2$

**49.** $y = 2\sqrt{\sqrt{x}e^{\sqrt{x}} - e^{\sqrt{x}} + 4}$

**57.** $y(x) = (4h/L^2)x^2 + a$

### EXERCISES 2.3  (PAGE 62)

**1.** $y = ce^{5x},\ (-\infty, \infty)$

**3.** $y = \frac{1}{4}e^{3x} + ce^{-x},\ (-\infty, \infty);\ ce^{-x}$ is transient

**5.** $y = \frac{1}{3} + ce^{-x^3},\ (-\infty, \infty);\ ce^{-x^3}$ is transient

**7.** $y = x^{-1}\ln x + cx^{-1},\ (0, \infty);$ solution is transient

**9.** $y = cx - x\cos x,\ (0, \infty)$

**11.** $y = \frac{1}{7}x^3 - \frac{1}{5}x + cx^{-4},\ (0, \infty);\ cx^{-4}$ is transient

**13.** $y = \frac{1}{2}x^{-2}e^x + cx^{-2}e^{-x},\ (0, \infty);\ cx^{-2}e^{-x}$ is transient

**15.** $x = 2y^6 + cy^4,\ (0, \infty)$

**17.** $y = \sin x + c\cos x,\ (-\pi/2, \pi/2)$

**19.** $(x + 1)e^x y = x^2 + c,\ (-1, \infty);$ solution is transient

**21.** $(\sec\theta + \tan\theta)r = \theta - \cos\theta + c,\ (-\pi/2, \pi/2)$

**23.** $y = e^{-3x} + cx^{-1}e^{-3x},\ (0, \infty);$ solution is transient

**25.** $y = -\frac{1}{5}x - \frac{1}{25} + \frac{76}{25}e^{5x};\ (-\infty, \infty)$

**27.** $y = x^{-1}e^x + (2 - e)x^{-1};\ (0, \infty)$

**29.** $i = \dfrac{E}{R} + \left(i_0 - \dfrac{E}{R}\right)e^{-Rt/L};\ (-\infty, \infty)$

**31.** $y = 2x + 1 + 5/x;\ (0, \infty)$

**33.** $(x + 1)y = x\ln x - x + 21;\ (0, \infty)$

**35.** $y = -2 + 3e^{-\cos x};\ (-\infty, \infty)$

**37.** $y = \begin{cases} \frac{1}{2}(1 - e^{-2x}), & 0 \le x \le 3 \\ \frac{1}{2}(e^6 - 1)e^{-2x}, & x > 3 \end{cases}$

**39.** $y = \begin{cases} \frac{1}{2} + \frac{3}{2}e^{-x^2}, & 0 \le x < 1 \\ \left(\frac{1}{2}e + \frac{3}{2}\right)e^{-x^2}, & x \ge 1 \end{cases}$

**41.** $y = \begin{cases} 2x - 1 + 4e^{-2x}, & 0 \le x \le 1 \\ 4x^2\ln x + (1 + 4e^{-2})x^2, & x > 1 \end{cases}$

**43.** $y = e^{x^2 - 1} + \frac{1}{2}\sqrt{\pi}e^{x^2}(\mathrm{erf}(x) - \mathrm{erf}(1))$

**45.** $y = e^{-e^x}\displaystyle\int_0^x e^{e^t}\,dt + e^{1 - e^x}$

**47.** $y = 10x^{-2}\big(\mathrm{Si}(x) - \mathrm{Si}(1)\big)$

**57.** $x = x_0 e^{-\lambda_1 t},\ y = \dfrac{x_0\lambda_1}{\lambda_2 - \lambda_1}e^{-\lambda_1 t} + \left(y_0 - \dfrac{x_0\lambda_1}{\lambda_2 - \lambda_1}\right)e^{-\lambda_2 t}$

### EXERCISES 2.4  (PAGE 70)

**1.** $x^2 - x + \frac{3}{2}y^2 + 7y = c$

**3.** $\frac{5}{2}x^2 + 4xy - 2y^4 = c$

**5.** $x^2 y^2 - 3x + 4y = c$

**7.** not exact

**9.** $xy^3 + y^2\cos x - \frac{1}{2}x^2 = c$

**11.** not exact

**13.** $xy - 2xe^x + 2e^x - 2x^3 = c$

**15.** $x^3 y^3 - \tan^{-1}3x = c$

**17.** $-\ln|\cos x| + \cos x\sin y = c$

**19.** $t^4 y - 5t^3 - ty + y^3 = c$

**21.** $\frac{1}{3}x^3 + x^2 y + xy^2 - y = \frac{4}{3}$

**23.** $4ty + t^2 - 5t + 3y^2 - y = 8$

**25.** $y^2\sin x - x^3 y - x^2 + y\ln y - y = 0$

**27.** $k = 10$

**29.** $x^2 y^2\cos x = c$

**31.** $x^2 y^2 + x^3 = c$

**33.** $3x^2 y^3 + y^4 = c$

**35.** $-2ye^{3x} + \frac{10}{3}e^{3x} + x = c$

**37.** $e^{y^2}(x^2 + 4) = 20$

**39. (c)** $y_1(x) = -x^2 - \sqrt{x^4 - x^3 + 4}$

$y_2(x) = -x^2 + \sqrt{x^4 - x^3 + 4}$

**45. (a)** $v(x) = 8\sqrt{\dfrac{x}{3} - \dfrac{9}{x^2}}$

**(b)** 12.7 ft/s

### EXERCISES 2.5  (PAGE 75)

**1.** $y + x\ln|x| = cx$

**3.** $(x - y)\ln|x - y| = y + c(x - y)$

**5.** $x + y\ln|x| = cy$

**7.** $\ln(x^2 + y^2) + 2\tan^{-1}(y/x) = c$

**9.** $4x = y(\ln|y| - c)^2$

**11.** $y^3 + 3x^3\ln|x| = 8x^3$

**13.** $\ln|x| = e^{y/x} - 1$

**15.** $y^3 = 1 + cx^{-3}$

**17.** $y^{-3} = x + \frac{1}{3} + ce^{3x}$

**19.** $e^{t/y} = ct$

**21.** $y^{-3} = -\frac{9}{5}x^{-1} + \frac{49}{5}x^{-6}$

**23.** $y = -x - 1 + \tan(x + c)$

**25.** $2y - 2x + \sin 2(x + y) = c$

**27.** $4(y - 2x + 3) = (x + c)^2$

**29.** $-\cot(x + y) + \csc(x + y) = x + \sqrt{2} - 1$

**35. (b)** $y = \dfrac{2}{x} + \left(-\frac{1}{4}x + cx^{-3}\right)^{-1}$

### EXERCISES 2.6  (PAGE 80)

**1.** $y_2 = 2.9800,\quad y_4 = 3.1151$

**3.** $y_{10} = 2.5937,\quad y_{20} = 2.6533;\ y = e^x$

**5.** $y_5 = 0.4198,\quad y_{10} = 0.4124$

**7.** $y_5 = 0.5639,\quad y_{10} = 0.5565$

**9.** $y_5 = 1.2194,\quad y_{10} = 1.2696$

**13.** Euler: $y_{10} = 3.8191,\quad y_{20} = 5.9363$

RK4: $y_{10} = 42.9931,\quad y_{20} = 84.0132$

## CHAPTER 2 IN REVIEW (PAGE 81)

**1.** $-A/k$, a repeller for $k > 0$, an attractor for $k < 0$

**3.** true

**5.** $\dfrac{d^3y}{dx^3} = x \sin y$

**7.** true

**9.** $y = c_1 e^{e^x}$

**11.** $\dfrac{dy}{dx} + (\sin x)y = x$

**13.** $\dfrac{dy}{dx} = (y-1)^2 (y-3)^2$

**15.** semi-stable for $n$ even and unstable for $n$ odd; semi-stable for $n$ even and asymptotically stable for $n$ odd.

**19.** $2x + \sin 2x = 2 \ln(y^2 + 1) + c$

**21.** $(6x + 1)y^3 = -3x^3 + c$

**23.** $Q = ct^{-1} + \frac{1}{25} t^4 (-1 + 5 \ln t)$

**25.** $y = \frac{1}{4} + c(x^2 + 4)^{-4}$

**27.** $y = e^{-2\sin x} + \frac{1}{2} e^{-2\sin x} \displaystyle\int_0^x t e^{2\sin t} dt$

**29.** $y = \dfrac{3}{x^2} + \dfrac{1}{x^2} \displaystyle\int_1^x t^2 e^{t^2} dt$

**31.** $y = \begin{cases} xe^{-x} + 5e^{-x}, & 0 \le x < 1 \\ 6e^{-x}, & x \ge 1 \end{cases}$

**33.** $y = \csc x, (\pi, 2\pi)$

**35. (b)** $y = \frac{1}{4}(x + 2\sqrt{y_0} - x_0)^2, (x_0 - 2\sqrt{y_0}, \infty)$

## EXERCISES 3.1 (PAGE 91)

**1.** 7.9 yr; 10 yr

**3.** 760; approximately 11 persons/yr

**5.** 11 h

**7.** 136.5 h

**9.** $I(15) = 0.00098 I_0$ or approximately 0.1% of $I_0$

**11.** approximately 15,963 years

**13.** $T(1) = 36.67°$ F; approximately 3.06 min

**15.** approximately 82.1 s; approximately 145.7 s

**17.** $390°$

**19.** about 1.6 hours prior to the discovery of the body

**21.** $A(t) = 200 - 170e^{-t/50}$

**23.** $A(t) = 1000 - 1000e^{-t/100}$

**25.** $A(t) = 1000 - 10t - \frac{1}{10}(100 - t)^2$;   100 min

**27.** 64.38 lb

**29.** $i(t) = \frac{3}{5} - \frac{3}{5} e^{-500t}; i \to \frac{3}{5}$ as $t \to \infty$

**31.** $q(t) = \frac{1}{100} - \frac{1}{100} e^{-50t}; i(t) = \frac{1}{2} e^{-50t}$

**33.** $i(t) = \begin{cases} 60 - 60e^{-t/10}, & 0 \le t \le 20 \\ 60(e^2 - 1)e^{-t/10}, & t > 20 \end{cases}$

**35. (a)** $v(t) = \dfrac{mg}{k} + \left( v_0 - \dfrac{mg}{k} \right) e^{-kt/m}$

**(b)** $v \to \dfrac{mg}{k}$ as $t \to \infty$

**(c)** $s(t) = \dfrac{mg}{k} t - \dfrac{m}{k}\left( v_0 - \dfrac{mg}{k} \right) e^{-kt/m}$
$+ \dfrac{m}{k}\left( v_0 - \dfrac{mg}{k} \right)$

**39. (a)** $v(t) = 0.024t^2 + 0.2t$   **(b)** $s(t) = 0.008t^3 + 0.1t^2$

**41. (a)** $v(t) = \dfrac{\rho g}{4k}\left( \dfrac{k}{\rho} t + r_0 \right) - \dfrac{\rho g r_0}{4k}\left( \dfrac{r_0}{\dfrac{k}{\rho} t + r_0} \right)^3$

**(c)** $33\frac{1}{3}$ seconds

**43. (a)** $P(t) = P_0 e^{(k_1 - k_2)t}$

**45. (a)** As $t \to \infty$, $x(t) \to r/k$.

**(b)** $x(t) = r/k - (r/k)e^{-kt}$; $(\ln 2)/k$

**49. (c)** 1.988 ft

## EXERCISES 3.2 (PAGE 101)

**1. (a)** $N = 2000$

**(b)** $N(t) = \dfrac{2000 e^t}{1999 + e^t}$; $N(10) = 1834$

**3.** 1,000,000;   52.9 mo

**5. (b)** $P(t) = \dfrac{4(P_0 - 1) - (P_0 - 4)e^{-3t}}{(P_0 - 1) - (P_0 - 4)e^{-3t}}$

**(c)** For $0 < P_0 < 1$, time of extinction is

$$t = -\frac{1}{3} \ln \frac{4(P_0 - 1)}{P_0 - 4}.$$

**7.** $P(t) = \dfrac{5}{2} + \dfrac{\sqrt{3}}{2} \tan\left[ -\dfrac{\sqrt{3}}{2} t + \tan^{-1}\left( \dfrac{2P_0 - 5}{\sqrt{3}} \right) \right]$;
time of extinction is

$$t = \frac{2}{\sqrt{3}}\left[ \tan^{-1} \frac{5}{\sqrt{3}} + \tan^{-1}\left( \frac{2P_0 - 5}{\sqrt{3}} \right) \right]$$

**9.** 29.3 g; $X \to 60$ as $t \to \infty$; 0 g of $A$ and 30 g of $B$

**11. (a)** $h(t) = \left( \sqrt{H} - \dfrac{4A_h}{A_w} t \right)^2$; $I$ is $0 \le t \le \sqrt{H} A_w / 4A_h$

**(b)** $576 \sqrt{10}$ s or 30.36 min

**13. (a)** approximately 858.65 s or 14.31 min

**(b)** 243 s or 4.05 min

**15. (a)** $v(t) = \sqrt{\dfrac{mg}{k}} \tanh\left( \sqrt{\dfrac{kg}{m}} t + c_1 \right)$

where $c_1 = \tanh^{-1}\left( \sqrt{\dfrac{k}{mg}} v_0 \right)$

**(b)** $\sqrt{\dfrac{mg}{k}}$

**(c)** $s(t) = \dfrac{m}{k} \ln \cosh\left( \sqrt{\dfrac{kg}{m}} t + c_1 \right) + c_2$,
where $c_2 = -(m/k)\ln(\cosh c_1)$

**17. (a)** $m\dfrac{dv}{dt} = mg - kv^2 - \rho V$,
where $\rho$ is the weight density of water

**(b)** $v(t) = \sqrt{\dfrac{mg - \rho V}{k}} \tanh\left( \dfrac{\sqrt{kmg - k\rho V}}{m} t + c_1 \right)$

**(c)** $\sqrt{\dfrac{mg - \rho V}{k}}$

**19. (a)** $W = 0$ and $W = 2$   **(b)** $W(x) = 2 \operatorname{sech}^2(x - c_1)$

**(c)** $W(x) = 2 \operatorname{sech}^2 x$

**21. (a)** $P(t) = \dfrac{1}{(-0.001350t + 10^{-0.01})^{100}}$

**(b)** approximately 724 months

**(c)** approximately 12,839 and 28,630,966

## EXERCISES 3.3 (PAGE 111)

**1.** $x(t) = x_0 e^{-\lambda_1 t}$

$y(t) = \dfrac{x_0 \lambda_1}{\lambda_2 - \lambda_1}(e^{-\lambda_1 t} - e^{-\lambda_2 t})$

$z(t) = x_0\left(1 - \dfrac{\lambda_2}{\lambda_2 - \lambda_1}e^{-\lambda_1 t} + \dfrac{\lambda_1}{\lambda_2 - \lambda_1}e^{-\lambda_2 t}\right)$

**3.** 5, 20, 147 days. The time when $y(t)$ and $z(t)$ are the same makes sense because most of $A$ and half of $B$ are gone, so half of $C$ should have been formed.

**5. (a)** $K(t) = K_0 e^{-(\lambda_1+\lambda_2)t}$, $C(t) = \dfrac{\lambda_1}{\lambda_1 + \lambda_2} K_0\left[1 - e^{-(\lambda_1+\lambda_2)t}\right]$,

$A(t) = \dfrac{\lambda_2}{\lambda_1 + \lambda_2} K_0\left[1 - e^{-(\lambda_1+\lambda_2)t}\right]$

**(b)** approximately $1.3 \times 10^9$ years

**(c)** 89%, 11%

**7.** $\dfrac{dx_1}{dt} = 6 - \dfrac{2}{25}x_1 + \dfrac{1}{50}x_2$

$\dfrac{dx_2}{dt} = \dfrac{2}{25}x_1 - \dfrac{2}{25}x_2$

**9. (a)** $\dfrac{dx_1}{dt} = 3\dfrac{x_2}{100-t} - 2\dfrac{x_1}{100+t}$

$\dfrac{dx_2}{dt} = 2\dfrac{x_1}{100+t} - 3\dfrac{x_2}{100-t}$

**(b)** $x_1(t) + x_2(t) = 150$;  $x_2(30) \approx 47.4$ lb

**15.** $L_1\dfrac{di_2}{dt} + (R_1 + R_2)i_2 + R_1 i_3 = E(t)$

$L_2\dfrac{di_3}{dt} + R_1 i_2 + (R_1 + R_3)i_3 = E(t)$

**17.** $i(0) = i_0$, $s(0) = n - i_0$, $r(0) = 0$

## CHAPTER 3 IN REVIEW (PAGE 114)

**1.** $dP/dt = 0.15P$

**3.** $P(45) = 8.99$ billion

**5.** approximately 3257 BCE

**7.** $x = 10 \ln\left(\dfrac{10 + \sqrt{100 - y^2}}{y}\right) - \sqrt{100 - y^2}$

**9. (a)** $\dfrac{BT_1 + T_2}{1 + B}$, $\dfrac{BT_1 + T_2}{1 + B}$

**(b)** $T(t) = \dfrac{BT_1 + T_2}{1 + B} + \dfrac{T_1 - T_2}{1 + B}e^{k(1+B)t}$

**13.** $x(t) = \dfrac{\alpha c_1 e^{\alpha k_1 t}}{1 + c_1 e^{\alpha k_1 t}}$,  $y(t) = c_2(1 + c_1 e^{\alpha k_1 t})^{k_2/k_1}$

**15.** $x^2 + y^2 = c_2$    **17.** $y^2 + 2x = c_2$    **21.** $t = \frac{1}{2}$

## EXERCISES 4.1 (PAGE 130)

**1.** $y = \frac{1}{2}e^x - \frac{1}{2}e^{-x}$

**3.** $y = 3x - 4x \ln x$

**9.** $(-\infty, 2)$

**11. (a)** $y = \dfrac{e}{e^2 - 1}(e^x - e^{-x})$    **(b)** $y = \dfrac{\sinh x}{\sinh 1}$

**13. (a)** $y = e^x \cos x - e^x \sin x$

**(b)** no solution

**(c)** $y = e^x \cos x + e^{-\pi/2}e^x \sin x$

**(d)** $y = c_2 e^x \sin x$, where $c_2$ is arbitrary

**15.** dependent    **17.** dependent

**19.** dependent    **21.** independent

**23.** The functions satisfy the DE and are linearly independent on the interval since $W(e^{-3x}, e^{4x}) = 7e^x \neq 0$; $y = c_1 e^{-3x} + c_2 e^{4x}$.

**25.** The functions satisfy the DE and are linearly independent on the interval since $W(e^x \cos 2x, e^x \sin 2x) = 2e^{2x} \neq 0$; $y = c_1 e^x \cos 2x + c_2 e^x \sin 2x$.

**27.** The functions satisfy the DE and are linearly independent on the interval since $W(x^3, x^4) = x^6 \neq 0$; $y = c_1 x^3 + c_2 x^4$.

**29.** The functions satisfy the DE and are linearly independent on the interval since $W(x, x^{-2}, x^{-2}\ln x) = 9x^{-6} \neq 0$; $y = c_1 x + c_2 x^{-2} + c_3 x^{-2}\ln x$.

**35. (b)** $y_p = x^2 + 3x + 3e^{2x}$;  $y_p = -2x^2 - 6x - \frac{1}{3}e^{2x}$

## EXERCISES 4.2 (PAGE 134)

**1.** $y_2 = xe^{2x}$            **3.** $y_2 = \sin 4x$

**5.** $y_2 = \sinh x$            **7.** $y_2 = xe^{2x/3}$

**9.** $y_2 = x^4 \ln|x|$          **11.** $y_2 = 1$

**13.** $y_2 = x \cos(\ln x)$      **15.** $y_2 = x^2 + x + 2$

**17.** $y_2 = e^{2x}$, $y_p = -\frac{1}{2}$    **19.** $y_2 = e^{2x}$, $y_p = \frac{5}{2}e^{3x}$

**21.** $y_2(x) = x\displaystyle\int_{x_0}^x \dfrac{e^{-t}}{t}dt$,  $x_0 > 0$

## EXERCISES 4.3 (PAGE 140)

**1.** $y = c_1 + c_2 e^{-x/4}$        **3.** $y = c_1 e^{3x} + c_2 e^{-2x}$

**5.** $y = c_1 e^{-4x} + c_2 x e^{-4x}$    **7.** $y = c_1 e^{2x/3} + c_2 e^{-x/4}$

**9.** $y = c_1 \cos 3x + c_2 \sin 3x$

**11.** $y = e^{2x}(c_1 \cos x + c_2 \sin x)$

**13.** $y = e^{-x/3}\left(c_1 \cos \frac{1}{3}\sqrt{2}x + c_2 \sin \frac{1}{3}\sqrt{2}x\right)$

**15.** $y = c_1 + c_2 e^{-x} + c_3 e^{5x}$

**17.** $y = c_1 e^{-x} + c_2 e^{3x} + c_3 x e^{3x}$

**19.** $u = c_1 e^t + e^{-t}(c_2 \cos t + c_3 \sin t)$

**21.** $y = c_1 e^{-x} + c_2 x e^{-x} + c_3 x^2 e^{-x}$

**23.** $y = c_1 + c_2 x + e^{-x/2}\left(c_3 \cos \frac{1}{2}\sqrt{3}x + c_4 \sin \frac{1}{2}\sqrt{3}x\right)$

**25.** $y = c_1 \cos \frac{1}{2}\sqrt{3}x + c_2 \sin \frac{1}{2}\sqrt{3}x$
     $+ c_3 x \cos \frac{1}{2}\sqrt{3}x + c_4 x \sin \frac{1}{2}\sqrt{3}x$

**27.** $u = c_1 e^r + c_2 r e^r + c_3 e^{-r} + c_4 r e^{-r} + c_5 e^{-5r}$

**29.** $y = 2\cos 4x - \frac{1}{2}\sin 4x$

**31.** $y = -\frac{1}{3}e^{-(t-1)} + \frac{1}{3}e^{5(t-1)}$

**33.** $y = 0$

**35.** $y = \frac{5}{36} - \frac{5}{36}e^{-6x} + \frac{1}{6}xe^{-6x}$

**37.** $y = e^{5x} - xe^{5x}$

**39.** $y = 0$

**41.** $y = \dfrac{1}{2}\left(1 - \dfrac{5}{\sqrt{3}}\right)e^{-\sqrt{3}x} + \dfrac{1}{2}\left(1 + \dfrac{5}{\sqrt{3}}\right)e^{\sqrt{3}x}$;

$y = \cosh\sqrt{3}x + \dfrac{5}{\sqrt{3}}\sinh\sqrt{3}x$

**49.** $y'' - 6y' + 5y = 0$   **51.** $y'' - 2y' = 0$

**53.** $y'' + 9y = 0$   **55.** $y'' + 2y' + 2y = 0$

**57.** $y''' - 8y'' = 0$

## EXERCISES 4.4 (PAGE 150)

**1.** $y = c_1 e^{-x} + c_2 e^{-2x} + 3$

**3.** $y = c_1 e^{5x} + c_2 x e^{5x} + \frac{6}{5}x + \frac{3}{5}$

**5.** $y = c_1 e^{-2x} + c_2 x e^{-2x} + x^2 - 4x + \frac{7}{2}$

**7.** $y = c_1 \cos\sqrt{3}x + c_2 \sin\sqrt{3}x + \left(-4x^2 + 4x - \frac{4}{3}\right)e^{3x}$

**9.** $y = c_1 + c_2 e^x + 3x$

**11.** $y = c_1 e^{x/2} + c_2 x e^{x/2} + 12 + \frac{1}{2}x^2 e^{x/2}$

**13.** $y = c_1 \cos 2x + c_2 \sin 2x - \frac{3}{4}x\cos 2x$

**15.** $y = c_1 \cos x + c_2 \sin x - \frac{1}{2}x^2 \cos x + \frac{1}{2}x\sin x$

**17.** $y = c_1 e^x \cos 2x + c_2 e^x \sin 2x + \frac{1}{4}x e^x \sin 2x$

**19.** $y = c_1 e^{-x} + c_2 x e^{-x} - \frac{1}{2}\cos x$

$+ \frac{12}{25}\sin 2x - \frac{9}{25}\cos 2x$

**21.** $y = c_1 + c_2 x + c_3 e^{6x} - \frac{1}{4}x^2 - \frac{6}{37}\cos x + \frac{1}{37}\sin x$

**23.** $y = c_1 e^x + c_2 x e^x + c_3 x^2 e^x - x - 3 - \frac{2}{3}x^3 e^x$

**25.** $y = c_1 \cos x + c_2 \sin x + c_3 x\cos x + c_4 x\sin x$

$+ x^2 - 2x - 3$

**27.** $y = \sqrt{2}\sin 2x - \frac{1}{2}$

**29.** $y = -200 + 200 e^{-x/5} - 3x^2 + 30x$

**31.** $y = -10 e^{-2x}\cos x + 9 e^{-2x}\sin x + 7 e^{-4x}$

**33.** $x = \dfrac{F_0}{2\omega^2}\sin\omega t - \dfrac{F_0}{2\omega}t\cos\omega t$

**35.** $y = 11 - 11 e^x + 9x e^x + 2x - 12x^2 e^x + \frac{1}{2}e^{5x}$

**37.** $y = 6\cos x - 6(\cot 1)\sin x + x^2 - 1$

**39.** $y = \dfrac{-4\sin\sqrt{3}x}{\sin\sqrt{3} + \sqrt{3}\cos\sqrt{3}} + 2x$

**41.** $y = \begin{cases} \cos 2x + \frac{5}{6}\sin 2x + \frac{1}{3}\sin x, & 0 \le x \le \pi/2 \\ \frac{2}{3}\cos 2x + \frac{5}{6}\sin 2x, & x > \pi/2 \end{cases}$

## EXERCISES 4.5 (PAGE 158)

**1.** $(3D - 2)(3D + 2)y = \sin x$

**3.** $(D - 6)(D + 2)y = x - 6$

**5.** $D(D + 5)^2 y = e^x$

**7.** $(D - 1)(D - 2)(D + 5)y = x e^{-x}$

**9.** $D(D + 2)(D^2 - 2D + 4)y = 4$

**15.** $D^4$   **17.** $D(D - 2)$

**19.** $D^2 + 4$   **21.** $D^3(D^2 + 16)$

**23.** $(D + 1)(D - 1)^3$   **25.** $D(D^2 - 2D + 5)$

**27.** $1, x, x^2, x^3, x^4$   **29.** $e^{6x}, e^{-3x/2}$

**31.** $\cos\sqrt{5}x, \sin\sqrt{5}x$   **33.** $1, e^{5x}, x e^{5x}$

**35.** $y = c_1 e^{-3x} + c_2 e^{3x} - 6$

**37.** $y = c_1 + c_2 e^{-x} + 3x$

**39.** $y = c_1 e^{-2x} + c_2 x e^{-2x} + \frac{1}{2}x + 1$

**41.** $y = c_1 + c_2 x + c_3 e^{-x} + \frac{2}{3}x^4 - \frac{8}{3}x^3 + 8x^2$

**43.** $y = c_1 e^{-3x} + c_2 e^{4x} + \frac{1}{7}x e^{4x}$

**45.** $y = c_1 e^{-x} + c_2 e^{3x} - e^x + 3$

**47.** $y = c_1 \cos 5x + c_2 \sin 5x + \frac{1}{4}\sin x$

**49.** $y = c_1 e^{-3x} + c_2 x e^{-3x} - \frac{1}{49}x e^{4x} + \frac{2}{343}e^{4x}$

**51.** $y = c_1 e^{-x} + c_2 e^x + \frac{1}{6}x^3 e^x - \frac{1}{4}x^2 e^x + \frac{1}{4}x e^x - 5$

**53.** $y = e^x(c_1 \cos 2x + c_2 \sin 2x) + \frac{1}{3}e^x \sin x$

**55.** $y = c_1 \cos 5x + c_2 \sin 5x - 2x\cos 5x$

**57.** $y = e^{-x/2}\left(c_1 \cos\dfrac{\sqrt{3}}{2}x + c_2 \sin\dfrac{\sqrt{3}}{2}x\right)$

$+ \sin x + 2\cos x - x\cos x$

**59.** $y = c_1 + c_2 x + c_3 e^{-8x} + \frac{11}{256}x^2 + \frac{7}{32}x^3 - \frac{1}{16}x^4$

**61.** $y = c_1 e^x + c_2 x e^x + c_3 x^2 e^x + \frac{1}{6}x^3 e^x + x - 13$

**63.** $y = c_1 + c_2 x + c_3 e^x + c_4 x e^x + \frac{1}{2}x^2 e^x + \frac{1}{2}x^2$

**65.** $y = \frac{5}{8}e^{-8x} + \frac{5}{8}e^{8x} - \frac{1}{4}$

**67.** $y = -\frac{41}{125} + \frac{41}{125}e^{5x} - \frac{1}{10}x^2 + \frac{9}{25}x$

**69.** $y = -\pi\cos x - \frac{11}{3}\sin x - \frac{8}{3}\cos 2x + 2x\cos x$

**71.** $y = 2e^{2x}\cos 2x - \frac{3}{64}e^{2x}\sin 2x + \frac{1}{8}x^3 + \frac{3}{16}x^2 + \frac{3}{32}x$

## EXERCISES 4.6 (PAGE 165)

**1.** $y = c_1 \cos x + c_2 \sin x + x\sin x + \cos x \ln|\cos x|$

**3.** $y = c_1 \cos x + c_2 \sin x - \frac{1}{2}x\cos x$

**5.** $y = c_1 \cos x + c_2 \sin x + \frac{1}{2} - \frac{1}{6}\cos 2x$

**7.** $y = c_1 e^x + c_2 e^{-x} + \frac{1}{2}x\sinh x$

**9.** $y = c_1 e^{-3x} + c_2 e^{3x} - \frac{3}{4}x^2 e^{-3x} - \frac{1}{4}x e^{-3x}$

**11.** $y = c_1 e^{-x} + c_2 e^{-2x} + (e^{-x} + e^{-2x})\ln(1 + e^x)$

**13.** $y = c_1 e^{-2x} + c_2 e^{-x} - e^{-2x}\sin e^x$

**15.** $y = c_1 e^{-t} + c_2 t e^{-t} + \frac{1}{2}t^2 e^{-t}\ln t - \frac{3}{4}t^2 e^{-t}$

**17.** $y = c_1 e^x \sin x + c_2 e^x \cos x + \frac{1}{3}x e^x \sin x$

$+ \frac{1}{3}e^x \cos x \ln|\cos x|$

**19.** $y = \frac{1}{4}e^{-x/2} + \frac{3}{4}e^{x/2} + \frac{1}{8}x^2 e^{x/2} - \frac{1}{4}x e^{x/2}$

**21.** $y = \frac{4}{9}e^{-4x} + \frac{25}{36}e^{2x} - \frac{1}{4}e^{-2x} + \frac{1}{9}e^{-x}$

**23.** $y = c_1 \cos x + c_2 \sin x - \cos x \displaystyle\int_{x_0}^x e^{t^2}\sin t\, dt + \sin x \displaystyle\int_{x_0}^x e^{t^2}\cos t\, dt$

**25.** $y = c_1 e^{-2x} + c_2 e^x - \frac{1}{3}e^{-2x}\displaystyle\int_{x_0}^x e^{2t}\ln t\, dt + \frac{1}{3}e^x \displaystyle\int_{x_0}^x e^{-t}\ln t\, dt, \; x_0 > 0$

**27.** $y = c_1 x^{-1/2}\cos x + c_2 x^{-1/2}\sin x + x^{-1/2}$

**29.** $y = c_1 + c_2 \cos x + c_3 \sin x - \ln|\cos x|$

$- \sin x \ln|\sec x + \tan x|$

**31.** $y = c_1 e^x - c_2 e^{-x} + c_3 e^{2x} + \frac{1}{30}e^{4x}$

## EXERCISES 4.7 (PAGE 171)

**1.** $y = c_1 x^{-1} + c_2 x^2$

**3.** $y = c_1 + c_2 \ln x$

**5.** $y = c_1 \cos(2\ln x) + c_2 \sin(2\ln x)$

**7.** $y = c_1 x^{(2-\sqrt{6})} + c_2 x^{(2+\sqrt{6})}$

**9.** $y = c_1 \cos\left(\frac{1}{5}\ln x\right) + c_2 \sin\left(\frac{1}{5}\ln x\right)$

**11.** $y = c_1 x^{-2} + c_2 x^{-2}\ln x$

**13.** $y = x^{-1/2}\left[c_1 \cos\left(\frac{1}{6}\sqrt{3}\ln x\right) + c_2 \sin\left(\frac{1}{6}\sqrt{3}\ln x\right)\right]$

**15.** $y = c_1 x^3 + c_2 \cos(\sqrt{2} \ln x) + c_3 \sin(\sqrt{2} \ln x)$

**17.** $y = c_1 + c_2 x + c_3 x^2 + c_4 x^{-3}$

**19.** $y = c_1 + c_2 x^5 + \frac{1}{5} x^5 \ln x$

**21.** $y = c_1 x + c_2 x \ln x + x(\ln x)^2$

**23.** $y = c_1 x^{-1} + c_2 x - \ln x$

**25.** $y = 2 - 2x^{-2}$

**27.** $y = \cos(\ln x) + 2 \sin(\ln x)$

**29.** $y = \frac{3}{4} - \ln x + \frac{1}{4} x^2$

**31.** $y = c_1 x^{-10} + c_2 x^2$

**33.** $y = c_1 x^{-1} + c_2 x^{-8} + \frac{1}{30} x^2$

**35.** $y = x^2 [c_1 \cos(3 \ln x) + c_2 \sin(3 \ln x)] + \frac{4}{13} + \frac{3}{10} x$

**37.** $y = 2(-x)^{1/2} - 5(-x)^{1/2} \ln(-x),\ x < 0$

**39.** $y = c_1 (x+3)^2 + c_2 (x+3)^7$

**41.** $y = c_1 \cos[\ln(x+2)] + c_2 \sin[\ln(x+2)]$

**47. (a)** $w(r) = c_1 + c_2 \ln r + c_3 r^2 + \dfrac{q}{64D} r^4$

   **(b)** $w(r) = \dfrac{q}{64D}(a^2 - r^2)^2$

## EXERCISES 4.8 (PAGE 182)

**1.** $y_p(x) = \frac{1}{4} \displaystyle\int_{x_0}^{x} \sinh 4(x - t) f(t)\, dt$

**3.** $y_p(x) = \displaystyle\int_{x_0}^{x} (x - t) e^{-(x-t)} f(t)\, dt$

**5.** $y_p(x) = \frac{1}{3} \displaystyle\int_{x_0}^{x} \sin 3(x - t) f(t)\, dt$

**7.** $y = c_1 e^{-4x} + c_2 e^{4x} + \frac{1}{4} \displaystyle\int_{x_0}^{x} \sinh 4(x - t) t e^{-2t}\, dt$

**9.** $y = c_1 e^{-x} + c_2 x e^{-x} + \displaystyle\int_{x_0}^{x} (x - t) e^{-(x-t)} e^{-t}\, dt$

**11.** $y = c_1 \cos 3x + c_2 \sin 3x + \frac{1}{3} \displaystyle\int_{x_0}^{x} \sin 3(x - t)(t + \sin t)\, dt$

**13.** $y_p(x) = \frac{1}{4} x e^{2x} - \frac{1}{16} e^{2x} + \frac{1}{16} e^{-2x}$

**15.** $y_p(x) = \frac{1}{2} x^2 e^{5x}$

**17.** $y_p(x) = -\cos x + \dfrac{\pi}{2} \sin x - x \sin x - \cos x \ln|\sin x|$

**19.** $y = \frac{25}{16} e^{-2x} - \frac{9}{16} e^{2x} + \frac{1}{4} x e^{2x}$

**21.** $y = -e^{5x} + 6x e^{5x} + \frac{1}{2} x^2 e^{5x}$

**23.** $y = -x \sin x - \cos x \ln|\sin x|$

**25.** $y = (\cos 1 - 2)e^{-x} + (1 + \sin 1 - \cos 1)e^{-2x} - e^{-2x}\sin e^{x}$

**27.** $y = 4x - 2x^2 - x \ln x$

**29.** $y = \frac{46}{45} x^3 - \frac{1}{20} x^{-2} + \frac{1}{36} - \frac{1}{6} \ln x$

**31.** $y(x) = 5e^x + 3e^{-x} + y_p(x),$

where $y_p(x) = \begin{cases} 1 - \cosh x, & x < 0 \\ -1 + \cosh x, & x \geq 0 \end{cases}$

**33.** $y = \cos x - \sin x + y_p(x),$

where $y_p(x) = \begin{cases} 0, & x < 0 \\ 10 - 10\cos x, & 0 \leq x \leq 3\pi \\ -20\cos x, & x > 3\pi \end{cases}$

**35.** $y_p(x) = (x - 1) \displaystyle\int_{0}^{x} t f(t)\, dt + x \displaystyle\int_{x}^{1} (t - 1) f(t)\, dt$

**37.** $y_p(x) = \frac{1}{2} x^2 - \frac{1}{2} x$

**39.** $y_p(x) = \dfrac{\sin(x - 1)}{\sin 1} - \dfrac{\sin x}{\sin 1} + 1$

**41.** $y_p(x) = -e^x \cos x - e^x \sin x + e^x$

**43.** $y_p(x) = \frac{1}{2}(\ln x)^2 + \frac{1}{2} \ln x$

## EXERCISES 4.9 (PAGE 187)

**1.** $x = c_1 e^t + c_2 t e^t$
   $y = (c_1 - c_2)e^t + c_2 t e^t$

**3.** $x = c_1 \cos t + c_2 \sin t + t + 1$
   $y = c_1 \sin t - c_2 \cos t + t - 1$

**5.** $x = \frac{1}{2} c_1 \sin t + \frac{1}{2} c_2 \cos t - 2c_3 \sin \sqrt{6} t - 2c_4 \cos \sqrt{6} t$
   $y = c_1 \sin t + c_2 \cos t + c_3 \sin \sqrt{6} t + c_4 \cos \sqrt{6} t$

**7.** $x = c_1 e^{2t} + c_2 e^{-2t} + c_3 \sin 2t + c_4 \cos 2t + \frac{1}{5} e^t$
   $y = c_1 e^{2t} + c_2 e^{-2t} - c_3 \sin 2t - c_4 \cos 2t - \frac{1}{5} e^t$

**9.** $x = c_1 - c_2 \cos t + c_3 \sin t + \frac{17}{15} e^{3t}$
   $y = c_1 + c_2 \sin t + c_3 \cos t - \frac{4}{15} e^{3t}$

**11.** $x = c_1 e^t + c_2 e^{-t/2} \cos \frac{1}{2} \sqrt{3} t + c_3 e^{-t/2} \sin \frac{1}{2} \sqrt{3} t$
   $y = \left(-\frac{3}{2} c_2 - \frac{1}{2} \sqrt{3} c_3\right) e^{-t/2} \cos \frac{1}{2} \sqrt{3} t$
   $\quad + \left(\frac{1}{2} \sqrt{3} c_2 - \frac{3}{2} c_3\right) e^{-t/2} \sin \frac{1}{2} \sqrt{3} t$

**13.** $x = c_1 e^{4t} + \frac{4}{3} e^t$
   $y = -\frac{3}{4} c_1 e^{4t} + c_2 + 5e^t$

**15.** $x = c_1 + c_2 t + c_3 e^t + c_4 e^{-t} - \frac{1}{2} t^2$
   $y = (c_1 - c_2 + 2) + (c_2 + 1)t + c_4 e^{-t} - \frac{1}{2} t^2$

**17.** $x = c_1 e^t + c_2 e^{-t/2} \sin \frac{1}{2} \sqrt{3} t + c_3 e^{-t/2} \cos \frac{1}{2} \sqrt{3} t$
   $y = c_1 e^t + \left(-\frac{1}{2} c_2 - \frac{1}{2} \sqrt{3} c_3\right) e^{-t/2} \sin \frac{1}{2} \sqrt{3} t$
   $\quad + \left(\frac{1}{2} \sqrt{3} c_2 - \frac{1}{2} c_3\right) e^{-t/2} \cos \frac{1}{2} \sqrt{3} t$
   $z = c_1 e^t + \left(-\frac{1}{2} c_2 + \frac{1}{2} \sqrt{3} c_3\right) e^{-t/2} \sin \frac{1}{2} \sqrt{3} t$
   $\quad + \left(-\frac{1}{2} \sqrt{3} c_2 - \frac{1}{2} c_3\right) e^{-t/2} \cos \frac{1}{2} \sqrt{3} t$

**19.** $x = -6c_1 e^{-t} - 3c_2 e^{-2t} + 2c_3 e^{3t}$
   $y = c_1 e^{-t} + c_2 e^{-2t} + c_3 e^{3t}$
   $z = 5c_1 e^{-t} + c_2 e^{-2t} + c_3 e^{3t}$

**21.** $x = e^{-3t+3} - t e^{-3t+3}$
   $y = -e^{-3t+3} + 2t e^{-3t+3}$

**23.** $mx'' = 0$
   $my'' = -mg;$
   $x = c_1 t + c_2$
   $y = -\frac{1}{2} g t^2 + c_3 t + c_4$

## EXERCISES 4.10 (PAGE 192)

**3.** $y = \ln|\cos(c_1 - x)| + c_2$

**5.** $y = \dfrac{1}{c_1^2} \ln|c_1 x + 1| - \dfrac{1}{c_1} x + c_2$

**7.** $(x + c_2)^2 + y^2 = c_1^2$

**9.** $\frac{1}{3} y^3 - c_1 y = x + c_2$

**11.** $y = \frac{2}{3}(x+1)^{3/2} + \frac{4}{3}$

**13.** $y = \tan\left(\frac{1}{4}\pi - \frac{1}{2}x\right),\ -\frac{1}{2}\pi < x < \frac{3}{2}\pi$

**15.** $y = -\dfrac{1}{c_1}\sqrt{1 - c_1^2 x^2} + c_2$

**17.** $y = 1 + x + \frac{1}{2}x^2 + \frac{1}{2}x^3 + \frac{1}{6}x^4 + \frac{1}{10}x^5 + \cdots$

**19.** $y = 1 + x - \frac{1}{2}x^2 + \frac{2}{3}x^3 - \frac{1}{4}x^4 + \frac{7}{60}x^5 + \cdots$

**21.** $y = -\sqrt{1 - x^2}$

## CHAPTER 4 IN REVIEW  (PAGE 193)

**1.** $y = 0$

**3.** false

**5.** $y = c_1 \cos 5x + c_2 \sin 5x$    **7.** $x^2 y'' - 3xy' + 4y = 0$

**9.** $y_p = x^2 + x - 2$    **11.** $(-\infty, \infty);\ (0, \infty)$

**13. (a)** $y = c_1 e^{3x} + c_2 e^{-5x} + c_3 x e^{-5x} + c_4 e^x + c_5 x e^x + c_6 x^2 e^x$

  **(b)** $y = c_1 x^3 + c_2 x^{-5} + c_3 x^{-5} \ln x + c_4 x + c_5 x \ln x$
  $+ c_6 x\,(\ln x)^2$

**15.** $x^2 y'' - 2xy' - 4y = 0$

**17.** $y = c_1 e^{(1+\sqrt{3})x} + c_2 e^{(1-\sqrt{3})x}$

**19.** $y = c_1 + c_2 e^{-5x} + c_3 x e^{-5x}$

**21.** $y = c_1 e^{-x/3} + e^{-3x/2}\left(c_2 \cos \frac{1}{2}\sqrt{7}x + c_3 \sin \frac{1}{2}\sqrt{7}x\right)$

**23.** $y = e^{3x/2}\left(c_2 \cos \frac{1}{2}\sqrt{11}x + c_3 \sin \frac{1}{2}\sqrt{11}x\right) + \frac{4}{5}x^3 + \frac{36}{25}x^2$
  $+ \frac{46}{125}x - \frac{222}{625}$

**25.** $y = c_1 + c_2 e^{2x} + c_3 e^{3x} + \frac{1}{5}\sin x - \frac{1}{5}\cos x + \frac{4}{3}x$

**27.** $y = e^x(c_1 \cos x + c_2 \sin x)$
  $- e^x \cos x \ln|\sec x + \tan x|$

**29.** $y = c_1 x^{-1/3} + c_2 x^{1/2}$

**31.** $y = c_1 x^2 + c_2 x^3 + x^4 - x^2 \ln x$

**33.** $y = c_1 \cos \omega x + c_2 \sin \omega x + A \cos \alpha x$
  $+ B \sin \alpha x,\ \ \omega \neq \alpha;$
  $y = c_1 \cos \omega x + c_2 \sin \omega x + Ax \cos \omega x$
  $+ Bx \sin \omega x,\ \ \omega = \alpha$

**35.** $y = c_1 \cos x + c_2 \sin x + e^{-x}(c_3 \cos 3x + c_4 \sin 3x)$

**37. (a)** $y = c_1 \cosh x + c_2 \sinh x + c_3 x \cosh x$
  $+ c_4 x \sinh x$

  **(b)** $y_p = Ax^2 \cosh x + Bx^2 \sinh x$

**39.** $y = e^{x-\pi} \cos x$

**41.** $y = \frac{13}{4}e^x - \frac{5}{4}e^{-x} - x - \frac{1}{2}\sin x$

**43.** $y = x^2 + 4$

**47.** $x = -c_1 e^t - \frac{3}{2}c_2 e^{2t} + \frac{5}{2}$
  $y = c_1 e^t + c_2 e^{2t} - 3$

**49.** $x = c_1 e^t + c_2 e^{5t} + t e^t$
  $y = -c_1 e^t + 3 c_2 e^{5t} - t e^t + 2 e^t$

## EXERCISES 5.1  (PAGE 209)

**1.** $\dfrac{\sqrt{2}\,\pi}{8}$    **3.** $x(t) = -\frac{1}{4}\cos 4\sqrt{6}\,t$

**5. (a)** $x\left(\frac{\pi}{12}\right) = -\frac{1}{4};\ x\left(\frac{\pi}{8}\right) = -\frac{1}{2};\ x\left(\frac{\pi}{6}\right) = -\frac{1}{4};$
  $x\left(\frac{\pi}{4}\right) = \frac{1}{2};\ x\left(\frac{9\pi}{32}\right) = \frac{\sqrt{2}}{4}$

**(b)** 4 ft/s; downward

**(c)** $t = \dfrac{(2n+1)\pi}{16},\ n = 0, 1, 2, \ldots$

**7. (a)** the 20-kg mass

**(b)** the 20-kg mass; the 50-kg mass

**(c)** $t = n\pi,\ n = 0, 1, 2, \ldots$ ; at the equilibrium position; the 50-kg mass is moving upward whereas the 20-kg mass is moving upward when $n$ is even and downward when $n$ is odd.

**9. (a)** $x(t) = \frac{1}{2}\cos 2t + \frac{3}{4}\sin 2t$

  **(b)** $x(t) = \frac{\sqrt{13}}{4}\sin(2t + 0.588)$

  **(c)** $x(t) = \frac{\sqrt{13}}{4}\cos(2t - 0.983)$

**11. (a)** $x(t) = -\frac{2}{3}\cos 10t + \frac{1}{2}\sin 10t$
  $= \frac{5}{6}\sin(10t - 0.927)$

  **(b)** $\dfrac{5}{6}$ ft; $\dfrac{\pi}{5}$

  **(c)** 15 cycles

  **(d)** 0.721 s

  **(e)** $\dfrac{(2n+1)\pi}{20} + 0.0927,\ n = 0, 1, 2, \ldots$

  **(f)** $x(3) = -0.597$ ft

  **(g)** $x'(3) = -5.814$ ft/s

  **(h)** $x''(3) = 59.702$ ft/s$^2$

  **(i)** $\pm 8\frac{1}{3}$ ft/s

  **(j)** $0.1451 + \dfrac{n\pi}{5};\ 0.3545 + \dfrac{n\pi}{5},\ n = 0, 1, 2, \ldots$

  **(k)** $0.3545 + \dfrac{n\pi}{5},\ n = 0, 1, 2, \ldots$

**13.** $k_{\text{eff}} = 160$ lb/ft; $x(t) = \frac{1}{8}\sin 16t$

**15.** $k_{\text{eff}} = 30$ lb/ft; $x(t) = \frac{\sqrt{3}}{6}\sin 4\sqrt{3}t$

**17.** $k_{\text{eff}} = 2k$

**21. (a)** above

  **(b)** heading upward

**23. (a)** below

  **(b)** heading upward

**25.** $\frac{1}{4}$ s; $\frac{1}{2}$ s, $x\left(\frac{1}{2}\right) = e^{-2}$; that is, the weight is approximately 0.14 ft below the equilibrium position.

**27. (a)** $x(t) = \frac{4}{3}e^{-2t} - \frac{1}{3}e^{-8t}$

  **(b)** $x(t) = -\frac{2}{3}e^{-2t} + \frac{5}{3}e^{-8t}$

**29. (a)** $x(t) = e^{-2t}\left(-\cos 4t - \frac{1}{2}\sin 4t\right)$

  **(b)** $x(t) = \frac{\sqrt{5}}{2}e^{-2t}\sin(4t + 4.249)$

  **(c)** $t = 1.294$ s

**31. (a)** $\beta > \frac{5}{2}$    **(b)** $\beta = \frac{5}{2}$    **(c)** $0 < \beta < \frac{5}{2}$

**33.** $x(t) = e^{-t/2}\left(-\frac{4}{3}\cos \frac{\sqrt{47}}{2}t - \frac{64}{3\sqrt{47}}\sin \frac{\sqrt{47}}{2}t\right)$
  $+ \frac{10}{3}(\cos 3t + \sin 3t)$

**35.** $x(t) = \frac{1}{4}e^{-4t} + te^{-4t} - \frac{1}{4}\cos 4t$

**37.** $x(t) = -\frac{1}{2}\cos 4t + \frac{9}{4}\sin 4t + \frac{1}{2}e^{-2t}\cos 4t$

$\qquad - 2e^{-2t}\sin 4t$

**39. (a)** $m\dfrac{d^2x}{dt^2} = -k(x - h) - \beta\dfrac{dx}{dt}$ or

$\qquad \dfrac{d^2x}{dt^2} + 2\lambda\dfrac{dx}{dt} + \omega^2 x = \omega^2 h(t),$

$\qquad$ where $2\lambda = \beta/m$ and $\omega^2 = k/m$

**(b)** $x(t) = e^{-2t}\left(-\frac{56}{13}\cos 2t - \frac{72}{13}\sin 2t\right) + \frac{56}{13}\cos t$

$\qquad + \frac{32}{13}\sin t$

**41.** $x(t) = -\cos 2t - \frac{1}{8}\sin 2t + \frac{3}{4}t\sin 2t + \frac{5}{4}t\cos 2t$

**43. (b)** $\dfrac{F_0}{2\omega}t\sin\omega t$

**49.** 4.568 C; 0.0509 s

**51.** $q(t) = 10 - 10e^{-3t}(\cos 3t + \sin 3t)$

$\qquad i(t) = 60e^{-3t}\sin 3t$; 10.432 C

**53.** $q_p = \frac{100}{13}\sin t + \frac{150}{13}\cos t$

$\qquad i_p = \frac{100}{13}\cos t - \frac{150}{13}\sin t$

**57.** $q(t) = -\frac{1}{2}e^{-10t}(\cos 10t + \sin 10t) + \frac{3}{2}; \frac{3}{2}$ C

**61.** $q(t) = \left(q_0 - \dfrac{E_0 C}{1 - \gamma^2 LC}\right)\cos\dfrac{t}{\sqrt{LC}}$

$\qquad + \sqrt{LC}\,i_0\sin\dfrac{t}{\sqrt{LC}} + \dfrac{E_0 C}{1 - \gamma^2 LC}\cos\gamma t$

$\qquad i(t) = i_0\cos\dfrac{t}{\sqrt{LC}} - \dfrac{1}{\sqrt{LC}}\left(q_0 - \dfrac{E_0 C}{1 - \gamma^2 LC}\right)\sin\dfrac{t}{\sqrt{LC}}$

$\qquad - \dfrac{E_0 C\gamma}{1 - \gamma^2 LC}\sin\gamma t$

## EXERCISES 5.2  (PAGE 219)

**1. (a)** $y(x) = \dfrac{w_0}{24EI}(6L^2 x^2 - 4Lx^3 + x^4)$

**3. (a)** $y(x) = \dfrac{w_0}{48EI}(3L^2 x^2 - 5Lx^3 + 2x^4)$

**5. (a)** $y(x) = \dfrac{w_0}{360EI}(7L^4 x - 10L^2 x^3 + 3x^5)$

**(c)** $x \approx 0.51933$, $y_{max} \approx 0.234799$

**7.** $y(x) = -\dfrac{w_0 EI}{P^2}\cosh\sqrt{\dfrac{P}{EI}}\,x$

$\qquad + \left(\dfrac{w_0 EI}{P^2}\sinh\sqrt{\dfrac{P}{EI}}L - \dfrac{w_0 L\sqrt{EI}}{P\sqrt{P}}\right)\dfrac{\sinh\sqrt{\dfrac{P}{EI}}\,x}{\cosh\sqrt{\dfrac{P}{EI}}L}$

$\qquad + \dfrac{w_0}{2P}x^2 + \dfrac{w_0 EI}{P^2}$

**9.** $\lambda_n = n^2, n = 1, 2, 3, \ldots;\quad y_n(x) = \sin nx$

**11.** $\lambda_n = \dfrac{(2n-1)^2\pi^2}{4L^2}, n = 1, 2, 3, \ldots;$

$\qquad y_n(x) = \cos\dfrac{(2n-1)\pi x}{2L}$

**13.** $\lambda_n = n^2, n = 0, 1, 2, \ldots;\quad y_n(x) = \cos nx$

**15.** $\lambda_n = \dfrac{n^2\pi^2}{25}, n = 1, 2, 3, \ldots;\quad y_n(x) = e^{-x}\sin\dfrac{n\pi x}{5}$

**17.** $\lambda_n = n^2, n = 1, 2, 3, \ldots;\quad y_n(x) = \sin(n\ln x)$

**19.** $\lambda_n = \dfrac{n^2\pi^2}{4}, n = 0, 1, 2, 3, \ldots;\quad y_n(x) = \cos\left(\dfrac{n\pi}{2}\ln x\right)$

**21.** $\lambda_n = n^4\pi^4, n = 1, 2, 3, \ldots;\quad y_n(x) = \sin n\pi x$

**23.** $x = L/4, x = L/2, x = 3L/4$

**27.** $\omega_n = \dfrac{n\pi\sqrt{T}}{L\sqrt{\rho}}, n = 1, 2, 3, \ldots;\quad y_n(x) = \sin\dfrac{n\pi x}{L}$

**29.** $u(r) = \left(\dfrac{u_0 - u_1}{b - a}\right)\dfrac{ab}{r} + \dfrac{u_1 b - u_0 a}{b - a}$

**31. (a)** $\lambda_n = \dfrac{n^4\pi^4}{L^4}, n = 1, 2, 3, \ldots;\quad y_n(x) = \sin\left(\dfrac{n\pi x}{L}\right)$

**(b)** $\omega_n = \dfrac{n^2\pi^2}{L^2}\sqrt{\dfrac{EI}{\rho}}, n = 1, 2, 3, \ldots$

## EXERCISE 5.3  (PAGE 228)

**7.** $\dfrac{d^2x}{dt^2} + x = 0$

**15. (a)** 5 ft      **(b)** $4\sqrt{10}$ ft/s      **(c)** $0 \le t \le \frac{3}{8}\sqrt{10}$; 7.5 ft

**17. (a)** $xy'' = r\sqrt{1 + (y')^2}$.

$\qquad$ When $t = 0, x = a, y = 0, dy/dx = 0$.

**(b)** When $r \neq 1$,

$\qquad y(x) = \dfrac{a}{2}\left[\dfrac{1}{1 + r}\left(\dfrac{x}{a}\right)^{1+r} - \dfrac{1}{1 - r}\left(\dfrac{x}{a}\right)^{1-r}\right] + \dfrac{ar}{1 - r^2}$

$\qquad$ When $r = 1$,

$\qquad y(x) = \dfrac{1}{2}\left[\dfrac{1}{2a}(x^2 - a^2) + \dfrac{1}{a}\ln\dfrac{a}{x}\right]$

**(c)** The paths intersect when $r < 1$.

**19. (a)** $\theta(t) = \omega_0\sqrt{\dfrac{l}{g}}\sin\sqrt{\dfrac{g}{l}}\,t$

**(b)** use at $\theta_{max}$, $\sin\sqrt{g/l}\,t = 1$

**(c)** use $\cos\theta_{max} \approx 1 - \frac{1}{2}\theta_{max}^2$

**(d)** $v_b \approx 21{,}797$ cm/s

## CHAPTER 5 IN REVIEW  (PAGE 232)

**1.** 8 ft      **3.** $\frac{5}{4}$ m

**5.** False; there could be an impressed force driving the system.

**7.** overdamped

**9.** $y = 0$ since $\lambda = 8$ is not an eigenvalue

**11.** 14.4 lb

**13.** $x(t) = -\frac{2}{3}e^{-2t} + \frac{1}{3}e^{-4t}$

**15.** $0 < m \le 2$

**17.** $\gamma = \frac{8}{3}\sqrt{3}$

**19.** $x(t) = e^{-4t}\left(\frac{26}{17}\cos 2\sqrt{2}\,t + \frac{28}{17}\sqrt{2}\sin 2\sqrt{2}\,t\right) + \frac{8}{17}e^{-t}$

**21. (a)** $q(t) = -\frac{1}{150}\sin 100t + \frac{1}{75}\sin 50t$

    **(b)** $i(t) = -\frac{2}{3}\cos 100t + \frac{2}{3}\cos 50t$

    **(c)** $t = \dfrac{n\pi}{50}, n = 0, 1, 2, \ldots$

**25.** $m\dfrac{d^2x}{dt^2} + kx = 0$

**27.** $mx'' + f_k\,\mathrm{sgn}(x') + kx = 0$

**31. (a)** $\theta_1(t) = \theta_0\cos\sqrt{\dfrac{g}{l}}\,t, \left[0, \dfrac{\pi}{2}\sqrt{\dfrac{l}{g}}\right]$

    **(b)** $\theta_2(t) = \frac{1}{2}\theta_0\sin 2\sqrt{\dfrac{g}{l}}\,t, \left[\dfrac{\pi}{2}\sqrt{\dfrac{l}{g}}, \pi\sqrt{\dfrac{l}{g}}\right]$

## EXERCISES 6.1 (PAGE 242)

**1.** $(-1, 1], R = 1$      **3.** $\left[-\frac{1}{2}, \frac{1}{2}\right), R = \frac{1}{2}$

**5.** $(-5, 15), R = 10$      **7.** $\left[0, \frac{2}{3}\right], R = \frac{1}{3}$

**9.** $\left(-\frac{75}{32}, \frac{75}{32}\right), R = \frac{75}{32}$      **11.** $\displaystyle\sum_{n=0}^{\infty} \dfrac{(-1)^n}{n!2^n}x^n$

**13.** $\displaystyle\sum_{n=0}^{\infty} \dfrac{(-1)^n}{2^{n+1}}x^n$      **15.** $\displaystyle\sum_{n=1}^{\infty} \dfrac{-1}{n}x^n$

**17.** $\displaystyle\sum_{n=0}^{\infty} \dfrac{(-1)^n}{(2n+1)!}(x - 2\pi)^{2n+1}$

**19.** $x - \frac{2}{3}x^3 + \frac{2}{15}x^5 - \frac{4}{315}x^7 + \cdots$

**21.** $1 + \frac{1}{2}x^2 + \frac{5}{24}x^4 + \frac{61}{720}x^6 + \cdots, (-\pi/2, \pi/2)$

**23.** $\displaystyle\sum_{k=3}^{\infty}(k - 2)c_{k-2}x^k$      **25.** $\displaystyle\sum_{k=0}^{\infty}[(k+1)c_{k+1} - c_k]x^k$

**27.** $2c_1 + \displaystyle\sum_{k=1}^{\infty}[2(k+1)c_{k+1} + 6c_{k-1}]x^k$

**29.** $c_0 + 2c_2 + \displaystyle\sum_{k=1}^{\infty}[(k+2)(k+1)c_{k+2} - (2k-1)c_k]x^k$

**35.** $y = c_0\displaystyle\sum_{k=0}^{\infty}\dfrac{1}{k!}(5x)^k$      **37.** $y = c_0\displaystyle\sum_{k=0}^{\infty}\dfrac{1}{k!}\left(\dfrac{x^2}{2}\right)^k$

## EXERCISES 6.2 (PAGE 251)

**1.** $5; 4$

**3.** $y_1(x) = c_0\left[1 - \dfrac{1}{2!}x^2 + \dfrac{1}{4!}x^4 - \dfrac{1}{6!}x^6 + \cdots\right]$

     $y_2(x) = c_1\left[x - \dfrac{1}{3!}x^3 + \dfrac{1}{5!}x^5 - \dfrac{1}{7!}x^7 + \cdots\right]$

**5.** $y_1(x) = c_0$

     $y_2(x) = c_1\left[x + \dfrac{1}{2!}x^2 + \dfrac{1}{3!}x^3 + \dfrac{1}{4!}x^4 + \cdots\right]$

**7.** $y_1(x) = c_0\left[1 - \dfrac{1}{2\cdot3}x^3 + \dfrac{1}{2\cdot3\cdot5\cdot6}x^6\right.$

         $\left. - \dfrac{1}{2\cdot3\cdot5\cdot6\cdot8\cdot9}x^9 + \cdots\right]$

     $y_2(x) = c_1\left[x - \dfrac{1}{3\cdot4}x^4 + \dfrac{1}{3\cdot4\cdot6\cdot7}x^7\right.$

         $\left. - \dfrac{1}{3\cdot4\cdot6\cdot7\cdot9\cdot10}x^{10} + \cdots\right]$

**9.** $y_1(x) = c_0\left[1 - \dfrac{1}{2!}x^2 - \dfrac{3}{4!}x^4 - \dfrac{21}{6!}x^6 - \cdots\right]$

     $y_2(x) = c_1\left[x + \dfrac{1}{3!}x^3 + \dfrac{5}{5!}x^5 + \dfrac{45}{7!}x^7 + \cdots\right]$

**11.** $y_1(x) = c_0\left[1 - \dfrac{1}{3!}x^3 + \dfrac{4^2}{6!}x^6 - \dfrac{7^2\cdot4^2}{9!}x^9 + \cdots\right]$

     $y_2(x) = c_1\left[x - \dfrac{2^2}{4!}x^4 + \dfrac{5^2\cdot2^2}{7!}x^7\right.$

         $\left. - \dfrac{8^2\cdot5^2\cdot2^2}{10!}x^{10} + \cdots\right]$

**13.** $y_1(x) = c_0; y_2(x) = c_1\displaystyle\sum_{n=1}^{\infty}\dfrac{1}{n}x^n$

**15.** $y_1(x) = c_0\left[1 + \frac{1}{2}x^2 + \frac{1}{6}x^3 + \frac{1}{6}x^4 + \cdots\right]$

     $y_2(x) = c_1\left[x + \frac{1}{2}x^2 + \frac{1}{2}x^3 + \frac{1}{4}x^4 + \cdots\right]$

**17.** $y_1(x) = c_0\left[1 + \dfrac{1}{4}x^2 - \dfrac{7}{4\cdot4!}x^4 + \dfrac{23\cdot7}{8\cdot6!}x^6 - \cdots\right]$

     $y_2(x) = c_1\left[x - \dfrac{1}{6}x^3 + \dfrac{14}{2\cdot5!}x^5 - \dfrac{34\cdot14}{4\cdot7!}x^7 - \cdots\right]$

**19.** $y(x) = -2\left[1 + \dfrac{1}{2!}x^2 + \dfrac{1}{3!}x^3 + \dfrac{1}{4!}x^4 + \cdots\right] + 6x$

     $= 8x - 2e^x$

**21.** $y(x) = 3 - 12x^2 + 4x^4$

**23.** $y_1(x) = c_0\left[1 - \frac{1}{6}x^3 + \frac{1}{120}x^5 + \cdots\right]$

     $y_2(x) = c_1\left[x - \frac{1}{12}x^4 + \frac{1}{180}x^6 + \cdots\right]$

## EXERCISES 6.3 (PAGE 260)

**1.** $x = 0$, irregular singular point

**3.** $x = -3$, regular singular point;
     $x = 3$, irregular singular point

**5.** $x = 0, 2i, -2i$, regular singular points

**7.** $x = -3, 2$, regular singular points

**9.** $x = 0$, irregular singular point;
     $x = -5, 5, 2$, regular singular points

**11.** for $x = 1$: $p(x) = 5, q(x) = \dfrac{x(x-1)^2}{x+1}$

     for $x = -1$: $p(x) = \dfrac{5(x+1)}{x-1}, q(x) = x^2 + x$

**13.** $r_1 = \frac{1}{3}, r_2 = -1$

**15.** $r_1 = \frac{3}{2}, r_2 = 0$

$$y(x) = C_1 x^{3/2}\left[1 - \frac{2}{5}x + \frac{2^2}{7 \cdot 5 \cdot 2}x^2\right.$$

$$\left. - \frac{2^3}{9 \cdot 7 \cdot 5 \cdot 3!}x^3 + \cdots\right]$$

$$+ C_2\left[1 + 2x - 2x^2 + \frac{2^3}{3 \cdot 3!}x^3 - \cdots\right]$$

**17.** $r_1 = \frac{7}{8}, r_2 = 0$

$$y(x) = c_1 x^{7/8}\left[1 - \frac{2}{15}x + \frac{2^2}{23 \cdot 15 \cdot 2}x^2\right.$$

$$\left. - \frac{2^3}{31 \cdot 23 \cdot 15 \cdot 3!}x^3 + \cdots\right]$$

$$+ c_2\left[1 - 2x + \frac{2^2}{9 \cdot 2}x^2\right.$$

$$\left. - \frac{2^3}{17 \cdot 9 \cdot 3!}x^3 + \cdots\right]$$

**19.** $r_1 = \frac{1}{3}, r_2 = 0$

$$y(x) = C_1 x^{1/3}\left[1 + \frac{1}{3}x + \frac{1}{3^2 \cdot 2}x^2\right.$$

$$\left. + \frac{1}{3^3 \cdot 3!}x^3 + \cdots\right]$$

$$+ C_2\left[1 + \frac{1}{2}x + \frac{1}{5 \cdot 2}x^2 + \frac{1}{8 \cdot 5 \cdot 2}x^3 + \cdots\right]$$

**21.** $r_1 = \frac{5}{2}, r_2 = 0$

$$y(x) = C_1 x^{5/2}\left[1 + \frac{2 \cdot 2}{7}x + \frac{2^2 \cdot 3}{9 \cdot 7}x^2\right.$$

$$\left. + \frac{2^3 \cdot 4}{11 \cdot 9 \cdot 7}x^3 + \cdots\right]$$

$$+ C_2\left[1 + \frac{1}{3}x - \frac{1}{6}x^2 - \frac{1}{6}x^3 - \cdots\right]$$

**23.** $r_1 = \frac{2}{3}, r_2 = \frac{1}{3}$

$$y(x) = C_1 x^{2/3}\left[1 - \frac{1}{2}x + \frac{5}{28}x^2 - \frac{1}{21}x^3 + \cdots\right]$$

$$+ C_2 x^{1/3}\left[1 - \frac{1}{2}x + \frac{1}{5}x^2 - \frac{7}{120}x^3 + \cdots\right]$$

**25.** $r_1 = 0, r_2 = -1$

$$y(x) = C_1 \sum_{n=0}^{\infty}\frac{1}{(2n+1)!}x^{2n} + C_2 x^{-1}\sum_{n=0}^{\infty}\frac{1}{(2n)!}x^{2n}$$

$$= C_1 x^{-1}\sum_{n=0}^{\infty}\frac{1}{(2n+1)!}x^{2n+1} + C_2 x^{-1}\sum_{n=0}^{\infty}\frac{1}{(2n)!}x^{2n}$$

$$= \frac{1}{x}[C_1 \sinh x + C_2 \cosh x]$$

**27.** $r_1 = 1, r_2 = 0$

$$y(x) = C_1 x + C_2\left[x \ln x - 1 + \frac{1}{2}x^2\right.$$

$$\left. + \frac{1}{12}x^3 + \frac{1}{72}x^4 + \cdots\right]$$

**29.** $r_1 = r_2 = 0$

$$y(x) = C_1 y(x) + C_2\left[y_1(x)\ln x + y_1(x)\left(-x + \frac{1}{4}x^2\right.\right.$$

$$\left.\left. - \frac{1}{3 \cdot 3!}x^3 + \frac{1}{4 \cdot 4!}x^4 - \cdots\right)\right]$$

where $y_1(x) = \sum_{n=0}^{\infty}\frac{1}{n!}x^n = e^x$

**33. (b)** $y_1(t) = \sum_{n=0}^{\infty}\frac{(-1)^n}{(2n+1)!}\left(\sqrt{\lambda}\, t\right)^{2n} = \frac{\sin\left(\sqrt{\lambda}\, t\right)}{\sqrt{\lambda}\, t}$

$y_2(t) = t^{-1}\sum_{n=0}^{\infty}\frac{(-1)^n}{(2n)!}\left(\sqrt{\lambda}\, t\right)^{2n} = \frac{\cos\left(\sqrt{\lambda}\, t\right)}{t}$

**(c)** $y = C_1 x \sin\left(\frac{\sqrt{\lambda}}{x}\right) + C_2 x \cos\left(\frac{\sqrt{\lambda}}{x}\right)$

### EXERCISES 6.4 (PAGE 273)

**1.** $y = c_1 J_{1/3}(x) + c_2 J_{-1/3}(x)$

**3.** $y = c_1 J_{5/2}(x) + c_2 J_{-5/2}(x)$

**5.** $y = c_1 J_0(x) + c_2 Y_0(x)$

**7.** $y = c_1 J_2(3x) + c_2 Y_2(3x)$

**9.** $y = c_1 I_{2/3}(4x) + c_2 I_{-2/3}(4x)$

**11.** $y = c_1 x^{-1/2}J_{1/2}(\alpha x) + c_2 x^{-1/2}J_{-1/2}(\alpha x)$

**13.** $y = x^{-1/2}\left[c_1 J_1(4x^{1/2}) + c_2 Y_1(4x^{1/2})\right]$

**15.** $y = x\left[c_1 J_1(x) + c_2 Y_1(x)\right]$

**17.** $y = x^{1/2}\left[c_1 J_{3/2}(x) + c_2 Y_{3/2}(x)\right]$

**19.** $y = x^{-1}\left[c_1 J_{1/2}\left(\frac{1}{2}x^2\right) + c_2 J_{-1/2}\left(\frac{1}{2}x^2\right)\right]$

**23.** $y = x^{1/2}\left[c_1 J_{1/2}(x) + c_2 J_{-1/2}(x)\right]$

$= C_1 \sin x + C_2 \cos x$

**25.** $y = x^{-1/2}\left[c_1 J_{1/2}\left(\frac{1}{8}x^2\right) + c_2 J_{-1/2}\left(\frac{1}{8}x^2\right)\right]$

$= C_1 x^{-3/2}\sin\left(\frac{1}{8}x^2\right) + C_2 x^{-3/2}\cos\left(\frac{1}{8}x^2\right)$

**33. (a)** $j_1(x) = \frac{\sin x}{x^2} - \frac{\cos x}{x}$, $j_2(x) = \left(\frac{3}{x^3} - \frac{1}{x}\right)\sin x - \frac{3\cos x}{x^2}$

**37.** $y = c_1 x^{1/2}J_{1/3}\left(\frac{2}{3}\alpha x^{3/2}\right) + c_2 x^{1/2}J_{-1/3}\left(\frac{2}{3}\alpha x^{3/2}\right)$,

$y = c_1 x^{1/2}I_{1/3}\left(\frac{2}{3}\alpha x^{3/2}\right) + c_2 x^{1/2}I_{-1/3}\left(\frac{2}{3}\alpha x^{3/2}\right)$

**47.** $P_2(x), P_3(x), P_4(x)$, and $P_5(x)$ are given in the text,

$P_6(x) = \frac{1}{16}(231x^6 - 315x^4 + 105x^2 - 5)$,

$P_7(x) = \frac{1}{16}(429x^7 - 693x^5 + 315x^3 - 35x)$

**49.** $\lambda_1 = 2, \lambda_2 = 12, \lambda_3 = 30$

**55.** $y = x - 4x^3 + \frac{16}{5}x^5$

### CHAPTER 6 IN REVIEW (PAGE 276)

**1.** False

**3.** $\left[-\frac{1}{2}, \frac{1}{2}\right]$

**7.** $x^2(x-1)y'' + y' + y = 0$

**9.** $r_1 = \frac{1}{2}, r_2 = 0$

$$y_1(x) = C_1 x^{1/2}\left[1 - \frac{1}{3}x + \frac{1}{30}x^2 - \frac{1}{630}x^3 + \cdots\right]$$

$$y_2(x) = C_2\left[1 - x + \frac{1}{6}x^2 - \frac{1}{90}x^3 + \cdots\right]$$

**11.** $y_1(x) = c_0\left[1 + \frac{3}{2}x^2 + \frac{1}{2}x^3 + \frac{5}{8}x^4 + \cdots\right]$

$$y_2(x) = c_1\left[x + \frac{1}{2}x^3 + \frac{1}{4}x^4 + \cdots\right]$$

**13.** $r_1 = 3, r_2 = 0$

$$y_1(x) = C_1 x^3\left[1 + \frac{1}{4}x + \frac{1}{20}x^2 + \frac{1}{120}x^3 + \cdots\right]$$

$$y_2(x) = C_2\left[1 + x + \frac{1}{2}x^2\right]$$

**15.** $y(x) = 3\left[1 - x^2 + \frac{1}{3}x^4 - \frac{1}{15}x^6 + \cdots\right]$

$$-2\left[x - \frac{1}{2}x^3 + \frac{1}{8}x^5 - \frac{1}{48}x^7 + \cdots\right]$$

**17.** $\frac{1}{6}\pi$

**19.** $x = 0$ is an ordinary point

**21.** $y(x) = c_0\left[1 - \frac{1}{3}x^3 + \frac{1}{3^2\cdot 2!}x^6 - \frac{1}{3^3\cdot 3!}x^9 + \cdots\right]$

$$+ c_1\left[x - \frac{1}{4}x^4 + \frac{1}{4\cdot 7}x^7\right.$$

$$\left. - \frac{1}{4\cdot 7\cdot 10}x^{10} + \cdots\right] + \left[\frac{5}{2}x^2 - \frac{1}{3}x^3\right.$$

$$\left. + \frac{1}{3^2\cdot 2!}x^6 - \frac{1}{3^3\cdot 3!}x^9 + \cdots\right]$$

## EXERCISES 7.1 (PAGE 285)

**1.** $\frac{2}{s}e^{-s} - \frac{1}{s}$

**3.** $\frac{1}{s^2} - \frac{1}{s^2}e^{-s}$

**5.** $\frac{1 + e^{-\pi s}}{s^2 + 1}$

**7.** $\frac{1}{s}e^{-s} + \frac{1}{s^2}e^{-s}$

**9.** $\frac{1}{s} - \frac{1}{s^2} + \frac{1}{s^2}e^{-s}$

**11.** $\frac{e^7}{s - 1}$

**13.** $\frac{1}{(s-4)^2}$

**15.** $\frac{1}{s^2 + 2s + 2}$

**17.** $\frac{s^2 - 1}{(s^2 + 1)^2}$

**19.** $\frac{48}{s^5}$

**21.** $\frac{4}{s^2} - \frac{10}{s}$

**23.** $\frac{2}{s^3} + \frac{6}{s^2} - \frac{3}{s}$

**25.** $\frac{6}{s^4} + \frac{6}{s^3} + \frac{3}{s^2} + \frac{1}{s}$

**27.** $\frac{1}{s} + \frac{1}{s - 4}$

**29.** $\frac{1}{s} + \frac{2}{s - 2} + \frac{1}{s - 4}$

**31.** $\frac{8}{s^3} - \frac{15}{s^2 + 9}$

**33.** Use $\sinh kt = \dfrac{e^{kt} - e^{-kt}}{2}$ and linearity to show that

$$\mathcal{L}\{\sinh kt\} = \frac{k}{s^2 - k^2}.$$

**35.** $\frac{1}{2(s - 2)} - \frac{1}{2s}$

**37.** $\frac{2}{s^2 + 16}$

**39.** $\frac{4\cos 5 + (\sin 5)s}{s^2 + 16}$

**43.** $\frac{\sqrt{\pi}}{s^{1/2}}$

**45.** $\frac{3\sqrt{\pi}}{4s^{5/2}}$

## EXERCISES 7.2 (PAGE 293)

**1.** $\frac{1}{2}t^2$

**3.** $t - 2t^4$

**5.** $1 + 3t + \frac{3}{2}t^2 + \frac{1}{6}t^3$

**7.** $t - 1 + e^{2t}$

**9.** $\frac{1}{4}e^{-t/4}$

**11.** $\frac{5}{7}\sin 7t$

**13.** $\cos\dfrac{t}{2}$

**15.** $2\cos 3t - 2\sin 3t$

**17.** $\frac{1}{3} - \frac{1}{3}e^{-3t}$

**19.** $\frac{3}{4}e^{-3t} + \frac{1}{4}e^t$

**21.** $0.3e^{0.1t} + 0.6e^{-0.2t}$

**23.** $\frac{1}{2}e^{2t} - e^{3t} + \frac{1}{2}e^{6t}$

**25.** $\frac{1}{5} - \frac{1}{5}\cos\sqrt{5}\,t$

**27.** $-4 + 3e^{-t} + \cos t + 3\sin t$

**29.** $\frac{1}{3}\sin t - \frac{1}{6}\sin 2t$

**35.** $y = -1 + e^t$

**37.** $y = \frac{1}{10}e^{4t} + \frac{19}{10}e^{-6t}$

**39.** $y = \frac{4}{3}e^{-t} - \frac{1}{3}e^{-4t}$

**41.** $y = 10\cos t + 2\sin t - \sqrt{2}\sin\sqrt{2}\,t$

**43.** $y = -\frac{8}{9}e^{-t/2} + \frac{1}{9}e^{-2t} + \frac{5}{18}e^t + \frac{1}{2}e^{-t}$

**45.** $y = \frac{1}{4}e^{-t} - \frac{1}{4}e^{-3t}\cos 2t + \frac{1}{4}e^{-3t}\sin 2t$

**47.** $y = \frac{10}{21}\cos 2t - \frac{10}{21}\cos 5t$

## EXERCISES 7.3 (PAGE 303)

**1.** $\frac{1}{(s - 10)^2}$

**3.** $\frac{6}{(s + 2)^4}$

**5.** $\frac{1}{(s - 2)^2} + \frac{2}{(s - 3)^2} + \frac{1}{(s - 4)^2}$

**7.** $\frac{3}{(s - 1)^2 + 9}$

**9.** $\frac{s}{s^2 + 25} - \frac{s - 1}{(s - 1)^2 + 25} + 3\frac{s + 4}{(s + 4)^2 + 25}$

**11.** $\frac{1}{2}t^2 e^{-2t}$

**13.** $e^{3t}\sin t$

**15.** $e^{-2t}\cos t - 2e^{-2t}\sin t$

**17.** $e^{-t} - te^{-t}$

**19.** $5 - t - 5e^{-t} - 4te^{-t} - \frac{3}{2}t^2 e^{-t}$

**21.** $y = te^{-4t} + 2e^{-4t}$

**23.** $y = e^{-t} + 2te^{-t}$

**25.** $y = \frac{1}{9}t + \frac{2}{27} - \frac{2}{27}e^{3t} + \frac{10}{9}te^{3t}$

**27.** $y = -\frac{3}{2}e^{3t}\sin 2t$

**29.** $y = \frac{1}{2} - \frac{1}{2}e^t\cos t + \frac{1}{2}e^t\sin t$

**31.** $y = (e + 1)te^{-t} + (e - 1)e^{-t}$

**33.** $x(t) = -\frac{3}{2}e^{-7t/2}\cos\frac{\sqrt{15}}{2}t - \frac{7\sqrt{15}}{10}e^{-7t/2}\sin\frac{\sqrt{15}}{2}t$

**37.** $\frac{e^{-s}}{s^2}$

**39.** $\frac{e^{-2s}}{s^2} + 2\frac{e^{-2s}}{s}$

**41.** $\frac{s}{s^2 + 4}e^{-\pi s}$

**43.** $\frac{1}{2}(t - 2)^2\,\mathcal{U}(t - 2)$

**45.** $-\sin t\,\mathcal{U}(t - \pi)$

**47.** $\mathcal{U}(t - 1) - e^{-(t-1)}\mathcal{U}(t - 1)$

**49.** (c)

**51.** (f)

**53.** (a)

**55.** $f(t) = 2 - 4\mathcal{U}(t - 3); \mathcal{L}\{f(t)\} = \dfrac{2}{s} - \dfrac{4}{s}e^{-3s}$

**57.** $f(t) = t^2\mathcal{U}(t - 1); \mathcal{L}\{f(t)\} = 2\dfrac{e^{-s}}{s^3} + 2\dfrac{e^{-s}}{s^2} + \dfrac{e^{-s}}{s}$

**59.** $f(t) = t - t\mathcal{U}(t - 2); \mathcal{L}\{f(t)\} = \dfrac{1}{s^2} - \dfrac{e^{-2s}}{s^2} - 2\dfrac{e^{-2s}}{s}$

**61.** $f(t) = \mathcal{U}(t - a) - \mathcal{U}(t - b); \mathcal{L}\{f(t)\} = \dfrac{e^{-as}}{s} - \dfrac{e^{-bs}}{s}$

**63.** $y = [5 - 5e^{-(t-1)}]\,\mathcal{U}(t - 1)$

**65.** $y = -\frac{1}{4} + \frac{1}{2}t + \frac{1}{4}e^{-2t} - \frac{1}{4}\mathcal{U}(t - 1)$

$$-\frac{1}{2}(t - 1)\mathcal{U}(t - 1) + \frac{1}{4}e^{-2(t-1)}\mathcal{U}(t - 1)$$

**67.** $y = \cos 2t - \frac{1}{6} \sin 2(t - 2\pi)\, \mathcal{U}(t - 2\pi)$
$+ \frac{1}{3} \sin(t - 2\pi)\, \mathcal{U}(t - 2\pi)$

**69.** $y = \sin t + [1 - \cos(t - \pi)]\mathcal{U}(t - \pi)$
$- [1 - \cos(t - 2\pi)]\, \mathcal{U}(t - 2\pi)$

**71.** $x(t) = \frac{5}{4} t - \frac{5}{16} \sin 4t - \frac{5}{4}(t - 5)\, \mathcal{U}(t - 5)$
$+ \frac{5}{16} \sin 4(t - 5)\, \mathcal{U}(t - 5) - \frac{25}{4} \mathcal{U}(t - 5)$
$+ \frac{25}{4} \cos 4(t - 5)\, \mathcal{U}(t - 5)$

**73.** $q(t) = \frac{2}{5} \mathcal{U}(t - 3) - \frac{2}{5} e^{-5(t-3)} \mathcal{U}(t - 3)$

**75. (a)** $i(t) = \dfrac{1}{101} e^{-10t} - \dfrac{1}{101} \cos t + \dfrac{10}{101} \sin t$

$- \dfrac{10}{101} e^{-10(t-3\pi/2)} \mathcal{U}\!\left(t - \dfrac{3\pi}{2}\right)$

$+ \dfrac{10}{101} \cos\!\left(t - \dfrac{3\pi}{2}\right) \mathcal{U}\!\left(t - \dfrac{3\pi}{2}\right)$

$+ \dfrac{1}{101} \sin\!\left(t - \dfrac{3\pi}{2}\right) \mathcal{U}\!\left(t - \dfrac{3\pi}{2}\right)$

**(b)** $i_{\max} \approx 0.1$ at $t \approx 1.7$, $i_{\min} \approx -0.1$ at $t \approx 4.7$

**77.** $y(x) = \dfrac{w_0 L^2}{16EI} x^2 - \dfrac{w_0 L}{12EI} x^3 + \dfrac{w_0}{24EI} x^4$

$- \dfrac{w_0}{24EI}\left(x - \dfrac{L}{2}\right)^4 \mathcal{U}\!\left(x - \dfrac{L}{2}\right)$

**79.** $y(x) = \dfrac{w_0 L^2}{48EI} x^2 - \dfrac{w_0 L}{24EI} x^3$

$+ \dfrac{w_0}{60EIL}\left[\dfrac{5L}{2} x^4 - x^5 + \left(x - \dfrac{L}{2}\right)^5 \mathcal{U}\!\left(x - \dfrac{L}{2}\right)\right]$

**81. (a)** $\dfrac{dT}{dt} = k[T - 70 - 57.5t - (230 - 57.5t)\mathcal{U}(t - 4)]$

## EXERCISES 7.4  (PAGE 315)

**1.** $\dfrac{1}{(s + 10)^2}$     **3.** $\dfrac{s^2 - 4}{(s^2 + 4)^2}$

**5.** $\dfrac{6s^2 + 2}{(s^2 - 1)^3}$     **7.** $\dfrac{12s - 24}{[(s - 2)^2 + 36]^2}$

**9.** $y = -\frac{1}{2} e^{-t} + \frac{1}{2} \cos t - \frac{1}{2} t \cos t + \frac{1}{2} t \sin t$

**11.** $y = 2 \cos 3t + \frac{5}{3} \sin 3t + \frac{1}{6} t \sin 3t$

**13.** $y = \frac{1}{4} \sin 4t + \frac{1}{8} t \sin 4t$
$- \frac{1}{8}(t - \pi) \sin 4(t - \pi)\mathcal{U}(t - \pi)$

**17.** $y = \frac{2}{3} t^3 + c_1 t^2$

**19.** $\dfrac{24}{s^5}$

**21.** $\dfrac{1}{s^2 - 1}$

**23.** $\dfrac{6}{s^5}$

**25.** $\dfrac{s - 1}{(s + 1)[(s - 1)^2 + 1]}$

**27.** $\dfrac{1}{s(s - 1)}$

**29.** $\dfrac{s + 1}{s[(s + 1)^2 + 1]}$

**31.** $\dfrac{1}{s^2(s - 1)}$

**33.** $\dfrac{3s^2 + 1}{s^2(s^2 + 1)^2}$

**35.** $e^t - 1$

**37.** $e^t - \frac{1}{2} t^2 - t - 1$

**41.** $f(t) = \sin t$

**43.** $f(t) = -\frac{1}{8} e^{-t} + \frac{1}{8} e^t + \frac{3}{4} t e^t + \frac{1}{4} t^2 e^t$     **45.** $f(t) = e^{-t}$

**47.** $f(t) = \frac{3}{8} e^{2t} + \frac{1}{8} e^{-2t} + \frac{1}{2} \cos 2t + \frac{1}{4} \sin 2t$

**49.** $y(t) = \sin t - \frac{1}{2} t \sin t$

**51.** $i(t) = 100[e^{-10(t-1)} - e^{-20(t-1)}]\, \mathcal{U}(t - 1)$
$- 100[e^{-10(t-2)} - e^{-20(t-2)}]\mathcal{U}(t - 2)$

**53.** $\dfrac{1 - e^{-as}}{s(1 + e^{-as})}$     **55.** $\dfrac{a}{s}\left(\dfrac{1}{bs} - \dfrac{1}{e^{bs} - 1}\right)$

**57.** $\dfrac{\coth(\pi s/2)}{s^2 + 1}$

**59.** $i(t) = \dfrac{1}{R}\left(1 - e^{-Rt/L}\right)$

$+ \dfrac{2}{R} \sum_{n=1}^{\infty} (-1)^n (1 - e^{-R(t-n)/L}) \mathcal{U}(t - n)$

**61.** $x(t) = 2(1 - e^{-t} \cos 3t - \frac{1}{3} e^{-t} \sin 3t)$

$+ 4 \sum_{n=1}^{\infty} (-1)^n \Big[1 - e^{-(t-n\pi)} \cos 3(t - n\pi)$

$- \frac{1}{3} e^{-(t-n\pi)} \sin 3(t - n\pi)\Big]\mathcal{U}(t - n\pi)$

## EXERCISES 7.5  (PAGE 321)

**1.** $y = e^{3(t-2)} \mathcal{U}(t - 2)$

**3.** $y = \sin t + \sin t\, \mathcal{U}(t - 2\pi)$

**5.** $y = -\cos t\, \mathcal{U}\!\left(t - \frac{\pi}{2}\right) + \cos t\, \mathcal{U}\!\left(t - \frac{3\pi}{2}\right)$

**7.** $y = \frac{1}{2} - \frac{1}{2} e^{-2t} + \left[\frac{1}{2} - \frac{1}{2} e^{-2(t-1)}\right] \mathcal{U}(t - 1)$

**9.** $y = e^{-2(t-2\pi)} \sin t\, \mathcal{U}(t - 2\pi)$

**11.** $y = e^{-2t} \cos 3t + \frac{2}{3} e^{-2t} \sin 3t$
$+ \frac{1}{3} e^{-2(t-\pi)} \sin 3(t - \pi)\, \mathcal{U}(t - \pi)$
$+ \frac{1}{3} e^{-2(t-3\pi)} \sin 3(t - 3\pi)\, \mathcal{U}(t - 3\pi)$

**13.** $y(t) = \sin t + \sin t \displaystyle\sum_{k=1}^{\infty} (-1)^k \mathcal{U}(t - k\pi)$

**15.** $y(x) = \begin{cases} \dfrac{w_0}{EI}\left(\dfrac{L}{4} x^2 - \dfrac{1}{6} x^3\right), & 0 \le x < \dfrac{L}{2} \\[2mm] \dfrac{w_0 L^2}{4EI}\left(\dfrac{1}{2} x - \dfrac{L}{12}\right), & \dfrac{L}{2} \le x \le L \end{cases}$

## EXERCISES 7.6  (PAGE 325)

**1.** $x = -\frac{1}{3} e^{-2t} + \frac{1}{3} e^t$     **3.** $x = -\cos 3t - \frac{5}{3} \sin 3t$
$y = \frac{1}{3} e^{-2t} + \frac{2}{3} e^t$          $y = 2 \cos 3t - \frac{7}{3} \sin 3t$

**5.** $x = -2e^{3t} + \frac{5}{2} e^{2t} - \frac{1}{2}$     **7.** $x = -\frac{1}{2} t - \frac{3}{4} \sqrt{2} \sin \sqrt{2} t$
$y = \frac{8}{3} e^{3t} - \frac{5}{2} e^{2t} - \frac{1}{6}$     $y = -\frac{1}{2} t + \frac{3}{4} \sqrt{2} \sin \sqrt{2} t$

**9.** $x = 8 + \dfrac{2}{3!} t^3 + \dfrac{1}{4!} t^4$

$y = -\dfrac{2}{3!} t^3 + \dfrac{1}{4!} t^4$

**11.** $x = \frac{1}{2} t^2 + t + 1 - e^{-t}$
$y = -\frac{1}{3} + \frac{1}{3} e^{-t} + \frac{1}{3} t e^{-t}$

**13.** $x_1 = \dfrac{1}{5} \sin t + \dfrac{2\sqrt{6}}{15} \sin \sqrt{6} t + \dfrac{2}{5} \cos t - \dfrac{2}{5} \cos \sqrt{6} t$

$x_2 = \dfrac{2}{5} \sin t - \dfrac{\sqrt{6}}{15} \sin \sqrt{6} t + \dfrac{4}{5} \cos t + \dfrac{1}{5} \cos \sqrt{6} t$

**15. (b)** $i_2 = \frac{100}{9} - \frac{100}{9} e^{-900t}$
$i_3 = \frac{80}{9} - \frac{80}{9} e^{-900t}$
**(c)** $i_1 = 20 - 20e^{-900t}$

**17.** $i_2 = -\frac{20}{13} e^{-2t} + \frac{375}{1469} e^{-15t} + \frac{145}{113} \cos t + \frac{85}{113} \sin t$

$i_3 = \frac{30}{13} e^{-2t} + \frac{250}{1469} e^{-15t} - \frac{280}{113} \cos t + \frac{810}{113} \sin t$

**19.** $i_1 = \frac{6}{5} - \frac{6}{5} e^{-100t} \cosh 50\sqrt{2}\,t - \frac{9\sqrt{2}}{10} e^{-100t} \sinh 50\sqrt{2}\,t$

$i_2 = \frac{6}{5} - \frac{6}{5} e^{-100t} \cosh 50\sqrt{2}\,t - \frac{6\sqrt{2}}{5} e^{-100t} \sinh 50\sqrt{2}\,t$

## CHAPTER 7 IN REVIEW  (PAGE 327)

**1.** $\dfrac{1}{s^2} - \dfrac{2}{s^2} e^{-s}$  **3.** false

**5.** true  **7.** $\dfrac{1}{s+7}$

**9.** $\dfrac{2}{s^2+4}$  **11.** $\dfrac{4s}{(s^2+4)^2}$

**13.** $\frac{1}{6} t^5$  **15.** $\frac{1}{2} t^2 e^{5t}$

**17.** $e^{5t} \cos 2t + \frac{5}{2} e^{5t} \sin 2t$

**19.** $\cos \pi(t-1)\,\mathcal{U}(t-1) + \sin \pi(t-1)\,\mathcal{U}(t-1)$

**21.** $-5$

**23.** $e^{-k(s-a)} F(s-a)$

**25.** $f(t)\mathcal{U}(t-t_0)$

**27.** $f(t-t_0)\mathcal{U}(t-t_0)$

**29.** $f(t) = t - (t-1)\mathcal{U}(t-1) - \mathcal{U}(t-4)$;

$\mathcal{L}\{f(t)\} = \dfrac{1}{s^2} - \dfrac{1}{s^2} e^{-s} - \dfrac{1}{s} e^{-4s}$;

$\mathcal{L}\{e^t f(t)\} = \dfrac{1}{(s-1)^2} - \dfrac{1}{(s-1)^2} e^{-(s-1)}$

$- \dfrac{1}{s-1} e^{-4(s-1)}$

**31.** $f(t) = 2 + (t-2)\,\mathcal{U}(t-2)$;

$\mathcal{L}\{f(t)\} = \dfrac{2}{s} + \dfrac{1}{s^2} e^{-2s}$;

$\mathcal{L}\{e^t f(t)\} = \dfrac{2}{s-1} + \dfrac{1}{(s-1)^2} e^{-2(s-1)}$

**33.** $\dfrac{e^{-s} - 1}{s(1 + e^{-s})}$

**35.** $y = 5te^t + \frac{1}{2} t^2 e^t$

**37.** $y = -\frac{6}{25} + \frac{1}{5} t + \frac{3}{2} e^{-t} - \frac{13}{50} e^{-5t} - \frac{4}{25}\mathcal{U}(t-2)$

$- \frac{1}{5}(t-2)\,\mathcal{U}(t-2) + \frac{1}{4} e^{-(t-2)}\,\mathcal{U}(t-2)$

$- \frac{9}{100} e^{-5(t-2)}\mathcal{U}(t-2)$

**39.** $y(t) = e^{-2t} + \left[-\frac{1}{4} + \frac{1}{2}(t-1) + \frac{1}{4} e^{-2(t-1)}\right]\mathcal{U}(t-1)$

$- 2\left[-\frac{1}{4} + \frac{1}{2}(t-2) + \frac{1}{4} e^{-2(t-2)}\right]\mathcal{U}(t-2)$

$+ \left[-\frac{1}{4} + \frac{1}{2}(t-3) + \frac{1}{4} e^{-2(t-3)}\right]\mathcal{U}(t-3)$

**41.** $y = 1 + t + \frac{1}{2} t^2$

**43.** $x = -\frac{1}{4} + \frac{9}{8} e^{-2t} + \frac{1}{8} e^{2t}$

$y = t + \frac{9}{4} e^{-2t} - \frac{1}{4} e^{2t}$

**45.** $i(t) = -9 + 2t + 9e^{-t/5}$

**47.** $y(x) = \dfrac{w_0}{12EIL}\left[-\dfrac{1}{5} x^5 + \dfrac{L}{2} x^4 - \dfrac{L^2}{2} x^3 + \dfrac{L^3}{4} x^2 \right.$

$\left. + \dfrac{1}{5}\left(x - \dfrac{L}{2}\right)^5 \mathcal{U}\left(x - \dfrac{L}{2}\right)\right]$

**49. (a)** $\theta_1(t) = \dfrac{\theta_0 + \psi_0}{2} \cos \omega t + \dfrac{\theta_0 - \psi_0}{2} \cos \sqrt{\omega^2 + 2K}\,t$

$\theta_2(t) = \dfrac{\theta_0 + \psi_0}{2} \cos \omega t - \dfrac{\theta_0 - \psi_0}{2} \cos \sqrt{\omega^2 + 2K}\,t$

**51. (a)** $x(t) = (v_0 \cos \theta)t,\quad y(t) = -\frac{1}{2} gt^2 + (v_0 \sin \theta)t$

**(b)** $y(x) = -\dfrac{g}{2v_0^2 \cos^2 \theta} x^2 + \dfrac{\sin \theta}{\cos \theta} x;\quad$ solve $y(x) = 0$

and use the double-angle formula for $\sin 2\theta$

**(d)** approximately 2729 ft; approximately 11.54 s

## EXERCISES 8.1  (PAGE 339)

**1.** $\mathbf{X}' = \begin{pmatrix} 3 & -5 \\ 4 & 8 \end{pmatrix}\mathbf{X}$, where $\mathbf{X} = \begin{pmatrix} x \\ y \end{pmatrix}$

**3.** $\mathbf{X}' = \begin{pmatrix} -3 & 4 & -9 \\ 6 & -1 & 0 \\ 10 & 4 & 3 \end{pmatrix}\mathbf{X}$, where $\mathbf{X} = \begin{pmatrix} x \\ y \\ z \end{pmatrix}$

**5.** $\mathbf{X}' = \begin{pmatrix} 1 & -1 & 1 \\ 2 & 1 & -1 \\ 1 & 1 & 1 \end{pmatrix}\mathbf{X} + \begin{pmatrix} 0 \\ -3t^2 \\ t^2 \end{pmatrix} + \begin{pmatrix} t \\ 0 \\ -t \end{pmatrix} + \begin{pmatrix} -1 \\ 0 \\ 2 \end{pmatrix}$,

where $\mathbf{X} = \begin{pmatrix} x \\ y \\ z \end{pmatrix}$

**7.** $\dfrac{dx}{dt} = 4x + 2y + e^t$

$\dfrac{dy}{dt} = -x + 3y - e^t$

**9.** $\dfrac{dx}{dt} = x - y + 2z + e^{-t} - 3t$

$\dfrac{dy}{dt} = 3x - 4y + z + 2e^{-t} + t$

$\dfrac{dz}{dt} = -2x + 5y + 6z + 2e^{-t} - t$

**17.** Yes; $W(\mathbf{X}_1, \mathbf{X}_2) = -2e^{-8t} \neq 0$ implies that $\mathbf{X}_1$ and $\mathbf{X}_2$ are linearly independent on $(-\infty, \infty)$.

**19.** No; $W(\mathbf{X}_1, \mathbf{X}_2, \mathbf{X}_3) = 0$ for every $t$. The solution vectors are linearly dependent on $(-\infty, \infty)$. Note that $\mathbf{X}_3 = 2\mathbf{X}_1 + \mathbf{X}_2$.

## EXERCISES 8.2  (PAGE 352)

**1.** $\mathbf{X} = c_1 \begin{pmatrix} 1 \\ 2 \end{pmatrix} e^{5t} + c_2 \begin{pmatrix} -1 \\ 1 \end{pmatrix} e^{-t}$

**3.** $\mathbf{X} = c_1 \begin{pmatrix} 2 \\ 1 \end{pmatrix} e^{-3t} + c_2 \begin{pmatrix} 2 \\ 5 \end{pmatrix} e^t$

**5.** $\mathbf{X} = c_1 \begin{pmatrix} 5 \\ 2 \end{pmatrix} e^{8t} + c_2 \begin{pmatrix} 1 \\ 4 \end{pmatrix} e^{-10t}$

**7.** $\mathbf{X} = c_1 \begin{pmatrix} 1 \\ 0 \\ 0 \end{pmatrix} e^t + c_2 \begin{pmatrix} 2 \\ 3 \\ 1 \end{pmatrix} e^{2t} + c_3 \begin{pmatrix} 1 \\ 0 \\ 2 \end{pmatrix} e^{-t}$

**9.** $\mathbf{X} = c_1 \begin{pmatrix} -1 \\ 0 \\ 1 \end{pmatrix} e^{-t} + c_2 \begin{pmatrix} 1 \\ 4 \\ 3 \end{pmatrix} e^{3t} + c_3 \begin{pmatrix} 1 \\ -1 \\ 3 \end{pmatrix} e^{-2t}$

**11.** $X = c_1 \begin{pmatrix} 4 \\ 0 \\ -1 \end{pmatrix} e^{-t} + c_2 \begin{pmatrix} -12 \\ 6 \\ 5 \end{pmatrix} e^{-t/2} + c_3 \begin{pmatrix} 4 \\ 2 \\ -1 \end{pmatrix} e^{-3t/2}$

**13.** $X = 3 \begin{pmatrix} 1 \\ 1 \end{pmatrix} e^{t/2} + 2 \begin{pmatrix} 0 \\ 1 \end{pmatrix} e^{-t/2}$

**15. (a)** $X' = \begin{pmatrix} -\frac{3}{100} & \frac{1}{100} \\ \frac{1}{50} & -\frac{1}{50} \end{pmatrix} X$

**(b)** $X(t) = -\frac{35}{3} \begin{pmatrix} -1 \\ 1 \end{pmatrix} e^{-t/25} + \frac{25}{3} \begin{pmatrix} 1 \\ 2 \end{pmatrix} e^{-t/100}$

**(d)** approximately 34.3 minutes

**21.** $X = c_1 \begin{pmatrix} 1 \\ 3 \end{pmatrix} + c_2 \left[ \begin{pmatrix} 1 \\ 3 \end{pmatrix} t + \begin{pmatrix} \frac{1}{4} \\ -\frac{1}{4} \end{pmatrix} \right]$

**23.** $X = c_1 \begin{pmatrix} 1 \\ 1 \end{pmatrix} e^{2t} + c_2 \left[ \begin{pmatrix} 1 \\ 1 \end{pmatrix} t e^{2t} + \begin{pmatrix} -\frac{1}{3} \\ 0 \end{pmatrix} e^{2t} \right]$

**25.** $X = c_1 \begin{pmatrix} 1 \\ 1 \\ 1 \end{pmatrix} e^{t} + c_2 \begin{pmatrix} 1 \\ 1 \\ 0 \end{pmatrix} e^{2t} + c_3 \begin{pmatrix} 1 \\ 0 \\ 1 \end{pmatrix} e^{2t}$

**27.** $X = c_1 \begin{pmatrix} -4 \\ -5 \\ 2 \end{pmatrix} + c_2 \begin{pmatrix} 2 \\ 0 \\ -1 \end{pmatrix} e^{5t}$
$+ c_3 \left[ \begin{pmatrix} 2 \\ 0 \\ -1 \end{pmatrix} t e^{5t} + \begin{pmatrix} -\frac{1}{2} \\ -\frac{1}{2} \\ -1 \end{pmatrix} e^{5t} \right]$

**29.** $X = c_1 \begin{pmatrix} 0 \\ 1 \\ 1 \end{pmatrix} e^{t} + c_2 \left[ \begin{pmatrix} 0 \\ 1 \\ 1 \end{pmatrix} t e^{t} + \begin{pmatrix} 0 \\ 1 \\ 0 \end{pmatrix} e^{t} \right]$
$+ c_3 \left[ \begin{pmatrix} 0 \\ 1 \\ 1 \end{pmatrix} \frac{t^2}{2} e^{t} + \begin{pmatrix} 0 \\ 1 \\ 0 \end{pmatrix} t e^{t} + \begin{pmatrix} \frac{1}{2} \\ 0 \\ 0 \end{pmatrix} e^{t} \right]$

**31.** $X = -7 \begin{pmatrix} 2 \\ 1 \end{pmatrix} e^{4t} + 13 \begin{pmatrix} 2t + 1 \\ t + 1 \end{pmatrix} e^{4t}$

**33.** Corresponding to the eigenvalue $\lambda_1 = 2$ of multiplicity five, the eigenvectors are

$$K_1 = \begin{pmatrix} 1 \\ 0 \\ 0 \\ 0 \\ 0 \end{pmatrix}, \quad K_2 = \begin{pmatrix} 0 \\ 0 \\ 1 \\ 0 \\ 0 \end{pmatrix}, \quad K_3 = \begin{pmatrix} 0 \\ 0 \\ 0 \\ 1 \\ 0 \end{pmatrix}.$$

**35.** $X = c_1 \begin{pmatrix} \cos t \\ 2\cos t + \sin t \end{pmatrix} e^{4t} + c_2 \begin{pmatrix} \sin t \\ 2\sin t - \cos t \end{pmatrix} e^{4t}$

**37.** $X = c_1 \begin{pmatrix} \cos t \\ -\cos t - \sin t \end{pmatrix} e^{4t} + c_2 \begin{pmatrix} \sin t \\ -\sin t + \cos t \end{pmatrix} e^{4t}$

**39.** $X = c_1 \begin{pmatrix} 5\cos 3t \\ 4\cos 3t + 3\sin 3t \end{pmatrix} + c_2 \begin{pmatrix} 5\sin 3t \\ 4\sin 3t - 3\cos 3t \end{pmatrix}$

**41.** $X = c_1 \begin{pmatrix} 1 \\ 0 \\ 0 \end{pmatrix} + c_2 \begin{pmatrix} -\cos t \\ \cos t \\ \sin t \end{pmatrix} + c_3 \begin{pmatrix} \sin t \\ -\sin t \\ \cos t \end{pmatrix}$

**43.** $X = c_1 \begin{pmatrix} 0 \\ 2 \\ 1 \end{pmatrix} e^{t} + c_2 \begin{pmatrix} \sin t \\ \cos t \\ \cos t \end{pmatrix} e^{t} + c_3 \begin{pmatrix} \cos t \\ -\sin t \\ -\sin t \end{pmatrix} e^{t}$

**45.** $X = c_1 \begin{pmatrix} 28 \\ -5 \\ 25 \end{pmatrix} e^{2t} + c_2 \begin{pmatrix} 4\cos 3t - 3\sin 3t \\ -5\cos 3t \\ 0 \end{pmatrix} e^{-2t}$
$+ c_3 \begin{pmatrix} 3\cos 3t + 4\sin 3t \\ -5\sin 3t \\ 0 \end{pmatrix} e^{-2t}$

**47.** $X = - \begin{pmatrix} 25 \\ -7 \\ 6 \end{pmatrix} e^{t} - \begin{pmatrix} \cos 5t - 5\sin 5t \\ \cos 5t \\ \cos 5t \end{pmatrix}$
$+ 6 \begin{pmatrix} 5\cos 5t + \sin 5t \\ \sin 5t \\ \sin 5t \end{pmatrix}$

**49. (a)** $X' = \begin{pmatrix} -\frac{1}{20} & 0 & \frac{1}{10} \\ \frac{1}{20} & -\frac{1}{20} & 0 \\ 0 & \frac{1}{20} & -\frac{1}{10} \end{pmatrix} X$

**(b)** $X(t) = -6 \begin{pmatrix} -\cos\frac{1}{20}t + \sin\frac{1}{20}t \\ -\sin\frac{1}{20}t \\ \cos\frac{1}{20}t \end{pmatrix} e^{-t/10}$
$-2 \begin{pmatrix} -\cos\frac{1}{20}t - \sin\frac{1}{20}t \\ \cos\frac{1}{20}t \\ \sin\frac{1}{20}t \end{pmatrix} e^{-t/10} + 11 \begin{pmatrix} 2 \\ 2 \\ 1 \end{pmatrix}$

### EXERCISES 8.3 (PAGE 360)

**1.** $X = c_1 \begin{pmatrix} -1 \\ 1 \end{pmatrix} e^{-t} + c_2 \begin{pmatrix} -3 \\ 1 \end{pmatrix} e^{t} + \begin{pmatrix} -1 \\ 3 \end{pmatrix}$

**3.** $X = c_1 \begin{pmatrix} 1 \\ -1 \end{pmatrix} e^{-2t} + c_2 \begin{pmatrix} 1 \\ 1 \end{pmatrix} e^{4t} + \begin{pmatrix} -\frac{1}{4} \\ \frac{3}{4} \end{pmatrix} t^2$
$+ \begin{pmatrix} \frac{1}{4} \\ -\frac{1}{4} \end{pmatrix} t + \begin{pmatrix} -2 \\ \frac{3}{4} \end{pmatrix}$

**5.** $X = c_1 \begin{pmatrix} 1 \\ -3 \end{pmatrix} e^{3t} + c_2 \begin{pmatrix} 1 \\ 9 \end{pmatrix} e^{7t} + \begin{pmatrix} \frac{55}{36} \\ -\frac{19}{4} \end{pmatrix} e^{t}$

**7.** $X = c_1 \begin{pmatrix} 1 \\ 0 \\ 0 \end{pmatrix} e^{t} + c_2 \begin{pmatrix} 1 \\ 1 \\ 0 \end{pmatrix} e^{2t} + c_3 \begin{pmatrix} 1 \\ 2 \\ 2 \end{pmatrix} e^{5t} - \begin{pmatrix} \frac{3}{2} \\ \frac{7}{2} \\ 2 \end{pmatrix} e^{4t}$

**9.** $X = 13 \begin{pmatrix} 1 \\ -1 \end{pmatrix} e^{t} + 2 \begin{pmatrix} -4 \\ 6 \end{pmatrix} e^{2t} + \begin{pmatrix} -9 \\ 6 \end{pmatrix}$

**11. (a)** $X' = \begin{pmatrix} -\frac{3}{100} & \frac{1}{100} \\ \frac{1}{50} & -\frac{1}{25} \end{pmatrix} X + \begin{pmatrix} 0 \\ 1 \end{pmatrix}$

**(b)** $X(t) = -\frac{70}{3} \begin{pmatrix} -1 \\ 2 \end{pmatrix} e^{-t/20} + \frac{80}{3} \begin{pmatrix} 1 \\ 1 \end{pmatrix} e^{-t/50} + \begin{pmatrix} 10 \\ 30 \end{pmatrix}$

**(c)** 10, 30

**13.** $X = c_1 \begin{pmatrix} 1 \\ 1 \end{pmatrix} + c_2 \begin{pmatrix} 3 \\ 2 \end{pmatrix} e^{t} - \begin{pmatrix} 11 \\ 11 \end{pmatrix} t - \begin{pmatrix} 15 \\ 10 \end{pmatrix}$

**15.** $X = c_1 \begin{pmatrix} 2 \\ 1 \end{pmatrix} e^{t/2} + c_2 \begin{pmatrix} 10 \\ 3 \end{pmatrix} e^{3t/2} - \begin{pmatrix} \frac{13}{2} \\ \frac{13}{4} \end{pmatrix} t e^{t/2} - \begin{pmatrix} \frac{15}{2} \\ \frac{9}{4} \end{pmatrix} e^{t/2}$

**17.** $X = c_1 \begin{pmatrix} 2 \\ 1 \end{pmatrix} e^{t} + c_2 \begin{pmatrix} 1 \\ 1 \end{pmatrix} e^{2t} + \begin{pmatrix} 3 \\ 3 \end{pmatrix} e^{t} + \begin{pmatrix} 4 \\ 2 \end{pmatrix} t e^{t}$

**19.** $X = c_1 \begin{pmatrix} 4 \\ 1 \end{pmatrix} e^{3t} + c_2 \begin{pmatrix} -2 \\ 1 \end{pmatrix} e^{-3t} + \begin{pmatrix} -12 \\ 0 \end{pmatrix} t - \begin{pmatrix} \frac{4}{3} \\ \frac{4}{3} \end{pmatrix}$

**21.** $X = c_1 \begin{pmatrix} 1 \\ -1 \end{pmatrix} e^{t} + c_2 \begin{pmatrix} t \\ \frac{1}{2} - t \end{pmatrix} e^{t} + \begin{pmatrix} \frac{1}{2} \\ -2 \end{pmatrix} e^{-t}$

**23.** $X = c_1 \begin{pmatrix} \cos t \\ \sin t \end{pmatrix} + c_2 \begin{pmatrix} \sin t \\ -\cos t \end{pmatrix} + \begin{pmatrix} \cos t \\ \sin t \end{pmatrix} t$

$\qquad + \begin{pmatrix} -\sin t \\ \cos t \end{pmatrix} \ln|\cos t|$

**25.** $X = c_1 \begin{pmatrix} \cos t \\ \sin t \end{pmatrix} e^{t} + c_2 \begin{pmatrix} \sin t \\ -\cos t \end{pmatrix} e^{t} + \begin{pmatrix} \cos t \\ \sin t \end{pmatrix} t e^{t}$

**27.** $X = c_1 \begin{pmatrix} \cos t \\ -\sin t \end{pmatrix} + c_2 \begin{pmatrix} \sin t \\ \cos t \end{pmatrix} + \begin{pmatrix} \cos t \\ -\sin t \end{pmatrix} t$

$\qquad + \begin{pmatrix} -\sin t \\ \sin t \tan t \end{pmatrix} - \begin{pmatrix} \sin t \\ \cos t \end{pmatrix} \ln|\cos t|$

**29.** $X = c_1 \begin{pmatrix} 2 \sin t \\ \cos t \end{pmatrix} e^{t} + c_2 \begin{pmatrix} 2 \cos t \\ -\sin t \end{pmatrix} e^{t} + \begin{pmatrix} 3 \sin t \\ \frac{3}{2} \cos t \end{pmatrix} t e^{t}$

$\qquad + \begin{pmatrix} \cos t \\ -\frac{1}{2} \sin t \end{pmatrix} e^{t} \ln|\sin t| + \begin{pmatrix} 2 \cos t \\ -\sin t \end{pmatrix} e^{t} \ln|\cos t|$

**31.** $X = c_1 \begin{pmatrix} 1 \\ -1 \\ 0 \end{pmatrix} + c_2 \begin{pmatrix} 1 \\ 1 \\ 0 \end{pmatrix} e^{2t} + c_3 \begin{pmatrix} 0 \\ 0 \\ 1 \end{pmatrix} e^{3t}$

$\qquad + \begin{pmatrix} -\frac{1}{4} e^{2t} + \frac{1}{2} t e^{2t} \\ -e^{t} + \frac{1}{4} e^{2t} + \frac{1}{2} t e^{2t} \\ \frac{1}{2} t^2 e^{3t} \end{pmatrix}$

**33.** $X = \begin{pmatrix} 2 \\ 2 \end{pmatrix} t e^{2t} + \begin{pmatrix} -1 \\ 1 \end{pmatrix} e^{2t} + \begin{pmatrix} -2 \\ 2 \end{pmatrix} t e^{4t} + \begin{pmatrix} 2 \\ 0 \end{pmatrix} e^{4t}$

**35.** $\begin{pmatrix} i_1 \\ i_2 \end{pmatrix} = 2 \begin{pmatrix} 1 \\ 3 \end{pmatrix} e^{-2t} + \frac{6}{29} \begin{pmatrix} 3 \\ -1 \end{pmatrix} e^{-12t} - \frac{4}{29} \begin{pmatrix} 19 \\ 42 \end{pmatrix} \cos t$

$\qquad + \frac{4}{29} \begin{pmatrix} 83 \\ 69 \end{pmatrix} \sin t$

### EXERCISES 8.4  (PAGE 365)

**1.** $e^{At} = \begin{pmatrix} e^{t} & 0 \\ 0 & e^{2t} \end{pmatrix}; \qquad e^{-At} = \begin{pmatrix} e^{-t} & 0 \\ 0 & e^{-2t} \end{pmatrix}$

**3.** $e^{At} = \begin{pmatrix} t+1 & t & t \\ t & t+1 & t \\ -2t & -2t & -2t+1 \end{pmatrix}$

**5.** $X = c_1 \begin{pmatrix} 1 \\ 0 \end{pmatrix} e^{t} + c_2 \begin{pmatrix} 0 \\ 1 \end{pmatrix} e^{2t}$

**7.** $X = c_1 \begin{pmatrix} t+1 \\ t \\ -2t \end{pmatrix} + c_2 \begin{pmatrix} t \\ t+1 \\ -2t \end{pmatrix} + c_3 \begin{pmatrix} t \\ t \\ -2t+1 \end{pmatrix}$

**9.** $X = c_3 \begin{pmatrix} 1 \\ 0 \end{pmatrix} e^{t} + c_4 \begin{pmatrix} 0 \\ 1 \end{pmatrix} e^{2t} + \begin{pmatrix} -3 \\ \frac{1}{2} \end{pmatrix}$

**11.** $X = c_1 \begin{pmatrix} \cosh t \\ \sinh t \end{pmatrix} + c_2 \begin{pmatrix} \sinh t \\ \cosh t \end{pmatrix} - \begin{pmatrix} 1 \\ 1 \end{pmatrix}$

**13.** $X = \begin{pmatrix} t+1 \\ t \\ -2t \end{pmatrix} - 4 \begin{pmatrix} t \\ t+1 \\ -2t \end{pmatrix} + 6 \begin{pmatrix} t \\ t \\ -2t+1 \end{pmatrix}$

**15.** $e^{At} = \begin{pmatrix} \frac{3}{2} e^{2t} - \frac{1}{2} e^{-2t} & \frac{3}{4} e^{2t} - \frac{3}{4} e^{-2t} \\ -e^{2t} + e^{-2t} & -\frac{1}{2} e^{2t} + \frac{3}{2} e^{-2t} \end{pmatrix};$

$\qquad X = c_1 \begin{pmatrix} \frac{3}{2} e^{2t} - \frac{1}{2} e^{-2t} \\ -e^{2t} + e^{-2t} \end{pmatrix} + c_2 \begin{pmatrix} \frac{3}{4} e^{2t} - \frac{3}{4} e^{-2t} \\ -\frac{1}{2} e^{2t} + \frac{3}{2} e^{-2t} \end{pmatrix}$  or

$\qquad X = c_3 \begin{pmatrix} 3 \\ -2 \end{pmatrix} e^{2t} + c_4 \begin{pmatrix} 1 \\ -2 \end{pmatrix} e^{-2t}$

**17.** $e^{At} = \begin{pmatrix} e^{2t} + 3t e^{2t} & -9t e^{2t} \\ t e^{2t} & e^{2t} - 3t e^{2t} \end{pmatrix};$

$\qquad X = c_1 \begin{pmatrix} 1 + 3t \\ t \end{pmatrix} e^{2t} + c_2 \begin{pmatrix} -9t \\ 1 - 3t \end{pmatrix} e^{2t}$

**23.** $X = c_1 \begin{pmatrix} \frac{3}{2} e^{3t} - \frac{1}{2} e^{5t} \\ \frac{3}{2} e^{3t} - \frac{3}{2} e^{5t} \end{pmatrix} + c_2 \begin{pmatrix} -\frac{1}{2} e^{3t} + \frac{1}{2} e^{5t} \\ -\frac{1}{2} e^{3t} + \frac{3}{2} e^{5t} \end{pmatrix}$  or

$\qquad X = c_3 \begin{pmatrix} 1 \\ 1 \end{pmatrix} e^{3t} + c_4 \begin{pmatrix} 1 \\ 3 \end{pmatrix} e^{5t}$

### CHAPTER 8 IN REVIEW  (PAGE 366)

**1.** $k = \frac{1}{3}$

**5.** $X = c_1 \begin{pmatrix} 1 \\ -1 \end{pmatrix} e^{t} + c_2 \left[ \begin{pmatrix} 1 \\ -1 \end{pmatrix} t e^{t} + \begin{pmatrix} 0 \\ 1 \end{pmatrix} e^{t} \right]$

**7.** $X = c_1 \begin{pmatrix} \cos 2t \\ -\sin 2t \end{pmatrix} e^{t} + c_2 \begin{pmatrix} \sin 2t \\ \cos 2t \end{pmatrix} e^{t}$

**9.** $X = c_1 \begin{pmatrix} -2 \\ 3 \\ 1 \end{pmatrix} e^{2t} + c_2 \begin{pmatrix} 0 \\ 1 \\ 1 \end{pmatrix} e^{4t} + c_3 \begin{pmatrix} 7 \\ 12 \\ -16 \end{pmatrix} e^{-3t}$

**11.** $X = c_1 \begin{pmatrix} 1 \\ 0 \end{pmatrix} e^{2t} + c_2 \begin{pmatrix} 4 \\ 1 \end{pmatrix} e^{4t} + \begin{pmatrix} 16 \\ -4 \end{pmatrix} t + \begin{pmatrix} 11 \\ -1 \end{pmatrix}$

**13.** $X = c_1 \begin{pmatrix} \cos t \\ \cos t - \sin t \end{pmatrix} + c_2 \begin{pmatrix} \sin t \\ \sin t + \cos t \end{pmatrix} - \begin{pmatrix} 1 \\ 1 \end{pmatrix}$

$\qquad + \begin{pmatrix} \sin t \\ \sin t + \cos t \end{pmatrix} \ln|\csc t - \cot t|$

**15. (b)** $X = c_1 \begin{pmatrix} -1 \\ 1 \\ 0 \end{pmatrix} + c_2 \begin{pmatrix} -1 \\ 0 \\ 1 \end{pmatrix} + c_3 \begin{pmatrix} 1 \\ 1 \\ 1 \end{pmatrix} e^{3t}$

### EXERCISES 9.1  (PAGE 373)

**1.** for $h = 0.1$, $y_5 = 2.0801$;   for $h = 0.05$, $y_{10} = 2.0592$

**3.** for $h = 0.1$, $y_5 = 0.5470$;   for $h = 0.05$, $y_{10} = 0.5465$

**5.** for $h = 0.1$, $y_5 = 0.4053$;   for $h = 0.05$, $y_{10} = 0.4054$

**7.** for $h = 0.1$, $y_5 = 0.5503$;   for $h = 0.05$, $y_{10} = 0.5495$

**9.** for $h = 0.1$, $y_5 = 1.3260$;   for $h = 0.05$, $y_{10} = 1.3315$

**11.** for $h = 0.1$, $y_5 = 3.8254$;   for $h = 0.05$, $y_{10} = 3.8840$;
at $x = 0.5$ the actual value is $y(0.5) = 3.9082$.

**13. (a)** $y_1 = 1.2$

**(b)** $y''(c)\dfrac{h^2}{2} = 4e^{2c}\dfrac{(0.1)^2}{2} = 0.02e^{2c} \le 0.02e^{0.2} = 0.0244$

**(c)** Actual value is $y(0.1) = 1.2214$. Error is $0.0214$.

**(d)** If $h = 0.05$, $y_2 = 1.21$.

**(e)** Error with $h = 0.1$ is $0.0214$. Error with $h = 0.05$ is $0.0114$.

**15. (a)** $y_1 = 0.8$

**(b)** $y''(c)\dfrac{h^2}{2} = 5e^{-2c}\dfrac{(0.1)^2}{2} = 0.025e^{-2c} \le 0.025$
for $0 \le c \le 0.1$.

**(c)** Actual value is $y(0.1) = 0.8234$. Error is $0.0234$.

**(d)** If $h = 0.05$, $y_2 = 0.8125$.

**(e)** Error with $h = 0.1$ is $0.0234$. Error with $h = 0.05$ is $0.0109$.

**17. (a)** Error is $19h^2e^{-3(c-1)}$.

**(b)** $y''(c)\dfrac{h^2}{2} \le 19(0.1)^2(1) = 0.19$

**(c)** If $h = 0.1$, $y_5 = 1.8207$.
If $h = 0.05$, $y_{10} = 1.9424$.

**(d)** Error with $h = 0.1$ is $0.2325$. Error with $h = 0.05$ is $0.1109$.

**19. (a)** Error is $\dfrac{1}{(c + 1)^2}\dfrac{h^2}{2}$.

**(b)** $\left| y''(c)\dfrac{h^2}{2}\right| \le (1)\dfrac{(0.1)^2}{2} = 0.005$

**(c)** If $h = 0.1$, $y_5 = 0.4198$. If $h = 0.05$, $y_{10} = 0.4124$.

**(d)** Error with $h = 0.1$ is $0.0143$. Error with $h = 0.05$ is $0.0069$.

**EXERCISES 9.2  (PAGE 377)**

**1.** $y_5 = 3.9078$; actual value is $y(0.5) = 3.9082$

**3.** $y_5 = 2.0533$          **5.** $y_5 = 0.5463$

**7.** $y_5 = 0.4055$          **9.** $y_5 = 0.5493$

**11.** $y_5 = 1.3333$

**13. (a)** $35.7130$

**(c)** $v(t) = \sqrt{\dfrac{mg}{k}}\tanh\sqrt{\dfrac{kg}{m}}\, t$;   $v(5) = 35.7678$

**15. (a)** for $h = 0.1$, $y_4 = 903.0282$;
for $h = 0.05$, $y_8 = 1.1 \times 10^{15}$

**17. (a)** $y_1 = 0.82341667$

**(b)** $y^{(5)}(c)\dfrac{h^5}{5!} = 40e^{-2c}\dfrac{h^5}{5!} \le 40e^{2(0)}\dfrac{(0.1)^5}{5!}$
$= 3.333 \times 10^{-6}$

**(c)** Actual value is $y(0.1) = 0.8234134413$. Error is
$3.225 \times 10^{-6} \le 3.333 \times 10^{-6}$.

**(d)** If $h = 0.05$, $y_2 = 0.82341363$.

**(e)** Error with $h = 0.1$ is $3.225 \times 10^{-6}$. Error with
$h = 0.05$ is $1.854 \times 10^{-7}$.

**19. (a)** $y^{(5)}(c)\dfrac{h^5}{5!} = \dfrac{24}{(c + 1)^5}\dfrac{h^5}{5!}$

**(b)** $\dfrac{24}{(c + 1)^5}\dfrac{h^5}{5!} \le 24\dfrac{(0.1)^5}{5!} = 2.0000 \times 10^{-6}$

**(c)** From calculation with $h = 0.1$, $y_5 = 0.40546517$.
From calculation with $h = 0.05$, $y_{10} = 0.40546511$.

**EXERCISES 9.3  (PAGE 381)**

**1.** $y(x) = -x + e^x$; actual values are $y(0.2) = 1.0214$,
$y(0.4) = 1.0918$, $y(0.6) = 1.2221$, $y(0.8) = 1.4255$;
approximations are given in Example 1.

**3.** $y_4 = 0.7232$

**5.** for $h = 0.2$, $y_5 = 1.5569$;   for $h = 0.1$, $y_{10} = 1.5576$

**7.** for $h = 0.2$, $y_5 = 0.2385$;   for $h = 0.1$, $y_{10} = 0.2384$

**EXERCISES 9.4  (PAGE 385)**

**1.** $y(x) = -2e^{2x} + 5xe^{2x}$; $y(0.2) = -1.4918$,
$y_2 = -1.6800$

**3.** $y_1 = -1.4928$, $y_2 = -1.4919$

**5.** $y_1 = 1.4640$, $y_2 = 1.4640$

**7.** $x_1 = 8.3055$, $y_1 = 3.4199$;
$x_2 = 8.3055$, $y_2 = 3.4199$

**9.** $x_1 = -3.9123$, $y_1 = 4.2857$;
$x_2 = -3.9123$, $y_2 = 4.2857$

**11.** $x_1 = 0.4179$, $y_1 = -2.1824$;
$x_2 = 0.4173$, $y_2 = -2.1821$

**EXERCISES 9.5  (PAGE 389)**

**1.** $y_1 = -5.6774$, $y_2 = -2.5807$, $y_3 = 6.3226$

**3.** $y_1 = -0.2259$, $y_2 = -0.3356$, $y_3 = -0.3308$,
$y_4 = -0.2167$

**5.** $y_1 = 3.3751$, $y_2 = 3.6306$, $y_3 = 3.6448$, $y_4 = 3.2355$,
$y_5 = 2.1411$

**7.** $y_1 = 3.8842$, $y_2 = 2.9640$, $y_3 = 2.2064$, $y_4 = 1.5826$,
$y_5 = 1.0681$, $y_6 = 0.6430$, $y_7 = 0.2913$

**9.** $y_1 = 0.2660$, $y_2 = 0.5097$, $y_3 = 0.7357$, $y_4 = 0.9471$,
$y_5 = 1.1465$, $y_6 = 1.3353$, $y_7 = 1.5149$, $y_8 = 1.6855$,
$y_9 = 1.8474$

**11.** $y_1 = 0.3492$, $y_2 = 0.7202$, $y_3 = 1.1363$, $y_4 = 1.6233$,
$y_5 = 2.2118$, $y_6 = 2.9386$, $y_7 = 3.8490$

**13. (c)** $y_0 = -2.2755$, $y_1 = -2.0755$, $y_2 = -1.8589$,
$y_3 = -1.6126$, $y_4 = -1.3275$

## CHAPTER 9 IN REVIEW (PAGE 389)

**1.** Comparison of numerical methods with $h = 0.1$:

| $x_n$ | Euler | Improved Euler | RK4 |
|---|---|---|---|
| 1.10 | 2.1386 | 2.1549 | 2.1556 |
| 1.20 | 2.3097 | 2.3439 | 2.3454 |
| 1.30 | 2.5136 | 2.5672 | 2.5695 |
| 1.40 | 2.7504 | 2.8246 | 2.8278 |
| 1.50 | 3.0201 | 3.1157 | 3.1197 |

Comparison of numerical methods with $h = 0.05$:

| $x_n$ | Euler | Improved Euler | RK4 |
|---|---|---|---|
| 1.10 | 2.1469 | 2.1554 | 2.1556 |
| 1.20 | 2.3272 | 2.3450 | 2.3454 |
| 1.30 | 2.5409 | 2.5689 | 2.5695 |
| 1.40 | 2.7883 | 2.8269 | 2.8278 |
| 1.50 | 3.0690 | 3.1187 | 3.1197 |

**3.** Comparison of numerical methods with $h = 0.1$:

| $x_n$ | Euler | Improved Euler | RK4 |
|---|---|---|---|
| 0.60 | 0.6000 | 0.6048 | 0.6049 |
| 0.70 | 0.7095 | 0.7191 | 0.7194 |
| 0.80 | 0.8283 | 0.8427 | 0.8431 |
| 0.90 | 0.9559 | 0.9752 | 0.9757 |
| 1.00 | 1.0921 | 1.1163 | 1.1169 |

Comparison of numerical methods with $h = 0.05$:

| $x_n$ | Euler | Improved Euler | RK4 |
|---|---|---|---|
| 0.60 | 0.6024 | 0.6049 | 0.6049 |
| 0.70 | 0.7144 | 0.7193 | 0.7194 |
| 0.80 | 0.8356 | 0.8430 | 0.8431 |
| 0.90 | 0.9657 | 0.9755 | 0.9757 |
| 1.00 | 1.1044 | 1.1168 | 1.1169 |

**5.** $h = 0.2$: $y(0.2) \approx 3.2$; $\quad h = 0.1$: $y(0.2) \approx 3.23$

**7.** $x(0.2) \approx 1.62$, $y(0.2) \approx 1.84$

## EXERCISES FOR APPENDIX A (PAGE APP-9)

**1.** 120

**3.** $\frac{3}{4}\sqrt{\pi}$

**5.** $\frac{105}{16}\sqrt{\pi}$

**7.** $\frac{4}{3}\sqrt{\pi}$

**9.** $\frac{16}{105}\sqrt{\pi}$

**11.** 0.9182

**13.** 0.9086

**15.** $(x + 3)(x + 2)(x + 1)x\,\Gamma(x)$

**17.** $\sqrt{\dfrac{\pi}{2}}$

**19.** $\frac{1}{324}$

**21.** $\frac{3}{128}$

**23.** $\sqrt{2\pi}$

**25.** 0.1836

**29.** $\pi$

**31.** $\frac{1}{4}\pi$

**33.** $\frac{5}{2}\pi$

**39.** $\frac{1}{12}$

**41.** $\frac{8}{15}$

**43.** $F'(x) = 6x^2 e^{-4x^9} - e^{-x^5} - 3x^2 \displaystyle\int_x^{2x^3} t^2 e^{-x^3 t^2}\,dt$

**45.** $F'(x) = -\sin x (x^2 + \cos^2 x)^{10} - 2x(x^2 + x^4)^{10}$
$$+ 20x \int_{x^2}^{\cos x} 10(x^2 + t^2)^9\,dt$$

**47.** $F'(x) = e^{\sin x}\sin(\cos x)$

**55. (a)** $\dfrac{\sqrt{\pi}}{2\sqrt{x}}$ **(b)** $\sqrt{\pi x} + C$ **(c)** $\sqrt{\pi x}$

## EXERCISES FOR APPENDIX B (PAGE APP-26)

**1. (a)** $\begin{pmatrix} 2 & 11 \\ 2 & -1 \end{pmatrix}$ **(b)** $\begin{pmatrix} -6 & 1 \\ 14 & -19 \end{pmatrix}$

**(c)** $\begin{pmatrix} 2 & 28 \\ 12 & -12 \end{pmatrix}$

**3. (a)** $\begin{pmatrix} -11 & 6 \\ 17 & -22 \end{pmatrix}$ **(b)** $\begin{pmatrix} -32 & 27 \\ -4 & -1 \end{pmatrix}$

**(c)** $\begin{pmatrix} 19 & -18 \\ -30 & 31 \end{pmatrix}$ **(d)** $\begin{pmatrix} 19 & 6 \\ 3 & 22 \end{pmatrix}$

**5. (a)** $\begin{pmatrix} 9 & 24 \\ 3 & 8 \end{pmatrix}$ **(b)** $\begin{pmatrix} 3 & 8 \\ -6 & -16 \end{pmatrix}$

**(c)** $\begin{pmatrix} 0 & 0 \\ 0 & 0 \end{pmatrix}$ **(d)** $\begin{pmatrix} -4 & -5 \\ 8 & 10 \end{pmatrix}$

**7. (a)** 180 **(b)** $\begin{pmatrix} 4 & 8 & 10 \\ 8 & 16 & 20 \\ 10 & 20 & 25 \end{pmatrix}$ **(c)** $\begin{pmatrix} 6 \\ 12 \\ -5 \end{pmatrix}$

**9. (a)** $\begin{pmatrix} 7 & 38 \\ 10 & 75 \end{pmatrix}$ **(b)** $\begin{pmatrix} 7 & 38 \\ 10 & 75 \end{pmatrix}$

**11.** $\begin{pmatrix} -14 \\ 1 \end{pmatrix}$

**13.** $\begin{pmatrix} -38 \\ -2 \end{pmatrix}$

**15.** singular

**17.** nonsingular; $\mathbf{A}^{-1} = \dfrac{1}{4}\begin{pmatrix} -5 & -8 \\ 3 & 4 \end{pmatrix}$

**19.** nonsingular; $\mathbf{A}^{-1} = \dfrac{1}{2}\begin{pmatrix} 0 & -1 & 1 \\ 2 & 2 & -2 \\ -4 & -3 & 5 \end{pmatrix}$

**21.** nonsingular; $\mathbf{A}^{-1} = -\dfrac{1}{9}\begin{pmatrix} -2 & -2 & -1 \\ -13 & 5 & 7 \\ 8 & -1 & -5 \end{pmatrix}$

**23.** $\mathbf{A}^{-1}(t) = \dfrac{1}{2e^{3t}}\begin{pmatrix} 3e^{4t} & -e^{4t} \\ -4e^{-t} & 2e^{-t} \end{pmatrix}$

**25.** $\dfrac{d\mathbf{X}}{dt} = \begin{pmatrix} -5e^{-t} \\ -2e^{-t} \\ 7e^{-t} \end{pmatrix}$

**27.** $\dfrac{d\mathbf{X}}{dt} = 4\begin{pmatrix} 1 \\ -1 \end{pmatrix}e^{2t} - 12\begin{pmatrix} 2 \\ 1 \end{pmatrix}e^{-3t}$

**29. (a)** $\begin{pmatrix} 4e^{4t} & -\pi \sin \pi t \\ 2 & 6t \end{pmatrix}$    **(b)** $\begin{pmatrix} \frac{1}{4}e^8 - \frac{1}{4} & 0 \\ 4 & 6 \end{pmatrix}$

**(c)** $\begin{pmatrix} \frac{1}{4}e^{4t} - \frac{1}{4} & (1/\pi) \sin \pi t \\ t^2 & t^3 - t \end{pmatrix}$

**31.** $x = 3, y = 1, z = -5$

**33.** $x = 2 + 4t, y = -5 - t, z = t$

**35.** $x = -\frac{1}{2}, y = \frac{3}{2}, z = \frac{7}{2}$

**37.** $x_1 = 1, x_2 = 0, x_3 = 2, x_4 = 0$

**41.** $\mathbf{A}^{-1} = \begin{pmatrix} 0 & \frac{2}{3} & \frac{1}{3} \\ 0 & -\frac{1}{3} & -\frac{2}{3} \\ \frac{1}{3} & -\frac{2}{3} & 0 \end{pmatrix}$

**43.** $\mathbf{A}^{-1} = \begin{pmatrix} 5 & 6 & -3 \\ 2 & 2 & -1 \\ -1 & -1 & 1 \end{pmatrix}$

**45.** $\mathbf{A}^{-1} = \begin{pmatrix} -\frac{1}{2} & -\frac{2}{3} & -\frac{1}{6} & \frac{7}{6} \\ 1 & \frac{1}{3} & \frac{1}{3} & -\frac{4}{3} \\ 0 & -\frac{1}{3} & -\frac{1}{3} & \frac{1}{3} \\ -\frac{1}{2} & 1 & \frac{1}{2} & \frac{1}{2} \end{pmatrix}$

**47.** $\lambda_1 = 6, \lambda_2 = 1, \mathbf{K}_1 = \begin{pmatrix} 2 \\ 7 \end{pmatrix}, \mathbf{K}_2 = \begin{pmatrix} 1 \\ 1 \end{pmatrix}$

**49.** $\lambda_1 = \lambda_2 = -4, \mathbf{K}_1 = \begin{pmatrix} 1 \\ -4 \end{pmatrix}$

**51.** $\lambda_1 = 0, \lambda_2 = 4, \lambda_3 = -4,$

$\mathbf{K}_1 = \begin{pmatrix} 9 \\ 45 \\ 25 \end{pmatrix}, \mathbf{K}_2 = \begin{pmatrix} 1 \\ 1 \\ 1 \end{pmatrix}, \mathbf{K}_3 = \begin{pmatrix} 1 \\ 9 \\ 1 \end{pmatrix}$

**53.** $\lambda_1 = \lambda_2 = \lambda_3 = -2,$

$\mathbf{K}_1 = \begin{pmatrix} 2 \\ -1 \\ 0 \end{pmatrix}, \mathbf{K}_2 = \begin{pmatrix} 0 \\ 0 \\ 1 \end{pmatrix}$

**55.** $\lambda_1 = 3i, \lambda_2 = -3i,$

$\mathbf{K}_1 = \begin{pmatrix} 1 - 3i \\ 5 \end{pmatrix}, \mathbf{K}_2 = \begin{pmatrix} 1 + 3i \\ 5 \end{pmatrix}$

# Index

Index

Index

Index

# Table of Laplace Transforms

| $f(t)$ | $\mathscr{L}\{f(t)\} = F(s)$ |
|---|---|
| **1.** $1$ | $\dfrac{1}{s}$ |
| **2.** $t$ | $\dfrac{1}{s^2}$ |
| **3.** $t^n$ | $\dfrac{n!}{s^{n+1}}$, $\quad n$ a positive integer |
| **4.** $t^{-1/2}$ | $\sqrt{\dfrac{\pi}{s}}$ |
| **5.** $t^{1/2}$ | $\dfrac{\sqrt{\pi}}{2s^{3/2}}$ |
| **6.** $t^\alpha$ | $\dfrac{\Gamma(\alpha+1)}{s^{\alpha+1}}$, $\quad \alpha > -1$ |
| **7.** $\sin kt$ | $\dfrac{k}{s^2 + k^2}$ |
| **8.** $\cos kt$ | $\dfrac{s}{s^2 + k^2}$ |
| **9.** $\sin^2 kt$ | $\dfrac{2k^2}{s(s^2 + 4k^2)}$ |
| **10.** $\cos^2 kt$ | $\dfrac{s^2 + 2k^2}{s(s^2 + 4k^2)}$ |
| **11.** $e^{at}$ | $\dfrac{1}{s - a}$ |
| **12.** $\sinh kt$ | $\dfrac{k}{s^2 - k^2}$ |
| **13.** $\cosh kt$ | $\dfrac{s}{s^2 - k^2}$ |
| **14.** $\sinh^2 kt$ | $\dfrac{2k^2}{s(s^2 - 4k^2)}$ |
| **15.** $\cosh^2 kt$ | $\dfrac{s^2 - 2k^2}{s(s^2 - 4k^2)}$ |
| **16.** $te^{at}$ | $\dfrac{1}{(s - a)^2}$ |
| **17.** $t^n e^{at}$ | $\dfrac{n!}{(s - a)^{n+1}}$, $\quad n$ a positive integer |
| **18.** $e^{at} \sin kt$ | $\dfrac{k}{(s - a)^2 + k^2}$ |
| **19.** $e^{at} \cos kt$ | $\dfrac{s - a}{(s - a)^2 + k^2}$ |
| **20.** $e^{at} \sinh kt$ | $\dfrac{k}{(s - a)^2 - k^2}$ |

| $f(t)$ | $\mathscr{L}\{f(t)\} = F(s)$ |
|---|---|
| **21.** $e^{at} \cosh kt$ | $\dfrac{s - a}{(s - a)^2 - k^2}$ |
| **22.** $t \sin kt$ | $\dfrac{2ks}{(s^2 + k^2)^2}$ |
| **23.** $t \cos kt$ | $\dfrac{s^2 - k^2}{(s^2 + k^2)^2}$ |
| **24.** $\sin kt + kt \cos kt$ | $\dfrac{2ks^2}{(s^2 + k^2)^2}$ |
| **25.** $\sin kt - kt \cos kt$ | $\dfrac{2k^3}{(s^2 + k^2)^2}$ |
| **26.** $t \sinh kt$ | $\dfrac{2ks}{(s^2 - k^2)^2}$ |
| **27.** $t \cosh kt$ | $\dfrac{s^2 + k^2}{(s^2 - k^2)^2}$ |
| **28.** $\dfrac{e^{at} - e^{bt}}{a - b}$ | $\dfrac{1}{(s - a)(s - b)}$ |
| **29.** $\dfrac{ae^{at} - be^{bt}}{a - b}$ | $\dfrac{s}{(s - a)(s - b)}$ |
| **30.** $1 - \cos kt$ | $\dfrac{k^2}{s(s^2 + k^2)}$ |
| **31.** $kt - \sin kt$ | $\dfrac{k^3}{s^2(s^2 + k^2)}$ |
| **32.** $\dfrac{a \sin bt - b \sin at}{ab(a^2 - b^2)}$ | $\dfrac{1}{(s^2 + a^2)(s^2 + b^2)}$ |
| **33.** $\dfrac{\cos bt - \cos at}{a^2 - b^2}$ | $\dfrac{s}{(s^2 + a^2)(s^2 + b^2)}$ |
| **34.** $\sin kt \sinh kt$ | $\dfrac{2k^2 s}{s^4 + 4k^4}$ |
| **35.** $\sin kt \cosh kt$ | $\dfrac{k(s^2 + 2k^2)}{s^4 + 4k^4}$ |
| **36.** $\cos kt \sinh kt$ | $\dfrac{k(s^2 - 2k^2)}{s^4 + 4k^4}$ |
| **37.** $\cos kt \cosh kt$ | $\dfrac{s^3}{s^4 + 4k^4}$ |
| **38.** $\sin kt \cosh kt + \cos kt \sinh kt$ | $\dfrac{2ks^2}{s^4 + 4k^4}$ |
| **39.** $\sin kt \cosh kt - \cos kt \sinh kt$ | $\dfrac{4k^3}{s^4 + 4k^4}$ |
| **40.** $\sinh kt - \sin kt$ | $\dfrac{2k^3}{s^4 - k^4}$ |